3. Energy Conversion Table

Three unusual but convenient "energy" quantities are temperature (K) representing the molecular energy kT at a given T, wavenumber (cm^{-1}) representing the molecular energy $hc\tilde{v}$ at a given frequency $\tilde{v}$ in cm^{-1} units (where c is the speed of light in cm s^{-1}), and electron volt (eV). All of these molecular energies are changed to molar energies on multiplication by Avogadro's number.

	K	cm^{-1}	kJ mol^{-1}	kcal mol^{-1}	eV
1 K =	1	0.69504	8.31451×10^{-3}	1.98722×10^{-3}	8.61739×10^{-5}
1 cm^{-1} =	1.43877	1	$1.19627 - 10^{-2}$	2.85914×10^{-3}	1.23984×10^{-4}
1 kJ mol^{-1} =	1.20272×10^{2}	8.35935×10^{1}	1	0.23901	1.03643×10^{-2}
1 kcal mol^{-1} =	5.03217×10^{2}	3.49755×10^{2}	4.18400	1	4.33641×10^{-2}
1 eV =	1.16044×10^{4}	8.06554×10^{3}	9.64853×10^{1}	2.30605×10^{1}	1

[a]These are "2002 CODATA recommended values"; see *Rev. Mod. Phys.* **77,** 1 (2005). The most recent values of physical constants can be obtained on the National Institute of Standards and Technology (NIST) website (http://physics.nist.gov/constants).

[b]F m^{-1} ≡ C^2 m^{-1} J^{-1} ≡ C m^{-1} V^{-1}.

[c]Note that $|c|$ is the *pure number* 29979245800 (i.e., it equals the magnitude of the speed of light in vacuum when it is expressed in cm s^{-1} units).

EXPERIMENTS IN PHYSICAL CHEMISTRY

EIGHTH EDITION

CARL W. GARLAND

Massachusetts Institute of Technology

JOSEPH W. NIBLER

Oregon State University

DAVID P. SHOEMAKER
(*deceased*)
Oregon State University

Boston   Burr Ridge, IL   Dubuque, IA   Madison, WI   New York   San Francisco   St. Louis
Bangkok   Bogotá   Caracas   Kuala Lumpur   Lisbon   London   Madrid   Mexico City
Milan   Montreal   New Delhi   Santiago   Seoul   Singapore   Sydney   Taipei   Toronto

EXPERIMENTS IN PHYSICAL CHEMISTRY, EIGHTH EDITION
International Edition 2009

Exclusive rights by McGraw-Hill Education (Asia), for manufacture and export. This book cannot be re-exported from the country to which it is sold by McGraw-Hill. This International Edition is not to be sold or purchased in North America and contains content that is different from its North American version.

10 09 08 07 06 05 04 03 02 01
20 09 08
CTF   ANL

When ordering this title, use ISBN: 978-007-126351-1 or MHID: 007-126351-9

Printed in Singapore

www.mhhe.com

Contents

Preface   ix

I. INTRODUCTION   1

Organization of the Book   2
Preparation for an Experiment   4
Execution of an Experiment   4
Safety   6
Recording of Experimental Data   7
Literature Work   9
Reports   10
Sample Report   13
Special Projects   26
Ethics   27

II. TREATMENT OF EXPERIMENTAL DATA   29

A. Calculations and Presentation of Data   29

Significant Figures   30
Precision of Calculations   31
Analytical Methods   32
Numerical Methods   33
Graphs and Graphical Methods   34
Exercises   37

B. Uncertainties in Data and Results   38

Errors   38
Rejection of Discordant Data   42
Statistical Treatment of Random Errors   43
Propagation of Errors   52
Case History of an Error Evaluation   59
Fundamental Limitations on Instrumental Precision   61
Summary   64
Exercises   65

III. USE OF COMPUTERS   68

Word Processors   68
Spreadsheets   69
Symbolic Mathematics Programs   79
Quantum Mechanical Programs   82
Interfacing with Experiments   85
Exercises   88

IV. GASES   91

1. Gas Thermometry   91
2. Joule–Thomson Effect   98
3. Heat-Capacity Ratios for Gases   106

V. TRANSPORT PROPERTIES OF GASES   119

Kinetic Theory of Transport Phenomena   119
4. Viscosity of Gases   128
5. Diffusion of Gases   136

VI. THERMOCHEMISTRY   145

Principles of Calorimetry   145
6. Heats of Combustion   152
7. Strain Energy of the Cyclopropane Ring   158
8. Heats of Ionic Reaction   167

VII. SOLUTIONS   172

9.   Partial Molar Volume   172
10. Cryoscopic Determination of Molar Mass   179
11. Freezing-Point Depression of Strong and Weak Electrolytes   188
12. Chemical Equilibrium in Solution   193

VIII. PHASE BEHAVIOR   199

13. Vapor Pressure of a Pure Liquid   199
14. Binary Liquid–Vapor Phase Diagram   207
15. Ordering in Nematic Liquid Crystals   215
16. Liquid–Vapor Coexistence Curve and the Critical Point   229

IX. ELECTROCHEMISTRY   235

17. Conductance of Solutions   235
18. Temperature Dependence of emf   245
19. Activity Coefficients from Cell Measurements   248

X. CHEMICAL KINETICS   254

20. Method of Initial Rates: Iodine Clock   254
21. NMR Study of a Reversible Hydrolysis Reaction   263
22. Enzyme Kinetics: Inversion of Sucrose   271
23. Kinetics of the Decomposition of Benzenediazonium Ion   283
24. Gas-Phase Kinetics   287

XI. SURFACE PHENOMENA   299

    25. Surface Tension of Solutions   299
    26. Physical Adsorption of Gases   308

XII. MACROMOLECULES   318

    27. Intrinsic Viscosity: Chain Linkage in Polyvinyl Alcohol   318
    28. Helix–Coil Transition in Polypeptides   327

XIII. ELECTRIC, MAGNETIC, AND OPTICAL PROPERTIES   336

    29. Dipole Moment of Polar Molecules in Solution   336
    30. Dipole Moment of HCl Molecules in the Gas Phase   347
    31. Magnetic Susceptibility   361
    32. NMR Determination of Paramagnetic Susceptibility   371
    33. Dynamic Light Scattering   379

XIV. SPECTROSCOPY   393

    34. Absorption Spectrum of a Conjugated Dye   393
    35. Raman Spectroscopy: Vibrational Spectrum of CCl_4 398
    36. Stimulated Raman Spectra of Benzene   407
    37. Vibrational–Rotational Spectra of HCl and DCl   416
    38. Vibrational–Rotational Spectra of Acetylenes   424
    39. Absorption and Emission Spectra of I_2 436
    40. Fluorescence Lifetime and Quenching in I_2 Vapor   446
    41. Electron Spin Resonance Spectroscopy   454
    42. NMR Determination of Keto–Enol Equilibrium Constants   466
    43. NMR Study of Gas-Phase DCl–HBr Isotopic Exchange Reaction   475
    44. Solid-State Lasers: Radiative Properties of Ruby Crystals   484
    45. Spectroscopic Properties of CdSe Nanocrystals   492

XV. SOLIDS   500

    46. Determination of Crystal Structure by X-Ray Diffraction   500
    47. Lattice Energy of Solid Argon   515
    48. Statistical Thermodynamics of Iodine Sublimation   523

XVI. ELECTRONIC DEVICES AND MEASUREMENTS   538

    Circuit Elements   538
    Operational Amplifiers   542
    Analog-to-Digital Conversion   547
    Digital Multimeters   550
    Potentiometer Circuits   552
    Wheatstone Bridge Circuits   554

XVII. TEMPERATURE   557

    Temperature Scales   557
    Triple-Point and Ice-Point Cell   561
    Thermometers   562
    Temperature Control   576

XVIII. VACUUM TECHNIQUES   587

Introduction   587
Pumps   587
Pressure Gauges   594
Safety Considerations   599

XIX. INSTRUMENTS   601

Balances   601
Barometers   605
Oscilloscopes   605
pH Meters   609
Polarimeters   611
Radiation Counters   612
Recording Devices and Printers   613
Refractometers   613
Signal-Averaging Devices   617
Spectroscopic Components   618
Spectroscopic Instruments   631
Timing Devices   636

XX. MISCELLANEOUS PROCEDURES   639

Volumetric Procedures   639
Purification Methods   644
Gas-Handling Procedures   644
Electrodes for Electrochemical Cells   651
Materials for Construction   652
Solders and Adhesives   658
Tubing Connections   660
Shopwork   661

XXI. LEAST-SQUARES FITTING PROCEDURES   663

Introduction   663
Foundations of Least Squares   664
Weights   669
Rejection of Discordant Data   671
Goodness of Fit   672
Comparison of Models   676
Uncertainties in the Parameters   678
Summary of Procedures   680
Sample Least-Squares Calculation   681

APPENDICES   **687**

    A.   Glossary of Symbols   687
    B.   International System of Units and Concentration Units   690
    C.   Safety   693
    D.   Literature Work   701
    E.   Research Journals   707
    F.   Numerical Methods of Analysis   709
    G.   Barometer Corrections   718
    H.   Ethical Conduct in Physical Chemistry   719

Index   **721**

Endpapers
    Physical Constants and Conversion Factors   Front
    Energy Conversion Table   Front
    Relative Atomic Masses of the Elements   Back

I often say that when you can measure what you are speaking about, and express it in numbers, you know something about it; but when you cannot express it in numbers, your knowledge is of a meagre and unsatisfactory kind; it may be the beginning of knowledge, but you have scarcely, in your thoughts, advanced to the stage of Science, whatever the matter may be.

Sir William Thomson
(Lord Kelvin)

It is much easier to make measurements than to know exactly what you are measuring.

J. W. N. Sullivan

It is a capital mistake to theorize before one has data. Insensibly one begins to twist facts to suit theories, instead of theories to suit facts.

Sir Arthur Conan Doyle

It is also a good rule not to put too much confidence in experimental results until they have been confirmed by theory.

Sir Arthur Eddington

Preface

This book is designed for use in a junior-level laboratory course in physical chemistry. It is assumed that the student will be taking concurrently (or has taken previously) a lecture course in physical chemistry. The book contains 48 selected experiments, which have been tested by extensive use.

Three of the experiments are new, and all of these involve optical measurements. One concerns a study of the birefringence of a liquid crystal to determine the evolution of nematic order near the nematic-isotropic phase transition (Exp. 15). The second is a study of the dynamic light scattering from an aqueous dispersion of polystyrene spheres in order to determine their particle size (Exp. 33). The third is a study of the absorption and fluorescence spectra of CdSe nanocrystals with an analysis based on predictions from a quantum model (Exp. 45). To conserve space, three experiments that appeared in the seventh edition have been deleted. These are the binary solid-liquid phase diagram, osmotic pressure, and single-crystal x-ray diffraction. All the other experiments from the seventh edition have been reviewed, and in some cases small changes have been made to either the theory or the experimental procedures. The most extensive of such changes occur in Experiments 1 (gas thermometry) and 4 (viscosity of gases), where the apparatus and procedure have been changed to eliminate the use of mercury. In numerous other experiments, mercury manometers and mercury-in-glass thermometers have been replaced by direct reading pressure gauges and temperature sensors.

Experiments. The 48 experiments provide a balance between traditional and modern topics in physical chemistry, with most of the classical topics in Chapters IV–X (Exps. 1–24) and most of the modern topics in Chapters XI–XV (Exps. 25–48). These experiments are not concerned primarily with "techniques" per se or with the analytical applications of physical chemistry. We believe that an experimental physical chemistry course should serve a dual purpose: (1) to illustrate and test theoretical principles, and (2) to develop a research orientation by providing basic experience with physical measurements that yield quantitative results of important chemical interest.

Each experiment is accompanied by a theoretical development in sufficient detail to provide a clear understanding of the method to be used, the calculations required, and the significance of the final results. The depth of coverage is frequently greater than that which is available in introductory physical chemistry textbooks. Experimental procedures are described in considerable detail as an aid to the efficient use of laboratory time and teaching staff. Emphasis is given to the reasons behind the design and procedure for each experiment, so that the student can learn the general principles of a variety of experimental techniques. Stimulation of individual resourcefulness through the use of special projects or variations on existing experiments is also desirable. We strongly urge that on some occasions the experiments presented here should be used as points of departure for work of a more independent nature.

Introductory and background material.   In addition to the experiments themselves, nine chapters contain material of a general nature. These chapters should be useful not only in an undergraduate laboratory course but also in special-project work, thesis research, and a broad range of general research in chemistry.

The first three chapters provide material that is valuable for all experimental work in physical chemistry, and it is recommended that these be read before beginning any laboratory work. Chapter I contains a wide range of introductory material, including advice on the preparation of laboratory reports, and Chapter II covers techniques for data analysis and the assessment of error limits. Chapter III, on the use of computer software and selected aspects of computer interfacing, has been revised and updated. The software aspect emphasizes the application of spreadsheets for the recording, plotting, and least-squares fitting of experimental data. Also discussed, and illustrated with examples, are more sophisticated programs such as Mathcad and Mathematica, along with the program Gaussian, which can be used to obtain molecular structures and properties from ab-initio quantum mechanics. Relevant theoretical calculations using such programs are suggested at the end of a number of experiments.

Chapters XVI–XX deal with basic experimental methods of broad value in many types of experimental work—electronic measurements, temperature measurement and control, vacuum techniques, diverse instruments that are widely used, and miscellaneous laboratory procedures. These chapters have been revised and updated in various ways. In the case of Chapters XVI and XVIII, the text has also been shortened from that which appeared in the seventh edition. Finally, Chapter XXI presents a thorough discussion of least-squares fitting procedures.

References cited at the end of each experiment and each of the general chapters have been updated. In cases where no new literature source could be found that covers a given topic as well as the source cited in the seventh edition, the original citation has been retained, since most libraries have available copies of older books and monographs.

In the development of the new experiments in this edition, we acknowledge expert advice from Prof. M. Bawendi (MIT), Prof. J. Thoen (Katholieke Universiteit Leuven), and Dr. B. Weiner (Brookhaven Instruments) as well as the assistance of faculty, teaching assistants, and undergraduate students (especially Nicole Baker, Matthew Martin, Colin Shear, Brain Theobald, and Robert Zaworski) at Oregon State University. Helpful comments also have come from a number of reviewers and from those who have used this book at other universities, and for these we are very appreciative. We encourage and welcome feedback from all who use this book, either as students or instructors.

Finally, we wish to acknowledge the suberb proofreading effort of Janelle Pregler, who has made the production of this edition go very smoothly.

Carl W. Garland
Joseph W. Nibler

I

Introduction

Physical chemistry deals with the physical principles underlying the properties of chemical substances. Like other branches of physical science, it contains a body of theory that has stood the test of experiment and is continually growing as a result of new experiments. In order to learn physical chemistry, you must become familiar with the experimental foundations on which the theoretical principles are based. Indeed, in many cases, the ability to apply the principles usefully requires an intimate knowledge of those methods and practical arts that are called "experimental technique."

For this reason, lecture courses in physical chemistry are usually accompanied by a program of laboratory work. Such experimental work should not just demonstrate established principles but should also develop research aptitudes by providing experience with the kind of measurements that can yield important new results. This book attempts to achieve that goal. Its aim is to provide a clear understanding of the principles of important experimental methods, the design of basic apparatus, the planning of experimental procedures, and the significance of the final results. In short, the aim is to help you become a productive research scientist.

Severe limitations of time and equipment must be faced in presenting a set of experiments as the basis of a laboratory course that will provide a reasonably broad coverage of the wide and varied field of physical chemistry. Although high-precision research measurements would require refinements in the methods described here and would often require more sophisticated and more elaborate equipment, each experiment in this book is designed so that meaningful results of reasonable accuracy can be obtained.

In the beginning, you will probably not have the skill and the time to plan detailed experimental procedures. Moreover, the trial-and-error method is not necessarily the best way to learn good experimental technique. Laboratory skills are developed slowly during a sort of apprenticeship. To enable you to make efficient use of available time, both the apparatus and the procedure for these experiments are described in considerable detail. *You should keep in mind the importance of understanding why the experiment is done in the way described.* This understanding is a vital part of the experience necessary for planning special or advanced experiments of a research character. As you become more experienced, it is desirable that there be opportunities to plan more of your own procedure. This can easily be accomplished by introducing variations in the experiments described here. A change in the chemical system to be studied, the use of equipment that purposefully differs from that described, or the choice of a different method for studying the same system will require modifications of the procedure. Finally, at the end of the course, it is recommended that, if possible, you carry out a special project that is completely independent of the experiments described in this book. A brief description of such special projects is given at the end of this chapter.

In addition to a general knowledge of laboratory techniques, creative research work requires the ability to apply two different kinds of theory. Many an experimental method is based on a special phenomenological theory of its own; this must be well understood in order to design the experiment properly and in order to calculate the desired physical property from the observed raw data. Once the desired result has been obtained, it is necessary to understand its significance and its interrelationship with other known facts. This requires a sound knowledge of the fundamental theories of physical chemistry (e.g., thermodynamics, statistical mechanics, quantum mechanics, and kinetics). Considerable emphasis has been placed on both kinds of theory in this book.

In the final analysis, however, research ability cannot be learned merely by performing experiments described in a textbook; it has to be acquired through contact with inspired teachers and through the accumulation of considerable experience. The goal of this book is to provide a solid frame of reference for future growth.

ORGANIZATION OF THE BOOK

Before undertaking any experiments, you should become familiar with the overall structure of this book. The material is divided into three blocks: essential introductory material is given in Chapters I–III, the experiments are given in Chapters IV–XV, and extensive reference material is given in Chapters XVI–XXI. The heart of the book is the collection of experiments, which are designed to provide significant "hands-on" laboratory experience. The general chapters present basic background information of value both for undergraduate laboratory courses and for independent research work.

INTRODUCTORY ESSENTIALS

Chapter I provides a general introduction to experimental work in physical chemistry. This includes advice on how to prepare for and carry out an experiment, how to record data in a laboratory notebook, and how to report results. Chapter II describes mathematical procedures for analyzing data. Part A of this chapter deals with calculations and the presentation of numerical results, and Part B is concerned with the quantitative assessment of random errors and their effect on the uncertainty in the final result. Chapter III deals with several types of computer hardware and software commonly used in physical chemistry. Commercial products for digital measurement of temperature, voltage, and pressure are now readily available, and the need for machine-level programming has been greatly reduced. Examples of some data collection systems are briefly described. Emphasis is placed on the use of spreadsheets, since these are very convenient for data handling, analysis, and presentation in both tabular and graphic forms. All three chapters should be read at the very beginning, since they provide general information of value for all experiments in physical chemistry.

It is important to mention briefly the issues of notation and units. A glossary of frequently used symbols is given in Appendix A. Note that the definition of thermodynamic work used here is the work done *on* a system (e.g., $dw = -p\,dV$). As a result, the first law of thermodynamics has the form $dE = dq + dw$. Note also that E is used for the thermodynamic internal energy and U for the potential energy instead of the IUPAC choices of

U and V respectively. Système International (SI) units, described in Appendix B, are used extensively but not slavishly. Chemically convenient quantities such as the gram (g), cubic centimeter (cm^3), and liter (L $\equiv$ dm^3 $=$ 10^3 cm^3) are still used where useful—densities in g cm^{-3}, concentrations in mol L^{-1}, molar masses in g. Conversions of such quantities into their SI equivalents is trivially easy. The situation with pressure is not so simple, since the SI pascal is a very awkward unit. Throughout the text, both bar and atmosphere are used. Generally bar ($\equiv 10^5$ Pa) is used when a precisely measured pressure is involved, and atmosphere ($\equiv 760$ Torr $\equiv 1.01325 \times 10^5$ Pa) is used to describe casually the ambient air pressure, which is usually closer to 1 atm than to 1 bar. Standard states for all chemical substances are officially defined at a pressure of 1 bar; normal boiling points for liquids are still understood to refer to 1-atm values. The conversion factors given inside the front cover will help in coping with non-SI pressures.

EXPERIMENTS

Chapters IV–XV contain the descriptions of the experiments, which are numbered 1 to 48. In addition, Chapters V and VI each contain some separate introductory material that is pertinent to all the experiments in the given chapter. Each figure, equation, and table in a given experiment is identified by a single number: e.g., Fig. 1, Eq. (8), Table 2. For cross references, double numbering is used: e.g., Fig. 38-1 refers to Fig. 1 in Exp. 38 and Eq. (V-8) refers to Eq. (8) in the introductory part of Chapter V.

Every attempt has been made to write each experiment in sufficient detail that it can be intelligently performed without the necessity of extensive outside reading. However, it is assumed that the student will refer frequently to a standard textbook in physical chemistry for any necessary review of elementary theory. Literature sources are cited explicitly for those topics that are beyond the scope of a typical undergraduate text-book, and these numbered references are listed together at the end of each experiment. In addition, a selected list of reading pertinent to the general topic of each experiment is given under the heading General Reading. It is hoped that you will do some reading in these books and journal articles, since this is an excellent way to broaden your scientific background.

A complete and very detailed list of equipment and chemicals is given at the end of each experiment.[1] The list is divided into two sections: Those items listed in the first para-graph are required for the exclusive use of a single team; those in the second paragraph are available for the common use of several teams. It is assumed that standardized stock solu-tions will be made up in advance and will be available for the students' use. The quantities indicated in parentheses† are for the use of the instructor and do *not* indicate the amounts of each chemical to be taken by a single team. In addition to the items included in these apparatus lists, it is assumed that the laboratory is equipped with analytical balances, a supply of distilled or deionized water, and a barometer as well as gas, water, and 110-V ac power lines. Also desirable, but not absolutely necessary, are gas-handling lines and a rough vacuum line.

BASIC REFERENCE MATERIAL

Chapters XVI–XX contain a variety of information on experimental procedures and devices. This includes the theory and practice of many types of electrical measurements,

---

†On the basis of the authors' experience, it is necessary to make available these amounts for each team that will do the experiment. They are scaled up from the amounts stated in the experiment to provide for possible wastage and to give a generous safety factor. It is hoped that they will be useful as a rough guide the first time an experi-ment is given.

the principles and devices used in measurement and control of temperature, vacuum techniques, detailed descriptions of frequently used instruments, and a description of many miscellaneous procedures such as gas handling and the construction of equipment.

Chapter XXI provides an introduction to least-squares fitting procedures and a discussion of how to assess the magnitude of random errors and how to judge the quality of any given fit to a set of data points.

PREPARATION FOR AN EXPERIMENT

Although most of the experiments in this book can be performed by a single person, they have been written under the assumption that a pair of students will work together as a team. Such teamwork is advantageous, since it provides an opportunity for valuable discussion of the experiment between partners. The amount of experimental work to be assigned will be based on the amount of laboratory time available for each experiment. Many of the experiments can be completed in full during a single 4-hour laboratory period. Many others are designed for 6 to 8 hours of laboratory work but may be abridged so that meaningful results can be obtained in a single 4-hour period. Some of the experiments require at least 6 hours and should not be attempted in a shorter time (in particular, Exps. 22, 24, 26, 28).

Before you arrive in the laboratory to perform a given experiment, it is essential that you study the experiment carefully, with special emphasis on the method, the apparatus design, and the procedure. It will usually be necessary to make changes in the procedure whenever the apparatus to be used or the system to be studied differs from that described in this book. Planning such changes or even successfully carrying out the experiment as described requires a clear understanding of the experimental method.

Experimental work in physical chemistry requires many complex and expensive pieces of apparatus; many of these have been constructed specially and cannot readily be replaced. Each team should accept responsibility for its equipment and should *check it over carefully before starting an experiment.*

EXECUTION OF AN EXPERIMENT

Making experimental measurements is like playing the piano or singing a song. You need some natural ability—bright students with no innate experimental skills often end up as theorists—but most people can learn to execute an experiment and have a lot of fun doing it. Like playing the piano and singing, you get better with practice. It is also useful to start with something reasonably simple and build up experience with a variety of techniques before taking on a really tough experiment.

GENERAL ADVICE

Some of the practice involved in developing experimental skills occurs within a single experiment and some occurs as the superposition of many experiments. In carrying out any given experiment, it may be necessary to make several "runs", i.e., sets of measurements. On some occasions, it is worthwhile to carry out a practice run as a warmup test of the apparatus and your own skills. Such a practice run can be made with a known standard material or standard set of conditions, or it can be a first try at a real set of measurements to get a feel for the operation of the equipment.

The design of the experiment will require a decision about the number of data points to be taken in each run and the number of runs needed; see the next subsection on data collection. Often it is desirable to have some indication of the reproducibility of a typical data point. In this case, multiple measurements of some initial value can be helpful for estimating the random errors as well as providing practice with the apparatus.

When carrying out an experiment, don't just record data and plan to think about them later during the analysis phase of the work. If you are measuring some quantity y as a function of a variable x—say, the pressure p of a fixed volume of gas as a function of the temperature T—make a rough plot of your data as they are acquired. Such a plot will quickly reveal any problems—whether one pressure reading is far "out of line" with the others or a series of pressure readings are erratic ("noisy"). It is better to stop and troubleshoot the apparatus rather than blindly march ahead with a long set of measurements that may prove very poor or even useless. Check some standard measurement or repeat a few data points. If a component of the apparatus is suspected to be malfunctioning, replace it if possible or borrow a properly functioning component for a quick test and then try to repair your component if its misbehavior is confirmed.

Be alert to possible external disturbances—such as high humidity, mechanical vibrations, stray electric fields, voltage fluctuations, unusual local heating—and internal misbehaviors—such as vacuum leaks, nonlinear meter readings, hysteresis. Some of these may cause equipment failure or result in noisy, erratic readings. Some can introduce subtler troubles in the form of systematic errors, where the data look fine but are not. See Chapter II for a detailed discussion of errors.

DATA COLLECTION

Data values are subject to both random and systematic errors. The effect of random errors can be reduced by increasing the number N of observations, but this is an inefficient process since, as shown in Chapter II, the error in the average of N values varies as $1/\sqrt{N}$. Thus one must quadruple the number of measurements in order to cut the random error in half. A much better approach is to improve the design and sensitivity of the apparatus. Even then, you must guard against systematic errors and personal bias in recording the data.

In most cases, data are collected sequentially rather than in random order. For example, chemical kinetics requires measuring the concentration of reactant or product sequentially as a function of time, and the measurement of pH during a titration must be made sequentially as a function of the volume of acid or base added. Thus you should be alert to the possibility that some uncontrolled variable (e.g., ambient pressure or temperature) may be undergoing a monotonic change that introduces a small but systematic effect into your data.

Even if you cannot influence the sequence of data collection, you can control the size of the interval between points. The general rule is: *Take points more closely spaced when the observed quantity is varying more rapidly.* Thus data points should be taken more frequently at early times for chemical kinetics and near the neutralization endpoint for an acid–base titration, as illustrated in Fig. 1.

In recording the absorption spectrum of a gas as a function of frequency v, you could in principle measure the intensity $I(v)$ at a set of randomly ordered frequency values. In practice, however, modern spectrometers have scanning motors that vary v in a monotonic fashion at variable programmable scan rates. With such a device, you would make a rapid overall scan to locate the interesting features and then repeat the measurement at a slow scan rate over the small frequency regions of interest. This scale expansion is equivalent to taking many discrete points in a region where $I(v)$ varies rapidly.

FIGURE 1

Optimal spacing of data points when the observed quantity varies rapidly in some regions and slowly in others: (*a*) concentration of reactant during a kinetics run; (*b*) pH of a solution during titration with a base.

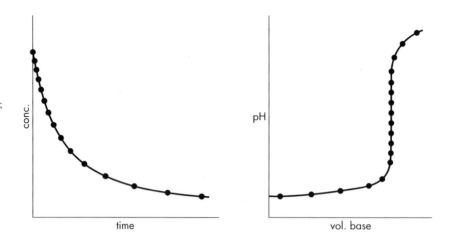

Another situation where you could take data in random order is in measuring the vapor pressure of a liquid as a function of temperature. However, it is better to make a series of measurements in a monotonic sequence—progressive heating or cooling—since equilibrium at each point is achieved more quickly. In this type of experiment, it is strongly recommended that you carry out both heating and cooling sequences. If the data points in both types of run agree, this demonstrates that equilibrium has been achieved and greatly strengthens the credibility of the data. The same technique can be used if you are measuring a property as a function of pressure p or electric field E.

BIAS

In collecting data, you must be alert to and avoid personal preconceptions and expectations that might influence the values recorded. Strive to be objective in recording experimental "readings." Put down in your notebook exactly what the device indicates and do not shade the reading toward what you feel it should be. Interpolate between the smallest divisions rather than rounding off to the nearest division. If adjusting a continuous dial to get some null condition, set the dial several times, approaching from different directions. Scientific objectivity is demanding. It means accepting an unattractive datum value unless there are valid experimental or statistical criteria for rejecting it. See Part B of Chapter II and Appendix H for a discussion of different aspects of this issue.

SAFETY

Experimental work can be subject to hazards of many kinds, and every person working in a laboratory should be alert to possible safety problems. Once you are aware of the particular hazards involved in an experimental procedure, the instinct for self-preservation usually provides sufficient motivation for finding ways of avoiding them. The principal danger lies in ignorance of specific hazards and in forgetfulness.

A detailed analysis of all types of laboratory hazards and the procedures for dealing with them is beyond the scope of this book. Specific hazards are pointed out in connection with individual experiments. It is also assumed that the instructing staff will provide specific warnings and reminders where needed. Some general remarks on the kinds of safety hazards that should be kept in mind are given in Appendix C, and very complete treatments can be found elsewhere.[2] At this point, we wish to stress a few basic principles that apply to all laboratory work.

1. Determine the potential hazards and review the safety procedures appropriate for any experiment before beginning the work.

2. Know the location and proper use of safety and emergency equipment such as fire extinguishers and alarms, first-aid kits, safety showers, eyewash fountains, emergency telephone numbers, and emergency exits.

3. Call attention to any unsafe conditions you observe. Someone else's accident can be dangerous to you as well as to him or her.

4. Check all electrical equipment carefully before plugging it into the power line; unplug equipment before making any changes in electrical connections. Make sure that no part of the equipment has exposed high-voltage components.

5. Do not use mouth suction to draw up chemicals into a pipette; use a pipetting bulb instead.

6. Safety glasses or other appropriate eye protection must be worn in the laboratory at all times.

7. Avoid safety hazards and environmental contamination by following accepted waste-disposal procedures.

8. *Never work in the laboratory alone.* (For graduate research work and professional work, it is sometimes necessary to work alone. If so, make sure that someone—a security guard or an individual working in a nearby laboratory—will check on you every few hours.)

RECORDING OF EXPERIMENTAL DATA

The laboratory notebook is the essential link between the laboratory and the outside world and is the ultimate reference concerning what took place in the laboratory.[3] As such, it is not only the source book for the production of reports and publications but also a permanent record, which may be consulted even after a great many years. It is standard practice in experimental research work to record *everything relevant* (data, calculations, notes and comments, literature surveys, and even some graphs) directly in a bound notebook with numbered pages. Spiral-bound notebooks are preferable, since they lie flat and stay open at the desired page. Such notebooks are available with pages that are ruled vertically as well as horizontally to give a $\frac{1}{4}$-inch grid; this facilitates tabulation of columns of figures and permits rough plots of the data to be made directly on the notebook pages during the experiment. In an undergraduate laboratory course, it is often convenient to make a carbon copy of the recorded data to include with the written report. (Notebooks with duplicate sets of numbered pages, in which alternate pages are perforated for removal, may be used; however, ordinary carbon paper and bond paper can be used with any style of research notebook.) In any case, the *original* is a part of a permanent notebook.

Whatever style of notebook is used, the principle is the same: *Record all data directly in your notebook.* Data may be copied into the notebook from a partner's notebook in those cases where it is clearly impossible for both partners to record data at the same time. Even then, a carbon copy or a photocopy of the original pages is better, since it avoids copying errors and saves time. In particular, *do not use odd scraps* of paper to record such incidental data as weights, barometer readings, and temperatures with the idea of copying them into the notebook at a later time. If anything must be copied from another source (calibration chart, reference book, etc.), identify it with an appropriate reference.

A ballpoint pen is best for recording data, especially if carbon copies are required; otherwise, any pen with permanent ink is satisfactory. Pencils are unsuitable for recording primary data. If a correction is necessary, draw a single line through the incorrect number so as to leave it legible and then write the correct number directly above or beside the old one. If something happens to vitiate the data on an entire page, cross out these data and record the circumstances. *No original pages should ever be removed from a laboratory notebook.* Never leave blank notebook pages or any significant number of blank lines on a notebook page, except at the end of an experiment. Always record your data in serial (chronological) fashion except when you are recording data in tabular form. Neatness and good organization are desirable, but legibility, proper labeling, and completeness are absolute necessities.

When you first obtain your notebook, put your name and the laboratory subject name on the cover. On the first page, record your name, your local address, your e-mail address, and telephone number plus a permanent home address and the date. Leave the next five or six pages blank so there will be room for a detailed table of contents. For each experiment undertaken, make an entry for it in this table of contents, indicating the page number and the date on which work began. If a long experiment has logical subdivisions, these should also be entered in the table of contents.

WHAT TO RECORD

Every data page for an experiment should have a clear heading that includes your name (and the name of any lab partner), the subject of the experiment, the date, and a page number.

The notebook should contain all the information that would be needed to permit someone else to perform the same experiment in the same way. In addition to the measurements and observations that constitute the results of the experiment, all other data that are relevant to the interpretation of the results should be recorded. It is not necessary to duplicate information that is conveniently available in some permanent record such as a journal article or laboratory text, provided that complete references are given.

The record of an experiment should begin with a brief statement of the experiment to be performed. The procedure being followed should be described in all essential detail. When the procedure is described elsewhere, the notebook entry may be abbreviated to a reference to the published description. However, be certain to record any variations from the published procedure and to specify the apparatus that was used (make and model for commercial equipment, otherwise a sketch). If various pieces of equipment have identification numbers or serial numbers, these should be recorded. Note also any date of modification for apparatus. Include any calibration constants or calibration curves that are provided. For each chemical substance used, record the name, formula, source, grade or stated purity, and concentration (in case of a solution). If you have developed a checklist or work plan for the experiment, record this in your notebook also.

Data should be entered directly in the notebook, in tabular form whenever possible. Use complete, explicit headings; do not rely on your memory for the meaning of unconventional symbols and abbreviations. Be sure that all numerical values are accompanied by the appropriate units. In many cases, it is necessary or at least wise to record such laboratory conditions as the ambient temperature, the atmospheric pressure, or the relative humidity.

As mentioned previously, it is strongly recommended that you plot data in your laboratory notebook as they are being acquired. In order to produce effective "rough" plots using the crosshatched grid in your notebook, choose convenient scales so that both variables can be quickly and accurately plotted. If data are being taken in two distinct runs, use different symbols to denote points on the separate runs.

Instrumental records—such as spectra, computer data plots, and printouts of computer files (see below)—should be dated and signed by the investigator. They may be taped or stapled into the notebook if they are of a convenient size or can be suitably reduced with a copying machine. This should not be done if there are many such records, as it makes the notebook clumsy and awkward to use. Records that are not fastened in the notebook should be numbered and placed in a laboratory file provided for the purpose (preferably a three-ring binder), and an appropriate reference should be made in the notebook.

COMPUTER FILES AND DISKS

If part or all of the data for an experiment were originally stored on the hard disk of a computer, it is necessary to back up this information with a hard (printed) copy or a personal compact disk (CD). If the volume of data is modest, printout is the most attractive format and such hard copy can be pasted or taped into the notebook. For more massive data sets, a clearly labeled CD is preferable to a thick stack of printout.

In any event, data files originally stored on a computer must be clearly named and dated. In the case of a hard disk, you should establish a personal directory named in a manner consistent with other users. Within this directory, you may wish to set up subdirectories for different experiments, although this is probably not necessary for an undergraduate physical chemistry course. Use extensions on your filenames to indicate run numbers and other analogous information: for example, Exp3.r01 or Exp6.cal. Record data file names in your notebook together with other notes on the experiment and also in the table of contents. Compact disks should be stored in a 6-in. $\times$ 8-in. manila envelope, glued securely inside the back cover of your notebook.

LITERATURE WORK

As a matter of policy, very few of the experiments contain a direct reference to published values of the final result that is to be reported. Students who wish to compare their results with accepted literature values are expected to do the necessary library work. As a general principle, it is best to refer directly to an original journal article rather than to some secondary source. It is commonly assumed that recent measurements are more precise than older ones; this assumption is based on the fact that methods and equipment are constantly being improved. But this does not mean that there is valid reason to reject or suspect a published result merely because it is old. The quality of research data depends strongly on the integrity, conscientious care, and patience of the research worker; much fine work done many years ago in certain areas of physical chemistry has not yet been improved upon. In evaluating results based on old but high-quality data, you must, however, be alert to the possible need for corrections necessitated by more recent theoretical developments or by improved values of physical constants. As an example, the international temperature scale was refined and modified slightly in 1990, and this has a small effect on previous absolute temperature values and thermodynamic properties, especially at high temperatures.[4]

Citing "literature values" in your report is attractive since it allows you to compare your result (with its estimated uncertainty) with accepted published results, which permits you to judge how well you have succeeded. A detailed discussion of literature work is given in Appendix D, but the essential approach needed for undergraduate laboratory reports is summarized in this section.

DATABASE AND ABSTRACT SEARCHES

Of the many compilations of physical properties described in Appendix D, the most convenient general source is the *CRC Handbook of Chemistry and Physics,* D. R. Lide (ed.), a large single volume published by CRC Press, Boca Raton, Florida. A new edition is published every year, but any recent edition will serve as well as the latest one. A very good general source of older physical data is *Landolt-Bornstein Numerical Data and Functional Relationships in Science and Technology, New Series.* This multivolume work is much more detailed but somewhat difficult to use.

Another resource is an Internet database search (see Appendix D). An introduction to the use of the Internet by chemists has been given in S. M. Bachrach (ed.), *The Internet: A Guide for Chemists.*[5] Approximately one half of this book provides a general coverage of many Internet issues, including the World Wide Web (WWW) and hypertext documents, the use of anonymous FTP (File Transfer Protocol), and database design. The remaining half is devoted to Internet resources of specific value to chemists, the Computational Chemistry List (CCL), which is a discussion forum, chemical Gopher sites, and web browsers (Uniform Resource Locators, URLs) for chemistry.

If you do not find the desired information in some encyclopedic compilation, then the next approach is to search for a scientific journal article giving this information. The first step is to use one of the two abstract journals, *Chemical Abstracts* or *Physics Abstracts,* to search for a useful journal article. Recommendations for their use are given in Appendix D. *Chemical Abstracts* is larger and more complete and is the preferred place to carry out a very thorough search. However, many physical quantities are also given in *Physics Abstracts,* which is faster and easier to use although not as comprehensive. Both of these abstract services maintain computer databases, which are described in Appendix D. Although web-based searches are increasingly available, the use of printed abstract journals is still a viable option for undergraduate laboratory work.

CHEMICAL NOMENCLATURE

In searching the chemical literature and in reading and writing research papers, a knowledge of systematic nomenclature for chemical compounds is indispensable. Since 1957 the International Union of Pure and Applied Chemistry (IUPAC) has recommended an international nomenclature. The more important nomenclature rules, together with recommended terminology, are available in several books:

G. J. Leigh, H. A. Favre, and W. V. Metanomski, *Principles of Chemical Nomenclature,* Blackwell, Oxford (1998).

B. P. Block, W. H. Powell, and W. C. Fernelias, *Inorganic Chemical Nomenclature,* American Chemical Society, Washington, DC (1990).

P. Fresenius, *Organic Chemical Nomenclature,* Prentice-Hall, Upper Saddle River, NJ (1989).

The most recent IUPAC recommendations for symbols to be used in physical chemistry are given in a pamphlet by Mills et al.,[6] which contains much other useful information.

REPORTS

The evaluation of any experimental work is based primarily on the contents of a written report. Indeed, the advancement of science depends heavily on the exchange of written information. Research work is not considered to be complete until the results have been properly reported. Such reports should be well organized and readable, so that anyone unfamiliar with the experiment can easily follow the presentation (with the aid of explicit references where necessary) and thereby obtain a clear idea as to what was actually done and what result was obtained.

An attempt should be made to use a scientific style comparable in quality to the literary style expected in an essay. Correct spelling and grammar should not be disregarded just because the report is to be read by a scientist instead of a general reader. The report should be as concise and factual as possible without sacrificing clarity. In particular, mathematical equations should be accompanied by enough verbal material to make their meaning clear.

Two general sources of information dealing with proper literary usage are

H. W. Fowler, *A Dictionary of Modern English Usage,* 2d ed., Oxford University Press, New York (1983) [Slightly dated but a classic that is still in print.]

W. Strunk, Jr., and E. B. White, *The Elements of Style,* 4th ed., Allyn and Bacon, Old Tappan, NJ (1999).

Four sources dealing specifically with technical writing are

M. Alley, *The Craft of Scientific Writing,* 3d ed., Springer-Verlag, New York (1996).

C. T. Brusaw, *Technical Writing,* 5th ed., St. Martin's Press, New York (1997).

J. G. Paradis and M. Zimmerman, *MIT Guide to Scientific and Engineering Communication,* MIT Press, Cambridge, MA (1997).

L. C. Perelman, E. C. Barrett, and J. G. Paradis, *The Mayfield Handbook of Scientific and Technical Writing,* Mayfield Publ. Co., Mountain View, CA (1997). [This is available both in hard copy as a book and in electronic forms. It can be used as a traditional print reference, but it was created primarily as a hypertext document. As the latter, it allows the user a great deal of freedom in choosing the sequence of topics and the level of detail.]

Stylistic details for preparing journal articles (such as recommended symbols, nomenclature, abbreviations, and the proper presentation of formulas, equations, tables, and literature citations) are discussed in

J. S. Dodd (ed.), *The ACS Style Guide,* 2d ed., American Chemical Society, Washington, DC (1997). [A general reference book on scientific writing with emphasis on manuscripts intended for one of the journals published by the American Chemical Society.]

J. T. Scott, *AIP Style Manual,* 4th ed., American Institute of Physics. New York (1990). [Provides guidance in the preparation of papers for AIP journals.]

The preparation of journal articles will not be discussed here. The recommendations given below concern student reports written as part of a laboratory course, although the same advice might well serve for technical reports of any kind.

Most important of all, the report must be an original piece of writing. Copying or even paraphrasing of material from textbooks, printed notes, or other reports is clearly dishonest and must be carefully avoided. Brief quotations, enclosed in quotation marks and accompanied by a complete reference, are permissible where a real advantage is to be gained. Certainly there is no point in giving more than a brief summary of the theory or the details of experimental procedure if these are adequately described in some readily available reference. In part, a report is likely to be judged on how clearly it states the essential points without oscillating between minute detail on one topic and vague generalities on another.

Except for general physical and numerical constants or well-known theoretical equations, any data or material taken from an outside source must be accompanied by a complete reference to that source.

The content and length of any given report will depend on the subject matter of the experiment and on the standards established by the instructor. It is our belief that at least in some cases the report should be quite complete and should include a quantitative analysis of the experimental uncertainties and a detailed discussion of the significance of the results (see the sample report given below). For many experiments a brief report with only a qualitative treatment of errors and a short discussion may be considered adequate. In either case, a clear presentation of the data, calculations, and results is essential to every report on experimental work.

STYLE

A report is a combination of text, tables, and figures that presents a continuous flow of information. Tables and figures should be numbered and referred to by number in the body of the text. One purpose of the text is to introduce the tables and figures and help the reader to understand their content. Do not start a paragraph or a sentence with a table or figure; this is too abrupt and difficult to follow. In the same spirit, one should not start a sentence with a symbol or arabic number. Adopt a logical and essentially chronological order—method used, data obtained, calculated results, discussion of results. See the Sample Report as an illustration. Insert short tables into the body of the text, and put long ones on a separate page. Generally, a figure should be placed on a separate page unless it is just a simple sketch rather than a quantitative plot.

Aim for a clear and simple writing style. Variation in sentence length and structure (begin some sentences with a subordinate clause or conjunctive phrase rather than the main clause) helps to hold the reader's interest. Avoid very long, complicated sentences. The usual practice in science writing is to avoid the use of the first person. Instead of "I carried out five runs," use "Five runs were carried out" or "Runs were carried out at five different temperatures." Try to balance this use of passive verbs with other sentences using active verbs such as "indicate" or "show." Instead of "The spectrum was recorded over the range 1000–3000 cm^{-1} and two strong peaks were observed," use "The spectrum showed two strong peaks in the 1000–3000 cm^{-1} region."

FORMAT

Unless otherwise instructed, all reports should be prepared on $8\frac{1}{2}$-in. $\times$ 11-in. paper with reasonable margins on all sides. The pages should be stapled together or bound in a folder. Legibility is absolutely essential. Double-spaced word-processed reports are best, but handwritten reports submitted in ink on wide-line ruled paper are acceptable unless you are cursed with illegible handwriting. Crossing out and the insertion of corrections are permissible, but try to keep the report as a whole reasonably neat.

TABLES

If only a few experimental data or calculated results are involved, they can be displayed in the body of the text as an unnumbered tabulation centered horizontally with wider margins than the text (see Sec. III of the Sample Report). In many cases, extensive data sets or multiple calculated results are obtained. These are best presented in formal tables with a number, a self-explanatory caption, and clearly labeled column heads with units given in parentheses. Footnotes may also be used at the bottom of the table to include references or comments on individual entries. See Tables 47-2 and XVII-1 as examples.

Any symbols used in a table should be defined previously in the text. In a column of numbers, align the decimal points and use a zero before the point for numbers less than 1 (i.e., use 0.39 rather than .39). If the numerical values are very large or very small, use an appropriate multiple of 10 in the heading: $10^3 p$ (bar) or p (millibar) are attractive for pressures such as $p = 0.00872$ bar, since the entry becomes 8.72. Or one would use $10^{-6} p$ (bar) or p (Mbar) as headings for pressures such as $p = 3,250,000$ bar, which results in the entry 3.25. It may be appropriate to include an estimate of the uncertainty value of each entry (say, 3.25 ± 0.05) or a general statement in the caption, such as "All pressure readings have an uncertainty of ± 0.05 Mbar."

FIGURES

A general discussion of the graphical treatment of experimental data is given in Chapter II. As part of that discussion, the proper technique for plotting experimental data is fully described for both computer-generated plots and any that may be prepared manually.

The vertical and horizontal axes must be labeled with the appropriate symbols or words and the units indicated in parentheses: for example, T (°C), t(s), A (cm^2), density (g cm^{-3}).† Often it is appropriate to add a smooth curve representing an equation arising from theory. The equation itself (if simple) or the number by which it is designated in the text must be given beside the theory curve. It is good practice to indicate directly on the figure the numerical values of any slopes, intercepts, areas, maxima, or other features that are important in the calculations.

If a figure has been computer generated, many of the above concerns may have been resolved automatically by the software plotting package. The user should, however, be aware that the "default" plot scales are often not ideal. Most plot programs allow the user to control the aspect ratio, the scale ranges, and the shape of the symbols. You must also make sure that the plotted variables are correctly specified and their units clearly indicated.

Each figure must have a figure number and a short legend prominently displayed, and it should be referred to by number in the body of the report.

SAMPLE REPORT

A sample undergraduate laboratory report on a very simple experiment is given here as a brief illustration of the typical structure to be used in technical reports. This example is intended to show the kind of material that each section should contain. It is not meant to provide a rigid model; the content of any given report will necessarily depend on the character of the experiment and the judgment of the individual student. In particular, it should be stressed that most reports will be longer and more complex than this example.

The sample report is displayed as a facsimile of a typed and handwritten report on the upper parts of the following pages. The lower parts of these pages contain a commentary on each section of the report. Although the given outline need not be followed exactly, the topics covered should be included somewhere in the report.

---

†Another convention is also widely used: e.g., $T/$°C, $t/$s, $p/$atm, $A/$cm^2, density/g cm^{-3}. Care must be taken in the way units are expressed in this style. For the style recommended in the above text, there is no confusion if density is written as ρ(g/cm^3) or ρ(g cm^{-3}), but $\rho/$g/cm^3 is not an acceptable alternative to $\rho/$g cm^{-3}.

DETERMINATION OF THE DENSITY OF

CRYSTALLINE GERMANIUM

Maria Smith   Sept. 25, 2008

(Partner: John Klein)

Abstract

A pycnometer has been used to determine the density of two samples of germanium at 25°C. The values obtained, with their 95% confidence limits, are ρ(sample I) = 5.310 ± 0.003 g cm^{-3} and ρ(sample II) = 5.332 ± 0.003 g cm^{-3}. Both samples consisted of several pieces of crystalline material, with the pieces in sample I being larger and more irregular. It seems likely that sample I had hidden defects and voids that introduced a systematic error since the x-ray value of the density is 5.327 ± 0.002 g cm^{-3}. Although the density of sample II is slightly high, it does agree with the x-ray value just within the sum of the cited uncertainties.

TITLE PAGE AND ABSTRACT

The front page of the report should display a title, your name, the name of any experimental partners, the date on which the report is submitted, and a brief abstract. An abstract is typically 50 to 100 words long; the example above contains about 90 words. It should summarize the results of the experiment and state any significant conclusions. *Numerical results with confidence limits should be included.*

I. Introduction

The purpose of this experiment is to measure the density
of crystals of germanium. Since the density ρ is defined by

$$\rho = W_S/V_S, \qquad\qquad (1)$$

it is desired to measure the volume V_S occupied by a known
weight W_S of the metal sample.

The method involves the use of a pycnometer of known
volume which is first weighed empty, then weighed
containing the solid sample to be studied. The difference
gives the weight of the solid, W_S. Finally the pycnometer
(containing the solid sample) is filled with a liquid of
known density and reweighed; the weight, and therefore the
volume, of the liquid can be found by difference. Since
the total volume of the pycnometer is known, one can then
calculate the volume V_S which is occupied by the solid.

II. Experimental Method

The experimental method was similar to that described
in the textbook (Aardvark and Zebra, 3rd ed., Exp. 13). The
design of the pycnometer used, which differs from that
described in the textbook, is shown in Fig. 1.

INTRODUCTION

The introduction should state the purpose of the experiment and give a *very brief*
outline of the necessary theory, which is often accomplished by citing pertinent equations
with references wherever appropriate. Each equation should appear on a separate line and
should be part of a complete sentence. Number all equations consecutively throughout the
entire report, and refer to them by number. All symbols should be clearly identified the
first time they appear. A very short description of the experimental method can also be
included. Use the present tense throughout this section.

The introduction should cover the above topics as concisely as possible; the sample
contains about 130 words. More complicated experiments will require longer introduc-
tions, but the normal length should be between 100 and 300 words. In the case of a journal
article, the introduction is often much longer, since pertinent recent work in the field is
cited and briefly summarized.

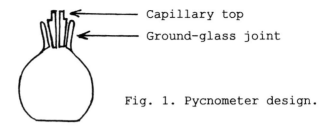

Fig. 1. Pycnometer design.

The procedure was modified as follows: After distilled water had been added to the pycnometer containing the sample, the pycnometer (with capillary top removed) was completely immersed in a flask of distilled water and boiled under low pressure for 15 minutes to remove air trapped by the solid or dissolved in the water [1]. After this boiling, the pycnometer was equilibrated for 30 min in a 25°C thermostat bath before the top was inserted.

The weight of the solid sample is given by

$$W_s = W_2 - W_1, \tag{2}$$

where W_1 is the weight of the empty pycnometer and W_2 is that of the pycnometer plus solid sample. The weight of water contained in the pycnometer, W_L, is

EXPERIMENTAL METHOD

The section on experimental method is usually *extremely brief* and merely cites the appropriate references that describe the details of the experimental procedure. If reference is made to the textbook and/or laboratory notes assigned for the course, an abbreviated title may be cited in lieu of a complete bibliographic entry. Any reference to other books or materials should be assigned footnote numbers and should be properly listed at the end of the experiment in the form illustrated by the references in this book. Description of experimental procedures should normally be given *only* for those features not described in or differing from the cited references. A simple sketch of apparatus is appropriate only when it differs from that described in the references. A summary statement or condensed derivation of the phenomenological equations used to analyze the raw data should be given. Except for text concerning such equations, this section should be written in the past tense and follow a chronological order. NOTE: A statement of the number of runs made and

$$W_L = W_3 - W_2, \qquad (3)$$

where W_3 is the weight of the pycnometer plus sample plus water. If the density of the liquid water is denoted by ρ_L, it follows from Eq. (3) that the volume of the solid sample is given by

$$V_S = V - V_L = (\rho_L V + W_2 - W_3)/\rho_L, \qquad (4)$$

where V is the total volume of the pycnometer. From Eqs. (1), (2), and (4), we obtain

$$\rho = W_S/V_S = \rho_L(W_2 - W_1)/(\rho_L V + W_2 - W_3). \qquad (5)$$

Since the values of V and ρ_L are known, it is only necessary to determine W_1, W_2, and W_3 in order to calculate the density of the solid.

Two duplicate runs, carried out using the same procedure, were made on each of two different germanium samples. Sample I consisted of larger and somewhat more irregular pieces than did Sample II.

the conditions under which they were carried out (concentration, temperature, etc.) should always be included at the end of this section.

RESULTS

The section on results should present the experimental results in full detail, making use of tables and figures where appropriate. No result should be excluded merely because it is unexpected or inconsistent with other data or theoretical models. The cause for discrepancies, if known, can be pointed out in the Discussion section. Primary measurements ("raw data") should be given, as well as derived quantities. It is essential that the units be completely specified.

It is undesirable to present detailed computations in the main body of the report; however, a typical sample calculation should be given in an appendix to illustrate how the data

III. Results

The average values of the measured weights W_1, W_2, and W_3 are listed below together with the stated value of the pycnometer volume V. The literature value for the density of water ρ_L is 0.99705 g cm^{-3} at 25°C (taken from the Handbook of Chemistry and Physics[2]).

$$V = 12.445 \pm 0.003 \text{ cm}^3 \text{ (given by instructor)}$$

$$W_1 = 8.6309 \text{ g}$$

$$\text{Sample I: } W_2 = 42.0301 \text{ g}, \quad W_3 = 48.1732 \text{ g}$$

$$\text{Sample II: } W_2 = 45.8479 \text{ g}, \quad W_3 = 51.3036 \text{ g}$$

The density of germanium can now be calculated by substitution of the above data into Eq. (5). The resulting densities are 5.315 g cm^{-3} for sample I and 5.337 g cm^{-3} for sample II. The weights used in these calculations have not been corrected for the effect of air buoyancy. Rather than correct each weight, we have used a simple formula given by Bauer and Lewin[1] for correcting the final result. This formula for the corrected density $\rho*$ is

$$\rho* = \rho + 0.0012[1 - (\rho/\rho_L)]. \tag{6}$$

analysis was carried out. If specialized computer programs have been used in processing the data, they should be cited in the Results section. If a program is not well known and documented elsewhere, provide a complete listing of the program in an appendix.

Tables should be numbered and given self-explanatory captions. The quantities displayed in a table should be clearly labeled with the units specified. Integrate the tables into the text so that the information presented flows smoothly for the reader.

Many reports will require a graphical presentation of data or calculated results. Each figure should be numbered and given a legend. Figures serve several purposes: to supplement or replace tables as a means of presenting results; to display relationships between two or more quantities; to find values needed in the calculation of results (see Chapter IIA). Some advantages of graphical display are that the relationship between two variables is shown more clearly by figures than by tables; deviations of individual results from expected trends are more readily apparent; "smoothing" of the data can be done better if it is necessary and appropriate; interpolation and extrapolation are easier. A disadvantage

When Eq. (6) is applied to our results, we obtain for ρ^* the following values:

Sample I: 5.310 g cm^{-3}

Sample II: 5.332 g cm^{-3} (7)

Average: 5.321 g cm^{-3}

According to Eq. (5) the uncertainty in ρ will depend on the uncertainty in each of five variables; however, the value of ρ_L is known to five significant figures and its uncertainty may be neglected in comparison to those in the other variables. We may take as reasonable 95% confidence limits for the weighings $\Delta W_1 = \Delta W_2 = 0.001$ g and $\Delta W_3 = 0.002$ g. The higher value for ΔW_3 includes the possible failure to attain an exact filling of the pycnometer with water. For ΔV we take 0.003 cm^3, the value given by the instructor, although such a value seems rather high. On the basis of these uncertainty values, it is clear that the major contributions to the limit of error $\Delta \rho$ are the uncertainty in V and, to a lesser extent, the uncertainty in the difference $W_2 - W_3$. The contribution to $\Delta \rho$ due to the uncertainty in the difference $W_2 - W_1$ is much less (since $W_2 - W_1$ is about 5.5 times larger than $W_2 - W_3$ and the uncertainty in $W_2 - W_1$ is less than that in $W_2 - W_3$), and it can be neglected in obtaining a good approximation for the

of figures is that graphical displays cannot always show the full precision of the results. For a report, it is normally advisable to include both figures and tables. Some advice on the preparation of figures is given on pp. 34–36. Indicate clearly the scales and units used. Error bars should be included when the uncertainties are greater than the size of the symbols enclosing the points.

An *error analysis* dealing with the uncertainty in the final result due to random errors in the measurements will normally be part of the Results section. The type of error analysis undertaken will depend a great deal on the nature of the experiment; see Chapters IIB and XXI for more details. The analysis given in the sample report is typical of a straightforward propagation-of-errors treatment. If a long and complex propagation-of-errors

value of $\Delta\rho$. With this simplification, the 95% confidence limit in ρ is given by

$$(\Delta\rho)^2/\rho^2 \approx [\rho_L^2(\Delta V)^2 + (\Delta W_2)^2 + (\Delta W_3)^2]/(\rho_L V + W_2 - W_3)^2 \quad (8)$$

The resulting error limits $\Delta\rho$ are ± 0.003 g cm^{-3} for both samples.

IV. Discussion

The values and limits of error obtained for the density of germanium at 25°C are

Sample I: 5.310 $\pm$ 0.003 g cm^{-3}

Sample II: 5.332 $\pm$ 0.003 g cm^{-3}

It is obvious that these two values deviate from each other by considerably more than the sum of the limits of error. Furthermore, this difference is even more disturbing when one considers the fact that any error in V will affect both densities in the same way and thus does not contribute to the uncertainty in the difference. Our results suggest

treatment is required, this should be given in an appendix and the resulting uncertainties should be stated, or shown as error bars on a plot, in the main body of the report. It is important to combine and simplify all expressions as much as possible in order to avoid obtaining unwieldy error equations. Since uncertainty values need not be calculated to better than about 15 percent accuracy, you should always try to find labor-saving approximations. Where the number of runs is so small that reliable limits of error cannot be deduced from statistical considerations, error limits must be assigned largely on the basis of experience and judgment. Try to make realistic assessments, avoiding excessive optimism (choosing error limits that are too small) or excessive pessimism (choosing limits that are unrealistically large). For a long and detailed report, a quantitative analysis of errors should always be derived and a numerical value of the uncertainty in the final result should be given. For a brief report, a qualitative discussion of the sources of error may suffice.

DISCUSSION

The Discussion is the most flexible section of the entire report, and you must depend heavily on your own judgment for the choice of topics for discussion. The final results of the experiment should be clearly presented. A comparison, often in a tabular or graphical

that the material examined may be somewhat inhomogeneous, yielding two samples of slightly different bulk densities. We suggest the possibility that voids inaccessible to the liquid are present in sample I. On this assumption, the greater confidence should be placed in the higher value.

The "literature" value given in the Handbook of Chemistry and Physics is 5.323 g cm^{-3} at room temperature [3], but no information about the precision or source of this value is given. A more accurate value can be calculated from the atomic mass (72.61 g mol^{-1} for Ge) and the volume of the unit cell as determined by x-ray crystallography. Germanium has a cubic crystal structure with 8 atoms per unit cell, and the cubic unit cell length is 5.65754 ± 0.00002 Angstrom at 25°C [4]. Thus the x-ray density is

$$\rho(\text{x-ray}) = (8 \times 72.61)/(6.022137\times10^{23})(5.65754\times10^{-8})^3$$
$$= 5.3266 \text{ g cm}^{-3} \text{ at } 25°C \qquad (9)$$

The uncertainty in this value is dominated by the uncertainty ±0.03 in the atomic mass of Ge [3]. The resulting uncertainty in the x-ray density is ±0.0022 g cm^{-3}.

Although these two literature values are in excellent agreement, the x-ray value is to be preferred since the temperature is specified and an uncertainty value is known. The value of ρ for sample I is lower than the x-ray value,

form, between these results and theoretical expectations or experimental values from the literature is usually appropriate. A comment should be made on any discrepancies with the accepted or expected values. In this sample discussion, comment is also made on "internal discrepancies," possible systematic errors and ways to reduce them, and the relative importance of various sources of random error. A brief suggestion should be made, if possible, for an improvement in the experimental method. Other possible topics include suitability of the method used compared with other methods, other applications of the method,

which could be explained by the presence of voids as suggested above. In contrast to this, the ρ value for sample II is slightly larger than the x-ray value (by an amount equal to the sum of the uncertainties in these two values). We have no good explanation for this difference, but it cannot be due to the presence of voids in the sample. One possible source of error is the volume V of the pycnometer. The stated value of 12.445 cm^3 has a rather large error limit of ±0.003 cm^3. Indeed, Appendix B shows that the largest contribution to the overall error in ρ comes from the uncertainty in V. No information is available about how V was determined or why it has such a large uncertainty. Our experimental precision indicates that a more reliable value for V could be obtained by measuring the weight of the pycnometer filled with water alone. Such a recalibration of the pycnometer volume would reduce the uncertainty in ρ due to random errors and might also shift the ρ values by a significant amount, which could perhaps improve the agreement between our ρ(sample II) and the x-ray value. However, a change in the value of V would not improve the agreement between the two samples.

Further work should be done to test the quality of the method (by studying a reference sample of well-known density) and the reproducibility of measurements on germanium samples prepared in different ways.

mention of any special circumstances or difficulties that might have influenced the results, discussion of any approximations that were made or could have been made, suggestions for changes or improvements in the calculations, mention of the theoretical significance of the result. At the end of several of the experiments in this book, there are questions that provide topics for discussion; however, you should usually go beyond these topics and include whatever other discussion you feel to be pertinent.

References

1. N. Bauer and S. Z. Lewin, "Determination of Density", in A. Weissberger and B. W. Rossiter (eds.), *Techniques of Chemistry*, Vol. I, Part IV, Chap. 2, esp. pp. 101–105, Wiley-Interscience, New York (1972).

2. D. R. Lide (ed.), *CRC Handbook of Chemistry and Physics*, 81th ed., p. 6-5, CRC Press, Boca Raton, FL (2000-2001).

3. *Ibid.*, p. 4-61.

4. A. Smakula and J. Kalnajs, Phys. Rev. **99**, 1737(1955); W. B. Pearson, *A Handbook of Lattice Spacings and Structures of Metals and Alloys*, Vol. 2, p. 971, Pergamon Press, Oxford (1967).

Appendix A: Data sheet

Appendix B: Sample calculations

REFERENCES

In citing a reference in the text of the report, use either a numerical superscript or a number in brackets to refer the reader to the appropriate entry in the reference list (e.g., . . . Smith[3] or . . . Smith [3]). The former style is used in this book, but the latter style is now more frequently used in scientific journals. An appropriate style for referring to a book is illustrated by entry 4. If the publisher's name is not well known, it should be given in full; if the city of publication is not well known, the state or country should also be given (see entry 2). The citation style for referring to a book containing chapters by several different authors is illustrated by entry 1. A recommended citation style for journal articles is shown in entry 4. Standard abbreviations for the titles of many important journals are given in Appendix E. For word-processed or handwritten reports, it is acceptable practice to underline the journal volume number (instead of using boldface type) and to use quotation marks to indicate book titles instead of italics.

APPENDICES

Important material that is so detailed that inclusion in the main body of the report would break the continuity of the text should be assembled in appendices. Examples are a long mathematical derivation, extensive tables of primary data (e.g., temperature–time values in calorimetry or composition–time values in chemical kinetics), extensive printout of computer data files, or a detailed listing of any nonstandard computer program.

Appendix A: Xerox copy of data for Exp. 13

24

Maria Smith   18 Sept. 2008
Partner: John Klein

Exp. 13: Density of Ge
Pycnometer #7 has $V = 12.445 \pm 0.003$ cm^3
Bath temperature $= 24.96°C$

	W_1 (empty)	W_2 (solid)	W_3 (solid + water)
Sample I			
Run 1	8.6313 g	42.0307 g	48.1749 g
Run 2	8.6308	42.0295 (0295 over 42.0259)	48.1715
Ave	—	42.0301 g	48.1732 g
Sample II			
Run 3	8.6316	45.8468	51.3055
Run 4	8.6299	45.8490	51.3017
Ave	—	45.8479 g	51.3036 g

Ave W_1 $= 8.6309$ g

estimate weighing uncertainty ± 0.001 g
temp variations of bath $\sim \pm 0.05°C$

    All experimental data must be recorded directly in a laboratory notebook along with any identifying numbers on special apparatus and all necessary calibration data; see pp. 7–9 for further details. For reports made in an undergraduate laboratory course, a carbon copy or a photocopy of all pertinent notebook pages should be included as an appendix to the report. The first page of this appendix should have a clear heading with the student's name, the name of any partner, and the dates on which the experimental work was performed.

Appendix B: Sample calculation for Exp. 13

The computation of ρ for germanium sample I is shown here in detail. Using Eq. (5) with weights in gram, V in cm^3, and ρ_L in $g\ cm^{-3}$, we obtain

$$\rho(I) = \frac{(0.99705)(42.0301 - 8.6309)}{(0.99707)(12.445) + 42.0301 - 48.1732}$$

$$= \frac{(0.99705)(33.3992)}{12.4085 - 6.1431} = \frac{33.3007}{6.2654}$$

$$= 5.3150\ g\ cm^{-3}$$

The correction for air buoyancy is given by Eq. (6), and we find

$$\rho^* = 5.3150 + 0.0012\left(1 - \frac{5.3151}{0.99707}\right)$$

$$= 5.3098\ gm\ cm^{-3}$$

for the corrected density ρ^* of sample I. The 95% confidence limit for the density of this sample is obtained from Eq. (8):

$$(\Delta\rho)^2 = \frac{(5.315)^2}{(6.265)^2}\left[(0.997)^2(.003)^2 + (.001)^2 + (.002)^2\right]$$

$$= 0.720\ [8.95 + 1 + 4] \times 10^{-6} = 10.0 \times 10^{-6}$$

$$\therefore \Delta\rho = 0.0032\ g\ cm^{-3}$$

A sample calculation should also be presented as an appendix to an undergraduate laboratory report. This appendix should show how one obtains the final results starting from the raw data. In general, the numbers used in the computations should have more significant figures than are justified by the precision of the final result, in order to avoid mathematical errors due to roundoff. Units should be included with each step of the calculation. Also specify the source of raw data used (e.g., run 5 on page 14 of notebook).

SPECIAL PROJECTS

In order to become a creative and independent research scientist, one must acquire a complex set of abilities. It is often necessary to invent new experimental methods or at least to adapt old ones to new needs. New apparatus must be designed, constructed, and fully tested. Most important of all, an intelligent procedure must be established for the use of this apparatus in making precise measurements.

Performing assigned experiments that are described in a book such as this one is the first step in developing research ability. Once some basic experience in carrying out physical measurements has been acquired, individually supervised experimental work on an original research problem is an excellent way to develop independence and experience with advanced research techniques in a specialized field. As a preparation for such research work, interested students should be encouraged to perform a special project in lieu of two or three regular experiments.

These special projects are intended to provide experience in choosing an interesting topic, in designing an experiment with the aid of literature references, in building apparatus, and in planning an appropriate experimental procedure. At least 20 hours of laboratory time should be available for carrying out such a project. Although certain limitations are imposed by the available time and equipment, challenging and feasible topics with a research flavor can be found in most branches of physical chemistry. Indeed, it is sometimes possible to make a significant start on an original research problem that will eventually lead to publishable results. The primary emphasis should, however, be placed on independent planning of the experimental work rather than on original proposals for new research. You will need to keep in mind that even well-planned original research takes much longer than 20 hours to complete.

The two (or possibly three) partners working on a given special project should plan the experiment together, starting 3 or 4 weeks in advance, and should discuss their ideas frequently with an instructor or a teaching assistant. All work done in the laboratory should be supervised by an experienced research worker in order to avoid safety hazards.

Some of the projects done in the laboratory courses at MIT and OSU are listed below as examples of the kind of work that might be attempted.

Rotational Raman spectra of N_2 and O_2

Resonance fluorescence spectrum of Br_2

Vibrational–rotational spectra of CD_3H and CH_3D

Spectrophotometric study of stability of metal ion–EDTA complexes

Kinetics of the $H_2 + I_2 = 2HI$ reaction in the gas phase

Weak-acid catalysis of BH_4^- decomposition

Photochemistry of the cis–trans azobenzene interconversion

Isotope effect on reaction-rate constants

Susceptibility of a paramagnetic solid as a function of temperature

Dielectric constant of polypropylene glycol

Determination of polymer molar masses by light scattering

X-ray study of short-range order in liquid mercury

Fluorescence and phosphorescence of complex ions in solution

Infrared study of hydrogen bonding of CH_3OD with various solvents

Light scattering near the critical point in ethane

Polanyi dilute flame reaction; e.g., $K + Br_2$

EPR study of gas-phase hydrogen and deuterium atoms

Photodissociation of NO_2

Fluorescence quenching of excited K atoms

Shock-tube kinetics: recombination of I atoms

Dielectric dispersion in high-polymer solutions

Mass spectrometry of fragmentation patterns of $CHCl_3$ and CH_2Cl_2

Many of these projects were quite ambitious and required hard work and enthusiasm on the part of both students and staff. Not all were completely successful in terms of precise numerical results, but each one was instructive and enjoyable. Frequently they resulted in an excellent scientific rapport between the students and the instructing staff.

SAFETY

It has already been emphasized that safe laboratory procedures require thoughtful awareness on the part of both students and instructors. This is especially important in the planning and execution of special projects, where new procedures need to be developed and often modified as the work progresses. Appendix C on safety hazards and safety equipment should be read before beginning a course of experimental work in physical chemistry and reviewed carefully before beginning any special project.

ETHICS

The proper handling of data acquired in the laboratory and the independent analysis of such data as well as the independent preparation of laboratory reports require adherence to a high ethical standard. Some ethical issues that arise in physical chemistry laboratory are described in Appendix H.

REFERENCES

1. A convenient source of information about the commercial availability of some 200,000 chemicals is M. Desing (ed.), *Chem Sources—USA,* published annually by Chemical Sources International, Clemson, SC 29633 in book format and as a CD-ROM. There is also a *Chem Sources—Online* search engine; see www.chemsources.com. Many suppliers of chemicals maintain online catalogues that can be accessed on the Web.
2. A. K. Furr (ed.), *CRC Handbook of Laboratory Safety,* 3d ed., CRC Press, Boca Raton, FL (1989); *Prudent Practices in the Laboratory: Handling and Disposal of Chemicals,* National Academy Press, Washington, DC (1995).
3. H. M. Kanare, *Writing the Laboratory Notebook,* American Chemical Society, Washington, DC (1985).
4. R. N. Goldberg and R. D. Wier, *Pure & Appl. Chem.* **64,** 1545 (1992).
5. S. M. Bachrach (ed.), *The Internet: A Guide for Chemists,* American Chemical Society, Washington, DC (1996).
6. I. Mills, T. Cvitas, K. Homann, N. Kallay, and K. Kuchitsu, *Quantities, Units and Symbols in Physical Chemistry,* 2d ed., International Union of Pure and Applied Chemistry, Blackwell, Oxford (1993).

GENERAL READING

Chemistry Writing Guide; available online at http://www.chem.orst.edu, which is the Oregon State University web page.

R. E. Maizell, *How to Find Chemical Information,* 3d ed., Wiley-Interscience, New York (1998).

D. C. Montgomery, *Design and Analysis of Experiments,* 4th ed., Wiley, New York (1996).

G. D. Wiggins, *Chemical Information Sources,* McGraw-Hill, New York (1990).

E. B. Wilson, Jr., *An Introduction to Scientific Research,* Dover, Mineola, NY (1998).

II

Treatment of Experimental Data

The usual objective of performing an experiment in physical chemistry is to obtain one or more numerical results. Between the recording of measured values and the reporting of results, there are processes of numerical calculations, some of which may involve averaging or smoothing the measured values but most of which involve the application of formulas derived from theory. Part A of this chapter is devoted to a discussion of general techniques for carrying out such calculations. The use of computer spreadsheet programs such as Excel, Quattro Pro, and Lotus 1-2-3 is particularly advantageous for data analysis. A discussion of the use of spreadsheets is given in Chapter III.

Our concern with the treatment of experimental data does not end when we have obtained a numerical result for the quantity of interest. We must also answer the question: "How good is the numerical result?" Without an answer to this question, the numerical result may be next to useless. The expression of how "good" the result may be is usually couched in terms of its *accuracy,* i.e., a statement of the degree of the uncertainty of the result. A related question, often to be asked before the experiment is begun, is: "How good does the result need to be?" The answer to this question may influence important decisions as to the experimental design, equipment, and degree of effort required to achieve the desired accuracy.

Indeed, it may be useful to discuss the matter also in economic terms. The "economic value" of the numerical result of an experiment often depends on its degree of accuracy. To claim too high an accuracy through ignorance, carelessness, or self-deception is to cheat the "consumer" who makes decisions on the basis of this result. To claim too low an accuracy through overconservatism or intellectual laziness lessens the value of the result to the "consumer" and wastes resources that have been employed to achieve the accuracy that could rightfully have been claimed. These issues of errors and the accuracy of a result are dealt with in Part B of this chapter.

A. Calculations and Presentation of Data

This part of Chapter II is concerned with the mathematical processing of data to obtain the desired quantitative results. At an initial stage, numerical methods are intrinsic, since data consist of a set of observed "points"—usually pairs x_i, y_i of a dependent quantity y_i measured for a specified value of the independent variable x_i. Of course, y could be a function of two variables x and z, in which case one often holds z constant and measures

y while varying x, then changes z to a new constant value and repeats the process. An example would be the measurement of the pressure of a gas as a function of volume at a series of constant temperatures. One can use numerical techniques to fit the data with a functional form: best line $y(x)$ through the points in an x–y plot or best surface $y(x, z)$ in a three-dimensional computer-graphics display. Then further processing of the data can be done with analytic methods. Or, in other cases, the data processing can be carried out completely with numerical methods.

SIGNIFICANT FIGURES

Many numerical data in the literature and in everyday use carry no explicit statement of uncertainty, such as an estimated standard deviation S or 95 percent confidence limit Δ. The definition and technique for evaluation of these error quantities are given in Part B of this chapter. Here we are concerned with more qualitative principles. Some idea of the uncertainty in the numerical value of a measurement ("reading") should be conveyed by the number of digits employed to express that numerical value. The digits that are essential to conveying the numerical value to its full accuracy or precision are called *significant figures*. Thus the numbers 4521 and 6.784 each have four significant figures, provided all of the digits are meaningful. So also do 0.006784 and 4,521,000, since the leading and trailing zeros generally determine only the magnitude of the number and have nothing to do with the relative precision. In the latter case, however, there is some ambiguity because one or more of the trailing zeros may be significant. The ambiguity may be removed by expressing the number in exponential notation: 4.521×10^6 or 4.5210×10^6, as the case may be.

This discussion applies to numerical values of variables whose possible range constitutes a continuum; in computer terminology these are called "real" numbers, as opposed to integers. Integers should be identified as such, and all of the digits necessary to express them should be given.

The use of too few digits in reporting numerical values robs the user of valuable information. The usual fault, however, is the reckless use of too many digits, which conveys a false sense of the accuracy or precision of the numerical value.

In reporting numerical values, certain rules regarding significant figures should be followed.

1. Numerical values are considered to be uncertain in the last digit by ± 3 or more, and perhaps slightly uncertain in the next-to-last digit. Ordinarily the next-to-last digit should not be uncertain by more than ± 2.

2. In rounding off numbers, (*a*) increase the last retained digit by 1 if the leftmost digit to be dropped is more than 5, or is 5 followed by any nonzero digits:

$$3.457 \rightarrow 3.46 \qquad 52.6502 \rightarrow 52.7$$

(*b*) leave the last retained digit unchanged if the leftmost digit to be dropped is less than 5:

$$0.34648 \rightarrow 0.346$$

(*c*) if the leftmost digit to be dropped is 5 followed by no digits except zero, then increase the last retained digit by 1 if it is odd, and leave it unchanged if it is even:

$$73.135 \rightarrow 73.14 \qquad 48.725 \rightarrow 48.72$$

For a discussion of the retention of digits in numbers during computation, see the next section.

PRECISION OF CALCULATIONS

The rules given above for determining the number of significant figures to be retained in a numerical result should be kept in mind. However, these rules do not apply strictly to the retention of digits in intermediate values obtained during a computation. It is necessary to retain *at least* one additional digit and recommended to retain two or more additional digits during computations to avoid unnecessary round-off errors in the final results. There are generally no problems with precision when the calculation is carried out on a computer; for example, spreadsheet programs typically have 15 to 17 digits and display power limits of 10^{-99} to 10^{+99}, although a power range of ± 308 is used for internal steps in a calculation. One must, of course, remember to adjust properly the number of significant figures when reporting the final results.

Computations, whether by hand calculator or by computer, should be done with awareness of certain principles concerning the effects of arithmetic operations. These follow directly from the principles underlying the propagation of errors, discussed in Part B of this chapter.

When a number of numerical values are *added,* the precision of the result can be no greater than that of the least precise numerical value involved. Thus, generally, the number of decimal places in the result should be the same as the number of decimal places in the component with the fewest:

$$32.7$$
$$+ \ \ 3.62$$
$$+ \ 10.008$$
$$46.328 \ \rightarrow \ 46.3$$

When the *difference* between two numbers yields a result that is relatively small, that difference not only has a precision limited to that of the less precise number but has also a relative precision much less than either number:

$$673.425$$
$$- \ 672.91$$
$$0.515 \ \rightarrow \ 0.52$$

The less precise of the two numbers has a relative precision of about 1 in 67,000 (0.0015 percent), while the difference has a relative precision of about 1 in 50 (2 percent). The effect can be particularly devastating in an intermediate calculation on a computer, where the user of the computer may be blissfully unaware of the loss of relative precision.

In *multiplication* and *division,* the relative precision of the result can be no greater than that of the least (relatively) precise number involved. Thus the number of significant figures in the result should be approximately that in the component having the fewest of them. In judging the appropriate number of significant figures in the result, bear in mind that the relative precision does not depend on the number of significant figures alone; 999, with three significant figures, is as precise as 1001, with four. When in doubt, employ the larger number of significant figures. For example,

$$346 \times 121 \times 900.0 = 37,679,400 \ \rightarrow \ 3.77 \times 10^7$$

The uncertainty in 121, the least relatively precise figure, is about 0.8 percent. One in 377 is about 0.27 percent, which implies a greater relative precision, but if the result is rounded to 3.8×10^7, with a relative precision of 2.6 percent, too much significance will be lost. The principles are no different when division is involved:

$$5.44 \times \frac{0.132}{8.092} = 0.0887395 \ \rightarrow \ 0.0887 \quad \text{or} \quad 0.089$$

The uncertainty in 0.132 is about 0.7 percent, that in 0.089 is about 1.1 percent, implying a small but tolerable loss of precision. If this is an intermediate calculation, keep the extra digit.

In the case of a calculation with many numbers, it should be kept in mind that round-off errors may accumulate to the extent that they reduce the precision significantly.

The above-mentioned considerations are not a substitute for a propagation-of-errors treatment to determine the uncertainty in the final result, especially when the calculations have been carried out on a computer and the result is given with an artificially large number of digits.

No matter what computational procedure has been used, there must be reliable assurance that mistakes in arithmetic and in the transcription of numerical data have not been made. Normally, an electronic calculator or computer can be trusted not to make mistakes in arithmetic, but mistakes in the manual transcription of input numbers and output numbers are all too easy to make. A printed record from a computer, which can be checked and double-checked, can provide satisfactory assurance against mistakes of this kind.

The best way to check a calculation is to repeat it, preferably in a different way, to the same precision. When an experiment has been performed by two persons as partners, the calculations should generally be done by each person independently, with cross-checking of intermediate and final numerical results.

ANALYTICAL METHODS

Much of the mathematical analysis required in physical chemistry can be handled by analytical methods. Throughout this book and in all physical chemistry textbooks, a variety of calculus techniques are used freely: differentiation and integration of functions of several variables; solution of ordinary and partial differential equations, including eigenvalue problems; some integral equations, mostly linear. There is occasional use of other tools such as vectors and vector analysis, coordinate transformations, matrices, determinants, and Fourier methods. Discussion of all these topics will be found in calculus textbooks and in other standard mathematical texts.

Accounts of the sort of applied mathematics useful in physics and chemistry can also be found in specialized monographs such as those listed below.

M. L. Boas, *Mathematical Methods in the Physical Sciences,* 2d ed., Wiley, New York (1983).

M. E. Starzak, *Mathematical Methods in Chemistry and Physics,* Kluwer Academic, Norwell, MA (1989).

F. Steiner, *The Chemistry Maths Book,* Oxford Univ. Press, Oxford (1996).

P. Tebbutt, *Basic Mathematics for Chemists,* 2d ed., Wiley, New York (1998).

G. Strang, *Introduction to Applied Mathematics,* Wellesley-Cambridge Press, Wellesley, MA 02482 (1986).

F. W. Byron, Jr., and R. W. Fuller, *Mathematics of Classical and Quantum Physics,* Dover, Mineola, NY (1992).

The first four of these are reasonably detailed presentations with a strongly applied flavor, whereas Strang and especially Byron and Fuller are on quite an advanced level.

It must be kept in mind that experimental situations do not yield directly an analytical function for the measured observable. Thus numerical methods are needed at least to establish the parameters of some theoretical form or to find some empirical approximation function. After a functional form $y = f(x)$ is found, one can often use analytical methods to evaluate quantities such as dy/dx, $\int y\, dx$ or to find the roots of $f(x) = 0$. However, some functions $f(x)$ are difficult to integrate analytically, and the roots of some equations are difficult to find. In such cases, numerical methods are very useful, especially since they can be implemented on computers with the use of very powerful algorithms.

NUMERICAL METHODS

Numerical calculations may be performed by any method that does not introduce round-off errors that are significant in comparison with the experimental errors.[1-5] A hand calculator will suffice for most of the experiments in this book. For lengthy or repetitive calculations, the use of a spreadsheet program is strongly recommended (see Chapter III). In some cases it may be advantageous to write special programs based on the mathematical procedures described in each experiment. As an aid to this, many useful algorithmic routines are available in Refs. 2–4. Reference 4 is particularly good because it contains 300 algorithms with a numerical-analysis text explaining the basic methods used. It is wise to carry out one typical calculation with a hand calculator in order to test any new program, and then if trouble is observed you can debug the program before using it.

Numerical methods can be applied to discrete (finite) data sets in order to carry out such procedures as differentiation, integration, solution of algebraic and differential equations, and data smoothing. Analytical methods, which deal with continuous functions, are exact or at least capable of being carried out to any arbitrary precision. Numerical methods applied to experimental data are necessarily approximate, being limited by the finite number of data points employed and their precision.

LEAST SQUARES

The *method of least squares* is a very powerful and widely used technique, and a detailed discussion of how to use it properly is given in Chapter XXI. Presented here is a brief commentary on this method.

Given a set of experimental data y_i measured at a series of values x_i for the independent variable x, it is often desired to construct a function $y(x)$ to represent them. If there is reason to believe that this function should conform to some analytic form depending on a few parameters, such as $y = (a/x) + b$, these parameters (a and b in the given case) can be adjusted to give the "best fit" to the data by using the method of least squares. The analytic form may be an equation arising from a theoretical model, which in principle may be exact. Alternatively, one may use some general "curve-fitting" function such as a polynomial that has no theoretical relationship to the experiment but will provide a useful empirical representation.

In general, the number of experimental points m is larger (often much larger) than the number of adjustable parameters. The resulting function is not an exact fit to each point but represents a best overall fit in the sense that the sum of the squares of the deviations $[y_i(\text{obs}) - y_{\text{calc}}(x_i)]$ is a minimum, hence the name. The fitting function $y(x)$ obtained by least squares can be differentiated, integrated, set equal to zero and solved to find roots, etc., by analytical methods. Thus there is no need for the application of other numerical methods unless the function is very awkward to handle analytically in closed form.

A particularly easy type of least-squares analysis called *multiple linear regression* is possible for fitting data with a low-order polynomial, and this technique can be used for many of the experiments in this book. The use of spreadsheet programs, as discussed in Chapter III, is strongly recommended in such cases. In the case of more complicated nonlinear fitting procedures, other techniques are described in Chapter XXI.

In some situations, no theoretically motivated analytic form $y = f(x)$ is apparent, and low-order series approximations do not provide a good fit. You should then be cautious about using high-order empirical series fitting forms, since they may yield very poor extrapolations, interpolations, and derivatives. In these cases, direct numerical (or graphical) methods are available.

OTHER METHODS

Many other numerical methods are available, and some of these are described in Appendix F: smoothing, series approximations, differentiation, integration, and root finding. It should be noted that spreadsheet programs can be utilized easily for all of these numerical operations. For example, Excel and Quattro Pro contain standard operations for linear regression, matrix multiplication and inversion, and also have procedures that will solve expressions such as $y = 10x^3 \sin x$ for the value of x corresponding to a given input value of y. These and other applications of spreadsheets are discussed in Chapter III.

Numerical methods beyond the scope of this book, such as the solution of linear equations, minimum–maximum problems, and Fourier transform methods, are also of value in some physical chemistry experiments. Algorithms for handling them with a computer or in some cases complete computer programs for the given task are described elsewhere.[2–4]

GRAPHS AND GRAPHICAL METHODS

By a *graph* we mean a representation of numerical values or functions by the positions of points and lines on a two-dimensional surface. A graph is inherently more limited in precision than a table of numerical values or an analytic equation, but it can contribute a "feel" for the behavior of data and functions that numerical tables and equations cannot. A graph reveals much more clearly such features as linearity or nonlinearity, maxima and minima, points of inflection, etc. Graphical methods of smoothing data and of differentiation and integration are sometimes easier than numerical methods. Graphs and graphical methods suffer, of course, from the limitations of a two-dimensional surface. Thus, normally, a plotted point has only two degrees of freedom, which we assume here to be represented by one independent variable x (increasing from left to right) and one dependent variable y (increasing from bottom to top). The word "normally" is required in the above sentence because three-dimensional plots, representing for example points with coordinates x, y, and z, can be represented in oblique projection on a two-dimensional surface by computer techniques.

Wherever possible, the data from an experiment should be plotted at an early stage, even if numerical methods will be used subsequently for greater accuracy. This is particularly true when functions are to be fitted by least-squares or other methods, since a graph may make evident special problems or requirements that might otherwise be missed.

COMPUTER-GENERATED PLOTS

In most cases, one can take advantage of the ability of computer spreadsheet programs to generate graphs. These may be viewed on the computer monitor for inspection and editing and then produced on external graphic devices such as ink-jet or laser printers. Since computer graphs can be produced quickly, it may be worthwhile to make several tries to obtain the best results. The first plot may be a "default plot" to see the overall appearance of the plotted data; then user-chosen scales, symbols, aspect ratio, and labels may be introduced as described below to create the final figure. If the computer output uses generic labels such as x and y, be sure to relabel with appropriate variables.

In preparing final plots, whether computer generated or prepared by hand, there are several important features to consider.

1. *Design an Overall Layout.* In planning the overall dimensions of the plot, determine the maximum and minimum values of *x* and *y* (remember, it is *not* necessary that they both be zero at the lower left corner!) so as to fit the points, lines, and curves into the available area. Be sure to make adequate allowance of space for borders all around, scale numbers and scale labels, at least at the bottom and at the left, and a legend telling what the graph is about. If the report is to be bound along the left edge, one must leave an extra wide left margin.

2. *Use Clear Scale Labels.* Choose suitable scale divisions with longer tick marks used for the major values along a scale that will be labeled with the appropriate numerical values. Under the scale numbers for the independent variable, enter the scale label, stating the quantity that is varying and its units: e.g.,

$$T(\text{K})$$

for absolute temperature in kelvin. To avoid scale numbers that are too large or too small for convenient use, multiply the quantity by a power of 10: e.g.,

$$\rho \times 10^4 (\text{g cm}^{-3}) \text{ or } \rho(\text{units of } 10^{-4} \text{ g cm}^{-3})$$

for the density of a gas. This is equivalent to multiplying all the scale numbers by 10^{-4} but is much more convenient. Similarly, enter the scale label for the dependent variable along the scale on the left side.

3. *Use Distinctive Symbols.* If data have been taken on two distinct runs—say a series of measurements on heating and then on subsequent cooling or merely duplicate runs made with the same protocol—use different symbols to denote the separate runs. If approximate estimates of the uncertainty in the variables is available, then add error bars to each point.

4. *Distinguish Smooth Curves.* If your plot contains several empirical smooth curves or several theoretical curves, it is wise to distinguish them by using dashed and dash–dot curves as well as solid lines.

5. *Add a Legend.* Somewhere on the figure (if possible, at the bottom) enter a legend with an identifying figure number. The legend should state the contents of the figure, identify the symbols and line types used, and provide any needed information that will not be provided in the text of the document in which the figure will appear.

For examples of well-prepared figures representing experimental data, consult current scientific journals such as those listed in Appendix E.

MANUAL PREPARATION OF GRAPHS

Although the use of computer-generated plots is extremely widespread, it is worth noting that good quality plots of data can be produced by hand using commercial graph paper. In addition to the five design considerations listed in the preceding section, two other issues play a role when producing figures by hand.

6. *Choose the Graph Paper and Indicate the Axes.* Graph paper is available with linear, semilog, and log–log scales. Most good graph paper is ruled in soft colors that do not distract the viewer from the plotted points and lines. Indeed, the presence of a lined grid makes it very easy to "read" values from the plot (say interpolate between points to find *y* at some desired value of *x*). In choosing the horizontal and vertical scales to be used, attention should be paid to making a commensurate choice where a major scale division corresponds to an attractive integral number of the smallest divisions.

7. *Plot the Points and Draw the Lines and Curves.* Using a sharp, hard pencil, plot the experimental points as small dots, as accurately as possible. Draw small circles (or squares, triangles, etc.) of uniform size (2 or 3 mm) around the points in ink to give

them greater visibility. If it is desired to draw a straight line "by eye" to provide a linear fit to the plotted points, draw a faint pencil test line with a good straightedge, such as a transparent ruler or draftsman's triangle. When you have achieved the best possible visual fit, go over the line with a pen, but do not allow the ink line to cross the circles or other symbols drawn about the experimental points. If a theoretical curve based on an analytic function is to be included, use a pencil to plot the value of the function at as many points as are reasonably needed; do not draw symbols around these plotted values. Carefully draw a smooth curve through the points, lightly at first and then "heavy it up".

GRAPHICAL DIFFERENTIATION AND INTEGRATION

The first derivative of a plotted function $y(x)$ with respect to x is given by the slope of a line drawn tangent to the curve at the point x, y concerned. Draw the tangent line long enough to maximize the accuracy of the slope determination. If two points x_1, y_1 and x_2, y_2 are accurately established at the two ends of the tangent line, the slope and thus the derivative is given by

$$y'(x) = \frac{y_2 - y_1}{x_2 - x_1} \tag{1}$$

One way to draw a tangent line is to use a compass to strike off arcs intersecting the curve on both sides and then draw a chord through the intersection points. The tangent can be drawn parallel to the chord if the curvature is uniform. If it is not, construct a second chord at a different distance in the same way and extend both lines to an intersection. Draw the (approximate) tangent line through this intersection point (see Fig. 1).

A definite integral, such as $\int_a^b y\,dx$, can be determined by measuring the area under the curve between the desired limits (see Fig. 1). The area can be measured by cutting out the area concerned with a pair of scissors and weighing it (having also weighed a rectangle of known area) or by approximating it as well as possible by a bar graph and adding the areas of the bars. Whatever the method used, the area to be measured should be minimized as much as possible; there is no need to include the rectangular area below $a'b'$ in Fig. 1, for instance, since it is given exactly by $(b - a)(a' - a)$.

FIGURE 1

An illustration of two graphical procedures: (*a*) the construction of a tangent by the method of chords; (*b*) the evaluation of a definite integral by approximating the curve by a bar graph. The top of each bar is drawn so that the two small areas thereby defined (e.g., the shaded and stippled areas on the first bar) appear equal.

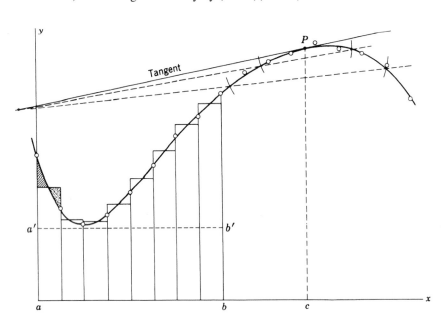

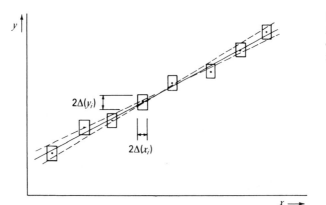

FIGURE 2
A graphical method of determining the limit of error in a slope.

GRAPHICAL UNCERTAINTIES

It frequently happens that an intermediate or final result of the calculations in a given experiment is obtained from the slope or intercept of a straight line on a graph—say a plot of y against x. In such a case it is desirable to evaluate the uncertainty in the slope or in the position of the intercept. A rough procedure for doing this is based on drawing a rectangle with width $2\Delta(x_i)$ and height $2\Delta(y_i)$ around each experimental point (x_i, y_i), with the point at its center. The assignments of the limits of error $\Delta(x_i)$ and $\Delta(y_i)$ are made as described in Part B. The significance of the rectangle is that any point contained in it represents a possible position of the "true" point (x_i, y_i) and all points outside are ruled out as possible positions. Having already drawn the best straight line through the experimental points, and having derived from this line the slope or intercept, draw two other (dashed) lines representing maximum and minimum values of the slope or intercept, consistent with the requirement that both lines pass through every rectangle, as shown in Fig. 2. Where there are a dozen or more experimental points, it may be justifiable to neglect partially or completely one or more obviously "bad" points in drawing the original straight line and the limiting lines, provided that good judgment is exercised. The difference between the two slopes or intercepts of the limiting lines can be taken as an estimate of twice the limit of error in the slope or intercept of the best straight line.

Where the number of points is sufficiently large, the limits of error of the position of plotted points can be inferred from their scatter. Thus an upper bound and a lower bound can be drawn, and the lines of limiting slope drawn so as to lie within these bounds. Since the theory of least squares can be applied not only to yield the equation for the best straight line but also to estimate the uncertainties in the parameters entering into the equation (see Chapter XXI), such graphical methods are justifiable only for rough estimates. In either case, the possibility of systematic error should be kept in mind.

EXERCISES

1. Each of the following quantities is uncertain (95 percent confidence) by 0.20 percent. Express the quantities with the proper number of significant figures and a specification of the (absolute) uncertainty.

   (*a*) 2.05675 cm

   (*b*) 388.982 K

   (*c*) 1.21335×10^3 Torr

2. The viscosity η of pure ethanol in centipoise was determined at eight different temperatures at 1 atm pressure with the following results:

t, °C	0	10	20	30	40	50	60	70
η, cP	1.78	1.45	1.17	0.98	0.83	0.71	0.60	0.49

Prepare a figure to display these data with a smooth curve drawn through the points, and then determine the temperature coefficient of viscosity $(\partial\eta/\partial t)_p$ at 25.0°C. It is recommended that this be done in two ways to allow a comparison: First, draw the smooth curve by eye and determine the temperature coefficient graphically; second, represent the data with an appropriate least-squares fitting function and differentiate analytically.

3. Bridgman obtained the following volume ratios V/V_0 for methanol, where $V_0 = 39.6$ cm^3 is the molar volume at 0°C and a pressure of 1 kg/cm^2.

p (kg cm^{-2})	1	500	1000	2000	3000	4000	5000
$V(20°C)/V_0$	1.0238	0.9823	0.9530	0.9087	0.8792	0.8551	0.8354
$V(50°C)/V_0$	1.0610	1.0096	0.9763	0.9271	0.8947	0.8687	0.8476

Use these data to evaluate the entropy change ΔS (in units of J K^{-1} mol^{-1}) when 1 mol of methanol at 35°C and 1 kg cm^{-2} is compressed isothermally to 5000 kg cm^{-2}. Recall that $\Delta S = -\int(\partial V/\partial T)_p\, dp \approx -\int(\Delta V/\Delta T)_p\, dp$. Thus you should determine values of $\Delta V/\Delta T$ and carry out the appropriate numerical integration.

B. Uncertainties in Data and Results

This part of the chapter is concerned with the evaluation of uncertainties in data and in calculated results. The concepts of random errors/precision and systematic errors/accuracy are discussed. Statistical theory for assessing random errors in finite data sets is summarized. Perhaps the most important topic is the propagation of errors, which shows how the error in an overall calculated result can be obtained from known or estimated errors in the input data. Examples are given throughout the text, the headings of key sections are marked by an asterisk, and a convenient summary is given at the end.

ERRORS

*RANDOM ERRORS AND PRECISION

We will assume that the numerical result to which our discussion applies is obtained with an instrument that measures a physical quantity for which the a priori range of possible values constitutes a continuum. This may be a continuous-reading instrument with a scale divided with numbered marks; customarily the scale reading is estimated to some fraction of the smallest interval ("scale division") between marks, typically one-fifth or one-tenth. The reading of the scale by a person with normal visual acuity is subject to a possible error (typically one- or two-tenths of a scale division). Even if the observer is capable of reading the scale with great precision, the measurement at any particular moment is subject to unpredictable, random fluctuations of the environmental conditions (temperature variations, voltage fluctuations in the electrical supply, electrical noise in the equipment, mechanical vibrations, etc.), which affect the instrumental response. Thus repetitions of the measurement generally yield different readings, distributed at random over a small range. The same may apply to a digital instrument when the smallest increment in the numerical output, corresponding to ± 1 in the final digit, is smaller than the differences expected between different readings of the same quantity. If the smallest increment of numerical output is *not* smaller than expected random fluctuations, the instrument itself is making a contribution to measurement error because of roundoff and is perhaps not suitable for its intended purpose. Whatever the measuring instrument, we may say that the measurements are subject to *random error.* A statement of the *precision* of a measurement concerns its reproducibility when the measurement is made repeatedly with the same instrumentation and is therefore an expression of the uncertainty due to random error.

Faced with the inevitability of random error, the observer customarily makes several measurements of a given physical quantity and then averages them to obtain a value that is expected to be more reliable than any single measurement. The average or mean of N measurements of an experimental variable x is

$$\bar{x} = \frac{1}{N} \sum_{i=1}^{N} x_i \tag{2}$$

where x_i is the result of the ith measurement.

The observer also customarily reports in some manner the scatter of the measurements that go into this expression in order to give an indication of the precision. A crude way is to state the *range* of measured values:

$$R = x_{\text{largest}} - x_{\text{smallest}} \tag{3}$$

For several reasons this does not give a very clear measure of the precision of the individual measurements but can be used as the basis of quick estimates of more sophisticated measures. Very commonly the average deviation is cited:

$$\text{ave. dev.} = \frac{1}{N} \sum_{i=1}^{N} |x_i - \bar{x}| \tag{4}$$

Traditionally this has been cherished for its arithmetic simplicity, a virtue of declining value in this age of computers. Its usefulness for quantitative comparisons is limited because under given experimental conditions its value depends sensitively on sample size (the value for a pair of measurements being, on the average, about 30 percent smaller than the value calculated with a large group of measurements).

A measure of precision that is unbiased by sample size is the *variance,* denoted by S^2 and defined by

$$S^2 \equiv \frac{1}{N-1} \sum_{i=1}^{N} (x_i - \bar{x})^2 \tag{5a}$$

This quantity has the property of additivity so that estimates from several sources of random error may be combined, and it is useful in probability calculations. An alternative form of S^2, algebraically equivalent and often more convenient for use with a calculator or computer, is

$$S^2 = \frac{1}{N-1}(\Sigma x_i^2 - N\bar{x}^2) = \frac{N}{N-1}(\overline{x^2} - \bar{x}^2) \tag{5b}$$

Particular attention should be paid to the divisor $(N-1)$, which is known as the number of "degrees of freedom" and is equal to the number of *independent* data on which the calculation of S^2 is based. The N values of $x_i - \bar{x}$ are at this stage of the game not all independent; one degree of freedom has been used up in calculating the mean $\bar{x}$. That is, one of the N deviations, say the Nth one $(x_N - \bar{x})$, is not an independent variable since it can be calculated from the other deviations:

$$(x_N - \bar{x}) = -\sum_{i=1}^{N-1}(x_i - \bar{x}) \tag{6}$$

The difference between a factor of $N - 1$ or N in the denominators of Eqs. (5a) and (5b) is trivial if one has 100 measurements but becomes very important when N is small.

The square root of the variance,

$$S = \frac{1}{\sqrt{N-1}} \left[\sum_{i=1}^{N} (x_i - \bar{x})^2 \right]^{1/2} \tag{7}$$

is often called the *estimated standard deviation*. The significance of this name will become clear later. This parameter is widely used to indicate the precision of individual measurements. More important is the precision of the mean of the measurements. It will be shown later that the *estimated standard deviation of the mean* of N values is

$$S_m = \frac{S}{\sqrt{N}} = \frac{1}{\sqrt{N(N-1)}} \left[\sum_{i=1}^{N} (x_i - \bar{x})^2 \right]^{1/2} \tag{8}$$

It is apparent that the precision of the mean can be increased by increasing the number of individual measurements. The gain in going from $N = 2$ to $N = 10$ is quite dramatic; that in going from $N = 10$ to $N = 20$ is less so; and going from $N = 90$ to $N = 100$ results in only a small increase in precision. Thus, in principle, the random error can be reduced to as small an amount as desired by repeated measurements. However, in practice, there comes rapidly a "point of diminishing returns" in continuing to repeat measurements. It is rare to make more than about 10 measurements of the same physical quantity, and often the number of measurements is considerably smaller.

SYSTEMATIC ERRORS AND ACCURACY

There is an important reason why excessive efforts should not be made to reduce the magnitude of random errors by making a very large number of measurements of the same quantity. This reason is that measurements are also subject to other kinds of error, chiefly *systematic errors,* the magnitude of which can all too easily exceed that of the random errors. A systematic error is one that cannot be reduced or eliminated by any number of repetitions of a measurement because it is inherent in the method, the instrumentation, or occasionally the interpretation of data. While the precision of a result, as already stated, is an expression of uncertainty due to random error, the *accuracy* of a result is an expression of *overall uncertainty including that due to systematic error.* Accuracy is very much more difficult to assess than precision.

Examples of systematic error are a calibration error in the instrument, failure to establish properly a "zero" reading of the instrument scale, improper graduation or alignment of an instrumental scale, uncompensated instrumental drift, leakage of material (e.g., of gas in a pressure or vacuum system) or of electricity (in an electrical circuit), and incomplete fulfillment of necessary experimental conditions (e.g., incomplete reaction in a calorimeter, incomplete dehydration of a sample prior to weighing). An example of another kind is faulty theoretical treatment of the results of the measurements to obtain the desired result, perhaps through a faulty approximation in the phenomenological theory involved.

A particular kind of systematic error might be called "personal error" because it results from subjective judgments or personal idiosyncrasies on the part of the observer. Examples are consistent parallax errors in reading scales and setting crosshairs, excessive reaction times in actuating a stopwatch or electronic timer, strong number prejudices that influence the reading of scales (tendency to favor 0s and 5s, or superstition in favor of 7 and against 13), and arbitrariness in rejecting or not rejecting readings that deviate somewhat from the mean of the others. Such errors can be particularly insidious if the observer has a personal stake in the outcome of the experiment (whether to confirm a theory and win a Nobel Prize or to get the result thought to be expected by the instructor and win an

"A" on the experiment). Persons of the highest character may not be altogether immune from this tendency, since biases are often subconscious.

Systematic errors are difficult to eliminate because they do not make their presence known as random errors do. They must be painstakingly sought out and eliminated through careful analysis of the entire experiment and of all assumptions on which the experimental method is founded. Some will be relatively easy to find, such as calibration and alignment errors; others, such as errors in the theory or the intervention of totally unexpected phenomena, can defy all attempts by ordinary means to identify them. All too often, important systematic errors are never found at all, and without doubt many are imbedded in the permanent scientific literature. Fortunately, some become apparent when new results, obtained by other methods, depart from the old ones by much more than the stated limits of error. As an example, R. A. Millikan's final value of Avogadro's number N_0, which he calculated as the ratio of the accurately known Faraday constant to his value of the electronic charge from the oil-drop experiment, was $(6.064 \pm 0.006) \times 10^{23}$ mol^{-1}. The currently recognized value is $(6.022137 \pm 0.000007) \times 10^{23}$, based on x-ray diffraction measurements of unit cell volume in a crystal together with the crystal density and formula weight. Millikan was a very careful experimenter; the systematic error in his value resided mainly in data reported by others that he used in his calculations.

As a rule, the best assurance of low systematic error is agreement among two or more entirely different methods of measurement. Further discussion of systematic errors can be found elsewhere.[6,7]

If all significant sources of systematic error cannot be eliminated, the experimentalist should at least seek to identify them and place limits on their possible magnitudes. However, to the maximum extent possible, you should aim to reduce the limits of systematic error to magnitudes that are small in comparison with the random errors of the experiment, so that statistical estimates made with equations in this chapter will be meaningful.

OTHER ERRORS

Another kind of error is important enough to mention. It is called by some "erratic error," which is a fancy name for "mistakes." Mistakes in arithmetic computation are an example, but these are inexcusable since all computations should be carefully checked. One-time reading errors of instrument scales in which digits are read improperly (the numeral 5 being mistaken for a 6, for example) or in which pairs of digits are transposed or a decimal point is misplaced in recording are examples of mistakes that can be very damaging, because they usually cannot be traced without repeating the experiment and because they may be very much larger than any conceivable random error. One of the most frequent examples is weighing errors, since a sample of material being prepared for an experiment is weighed only once. Other mistakes are failure to execute some essential step in the procedure (such as turning on a heater or a circulating pump) or faulty execution of the procedure (such as use of a wrong optical filter or even a wrong chemical substance for investigation). These mistakes may be untraceable, since clues may not exist in the laboratory notebook if the experimenter has been less than meticulous. Serious mistakes may be revealed only when the consequent numerical result appears to be impossible or ridiculous or is in wide disagreement with a theoretical value or prior experimental result.

Any measurement that of its nature will be done only once should provoke in you a strong sense of insecurity that will motivate you to be very deliberate in recording the data in the notebook and checking them thoroughly before going on to something else. Wherever possible, measurements should be done at least in duplicate. Duplicate measurements are NOT, as many suppose, for the purpose of enabling statistical treatment of random error! For that purpose, at least four measurements should be made, preferably more. Mistakes in procedure can be avoided only by exercise of due care in planning and executing the experiment. If the procedure is excessively complicated, prepare a checklist ahead of time. Make sure that illumination is adequate, labels are clear, scales are clean. Record in the notebook the identification numbers of optical filters, tared vessels, and all relevant data from labels of sample containers. Keep the experimental setup orderly and as uncomplicated as possible.

*REJECTION OF DISCORDANT DATA

It occasionally happens in making multiple measurements that one value differs from the rest considerably more than they differ from one another. Should the discordant value be rejected before an average is taken? This question provides one of the most severe tests of your scientific objectivity, for you may (indeed, you should) reject any measurement, whether concordant or discordant, if there is a valid reason to believe that it is defective.

The first step should always be to check for evidence of a determinate error (used the wrong pipette, added the wrong reagent, inverted the digits in writing down the number, etc.) or poorly controlled conditions (temperature control erratic during that measurement, forgot to check instrument zero setting, etc.). If no objective mistake or operating difficulty can be identified, the suspect value must be included unless valid statistical arguments can be presented to show that such a large deviation on the part of a member of the hypothetical ideal population of similar measurements is highly improbable. There are a number of ways of going about this. Most of them—such as the traditional rule about discarding values that differ from the mean of the others by more than four times the average deviation of the others—are based on faulty application of large-number statistics to small-sample problems and are not to be trusted. There is, however, a very simple approach, known as the Q test, which is both statistically sound and straightforward to use.

Q Test. When one in a series of 3 to 10 measurements appears to deviate from the mean by more than seems reasonable, calculate the quantity Q, defined by

$$Q \equiv \frac{|(\text{suspect value}) - (\text{value closest to it})|}{(\text{highest value}) - (\text{lowest value})} \tag{9}$$

Compare this value of Q with the critical value Q_c in Table 1 corresponding to the number of observations in the series. If Q is equal to or larger than Q_c, the suspect measurement should be rejected. If Q is less than Q_c, this measurement must be retained.

EXAMPLE. Five determinations of the baseline reading of a spectrophotometer at a standardizing wavelength were 0.32, 0.38, 0.21, 0.35, and 0.34 absorbance units. May the 0.21 be discarded?

$$Q = \frac{|0.21 - 0.32|}{0.38 - 0.21} = 0.65$$

From Table 1, the critical value Q_c for $N = 5$ is 0.64. The value of Q is higher, hence the 0.21 value may be discarded with 90 percent confidence.

TABLE 1 Critical Q values for rejection of a discordant value at 90 percent confidence level[8]

N	3	4	5	6	7	8	9	10
Q_c	0.94	0.76	0.64	0.56	0.51	0.47	0.44	0.41

It should be noted that only *one* value in a series may be discarded in this manner. If there is more than one divergent value in a small series, it means that the data really do show a lot of scatter. When one value in a small series is considerably off but not sufficiently to justify rejection by the Q test, there is good reason to consider reporting the median value (i.e., the middle value) instead of the mean. As an illustration of this, consider the situation if the five spectrophotometer readings had been 0.32, 0.38, 0.23, 0.35, and 0.34. The value of Q for the suspect member of this set is 0.60, which does not quite exceed the critical value Q_c; and the 0.23 value cannot be rejected. However, reporting the median value 0.34 avoids the large effect that 0.23 would have if included in calculating the mean, while still allowing it a "vote." The mean of all five values is 0.32, whereas the mean without the suspect value is 0.35.

When N exceeds 10 it is suggested that an observation be rejected if its deviation from the mean of the others exceeds $2.6S$, where S is the estimated standard deviation of the others from their mean; this corresponds to about a 1 percent probability that the suspect observation is valid.

Whether the criteria for rejection are those recommended above or alternative criteria, *it is important that the chosen criteria be used consistently*. The criteria must be decided upon ahead of time; selection of criteria after the measurements have been made runs the grave risk of introducing bias. If a series of measurements presents a problem that cannot be treated objectively with previously selected criteria, then if at all possible the entire series should be rejected and a new series of measurements made after careful review of the experimental procedure to minimize the possibility of large errors, mistakes, or bias.

STATISTICAL TREATMENT OF RANDOM ERRORS

Before presenting the results of statistical analysis of random errors, let us recall the definition of a normalized probability P. As an example, consider flipping a two-sided coin (heads or tails) 10 times and recording the number h of heads. Clearly, the variable h observed in that measurement must have one of the 11 values between $h_1 = 0$ and $h_{11} = 10$. Now carry out many measurements (each a set of 10 flips) and keep track of the number of times n_i that a given value h_i is observed. The probability P_i of observing the value h_i is defined by $P_i = n_i/\Sigma n_i = n_i/N$, where N is the total number of observations. It is clear that $\Sigma P_i = 1$, which corresponds to the *certainty* that all h_i values must lie somewhere between 0 and 10. In fact, we expect the most likely average of many measurements to be 5; that is, P_i should be sharply peaked at 5. But P is not a delta function; there is some spread about 5, and any single measurement may yield an h value of 3 or 6 or even 10!

The next three subsections describe the background and principles of random error treatment, and they introduce two important quantities: standard deviation σ and 95 percent confidence limits.[9] The four subsections following these—Uncertainty in Mean Value, Small Samples, Estimation of Limits of Error, and Presentation of Numerical Results—are essential for the kind of random error analysis most frequently required in the experiments given in this book. The Student t distribution is particularly important and useful.

ERROR FREQUENCY DISTRIBUTION

Let it be supposed that a very large number of measurements x_i ($i = 1, 2, \ldots, N$) are made of a physical quantity x and that these are subject to random errors ϵ_i. For simplicity we shall assume that the true value x_0 of this quantity is known and therefore that the errors are known. We are here concerned with the frequency $n(\epsilon)$ of occurrence of errors of size ϵ. This can be shown by means of a bar graph, like that of Fig. 3, in which the error scale is divided into ranges of equal width and the height of each bar represents the number of

FIGURE 3

A typical distribution of errors. The bar graph represents the actual error frequency distribution $n(\epsilon)$ for 376 measurements; the estimated normal error probability function $P(\epsilon)$ is given by the dashed curve. Estimated values of the standard deviation σ and the 95 percent confidence limit Δ are indicated in relation to the normal error curve.

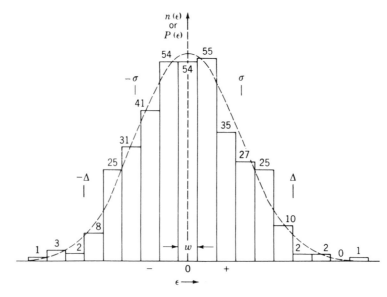

measurements yielding errors that fall within the respective range. The width w is chosen as a compromise between the desirability of having the numbers in each bar as large as possible and the desirability of having the number of bars as large as possible.

It is seen that even with as many as 376 measurements the graph shows irregularities, owing to the fact that the number of measurements represented by each bar is subject to statistical fluctuations that are not small in comparison with the number itself. From this graph we can make rough predictions concerning the probability that a measurement will yield an error of a given size. If we greatly increase the number of measurements represented (while perhaps decreasing the width w of the range more slowly than the rate of increase in the total number of measurements), the statistical fluctuations will become smaller in relation to the heights of the bars and our probability predictions are improved; we may even draw a smooth curve through the tops of the bars and assume it to represent an *error probability function* $P(\epsilon)$. The vertical scale of the $P(\epsilon)$ distribution should be adjusted by multiplication with an appropriate factor so that the function $P(\epsilon)$ is normalized: i.e., so that

$$\int_{-\infty}^{\infty} P(\epsilon)\, d\epsilon = 1 \tag{10}$$

Its significance is that the probability that a single measurement will be in error by an amount lying in the range between ϵ and $\epsilon + d\epsilon$ is equal to $P(\epsilon)\, d\epsilon$.

A probability function derived in this way is approximate; the true probability function cannot be inferred from any finite number of measurements. However, it can often be assumed that the probability function is represented by a Gaussian distribution called the *normal error probability function,*

$$P(\epsilon) = \frac{1}{\sqrt{2\pi}\sigma}\, e^{-\epsilon^2/2\sigma^2} \tag{11}$$

where the *standard deviation* σ is a parameter that characterizes the width of the distribution.[1,7,9] It is the root-mean-square error expected with this probability function:

$$\sigma \equiv (\overline{\epsilon^2})^{1/2} = \left(\frac{1}{\sqrt{2\pi}\sigma} \int_{-\infty}^{\infty} \epsilon^2 e^{-\epsilon^2/2\sigma^2}\, d\epsilon \right)^{1/2} \tag{12}$$

If the true value x_0 and thus the errors ϵ_i themselves are known, σ can be estimated from

$$\sigma = \left(\frac{1}{N} \sum_{i=1}^{N} \epsilon_i^2 \right)^{1/2} \tag{13}$$

The dashed curve in Fig. 3 represents a normal error probability function, with a value of σ calculated with Eq. (13) from the 376 errors ϵ_i.

The usual assumptions leading to the normal error probability function are those required for the validity of the *central limit theorem*.[10,11] The assumptions leading to this theorem are sufficient but not always altogether necessary: the normal error probability function may arise at least in part from circumstances different from those associated with the theorem. The factors that in fact determine the distribution are seldom known in detail. Thus it is common practice to *assume* that the normal error probability function is applicable even in the absence of valid a priori reasons. For example, the normal error probability function appears to describe the 376 measurements of Fig. 3 quite well. However, a much larger number of measurements might make it apparent that the true probability function is slightly skewed or flat topped or double peaked (bimodal), etc.

INFINITELY LARGE SAMPLE

So far the discussion has dealt with the errors themselves, as if we knew their magnitudes. In actual circumstances we cannot know the errors ϵ_i by which the measurements x_i deviate from the true value x_0, but only the deviations $(x_i - \bar{x})$ from the mean $\bar{x}$ of a given set of measurements. If the random errors follow a Gaussian distribution and the systematic errors are negligible, the best estimate of the true value x_0 of an experimentally measured quantity is the arithmetic mean $\bar{x}$. If you as an experimenter were able to make a very large (theoretically infinite) number of measurements, you could determine the *true mean* μ exactly, and the spread of the data points about this mean would indicate the precision of the observation. Indeed, the probability function for the deviations would be

$$P(x - \mu) = \frac{1}{\sqrt{2\pi}\sigma} \exp\left[-\frac{(x-\mu)^2}{2\sigma^2} \right] \tag{14}$$

where μ is the mean and σ the standard deviation of the distribution of the hypothetical *infinite population* of all possible observations, given by

$$\sigma = \lim_{N\to\infty} \left[\frac{1}{N} \sum_{i=1}^{N} (x_i - \mu)^2 \right]^{1/2} \tag{15}$$

In the absence of systematic errors, μ should be equal to the true value x_0.

The normal probability function as expressed by Eq. (14) is useful in theoretical treatments of random errors. For example, the normal probability distribution function is used to establish the probability P that an error is less than a certain magnitude δ, or conversely to establish the limiting width of the range, $-\delta$ to δ, within which the integrated probability P, given by

$$P = \int_{-\delta}^{\delta} \frac{1}{\sqrt{2\pi}\sigma} \exp\left[-\frac{\epsilon^2}{2\sigma^2} \right] d\epsilon \tag{16}$$

has a certain value. If δ is specified as equal to σ, it is found that $P = 0.6826$. That is, 68.26 percent of all errors are less than the standard deviation in magnitude; see Fig. 4a. If P is specified as 0.95 (95 percent), then

$$\delta_{0.95} = 1.96\sigma \simeq 2\sigma \tag{17}$$

FIGURE 4

The integrated probability P for a normal error distribution: i.e., the statistical probability that the error lies between the specified limits. The value of P is given by the shaded area: (a) standard deviation error limits, $\pm\sigma$, (b) 95 percent confidence limits, $\pm1.96\sigma$.

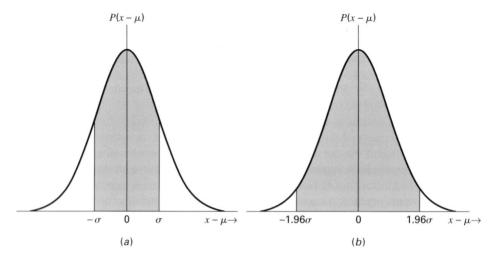

(a)   (b)

This may be described as a "95 percent confidence limit" for an individual measurement and is illustrated in Fig. 4b. The relationship between a specified uncertainty and the corresponding confidence limit is shown in Table 2. There is a 68.26 percent probability that the x value obtained in a single measurement will lie between $\mu - \sigma$ and $\mu + \sigma$. In contrast, the level of confidence that a single x value will lie between $\mu - 1.96\sigma$ and $\mu + 1.96\sigma$ is 95 percent.

Confidence limits will be discussed more fully in the section on uncertainties in mean values. It must be emphasized that Eq. (17) can be used only when σ is known. For σ to be known to a satisfactory approximation from the scatter of N measurements alone, N should be at least 20, preferably more.

LARGE FINITE SAMPLE

If a large enough number of data points are available (at least 20 and preferably 100), classical probability calculations provide a very satisfactory description of the precision of the measurements in question. In practice, one rarely has anything like 20 measurements of the same quantity; most experiments are based on averages of 6 or less. An important exception to this is computer-controlled automated experiments, where a digital reading (say a voltage drop or the resistance of a thermistor as given by a high-resolution multimeter) can be made rapidly—say within 100 ms. In this case, it is feasible and attractive to make 20 or more sequential readings and average them.

Classical probability statistics are inadequate for the treatment of small numbers of observations, and techniques developed only within recent years are necessary to avoid large errors in the estimates of error. For a finite (and usually small) number of observations of a quantity, one obtains data that show a certain amount of spread. The true mean μ and the true spread of the hypothetical infinite population of measurements are what one wishes to have. Usually the best that we can actually have with a finite number N of measurements are *estimates* of μ and σ. These are, respectively, the mean $\bar{x}$ and the spread

TABLE 2 Correspondence between uncertainty value and level of confidence

Uncertainty	$\pm\sigma$	$\pm1.64\sigma$	$\pm1.96\sigma$	$\pm2.58\sigma$	$\pm3.29\sigma$
Confidence level	68.26	90	95	99	99.9

of the measurements for which we use the quantity S, the estimated standard deviation, defined in Eq. (7). Clearly, as $N \rightarrow \infty$,

$$\bar{x} \rightarrow \mu \quad \text{and} \quad S \rightarrow \sigma$$

If N is larger than 20, S provides a satisfactory estimate of σ. Note that the estimated standard deviation S is a finite sample standard deviation and the so-called standard deviation σ is more precisely the standard deviation of an infinite sample set.

UNCERTAINTY IN MEAN VALUE

The error ϵ_m in the mean of N observations [Eq. (2)] is the mean of the individual errors ϵ_i:

$$\epsilon_m = \frac{1}{N} \sum_{i=1}^{N} \epsilon_i \tag{18}$$

The mean square error is obtained by squaring both sides of this expression and averaging over an infinite population of sets of N measurements. The averaging is done by multiplying the square of each side of Eq. (18) by the probability distribution function $P(\epsilon)$ and integrating from $-\infty$ to ∞:

$$S_m^2 = \overline{\epsilon_m^2} = \frac{1}{N^2} \left[\sum_{i=1}^{N} \overline{\epsilon_i^2} + \sum_{i=1}^{N} \sum_{j=1}^{N} {}' \overline{\epsilon_i \epsilon_j} \right] \tag{19}$$

where the prime indicates a sum over j values that are not equal to i. The mean of $\epsilon_i \epsilon_j$ vanishes because $\epsilon_i \epsilon_j$ is as often negative as positive. Therefore

$$S_m^2 = \frac{1}{N^2} N S^2 \quad \text{or} \quad S_m = \frac{S}{\sqrt{N}} \tag{20}$$

where S_m is the estimated standard deviation of the mean, in agreement with Eq. (8). This is a quantitative statement of the intuitively obvious conviction that the mean of a group of N independent measurements of equal weight has a higher precision than any single one of the measurements.

Previously we were concerned with estimating μ and σ, which are parameters of an infinite sample set. Now we are dealing with a large but finite set—say 20 or more measurements of x. Consider several such finite sets, each with its average $\bar{x}$. The distribution of these *averages* follows a normal distribution characterized by a sample standard deviation S_m. That is, there is a 68.26 percent probability that the true value of x lies between $\bar{x} - S_m$ and $\bar{x} + S_m$ for a sample of $N \geq 20$ measurements.

As before, we can also determine a 95 percent confidence limit *in the mean*, to be denoted as Δ:

$$\Delta \equiv \delta_{m,0.95} = \frac{1.96S}{\sqrt{N}} \approx \frac{2S}{\sqrt{N}} \tag{21}$$

When N is less than about 20, small sample statistics are needed and the Student t distribution described below should be used instead of the normal distribution.

The general results given above can be proved through the use of the normal probability distribution, Eq. (11).[8–10] The *joint probability* $P_{\text{joint}}(x_1, x_2, \ldots, x_N)$ that of N observations the first lies in the range x_1 to $x_1 + dx_1$, the second lies in the range x_2 to $x_2 + dx_2$, and so on, can be defined. Then changing from $x_1, x_2, \ldots, x_N$ to a new set of N variables and integrating over all of them except $\bar{x}$, one can obtain[8]

$$P_m(\bar{x}) = \frac{1}{\sqrt{2\pi}(\sigma/\sqrt{N})} \exp\left[-\frac{(\bar{x} - \mu)^2}{2(\sigma/\sqrt{N})^2} \right] \tag{22}$$

as the probability distribution function for the distribution of mean values $\bar{x}$ around the true mean of the distribution μ. Note that this equation resembles Eq. (11) except that $\sigma/\sqrt{N}$ replaces σ. Therefore, in the case where σ is known, we may write

$$\sigma_m = \frac{\sigma}{\sqrt{N}} \tag{23}$$

*SMALL SAMPLES

In the analysis presented above, we discussed infinite ($N \to \infty$) and large finite ($N \geq 20$) samples of multiple measurements of some observable quantity x. Now we need to consider small samples ($1 < N < 20$), since in many cases N is as small as 3 or 4. As before, the mean $\bar{x}$ given by Eq. (2) is still our best estimate of the true value. The sample standard deviation S and the standard deviation of the mean S_m can still be calculated with Eqs. (7) and (20), but their usefulness is not clear since the probability distribution is not known. The problem is that the standard deviation σ is not known a priori, and we need to find some kind of distribution function that will apply for a small sample with unknown σ.

Student t Distribution. The way to proceed for small sample sets is to consider the joint probability P_{joint} and integrate it after a different change of variables[7] than that used previously. In the present case, the only variable that we do not integrate over is

$$\tau \equiv \frac{\bar{x} - \mu}{(S/\sqrt{N})} = \frac{\bar{x} - \mu}{S_m} \tag{24}$$

After integration over all new variables except τ we obtain

$$dP = P(\tau)\, d\tau \tag{25}$$

where

$$P(\tau) = \left[\frac{\Gamma(N/2)}{\sqrt{(N-1)\pi}\,\Gamma([N-1]/2)} \right] \left(1 + \frac{\tau^2}{N-1} \right)^{-N/2} \tag{26}$$

The factor in large square brackets, containing the gamma functions, is a normalization constant, with a value which assures that

$$\int_{-\infty}^{\infty} P(\tau)\, d\tau = 1 \tag{27}$$

The distribution represented in Eqs. (25) and (26) and is known as the *Student t distribution;* "Student" is a pseudonym for W. S. Gosset, who developed this distribution for application to biometrics. Note that σ does not appear in the expression. The shape of the Student t distribution, unlike that of the normal distribution, is variable, depending on N. When N is large, it rapidly approaches the normal distribution. The distribution functions for several values of the number of degrees of freedom ν ($\nu = N - 1$ for determining the mean $\bar{x}$) are shown in Fig. 5.

We may use this distribution in much the same way we use the normal distribution. Suppose we wish to find the (symmetrical) range of values of τ over which the integral of the Student probability function is a fraction P (say 0.95, or 95 percent). Let us define this range to be from $-t$ to t:

$$\int_{-t}^{t} P(\tau)\, d\tau = P \tag{28}$$

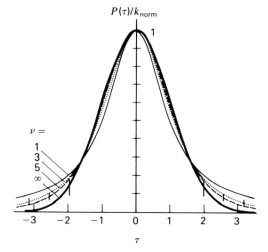

$P(\tau)/k_{norm}$

$\nu =$
1
3
5
∞

−3   −2   −1   0   1   2   3

τ

FIGURE 5

Student t distribution functions $P(\tau)$ for $\nu = 1, 3, 5, \infty$, where ν = number of degrees of freedom. The quantities actually plotted are

$$P(\tau)/k_{norm} = \left[1 + \frac{\tau^2}{N-1}\right]^{-N/2}$$
$$= [1 + \tau^2/\nu]^{-(\nu+1)/2}$$

which are not normalized; they need to be multiplied by the normalizing constant k_{norm}, which is the bracketed quantity in Eq. (26) with $N - 1 = \nu$. For $\nu = 1$ this is 0.3183, for $\nu = 3$ it is 0.3676, for $\nu = 5$ it is 0.3796, and for $\nu = \infty$ it is $1/\sqrt{2\pi} = 0.3989$. The curve for $\nu = \infty$ is the normal error probability curve. The short vertical bars mark the 95 percent confidence limits. These are not indicated for $\nu = 1$, as they are at $\tau = \pm12.7$, far outside the plotting range.

"Critical" values of t for a given number of degrees of freedom ν and a given P are given in Table 3 and in published tables.[12] Let us define a limit of error δ as the value of $\bar{x} - \mu$ that corresponds to the limit of integration t. Then, from Eq. (24),

$$\delta = tS_m = t\frac{S}{\sqrt{N}} \qquad \Delta = t_{0.95}S_m = t_{0.95}\frac{S}{\sqrt{N}} \qquad (29)$$

where again we use Δ to mean a 95 percent confidence limit in the mean. We emphasize the use of ν rather than $N - 1$ in Fig. 5 and Table 3 because the figure and the table are then relevant to cases in which several parameters are determined from the N measurements,

TABLE 3 Critical values of t (taken from Ref. 12)[a]

ν	P	0.50	0.80	0.90	0.95	0.98	0.99	0.999
	P'	0.75	0.90	0.95	0.975	0.99	0.995	0.9995
1		1.00	3.08	6.31	12.7	31.8	63.7	637.0
2		0.816	1.89	2.92	4.30	6.96	9.92	31.6
3		0.765	1.64	2.35	3.18	4.54	5.84	12.9
4		0.741	1.53	2.13	2.78	3.75	4.60	8.61
5		0.727	1.48	2.02	2.57	3.36	4.03	6.87
6		0.718	1.44	1.94	2.45	3.14	3.71	5.96
7		0.711	1.41	1.89	2.36	3.00	3.50	5.41
8		0.706	1.40	1.86	2.31	2.90	3.36	5.04
9		0.703	1.38	1.83	2.26	2.82	3.25	4.78
10		0.700	1.37	1.81	2.23	2.76	3.17	4.59
15		0.691	1.34	1.75	2.13	2.60	2.95	4.07
20		0.687	1.33	1.72	2.09	2.53	2.85	3.85
30		0.683	1.31	1.70	2.04	2.46	2.75	3.65
∞		0.674	1.28	1.64	1.96	2.33	2.58	3.29

[a]P is the probability that the mean μ of the population does not *differ from* the sample mean $\bar{x}$ by a factor of more than t, the tabular entry for a given number of degrees of freedom ν. P' is the probability that μ does not *exceed* $\bar{x}$ (or, alternatively, the probability that $\bar{x}$ does not *exceed* μ) by a factor of more than t for a given number of degrees of freedom ν.

TABLE 4 Factors for small-sample estimates based on the range R

N	K_2	$J(95\%)$
3	0.59	1.3
4	0.49	0.72
5	0.43	0.51
6	0.40	0.40
7	0.37	0.33
8	0.35	0.29
9	0.34	0.26
10	0.33	0.23

rather than only one (in the present case, the mean $\bar{x}$). Such a case is least-squares curve fitting (Chapter XXI). If the number of parameters to be determined by least squares is n and the number of data points is N, then $\nu = N - n$.

Caution is required when using tables from some sources,[12] since for those tables the integration of Eq. (28) is from $-\infty$ to t rather than from $-t$ to t. If we designate the corresponding integrated probability as P', then

$$P' = (P + 1)/2 \qquad \text{or} \qquad P = 2P' - 1 \tag{30}$$

Values of P' are also given in Table 3, and are useful in significance tests under certain circumstances to be described later.

Short-Cut Methods Based on the Range.[13] For small samples (N from 3 to about 10) the range, $R = x(\text{largest}) - x(\text{smallest})$, can be used to obtain rough uncertainty estimates. The range, when multiplied by the appropriate value of K_2 from Table 4, gives an approximate but useful estimate of the standard deviation. The factor J in Table 4 is equivalent to $tK_2/\sqrt{N}$, where t is the critical value from the $P = 95$ percent column in Table 3 for the appropriate degree of freedom. Hence, to get a quick estimate of the 95 percent confidence limits of the mean of a few observations, simply calculate $\pm JR$. In summary,

$$\sigma \simeq S \simeq K_2 R \qquad \text{and} \qquad \Delta\,(95\%\ \text{confidence}) \simeq JR \tag{31}$$

These techniques, which have developed out of the statistical methodology known as quality control, are becoming widely adopted because of their convenience. They are however decidedly inferior to the methods already described.

*ESTIMATION OF LIMITS OF ERROR

It must be kept in mind that the above discussion of random errors is based on the notion that one has no prior information about the precision of the measurements, in which case the uncertainty must be estimated from the scatter of the data in a given sample. In practice the number of independent measurements of an experimental quantity is often very small (sometimes two or only one) owing to time limitations, and it is unrealistic or impossible to estimate the confidence limits from the scatter or the range. However, the experimenter frequently has a good idea of the precision of the instrument or method being used. In this case, and also in cases where systematic errors must be estimated as well, the 95 percent confidence limits must be assigned on the basis of the individual judgment and experience of the experimenter. The ability to assign realistic limits of error, large enough to be safe but not so large as to detract unnecessarily from the value of the measurement, is one of the marks of a good experimentalist. It would be very difficult to formulate a set of rules that would be applicable in all cases; the confidence limits assigned in any given case will depend on the characteristics and capabilities of the instrument, its age and condition, the reproducibility of its reading, the quality of its calibration, and the user's experience and familiarity with it. A reasonable 95 percent confidence limit in weighing a small solid object on an analytical balance with weights calibrated to 0.05 mg might be 0.3 mg; in determining a short time interval with a manually actuated timer it might be 0.4 s; in reading an instrumental scale it might be 0.2 of the smallest scale division or more, depending on the instrument; etc.

Confidence limits assigned in this way will be represented in later sections by Δ on the rough assumption that they are 95 percent, based on long-term experience. However, one person's 95 percent estimate might be another's 90 percent, or 99 percent; this is unavoidable, and you as the experimenter simply have to do the best you can. Estimation of meaningful standard deviations in this manner is more difficult; if these are needed, rough values can be obtained as the estimated 95 percent confidence limits divided by 2.

*PRESENTATION OF NUMERICAL RESULTS

At some point the final numerical result or results of an experiment and their uncertainties must be presented in a report or a publication, and it is important that this be done in a clear manner. Uncertainties are often presented (usually with "±") without making clear what kind of uncertainty estimates are being employed. Some conventions have arisen, which must be used with caution since they are by no means universally accepted; see Table 5.

Whatever the uncertainty figure, it is important that the number of decimal places given or implied by the uncertainty figure be the same as that of the numerical value to which it applies. Obviously, the same units apply to the uncertainty figure as to the numerical value. However, it is usually not necessary to state the units twice; it is sufficient to state them following the uncertainty figure. To be sure to avoid confusion, even with the confidence level at the usual 95 percent, it is becoming increasingly common to state the level of confidence and the number of measurements N or the number of degrees of freedom v: e.g.,

$$w = 142.8 \pm 0.3 \text{ mg} \qquad (95 \text{ percent}, N = 7)$$

or

$$w = 142.8 \pm 0.3 \text{ mg} \qquad (95 \text{ percent}, v = 6)$$

when the result presented is a mean of several independent measurements. Otherwise, when the limits are estimated on the basis of judgment and experience,

$$w = 142.8 \pm 0.3 \text{ mg} \qquad (95 \text{ percent})$$

is better than a presentation with no confidence limit explicitly stated.

TABLE 5 Most common modes of uncertainty presentation

Result	Uncertainty	Presentation
In terms of the estimated standard deviation S with parentheses around the uncertain digits		
3.42 bar	$S = 0.04$ bar	3.42(4) bar
0.451 nm	$S = 0.012$ nm	0.451(12) nm
In terms of the 95 percent confidence limit Δ with ±		
66.32 J K^{-1}	$\Delta = 0.21$ J K^{-1}	66.32 ± 0.21 J K^{-1}
5.02×10^{-4} V	$\Delta = 5 \times 10^{-6}$ V	$(5.02 \pm 0.05) \times 10^{-4}$ V

As results are prepared for presentation the relationship between the 95 percent confidence limit and the number of significant figures retained in the final result should be recognized: Generally, the confidence limit should be expressed as a single digit equal to or greater than 3 or as a two-digit number not greater than 25. Correspondingly, if the uncertainty is expressed as an estimated standard deviation S, it should be expressed as a single digit equal to or greater than 2 or a two-digit number not greater than 15.

PROPAGATION OF ERRORS

An experiment has been performed; a variety of direct measurements have been made—weights, volumes, temperatures, measured electromotive forces, spectral absorbances, etc.—and uncertainties in all of them have been estimated, either from statistical data obtained by repeated measurements or from judgment and experience. From the values obtained by direct measurement and with the aid of a phenomenological theory, a final numerical result is calculated. Let the desired numerical result be designated by F and the directly measured quantities be designated by $x, y, z, \ldots$. *The latter quantities are assumed to be mutually independent.* Let their uncertainties, usually in the form of 95 percent confidence limits, be designated $\Delta(x), \Delta(y), \Delta(z), \ldots$. The value of F is determined by substituting the experimentally determined values of $x, y, z, \ldots$ into a mathematical formula, which we write symbolically as

$$F = f(x, y, z, \ldots) \tag{32}$$

The issue to be discussed here is how one can estimate the uncertainty of the final result F, usually in the form of a 95 percent confidence limit.

Infinitesimal changes dx, dy, etc., in the experimentally determined values will produce in F the infinitesimal change dF, where

$$dF = \frac{\partial F}{\partial x} dx + \frac{\partial F}{\partial y} dy + \frac{\delta F}{\partial z} dz + \ldots \tag{33}$$

If the changes are finite rather than infinitesimal, but are small enough that the values of the partial derivatives are not appreciably affected by the changes, we have approximately

$$\Delta F = \frac{\partial F}{\partial x} \Delta x + \frac{\partial F}{\partial y} \Delta y + \frac{\delta F}{\partial z} \Delta z + \ldots \tag{34}$$

This is equivalent to a Taylor expansion in which only the first-power terms have been retained. Now suppose that Δx represents the experimental error $\epsilon(x)$ in the quantity x:

$$\Delta x = \epsilon(x) \equiv x \text{ (measured)} - x \text{ (true)} \tag{35}$$

Such errors will produce an error in F,

$$\Delta F = \epsilon(F) \equiv F \text{ (calculated from measured } x, y, z, \ldots) - F \text{ (true)} \tag{36}$$

the value of which is given by

$$\epsilon(F) = \frac{\partial F}{\partial x} \epsilon(x) + \frac{\partial F}{\partial y} \epsilon(y) + \frac{\partial F}{\partial z} \epsilon(z) + \ldots \tag{37}$$

PROPAGATION OF SYSTEMATIC ERRORS

Systematic or "determinate" errors are not governed by probability statistics and do not involve such concepts as the variance or standard deviation. If a systematic error in a quantity x is established, the sign and magnitude of $\epsilon(x)$ is known, and Eq. (37) can be used for the propagation of the systematic error in F. Two important examples of this are

$$F = x + y - z \qquad \epsilon(F) = \epsilon(x) + \epsilon(y) - \epsilon(z) \tag{38}$$

$$F = \frac{xy}{z} \qquad \frac{\epsilon(F)}{F} = \frac{\epsilon(x)}{x} + \frac{\epsilon(y)}{y} - \frac{\epsilon(z)}{z} \tag{39}$$

Of course, if systematic errors are known, they should be corrected. Equation (37) provides one way of calculating the correction; normally it would be better simply to recalculate the result. Much more often, systematic errors are unknown; if suspected and of unknown sign, allowance for their possible magnitudes can be included in the estimated limits of "random" error.

*PROPAGATION OF RANDOM ERRORS

In the case of random errors, we do not know the actual values of $\epsilon(x)$, $\epsilon(y)$, etc., and we cannot determine the actual value of $\epsilon(F)$. However, we have already assigned to each experimental variable an estimated standard deviation S or a confidence limit Δ. We now wish to deduce the corresponding uncertainty in the final result F, taking into consideration the high probability that the errors in the several variables will tend somewhat to cancel one another out. Let us square both sides of Eq. (37):

$$[\epsilon(F)]^2 = \left(\frac{\partial F}{\partial x}\right)^2 [\epsilon(x)]^2 + \left(\frac{\partial F}{\partial y}\right)^2 [\epsilon(y)]^2 + \ldots + 2\left(\frac{\partial F}{\partial x}\right)^2\left(\frac{\partial F}{\partial y}\right)^2 \epsilon(x)\,\epsilon(y) + \ldots \tag{40}$$

Now let us average this expression over all values expected for $\epsilon(x)$, $\epsilon(y)$, $\ldots$, in accordance with the normal frequency distribution. Since the $\epsilon(x)$ independently have average value zero, we expect the cross-terms to vanish. However, the squared terms are always positive and will not vanish. If we replace the average of each squared error by the variance, we obtain

$$S^2(F) = \left(\frac{\partial F}{\partial x}\right)^2 S^2(x) + \left(\frac{\partial F}{\partial y}\right)^2 S^2(y) + \ldots \tag{41}$$

It can be seen immediately that the variance in the mean of N measurements $S^2(\bar{x}) = S^2(x)/N$, which was discussed previously, can be obtained as a trivial example of Eq. (41) by taking

$$F = \bar{x} = \frac{1}{N}\sum_{i=1}^{N} x_i$$

Equation (41), which is not limited to normal distributions, is an equation that can be used when all or substantially all of the experimental variables are means of sets of measurements from which statistical measures of variance are available. From $S(F)$ a confidence limit $\Delta(F)$ can be derived if the total number of degrees of freedom can be ascertained. In some cases this may be the sum of the number of degrees of freedom from all of the experimental variables; in general it will be larger than the number of degrees of freedom from any one variable. This complicated problem may be bypassed by assuming that the central limit theorem is at work: the error probability distribution in F is more closely normal than are the error probability distributions in the experimental variables. Thus, to a satisfactory approximation, the value of t to be used can be taken to be that for $\nu = \infty$: For $P = 95$ percent, $t = 1.96$.

However, we are usually working in situations where many or most of the uncertainties in the experimental variables are estimated on the basis of judgment and experience. On the assumption that the errors in such variables are distributed in a roughly normal manner, we may resort to the following approximate expression in terms of confidence limits:†

$$\Delta^2(F) = \left(\frac{\partial F}{\partial x}\right)^2 \Delta^2(x) + \left(\frac{\partial F}{\partial y}\right)^2 \Delta^2(y) + \left(\frac{\partial F}{\partial z}\right)^2 \Delta^2(z) + \dots \tag{42}$$

In certain cases, the propagation of random errors can be carried out very simply:

1. For $F = ax \pm by \pm cz$,

$$\Delta^2(F) = a^2\Delta^2(x) + b^2\Delta^2(y) + c^2\Delta^2(z) \tag{43}$$

2. For $F = axyz$ (or axy/z or ax/yz or a/xyz),

$$\frac{\Delta^2(F)}{F^2} = \frac{\Delta^2(x)}{x^2} + \frac{\Delta^2(y)}{y^2} + \frac{\Delta^2(z)}{z^2} \tag{44}$$

3. For $F = ax^n$,

$$\frac{\Delta^2(F)}{F^2} = n^2 \frac{\Delta^2(x)}{x^2} \rightarrow \frac{\Delta(F)}{F} = n\frac{\Delta(x)}{x} \tag{45}$$

4. For $F = ae^x$,

$$\Delta^2(F) = a^2 e^{2x}\Delta^2(x) \rightarrow \frac{\Delta(F)}{F} = \Delta(x) \tag{46}$$

5. For $F = a \ln x$,

$$\Delta^2(F) = \frac{a^2}{x^2} \Delta^2(x) \rightarrow \Delta(F) = a\frac{\Delta(x)}{x} \tag{47}$$

Note the differences between Eqs. (43) and (44) for random errors and the comparable expressions in Eqs. (38) and (39) for systematic errors.

Equation (43) includes the important special case $(x - y)$, which is applicable, for example, to differences between two readings of the same variable (e.g., the weight of a liquid plus container and the weight of the empty container, the initial and final readings of a thermometer). In many cases the limits of error in the initial and final readings may be taken as equal, in which case the limit of error in the difference between the two readings is equal to $\sqrt{2}$ times the limit of error in a single reading. Equation (44) can be simplified in the special case that the fractional error in each variable is approximately the same [i.e., $\Delta(x)/x \simeq \Delta(y)/y \simeq \Delta(z)/z$]. In this special case, the fractional error in F is roughly equal to the fractional error in an individual variable multiplied by the square root of the number of variables [$\Delta F/F \simeq \sqrt{n}(\Delta x/x)$].

Many propagation-of-error treatments are not so simple as those illustrated by Eqs. (43) to (47). The expression for F is often complicated enough or the functional form sufficiently awkward to justify a breakdown of the procedure into steps:

$$F = f_0(A, B, \dots) \tag{48}$$

where

$$A = f_1(x_1, x_2, \dots), \quad B = f_2(y_1, y_2, \dots), \quad \text{etc.}$$

We can then write

$$\Delta^2(F) = \left(\frac{\partial F}{\partial A}\right)^2 \Delta^2(A) + \left(\frac{\partial F}{\partial B}\right)^2 \Delta^2(B) + \dots$$

$$\Delta^2(A) = \left(\frac{\partial A}{\partial x_1}\right)^2 \Delta^2(x_1) + \left(\frac{\partial A}{\partial x_2}\right)^2 \Delta^2(x_2) + \dots \tag{49}$$

$$\Delta^2(B) = \left(\frac{\partial B}{\partial y_1}\right)^2 \Delta^2(y_1) + \left(\frac{\partial B}{\partial y_2}\right)^2 \Delta^2(y_2) + \dots, \quad \text{etc.}$$

†The use of Δ as a limit of error, as in $\Delta(x)$, is not to be confused with its use as a difference, as in Δx.

provided that A, B, . . . are *independent*. If the quantities A, B, etc., are not independent (that is, in the event that some of the variables x_i are identical with some of the y_i), the limit of error calculated in this manner will be incorrect. For example, if we let

$$F = a(e^{kx} - 1) + b(y - cx) = A + B$$

it would be improper to write

$$\Delta^2(F) = \Delta^2(A) + \Delta^2(B)$$

$$\Delta^2(A) = (ake^{kx})^2\Delta^2(x) \qquad \Delta^2(B) = b^2\Delta^2(y) + b^2c^2\Delta^2(x)$$

Treatment of the function as a whole gives the correct result

$$\Delta^2(F) = (ake^{kx} - bc)^2\Delta^2(x) + b^2\Delta^2(y)$$

which differs from the result first given by a term $-2abcke^{kx}\Delta^2(x)$.

When the expression for F is a complicated one, it is advisable to watch for opportunities for transforming the expression or parts of the expression so that single symbols representing known quantities can be substituted for groupings of several symbols. The most advantageous simplifications are usually obtained when the quantities substituted are the end results of the calculations, such as F itself, or at least quantities that represent the later stages of the calculation of F. This is advantageous not only because it results in the greatest economy of symbols but also because it presents the most frequent opportunities for cancellation and simplification. For example, the limit of error of

$$F = C\frac{x^2}{y}$$

can be obtained from

$$\Delta^2(F) = \left(\frac{2Cx}{y}\right)^2\Delta^2(x) + \left(\frac{Cx^2}{y^2}\right)^2\Delta^2(y)$$

but for computational purposes it is more convenient to divide this equation by F^2, since the value of F will already have been calculated; we obtain

$$\frac{\Delta^2(F)}{F^2} = 4\frac{\Delta^2(x)}{x^2} + \frac{\Delta^2(y)}{y^2}$$

Estimation of uncertainties is often more complicated than in the above instances because the functional form of F may be more complicated. Before one begins to differentiate F, which is always in principle possible but often yields masses of unwieldy expressions with the ever-present possibility of mistakes, it is well to examine F carefully with two purposes in mind: (1) to see if the expression for F can be simplified to make differentiation easier; and (2) to see if there are any terms that can be neglected because their effect on the uncertainty is small. While F must be evaluated in its precise form to yield the desired numerical result to the maximum possible precision, it should be remembered that the value of the limit of error need not be known to great precision; thus a quantity can often be dropped in a limit-of-error treatment even if it cannot be neglected in the calculation of F itself. A term can ordinarily be dropped in the limit-of-error treatment if the effect produced in $\Delta(F)$ is no more than about 5 percent or even 10 percent. An overgenerous (overconservative) estimate of the limit of error is to be preferred to an insufficient (overoptimistic) estimate; if a term is to be dropped, it is better that dropping it make $\Delta(F)$ a little too large than a little too small. For example, in the cryoscopic determination of the molar mass of an unknown substance, the molar mass is given by Eq. (10-21) as

$$M = \frac{1000gK_f}{G\Delta T_f}(1 - k_f\Delta T_f)$$

Since k_f is of the order of 0.01 K^{-1}, the term $k_f\Delta T_f$ is itself very small compared with unity and will contribute negligibly to the uncertainty in comparison to ΔT_f itself. Therefore, the treatment of uncertainties, but not the calculation of M itself, may be based on the simpler, approximate expression

$$M = \frac{1000gK_f}{G\Delta T_f}$$

The value of $\Delta(M)/M$ is then easily obtained by using the form given in Eq. (44).

Another technique that may be used in cases where the formula for F is very complicated and difficult to differentiate is "computational differentiation," which is extremely tedious to do by hand or with a calculator but reasonably easy to do on a programmable calculator or a computer that has already been programmed to compute the result F. This consists of first computing the result F by substituting in the values of the x_i, then calculating the result $F + \Delta_1 F$ obtained by changing x_1 to $x_1 + \Delta x_1$, where Δx_1 is a very small increment, leaving the other x_i unchanged. Then the change $\Delta_1 F$ is obtained by subtraction. This is done successively with $x_2, x_3, \ldots$, yielding $\Delta_2 F, \Delta_3 F, \ldots$. Then

$$\frac{\partial F}{\partial x_i} \simeq \frac{\Delta_i F}{\Delta x_i} \tag{50}$$

These values of partial derivatives may then be substituted into Eq. (42) to yield the desired limit-of-error value.

*EXAMPLES

Given below are three numerical examples of a propagation of errors treatment.

Example 1. The first example is related to the optical rotation measurement described as a case study in the next section. The formula for calculating the specific rotation $[\alpha]$ is

$$[\alpha] = \frac{V}{Lm}\alpha \tag{51}$$

where V is the volume of the solution in cubic centimeters, L is the length of the polarimeter tube in decimeters, m is the mass of solute contained in volume V, and α is the rotation angle in degrees. Values of these four independent variables together with their 95 percent confidence limits are as follows:

V	L	m	$\bar{\alpha}$
25.00 cm^3	2.000 dm	1.7160 g	+20.950°
$\Delta = 0.02$ cm^3	0.002 dm	0.0003 g	0.016°

The confidence limits for the first three variables were estimated on the basis of judgment and experience. The last was determined from a statistical analysis of data from 10 individual measurements with Eqs. (8) and (29). From Eq. (51) we calculate

$$F = [\alpha]_D^{25} = \frac{25.00}{2.000 \times 1.7160} \times 20.950 = 152.61 \text{ deg dm}^{-1}(\text{g cm}^{-3})^{-1}$$

We observe that Eq. (51) is of the type for which Eq. (44) is applicable. Therefore

$$\Delta^2([\alpha]) = [\alpha]^2\left[\frac{\Delta^2(V)}{V^2} + \frac{\Delta^2(L)}{L^2} + \frac{\Delta^2(m)}{m^2} + \frac{\Delta^2(\alpha)}{\alpha^2}\right]$$

$$= 152.61^2\left(\frac{0.02^2}{25.00^2} + \frac{0.002^2}{2.000^2} + \frac{0.0003^2}{1.7160^2} + \frac{0.016^2}{20.950^2}\right)$$

$$= 2.33 \times 10^4(64 + 100 + 3 + 58) \times 10^{-8} = 0.0524$$

$$\Delta([\alpha]) = 0.23 \text{ deg dm}^{-1} (\text{g cm}^{-3})^{-1}$$

One would report this result as $[\alpha] = 152.61 \pm 0.23$ deg dm^{-1} (g cm^{-3})$^{-1}$ or 1.5261 ± 0.0023 deg m^2 kg^{-1} in SI units.

Example 2. The second example is based on Exp. 3 (method B) and involves the determination of $C_p/C_v = \gamma$ for argon gas from the measurement of the sound velocity c. The formula for calculating γ for an ideal gas, which is a very good approximation for Ar, is

$$\frac{C_p}{C_v} \equiv \gamma = \frac{Mc^2}{RT} \tag{52}$$

where M is the molar mass in kg mol^{-1} (0.039948 for Ar), c is the sound velocity in m s^{-1}, R is the gas constant (8.31447 J K^{-1} mol^{-1}), and T is the temperature in kelvin.

The experiment was carried out at an average temperature of $25.1 \pm 0.4°$C $=$ 298.25 ± 0.4 K. The estimated 95 percent confidence limit $\Delta(T) = 0.4$ K is fairly large since there was no thermostat jacket to maintain better temperature control. The sound velocity is given by $c = \lambda f$, where λ is the measured wavelength of an acoustic standing wave of frequency f. This frequency was very precisely known: $f = 1023.0$ Hz (1 Hz = 1 s^{-1}) with negligible uncertainty. Since the average of many wavelength measurements gave $\lambda = 31.75 \pm 0.35$ cm (95 percent), we find $c = 324.8$ m s^{-1} and from Eq. (43), which describes this simple case, we obtain $\Delta(c) = f\Delta(\lambda) = 3.58$ m s^{-1}.

Inserting the data given above into Eq. (52), we obtain the dimensionless value

$$\gamma = \frac{(0.039948)(324.8)^2}{(8.31447)(298.25)} = 1.699$$

To evaluate the confidence limit in γ, we use Eq. (42):

$$\Delta^2(\gamma) = \left(\frac{2Mc}{RT}\right)^2 \Delta^2(c) + \left(-\frac{Mc^2}{RT^2}\right)^2 \Delta^2(T)$$

$$= \frac{4\gamma^2}{c^2}\Delta^2(c) + \frac{\gamma^2}{T^2}\Delta^2(T)$$

$$\frac{\Delta^2(\gamma)}{\gamma^2} = 4\frac{\Delta^2(c)}{c^2} + \frac{\Delta^2(T)}{T^2} = 4\frac{(3.58)^2}{(324.8)^2} + \frac{(0.4)^2}{(298.25)^2}$$

$$= 4.86 \times 10^{-4} + 1.8 \times 10^{-6} = 4.88 \times 10^{-4}$$

or $\Delta(\gamma) = 0.0221\,\gamma = 0.0375$

The results of this experiment should be reported as $\gamma = 1.70 \pm 0.04$ (95 percent) for Ar at 298.25 K.

A literature search indicated that the ideal gas theoretical value for γ of a monatomic gas is $5/3 = 1.6667$ and the best experimental value for Ar is $\gamma = 1.6677$. Happily, both these values lie within our 95 percent confidence limits.

Example 3. As a third and more complicated example, consider the measurement of the temperature of a boiling organic liquid with the gas thermometer described in Exp. 1. The absolute temperature is given by

$$T = T_0 \frac{p(1 + pv/p_rV + 3\alpha t - B/\widetilde{V})}{p_0(1 + p_0v/p_rV - B_0/\widetilde{V})} \tag{53}$$

This equation has been obtained by combining Eq. (1-8) with a similar equation in which T is replaced by T_0, which is the temperature at the ice point (273.15 K). For definitions of the quantities in this expression, see Exp. 1; quantities with subscript zero are those corresponding to T_0. The volume V of the thermometer bulb is not to be confused with $\widetilde{V} \simeq RT_0/p_0$, which is the molar volume of the gas employed. The direct partial differentiation of this equation with respect to p or p_0 would yield a very messy expression, so we are motivated to look for ways to simplify the equation. Since the last two terms in parentheses in the denominator are small compared to unity, we can apply the series expansion expression for the reciprocal of the binomial $1 + x$,

$$\frac{1}{1 + x} = 1 - x + x^2 - \dots$$

keeping only the first two terms on the right. Thus we may write the expression for T, ignoring terms of second order in small quantities, as [cf. Eqs. (1-12), (1-13)]

$$T = T_0 \frac{p}{p_0}\left[1 + \frac{(p - p_0)}{p_r} \frac{v}{V} + 3\alpha t - \frac{(B - B_0)}{\tilde{V}}\right]$$

$$= T_0\left[\frac{p}{p_0}\left(1 + 3\alpha t - \frac{p_0}{RT_0}(B - B_0)\right) + \frac{(p^2 - pp_0)}{p_0 p_r} \frac{v}{V}\right]$$

This may be differentiated with respect to p, p_0, p_r, and v/V without much difficulty. Before going to the trouble of differentiating, we should examine the terms for magnitude. Using values of α, B, and B_0 given in Exp. 1 and noting that v/V is of the order of 0.01, we may surmise that, for the purpose of deriving a limit-of-error estimate, all terms except the leading one may be neglected:

$$T = T_0 \frac{p}{p_0}$$

Let us put in some numbers. Assume that the gas in the bulb is helium; then $B - B_0$ is certainly negligible (see Table 1-1). For $p = 510.1 \pm 1.0$ Torr and $p_0 = 418.2 \pm 1.0$ Torr (where the cited uncertainties are 95 percent confidence limits) and $T_0 = 273.15$ K (regarded as exact), T by the above simple equation is 333.2 K; $t = 60.0°C$. With, in addition, $p_r = 456.5 \pm 1.0$ Torr, $3\alpha t = 3(3.2 \times 10^{-6})60 = 0.00058$, and $v/V = 0.011 \pm 0.003$, the original precise expression gives $T = 333.94$ K. Using the simple expression for our limit-of-error treatment we have

$$\frac{\Delta^2(T)}{333.94^2} = \frac{1.0^2}{510.1^2} + \frac{1.0^2}{418.2^2}$$

or $\Delta(T) = 1.06$ K. Thus

$$T = 333.9 \pm 1.1\,\text{K} \qquad \text{or} \qquad t = 60.8 \pm 1.1°C$$

Suppose, however, that we are slightly unsure about the effect of the neglect of the v/V term on the limit of error. Let us disregard the extremely small terms involving α and $(B - B_0)$ and otherwise carry out a full-fledged limit-of-error treatment. The partial derivatives and their values are

$$\frac{\partial T}{\partial p} = \frac{T_0}{p_0}\left[1 + \frac{(2p - p_0)}{p_r} \frac{v}{V}\right] = 0.662\ \text{K Torr}^{-1}$$

(instead of 0.653 K Torr^{-1} from the simple approximate expression),

$$\frac{\partial T}{\partial p_0} = -T_0\left(\frac{p}{p_0^2} + \frac{p^2}{p_0^2 p_r} \frac{v}{V}\right) = -0.807\ \text{K Torr}^{-1}$$

(instead of -0.798 K Torr^{-1} from the simple approximate expression),

$$\frac{\partial T}{\partial p_r} = -T_0 \frac{p(p - p_0)}{p_0 p_r^2} \frac{v}{V} = -0.0016\ \text{K Torr}^{-1}$$

(instead of being neglected altogether), and

$$\frac{\partial T}{\partial (v/V)} = T_0 \frac{p(p - p_0)}{p_0 p_r} = 67.1\ \text{K}$$

(instead of being neglected altogether). Then, we obtain from Eq. (42)

$$\Delta(T) = [1.0^2(0.662^2 + 0.807^2 + 0.0016^2) + 0.003^2(67.1)^2]^{1/2}$$
$$= (1.0895 + 0.0405)^{1/2}$$
$$= 1.063 \text{ K}$$

which is not significantly different from the result of the earlier calculation.

None of the experiments described in this book will involve mathematical treatments for limit-of-error estimation more complicated than this example. With careful judgment, most limit-of-error treatments can be made much simpler.

CASE HISTORY OF AN ERROR EVALUATION

A professor once received from a chemical supply house a specimen of a crystalline organic compound needed in her research. This compound was optically active, with a positive rotation (i.e., it was dextrorotatory). The professor wanted to check that the sample was not contaminated by the other (levorotatory) optical isomer. Such contamination would decrease the magnitude of the positive rotation. Therefore she gave a small amount of the material to a graduate student with instructions to determine its *specific rotation* at 25°C with the highly precise polarimeter available in the laboratory (see Exp. 28 and Chapter XIX).

The next day, the student came back to the professor bearing his laboratory notebook. This showed that he had first adjusted the thermostat bath that supplies water to the jacket of the polarimeter so that it would cycle nicely between 24.8 and 25.3°C. He then weighed out 1.5220 g of the material (which he had stored overnight in a desiccator) and dissolved it in distilled water in a volumetric flask to make 25.00 mL of solution, being careful to equilibrate the flask in the thermostat bath before making the solution up to the mark. He then filled a 20.00-cm (2.000-dm) polarimeter tube with the solution and obtained the following values for the optical rotation with sodium-D light, taking care to approach the extinction point from alternate sides:

α (deg)	dev $= \alpha - \bar\alpha$
+20.04	−0.004
+20.07	+0.026
+20.05	+0.006
(+20.09)	rejected
+20.04	−0.004
+20.02	−0.024
+20.04	−0.004
+20.03	−0.014
+20.06	+0.016
+20.05	+0.006

$$\bar\alpha = +20.044 \qquad \Sigma \, \text{dev}^2 = 0.001824$$

$$S = \left(\frac{0.001824}{9 - 1} \right)^{1/2} = 0.015° \qquad S_m = \frac{0.015°}{\sqrt{9}} = 0.005°$$

In order to evaluate a 95 percent confidence limit Δ for this mean value $\bar\alpha$, the Student t distribution can be used. In the present case, we have $N = 9$, $v = 8$ and should look in Table 3 under the $P = 0.95$ column to the row corresponding to $\nu = 8$. This yields a $t_{0.95}$ value of 2.31. Equation (29) then gives

$$\Delta = t_{0.95}S_m = 2.31 \times 0.005° = 0.012°$$

Thus

$$\bar{\alpha} = +20.044° \pm 0.012° \text{ (95 percent, } \nu = 8)$$

which means that, if the student were to repeat this procedure many times, he would find that the mean value of α would differ from 20.044° by no more than 0.012° in 95 percent of the cases.

Using the formula for specific rotation given by Eq. (51), which is equivalent to Eq. (28-14), the student calculated for the specific rotation at 25°C with sodium-D light

$$[\alpha]_D^{25} = \frac{25.00}{2.0000 \times 1.5220} \times (+20.044) = +164.62 \text{ deg dm}^{-1} \text{ (g cm}^{-3})^{-1}$$

Estimating from his general experience 95 percent confidence error limits of 0.02 cm³ in V, 0.002 dm in L, and 0.0003 g in m, and using his 95 percent confidence limits of 0.012° in the mean value of α, the student calculated by propagation-of-errors techniques 95 percent confidence limits of ± 0.22 in the $[\alpha]_D^{25}$ result.

When the professor saw the final result for the specific rotation, $+164.62 \pm 0.22$, she was bewildered. The reported value *exceeded* the literature value of the pure dextrarotatory compound, $+152.70$, by 11.92, about 50 times the claimed limit of error! This ridiculous result could mean only that a serious error or mistake had been made in this determination, unless the literature value itself were seriously in error, which seemed unlikely. She looked for, but could not find, a mistake in arithmetic. She examined the data and asked the student why the reading of 20.09° had been rejected. The student replied that the reading seemed too far out of line. In order to check on this, the professor suggested that they carry out a Q test. This test showed that the reading had been wrongfully rejected (calculated $Q = 0.29$; for $N = 10$, $Q_c = 0.41$; see Table 1 and accompanying discussion). With a revised average of $+20.049°$ ($S = 0.020°$, $S_m = 0.006°$), the professor quickly recalculated the specific rotation, obtaining

$$[\alpha]_D^{25} = +164.66 \pm 0.23 \text{ deg dm}^{-1} \text{ (g cm}^{-3})^{-1}$$

Clearly, the real problem had not yet been found.

She asked to be shown the experimental setup. On seeing the balance on which the student had weighed out the material she noticed that the numerical scale on one of the weight dials was smudged in a few places; one of the 6s looked a little like a 5. Could the student have recorded 1.5220 g when the true weight might have been 1.6220 g? That would be enough to account for most of the difference between the experimental and literature values. Under pressure, the student conceded that this was possible, even likely. The professor calculated, tentatively,

$$[\alpha]_D^{25} = +164.66 \times \frac{1.5220}{1.6220} = +154.51$$

This was now within shooting distance of the literature value, although the occurrence of the weighing mistake could not be considered a certainty. (The professor did point out that had the error gone the other way, she might have thrown out the sample and tried to find new material.)

The professor then inspected the polarimeter setup. She noticed that the circulating pump, which was to supply constant-temperature water from the thermostat bath to the water jacket, was not running. "Did you have the circulating pump on during your measurements?" she asked. "I thought that it went on automatically," replied the student, who had used the prominent on–off switch for the sodium-vapor lamp but had not noticed the rather inconspicuous pump switch. Unfortunately, he had also not bothered to read the thermometer that was attached to the polarimeter jacket, which now read 18.0°C rather than 25.0°C. The professor was momentarily stumped, for there were no available data on the temperature coefficient of rotation for this substance. At length she reasoned that the principal effect of temperature on the rotation would be through the temperature dependence of the density of the solution. Using density values for water at the two temperatures, the professor estimated

$$[\alpha]_D^{25} = +154.51 \times \frac{0.998625(@18°C)}{0.997075(@25°C)} = +154.75$$

"Wait a minute," said the student. "If the polarimeter tube were at 18°C instead of 25°C, the solution density would be greater than the value used; thus the corrected specific rotation should be lower, not higher." Nodding in agreement, the professor inverted the ratio and obtained +154.27, which did not help much since the result was still more than a degree higher than the literature value.

One more thought occurred to her. Something her experience told her should be in the notebook was not there. She took the polarimeter tube, rinsed it thoroughly and filled it with distilled water, and placed it in the instrument (after turning the circulating pump on). Instead of a rotation reading of 0.00° she obtained a reading of +0.26°. The student had failed to check the zero adjustment on the polarimeter scale! The average of 10 readings gave +0.258° ($S_m = 0.006°$). Making the needed correction, the professor obtained $[\alpha]_D^{25} = +152.52 \deg \mathrm{dm}^{-1} (\mathrm{g\ cm}^{-3})^{-1}$, which was now slightly *below* the literature value of 152.70.

Because of doubt about the weighing error and the temperature correction, it was necessary to repeat the determination; see the results given in Example 1 on p. 56. The new value of $[\alpha]_D^{25}$ was $+152.61 \pm 0.23 \deg \mathrm{dm}^{-1} (\mathrm{g\ cm}^{-3})^{-1}$. Evidently the optical purity of the specimen was quite high; the difference from the literature value of +152.70 is not significant at the 95 percent confidence level.

Final Caution. The Student t distribution is to be used when the numerical value to which it attaches is a mean of a definite number of direct observations or is a numerical result calculated from such a mean by a procedure that introduces no uncertainties comparable in magnitude to the random errors of the direct observations. This is *not* the case for the specific rotation $[\alpha]_D^{25}$ discussed above. The uncertainty contribution to the final result that is due to random errors in the raw data on optical rotations α_i is *not* large compared to the contributions due to the *estimated* uncertainties in the other variables (particularly V and L) that are required for calculating the final result $[\alpha]_D^{25}$. The number of degrees of freedom applicable to the final result cannot be known precisely, or at least it would be an extremely difficult matter to determine. In addition we must reiterate that the statistical treatment can mean little or nothing in the face of large systematic errors, except to indicate the *precision* of the measurement method.

FUNDAMENTAL LIMITATIONS ON INSTRUMENTAL PRECISION

In all experiments given in this book the precision of the measurements is governed by practical limitations of apparatus construction and operation; the precision is presumably capable of being increased by further refinement of the apparatus or the technique for using it, and it might well be imagined that there is no limit to the possible improvement. However, there are certain definite limitations imposed by physics. Although these limitations are not likely to be encountered by the beginning laboratory student, it is well to be aware of them.

LIMITATIONS DUE TO THERMAL AGITATION

The classical law of equipartition of energy states that every degree of freedom of a system has an average kinetic energy equal to $kT/2$, where k is the Boltzmann constant. The student is used to thinking of this amount of energy (about 2×10^{-21} J at room temperature) as being of significance only for individual atoms and molecules or at most for microscopic particles undergoing Brownian motion. However, Brownian motion may be observed in systems as large as laboratory instruments. An example of a phenomenon analogous to this is observed in electrical circuits in the form of thermal noise (called Johnson noise). This may, for example, impose an important limitation on the sensitivity of a very sensitive voltmeter or electrometer. It constitutes a very severe limitation on the reception of extremely faint radio waves, especially in radio astronomy, where increasing the amplification is futile because the electrical noise is amplified to the same degree as the desired signal.

An electrical measuring instrument contains electrical circuits incorporating capacitance, inductance, and resistance. In the absence of resistance, a circuit tends to oscillate with a definite frequency f when disturbed. For optimum performance an amount of resistance is incorporated that is barely sufficient to damp the oscillations resulting from transient inputs; the circuit is then said to be "critically damped." For a critically damped circuit it can be shown that the root-mean-square (rms) fluctuations in voltage V and in current I are given by

$$V_{\text{rms}} \equiv (\overline{V^2})^{1/2} = (\pi kTRf)^{1/2} \tag{54}$$

$$I_{\text{rms}} \equiv (\overline{I^2})^{1/2} = (\pi kTf/R)^{1/2} \tag{55}$$

where R is the total resistance of the circuit, including damping resistance. If the units used in the above equations are volts, amperes, ohms, and hertz (cycles per second), k must be given in joules per kelvin (1.38×10^{-23}).

The rms values given above effectively constitute estimated standard deviations S for *individual* measurements of V or I. These limitations on precision can be overcome to some degree by making many measurements and taking the mean, provided the conditions of measurement do not change during the time required. If a voltage V is automatically and continuously recorded on a chart recorder, a straight line or a smooth curve can be drawn through the wiggly curve traced by the recorder pen. Or a voltage V can be measured multiple times with a computer-controlled digital voltmeter and averaged (beware of round-off effects due to D-to-A finite bit size). However, as in cases discussed earlier, the improvement in precision varies as the square root of the number of measurements and is therefore limited in practice. Since the fluctuations are thermal in origin, they can in principle, and also in practice in some cases, be reduced by reducing the temperature of the detection device.

LIMITATIONS DUE TO PARTICULATE NATURE

An important limitation that is sometimes encountered is due to the particulate nature of electricity (electrons, ions) and of radiation (photons). The measurement of radiation intensities is in certain cases (e.g., X rays) performed by counting particles or photons one at a time. The number N counted in a time interval of given magnitude is subject to statistical fluctuations; a count of N is subject to an estimated standard error given by

$$S = \sqrt{N} \tag{56}$$

The only way to overcome limitations imposed by the particulate nature of electricity and radiation is to scale up the experiment in physical size, time span, or intensity, all of which are subject to severe practical limitations. Doubling the source intensity in an x-ray experiment or doubling the counting time will (on the average) result in the number of counts being doubled, while the standard error is increased by a factor of $\sqrt{2}$ and the relative error is decreased in the same proportion. To improve the relative precision by a factor of 10 requires a scale-up factor of 100; thus the square-root relationship imposes a severe limitation to improvement of precision.

LIMITATIONS DUE TO THE UNCERTAINTY PRINCIPLE

According to quantum mechanics, one cannot measure simultaneously both of two conjugate variables (e.g., position and momentum or energy and time) to infinite precision; the product of the standard errors in the two variables must be greater than $h/2\pi$, where h is the Planck constant (6.64×10^{-34} J s). For instance the energy of an excited state of an atom relative to the ground state, as determined by the wavelength of an emitted photon, is somewhat uncertain because the lifetime of the excited state is limited by collisions or other perturbations if not by the finite probability of spontaneous emission. The uncertainty principle applies to the results of any *one* experiment comprising simultaneous measurements of a pair of conjugate variables. Suppose that a single photon emitted by a system under study (say a gas undergoing an electrical discharge) were to pass between the slits of a very-high-resolution spectrometer, pass through the dispersing element (prism or grating), pass between the very narrow slits of a detector, and then evoke from a photomultiplier tube a response that is recorded as a single event. The wavelength λ of the photon is determined by the angular position of the detector slits with respect to the dispersing element, and the energy of the photon is therefore determined ($E = h\nu = hc/\lambda$, where c is the velocity of light). The conjugate variable t is known, say from the pressure of the light-emitting gas, which determines the rms time between deactivating molecular collisions. That is just the one photon; many others from the same source possess slightly different wavelengths and therefore miss the detector slit opening. If the experiment were to stop right there, the determination of the wavelength that characterizes the spectral transition under study would be truly limited in precision due to the uncertainty principle.

However, that is not the way we would conduct the wavelength determination in practice. Let the detector slit scan slowly over a narrow range bracketing the position corresponding to the expected wavelength; photons will begin to enter the slit opening at a very low rate at first, and then at a gradually increasing rate, which goes through a maximum and declines; if the counting rate is automatically recorded, a more-or-less Gaussian peak will be obtained. This constitutes in effect a large number (perhaps several thousand) of "individual experiments." The width of the peak is related to the estimated standard deviation S; the center of the peak can be measured with a precision corresponding to S_m. In any practical spectroscopic instrument, a peak of finite width is obtained even with spectral lines of essentially zero width, owing to the finite width of the detector slit and other geometrical factors; in fact,

$$S = (S_{qm}^2 + S_{inst}^2)^{1/2} \tag{57}$$

where S_{qm} characterizes the quantum mechanical estimated standard deviation and S_{inst} characterizes the instrumental estimated standard deviation. In this kind of scanning spectrophotometer, the number of "individual experiments" is so large that the square-root relationship between precision and number of measurements almost never provides a severe limitation; in the experiments described above, the precision of the determination of the center of the spectral line is limited more by uncertainties in the geometric dimensions of the apparatus than by quantum mechanics.

SUMMARY

The principal equations in Part B of this chapter and key references to the text are cited here as a practical aid in the treatment of experimental data. This summary is not intended to be a substitute for thoughtful judgment. Use the equations with care; if there is any doubt concerning the applicability of an equation, review the associated text material.

Rejection of a Discordant Value with Q Test. For $3 \leq N \leq 10$, where N is the number of measurements of the same quantity, calculate

$$Q = \frac{|(\text{suspect value}) - (\text{value closest to it})|}{(\text{highest value}) - (\text{lowest value})} \tag{9}$$

and compare it with Q_c from Table 1. If $Q > Q_c$ then reject the suspect value; otherwise retain it. If $N > 10$, reject the suspect value if and only if it differs from the mean of the others by more than $2.6 S_m$, where S_m is the estimated standard deviation of the mean of the others.

Mean, Variance, Estimated Standard Deviation S. For a data sample of N values of x, denoted x_i where $i = 1, 2, \ldots, N$, the unweighted mean is

$$\bar{x} = \frac{1}{N} \sum_{i=1}^{N} x_i \tag{2}$$

The variance S^2 is given by

$$S^2 = \frac{1}{N-1} \sum_{i=1}^{N} (x_i - \bar{x})^2 = \frac{N}{N-1}(\overline{x^2} - \bar{x}^2) \tag{5}$$

The estimated standard deviation S is the square root of the variance:

$$S = \frac{1}{\sqrt{N-1}} \left[\sum_{i=1}^{N} (x_i - \bar{x})^2 \right]^{1/2} \tag{7}$$

Estimated Standard Deviation of the Mean S_m; Confidence Limit for the Mean. The value of S_m is given by

$$S_m = \frac{S}{\sqrt{N}} = \frac{1}{\sqrt{N(N-1)}} \left[\sum_{i=1}^{N} (x_i - \bar{x})^2 \right]^{1/2} \tag{8}$$

The 95 percent confidence limit of error Δ is

$$\Delta = t_{0.95} S_m = t_{0.95} \frac{S}{\sqrt{N}} \tag{29}$$

The value of $t_{0.95}$ is given in Table 3 under $P = 0.95$ for the applicable number of degrees of freedom

$$v = N - 1$$

Estimates Based on the Range. See Table 4 and Eq. (31).

Estimates Based on Experience. See the text on p. 50–51.

Presentation of Results with Uncertainties. See Table 5.

Propagation of Random Errors. Let x, y, z, ... be independent, directly measured quantities; let F be a result calculated from them with an equation represented symbolically by

$$F = f(x, y, z, \ldots) \tag{32}$$

$$\Delta^2(F) = \left(\frac{\partial F}{\partial x}\right)^2 \Delta^2(x) + \left(\frac{\partial F}{\partial y}\right)^2 \Delta^2(y) + \left(\frac{\partial F}{\partial z}\right)^2 \Delta^2(z) + \ldots \tag{42}$$

where $\Delta(x)$ represents the 95 percent confidence limit for x, etc. In many cases, the expression for $\Delta(F)$ or $\Delta^2(F)$ can be greatly simplified. See p. 54 for five common simple examples and pp. 54–59 for a discussion of how to handle complicated cases.

EXERCISES

1. For each of the following series of measurement values, determine whether any measurements should be rejected at the 90 percent confidence level. After rejecting values where appropriate, determine S, S_m, and Δ (95 percent confidence limit) for each series.

   (*a*) 2.8   2.7   2.7   2.5   2.9   2.6   3.0   2.6
   (*b*) 97.13   97.10   97.20   97.35   97.10   97.19
   (*c*) 0.134   0.120   0.109   0.124   0.131   0.119   0.135   0.132   0.132

2. A chemist dispenses a titrant, 0.1000 M HCl, from a burette at 25.0°C. The initial reading is taken six times; the values (in milliliters) are

         6.79   6.78   6.79   6.77   6.76   6.78

   The final reading is also taken six times; the values are

         28.02   28.03   28.01   28.02   28.02   28.03

   Calculate the number n of moles of HCl in the solution dispensed. Give S and S_m for the initial and final volumes, and give a limit of error (95 percent confidence) for n.

3. The heat of vaporization of a liquid may be obtained from the approximate integrated form of the Clausius-Clapeyron equation,

   $$\Delta \tilde{H}_v = -R \frac{\ln(p_2/p_1)}{1/T_2 - 1/T_1} = R \frac{T_1 T_2}{T_2 - T_1} \ln \frac{p_2}{p_1}$$

   (See Exp. 13 for the exact form of the Clapeyron equation. The integrated form given here involves certain reasonable approximations.) For water, the vapor pressure is measured to be 9.2 Torr at 10.0°C (283.15 K) and 55.3 Torr at 40.0°C (313.15 K). Taking R to be 8.3145 J K^{-1} mol^{-1}, calculate $\Delta \tilde{H}_v$. Taking the limit of error (95 percent confidence) in a temperature measurement to be 0.1 K and that in a pressure measurement to be 0.1 Torr, calculate the uncertainty in $\Delta \tilde{H}_v$. A handbook gives 43,893 J mol^{-1} for $\Delta \tilde{H}_v$ at 25°C.

4. A student determines the density of a solid with a pycnometer (see Sample Report, Chapter I) at 25.0°C. The weight of the empty, dry pycnometer is 6.2330 g. Some of the solid material is introduced; the weight of pycnometer plus solid is 39.4156 g. Water (density $\rho = 0.997048$ g cm^{-3} at 25.0°C) is added to fill all space not occupied by the solid; the weight of pycnometer plus solid plus water is 46.1357 g. The pycnometer is emptied and filled with water only; the weight is 19.7531 g. The student's estimated 95 percent confidence limit of error in any single weighing is 0.0020 g. Calculate the density and its 95 percent confidence limit of error, and present it and its limit of error in proper form with the correct number of significant figures and proper units.

REFERENCES

1. P. R. Bevington and D. K. Robinson, *Data Reduction and Error Analysis for the Physical Sciences,* 3d ed., McGraw-Hill, New York (2003).

2. L. F. Shampine, R. C. Allen, Jr., and S. Pruess, *Fundamentals of Numerical Computing,* Wiley, New York (1997).

3. J. Murphy and B. McShane, *Computational Mathematics Applied to Numerical Methods,* Prentice-Hall, Upper Saddle River, NJ (1998); J. L. Mohamed and J. E. Walsh, *Numerical Algorithms,* Oxford Univ. Press, New York (1986).

4. W. H. Press, S. A. Teukolsky, W. T. Vetterling, and B. P. Flannery, *Numerical Recipes in C—The Art of Scientific Computing,* 2d ed., Cambridge Univ. Press, New York (1992). [Available in C, Pascal, and FORTRAN versions. CD-ROM disk versions of the text are available for C and FORTRAN 90. *Numerical Recipes Example Book,* 2d ed., accompanied by disks containing the programs given in the books on C, Pascal, and FORTRAN 77 are also available.]

5. R. W. Hamming, *Numerical Methods for Scientists and Engineers,* 2d ed., pp. 567–573, 597–599, Dover, Mineola, NY (1987).

6. D. A. Skoog, D. M. West, and F. J. Holler, *Fundamentals of Analytical Chemistry,* 7th ed., Harcourt College, Fort Worth, TX (1996); D. A. Skoog, *Analytical Chemistry: An Introduction,* 7th ed., Harcourt College, Fort Worth, TX (2000).

7. E. B. Wilson, Jr., *An Introduction to Scientific Research,* pp. 232–242, Dover, Mineola, NY (1998).

8. W. J. Dixon, *Ann. Math Stat.,* **22,** 68 (1951); R. B. Dean and W. J. Dixon, *Anal. Chem.* **23,** 636 (1951).

9. J. R. Taylor, *An Introduction to Error Analysis,* 2d ed., University Science Books, Sausalito, CA (1997); J. Pitman, *Probability,* Springer-Verlag, New York (1993).

10. F. Reif, *Fundamentals of Statistical and Thermal Physics,* pp. 37–39, McGraw-Hill, New York (1965).

11. A. I. Khinchin, *Mathematical Foundations of Statistical Mechanics,* p. 166, Dover, Mineola, NY (1998).

12. W. H. Beyer (ed.), *CRC Handbook of Tables for Probability and Statistics,* 2d ed., pp. 296, 306, CRC Press, Boca Raton, FL (1968); W. H. Beyer (ed.), *CRC Handbook of Mathematical Sciences,* 6th ed., CRC Press, Boca Raton, FL (1987).

13. R. S. Burlington and D. C. May, *Handbook of Probability and Statistics with Tables,* 2d ed., p. 325, McGraw-Hill, New York (1970).

GENERAL READING

D. C. Baird, *Experimentation: An Introduction to Measurement Theory and Experimental Design,* 3d ed., Prentice-Hall, Upper Saddle River, NJ (1994).

P. R. Bevington and D. K. Robinson, *Data Reduction and Error Analysis for the Physical Sciences,* 3d ed., McGraw-Hill, New York (2003).

W. H. Beyer (ed.), *CRC Standard Probability and Statistics: Tables and Formulae,* 29th ed., CRC Press, Boca Raton, FL (1991).

E. O. Doebelin, *Measurement Systems,* 4th ed., McGraw-Hill, New York (1989).

S. Ghahramani, *Fundamentals of Probability,* 2d ed., Prentice-Hall, Upper Saddle River, NJ (2000).

N. C. Giri, *Introduction to Probability and Statistics,* 2d ed., Dekker, New York (1993).

R. W. Hamming, *Numerical Methods for Scientists and Engineers,* 2d ed. Dover, Mineola, NY (1987).

R. G. Mortimer, *Mathematics for Physical Chemistry,* 2d ed., Academic Press, San Diego, CA (1999).

J. R. Taylor, *An Introduction to Error Analysis,* 2d ed., University Science Books, Sausalito, CA.

E. Whittaker and G. Robinson, *The Calculus of Observations,* 4th ed., Blackie, Glasgow (1944).

E. B. Wilson, Jr., *An Introduction to Scientific Research.* Dover, Mineola, NY (1998).

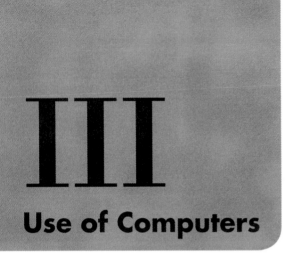

III

Use of Computers

In most fields of physical chemistry, the use of digital computers is considered indispensable. Many things are done today that would be impossible without modern computers. These include Hartree–Fock ab initio quantum mechanical calculations, least-squares refinement of x-ray crystal structures with hundreds of adjustable parameters and many thousands of observational equations, and Monte Carlo calculations of statistical mechanics, to mention only a few. Moreover computers are now commonly used to control commercial instruments such as Fourier transform infrared (FTIR) and nuclear magnetic resonance (FT-NMR) spectrometers, mass spectrometers, and x-ray single-crystal diffractometers, as well as to control specialized devices that are part of an independently designed experimental apparatus. In this role a computer may give all necessary instructions to the apparatus and record and process the experimental data produced, with relatively little human intervention.

Accordingly, familiarity with computers and their applications is essential to a sound foundation in chemistry, particularly in physical chemistry. Software programs are now widely available for sophisticated mathematical calculations of the structures and physical properties of molecules.[1–3] Some of these are considered in this chapter; however, the emphasis will be on the use of computer software for the organization, display, analysis, and discussion of data that you might record in experiments like those in this book. To a large extent, this task can be achieved with commercial spreadsheet and word processing programs,[4] and no detailed knowledge of computer programming is necessary. This is not the case as applications become more complex, and ultimately some knowledge of programming structure and language is very valuable for a physical chemist. This is especially true when you wish to interface a computer to an experiment, a topic treated briefly at the end of this chapter. As discussed there, many of the experiments in this book lend themselves to computerized data acquisition, and this can provide valuable experience with this important application of computers.

WORD PROCESSORS

It is unlikely that many students taking a physical chemistry laboratory course at the junior or senior level have not already used a word processor in writing reports and compositions in previous courses. Examples of such word processing programs are Microsoft

Word and WordPerfect. Although these and other commercial word processors vary somewhat in the details of their operation, all have the basic capabilities for drafting quality documents and allow the insertion of tables and figures from spreadsheets and other programs so that technical reports can be constructed with relatively little effort. This chapter assumes that you are familiar with keyboard entry, with the setting of margins and font types and sizes, with the use of subscripts and superscripts, and with the insertion and placement of graphs and tables. For scientific writing, use of a simple font such as Times Roman, used in this book, in a point size of 12 is recommended. The use of boldface and different font sizes for headings in a report can make it more attractive and readable. Most word processors provide Greek and other special characters of use in chemistry and many allow one, with some practice, to construct and display complicated equations and chemical formulas. Spelling and grammar checkers are usually built into word processors, as are thesaurus and dictionary tools, and these aids can be helpful.

If you have not already mastered the use of word processing tools, it is strongly advised that you develop these skills. Of course, the attractive appearance of a report or paper does not determine its intellectual content, and the old computer programming adage—"garbage in, garbage out"—applies here too. Neat garbage will not impress a laboratory instructor, employer, journal editor, or granting agency. Nonetheless, efficient use of analysis and presentation software can make easier the tasks of extracting physical meaning from a set of data and of assessing its value. It can aid significantly in communicating the important results in a report or paper. The development of good habits in assembling clear and logical laboratory reports will pay off throughout your scientific career.

SPREADSHEETS

Next to a word processor, the software package of greatest use to most students and scientists is likely to be a spreadsheet. Originally designed to resemble an accountant's entry ledger and to serve the large-business market, spreadsheet programs have grown increasingly sophisticated and have become very useful to scientists in displaying and analyzing data of all types. Because of their mass market, spreadsheet programs are relatively inexpensive, and most offer impressive graphing and display capabilities along with a remarkable number of mathematical operations of use in physical chemistry and other fields. Examples of spreadsheet programs are Excel, Lotus 1-2-3, and Quattro Pro. Most of the specific features and functions of these three programs are very similar, and a person familiar with one can quickly learn another. Each program comes with extensive tutorial and Help features, which can be utilized in learning the basics of spreadsheet manipulations and in making use of some of the more advanced options.[4]

GENERAL STRUCTURE

A typical spreadsheet consists of pages of worksheets, each with a grid of rows and columns, as shown in Fig. 1. The columns are labeled alphabetically and the rows numerically, and the intersection of a row and column is a *cell*. Reference to the contents of a cell is made by indicating its unique column/row address, e.g., B5, Q77, etc.; and the active or current cell (B14 in Fig. 1) is identified by outlining and is also indicated at the top left of the worksheet. The contents of a cell can be text, a number, or a formula whose numerical value is displayed in the worksheet. Movement from one cell to the next can be done with arrow, Page Up/Down, or Home/End keys, or by use of a mouse to reposition a cursor. Clicking the left key of the mouse makes the cursor cell active, and by holding the key down one can select a block of cells for copying, moving, deleting, etc. The right mouse key gives access to the cell or block properties such as the numerical format, text font and

FIGURE 1

An Excel spreadsheet comparing potential energy curves calculated for HCl for Morse and harmonic oscillator models with ab initio quantum mechanical results obtained with the program Gaussian. The example illustrates the use of cell formulas and some of the text Format options, such as bold and italic fonts of various sizes, subscripts and superscripts, and Greek and other special characters.

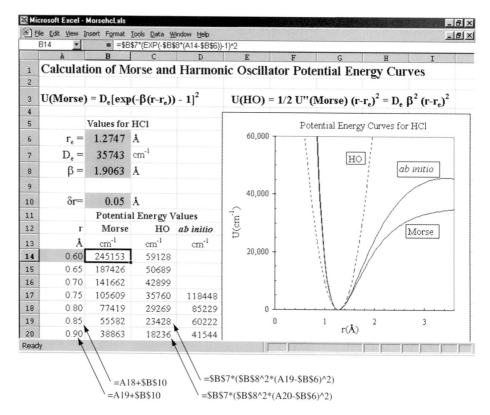

size, color, outlining, and so on. The Open and Save operations in the File menu can be used to open or save a spreadsheet, and the Print operation in this menu allows one to print all or a portion of the worksheets, including graphs, as in Fig. 1. You should experiment freely in order to become comfortable with these various manipulations.

DATA ENTRY AND MANIPULATION

Text and numerical data are entered into a cell by simply moving to the desired location and typing it. In the Excel worksheet example of Fig. 1, potential energy curves are calculated for Morse and harmonic oscillator models of a diatomic molecule from parameters entered into cells B6, B7, and B8. Data from a calculation using the ab initio quantum mechanical program Gaussian, described later, have been entered into column D. Under an operating system such as Windows, this transfer is easily effected by Copying the desired data from one Windows program to an intermediate Clipboard and then Pasting it into the spreadsheet or, in some cases, by directly "dragging" a selected block between two active programs. Alternatively, data files in standard ASCII form can be directly read into a spreadsheet and parsed as desired to separate into columns x, y, and other text or numerical quantities.

Text can be displayed, and printed, in various fonts and font sizes, as subscripts or superscripts, or as special math or Greek characters by modifying the characteristics of cell entries under the Format options at the top of the worksheet (also accessible via the right mouse button). Similarly, one can choose centering or left–right adjustment (alignment) of the text or numbers, bold or italic text, the widths of columns, and the format of numerical entries (0.00560 or 5.60E2, $27.37, etc.). Numerical data typically are stored and generated with at least 15 digits of accuracy, regardless of the number entered or displayed. Other spreadsheet niceties include color choices for text and shading and outlining of selected areas to distinguish, for example, cells for input data from those for calculated quantities. While many of these features are unnecessary for quick calculations and most student reports, it is useful to be aware of their existence, especially since they create a logical structure for spreadsheets, making them easier to understand by others and by yourself at a later time.

FORMULA ENTRY AND ADDRESSING

One of the most useful features of spreadsheets is the ease with which repetitive calculations can be done by copying the formula in one cell to others. A formula in a cell is a mathematical operation that can utilize values contained in other cells, as shown by the formula content of the active cell B14. This formula is displayed in the information bar at the top of the spreadsheet of Fig. 1 and is the equation for the Morse potential function, with the cell addresses of the constants r_e, D_e, β, and the variable r entered instead of numerical values. By using the Edit menu (or the right mouse button), the formula operation of B14 can be Copied and then Pasted into cells B15, B16, . . . , B73, thus giving a series of 60 points in this column. In doing this, the *relative address* of the r variable changes to A15, A16, . . . , but the *absolute addresses* of the constants ($B$6, $B$7, $B$8) are fixed by using the $ prefix. In a similar way, the formula contents indicated at the bottom of the figure for cells C19 and C20 illustrate the address changes resulting from the Copy and Paste operations for the harmonic oscillator formula of cell C14.

Also shown at the bottom of the figure are the formulas of cells A19 and A20, which reveal that each cell consists of the value in the cell above plus the r variable increment δr specified in B10. The initial value of r is entered into cell A14 (which is shaded to mark it as an input cell). In this way, the starting point and r increment of the calculation can be easily changed without changing all the cell formulas. Similarly, the constants r_e, D_e, and β entered into B6, . . . , B8 for HCl can be changed for calculations for another molecule, such as I_2 studied in Exp. 39.

FUNCTIONS

Many mathematical, logical, and statistical operations are provided as *functions* in spreadsheets. In the example of Fig. 1, the exponentiation operation is invoked in Excel by typing EXP (or exp) followed by an argument in parentheses. With Lotus 1-2-3 and Quattro Pro, functions are preceded with an @ symbol, e.g., @ EXP($-1.43*$A7), @SQRT(B9), @LN(27), @LOG($27+$$A4), @SIN(THETA*@PI/180), etc. In the last expression, THETA is a name that has been assigned to a particular cell location, a procedure that can make formulas involving it more readable. Logical operations such as the @IF(Cond,TrueExpr, FalseExpr) function can be used to provide conditional values in calculations: e.g., the entry @IF(A1>B1,A1,B1) in cell C1 would place the larger of the values in cells A1 and B1 into C1.

Many of the statistical quantities discussed in Chapter II and in Chapter XXI on least-squares methods are available as functions. For example, the average of entries in a range of cells, say B7, . . . , B23, is easily obtained by typing @AVG(B7, . . . , B23) in a cell where the result is desired. Likewise, the estimated standard deviation S, defined in Eq. (II-7), for this sample set is given by @STDS(B7, . . . , B23) in Lotus 1-2-3 and Quattro Pro and by STDEV(B7, . . . , B23) in Excel. The maximum and minimum values in this range are @MAX(B7, . . . , B23) and @MIN(B7, . . . , B23).

Of special interest is the normal probability function discussed in Chapter II [see Eq. (II-14)]:

$$P(x - \mu) = \frac{1}{\sqrt{2\pi}\sigma} \exp\left[-\frac{(x - \mu)^2}{2\sigma^2}\right] = @\text{NORMDIST}(x, \mu, \sigma, 0) \qquad (1)$$

The value of this is obtained with the indicated command and the integral from $-\infty$ to x is given by @NORMDIST($x, \mu, \sigma, 1$). As noted in Chapter II, for small sample sets with v degrees of freedom, the Student t distribution function is more appropriate and the critical values of t listed in Table II-3 can be obtained directly as @TINV($1 - P, v$), e.g., $t_{0.95}$ = @TINV($1 - 0.95, 6$) = 2.45. This is used with S to calculate the conventional 95 percent confidence interval Δ, Eq. (II-29), for the average of a series of N measurements. Even more directly, Δ for the average of the B7-to-B23 cell range (which contains $N =$ @COUNT(B7, ..., B23) = 17 data points) can be simply calculated as

$$\Delta = t_{0.95}\frac{S}{\sqrt{N}} = @\mathrm{CONFIDENCE}(1 - 0.95, @\mathrm{STDS}(\mathrm{B7}, ..., \mathrm{B23}), 17) \qquad (2)$$

GRAPHS

The plotting capabilities of spreadsheets are extremely convenient in presenting data and in making comparisons with values calculated from a theoretical or empirical expression. A wide variety of two- and three-dimensional graph modes are offered, but the most generally useful one for scientific presentations is the simple Y-versus-X plot. For example the graph inserted in the Excel spreadsheet of Fig. 1 is an $X-Y$ plot of the three Y series B14 ... B73, C14 ... C73, D14 ... D73 versus the X variable A14 ... A73. Similarly, Fig. 2 shows a Quattro Pro spreadsheet with plots of the observed and calculated absorption coefficients of iodine vapor versus temperature, using data from Table 48-2 on the sublimation of this molecule. The inset shows the Source Data menu, where the appropriate X and Y spreadsheet data ranges are entered (alternatively, the Y series can be selected and "dragged" directly onto a graph insert). By positioning the mouse cursor on various elements of a graph and double clicking the left mouse button, one can add titles, adjust the X and Y scales, and choose separate left or right Y scales for each Y series. Different plotting symbols, termed *markers* in spreadsheets, can be selected for each Y series, as can the choice of lines (solid, dashed, dotted, etc.) connecting the points. It should be noted that the multiple Y series in Fig. 1 have the same X data values. Often one wishes to plot separate series $x_1 ... x_m, y_1 ... y_m,$ and $x'_1 ... x'_n, y'_1 ... y'_n$. Excel and Quattro Pro have graphic options that provide this capability.

In general, experimental data should be presented as points with markers but without connecting lines. Values calculated from a model for comparison should be displayed without markers and should be connected with a line. Thus, for example, in Fig. 2 the experimental data in column B are plotted with markers only, while the calculated values in column C are plotted without markers and with straight-line segments between each point. If the experimental data are too closely spaced to give a reasonable display with markers, a line can be used and any theoretical fit should then be shown with a dashed line. Frequently, as in Fig. 2, it will be difficult to distinguish the difference between observed and calculated values, and it is recommended that a curve of residuals (obs. – calc.) then be displayed using the secondary Y axis scale on the right-hand side of the figure. By choosing an appropriate manual (as opposed to automatic) scaling of the right axis, the residuals can be positioned at the bottom of the figure, without confusing overlap with the main curve. Such a display of residuals is very helpful in judging whether systematic trends exist in the data that might warrant extensions in a theoretical model, as illustrated in the following section.

LINEAR REGRESSIONS

For almost all measurements in physical chemistry, the objective is to compare experimental results with values predicted from a theoretical relationship derived from a physical model, with the goal of assessing the validity of the model and of perhaps improving

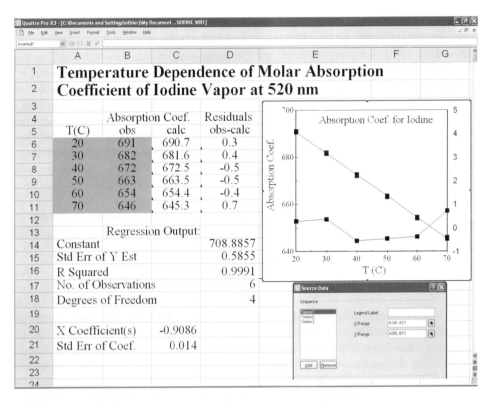

FIGURE 2
A Quattro Pro spreadsheet plot showing the I_2 vapor absorption coefficient at 520 nm as a function of temperature. Comparison is made with values obtained from a linear fit using the Regression operation under the Tools menu, and the obs.−calc. residuals are plotted on the secondary Y axis on the right of the figure. The inset shows the Graph menu in which the series plotted are indicated. The A in the range references A:A6 ... A11, and A: B6 ... B11 refers to sheet A of the spreadsheet; up to 256 sheets can be used in a spreadsheet notebook.

it. In the example of Fig. 2, the molar absorption coefficient ε of iodine vapor at the peak of an unresolved visible spectrum is found to have a slight temperature dependence, and this must be taken into account in Exp. 48 in order to determine accurately the sublimation vapor pressure from the absorbance and Beer's law. This temperature dependence comes from changes in the Boltzmann populations of different vibrational–rotational levels as temperature is altered, and a proper theoretical modeling of this would be complex. Since the measured variation is small and smooth, a simple linear interpolation would be adequate to obtain ε at intermediate temperatures between those listed. Alternatively, one could fit the entire data set to an empirical relation

$$\varepsilon = a + bT \tag{3}$$

and the least-squares values for a and b could then be used to calculate ε at any desired temperature.

A least-squares fit of Eq. (3) to the data is easily done using the spreadsheet linear Regression Tool. In the menu for this, the ranges of the independent X variable T and the dependent Y variable ε are entered, as is the output location. An option is offered to constrain the Y data to be zero at X equal zero, but this would not be appropriate in the present case. The Regression output, shown in Fig. 2, gives the a and b constants (708.886 and –0.9086) that are calculated from the relations

$$a = \frac{1}{D}\left(\sum x_i^2 \sum y_i - \sum x_i \sum x_i y_i\right) \tag{4}$$

$$b = \frac{1}{D}\left(m\sum x_i y_i - \sum x_i \sum y_i\right) \tag{5}$$

$$D = m\sum x_i^2 - \left(\sum x_i\right)^2 \tag{6}$$

In these relations, m is the number of observations. For the derivation of these expressions, see Chapter XXI and Eqs. (XXI-20a, b), (XXI-21). In comparing these equations, it should be noted that Eqs. (4) to (6) are the *equal-weights* versions of (XXI-20a, b), (XXI-21); i.e., if the weights w_i for each measurement y_i are all assigned the same value w, this value cancels out in Eqs. (4) to (6). Such a choice of equal weights is reasonable for the absorption coefficient data, since no basis such as standard deviations of each datum is available to estimate w_i and there is no reason to favor one measurement value over another. Basic spreadsheet regression operations do not offer weighting options and thus implicitly assume equal weights with $w = 1$. We adopt this choice throughout this chapter. The preferred, but more involved, procedure in which other weighting choices are used is discussed in Chapter XXI, and an example is given there of a weighted least-squares regression using spreadsheets.

The Regression output gives three quantities that are helpful in judging the quality of the fit of the model to the data. The first of these is the correlation coefficient R Squared, defined[5] as

$$R^2 = \frac{\left(m\sum x_i y_i - \sum x_i \sum y_i\right)^2}{\left[m\sum x_i^2 - \left(\sum x_i\right)^2\right]\left[m\sum y_i^2 - \left(\sum y_i\right)^2\right]} \tag{7}$$

This can range from 0 to 1, with 1 indicating a perfect fit and low values suggesting that the relationship between the X and Y variables is weak or nonexistent. An R^2 value of 0.912 means that about 91 percent of the variation in Y is explained by the change in X. Although it depends on the application and the number of samples, fits with values of R less than about 0.9 ($R^2 < 0.81$) would generally be considered poor.

A second quantity related to the quality of the fit is the estimated standard error in Y (Std. Err. of Y Est.), which is an estimate of the standard deviation of an observation of unit weight. This quantity, which we denote as ESE, is defined by the equation

$$\text{ESE} = \left(\sum \frac{(y_i - y_{i\text{calc}})^2}{\nu}\right)^{1/2} \tag{8}$$

where the degrees of freedom ν is m minus the number of adjustable parameters n ($\nu = 6 - 2 = 4$ in this example). For the equal-weights case, ESE is proportional to the "reduced chi-square" quantity χ_ν^2 discussed in Chapter XXI, the relation being $\chi_\nu^2 \simeq w\,\text{ESE}$ [see Eq. (XXI-31b)]. Since for equal-weights fitting, ESE and χ_ν^2 are both proportional to the sum of residuals squared, $(y_i - y_{i\text{calc}})^2$, all three quantities are minimized by the least-squares operation. In the case where accurate absolute weights w_i are known and used in the least-squares procedure, the value of χ_ν^2 can be used in statistical tests of the goodness of fit (see this section in Chapter XXI). A value of zero would correspond to a perfect fit, but one doesn't expect better than random scatter of the experimental points about the values predicted by the model. If all deviations are random and the theoretical model is correct, a value of 1 is expected for χ_ν^2. If only relative weights are known, as in the present case, the magnitude of ESE is less useful, but changes in its relative value still can be helpful in judging the relative merits of two different models that might be fit to the data.

Finally, the standard errors, $S(a)$ and $S(b)$, in the model parameters are useful in evaluating the fit of a model to the data. Expressions for these are derived in Chapter XXI [see Eqs. (XXI-42a) and (XXI-42b)] and are

$$S(a) = \text{ESE}\left(\sum \frac{x_i^2}{D}\right)^{1/2} \tag{9}$$

$$S(b) = \text{ESE}\left(\frac{m}{D}\right)^{1/2} \tag{10}$$

Note that $S(b)$ is given directly below the constant b (i.e., the "X coefficient") in the output of Fig. 2. Curiously, $S(a)$ is usually *not* given in the simplest Lotus 1-2-3 and Quattro Pro Regression formats. For the simple two-parameter case, it can be calculated from

$$S(a) = S(b) \left(\sum \frac{x_i^2}{m} \right)^{1/2} = S(b) \left[\frac{@SUMSQ(A6\ldots A11)}{@COUNT(A6\ldots A11)} \right]^{1/2} \tag{11}$$

If the model has more parameters, a convenient procedure for obtaining $S(a)$ is given in the next section. Alternatively, the Advanced Regression tool of Quattro Pro or its counterpart in Excel can be used to obtain all parameters and their standard errors directly, along with other statistical quantities.

The standard errors $S(a)$ and $S(b)$ define plus and minus uncertainties in a and $b;$ they would be zero for a perfect fit and generally should be small relative to the coefficient values. Should the error in a parameter times $t_{0.95}$ exceed the parameter value, inclusion of the parameter in the model is dubious and the model should be subject to further examination, as illustrated later.

MULTIPLE LINEAR REGRESSIONS

Another general way of obtaining $S(a)$ directly is by doing the regression of the equation $y = 0 + ax^0 + bx^1$: i.e., do a *multiple* linear regression with the X array expanded to include x^0 (a column of 1s) and x^1, and with the intercept option set *equal to zero*. The lower left part of the Lotus 1-2-3 spreadsheet in Fig. 3 shows the result of such a regression, and by comparison with Fig. 2, it is seen that the same a and b values result but now standard errors are given for both. $S(a)$ is 0.6737, a value different from the value 0.5855 for ESE. This "trick" for obtaining $S(a)$ has the advantage that it can be used for more complicated multiple linear regressions, such as $f(x, y) = 0 + ax^0 + bx^1 + cx^2 + dy^1 + ex^1y^1$. Note that, although a term such as cx^2 is *quadratic* in the x variable, entry of it as x^2 in the spreadsheet makes the dependence a *linear* one so that the usual linear least-squares equations apply. In other cases, discussed later, the parameters cannot be isolated as linear coefficients, e.g., $f(x) = a/(b - cx^2)$, and *nonlinear* least-squares methods must be used.

As a further example of a multiple regression, consider the case where a quadratic term cT^2 is added to Eq. (3) in fitting the absorption coefficient data. The basis for considering such a term might be the apparent nonrandom distribution of the residuals that were calculated and plotted in Fig. 2 for the linear fit (Model 1). The bottom right side of Fig. 3 shows the corresponding results for Model 2, and it is seen that the constants a and b change by more than their previous standard errors when c is allowed to have a nonzero value (0.00161 in this case). The standard errors in a and b are larger for Model 2 but these, and that for c, are still small relative to a, b, and c. The correlation coefficient improves, as it must since there are more parameters to fit the data (although it could remain the same should the least-squares value be exactly $c = 0$). In judging the two models, one might speculate that the residuals for Model 2 in Fig. 3 have a more random appearance than those for Model 1 in Fig. 2, suggesting that the addition of the quadratic term might be warranted.

SIGNIFICANCE TESTS

An objective judgment of the relative merits of Models 1 and 2 can be made by means of an F test, as discussed in Chapter XXI. This is a statistical test to determine if a decision to include the cT^2 term is justified at the conventional 95 percent confidence level. For the present case, where the degrees of freedom ν_1 and ν_2 are 4 and 3 for the linear and quadratic models, respectively, the F_{obs} value is calculated as [see Eq. (XXI-38)]

$$F_{obs} = \frac{\nu_1(S_{(1)1})^2 - \nu_2(S_{(1)2})^2}{(S_{(1)2})^2} = \frac{4(0.5855)^2 - 3(0.3684)^2}{(0.3684)^2} = 7.105 \tag{12}$$

FIGURE 3

A Lotus 1-2-3 spreadsheet showing multiple linear regression results for two expressions describing the temperature dependence of the I_2 vapor absorption coefficient at 520 nm. The X-array block was chosen as A7 . . . B12 and as A7 . . . C12 for Models 1 and 2, respectively. By including the A7 . . . A12 column of ones in the X-array range and choosing the Regression option to constrain the intercept to be zero, the standard error $S(a)$ is obtained (see discussion in text). Plots of the calculated curve and residuals for Model 2 are shown, and these can be compared with plots for Model 1 in Fig. 2. The P and F tests at the bottom show that the neglect of the quadratic cT^2 term is justified at the 95 percent confidence level: i.e., Model 2 should not be used.

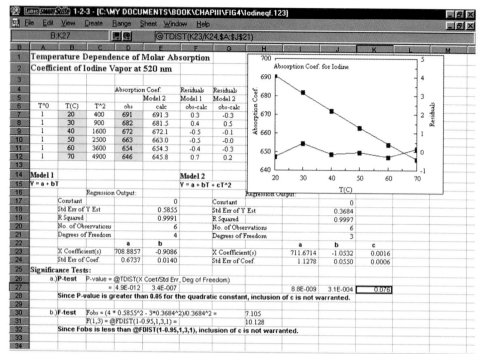

This value can then be compared with a limiting $F(\nu_1 - \nu_2, \nu_1) = F(1, 3)$ value at the 95 percent confidence level from Table XXI-1 or, more conveniently, from the Lotus 1-2-3 function @FDIST(1-0.95,1,3,1) = 10.128 [this function is @FINV(1-0.95,1,3) in Quattro Pro]. Since F_{obs} is *less* than the limiting value, the "better" fit of Model 2 is not statistically significant and the simpler linear model should be used. Put another way, if this test is done on an infinite number of new sets of similar absorption coefficient measurements, the decision to leave out the quadratic term will be correct 95 percent of the time.

A simpler method, related to the F test but more direct, to answer the question of which parameters a, b, c are significant is the calculation of the P value for each, using the spreadsheet T distribution function:

$$P \text{ value } = \text{TDIST}(X \text{ Coef}/\text{Std Err, Degrees of Freedom}) \tag{13}$$

If the P value for a parameter is greater than 0.05, the parameter is insignificant and should be dropped from the model. The calculation is done first for the model with the greatest number of parameters and the following "rule" is applied: *if any* P *value is greater than 0.05, repeat the regression, leaving the least significant parameter (largest* P *value) out of the fitting function.* Continue this procedure, successively eliminating all insignificant parameters until all P values are less than 0.05. This final fit gives the set of parameters that are significant at the 95 percent confidence level. An example of the P value calculation using Lotus 1-2-3 is shown in Fig. 3; in Excel, the P value is automatically given as part of the Regression procedure and in Quattro Pro it is calculated in the *advanced regression option.*

Use of this P test (or F test) procedure to judge the value of added parameters in fitting experimental data with a model is strongly encouraged, and it can be employed in a number of experiments in this book. For example, in Exps. 37 and 38 the possible inclusion of a centrifugal distortion term can be explored in fitting vibrational–rotational data for DCl and acetylene. Similarly, in Exp. 39 one can test the significance of a higher anharmonicity term $\nu_e y_e$ in fitting vibrational data for I_2. By using such tests, one can avoid the normal impulse to "do a better job" of fitting data by adding more, but statistically unsupported and perhaps physically meaningless, parameters to a theoretical model.

ADVANCED SPREADSHEET TOOLS

Many other operations can be carried out with spreadsheets, including matrix multiplication, operations with complex variables, and fast Fourier transforms. Some of these tools can be very useful in organizing and analyzing data of the type a student might record in experiments in this book. For example, the Sort operation allows one to rearrange a selected block of columns alphabetically or numerically, using one (or more, in order of precedence) of the columns for the ordering basis. An instructor can use such a procedure to sort a class list by name, Social Security number, or final score. A student doing Exp. 37 on the vibrational–rotational spectrum of DCl might employ Sort to rearrange the FTIR instrument output, which will consist of a file of interleaved peak wavenumbers for the $D^{35}Cl/D^{37}Cl$ isotopic molecules. After importing the peak wavenumbers into, say, column A of a spreadsheet, a value of 35 or 37 can be entered in column B beside each peak, using the fact that the $D^{35}Cl$ peaks will be more intense than the $D^{37}Cl$ features. A Sort operation on both columns, with B having precedence over A, can then be used to isolate each isotopic data set in increasing (or decreasing) wavenumber order for subsequent analysis. By making no entry for questionable or "noise" peaks (often given by FTIR peak-pick routines), these will be isolated and can be excluded. A similar sorting operation can be used in Exp. 39 to sort the electronic transition data for I_2 according to quantum numbers assigned for upper and lower states.

The Solver tool of spreadsheets can be used to obtain numerical solutions to a complex equation. For example, to obtain an x solution to the equation $y - \cos(x)\exp(-x) = 0$ for a particular y value, Solver uses methods such as the Newton–Raphson procedure (App. F) to iteratively vary a cell containing an initial guess of x until the value of a cell containing the formula is zero. The result will be the root closest to the initial guess; to determine other roots, a plot of the equation versus x should be made to identify approximate zero-crossing points.

Solver (or Optimizer in Quattro Pro) can also be used to do least-squares fitting of data with a theoretical function that cannot be cast in the form required for the usual linear regression. As an illustration, Fig. 4 shows a radiative decay curve measured by a student for a ruby crystal as part of Exp. 44. After laser excitation the digitized emission signal I decreases exponentially and can be analyzed for k (the radiative rate constant, whose reciprocal is the lifetime) according to the relation

$$\ln(I - I_\infty) = \ln I_0 - kt \tag{14}$$

Note that a linear regression can be done only if a value for I_∞ is determined experimentally or, as is normally the case, if it is experimentally adjusted to zero while blocking the excitation source. In this example, the background adjustment was neglected and the analysis must compensate for this oversight. The obvious curvature of the plot of $\ln I$ in the figure makes it clear that the student must consider ways to estimate I_∞. By examining the data at long times, a reasonable value of I_∞ could have been chosen but unfortunately data beyond 5 ms had been discarded because the signal had dropped so low that the resolution of the analog-to-digital converter began to limit the precision of the measurements. Rather than simply make a series of guesses at I_∞, the student decided to use the nonlinear fitting capability of Solver in Excel to obtain optimum values of the parameters I_∞, I_0, and k. This was done by entering initial guesses for these parameters in cells C10, C11, and C12 and computing the sum of squares of the residuals $(I - I_{calc})$ in cell C14 of the worksheet

FIGURE 4

An Excel spreadsheet illustrating the use of the Solver tool for nonlinear least-squares analysis of a fluorescent decay curve of a ruby crystal. The sum of the squares of residuals is calculated in cell C14 and is minimized in Solver by iterative variation of the parameters in cells C10, C11, and C12.

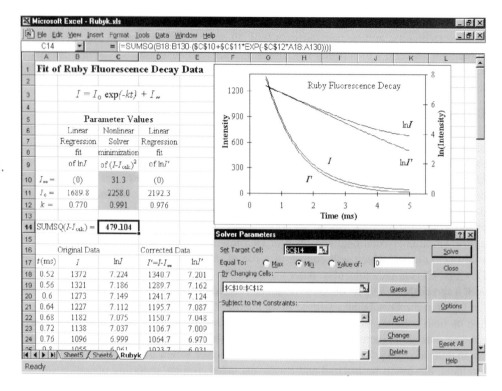

shown in Fig. 4. Note in the formula line at the top of the spreadsheet the compact equation for doing this in terms of array ranges, denoted by { } brackets in Excel. Solver was next used to minimize the value in this cell by iteratively varying the parameters, thereby yielding a value of 31.3 for I_∞. This value was then subtracted from I and the resultant intensity I' gave a satisfyingly linear $\ln I'$ plot, thereby salvaging the measurements and earning the student bonus points for ingenuity.†

More generally, this least-squares procedure can be useful in fitting with a variety of nonlinear relations, such as double exponential decays, $\ln(I - I_\infty) = \ln[A \exp(-k_A t) + B \exp(-k_B t)]$, or in Exp. 24, the expression $\ln(np^0 - p) = a + kt$ for the pressure changes in the thermal decomposition reactions of cyclopentene or *tert*-butyl peroxide. In the latter case, the parameters to be determined are p^0, a, and k since n is expected to be 2 or 3, depending on the stoichiometry of the reaction. If n were not known, it could also be determined using Solver, which in some spreadsheets can even add the constraint that n be an integer. Alternatively, other analysis and plotting packages, such as Curvefit, GRAMS, IGOR PRO, and Origin, exist with nonlinear fitting options, and many of these have the added benefit of giving standard errors for the parameters determined. A method for obtaining standard errors for nonlinear fit parameters generated by Solver is given in Ref. 6.

Finally, we note that programming of a sort is possible using the spreadsheet Macro feature. This allows one to record a sequence of operation (Copy, Move, Sort, Regress, etc.) carried out in the spreadsheet. This sequence is given a name and can then be repeated later by Playing this macro. Macros can be fairly sophisticated since it is possible to use them to branch to different locations, to run other Macros, and to perform "Do loops" in a manner as done in Visual Basic, C, or other programming languages.

---

†It should be noted that slightly different parameters are obtained when the least-squares quantity minimized is $[I - I_{calc}]^2$ rather than $[\ln I' - \ln I'_{calc}]^2$. Minimization of the latter is more common for first-order decay measurements such as this.

SYMBOLIC MATHEMATICS PROGRAMS

In addition to spreadsheet programs, a number of other software packages offer powerful mathematical capabilities of use in physical chemistry. Among these are symbolic mathematics programs such as Maple, Mathcad, and Mathematica, which allow solutions of algebraic or differential equations in terms of defined symbols. With these, one can obtain algebraic expressions for integrals, differentials, matrix products, or roots of equations, as well as numerical values for a given choice of all symbols and easy display of graphs of functions. The ease with which one can use such programs to solve complex problems, and to visualize the results, greatly reduces the "activation barrier" to applying mathematics to chemical problems.[3]

As an example of a symbolic calculation, Fig. 5 displays a Mathcad solution of one 4×4 block of the secular determinant of a Hückel molecular orbital calculation done in Exp. 41 for *ortho*-benzosemiquinone (compare with Table 41-2). Mathcad is a software package for numerical analysis but also makes use of a subset of the symbolic routines of Maple. The algebraic expansion of the determinant is generated and solved with two

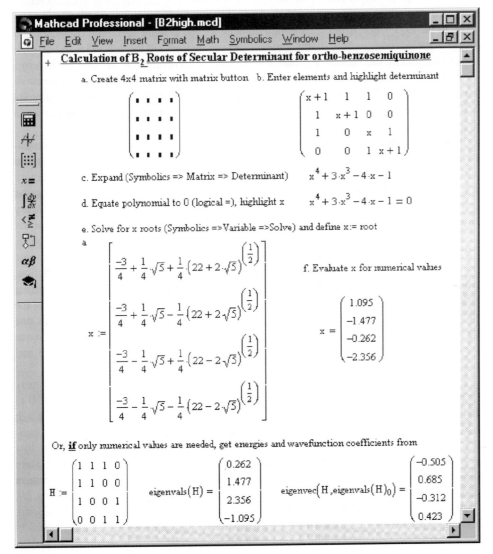

FIGURE 5

A display of the Hückel molecular orbital calculation of the energies of the B_2 block of *ortho*-benzosemiquinone. The figure illustrates the calculation of the determinant and its roots in symbolic form using Mathcad. In a separate calculation, the bottom portion shows the corresponding H matrix of coefficients and Mathcad operations giving the four $H_{0\ldots3}$ eigenvalues and the eigenvector of the first of these [$(H)_0 = 2.356$]. Note that the energies are given by $-H_{0\ldots3}$ and that the sign of the eigenvector elements can be multiplied by -1 without changing the relative values of the coefficients in the wavefunction.

Symbolics menu commands. The resultant exact expressions can then be evaluated for numerical values. Alternatively, as shown at the bottom of the figure, numerical values could have been obtained as minus the eigenvalues of the Hamiltonian matrix H, which is diagonalized via the *eigenvals* command. Similarly, the *eigenvec* command can be used to obtain eigenvector (wavefunction) coefficients for any one of the eigenvalues. This linear algebra procedure of diagonalizing matrices is widely used in physical chemistry.[7] Other applications of Mathcad in physical chemistry can be found in Ref. 8.

As one other example illustrating the power of symbolic mathematics programs, Fig. 6 shows a Mathematica calculation of Franck–Condon factors for the I_2 $B \leftarrow X$ electronic absorption spectrum studied in Exp. 39. These factors are the squares of the overlap integrals of the vibrational wavefunctions for the lower (v'') and upper (v') vibrational levels involved in a transition:

$$\text{FCF} = \left| \int \psi_{v'}(r)\psi_{v''}(r)dr \right|^2 \tag{15}$$

The absorption spectrum consists of sequences of transitions from $v'' = 0, 1, 2$ to various v' levels in the upper state, and the relative intensities of the vibration–rotation bands are given primarily by the product of the FCF value and a Boltzmann term, which can be taken to be $\exp(-hc\tilde{v}_e v''/kT)$. Common choices for the ψ's are harmonic oscillator and Morse wavefunctions, whose mathematical form can be found in Refs. 7 and 9 and in other books on quantum mechanics. The harmonic oscillator wavefunctions are defined in terms of the Hermite functions,[7,9] while the Morse counterparts are usually written in terms of hypergeometric[9] or Laguerre functions.[10] All three types of functions are polynomial series defined with a single statement in Mathematica, and they can be easily manipulated even though they become quite complicated for higher v values.

In Fig. 6, section 1 defines the constants for the I_2 molecule and calculates parameters to make the arguments of the wavefunctions unitless. One need only change these numbers to perform the calculations for some other molecule, such as N_2 or Br_2. Section 2 gives the two equations necessary to define the two types of wavefunctions, and section 3 shows the statements necessary to calculate and plot these functions for an arbitrary choice of v for X and B states. The curves shown are the squared wavefunctions for the $v'' = 2$ and $v' = 20$ levels of the X and B electronic states and, at the top of each figure, the $\psi''\psi'$ product whose integral squared is the Franck–Condon factor. The harmonic oscillator function shown at the left is similar but not identical to the Morse function, which has a higher amplitude (probability) at large r, near the so-called classical turning point. Note that the overlap region shifts to the right for the Morse function because of the steeper potential of the latter at the inner turning point. These differences increase as one goes to higher v' levels, where anharmonicity is increasingly important. This behavior is easily explored by simply changing the quantum number arguments in section 3 and viewing the resultant plots.

Section 4 gives the statements necessary to calculate and plot the relative intensity factors for the $v'' = 0, 1, 2$ vibrational sequences to upper levels ranging from $v' = 0$ to 70. Because of the Boltzmann factor, the intensities decrease in the order $v'' = 0 > 1 > 2$; and by changing T in section 4, the effect of temperature on these intensities can be examined. For $v'' = 1, 2$, the harmonic oscillator distributions show multiple maxima at low v' values, whereas the intensity distributions are quite different for the Morse case. Here the distribution is broader, with a single dominant maximum at higher v' values, a pattern conforming much more closely to the I_2 experimental results you will obtain in Exp. 39. It should be said that, although section 4 is compact in form, the calculation is lengthy and may take several minutes, since it involves numerical integrations of many terms in the higher v' wavefunctions. For accuracy in these integrations, it is necessary to set the

1. Generation of unitless parameters for ground (X) and excited (B) states; distance in angstroms, energy in cm-1, mass in kg/mol

re[X] = 2.6663`20.; we[X] = 214.502`20.; De[X] = 12547.`20.;
re[B] = 3.0247`20.; we[B] = 125.697`20.; De[B] = 4391.`20.; mu = (0.1269`20./2)/6.022`20.*10^23;
c=2.998`20.*10^10; h=6.626`20./10^34; pi=3.1416`20.; constant=(2*pi*Sqrt[(c*mu)/h])/10^10;
d[X] = (2*De[X])/we[X]; beta[X] = constant*re[X]*Sqrt[we[X]]; alpha[X] = beta[X]/Sqrt[d[X]];
d[B] = (2*De[B])/we[B]; beta[B] = constant*re[B]*Sqrt[we[B]]; alpha[B] = beta[B]/Sqrt[d[B]];

2. Normalized harmonic oscillator (HO) and Morse oscillator wavefunctions

$$H[r\_?\,NumberQ, e\_, v\_] := \frac{\text{HermiteH}\left[v, z = \text{beta}[e]\left(\frac{r}{re[e]} - 1\right)\right]e^{-\frac{z^2}{2}}}{\sqrt{\frac{\sqrt{\pi}\,2^v\,v!\,re[e]}{\text{beta}[e]}}};$$

$$M[r\_?\,NumberQ, e\_, v\_] := \frac{\text{LaguerreL}\left[v, a = 2\,d[e] - 2\,v - 1, y = \frac{2\,d[e]}{e^{\text{alpha}[e]\left(\frac{r}{re[e]}-1\right)}}\right]\sqrt{\frac{y^a}{e^y}}}{\sqrt{\frac{\left(\sum_{s=0}^{v}\frac{\text{Gamma}[s+a]}{s!}\right)re[e]}{\text{alpha}[e]}}}$$

3. Plots of HO and Morse squared functions and overlap functions for levels vX, vB

vX=1; vB=20; plot[f_]:= Plot[{f[r,X,vX]^2, 10+f[r,B,vB]^2, 18-f[r,X,vX]*f[r,B,vB]}, {r,2.4`20.,3.8`20.},
PlotRange -> {0, 24}, PlotPoints -> 100, PlotStyle → {Black}, FrameTicks -> {Automatic, None}, Frame -> True,
FrameLabel -> {"r (angstroms)", " X*X   B*B   X*B"}, LabelStyle-> Directive[Large,Bold],
DisplayFunction -> Identity];Show[GraphicsArray[{plot[H], plot[M]}]]

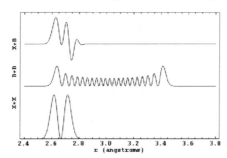 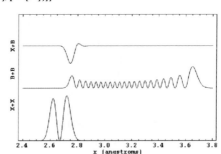

4. Plots of HO and Morse relative intensity (Boltzmann × Franck-Condon) factors

T = 298.15`20.; k=1.381`20./10^23; BoltzmannFactor[v_] := E^(-((v*we[X]*h*c)/(k*T)));
FCFactor[vX_,vB_,f_]:= NIntegrate[f[r,X,vX]*f[r,B,vB],{r,2.4`20.,2.9`20.},WorkingPrecision->20]^2;
RelativeIntensity[vX_,vB_,f_]:= BoltzmannFactor[vX]*FCFactor[vX,vB,f];

plot[f_] := Show[Table[ListPlot[Table[{vB, RelativeIntensity[vX, vB, f]}, {vB, 0, 70}], Joined -> True,
PlotRange -> All, PlotStyle → {Black},DisplayFunction -> Identity], {vX, 0, 2}], FrameTicks -> {Automatic, None},
Frame -> True, FrameLabel -> {"v for B state", "Relative Intensity"}, LabelStyle-> Directive[Large,Bold]];
Show[GraphicsArray[{plot[H], plot[M]}]], DisplayFunction → $DisplayFunction]

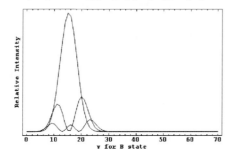 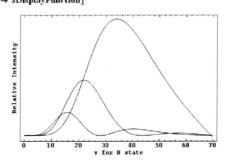

FIGURE 6

Direct printout of Mathematica commands to calculate and plot wavefunctions and electronic transition intensity factors for a diatomic molecule (I_2) using harmonic and Morse oscillator wavefunctions. See text for discussion. A Maple version of this calculation can be found on the Maple applications website.[11]

precision of all numerical quantities at a high value (achieved with the `20 entry for numbers in this example).

The expressions can also be used to calculate relative intensities for emission from, for example, the $v' = 43$ level (excited by an argon ion laser in Exp. 39) to various v'' levels. In this case, the Boltzmann factor in section 5 can be set equal to 1, since only one upper state is involved. The emission experiment shows dramatic intensity variations for different v'' values, and quantitative relative peak intensity values can be measured from the few narrow rotational lines seen for each vibrational transition.

QUANTUM MECHANICAL PROGRAMS

Also of considerable use in chemistry are ab initio and semiempirical programs that employ quantum mechanical principles to calculate structures, orbitals, charge distributions, and other molecular properties. Examples of ab initio programs useful for molecules of moderate size (less than about 20 atoms) are Gaussian and Spartan. For larger molecules, Gaussian, Spartan, and other programs such as HyperChem can make use of empirical parameters and less rigorous theory. The results are nonetheless of considerable value in modeling the structures of complex molecules and in visualizing their electron distributions. In all such programs, the user is "insulated" from the rather complex details of the calculations, a situation a quantum theoretician might bemoan but one that has made quantum mechanical model results available to many chemists.[1] A person with one year of physical chemistry background has been exposed to the fundamental relations used in such programs, i.e., the Schrödinger equation (with its coulombic form for the potential of interaction of charged nuclei and electrons) and the variational principle. The solution for energies and wavefunctions of the hydrogen atom are familiar, and the concepts of molecular orbitals, Slater determinants, configuration interaction, and self-consistent field calculations are usually discussed, albeit briefly, in most physical chemistry texts.

QUANTUM MECHANICAL "THEORY" EXPERIMENTS

A number of experiments in this book can benefit from an accompanying theoretical computation using available programs. For example, electron spin densities deduced from ESR measurements in Exp. 41 can be calculated using semiempirical programs, as can electronic transition frequencies of the dye molecules studied in Exp. 34. Bond energies and reaction enthalpies from energy differences between reactants and products can be compared with values measured in Exps. 6 and 7. Calculations of NMR chemical shifts, potential energy surfaces, reaction coordinates, and transition states are also possible.[2,12]

Because of the dramatic increases in the power of inexpensive computers and workstations, it is now possible to do sophisticated ab initio calculations even on a personal computer, using a program such as Gaussian for Windows.[13] The ease of this process is illustrated below by the few input file statements necessary for a Gaussian calculation of the HCl potential energy curve shown earlier in Fig. 1. Descriptive comments have been added to the right.

```
#HF STO-3G SCAN              (Sets Level as HF, Basis as STO-3G, and scans Re)
                             (Blank line)
HCl POTENTIAL CURVE          (Title)
                             (Blank line)
0 1                          (Molecular Charge and Spin Multiplicity)
H                            (Atom 1 Identifier)
Cl 1 Re                      (Atom 2 connected to Atom 1 by bond of length Re)
                             (Blank line)
Re=0.75 135 0.05             (Re to be varied from 0.75 Å in 135 steps of 0.05 Å)
```

The notation HF chooses a Hartree–Fock self-consistent-field (SCF) *level* of calculation which involves finding linear coefficients in the chosen basis set that give the lowest energy and which assigns each electron to a single spatial orbital. The shorthand notation STO-3G defines the wavefunction *basis set* to be used; here the STO means that hydrogenlike Slater-type orbitals are used for the radial part of the atomic wavefunctions and the 3G means that these are approximated by three Gaussian functions. Gaussian functions have exponential factors involving r^2 rather than r, a feature that makes evaluation of multicenter integrals easier in the Hartree–Fock calculation but at the expense of having to use more basis functions to obtain accurate results. Explicit "standardized" linear combinations of Gaussian functions that reproduce well the atomic properties have been determined and are described in Refs. 2 and 12. Examination of the potential energy curves in Fig. 1 shows that this relatively low-level calculation for HCl agrees well with the "experimental" Morse curve near the minimum position but less well at larger atomic separations. Better values of the limiting bond dissociation energy are obtained as one adds more basis functions and accounts more correctly for electron repulsions (see below).

Ab initio calculations can give a variety of molecular properties besides energy. By changing the SCAN command to OPT in the above list of input statements, the R_e value of the HCl bond length can be optimized to give the lowest energy and the bond length can then be compared with the value determined experimentally in Exp. 37. Addition of FREQ and POLAR commands to the second line yields properties such as the harmonic vibrational frequency ($\tilde{v}_e$), the dipole moment μ, and the polarizability tensor elements α_{xx}, α_{zz}, α_{xy}, etc. The calculated dipole moment can be contrasted with that determined in Exp. 30 and the mean polarizability $\alpha = (\alpha_{xx} + \alpha_{yy} + \alpha_{zz})/3$ can be compared with the value obtained from the refractive index measurement in the same experiment. Comparisons can also be made of $\tilde{v}_e$ with experimental values determined for I_2 (Exp. 39), as well as for HCl (Exp. 37) if $\tilde{v}_e$ is deduced from overtone measurements.

Similar calculations of structures and properties can of course be done for polyatomic molecules, such as CCl_4, C_6H_6, and C_2H_2, studied in Exps. 35, 36, and 38. Examples of input file forms for polyatomics can be found in program manuals and, for Gaussian, in the clear and readable text of Foresman and Frisch.[2] GaussView, an accessory program for Gaussian for Windows, provides a convenient means for preparing input files and for viewing ab initio output such as structures, molecular orbitals, vibrational modes, and spectra. Also given in Ref. 2 are examples of other calculations of chemical interest, such as reaction enthalpies of the type measured in combustion experiments; see Exps. 6 and 7.

In all such computations, it can be expected that the accuracy of the calculated results will generally improve with an increase in the size of the basis set and in the level of the calculation.[2,12] However, this improvement often comes at considerable increase in computation time, so explorations of basis set and level effects are probably best done as class projects, with each student responsible for one or two calculations. The grid below gives a

Basis Set	# Basis Functions	Level			
		HF	MP2	QCISD (T)	B3LYP
STO-3G	10				
3-21G	15				
6-31G (d)	21				
6-311+G (d, p)	36				
6-311+G (2d, 2p)	44				
6-311++G (3df, 3pd)	65				

specific example of such a project, with the basis set versus level that might be explored in calculating r_e, ν_e, μ, and α values of HCl, using in each case the experimental bond length of 1.2747 Å as the initial "guess."

The detailed meaning of the shorthand notation for the basis sets and the explicit linear combinations of Gaussian functions can be found in Refs. 2 and 12. The 6-31G(d) basis set has more than one size (exponential charge coefficient) for each orbital, and the symbol d means that d-like functions are added, for example, to C atoms, thereby allowing for so-called *polarization* effects. This basis set is also known as 6-31G* and is a standard for calculations involving small- to medium-sized molecules. A single + symbol indicates that orbitals with very large sizes (*diffuse* functions) have been added to atoms other than hydrogen, to account for the spatial extent of lone pair electrons. The ++ notation in the basis 6-311++G(3df,3pd) means that diffuse functions have been added to H as well, and the terms in parentheses mean that d and f functions have been added to the Cl atom, while p and d functions have been added to the H atom.

Also discussed in Refs. 2, 7, and 9 are the different levels of calculation, denoted by HF = Hartree−Fock, MP = Møller–Plesset, and CI = Configuration Interaction. The MP2 level calculation starts with the Hartree–Fock result and uses second-order perturbation theory to try to account for instantaneous (rather than average) interelectron repulsions. Such electron *correlation* effects can also be achieved by allowing one or more of the electrons to occupy HF-SCF orbitals, which correspond to energies above the lowest level. This procedure, termed *configuration interaction,* would be analogous to allowing the electron in the hydrogen atom to be in 2s, 2p, and higher levels. The result of mixing in excited orbital contributions of this type is to allow the electrons more spatial freedom and hence to reduce their repulsion energy. The designation QCISD(T) stands for quadratic configuration interaction including single, double, and (partially) triple excitations of the electrons.

Another method which takes into account electron correlation is *density functional theory* (DFT), which is implemented in Gaussian with the label B3LYP. The basic idea of DFT is to calculate the energy of an atom or molecule in terms of electron density rather than the wavefunction ψ.[14] The method has become increasingly popular because it can do reasonably accurate calculations on large molecules in significantly less time than that for the HF method.

It should be noted that only the HF and CI methods satisfy the variational principle and hence give an upper bound to the ground state energy. The DFT and MP perturbation methods do not satisfy this principle and can give an energy *lower* than the true energy. CI methods generally give the most accurate results for most properties but are also the most computationally intensive. In the table shown above, the total number of basis functions for H and Cl are indicated, and as part of the project, you should record the time for the calculation as the basis size increases. Using the Gaussian for Windows program on a modern PC, computation times for HCl should range from a few seconds to several hours in going from the top left to bottom right limits.

Although the emphasis in this book is on the *measurement* of molecular properties, the above theoretical "experiment" on HCl is strongly recommended to enable you to judge, by comparison with your laboratory results, the quality of such calculations in at least this relatively simple case. For such small gas-phase molecules, one might well question today the validity of the famous 1929 declaration of P. A. M. Dirac, Nobel laureate in physics:

> The underlying physical laws necessary for the mathematical theory of a large part of physics and the whole of chemistry are thus completely known, and the difficulty is only that the exact application of these laws leads to equations much too complicated to be soluble.

However, the whole of chemistry is much more than just small, noninteracting gas-phase molecules, and much remains to be done before quantum theory can accurately describe complex chemical reactions and the properties of condensed matter such as hydrogen-bonded liquids, superconductors, liquid crystals, and glasses. Thus the experimental skills you will hone in the experiments in this book can be expected to serve you well in the foreseeable future.

INTERFACING WITH EXPERIMENTS

In addition to their use in the analysis of data obtained in laboratory measurements, computers also play a critical role in the efficient and accurate acquisition of such data. For instance, a computer might be used to control the scanning of a spectrometer over a wavelength range while at the same time converting the spectral intensity into a series of data points that can be plotted and analyzed on the computer screen. Another common use in physical chemistry is the acquisition of temperature or pressure values in an experiment using appropriate transducers interfaced to a computer. In the past, such operations usually involved specialized hardware and detailed programming by the user, but in recent years much of the necessary effort has been reduced. A variety of commercial measurement systems is now available at reasonable cost, and these are easily implemented for many of the experiments done in this book. In this section, we consider some of the general requirements for interfacing of experiments to computers and then present two examples in some detail.

The process of measurement of a physical property using a computer involves four steps. Step one is conversion of the property, such as temperature, pressure, pH, and light intensity, to a voltage by a suitable transducer. Such transducers include thermistors, capacitance manometers, pH probes, photomultipliers, and photodiodes; and the characteristics of some of these are discussed in Chapters XVI to XIX. Step two involves conversion of the analog voltage from the transducer to a digital form recognizable by the computer. Typically, this involves 8- to 12-bit signals, corresponding to a precision of 0.8 percent or 0.05 percent, respectively, although analog-to-digital (A/D) converters with 16- to 24-bit precision are not uncommon. It is often necessary to amplify low-level (mV) voltages of transducers to values in the range -5 to 5 V to make full use of the digitizing capability of the A/D converters; frequently such amplification is provided as a programmable part of the A/D device.

Step three is the transfer of the measurement information to the computer. If the A/D board is in the computer itself, this involves a simple connection of transducer/amplifier leads to inputs on the back of the computer. However, to reduce electrical noise picked up by connecting lines and generated by the computer, it is frequently better to convert the low-level signals to digital form remotely, closer to the transducer itself. Transfer of this digital information can then be made reliably by means of parallel (8 bits simultaneously) or serial (sequential bit by bit) methods. For parallel transmission, an early development was the IEEE-488 bus, also known as the HP-IB (Hewlett-Packard Instrument Bus) or GPIB (General Purpose Interface Bus). This bus is still in wide use for connection of up to 15 devices to a single computer and consists of a bundle of 24 wires capable of carrying 8 bits (one byte) per transfer at rates of about 64 Mbits/s. For serial transfer, fewer wires are used and one of the first Recommended Standards was termed RS-232 and utilized a port present on all computers even though it was nominally limited to rates of 20 Kbits/s or less. This "legacy" serial bus has now been largely supplanted by much faster connections such as Ethernet, Firewire, and USB (Universal Serial Bus). The most commonly used is now the USB connection, which can achieve data transfer rates ranging from 1.5 to 480 Mbits/s. Most modern computers have several USB ports for cable connection to devices, and many also make use of wireless transfer, which can achieve rates up to 480 Mbits/s for distances less than 3 meters and 110 Mbits/s at 10 meters.

Step four in the interfacing operation is the use of computer software to communicate with a device and to capture the information it sends for analysis and display. This can be done at a basic level using machine language or higher programming tools such as Visual Basic, but increasingly this capability is provided by manufacturers of instruments, measurement systems, or interface boards. This relieves the user of much of the effort in interfacing devices and allows efficient and easy set up of different experiments typically done in physical chemistry laboratories. As illustration, we consider two examples of measurement approaches that can be used in a number of the experiments in this book.

Example One. The first of these examples involves the use of inexpensive measurement systems such as those produced by MeasureNet Technology, Ltd., and Vernier Software & Technology. The LabPro system produced by the latter company consists of an external 12-bit A/D signal process module to which up to four sensor devices can be connected. A wide variety of sensors is available, including pressure sensors, thermistor temperature probes, and simple voltage probes. Digital data transfer to the computer is done over a serial USB cable at rates up to 50,000 measurements/s. Software is provided to permit easy collection and display of data and to generate files for processing in spreadsheet and other common programs. Similar capabilities are provided by the MeasureNet system, which also allows multiple stations to connect to a single computer.

Figure 7 shows the screen display of measurements made using the Vernier LabPro system in the development of a liquid crystal experiment for this text (Exp. 15). Tabulated and plotted versus time are the temperature and the (amplified) photomultiplier voltage as a liquid crystal sample was heated through the nematic to isotropic phase change and then cooled back down. The temperature probe is a calibrated thermistor contained in a metal

FIGURE 7

Screen display of data collected with the LabPro measurement system.

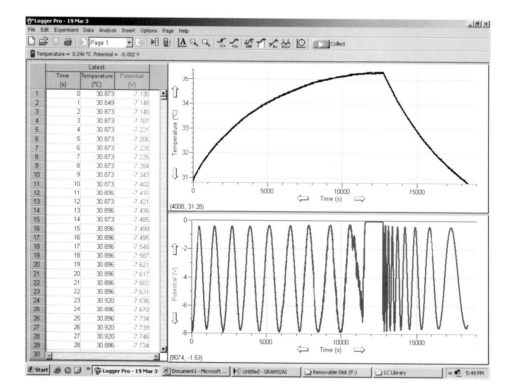

sheath, and the photomultiplier signal shows the temperature effect on the optical rotation of a polarized laser beam as it passes through the liquid crystal cell. In this test run, heating and cooling were done very slowly, over 5 hours, with measurements made every second, so the advantage of computer data collection is easily appreciated.

Example Two. The second example illustrates the interfacing of an instrument, a Tektronix TDS 350 oscilloscope in this case, connected by a GPIB parallel cable to a National Instruments digital input/output board contained in a PC computer. Control of the instrument and transfer of the data produced by it are done using LabVIEW, a popular visual interfacing software tool produced by National Instruments. The application shown is to transfer a waveform captured by the oscilloscope to the computer, where it is displayed and also written in numerical form to a spreadsheet file. The waveform might be a single sweep or an average of many sweeps collected internally in the oscilloscope. The use of such a measurement could occur in the study of radiative lifetimes such as those determined in Exps. 40 and 44, and LabVIEW is used in a different but essential way to determine an autocorrelation function from dynamic light scattering data in Exp. 33.

LabVIEW allows a user to design a Virtual Instrument (VI) for controlling a supported instrument by "drag and drop" manipulation of icons or objects representing various components needed in an interfacing application. These are provided by an extensive array of cascading windows containing icons that represent various functions, operations, displays such as meters and graphs, and statistical and mathematical treatments such as curve fitting and fast Fourier transforms. The icons can represent a single function operation or, more powerfully, can be a combination of operations termed a subVI. Connections can be made between the icons by drawing "wires" with the mouse between "nodes" that light up when the mouse is passed over the icon.

In operation, the LabVIEW program has two windows of concern to the programmer: Front Panel and Block Diagram, shown in Fig. 8. The Front Panel is the interface that the program user will usually see and contains chosen data inputs and outputs displayed in

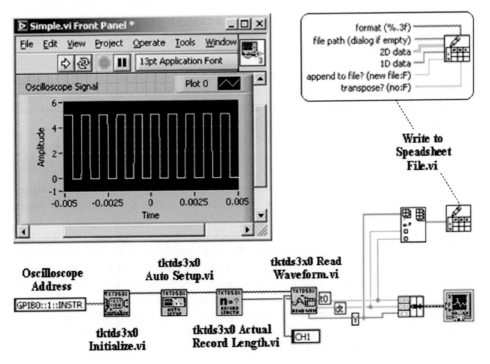

FIGURE 8
LabVIEW Front Panel (top left) and Block Diagram (bottom) for a Virtual Instrument (VI) for transfer of oscilloscope data to a computer for display and generation of a spreadsheet file. The content of the subVI for the latter operation is shown at the top right.

text boxes, graphs, charts, etc. The Block Diagram window is a visual representation of the manipulations required to produce the information on the Front Panel window, and, in this example, it shows the flow of data from the oscilloscope to boxes that process the waveform for display and write it to a file that can be read by a spreadsheet program. The Initialize.vi box on the left of the figure and the three following VIs are instrument-specific routines obtained from a large library on the National Instruments website. The Initialize.vi serves to alert the oscilloscope to a pending transfer of data via a GPIB0 line address. Auto Setup.vi starts the transfer of data previously captured by the oscilloscope and this flows through Actual Record Length.vi, which ensures that the full sweep is transferred. This information, coming from channel 1 of the oscilloscope, is in binary form and is converted to time zero ($t0$), time increment (dt), and amplitude (Y) values by Read Waveform.vi. This information is further processed by single-operation icons before being displayed (on Front Panel) or written to a spreadsheet by Write to Spreadsheet File.vi. Each subVI is a set of instructions that can be revealed by double clicking the mouse on the corresponding icon. For example, the details of the spreadsheet VI, shown in Fig. 8, indicate that it is possible to change the format of the data, to choose a filename, to transpose the data from row to column form, and so on.

Learning to use LabVIEW effectively can be challenging, but many examples are provided by National Instruments, and its website[15] lists a number of texts that can be useful. Several articles describing LabVIEW applications in chemistry are given in Ref. 16. LabVIEW and similar programs and measurement systems from other vendors are powerful tools that make accurate and repetitive measurements much more practical than in the past. The specific use of such tools will depend upon the equipment available in a particular laboratory and also on the ingenuity of the students and teaching staff. It is clear, however, that some experience in interfacing instruments to computers is an important component in the laboratory training of a modern physical chemist.

EXERCISES

Most of the exercises in Chapter II can be done using a spreadsheet program, and you are encouraged to do so. It would also be worthwhile to duplicate the examples in Chapter II, and in Chapter XXI on least-squares procedures, using spreadsheet functions.

1. With any available spreadsheet program, duplicate as closely as you can the spreadsheet shown in Fig. 1 in the text. (You will not have the ab initio values in column D, and not all spreadsheet programs allow mixtures of font types and subscripts and superscripts within a cell and on a graph.)
2. Use any available spreadsheet program to duplicate, as much as possible, Fig. 3.
3. Use the data shown in Fig. 2 and Eqs. (4) to (11) to confirm the least-squares results from the spreadsheet linear regression of Fig. 2.
4. Use the Solver tool in a spreadsheet to obtain the roots of the equation $x^4 + 3x^3 - 4x - 1 = 0$. Make a plot of this function versus x. Compare your results with those in Fig. 5.
5. The rotational transitions of a diatomic molecule fit the relation $\nu\,(J \rightarrow J + 1) = 2(J + 1)B_0 - 4(J + 1)^3 D_e$, where B_0 is the rotational constant for the $\upsilon = 0$ ground vibrational state and D_e is the centrifugal distortion constant. For CO, rotational transitions originating in J levels 0 to 4 have been observed at 38.45, 76.90, 115.34, 153.79, and 192.22 cm^{-1}, respectively. Use the linear regression procedure to fit these data with the equation both with and without inclusion of the D_e term. Give your values of the parameters and their standard errors. Note that these are most easily obtained if you choose the X column to be $2(J + 1)$ and $-4(J + 1)^3$: i.e., if you isolate the parameters

before the regression. Make plots of the residuals in both cases and use the 95 percent P and F tests to judge whether D_e should be included in the analysis.

6. Use a spreadsheet program to generate a random-walk display on an $X - Y$ graph. *Hint:* Create two columns of x, y points, starting with 0, 0 entries at the top in cells A5 and B5. In cell A6, enter the formula $+A5 + @IF@RAND> = 0.5, -1, +1)$, which serves to add or subtract one unit in x depending on @RAND, a random number between 0 and 1. Copy the A6 formula into say the A7 . . . B105 block and plot this xy array, using connected lines, not markers. Note that the walk changes every time a recalculation is done in the spreadsheet. *Optional:* Write a macro to repeat the walk 1000 times and to store the final x, y coordinates from each walk in an array. Plot the array (using markers only) to see if the distribution is really random. For each of the repetitions, compute the beginning-to-end distance and then compute the mean value. Use the Data/Frequency tool in the spreadsheet to bin the data and to plot the distribution about the mean. (Such random-walk processes emulate the random coiling of polymers.)

7. Use Mathcad or some other symbolic algebra program to solve the A_2 secular determinant of Table 41-2 in a manner similar to that shown in Fig. 5 for the B_2 determinant.

8. Modify the Mathematica commands shown in Fig. 6 to see the square of the I_2 harmonic oscillator and Morse wavefunctions and their overlap product for $v'' = 2$ and $v' = 0, 5, 10, 15, 20,$ and 25. Obtain plots of these results and discuss the trends that you see. Repeat the exercise for $v' = 40$ and $v'' = 0, 1, 2, 3, 4,$ and 5 and note the dramatic intensity variations for the Morse oscillator. Emission from this state, which can be populated by the 520.8-nm krypton ion laser line, is strong to even v'' levels but is very weak to odd v'' levels (up to about $v'' = 30$).

REFERENCES

1. Applications and reviews of commercial software packages often appear in *J. Chem. Educ.* and in *Reviews of Computational Chemistry.* A variety of programs are available at modest cost from *J. Chem. Educ.: Software* and from the Quantum Chemistry Program Exchange, Department of Chemistry, Indiana Univ., Bloomington, IN 47405.

2. J. B. Foresman and A. Frisch, *Exploring Chemistry with Electronic Structure,* 2d ed. Gaussian, Pittsburgh, PA (1996).

3. Descriptions of symbolic mathematics programs and some of their applications can be found, for example, in W. H. Cropper, *Mathematica Computer Programs for Physical Chemistry,* Springer-Verlag, New York (1998) and in J. F. Ogilvie, *Mathematics for Chemistry with Symbolic Computation,* an electronic text containing Maple worksheets, available at http://www.maplesoft.com.

4. Word processing and spreadsheet programs generally come with extensive online Help and tutorial files as learning and reference aids. Since new program versions are introduced frequently, reviews of current books that give further details and useful examples regarding spreadsheets are best accessed on the Web.

5. P. R. Bevington and D. K. Robinson, *Data Reduction and Error Analysis for the Physical Sciences,* 3d ed., McGraw-Hill, New York (2003).

6. R. de Levie, *J. Chem. Educ.* **76,** 1594 (1999).

7. I. N. Levine, *Quantum Chemistry,* 5th ed., Prentice-Hall, Upper Saddle River, NJ (2000).

8. M. P. Cady and C. A. Trapp, *A Mathcad Primer for Physical Chemistry,* Freeman, New York (2000); T. J. Zielinski, *J. Chem. Educ.* **77,** 668 (2000); J. H. Noggle, *Physical Chemistry Using Mathcad,* Pike Creek, Newark, DE (1997).

9. S. Flugge, *Practical Quantum Mechanics,* Springer-Verlag, Berlin (1971).

10. P. M. Morse, *Phys. Rev.,* **34,** 57 (1929); H. Halmann and I. Laulicht, *J. Chem. Phys.,* **43,** 438 (1965).

11. A Maple worksheet for a similar calculation of Franck–Condon factors for harmonic and Morse oscillators is available from the Maple Application Center at www.maple-apps.com; see *Estimation of Franck–Condon Factors with Model Wave Functions* by G. J. Fee, J. W. Nibler, and J. F. Ogilvie (2001). A Mathcad calculation for a harmonic oscillator is described by T. J. Zielinski, *J. Chem. Educ.* **75,** 1189 (1998).

12. W. J. Hehre, L. Radom, P. von R. Schleyer, and J. A. Pople, *Ab Initio Molecular Orbital Theory,* Wiley, New York (1986).

13. D. L. Williams, P. R. Minarik, and J. W. Nibler, *J. Chem. Educ.* **73,** 608 (1996).

14. Discussions of ab initio and density functional theory can be found in Refs. 2 and 6 and in P. W. Atkins and R. S. Friedman, *Molecular Quantum Mechanics,* 3d ed., Oxford Univ. Press, Oxford (1997).

15. http://www.ni.com/academic/resources_textbooks.htm.

16. S. M. Drew, *J. Chem. Educ.* **73,** 1107 (1996); M. A. Muyskens et al., *J. Chem. Educ.* **73,** 1112 (1996); P. J. Ogren and T. P. Jones, *J. Chem. Educ.* **73,** 1115 (1996).

GENERAL READING

S. H. Young, J. D. Madura, F. Rioux, "Software for Teaching and Using Numerical Methods in Physical Chemistry," Chapter 10 in *Using Computers in Chemistry and Chemical Education,* T. J. Zielinski and M. L. Swift (eds.), American Chemical Society, Washington, DC (1997).

R. G. Mortimer, *Mathematics for Physical Chemistry,* 3d ed., Elsevier, Amsterdam (2005).

P. R. Bevington and D. K. Robinson, *Data Reduction and Error Analysis for the Physical Sciences,* 3d ed., McGraw-Hill, New York (2003).

I. N. Levine, *Quantum Chemistry,* 5th ed., Prentice-Hall, Upper Saddle River, NJ (2000).

IV

Gases

EXPERIMENTS

1. Gas Thermometry
2. Joule–Thomson Effect
3. Heat-capacity Ratios for Gases

EXPERIMENT 1

Gas Thermometry

A fundamental attribute of temperature is that for any body in a state of equilibrium the temperature may be expressed by a number on a *temperature scale,* defined without particular reference to that body. The applicability of a universal temperature scale to all physical bodies at equilibrium is a consequence of an empirical law (sometimes called the "zeroth law of thermodynamics"), which states that if a body is in thermal equilibrium separately with each of two other bodies these two will be also in thermal equilibrium with each other.

However, a temperature scale must be somehow defined, in order that each attainable temperature will have a unique numerical value. A temperature scale may be defined over a certain range by the readings of a *thermometer,* which is a body possessing some easily measurable physical property that is for all practical purposes a sensitive function of the temperature alone. The specific volume of a fluid, the electrical resistivity of a metal, and the thermoelectric potential at the junction of a pair of different metals are examples of properties that are frequently used. Since these properties can be measured much more easily and precisely on a relative basis than on an absolute one, it is convenient to base a temperature scale in large part on certain reproducible "fixed points" defined by systems whose temperatures are fixed by nature. Examples of these are the triple point of water (the temperature at which ice, liquid water, and water vapor coexist in equilibrium) and the melting or boiling points of various pure substances under 1 atm pressure. Once the temperature has been fixed at two or more points, the temperatures at other points in the range of the thermometer can be defined in terms of the value of the physical property concerned. Thus the thermometer may serve as a device for interpolating among two or more fixed points.

In the original centigrade scale of temperature the two fixed points were taken as the "ice point" (the temperature at which ice and air-saturated water are in equilibrium under a total pressure of 1 atm), assigned a value of 0°C, and the "steam point" (the temperature at which pure water and water vapor are in equilibrium under a pressure of 1 atm), assigned a value of 100°C. A mercury thermometer with a capillary of uniform bore could be marked at 0°C and 100°C on the capillary stem by use of these fixed points, and the intervening range could then be marked off into 100 equal subdivisions. It is important to recognize that the scale thus defined is not identical with one similarly defined with alcohol (for example) as the thermometric fluid; in general, the two thermometers would give different scale readings at the same temperature because the thermal expansion coefficients of the fluids will have different temperature dependences.

Thus any one physical property of any arbitrarily specified substance would seem to define a temperature scale of a rather arbitrary kind. It would seem clearly preferable to define the temperature on the basis of some fundamental law. In the middle of the nineteenth century, Clausius and Kelvin stated the second law of thermodynamics and proposed the *thermodynamic temperature scale,* which is based on that law.[1] This scale is fixed at its lower end at the absolute zero of temperature. A scale factor, corresponding to the size of the degree, must be specified to complete the definition of the temperature scale; this can be accomplished by specifying the numerical value of the temperature of a reproducible fixed point or by specifying the numerical width of the interval between two fixed points. Kelvin adopted the latter procedure in order to make the degree coincide with the (mean) centigrade degree; accordingly the interval between the ice point and the steam point was fixed at 100 K, and the absolute (or Kelvin) temperature of the ice point has been found by experiment to be approximately 273.15 K. However, by international agreement (1948, 1954), the former procedure is now used. The triple point of pure water is defined as exactly 273.16 K (Kelvin) and as exactly 0.01°C (degrees Celsius). Thus the relation of the Celsius (formerly centigrade) temperature t to the Kelvin (or absolute) temperature T is†

$$t(°C) \equiv T(K) - 273.15 \tag{1}$$

Practical difficulties arise in making very precise determinations of temperature on the thermodynamic scale; the precision of the more refined thermometric techniques considerably exceeds the accuracy with which the experimental thermometer scale may be related to the thermodynamic scale. For this reason, a scale known as the *International Temperature Scale* has been devised, with several fixed points and with interpolation formulas based on practical thermometers (e.g., the platinum resistance thermometer between 13.803 K and 1234.93 K). This scale is intended to correspond as closely as possible to the thermodynamic scale but to permit more precision in the measurement of temperatures. Further details about this scale are given in Chapter XVII.

METHOD

The establishment of the International Temperature Scale has required that the thermodynamic temperatures of the fixed points be determined with as much accuracy as possible. For this purpose a device was needed that measures essentially the thermodynamic temperature and does not depend on any particular thermometric substance. On the other

---

†The differences between the original and the present Kelvin scales and between the old centigrade and the thermodynamic Celsius scales are small (on the order of 10^{-3} K) and of importance only for refined measurements.

hand, since it was needed only for a few highly accurate measurements, it did not need to have the convenience of such instruments as resistance thermometers or thermocouples. A device that has filled this need is the *gas thermometer*. It is based on the perfect-gas law, expressed by

$$pV = nRT \tag{2}$$

where n is the number of moles, R is a universal constant called the *gas constant,* and T is the "perfect-gas" temperature. At ordinary gas densities there are deviations from the perfect-gas law, but it is exact in the limit of zero gas density, where it is applicable to all gases. The "perfect-gas temperature scale" defined by Eq. (2) can be shown by thermodynamics to be identical with a thermodynamic temperature scale,† defined in terms of the second law of thermodynamics.

Gas thermometry measurements must of course be made with a real gas at ordinary pressures. However, it is possible to estimate the deviations from perfect-gas behavior and to convert measured pV values to "perfect-gas" pV values. For this purpose a virial equation of state for a real gas is often used:

$$\frac{p\widetilde{V}}{RT} = 1 + \frac{B}{\widetilde{V}} + \frac{C}{\widetilde{V}^2} + \dots \tag{3}$$

where $B, C, \dots$ are known as the second, third, $\dots$ *virial coefficients* and $\widetilde{V}$ is the molar volume of the gas at p and T. At ordinary pressures the series converges rapidly, and for many purposes terms beyond the one containing B can be neglected. Values of second (and in some cases third and fourth) virial coefficients are known over a wide range of temperatures from gas compressibility measurements; these values of B are useful in correcting the readings of a gas thermometer.

The establishment of the International Temperature Scale has been accomplished largely with the aid of measurements made with the helium gas thermometer.[2] The most precise gas thermometry method is the constant-volume method, in which a definite quantity of the gas is confined in a bulb of constant volume V at the temperature T to be determined and the pressure p of the gas is measured. A problem is encountered however in measuring the pressure; a way must be found to communicate between the bulb and the pressure gauge. This is usually accomplished by connecting the bulb to the room-temperature part of the system by a slender tube and allowing a portion of the gas to occupy a relatively small, constant "dead-space" volume at room temperature. Thus, it is important that the gas volume in the pressure manometer be as small as possible.

THEORY

The number of moles of gas n in the bulb of volume V at temperature T and in the dead space of volume v at room temperature is constant. Assuming the perfect-gas law, we have

$$\frac{pV}{RT} + \frac{pv}{RT_r} = n \tag{4}$$

†The perfect-gas properties required for this identity are (1) Boyle's law is obeyed: $p\widetilde{V} = f(T)$; and (2) the internal energy per mole is a function of temperature only: $\widetilde{E} = g(T)$.

where T_r is room temperature. The small region of large temperature gradient actually existing between the bulb and the dead space is replaced, without appreciable error, by a sharp division between two uniform temperatures T and T_r. Since the second term is small in comparison with the first, we shall introduce only a very small error ($\sim$0.05 percent with the present apparatus) by making the substitution

$$T_r = \frac{p_r}{p} T \tag{5}$$

where p_r is the pressure measured by the manometer when the bulb is at room temperature. We then obtain

$$\frac{pV}{RT}\left(1 + \frac{pv}{p_r V}\right) = n \tag{6}$$

Since only the pressure is being measured directly, the volume V must remain constant or else vary in a known way with the temperature (and pressure). If α is the coefficient of linear thermal expansion of the material from which the bulb is constructed, the coefficient of volume expansion is 3α. If this is assumed constant with temperature, we can write

$$\frac{pV_0}{RT}\left(1 + \frac{pv}{p_r V} + 3\alpha t\right) = n \tag{7}$$

where V_0 is the bulb volume at 0°C and t is the Celsius temperature of the gas thermometer bulb. For very precise work, other variables, including the dependence of the bulb volume on the difference between internal and external pressure, must be taken into account. To the precision of the present experiment, this effect of pressure is inconsequential.

Finally, if we allow for gas imperfections by including the second virial term, we can write

$$\frac{pV_0}{RT}\left(1 + \frac{pv}{p_r V} + 3\alpha t - \frac{B}{\widetilde{V}}\right) = n \tag{8}$$

where

$$\frac{1}{\widetilde{V}} = \frac{n}{V} \cong \frac{p_0}{RT_0} \tag{9}$$

and p_0 and T_0 are the values of p and T at (say) the ice point.

The above equations contain approximations that are acceptable for the present experiment. For high-precision work a much more detailed treatment is required.[2]

EXPERIMENTAL

The object of the experiment described below is to set up a gas thermometer, calibrate it at the ice point (in lieu of the experimentally more difficult triple point), and use it to determine the temperatures of one or more other fixed points. These may include the steam point, the boiling point of liquid nitrogen, the sublimation temperature of solid carbon dioxide (Dry Ice), the transition temperature of sodium sulfate decahydrate to the monohydrate and saturated solution, etc. The experiment will be performed with an apparatus that

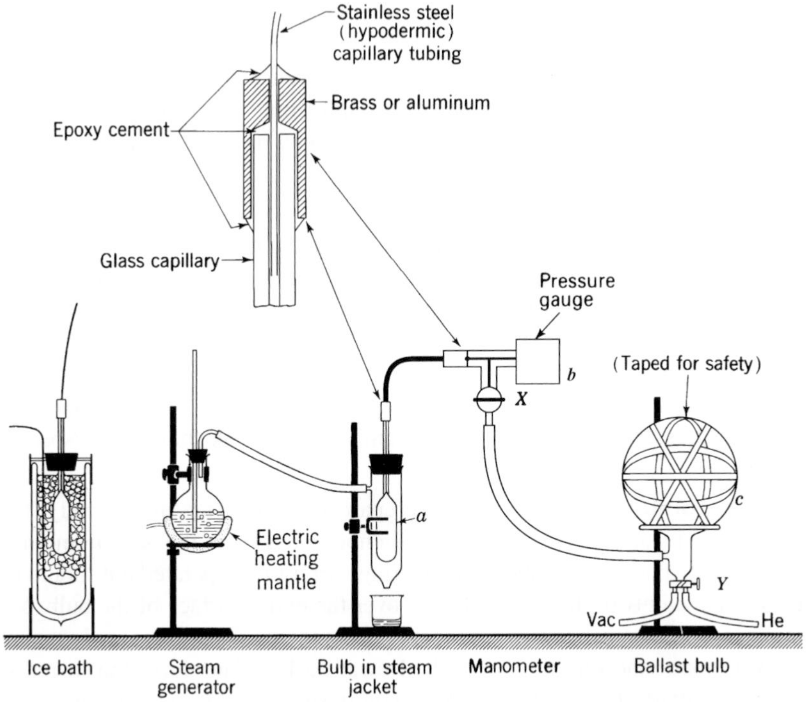

FIGURE 1

Gas thermometry apparatus.

resembles in principle a research gas thermometer but is very much simpler and somewhat less precise. The apparatus is shown in Fig. 1.

Gas thermometer bulb *a*. This is a Pyrex bulb with a volume V of about 100 cm^3, with an attached glass capillary passing through a large rubber stopper. The upper end of this capillary tube is connected, by a length of flexible stainless-steel capillary tubing, to the manometer.

Manometer *b*. The pressure manometer is connected to the stainless-steel capillary by a capillary *T*-tube containing a capillary stopcock, *X*, which can be opened to admit or remove gas from the bulb. During measurements this stopcock is kept closed and the quantity of gas in the bulb and dead space is thereby fixed. The pressure manometer should have a small internal volume and be capable of measuring absolute pressures with a resolution of ± 1 Torr or better. As discussed in Chapter XVIII, either capacitance or strain-gauge manometers can be used. The latter are generally less expensive, and high-precision versions of these gauges are available.

Ballast bulb *c*. This is a bulb of large volume (at least 5 L), which permits small adjustments in pressure to be made by the addition or removal of small quantities of gas or air through the three-way stopcock *Y* at the bottom. One of the two lower arms of this stopcock is connected to the source of gas to be used (or to room air); the other is connected to a laboratory vacuum line or a vacuum pump.

The gas thermometer bulb is first filled with the gas to be used, preferably helium or nitrogen (free of water contamination). With the stopcock *X* open, the entire system is evacuated and then filled to the desired pressure with the gas. If complete evacuation is not possible (as is the case when a laboratory vacuum line is used), the filling must be done several times to flush out residual air. When the bulb has been filled with the gas to the desired pressure, stopcock X is closed.

When a temperature value is being measured with the thermometer, pressure readings should be taken over a sufficient length of time to ensure that equilibrium has been attained. Then obtain four pressure readings taken over a 5 minute period and record the average of these values.

Procedure. Assemble the apparatus as indicated in Fig. 1. Make sure that stopcock X is properly greased. Then fill the apparatus with the gas to be used, as described above. At the final filling, adjust the pressure to about 600 Torr, and after 1 or 2 min, close stopcock X. After drift has ceased, make a reading of the pressure of gas in the bulb at room temperature, p_r.

To check for possible leaks, it is advisable to wait for 10 or 15 min and then redetermine the pressure. If p_r has increased significantly, a leak should be suspected.

Ice point. In the Dewar flask, prepare a "slushy" mixture of finely shaved clean ice and distilled water. This should be fluid enough to permit the gas thermometer bulb to be lowered into place and to allow a metal ring stirrer to be operated but should have sufficient ice to maintain two-phase equilibrium over the entire surface of the bulb. Mount the bulb in place, and commence stirring the mixture, moving the stirrer slowly up and down from the very top to the very bottom of the Dewar flask. Do not force the stirrer, since the bulb and its capillary stem are fragile. After equilibrium is achieved, take the ice-point pressure readings and record p_0.

Steam point. While the ice-point readings are being made, the water in the steam generator should be heated to boiling. When ready, the gas thermometer bulb should be positioned in the steam jacket so that it does not touch the wall at any point. Steam should be passed through the jacket at a rate slow enough to avoid overpressure in the jacket; steam should emerge from the bottom at low velocity. The rubber tubing connecting the steam generator to the steam jacket should run downhill all the way, to permit steady drainage of any water condensed in the tubing. When drift ceases, the steam-point pressure readings may be taken. During the steam-point measurements, record the barometric pressure. (The barometer reading must be corrected for temperature. A discussion of the use of precision barometers is given in Chapter XIX, and the necessary corrections are given in Appendix G.)

Repeat the ice-point measurements. The results should agree with those obtained previously to within experimental error; if the differences are larger, a leak must be suspected. Measure the other fixed points assigned. If liquid nitrogen is used, insert the bulb into the Dewar *slowly* to prevent violent boiling and excessive loss of liquid.

If time permits, refill the gas thermometer bulb with another gas or reduce p_r to 300 Torr and repeat some or all of the above measurements.

Also obtain from the instructor the values of V and v for the apparatus used in the experiment.

CALCULATIONS

Equation (8) can be written in the form

$$T = Ap \qquad (10)$$

TABLE I   Second virial coefficients (in cm^3 mol^{-1})

T, °C	He	Ar	N$_2$	CO$_2$
−250	~0			
−200	+10.4			
−150	11.4			
−100	11.7	−64.3	−51.9	
−50	11.9	−37.4	−26.4	
0	11.8	−21.5	−10.4	−154
50	11.6	−11.2	−0.4	−103
100	11.4	−4.2	+6.3	−73
150	11.0	+1.1	11.9	−51

where the proportionality factor

$$A = \frac{V_0}{nR}\left(1 + \frac{pv}{p_r V} + 3\alpha t - \frac{B}{\widetilde{V}}\right) \tag{11}$$

is nearly constant with temperature; but for the precision attainable with this experiment, the proportionality factor should be evaluated at each temperature. Making use of the known thermodynamic temperature of the ice point, we can write

$$T = \frac{273.15}{p_0}\frac{A}{A_0}p \tag{12}$$

$$\frac{A}{A_0} = 1 + \frac{p - p_0}{p_r}\frac{v}{V} + 3\alpha t - \frac{1}{\widetilde{V}}(B - B_0) \tag{13}$$

where p_0, A_0, and B_0 pertain to the ice point. Since the last two terms depend on the temperature, an approximate value of the temperature must be known before they can be evaluated; this can be obtained by setting A/A_0 equal to unity in Eq. (12). For Pyrex glass, $\alpha = 3.2 \times 10^{-6}$ K^{-1}. Second virial coefficients for various gases are given in Table 1, and $\widetilde{V}$ can be calculated from Eq. (9).

The student should report the temperature determined for each fixed point on both the Kelvin and Celsius scales. In cases where the temperature is a boiling point or sublimation point, calculate and report also the temperature on both scales corrected to a pressure of 1 atm. In the neighborhood of 1 atm, the boiling point of water increases with pressure by 0.037 K Torr^{-1}. For liquid nitrogen, the increase is 0.013 K Torr^{-1}.

DISCUSSION

What property of helium makes it particularly suitable for gas thermometry over the temperature range covered by this experiment? Is the correction for gas imperfection in this experiment of significant magnitude in relation to the experimental uncertainty? If not by how much must the precision of the pressure measurements be improved before gas imperfection corrections become significant? How does this depend on the choice of gas to be used?

SAFETY ISSUES

The ballast bulb must be taped to prevent flying glass fragments in the unlikely event of breakage. Safety glasses should be worn for all laboratory work. Gas cylinders must be chained securely to the wall or laboratory bench (see pp. 644–646 and Appendix C). Liquid nitrogen must be handled properly (see Appendix C).

APPARATUS

Pressure manometer (such as a Honeywell stain-gauge device) and digital voltmeter for readout; properly taped and mounted ballast bulb; heavy-wall pressure tubing; Dewar flask; large ring stirrer; notched cover plate for Dewar with hole for mounting gas thermometer bulb; electrical heating mantle; steam generator with rubber connecting tubing; steam jacket; two ring stands; ring clamp; two clamp holders; one large and one medium clamp.

Cylinder of helium or dry nitrogen; pure ice (1 kg); ice grinder; liquid nitrogen (1 L); boiling chips; stopcock grease; vacuum pump or water aspirator.

REFERENCES

1. R. J. Silbey, R. A. Alberty, and M. G. Bawendi, *Physical Chemistry,* 4th ed., pp. 7–8, 97, Wiley, New York (2005).

2. J. A. Beattie and coworkers, *Proc. Am. Acad. Arts Sci.* **74,** 327 (1941); **77,** 255 (1949).

GENERAL READING

J. R. Leigh, *Temperature Measurement and Control,* INSPEC, Edison, NJ (1988).

R. P. Benedict, *Fundamentals of Temperature, Pressure, and Flow Measurements,* 3d ed., Wiley-Interscience, New York (1984).

J. F. Schooley, *Thermometry,* CRC Reprint, Franklin, Elkins Park, PA (1986).

EXPERIMENT 2
Joule–Thomson Effect

The Joule–Thomson effect is a measure of the deviation of the behavior of a real gas from what is defined to be ideal-gas behavior. In this experiment a simple technique for measuring this effect will be applied to a few common gases.

THEORY

An ideal gas may be defined as one for which the following two conditions apply at all temperatures for a fixed quantity of the gas: (1) Boyle's law is obeyed; i.e.,

$$pV = f(T)$$

and (2) the internal energy E is independent of volume. Accordingly, E is independent of pressure as well, and in the absence of other pertinent variables (such as applied fields), E of an ideal gas is therefore a function of the temperature alone:

$$E = g(T)$$

It is apparent that the enthalpy H of an ideal gas is also a function of temperature alone:

$$H \equiv E + pV = h(T)$$

Accordingly, we can write for a definite quantity of an ideal gas at all temperatures

$$\left(\frac{\partial E}{\partial V}\right)_T = \left(\frac{\partial E}{\partial p}\right)_T = \left(\frac{\partial H}{\partial V}\right)_T = \left(\frac{\partial H}{\partial p}\right)_T = 0 \tag{1}$$

The absence of any dependence of the internal energy of a gas on volume was suggested by the early experiments of Gay-Lussac and Joule. They found that, when a quantity of gas in a container initially at a given temperature was allowed to expand into another previously evacuated container without work or heat flow to or from the surroundings ($\Delta E = 0$), the final temperature (after the two containers came into equilibrium with each other) was the same as the initial temperature. However, that kind of experiment (known as the *Joule experiment*) is of limited sensitivity, because the heat capacity of the containers is large in comparison with that of the gases studied. Subsequently, Joule and Thomson[1] showed, in a different kind of experiment, that real gases do undergo small temperature changes upon free expansion. This experiment utilized continuous gas flow through a porous plug under adiabatic conditions. Because of the continuous flow, the solid parts of the apparatus come into thermal equilibrium with the flowing gas, and their heat capacities impose a much less serious limitation than in the case of the Joule experiment.

Let it be imagined that gas is flowing slowly from left to right through the porous plug in Fig. 1. To the left of the plug, the temperature and pressure of the gas are T_1 and p_1; and to the right of the plug, they are T_2 and p_2. The volume of a definite quantity of gas (say 1 mol) is V_1 on the left and V_2 on the right, and the internal energy is E_1 and E_2, respectively. When 1 mol of gas flows through the plug, the work done on the system by the surroundings is

$$w = p_1V_1 - p_2V_2$$

Since the process is adiabatic, the change in internal energy is

$$\Delta E = E_2 - E_1 = q + w = w$$

Combining these two equations we obtain

$$E_1 + p_1V_1 = E_2 + p_2V_2$$

or

$$H_1 = H_2 \tag{2}$$

Thus this process takes place at constant enthalpy.

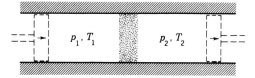

FIGURE 1

Schematic diagram of the Joule–Thomson experiment. The stippled area represents a porous plug.

For a process involving arbitrary infinitesimal changes in pressure and temperature, the change in enthalpy is

$$dH = \left(\frac{\partial H}{\partial p}\right)_T dp + \left(\frac{\partial H}{\partial T}\right)_p dT \tag{3}$$

In the present experiment dH is zero and dT and dp cannot be arbitrary but are related by

$$\mu \equiv \left(\frac{\partial T}{\partial p}\right)_H = -\frac{(\partial H/\partial p)_T}{(\partial H/\partial T)_p} \tag{4}$$

The quantity μ defined by this equation is known as the *Joule–Thomson coefficient*. It represents the limiting value of the experimental ratio of temperature difference to pressure difference as the pressure difference approaches zero:

$$\mu = \lim_{\Delta p \to 0} \left(\frac{\Delta T}{\Delta p}\right)_H \tag{5}$$

Experimentally, ΔT is found to be very nearly linear with Δp over a considerable range; this is in accord with expectations based on the theory given below.

The denominator on the right side of Eq. (4) is the heat capacity at constant pressure C_p. The numerator is zero for an ideal gas [see Eq. (1)]. Accordingly, for an ideal gas the Joule–Thomson coefficient is zero, and there should be no temperature difference across the porous plug. For a real gas, the Joule–Thomson coefficient is a measure of the quantity $(\partial H/\partial p)_T$ [which can be related thermodynamically to the quantity involved in the Joule experiment, $(\partial E/\partial V)_T$]. Using the general thermodynamic relation[2]

$$\left(\frac{\partial H}{\partial p}\right)_T = -T\left(\frac{\partial V}{\partial T}\right)_p + V \tag{6}$$

it can be shown that, for an ideal gas satisfying the criteria already given,

$$pV = \text{const} \times T \tag{7}$$

where T is the absolute thermodynamic temperature. The coefficient $(\partial H/\partial p)_T$ is therefore a measure of the deviation from the behavior predicted by Eq. (7). On combining Eqs. (4) and (6), we obtain

$$\mu = \frac{T(\partial V/\partial T)_p - V}{C_p} \tag{8}$$

In order to predict the magnitude and behavior of the Joule–Thomson coefficient for a real gas, we can use the van der Waals equation of state,[2] which is

$$\left(p + \frac{a}{\widetilde{V}^2}\right)(\widetilde{V} - b) = RT \tag{9}$$

where $\widetilde{V}$ is the molar volume. We can rearrange this equation (neglecting the very small second-order term $ab/\widetilde{V}^2$ and substituting p/RT for $1/\widetilde{V}$ in a first-order term) to obtain

$$p\widetilde{V} = RT - \frac{ap}{RT} + bp$$

Thus,

$$\left(\frac{\partial \widetilde{V}}{\partial T}\right)_p = \frac{R}{p} + \frac{a}{RT^2}$$

Combination of these two equations yields

$$\left(\frac{\partial \widetilde{V}}{\partial T}\right)_p = \frac{\widetilde{V} - b}{T} + \frac{2a}{RT^2} \tag{10}$$

which on substitution into Eq. (8) gives the expression

$$\mu = \frac{(2a/RT) - b}{\widetilde{C}_p} \quad \text{(van der Waals)} \tag{11}$$

This expression does not contain p or $\widetilde{V}$ explicitly, and the molar heat capacity $\widetilde{C}_p$ may be considered essentially independent of these variables. The temperature dependence of $\widetilde{C}_p$ is small, and accordingly that of μ is also small enough to be neglected over the ΔT obtainable with a Δp of about 1–5 bar (namely about 4 K or less for the gases considered here). Accordingly, we may expect that μ will be approximately independent of Δp over a wide range, as stated previously.

For most gases under ordinary conditions, $2a/RT > b$ (the attractive forces predominate over the repulsive forces in determining the nonideal behavior) and the Joule–Thomson coefficient is therefore positive (gas cools on expansion). At a sufficiently high temperature, the inequality is reversed, and the gas warms on expansion. The temperature at which the Joule–Thomson coefficient changes sign is called the *inversion temperature* T_I. For a van der Waals gas,

$$T_I = \frac{2a}{Rb} \tag{12}$$

This temperature is usually several hundred degrees above room temperature. However, hydrogen and helium are exceptional in having inversion temperatures that are well below room temperatures. This results from the very small attractive forces in these gases (see Table 1 for values of the van der Waals constant a).

TABLE 1 Values of constants in equations of state[a] and the Lennard–Jones potential

	He	H$_2$	N$_2$	CO$_2$
van der Waals[3]:				
a	0.03457	0.2476	1.408	3.640
b	0.02370	0.02661	0.03913	0.04267
Beattie–Bridgeman[4,5]:				
A_0	0.0219	0.2001	1.3623	5.0728
a	0.05984	−0.00506	0.02617	0.07132
B_0	0.01400	0.02096	0.05046	0.10476
b	0.0	−0.04359	−0.00691	0.07235
$10^{-4}c$	0.0040	0.0504	4.20	66.00
Lennard–Jones[6]:				
ϵ/k (K)	6.03	29.2	95.0	189
σ (nm)	0.263	0.287	0.370	0.449

[a] Units assumed are V in dm^3 mol^{-1} ≡ L mol^{-1}, p in bar ≡ 10^5 Pa, T in K. ($R = 0.083145$ bar dm^3 K^{-1} mol^{-1}.)

Other semiempirical equations of state can be used to predict Joule–Thomson coefficients. Perhaps the best of these is the Beattie–Bridgeman equation,[4,5] which can be written (for 1 mol) as

$$p = \frac{RT(1 - \epsilon)}{\widetilde{V}^2}(\widetilde{V} + B) - \frac{A}{\widetilde{V}^2} \tag{13}$$

where $A = A_0(1 - a/\widetilde{V})$, $B = B_0(1 - b/\widetilde{V})$, and $\epsilon = c/\widetilde{V}T^3$. In this equation of state, there are five constants which are characteristic of the particular gas: A_0, B_0, a, b, and c. In terms of these constants and the pressure and temperature, the Joule–Thomson coefficient is given[4] by

$$\mu = \frac{1}{\widetilde{C}_p}\left\{-B_0 + \frac{2A_0}{RT} + \frac{4c}{T^3} + \left[\frac{2B_0b}{RT} - \frac{3A_0a}{(RT)^2} + \frac{5B_0c}{RT^4}\right]p\right\} \tag{14}$$

This equation predicts a small dependence on pressure not shown by Eq. (11), which is based on the van der Waals equation.

The most general of the equations of state is the *virial equation*, which is also the most fundamental since it has a direct theoretical connection to the intermolecular potential function. The virial equation of state expresses the deviation from ideality as a series expansion in density and, in terms of molar volume, can be written

$$\frac{p\widetilde{V}}{RT} = 1 + \frac{B_2(T)}{\widetilde{V}} + \frac{B_3(T)}{\widetilde{V}^2} + \dots \tag{15}$$

The virial coefficients B_2 and B_3 depend only on temperature and are determined by two- and three-body interactions between molecules, respectively. For pressures below about 10 bar, the B_3 term is very small and can be neglected. Solving Eq. (15) for $\widetilde{V}$ and $(\partial \widetilde{V}/\partial T)_p$ in a manner similar to that for the van der Waals case above gives

$$\mu = \frac{T(\partial B_2/\partial T)_p - B_2}{\widetilde{C}_p} \tag{16}$$

From statistical mechanics,[6] $B_2(T)$ is given by

$$B_2(T) = N_0 \int_0^\infty [1 - e^{-U(r)/kT}]2\pi r^2 \, dr \tag{17}$$

and $(\partial B_2/\partial T)_p$ can be obtained by differentiation. $U(r)$ is the potential energy as a function of the separation of the molecules, taken to be spherical, and is important because it can be used to predict many of the transport and collisional properties of a molecule. One common choice for $U(r)$ is the so-called Lennard–Jones 6-12 potential, which has the form

$$U(r) = 4\varepsilon\left[\left(\frac{\sigma}{r}\right)^{12} - \left(\frac{\sigma}{r}\right)^6\right] \tag{18}$$

where ε is the well depth corresponding to the minimum in the potential and σ is the separation corresponding to $U(r) = 0$; see Fig. 47-1. Values for these parameters are included in Table 1 for the gases of interest in this experiment.

EXPERIMENTAL

The experimental apparatus shown in Fig. 2 is patterned after a design given in Ref. 7. The "porous plug" is a $\frac{3}{8}$-in.-OD stainless steel frit of 2 μm pore size and $\frac{1}{16}$-in. thickness pressed into a $\frac{3}{8}$-in. Swagelok tee made of nylon for reduced thermal conductivity.

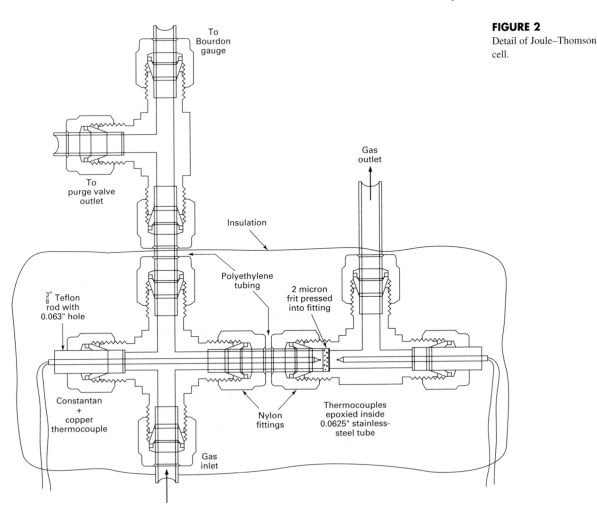

FIGURE 2
Detail of Joule–Thomson cell.

The high-pressure inlet is attached to a $\frac{3}{8}$-in. cross to provide ports for gas introduction, pressure measurement, and thermocouple placement just in front of the frit. The Bourdon gauge (0–10 bar) should be connected via a tee to a purge valve to facilitate gas changes. Before use the assembly should be tested at 10 bar for leaks. Thermal insulation such as glass wool should be wrapped around the frit assembly to keep the expansion as adiabatic as possible.

The temperature difference ΔT across the frit is measured with two copper-Constantan (type T) thermocouples with wires of 0.010-in. diameter or less for reduced thermal conductivity. The thermocouples can be sealed with epoxy into a $\frac{1}{16}$-in. stainless-steel sheathing tube, which can be connected to the cross fitting by a $\frac{1}{16}$-in. to $\frac{3}{8}$-in. Swagelok adaptor, or more simply by swaging a $\frac{3}{8}$-in. Teflon rod with a 0.063-in. feedthrough hole for the thermocouple tube, as shown in Fig. 2. A convenient 6-in., $\frac{1}{16}$-in.-OD sealed subminiature probe with an exposed thermocouple junction and external strain relief is available from Omega (e.g., probe TMTSS-062E-6). The 6-in. length is sufficient to allow the thermocouples to be positioned adjacent to the center of the frit, as shown in the figure.

Because the maximum temperature change will be only 0.5 to 4 K, a sensitive digital voltmeter (0.1 μV), null voltmeter, or potentiometer is desirable for accurate measurements. To obtain the temperature *difference* directly, the two Constantan leads of the thermocouples

should be clamped together and the copper leads should be attached to the measuring device.†
For best absolute accuracy, the two thermocouples should be calibrated (e.g., using a standard
thermometer) to determine their temperature coefficients (Seebeck coefficient) α. However, for
a copper–Constantan thermocouple, α varies only slightly with temperature, from 39 to 43 μV
K^{-1} from 0 to 50°C. At 25°C, α is 40.6 μV K^{-1} and the variation is small from one thermocouple to another. For small temperature differences, a linear relation $\Delta V_{TC} = \alpha\Delta T = \alpha(dT/dP)\Delta p$
is a good approximation for the thermocouple potential difference between the two junctions.‡
Thus, to the accuracy needed for this experiment, the slope of a plot of ΔV_{TC} versus Δp can be
combined with an assumed value of $\alpha = 40.6$ μV K^{-1} to yield (dT/dp) and hence μ.

Procedure. Set up the apparatus shown in Fig. 2. The gas supply should be a cylinder or supply line equipped with a pressure regulator and a control valve. The supply pressure should be constant during the measurements. Because a significant temperature change
occurs as gases go from high to low pressure through the pressure regulator itself, the gas
should be passed through about 50 ft of $\frac{1}{4}$-in. coiled copper tubing contained in a water
bath at 25 ± 1°C. A $\frac{1}{4}$-in.-to-$\frac{3}{8}$-in. adaptor can be used for a short, insulated polyethylene
tubing connection to the expansion apparatus. Before initiating gas flow, record the bath
temperature and determine any offset voltage between the two thermocouples.

Start the measurements with CO_2 with the pressure regulator set to minimum pressure.
Open the control valve and purge the copper line and pressure gauge of air or any other
gases with the purge valve open. Then close the purge valve and slowly increase the pressure to 4 bar. After this pressure is reached, record the thermocouple reading every 30 s,
until the values become constant (typically a few minutes). Lower the regulator pressure by
about 0.5 bar and again take readings every 30 s until a constant value is obtained. Continue
this procedure down to a final pressure of 0.5 bar. Note that this is the excess pressure over
the discharge pressure into the room (assumed to be at 1 bar).

Change the gas supply to N_2 and again purge the copper coil and pressure gauge with
the purge valve open. Close this valve and bring the pressure slowly to 10 bar, a higher
value than for CO_2 since the cooling is less. After the temperature has stabilized, repeat the
sequence of measurements as for CO_2 but at 1-bar intervals. Finally, repeat the N_2 procedure using He gas. In this case, the temperature change will be much smaller and positive:
i.e., the gas heats on expansion because it is above the so-called Joule–Thomson inversion
point, the temperature at which the coefficient μ is zero. After completion of the experiment, make sure that all cylinder valves are closed.

CALCULATIONS

For each gas studied, do a linear regression to fit ΔV_{TC} (or ΔT) versus Δp so as to
obtain the slope along with its standard error. On a single graph, show for each of the three
gases the best-fit straight line along with the experimental data points. From the slopes,
evaluate the Joule–Thomson coefficient μ in units of K bar^{-1}. Compare your results with
literature values given in Ref. 7. Calculate μ for these gases at 25°C from the van der

---

†As an alternative to a thermocouple, one can use two sensitive thermistor probes and an appropriate resistance
bridge circuit (see Chapters XVI and XVII). A calibration to convert the bridge measurement to ΔT is required
in this case.

‡In practice, one often finds that $\Delta V_{TC} = \alpha\Delta T + \delta V_{TC}$, where δV_{TC} is a small offset voltage ($\sim$1–3 μV) observed
when both the reference and the measuring junction are at the same temperature. This "nonthermodynamic"
result can occur if the thermocouple wire has regions of compositional variation or strain (e.g., from kinking) that
are subject to a temperature gradient. δV_{TC} can be ignored in this experiment, since it affects only the intercept of
the plot of ΔV_{TC} versus Δp and not the slope.

Waals and Beattie–Bridgeman constants given in Table 1. $\widetilde{C}_p$ values for He, N_2, and CO_2 at 25°C are 20.79, 29.12, and 37.11 J K^{-1} mol^{-1}, respectively.

Plot the Lennard–Jones potentials for each of the gases studied. Obtain μ from Eqs. (16)–(18) by numerical integration and compare the values from this two-parameter potential with those from the van der Waals and Beattie–Bridgeman equations of state. [*Optional:* A simple square-well potential model can also be used to crudely represent the interaction of two molecules. In place of Eq. (18), use the square-well potential and parameters of Ref. 6 to calculate μ. Contrast with the results from the Lennard–Jones potential and comment on the sensitivity of the calculations to the form of the potential.]

DISCUSSION

The Joule–Thomson coefficient gives a measure of how much potential energy is converted into kinetic energy or vice versa as molecules in a dense gas change their average separation during an adiabatic expansion. As mentioned earlier, the magnitude and sign of μ are determined by the balance of attractive and repulsive interactions and, for most gases at room temperature, cooling occurs as molecules work against a net attractive force as they move apart. The exceptions are the weakly interacting species He and H_2, where μ is negative at 300 K and precooling below the inversion temperature is first necessary before cooling can occur on expansion. Calculate the inversion temperature for the gases of Table 1 using Eqs. (11), (14), and (16), neglecting the last, small pressure-dependent term in (14), and compare your values with experimental ones you find in the literature. Equations (14) and (16) can be most easily solved for T_I by iteration, using for example the Solve For function of spreadsheet programs, as discussed in Chapter III.

Joule–Thomson cooling is the basis for the Linde method of gas liquefaction, in which a gas is compressed, allowed to cool by heat exchange, and is then expanded to cool sufficiently that the gas liquefies. This effect is also important in the operation of refrigerators and heat pumps. Using cylinders of high-pressure gas, cooling can be achieved without power input in a device without moving parts, and hence the Joule–Thomson process has been used in cooling of small infrared and optical detectors on space probes. Discuss some of the design factors that might be important in achieving maximum cooling efficiency in the latter kind of a device.

For the more difficult Joule experiment, we can write

$$\eta \equiv -\left(\frac{\partial T}{\partial V}\right)_E = \frac{(\partial E/\partial V)_T}{(\partial E/\partial T)_V} = \frac{T(\partial p/\partial T)_V - p}{C_v} \tag{19}$$

This quantity is called the *Joule coefficient*. It is the limit of $-(\Delta T/\Delta V)_E$, corrected for the heat capacity of the containers as ΔV approaches zero. With the van der Waals equation of state, we obtain $\eta = a/\widetilde{V}^2\widetilde{C}_v$. The corrected temperature change when the two containers are of equal volume is found by integration to be $\Delta T = -a/2\widetilde{V}\widetilde{C}_v$, where $\widetilde{V}$ is the initial molar volume and $\widetilde{C}_v$ is the molar constant-volume heat capacity. It is instructive to calculate this ΔT for a gas such as CO_2. In addition, the student may consider the relative heat capacities of 10 L of the gas at a pressure of 1 bar and that of the quantity of copper required to construct two spheres of this volume with walls (say) 1 mm thick and then calculate the ΔT expected to be observed with such an experimental arrangement.

SAFETY ISSUES

Gas cylinders must be chained securely to the wall or laboratory bench (see pp. 644–646 and Appendix C).

APPARATUS

Insulated Joule–Thomson cell similar to that of Fig. 2 (suitable stainless steel frits can be obtained from chromatographic parts suppliers, e.g., Upchurch Scientific part C-414); metal or nylon tees, crosses, and reducers (available from Swagelok and other manufacturers); $\frac{3}{8}$-in. Teflon rod; type T insulated copper–Constantan thermocouples with 0.010-in.-diameter wires; voltmeter with 0.1-μV resolution (e.g., Keithley 196), null voltmeter (e.g., Hewlett Packard 419A or Keithley 155), or sensitive potentiometer (e.g., Keithley K-3). Cylinders of CO_2, N_2, and He with regulators and control valves; 50 ft of $\frac{1}{4}$-in. copper coil, $\frac{3}{8}$-in. and $\frac{1}{4}$-in. polyethylene tubing; 0- to 10-bar Bourdon gauge; 25°C water bath.

REFERENCES

1. J. P. Joule and W. Thomson (Lord Kelvin), *Phil. Trans.* **143**, 357 (1853); **144**, 321 (1854). [Reprinted in *Harper's Scientific Memoirs I, The Free Expansion of Gases,* Harper, New York (1898).]

2. R. J. Silbey, R. A. Alberty, and M. G. Bawendi, *Physical Chemistry,* 4th ed., p. 127, Wiley, New York (2005).

3. *Landolt-Börnstein Physikalisch-chimische Tabellen,* 5th ed., p. 254, Springer, Berlin (1923). [Reprinted by Edwards, Ann Arbor, MI (1943).] [This is the source of van der Waals constants cited in the *CRC Handbook of Chemistry and Physics.*]

4. J. A. Beattie and W. H. Stockmayer, "The Thermodynamics and Statistical Mechanics of Real Gases," in H. S. Taylor and S. Glasstone (eds.), *A Treatise on Physical Chemistry,* Vol. II, pp. 187ff., esp. pp. 206, 234, Van Nostrand, Princeton, NJ (1951).

5. J. A. Beattie and O. C. Bridgeman, *J. Am. Chem. Soc.* **49**, 1665 (1927); *Proc. Am. Acad. Arts Sci.* **63**, 229 (1928).

6. J. O. Hirschfelder, C. F. Curtiss, and R. B. Bird, *Molecular Theory of Gases and Liquids,* chap. 3 and table I-A, Wiley, New York (1964).

7. A. M. Halpern and S. Gozashti, *J. Chem. Educ.* **63**, 1001 (1986).

EXPERIMENT 3
Heat-Capacity Ratios for Gases

The ratio C_p/C_v of the heat capacity of a gas at constant pressure to that at constant volume will be determined by either the method of adiabatic expansion or the sound velocity method. Several gases will be studied, and the results will be interpreted in terms of the contribution made to the specific heat by various molecular degrees of freedom.

THEORY

In considering the theoretical calculation of the heat capacities of gases, we shall be concerned only with perfect gases. Since $\widetilde{C}_p = \widetilde{C}_v + R$ for an ideal gas (where $\widetilde{C}_p$ and $\widetilde{C}_v$ are the molar quantities C_p/n and C_v/n), our discussion can be restricted to C_v.

The number of *degrees of freedom* for a molecule is the number of independent coordinates needed to specify its position and configuration. Hence a molecule of N atoms has $3N$ degrees of freedom. These could be taken as the three Cartesian coordinates of the N individual atoms, but it is more convenient to classify them as follows.

1. *Translational degrees of freedom:* Three independent coordinates are needed to specify the position of the center of mass of the molecule.
2. *Rotational degrees of freedom:* All molecules containing more than one atom require a specification of their orientation in space. As an example, consider a rigid diatomic molecule; such a model consists of two point masses (the atoms) connected by a rigid massless bar (the chemical bond). Through the center of mass, which lies on the rigid bar, independent rotation can take place about two axes mutually perpendicular to each other and to the rigid bar. (The rigid bar itself does not constitute a third axis of rotation under ordinary circumstances for reasons based on quantum theory, there being no appreciable moment of inertia about this axis.) Rotation of a *diatomic* molecule or any linear molecule can thus be described in terms of *two* rotational degrees of freedom. *Nonlinear* molecules for which the third axis has a moment of inertia of appreciable magnitude and constitutes another axis of rotation require *three* rotational degrees of freedom.
3. *Vibrational degrees of freedom:* One must also specify the displacements of the atoms from their equilibrium positions (vibrations). The number of vibrational degrees of freedom is $3N - 5$ for linear molecules and $3N - 6$ for nonlinear molecules. These values are determined by the fact that the total number of degrees of freedom must be $3N$. For each vibrational degree of freedom, there is a "normal mode" of vibration of the molecule, with characteristic symmetry properties and a characteristic harmonic frequency. The vibrational normal modes for CO_2 and H_2O are illustrated schematically in Fig. 1.

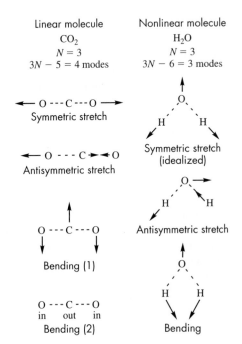

Linear molecule
CO_2
$N = 3$
$3N - 5 = 4$ modes

Symmetric stretch

Antisymmetric stretch

Bending (1)

Bending (2)

Nonlinear molecule
H_2O
$N = 3$
$3N - 6 = 3$ modes

Symmetric stretch (idealized)

Antisymmetric stretch

Bending

FIGURE 1
Schematic diagrams of the vibrational normal modes for CO_2, a linear molecule, and H_2O, a bent molecule.

Using classical statistical mechanics, one can derive the theorem of the equipartition of energy. According to this theorem, $kT/2$ of energy is associated with each quadratic term in the expression for the energy.[1] Thus there is associated with each translational or rotational degree of freedom for a molecule a contribution to the energy of $kT/2$ of kinetic energy and for each vibrational degree of freedom a contribution of $kT/2$ of kinetic energy and $kT/2$ of potential energy. (The corresponding contributions to the energy per *mole* of gas are $RT/2$.)

Clearly, a monatomic gas has no rotational or vibrational energy but does have a translational energy of $\frac{3}{2}RT$ per mole. The constant-volume heat capacity of a *monatomic* perfect gas is thus

$$\widetilde{C}_v = \left(\frac{\partial \widetilde{E}}{\partial T} \right)_V = \tfrac{3}{2}R \tag{1}$$

For diatomic or polyatomic molecules, we can write

$$\widetilde{E} = \widetilde{E}(\text{trans}) + \widetilde{E}(\text{rot}) + \widetilde{E}(\text{vib}) \tag{2}$$

In Eq. (2), contributions to the energy from electronic states have been neglected, since they are not significant at room temperature for most molecules. Also, any small intermolecular energies that occur for imperfect gases are not considered.

The equipartition theorem is based on classical mechanics. Its application to translational motion is in accord with quantum mechanics as well. At ordinary temperatures the rotational results are also in accord with quantum mechanics. (The greatest deviation from the classical result is in the case of hydrogen, H_2. At temperatures below 100 K the rotational energy of H_2 is significantly below the equipartition value, as predicted by quantum mechanics.)

The vibrational energy is however highly quantized and depends strongly on temperature: the various vibrational modes are at ordinary temperatures only partially "active," and the degree of activity depends strongly on the temperature. As a general rule, the heavier the atoms or the smaller the force constant of the bond (i.e., the lower the vibrational frequency), the more "active" is a given degree of freedom at a given temperature and the greater is the contribution to the heat capacity. Moreover, the frequencies of modes that are predominantly bending of bonds tend to be much lower than those that are predominantly stretching of bonds. In the case of most gaseous *diatomic* molecules (where the one vibrational mode is a pure stretch), the vibrational contribution to $\widetilde{C}_v$ is very small: for example, N_2 would have its classical equipartition value for $\widetilde{C}_v$ only above about 4000 K. Many polyatomic molecules, especially those containing heavy atoms, will at room temperature have significant *partial* vibrational contributions to $\widetilde{C}_v$.

For polyatomic molecules, theory based on the equipartition theorem allows one to calculate only limiting values for $\widetilde{C}_v$ by either completely ignoring all vibrational contributions or assuming that the vibrational contributions achieve their full classical value. For monatomic gases and all ordinary diatomic molecules (where the vibrational contribution is not important at room temperature and can be ignored), definite $\widetilde{C}_v$ values can be calculated. For a brief discussion of a more accurate calculation of $\widetilde{C}_v(\text{vib})$, see Exp. 37.

For a given $\widetilde{C}_v$ value, the ratio $\widetilde{C}_p / \widetilde{C}_v$ for a perfect gas is given by

$$\gamma \equiv \frac{\widetilde{C}_p}{\widetilde{C}_v} = 1 + \frac{R}{\widetilde{C}_v} \tag{3}$$

A.   Adiabatic Expansion Method

For the reversible adiabatic expansion of a perfect gas, the change in energy content is related to the change in volume by

$$dE = -p \, dV = -\frac{nRT}{V} dV = -nRT \, d \ln V \tag{4}$$

Moreover, since E for a perfect gas is a function of temperature only, we can also write $dE = C_v \, dT$. Substituting this expression into Eq. (4) and integrating, we find that

$$\tilde{C}_v \ln \frac{T_2}{T_1} = -R \ln \frac{\tilde{V}_2}{\tilde{V}_1} \tag{5}$$

where $\tilde{C}_v$ and $\tilde{V}$ are molar quantities (that is, C_v/n, V/n). It has been assumed that C_v is constant over the temperature range involved. This equation predicts the decrease in temperature resulting from a reversible adiabatic expansion of a perfect gas.

Consider the following two-step process involving a perfect gas denoted by A:

Step I: Allow the gas to expand adiabatically and reversibly until the pressure has dropped from p_1 to p_2:

$$A(p_1, \, \tilde{V}_1, \, T_1) \rightarrow A(p_2, \, \tilde{V}_2, \, T_2) \tag{6}$$

Step II: At constant volume, restore the temperature of the gas to T_1:

$$A(p_2, \, \tilde{V}_2, \, T_2) \rightarrow A(p_3, \, \tilde{V}_2, \, T_1) \tag{7}$$

For step I, we can use the perfect-gas law to obtain

$$\frac{T_2}{T_1} = \frac{p_2 \tilde{V}_2}{p_1 \tilde{V}_1} \tag{8}$$

Substituting Eq. (8) into Eq. (5) and combining terms in $\tilde{V}_2 / \tilde{V}_1$, we write

$$\ln \frac{p_2}{p_1} = \frac{-(\tilde{C}_v + R)}{\tilde{C}_v} \ln \frac{\tilde{V}_2}{\tilde{V}_1} = -\frac{\tilde{C}_p}{\tilde{C}_v} \ln \frac{\tilde{V}_2}{\tilde{V}_1} \tag{9}$$

since for a perfect gas

$$\tilde{C}_p = \tilde{C}_v + R \tag{10}$$

For step II, which restores the temperature to T_1,

$$\frac{\tilde{V}_2}{\tilde{V}_1} = \frac{p_1}{p_2} \tag{11}$$

Thus

$$\ln \frac{p_1}{p_2} = \frac{\tilde{C}_p}{\tilde{C}_v} \ln \frac{p_1}{p_3} \tag{12}$$

This can be rewritten in the form

$$\gamma \equiv \frac{\tilde{C}_p}{\tilde{C}_v} = \frac{\ln (p_1/p_2)}{\ln (p_1/p_3)} \tag{13}$$

Although in theoretical calculations of $\widetilde{C}_v$ we are concerned only with perfect gases, equations applying to reversible adiabatic expansion can be derived that are not limited to perfect gases. We shall here derive a general expression for $(\partial p/\partial \widetilde{V})_S$ that we will apply to a perfect gas to obtain Eq. (9); we shall also apply it to a van der Waals gas. The adiabatic expansion method may be incapable of sufficient precision to justify a van der Waals treatment, but the expressions obtained may be used with the more precise sound velocity method.

We may write

$$\left(\frac{\partial p}{\partial \widetilde{V}}\right)_S = \left(\frac{\partial T}{\partial \widetilde{V}}\right)_S \bigg/ \left(\frac{\partial T}{\partial p}\right)_S \tag{14}$$

From

$$d\widetilde{S} = \left(\frac{\partial \widetilde{S}}{\partial \widetilde{V}}\right)_T d\widetilde{V} + \left(\frac{\partial \widetilde{S}}{\partial T}\right)_V dT \tag{15a}$$

and

$$d\widetilde{S} = \left(\frac{\partial \widetilde{S}}{\partial p}\right)_T dp + \left(\frac{\partial \widetilde{S}}{\partial T}\right)_p dT \tag{15b}$$

we obtain, setting $d\widetilde{S} = 0$ in both cases,

$$\left(\frac{\partial T}{\partial \widetilde{V}}\right)_S = -\left(\frac{\partial \widetilde{S}}{\partial \widetilde{V}}\right)_T \bigg/ \left(\frac{\partial \widetilde{S}}{\partial T}\right)_V \tag{16a}$$

$$\left(\frac{\partial T}{\partial p}\right)_S = -\left(\frac{\partial \widetilde{S}}{\partial p}\right)_T \bigg/ \left(\frac{\partial \widetilde{S}}{\partial T}\right)_p \tag{16b}$$

Now, $(\partial \widetilde{S}/\partial T)_V = \widetilde{C}_v/T$ and $(\partial \widetilde{S}/\partial T)_p = \widetilde{C}_p/T$. On applying the appropriate Maxwell relations (reciprocity relations arising from the perfect differentials $d\widetilde{A} = -\widetilde{S}\,dT - p\,d\widetilde{V}$ and $d\widetilde{S} = -\widetilde{S}\,dT + \widetilde{V}\,dp$, respectively), namely,

$$\left(\frac{\partial \widetilde{S}}{\partial \widetilde{V}}\right)_T = \left(\frac{\partial p}{\partial T}\right)_V \quad \text{and} \quad \left(\frac{\partial \widetilde{S}}{\partial p}\right)_T = -\left(\frac{\partial \widetilde{V}}{\partial T}\right)_p \tag{17a,b}$$

we obtain the relation

$$\left(\frac{\partial p}{\partial \widetilde{V}}\right)_S = -\frac{\widetilde{C}_p}{\widetilde{C}_v}\left(\frac{\partial p}{\partial T}\right)_V \bigg/ \left(\frac{\partial \widetilde{V}}{\partial T}\right)_p \tag{18}$$

The treatment is perfectly general up to this point: Eq. (18) can be applied to any gas law $f(p, \widetilde{V}, T) = 0$. For the perfect-gas law, we easily obtain

$$\left(\frac{\partial p}{\partial \widetilde{V}}\right)_S = -\gamma \frac{p}{\widetilde{V}} \tag{19}$$

Integration of this equation at constant entropy yields Eq. (9). For a van der Waals gas, we have

$$\left(p + \frac{a}{\widetilde{V}^2}\right)(\widetilde{V} - b) = RT \tag{20}$$

from which we obtain (using the perfect-gas approximation in the small term involving a)

$$\left(\frac{\partial p}{\partial T}\right)_V = \frac{R}{\widetilde{V}-b} \tag{21a}$$

$$\left(\frac{\partial \widetilde{V}}{\partial T}\right)_p = \frac{\widetilde{V}-b}{T} + \frac{2a}{RT^2} \tag{21b}$$

Combining these with Eq. (18) yields, on simplification (keeping only terms to first order in the small quantities a and b),

$$\left(\frac{\partial p}{\partial \widetilde{V}}\right)_S = -\gamma\frac{p}{\widetilde{V}}\left(1 - \frac{a}{p\widetilde{V}^2} + \frac{b}{\widetilde{V}}\right) \tag{22}$$

The adiabatic expansion method, due to Clement and Desormes,[2] uses the very simple apparatus shown in Fig. 2. The change in state (6) is carried out by quickly removing and replacing the stopper of a large carboy containing the desired gas at a pressure initially somewhat higher than 1 atm pressure, so that the pressure of gas in the carboy momentarily drops to atmospheric pressure p_2. The change in state (7) consists of allowing the gas remaining in the carboy to return to its initial temperature. The initial pressure p_1 and the final pressure p_3 are read from an open-tube manometer.

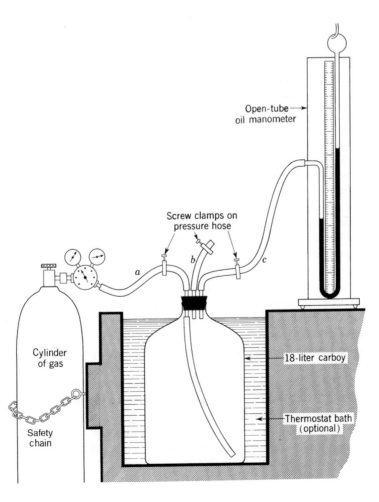

FIGURE 2
Apparatus for the adiabatic expansion of gases.

The thermodynamic equations, Eqs. (4) to (13), apply only to that part of the gas that remains in the carboy after the stopper is replaced. We may imagine the gas initially in the carboy to be divided into two parts by an imaginary surface; the part above the surface leaves the carboy when the stopper is removed and presumably interacts irreversibly with the surroundings, but the part below the surface expands *reversibly* against this imaginary surface, doing work in pushing the upper gas out. The process is approximately adiabatic only because it is rapid; within a few seconds the gas near the walls will have received an appreciable quantity of heat by direct conduction from the walls, and the pressure can be seen to rise almost as soon as the stopper is replaced.

The approximate assumptions on which this experiment is based are not without controversy. An alternative assumption,[3] with which we do not agree, is that the whole of the gas in the container, initially at pressure p_1, upon removal of the stopper expands *irreversibly* against the surroundings at a *constant* external pressure p_2 ($\sim$1 atm) until the internal pressure has been reduced to p_2. This assumption raises the question of where the boundary of the system is to be defined, since the external pressure p_2 is exerted only *outside* the neck of the carboy and since the treatment requires a closed system of uniform internal pressure and temperature. In any case, the alternative assumption results in the final equation

$$\gamma = \frac{(p_1/p_2) - 1}{(p_1/p_3) - 1} \qquad (23)$$

The numerator and the denominator in this equation are the leading terms in the power-series expansions of the logarithms in the numerator and the denominator in Eq. (13). Under the conditions of the experiment, the difference between the values of γ given by Eqs. (13) and (23) is only about 1 or 2 percent.

The experiment will presumably give a somewhat *low* result (low p_3 and therefore a low $\widetilde{C}_p / \widetilde{C}_v$ ratio) if the expansion is to an appreciable degree irreversible, a somewhat *low* result if the stopper is left open so long that the conditions are not sufficiently adiabatic, and a somewhat *high* result if the stopper has not been removed for a long enough time to permit the pressure to drop momentarily to atmospheric. There should be no significant irreversibility if during the expansion there are no significant pressure gradients in the gas below the imaginary surface mentioned above, and such pressure gradients should not be expected if the throat area is small in comparison with the effective area of this imaginary surface. With an 18-L carboy and a p_1 about 50 Torr above 1 atm, the volume of air forced out should be about 1 L, which should provide a large enough surface area to fulfill this condition approximately if the throat diameter is not more than about 2 or 3 cm. The effect of the length of time the stopper is removed is much more difficult to estimate by calculation; some idea can be obtained from the duration of the sound produced when the stopper is removed and from the rate of rise of the manometer reading immediately after the stopper is replaced. For the purpose of this experiment, the student may assume that, if the stopper is removed completely from the carboy to a distance of 2 or 3 in. away and replaced tightly as soon as physically possible, the desired experimental conditions are approximately fulfilled. Some additional assurance may be gained from the reproducibility obtained in duplicate runs.

The adiabatic expansion method is not the best method of determining the heat capacity ratio. Much better methods are based on measurements of the velocity of sound in gases. One such method, described in Part B of this experiment, consists of measuring the wavelength of sound of an accurately known frequency by measuring the distance between nodes in a sonic resonance set up in a Kundt's tube. Methods also exist for determining the heat capacities directly,[4] although the measurements are not easy.

EXPERIMENTAL

The apparatus should be assembled as shown in Fig. 2. If desired, the carboy may be mounted in a thermostat bath; if so it must be clamped securely to overcome buoyancy. The manometer is an *open-tube* manometer, one side of which is open to the atmosphere; the pressure that it measures is therefore the difference of pressure from atmospheric pressure. A suitable liquid for the manometer is dibutyl phthalate, which has a density of 1.046 g cm^{-3} at room temperature (20°C). To convert manometer readings (millimeters of dibutyl phthalate) to equivalent readings in millimeters of mercury (Torr), multiply by the ratio of this density to the density of mercury, which is 13.55 g cm^{-3} at 20°C. To find the total pressure in the carboy, the converted manometer readings should be added to atmospheric pressure as given by a barometer. It is unnecessary to correct all readings to 0°C, as all pressures enter the calculations as ratios.

Seat the rubber stopper firmly in the carboy and open the clamps on tubes a and b. Clamp off the tube c. The connections shown in Fig. 2 are based on the assumption that the gas to be studied is heavier than air (or the previous gas in the carboy) and therefore should be introduced at the bottom in order to force the lighter gas out at the top; in the event that the gas to be studied is lighter, the connections a and b should be reversed.

Allow the gas to be studied to sweep through the carboy for 15 min. The rate of gas flow should be about 6 L min^{-1}, or 100 mL s^{-1} (measure roughly in an inverted beaker held under water in the thermostat bath). Thus five volumes of gas (90 L) will pass through the carboy.

Retard the gas flow to a fraction of the flushing rate by partly closing the clamp on tube a. *Carefully* open the clamp on tube c, and then cautiously (to avoid blowing liquid out of the manometer) clamp off the exit tube b, keeping a close watch on the manometer. When the manometer has attained a reading of about 600 mm of oil, clamp off tube a. Allow the gas to come to the temperature of the thermostat bath (about 15 min), as shown by a constant manometer reading. Record this reading; when it is converted to an equivalent mercury reading and added to the barometer reading, p_1 is obtained.

Remove the stopper *entirely* (a distance of 2 or 3 in.) from the carboy, and replace it *in the shortest possible time*, making sure that it is tight. As the gas warms back up to the bath temperature, the pressure will increase and finally (in about 15 min) reach a new constant value p_3, which can be determined from the manometer reading and the barometer reading. At some point in the procedure, a barometer reading (p_2) should be taken as well as a bath-temperature reading.

Repeat the steps above to obtain two more determinations with the same gas. For these repeat runs, long flushing is not necessary; one additional volume of gas should suffice to check the effectiveness of the original flushing.

Measurements are to be made on both helium and nitrogen. If there is sufficient time, study carbon dioxide also.

CALCULATIONS

For each of the three runs on He and N$_2$ (and perhaps CO$_2$), calculate $\widetilde{C}_p/\widetilde{C}_v$ using Eq. (13). Also calculate the theoretical value of $\widetilde{C}_p/\widetilde{C}_v$ predicted by the equipartition theorem. In the case of N$_2$ and CO$_2$, calculate the ratio both with and without a vibrational contribution to $\widetilde{C}_v$ of R per vibrational degree of freedom.

Assuming that all the gases can be treated as ideal and thus $\widetilde{C}_p = \widetilde{C}_v + R$, obtain approximate values for the molar constant-volume heat capacity $\widetilde{C}_v$ for He and N$_2$ (and CO$_2$).

DISCUSSION

Compare your experimental ratios with those calculated theoretically, and make any deductions you can about the rotational and vibrational contributions, taking due account of the uncertainties in the experimental values. For CO_2, how would the theoretical ratio be affected if the molecule were nonlinear (such as SO_2) instead of linear? Could you decide between these two structures from the $\widetilde{C}_p / \widetilde{C}_v$ ratio alone?

With a knowledge of the vibrational frequencies for the normal modes of N_2 and CO_2 and the appropriate statistical thermodynamic formulae (see Exp. 37), one can calculate quite accurate $\widetilde{C}_v$ values for N_2 and CO_2.

B. Sound Velocity Method

The heat-capacity ratio $\gamma \equiv C_p/C_v$ of a gas can be determined with good accuracy by measuring the speed of sound c. For an ideal gas,

$$\gamma = \frac{Mc^2}{RT} \tag{24}$$

where M is the molar mass. A brief derivation[5] of this equation will be sketched below.

For longitudinal plane waves propagating in the x direction through a homogeneous medium of mass density ρ, the wave equation is

$$\frac{\partial^2 \xi}{\partial t^2} = c^2 \frac{\partial^2 \xi}{\partial x^2} \tag{25}$$

where ξ is the particle displacement. Consider a layer AB of the fluid, of thickness δx and unit cross-sectional area, which is oriented normal to the direction of propagation. The mechanical strain on AB due to the sound wave is $\partial \xi/\partial x$:

$$\frac{\partial \xi}{\partial x} = \frac{\Delta V}{V} = -\frac{\Delta \rho}{\rho} \tag{26}$$

where ΔV and $\Delta \rho$ are, respectively, the volume and density changes caused by the acoustic pressure (mechanical stress) change Δp in the ambient pressure p. From Hooke's law of elasticity,

$$-\Delta p = B_S \frac{\partial \xi}{\partial x} \tag{27}$$

where B_S is the adiabatic elastic modulus.† The net acoustic force acting on AB due to the sound wave is $-(\partial \Delta p/\partial x)\, \delta x$. According to Newton's second law of motion, this force is also equal to $(\rho\, \delta x)\, \partial^2 \xi/\partial t^2$. Thus

$$\frac{\partial \Delta p}{\partial x} = -\rho \frac{\partial^2 \xi}{\partial t^2} \tag{28}$$

or

$$\frac{\partial^2 \xi}{\partial t^2} = \frac{B_S}{\rho} \frac{\partial^2 \xi}{\partial x^2} \tag{29}$$

---

†Since the strain variations occur so rapidly that there is not sufficient time for heat flow from adjacent regions of compression and rarefaction, the adiabatic modulus is the appropriate one for sound propagation.[5]

Comparison of Eqs. (25) and (29) leads to

$$\rho c^2 = B_S \tag{30}$$

and one can now rewrite Eq. (27) as $\Delta p = \rho c^2 (\Delta \rho / \rho) = c^2 \, \Delta \rho$, which is equivalent to

$$c^2 = \left(\frac{\partial p}{\partial \rho} \right)_S \tag{31}$$

Now

$$\rho = \frac{M}{\widetilde{V}} \tag{32}$$

Therefore

$$c^2 = -\frac{\widetilde{V}}{\rho} \left(\frac{\partial p}{\partial \widetilde{V}} \right)_S = -\frac{\widetilde{V}^2}{M} \left(\frac{\partial p}{\partial \widetilde{V}} \right)_S \tag{33}$$

If sound propagation is considered as a reversible adiabatic process, we can use Eq. (19) to obtain for a perfect gas

$$c^2 = \gamma \frac{p}{\rho} = \gamma \frac{RT}{M} \tag{34}$$

which can be rewritten as Eq. (24). For a van der Waals gas, we obtain (to first order in a and b)

$$\begin{aligned} c^2 &= \gamma \frac{p}{\rho} \left(1 - \frac{a}{p \widetilde{V}^2} + \frac{b}{\widetilde{V}} \right) \\ &= \gamma \frac{RT}{M} \left(1 - \frac{2a}{p \widetilde{V}^2} + \frac{2b}{\widetilde{V}} \right) \end{aligned} \tag{35}$$

which may be readily solved for r. For comparison the value of γ for a van der Waals gas is (to first order in a)

$$\gamma = 1 + \frac{R}{\widetilde{C}_v} \left(1 + \frac{2a}{p \widetilde{V}^2} \right) \tag{36}$$

assuming that we are given a theoretical value for $\widetilde{C}_v$. (If this is based on a perfect gas, the translational part may be slightly in error because of gas imperfections, but that error is expected to be much less for $\widetilde{C}_v$ than for $\widetilde{C}_p$.)

METHOD

This experiment is based on a modified version of Kundt's tube, in which the wavelength λ of standing waves of frequency f are determined electronically. Figure 3 shows the experimental apparatus, which utilizes an audio oscillator driving a miniature speaker as the source and a microphone as the detector. Also, an alternative apparatus based on a small spherical resonator has been described by Colgate et al.[6]

It is possible to connect the microphone directly to a sensitive voltmeter and watch the variation in the intensity of the received signal as the detector is moved along the tube. The intensity will decrease to a minimum when an antinode is reached and then increase to a maximum at a node. However, it is more precise and more convenient to use an oscilloscope to indicate the phase relationship between the speaker input and the microphone signal. The output of the audio oscillator should drive the horizontal sweep of the oscilloscope. The signal received by the microphone after propagation through the gas is applied (with amplification if necessary) to the vertical sweep of the oscilloscope. The phase difference is determined from the Lissajous figures that appear on the screen (see Chapter XIX). When the pattern changes from a straight line tilted at 45° to the right (in phase, 0° phase angle) to another straight line tilted at 45° to the left (out of phase, 180° phase angle), the piston has moved one-half the wavelength of sound.

EXPERIMENTAL

The apparatus should be assembled as shown in Fig. 3. Details of the procedure will depend on the particular electronic components used, especially the type of oscilloscope. Additional instructions may be provided by the instructor.

If the oscillator is not calibrated accurately, this can be done with a calibrated frequency meter or by (1) displaying the waveform on the screen if the oscilloscope has an accurate timebase; (2) obtaining a Lissajous figure with a standard frequency source (internal or external); or (3) holding a vibrating tuning fork near the microphone and obtaining Lissajous figures at the fundamental and several harmonics (disconnect the speaker unit during this calibration). The oscillator should be calibrated at several frequencies between 1 and 2 kHz.†

Move the piston to a position near the far end of its travel. Allow the gas to be studied to sweep through the Kundt's tube for about 10 min in order to displace all the air. Then reduce the flow rate until there is a very slow stream of gas through the tube; this will prevent air from diffusing in during the run. Adjust the gain controls so that the two signals are about equal in magnitude (i.e., a circle is seen when the signals differ in phase by 90°).

Set the oscillator at the desired frequency (1 kHz for N_2 and CO_2, 2 kHz for He), and slowly move the piston toward the speaker until a straight-line pattern is observed on the screen. Record the position of the piston and note whether the phase shift is 0° or 180°. Then move the piston inward again until the next straight-line pattern is observed, and again record the position. Continue to record such readings as long as a satisfactory pattern can be observed. Record the temperature of the outlet gas several times during the run.

†The frequency unit cycles per second is called a hertz (Hz). Thus 1 kHz = 1000 Hz = 1000 cps.

FIGURE 3

Apparatus for measuring the speed of sound in a gas. The Kundt's tube should be a glass tube ~150 cm long and ~5 cm in diameter. The close-fitting Teflon piston should move over a range of at least 100 cm. If a constant-temperature jacket is provided for the Kundt's tube, the position of the piston can be determined from the position of a fiducial mark on the rod.

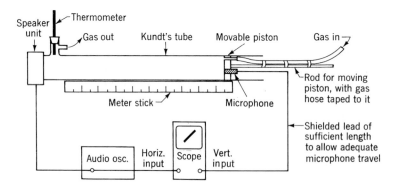

If necessary, repeat the entire procedure for a given gas until consistent results are obtained. If time permits, repeat the procedure using a different frequency. Measurements are to be made on helium, nitrogen, and carbon dioxide. At some point, obtain a reading of the barometric pressure.

CALCULATIONS

From the average value of the spacing $\lambda/2$ between adjacent nodes and the known frequency f, calculate the speed of sound c in each gas. Use Eq. (24) to calculate γ, and compare these experimental values with the theoretical values predicted by the equipartition theorem. In the case of N_2 and CO_2, calculate the theoretical ratio both with and without a vibrational contribution to $\widetilde{C}_v$ of R per vibrational degree of freedom.

Using the van der Waals constants given for N_2 and CO_2 in Table 2-1, recalculate γ for those gases using Eq. (35) and compare them with the theoretical value given by Eq. (36). For this purpose, the quantity $\widetilde{V}$ can be approximated by the ideal gas value RT/p.

Assuming that all the gases can be treated as ideal and thus $\widetilde{C}_p = \widetilde{C}_v + R$, obtain approximate values for the molar constant-volume heat capacity $\widetilde{C}_v$ for He and N_2 (and CO_2).

DISCUSSION

Compare your experimental ratios with those calculated theoretically, and make any deductions you can about the presence or absence of rotational and vibrational contributions, taking due account of the uncertainties in the experimental values. For CO_2 how would the theoretical ratio be affected if the molecule were nonlinear (like SO_2) instead of linear? Could you decide between these two structures from the $\widetilde{C}_p / \widetilde{C}_v$ ratio alone?

With a knowledge of the vibrational frequencies for the normal modes of N_2 and CO_2 and the appropriate statistical thermodynamic formulae (see Exp. 37), one can calculate accurate $\widetilde{C}_v$ values for these molecules.

Compare the speed of sound in each gas with the average speed of the gas molecules and with the average velocity component in a given direction. Explain why the speed of sound is independent of the pressure.

Does the precision of the measurements justify the van der Waals expression, Eq. (35), instead of the perfect-gas expression, Eq. (34)? Justify your answer in terms of your estimates of uncertainty.

SAFETY ISSUES

Gas cylinders must be chained securely to the wall or laboratory bench (see pp. 644–646 and Appendix C).

APPARATUS

Adiabatic Expansion Method. Large-volume vessel (such as an 18-L glass carboy); three-hole stopper fitted with three glass tubes; open-tube manometer, with dibutyl phthalate as indicating fluid (may contain a small amount of dye for ease of reading); three long lengths and one short length of rubber pressure tubing; three screw clamps; cork ring and brackets, needed if vessel is to be mounted in a water bath; thermometer; 500-mL beaker.

Thermostat bath, set at 25°C (unless each vessel has an insulating jacket); cylinders of helium, nitrogen, and carbon dioxide.

Sound Velocity Method. Kundt's tube, complete with speaker unit (miniature cone or horn driver) and movable piston holding microphone; stable audio oscillator, preferably with calibrated dial; small crystal microphone; oscilloscope; audio-frequency amplifier, to be used if scope gain is inadequate; electrical leads; rubber tubing; stopper; thermometer.

Cylinders of helium, nitrogen, and carbon dioxide.

REFERENCES

1. R. J. Silbey, R. A. Alberty, and M. G. Bawendi, *Physical Chemistry,* 4th ed., Sec. 16.9, Wiley, New York (2005).

2. Lord Rayleigh, *The Theory of Sound,* 2d ed., Vol. II, pp. 15–23, Dover, Mineola, NY (1945).

3. G. L. Bertrand and H. O. McDonald, *J. Chem. Educ.* **63,** 252 (1986).

4. K. Schell and W. Heuse, *Ann. Physik* **37,** 79 (1912); **40,** 473 (1913); **59,** 86 (1919).

5. J. Blitz, *Fundamentals of Ultrasonics,* 2d ed., pp. 10–13, Butterworth, London (1967); V. A. Shutilov, *Fundamental Physics of Ultrasound,* Gordon and Breach, Newark, NJ (1988).

6. S. O. Colgate, K. R. Williams, K. Reed, and C. A. Hart, *J. Chem. Educ.* **64,** 553 (1987).

GENERAL READING

V. A. Shutilov, *Fundamental Physics of Ultrasound,* Gordon and Breach, Newark, NJ (1988).

J. R. Partington, *An Advanced Treatise on Physical Chemistry,* Vol. I, pp. 792ff., Books on Demand, Ann Arbor, MI (1949).

Transport Properties of Gases

EXPERIMENTS

4. Viscosity of Gases
5. Diffusion of Gases

KINETIC THEORY OF TRANSPORT PHENOMENA

In this section we shall be concerned with a molecular theory of the transport properties of gases. The molecules of a gas collide with each other frequently, and the velocity of a given molecule is usually changed by each collision that the molecule undergoes. However, when a one-component gas is in thermal and statistical equilibrium, there is a definite distribution of molecular velocities—the well-known Maxwellian distribution.[1] Figure 1 shows how the molecular velocities are distributed in such a gas. This distribution is isotropic (the same in all directions) and can be characterized by a *root-mean-square* (*rms*) *speed u,* which is given by

$$u = \left(\frac{3kT}{m} \right)^{1/2} = \left(\frac{3RT}{M} \right)^{1/2} \tag{1}$$

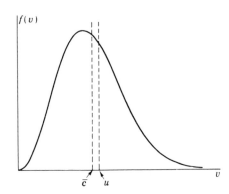

FIGURE 1
Schematic Maxwellian velocity distribution: $f(v)\, dv$ is the fraction of the molecules with velocities between v and $v + dv$. Values of the rms speed u and the mean speed $\bar{c}$ are shown.

where k is the Boltzmann constant, R is the gas constant, m is the mass of one molecule, and M is the molar mass. Sometimes it is more convenient to use the *mean speed* $\bar{c}$:

$$\bar{c} = \left(\frac{8kT}{\pi m}\right)^{1/2} = \left(\frac{8}{3\pi}\right)^{1/2} u \tag{2}$$

Another very important concept in kinetic theory is the average distance a molecule travels between collisions—the so-called *mean free path*. On the basis of a very simple conception of molecular collisions, the following equation for the mean free path λ can be derived:

$$\lambda = \frac{1}{\pi d^2 \bar{N}}$$

where d is the "molecular diameter," or center-to-center collision distance, and $\bar{N}$ is the number of molecules per unit volume. When the fact is properly taken into account that all molecules do not have the same velocities, a different numerical coefficient is obtained:[1]

$$\lambda = \frac{1}{\sqrt{2}\pi d^2 \bar{N}} = \frac{1}{\sqrt{2}\pi d^2}\frac{kT}{p} = \left(\frac{1}{\sqrt{2}\pi N_0}\right)\frac{RT}{pd^2} = 3.738 \times 10^{-25}\frac{RT}{pd^2} \tag{3}$$

where N_0 is Avogadro's number. Note that the ideal-gas law has been used for $\bar{N}$ and that the numerical coefficient in the final expression has units of mole. If SI units are used for all quantities, which means expressing the pressure in pascals, the resulting value of λ will be in meters. However, for obtaining a feeling for the magnitude of λ compared to the scale of typical apparatus, centimeter units are more convenient. Some typical values of λ are given in Table 1.

We shall consider in detail the predictions of the hard-sphere model for the viscosity, thermal conductivity, and diffusion of gases; indeed, the kinetic theory treatment of these three transport properties is very similar. But first let us consider the simpler problem of molecular effusion.

EFFUSION OF GASES

The theory of effusion (Knudsen flow) is quite straightforward since only molecular flow is involved: i.e., in the process of effusing, the molecules act independently of each other. For a Maxwellian distribution of velocities it can be shown[1,2] that the number Z of molecular impacts on a unit area of wall surface in unit time is

$$Z = \frac{1}{4}\bar{c}\bar{N} = \frac{1}{\sqrt{6\pi}}u\bar{N} \tag{4}$$

TABLE 1   Mean free paths of various gases at 25°C

Gas	Mean free path, cm		
	at 760 Torr	at 0.1 Torr	at 10^{-4} Torr
He	19.0×10^{-6}	14.5×10^{-2}	145
H_2	12.1×10^{-6}	9.2×10^{-2}	92
Ar	6.8×10^{-6}	5.2×10^{-2}	52
O_2	6.9×10^{-6}	5.3×10^{-2}	53
N_2	6.4×10^{-6}	4.9×10^{-2}	49
CH_4	5.2×10^{-6}	4.0×10^{-2}	40
CO_2	4.2×10^{-6}	3.2×10^{-2}	32
SO_2	3.0×10^{-6}	2.3×10^{-2}	23

Combining Eq. (1) or Eq. (2) with Eq. (4), we obtain

$$Z = \left(\frac{RT}{2\pi M} \right)^{1/2} \bar{N}$$

The number z of *moles* colliding with a unit area of wall surface per second is

$$z = \left(\frac{RT}{2\pi M} \right)^{1/2} \frac{n}{V} \tag{5}$$

where n/V is the number of moles per unit volume. For a perfect gas Eq. (5) becomes

$$z = \frac{p}{\sqrt{2\pi MRT}} \tag{6}$$

Let two bodies of the same gas, at the same temperature but at two different pressures (p and p'), communicate through a small hole of area A, all dimensions of which are *small* in comparison with the mean free path of the gas molecules. It can then be assumed that the Maxwellian distribution of velocities is essentially undisturbed and that the same number of molecules will enter the hole as would otherwise have collided with the corresponding area of wall surface. It will also be assumed that all molecules entering the hole will pass through to the other side. The *net* rate of effusion through the hole will be the difference between the numbers of moles flowing through the hole in unit time in the two directions:

$$\frac{dn_{\text{eff}}}{dt} = \frac{A(p - p')}{\sqrt{2\pi MRT}} \tag{7}$$

This equation assumes that the gas is perfect, but it should be emphasized that the rate of effusion calculated from it is independent of the size and shape of the molecules provided only that these are such as to give a sufficiently long mean free path at the given temperature and pressure.

GAS VISCOSITY

Consider a body of gas under shear as shown in Fig. 4-1 and discussed in the introduction to Exp. 4. Assume that all the molecules in the gas are of one kind and that the gas undergoes laminar flow. By the mass velocity of the gas at each point, we understand the (vectorial) average velocity of all molecules passing through an infinitesimal region at that point. If the velocity gradient is substantially constant within one or a few mean free paths, we may also say that the mass velocity of the gas at a given point is the (vectorial) average velocity of all molecules that have suffered their most recent collision in an infinitesimal region at that point. We shall assume that, with respect to a system of coordinates moving at the mass velocity at a given point, the distribution of velocities of gas molecules that underwent their most recent collisions at that point is the same as it would be if the gas were not under shear (i.e., that the distribution is Maxwellian and therefore isotropic). We shall also assume for simplicity that all molecules travel the same distance λ between collisions.

Let us focus our attention on a horizontal laminar plane $x = x'$, with $v = v'$. Molecules coming from below and colliding with molecules on this layer may be assumed (in a rough approximation) to have suffered their most recent previous collision on a plane at a distance λ below the plane in question. The average momentum component in the shear direction parallel to the lamina (call it the y direction) for one of these molecules is therefore

$$mv = mv' - m\frac{dv}{dx}\lambda$$

where m is the mass of a molecule. Since by hypothesis the molecules have average momentum mv' after collision, an average amount of momentum

$$-m\frac{dv}{dx}\lambda$$

is transferred to the laminar plane x' from below, per collision. The number of molecules reaching a unit area of the plane from below per unit time is approximately[3]

$$\frac{1}{6}\bar{N}\bar{c}$$

where $\bar{N}$ is the number of molecules per unit volume and $\bar{c}$ is their mean molecular speed. Therefore, the rate of upward transfer of momentum to a unit area of the plane from below is

$$-\frac{1}{6}m\bar{N}\bar{c}\lambda\frac{dv}{dx}$$

Similarly, it can be shown that the rate of downward transfer of momentum from a unit area of the plane x' to the plane at $(x' - \lambda)$ is

$$+\frac{1}{6}m\bar{N}\bar{c}\lambda\frac{dv}{dx}$$

A downward flow of momentum is equivalent to an upward flow of momentum of the opposite sign. Thus the net rate of *upward* flow of momentum through any given plane per unit area is

$$-\frac{1}{3}m\bar{N}\bar{c}\lambda\frac{dv}{dx}$$

This rate of change of momentum must be balanced by a force in accordance with Newton's second law:

$$\frac{f}{A} = \frac{1}{3}m\bar{N}\bar{c}\lambda\frac{dv}{dx} \tag{8}$$

We can write Eq. (4-1), the defining equation for the viscosity coefficient η, as

$$\frac{f}{A} = \eta\frac{dv}{dx} \tag{9}$$

Combining Eqs. (8) and (9), we obtain an expression for η:

$$\eta \cong \frac{1}{3}m\bar{N}\bar{c}\lambda \tag{10}$$

Making use of Eq. (2) for $\bar{c}$ and Eq. (3) for λ, we find

$$\eta = \frac{2}{3\pi^{3/2}}\frac{(mkT)^{1/2}}{d^2} = \frac{2}{3\pi^{3/2}N_0}\frac{(MRT)^{1/2}}{d^2} \tag{11}$$

This *mean-free-path treatment* involves several simplifying and somewhat crude approximations; a more sophisticated and rigorous hard-sphere theory[1] gives results identical with Eqs. (10) and (11) except for the numerical factors:

$$\eta = \frac{5\pi}{32}m\bar{N}\bar{c}\lambda \tag{12}$$

and

$$\eta = \frac{5}{16\pi^{1/2}N_0}\frac{(MRT)^{1/2}}{d^2} = 2.928 \times 10^{-25}\frac{(MRT)^{1/2}}{d^2} \tag{13}$$

It should be noted again that the numerical coefficient above has units of mole. According to this equation, the gas viscosity coefficient should be independent of pressure and should increase with the square root of the absolute temperature. The viscosities of gases are in fact found to be substantially independent of pressure over a wide range. The temperature dependence generally differs to some extent from $T^{1/2}$ because the effective molecular diameter is dependent on how hard the molecules collide and therefore depends somewhat on temperature. Deviation from hard-sphere behavior in the case of air (diatomic molecules, N_2 and O_2) is demonstrated by Eq. (4-19).

Equation (13) can also be used for the calculation of molecular diameters from experimental viscosity data. These can be compared with molecular diameters as determined by other experimental methods (molecular beams, diffusion, thermal conductivity, x-ray crystallographic determination of molecular packing in the solid state, etc.).

The SI unit for viscosity is $N\ m^{-2}\ s = Pa\ s = kg\ m^{-1}\ s^{-1}$, which has no special name. The cgs unit is the *poise* (P), which equals $1\ dyn\ cm^{-2}\ s = 0.1\ N\ m^{-2}\ s$; this unit is widely used and will probably continue to be used for the foreseeable future. Remember to convert poise into SI units before carrying out calculations involving viscosities.

THERMAL CONDUCTIVITY

The mechanism of thermal conduction is analogous to that of viscous resistance to fluid flow. In the case of fluid flow, where a velocity gradient exists, momentum is transported from point to point by gas molecules; in thermal conduction, where a temperature gradient exists, it is kinetic energy that is transported from point to point by gas molecules. If, in the mean-free-path treatment of viscosity, the momentum is replaced by the average kinetic energy $\bar{\varepsilon}$ of a molecule, we obtain for one-dimensional heat flow in the x direction an equation analogous to Eq. (8):

$$\dot{q} = \frac{1}{3}\bar{N}\bar{c}\lambda\left(-\frac{d\bar{\varepsilon}}{dx}\right) \tag{14}$$

where $\dot{q}$ is rate of heat flow per unit area, $\bar{N}$ is the number of molecules per unit volume, $\bar{c}$ is the mean speed, and λ is the mean free path. Now

$$\frac{d\bar{\varepsilon}}{dx} = \frac{d\bar{\varepsilon}}{dT}\frac{dT}{dx} = m\bar{C}_v\frac{dT}{dx} \tag{15}$$

where m is the mass of a molecule and $\bar{C}_v$ is the constant-volume specific-heat capacity (i.e., per unit mass, which means per kilogram using SI units) of the gas; thus we obtain

$$\dot{q} = \frac{1}{3}m\bar{N}\bar{c}\lambda\bar{C}_v\left(-\frac{dT}{dx}\right) \tag{16}$$

By comparison with $\dot{q} = -K(dT/dx)$, the defining equation for the coefficient of thermal conductivity K, we find that

$$K = \frac{1}{3}m\bar{N}\bar{c}\lambda\bar{C}_v \tag{17}$$

Using Eq. (10) for the coefficient of viscosity of a gas, we obtain the following relationship between coefficients of thermal conductivity and of viscosity:

$$\frac{K}{\eta} = \bar{C}_v \tag{18}$$

This simple expression, which is derived using the mean-free-path treatment, is correct in form, but the numerical coefficient is wrong. More advanced treatments[1] for a hard-sphere monatomic gas yield

$$K = \frac{25\pi}{64} m\bar{N}\bar{c}\lambda \bar{C}_v \quad \text{monatomic} \tag{19}$$

From this result and Eq. (12), we find for monatomic gases

$$\frac{K}{\eta} = \frac{5}{2}\bar{C}_v = \frac{5}{2}\frac{\tilde{C}_v}{M} = \frac{15R}{4M} \tag{20}$$

where the statistical mechanical value $3R/2$ has been used for the molar heat capacity $\tilde{C}_v$. Interactions between diatomic and polyatomic molecules, involving transfer of rotational and vibrational energy, are much more difficult to treat theoretically[1,2] and will not be discussed here.

The SI unit for the thermal conductivity coefficient K is $J\ m^{-1}\ K^{-1}\ s^{-1}$, although the values are often cited in units of $J\ cm^{-1}\ K^{-1}\ s^{-1}$. Use of SI units for both K and η yields units of $J\ kg^{-1}\ K^{-1}$ for K/η as expected from R/M.

DIFFUSION

The theory of the diffusion of a gas is related to the theory of viscosity and of thermal conductivity in the sense that all three are concerned with a transport of some quantity from one point to another by the thermal motion of gas molecules. For diffusion, where a concentration gradient exists, chemical identity is transported.

Let there be two kinds of perfect gases, designated 1 and 2, present in a closed system of fixed total volume at concentrations of $\bar{N}_1$ and $\bar{N}_2$ molecules per unit volume or $\bar{n}_1$ and $\bar{n}_2$ moles per unit volume, respectively. These gases will have mean free paths of λ_1 and λ_2.

Let it be assumed that there is a concentration gradient in the vertical x direction only. Consider a horizontal plane x_0 and consider for a moment those molecules of kind 1 that pass through a small area $dA = dy\ dz$ on that plane. Assume for simplicity that these molecules suffered their most recent collision at or near the surface of an imaginary sphere of radius λ_1 with its center at dA. Now the mean distance from the plane x_0 up or down to the sphere is $2\lambda_1/3$. Therefore the mean concentration of these molecules, in the regions below the plane where they suffered their most recent collisions, is

$$\bar{N}_- = \bar{N}_1(x_0) - \frac{2\lambda_1}{3}\frac{d\bar{N}_1}{dx}$$

and the mean concentration similarly defined above the plane is

$$\bar{N}_+ = \bar{N}_1(x_0) + \frac{2\lambda_1}{3}\frac{d\bar{N}_1}{dx}$$

The number of molecules passing through a unit area of a given plane surface depends on the concentration of the molecules:

$$Z = \frac{1}{4}\bar{c}\bar{N} \tag{21}$$

where $\bar{c}$ is the mean speed. The *net* number of molecules of kind 1 crossing the x_0 plane in the upward direction is therefore

$$Z_1 = \frac{1}{4}\bar{c}_1(\bar{N}_- - \bar{N}_+) = -\frac{1}{3}\bar{c}_1\lambda_1\frac{d\bar{N}_1}{dx} \tag{22}$$

So far we have neglected the mass flow, if any, that is needed for the maintenance of constant and uniform pressure. If there is mass flow with a given velocity w_0 (which will be taken as positive when the mass flow is upward), the molecular velocities will be Maxwellian only with respect to a set of coordinate axes moving with that velocity, since the diffusing molecules suffered their recent collisions with molecules moving upward at the *mean* velocity. Equation (22) therefore represents the number of molecules per second crossing a unit area on a plane (parallel to the y and z axes) that is moving upward at velocity w_0. The number of molecules per second crossing a unit area on a *fixed* plane is then

$$Z_1 = \bar{N}_1 w_0 - \frac{1}{3}\bar{c}_1 \lambda_1 \frac{d\bar{N}_1}{dx}$$
$$Z_2 = \bar{N}_2 w_0 - \frac{1}{3}\bar{c}_2 \lambda_2 \frac{d\bar{N}_2}{dx} \tag{23}$$

Now, for maintenance of constant pressure, we must have $Z_1 = -Z_2$; thus we obtain

$$w_0 = \frac{1}{3(\bar{N}_1 + \bar{N}_2)}\left(\bar{c}_1 \lambda_1 \frac{d\bar{N}_1}{dx} + \bar{c}_2 \lambda_2 \frac{d\bar{N}_2}{dx}\right)$$
$$= \frac{1}{3\bar{N}}\frac{d\bar{N}_1}{dx}(\bar{c}_1 \lambda_1 - \bar{c}_2 \lambda_2) \tag{24}$$

and

$$Z_1 = -\frac{1}{3\bar{N}}(\bar{N}_1 \lambda_2 \bar{c}_2 + \bar{N}_2 \lambda_1 \bar{c}_1)\frac{d\bar{N}_1}{dx} = -Z_2 \tag{25}$$

where, since the pressure is constant, we have used the relation

$$-\frac{d\bar{N}_1}{dx} = \frac{d\bar{N}_2}{dx} \tag{26}$$

We can also write, in terms of moles,

$$z_1 = -\frac{1}{3\bar{n}}(\bar{n}_1 \lambda_2 \bar{c}_2 + \bar{n}_2 \lambda_1 \bar{c}_1)\frac{d\bar{n}_1}{dx} = -z_2$$
$$= -\frac{1}{3}(X_1 \lambda_2 \bar{c}_2 + X_2 \lambda_1 \bar{c}_1)\frac{d\bar{n}_1}{dx} \tag{27}$$

where X_1 and X_2 are mole fractions ($X_1 = \bar{n}_1/\bar{n}$, $X_2 = \bar{n}_2/\bar{n}$; $\bar{n} = \bar{n}_1 + \bar{n}_2$). By comparison with Eq. (5-2), the defining equation for the diffusion constant D_{12}, we find that

$$D_{12} = \frac{1}{3}(X_1 \lambda_2 \bar{c}_2 + X_2 \lambda_1 \bar{c}_1) \tag{28}$$

This equation appears to predict that D_{12} will be a function of the composition of the gas, although experimentally the diffusion constant is almost independent of composition. However, we must be careful in our definition of λ_1 and λ_2. We must take account of the fact that collisions of molecules of one species with one another can have no significant effect on the diffusion; such collisions do not affect the total momentum possessed by all the molecules of that species and thus do not affect the mean mass velocity of the species in its diffusion. Thus the total number of molecules of that species crossing the reference plane in a given period of time is not affected by such collisions and is the same as if such collisions did not take place at all. We should define λ_1 for our present purpose as the mean free path of molecules of species 1 between successive collisions with molecules of species 2 and vice versa. Accordingly, we write [see Eq. (3)]

$$\lambda_1 = \frac{1}{\sqrt{2}\pi \bar{N}_2 \, d_{12}^2} \tag{29}$$

where by d_{12} we mean the center-to-center collision distance (assuming the molecules to be "hard spheres"). Equation (28) then becomes

$$D_{12} = \frac{1}{3\sqrt{2}\pi d_{12}^2 \bar{N}}(\bar{c}_1 + \bar{c}_2) \tag{30}$$

which is independent of the composition. If we now introduce the kinetic theory expressions for $\bar{c}_1$ and $\bar{c}_2$ as given by Eq. (2) and make use of the perfect-gas expression for $\bar{N}$, namely,

$$\bar{N} = N_0 \bar{n} = N_0 \frac{p}{RT} \tag{31}$$

where N_0 is Avogadro's number, we obtain for the diffusion constant

$$D_{12} = \frac{2}{3\pi^{3/2} N_0} \frac{(RT)^{3/2}}{p} \frac{\sqrt{1/M_1} + \sqrt{1/M_2}}{d_{12}^2} \tag{32}$$

and for the *self-diffusion constant* (see Exp. 5)

$$D_1 = \frac{4}{3\pi^{3/2} N_0} \frac{(RT)^{3/2}}{p} \frac{\sqrt{1/M_1}}{d_1^2} \tag{33}$$

More advanced treatments[1] lead to a slightly different form of expression for D_{12} and different numerical coefficients for both D_{12} and D_1:

$$D_{12} = \frac{3}{8\sqrt{2}\pi N_0} \frac{(RT)^{3/2}}{p} \frac{\sqrt{1/M_1 + 1/M_2}}{d_{12}^2} \tag{34}$$

and

$$D_1 = \frac{3\pi}{16}\bar{c}_1 \lambda_1 = \frac{3}{8\sqrt{\pi} N_0} \frac{(RT)^{3/2}}{p} \frac{\sqrt{1/M_1}}{d_1^2} \tag{35}$$

where λ_1 is the mean free path as usually defined for a pure gas.

Equations (32) and (34), which are based on a "hard-sphere" model, are in agreement in predicting no dependence of the diffusion constant on gas composition. However, for real molecules, a slight composition dependence should exist, which depends on the form of the intermolecular potential.[1]

That Eqs. (32) and (34) differ in the dependence of D_{12} on the molecular weights should not be altogether surprising, since in our simple treatment we have assumed that the velocities of molecules after collisions are uncorrelated with their velocities before collisions, whereas in fact such correlations in general exist and are functions of the ratio of the masses of the colliding particles. To take this and other remaining factors properly into account would require a treatment that is beyond the scope of this book. It may suffice to point out here that the square-root quantity in Eq. (34) is equivalent to $\sqrt{1/\mu_{12}N_0}$, where

$$\mu_{12} = \frac{m_1 m_2}{m_1 + m_2}$$

is the "reduced mass" of a system comprising a molecule of kind 1 and a molecule of kind 2.

The diffusion constant is predicted by Eqs. (32) to (35) to be inversely proportional to the total pressure. Experimentally, this is the case to roughly the degree to which the perfect-gas law applies. The equations appear to predict that the diffusion constant will be proportional to the three-halves power of the temperature; however, as in the case of viscosity, significant deviations from this behavior occur, as actual molecules are not truly "hard spheres" and have collision diameters that depend on the relative speeds with which molecules collide with one another.

By means of Eq. (34) it is possible to calculate the center-to-center collision distance d_{12} from a measured diffusion constant D_{12}. If three diffusion constants D_{12}, D_{13}, and D_{23} can be measured, individual molecular diameters d_1, d_2, and d_3 can be obtained if it can be assumed that the collision distances are the arithmetic averages of the molecular diameters involved, as would be the case with hard spheres:

$$d_{ij} = \frac{1}{2}(d_i + d_j) \tag{36}$$

If a self-diffusion constant D_1 can be measured approximately through one of the approaches discussed in Exp. 5, a molecular diameter d_1 can be obtained directly from Eq. (35). Conversely, Eqs. (34) and (35) can be used to estimate a diffusion constant if molecular diameters are known from some other source or to calculate self-diffusion constants from binary diffusion constants (or vice versa) by calculating first the molecular diameters. In the latter case, it can be argued that any remaining approximations in the treatment will largely cancel out.

As expected the self-diffusion constant can be related to other transport coefficients. From Eqs. (12) and (35), we find

$$\frac{D}{\eta} = \frac{6}{5\rho} \tag{37}$$

where ρ is the mass density of the gas. This is a hard-sphere model result, and more advanced treatments[1] show that the coefficient of $1/\rho$ is larger for soft-sphere models. For spherical molecules with a somewhat artificial inverse-fifth-power repulsion, the coefficient is about 1.5. Clearly D can also be related to K for monatomic gases. From Eqs. (20) and (37),

$$\frac{D}{K} = \frac{8M}{25\rho R} \tag{38}$$

The SI unit for D is $m^2\ s^{-1}$, but again it is common to cite values in $cm^2\ s^{-1}$. To avoid problems associated with calculation of ratios such as D/η and D/K, consistent use of SI units is recommended.

REFERENCES

1. R. D. Present, *Kinetic Theory of Gases,* McGraw-Hill, New York (1958); P. P. Schram, *Kinetic Theory of Gases and Plasmas,* Kluwer, Norwell, MA (1991).

2. I. N. Levine, *Physical Chemistry,* 6th ed., chaps. 14 and 15, McGraw-Hill, New York (2009).

3. G. W. Castellan, *Physical Chemistry,* 3d ed., chap. 30, Addison-Wesley, Reading, MA (1983).

GENERAL READING

S. Chapman and T. G. Cowling, *The Mathematical Theory of Non-uniform Gases,* 3d ed., Cambridge Univ. Press, Cambridge (1970).

L. C. Woods, *An Introduction to the Kinetic Theory of Gases and Magnetoplasmas,* Oxford Univ. Press, New York (1993).

EXPERIMENT 4

Viscosity of Gases

It is a general property of fluids (liquids and gases) that an applied shearing force that produces flow in the fluid is resisted by a force that is proportional to the gradient of flow velocity in the fluid. This is the phenomenon known as *viscosity.*

Consider two parallel plates of area A, at distance D apart, as in Fig. 1. It is convenient to imagine that D is small in comparison with any dimension of the plates in order to avoid edge effects. Let there be a uniform fluid substance between the two plates. If one of the plates is held at rest while the other moves with uniform velocity v_0 in a direction parallel to its own plane, under ideal conditions the fluid undergoes a pure shearing motion and a flow velocity gradient of magnitude v_0/D exists throughout the fluid. This is the simplest example of *laminar flow,* or pure viscous flow, which takes place under such conditions that the inertia of the fluid plays no significant role in determining the nature of the fluid motion. The most important of these conditions is that the flow velocities be small. In laminar flow in a system with stationary solid boundaries, the paths of infinitesimal mass elements of the fluid do not cross any of an infinite family of stationary laminar surfaces that may be defined in the system. In the simple example above, these laminar surfaces are planes parallel to the plates. When fluid velocities become high, the flow becomes *turbulent* and the momentum of the fluid carries it across such laminar surfaces so that eddies or vortices form.

In the above example, with laminar flow the force f resisting the relative motion of the plates is proportional to the area A and to the velocity gradient v_0/D:

$$f = \eta A \frac{v_0}{D} \tag{1}$$

The constant of proportionality η is called the *coefficient of viscosity* of the fluid, or simply the *viscosity* of the fluid. A convenient unit for viscosity is the *poise,* which is the cgs unit equivalent to 10^{-1} N s m^{-2} = 10^{-1} Pa s.

The viscosities of common liquids are of the order of a centipoise (cP); at 20.20°C, the viscosity of water is 1.000 cP while that of ethyl ether is 0.23 cP and that of glycerin is 830 cP (8.3 P). The viscosity of pitch or asphalt at room temperature is of the order of

FIGURE 1
Ideal plane–parallel laminar flow.

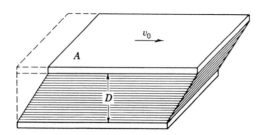

10^{10} P, and the viscosity of glass (a supercooled liquid) is many powers of 10 higher. The viscosities of gases are of the order of 100 or 200 μP. The viscosities of liquids and soft solids decrease with increasing temperature, while those of gases increase with increasing temperature. Viscosities of all substances are substantially independent of pressure at ordinary pressures but show change at very high pressures and, apparently in the case of gases alone, at very low pressures also.

METHOD

Among the various methods[1,2] that have been used for determining the viscosities of liquids and gases are several that involve the measurement of viscous drag on a rotating disk or cylinder immersed in the fluid or a sphere falling through it. The simplest and most commonly used methods however depend on measurement of the rate of viscous flow through a cylindrical capillary tube in relation to the pressure gradient along the capillary.

For absolute viscosity measurements, the pressure should be constant at each end of the capillary. A simple but excellent method for gases has been described by A. O. Rankine,[2] in which a constant pressure difference is maintained across the ends of a capillary by a pellet of mercury falling in a parallel tube of considerably larger diameter (up to 3.5 mm). However, it is rarely necessary to make absolute determinations. Most viscosity measurements are made in "relative viscosimeters," in which the viscosity is proportional to the time required for the flow of a fixed quantity of fluid through the capillary under a pressure differential that varies during the experiment *but always in the same manner.* If the pressure differential as a function of the volume of fluid passed through the capillary is known accurately, the viscosimeter can be used for absolute determinations. More commonly the unknown factors are lumped into a single apparatus constant, the value of which is determined by time-of-flow measurements on a reference substance of known viscosity. This is the principle of the Ostwald viscosimeter routinely used for measurements on liquids and solutions, in which capillary flow is induced by gravity (and accordingly the time of flow is proportional to the viscosity divided by the liquid density). It is also the principle of the method used for measurements on gases in this experiment.

THEORY

Let us imagine a long cylindrical capillary tube, of length L and radius r. Let x denote the distance outward radially from the axis of the tube and z denote the distance along the axis. Under conditions of laminar flow, the laminar surfaces are a family of right circular cylinders coaxial with the tube.

Consider two such cylindrical surfaces, of radius x and $x + dx$ and length dz as shown in Fig. 2a. What is the difference in velocity dv between the two cylinders when the flow rate is such as to produce a pressure gradient dp/dz along the tube?

The force tending to cause the inner cylinder to slide past the outer one is the same as it would be for a *solid* inner piston of radius x sliding with the same relative velocity in a *solid* cylinder of radius $x + dx$, with the fluid playing the role of a "lubricant" between. Since the spacing dx is infinitesimal in comparison with the radius of curvature, we may imagine that the conditions existing between the two surfaces resemble those of Fig. 1. The force acting on an element of length dz of the tube is

$$df = \frac{dp}{dz} \pi x^2 dz$$

FIGURE 2
(*a*) Definition of variables.
(*b*) Parabolic velocity
distribution. (*c*) Abnormal
(approximately uniform)
velocity distribution existing
near inlet.

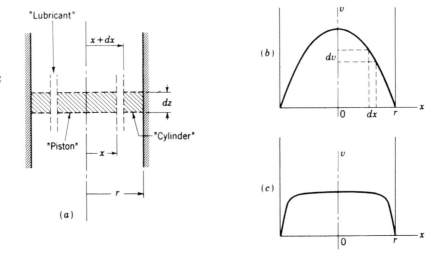

The corresponding cross-sectional area is

$$dA = 2\pi x\, dz$$

and the spacing D is dx. In order to determine the corresponding velocity difference dv, we substitute for f, A, D, and v_0 in Eq. (1) and rearrange to obtain

$$dv = \frac{\pi x^2}{\eta 2\pi x}\frac{dp}{dz}dx = -\frac{1}{2\eta}\left(-\frac{dp}{dz}\right)x\, dx \qquad (2)$$

To find the velocity at any value of x, we integrate from $x = r$, where the velocity vanishes, to $x = x$:

$$v(x) = \int_r^x dv = \frac{1}{4\eta}\left(-\frac{dp}{dz}\right)(r^2 - x^2) \qquad (3)$$

This equation predicts that the distribution of velocities in the tube can be represented by a parabola, as in Fig. 2*b*. The volume rate of flow past a given point (volume per unit time) is

$$\phi_V = \int_0^r v(x)2\pi x\, dx = \frac{\pi r^4}{8\eta}\left(-\frac{dp}{dz}\right) \qquad (4)$$

and the mean flow velocity $\bar{v}$ is†

$$\bar{v} = \frac{\phi_V}{\pi r^2} = \frac{1}{8\eta}\left(-\frac{dp}{dz}\right)r^2 = \frac{1}{2}v_{max} \qquad (5)$$

If the fluid is incompressible, the volume rate of fluid flow must be constant along the tube; and we find, where p_1 and p_2 are the pressures at the inlet and outlet of the tube, respectively, that

$$\phi_V = \frac{\pi r^4(p_1 - p_2)}{8\eta L}$$

This equation, known as *Poiseuille's law,* can be applied under ordinary laminar-flow conditions to liquids and also to gases in the limiting case where $(p_1 - p_2)$ is negligible in

†It should be emphasized that $\bar{v}$ is *not* the mean molecular speed $\bar{c}$ defined in Eq. (V-2).

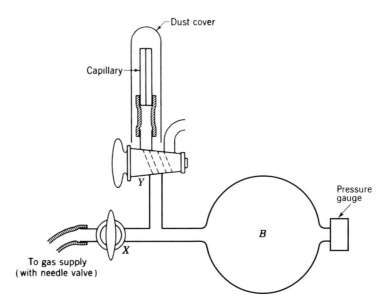

FIGURE 3
Gas viscosity apparatus.

comparison with p_1 or p_2. For a more general treatment of gases, their compressibility must be taken into account. If we assume the perfect-gas law, we can write for the *molar* rate of flow (moles per unit time)

$$\phi_n = \frac{p}{RT}\phi_V = \frac{\pi r^4}{8\eta RT}p\left(-\frac{dp}{dz}\right) = \frac{\pi r^4}{16\eta RT}\left[-\frac{d(p^2)}{dz}\right] \tag{6}$$

Since conservation of matter requires that the molar rate of flow be constant along the tube, $d(p^2)/dz$ must be constant, and accordingly we can write

$$-\frac{d(p^2)}{dz} = \frac{p_1^2 - p_2^2}{L} \tag{7}$$

We now obtain

$$\phi_n = \frac{\pi r^4(p_1^2 - p_2^2)}{16\eta LRT} \tag{8}$$

In the apparatus to be used in this experiment (see Fig. 3), the gas contained in bulb B flows through the capillary tube over a time period t_f during which the inlet pressure drops from a predetermined initial value p_1^o to a predetermined final pressure p_1^f. The outlet pressure p_2 is constant at the ambient atmospheric pressure during this period. The volume V occupied by gas below the capillary (bulb B plus the pressure gauge and associated tubing) is constant, but the number of moles n of gas contained in volume V varies according to the ideal gas law:

$$\frac{dn}{dt} = \frac{1}{RT}\frac{d(p_1V)}{dt} = \frac{V}{RT}\frac{dp_1}{dt}$$

Setting $(-dn/dt)$ equal to ϕ_n,† we obtain

$$-\frac{dp_1}{dt} = \frac{RT}{V}\phi_n = \frac{\pi r^4(p_1^2 - p_2^2)}{16\eta LV} \tag{9}$$

†The minus sign is necessary, since dn/dt is negative while ϕ_n is positive.

On inverting and integrating, we find that the time required for the inlet pressure to drop from p_1^0 to p_1^f is

$$t_f = \eta \frac{16L}{\pi r^4} V \int_{p_1^f}^{p_1^0} \frac{dp_1}{(p_1^2 - p_2^2)}$$

$$= \eta \frac{16L}{\pi r^4} \frac{V}{2p_2} \left[\ln \frac{p_1 - p_2}{p_1 + p_2} \right]_{p_1^f}^{p_1^0} = K\eta \qquad (10)$$

where the "apparatus constant" K contains a part $(16L/\pi r^4)$ determined by the capillary tube and a part (V times the integral over p_1) determined by the dimensions of the bulb and the chosen initial and final inlet pressures. While it is possible to determine K from the known capillary dimensions and a careful determination of the volume V, more satisfactory results can be obtained by determining K from the time of flow for a gas of known viscosity.

Factors Influencing Apparatus Design. Practical application of the theory presented above depends on the validity of a number of assumptions. One, which has already been mentioned, is the assumption that the flow is laminar. It has been shown by dimensional analysis that, for any given type of hydrodynamic experiment, the conditions for onset of turbulent flow depend on the magnitude of a certain combination of pertinent experimental variables that is a pure number, called the Reynolds number Re. For flow through a long, round, straight tube,

$$Re = r\bar{v}\frac{\rho}{\eta} \qquad (11)$$

where ρ is the density of the fluid. It is found empirically that laminar flow is always obtained in such a tube when Re is less than 1000, regardless of the magnitude of any of the individual variables r, $\bar{v}$, ρ, η. Laminar flow can be obtained with a Reynolds number as high as 10,000 on even 25,000 if sufficiently careful attention is given to the smoothness of the walls and to the shape of the inlet, but ordinarily turbulent flow is obtained with Reynolds numbers in excess of a few thousand.[3]

Another important factor concerns the transition from a uniform velocity distribution at the inlet of the tube to the parabolic velocity distribution inside the tube (see Figs. 2*b* and 2*c*). The region in which this transition takes place may be of considerable length, and Eq. (8) will not apply very well over this region. Thus it is important that the length of the region be small in comparison with the length of the tube. Experiments have shown[3] that the length L' of the transition region is given approximately by the equation

$$L' = \frac{1}{4}(Re)r \qquad (12)$$

where r is the radius of the tube and Re is the Reynolds number.

It is important that the tube be very straight; the critical Reynolds number for onset of turbulence will be greatly reduced if the tube is appreciably curved. As a suitable criterion, it might be specified that deviations from straightness should be negligible in comparison with r over distances of the order of L'.

Kinetic-Energy Correction. In deriving Eq. (10) it was assumed that "end effects" are negligible. Such end effects are most likely to be important at the inlet of the capillary, where a pressure drop occurs as a result of the necessity of imparting kinetic energy to the fluid.

By the law of Bernoulli, the pressure drop is equal to the "flow-average" *kinetic-energy density* of the fluid (that is, the rate of flow of kinetic energy past a given fixed point divided by the volume rate of flow of fluid past this point). For flow through a pipe or capillary,

$$\Delta p = \frac{1}{\pi r^2 \bar{v}} \int_0^r \rho \frac{v^2}{2} v 2\pi x \, dx \tag{13}$$

Beyond the transitional region of length L', we may assume the parabolic velocity distribution given by Eqs. (3) and (5) is

$$v = 2\bar{v}\left(1 - \frac{x^2}{r^2}\right) \tag{14}$$

The overall kinetic-energy pressure drop from the inlet to the point where the distribution has become parabolic is obtained by substituting Eq. (14) into Eq. (13) and is

$$\Delta p = \rho(\bar{v})^2 = \frac{\rho}{\pi^2 r^4} \phi_V^2 \tag{15}$$

This quantity is called the *kinetic-energy correction*. In practice, it is subtracted from the overall pressure difference $p_1 - p_2$, since no comparable pressure change need ordinarily be considered at the outlet end. This results from the fact that the kinetic energy of the fluid carries it in a jet stream far out into the body of fluid beyond the outlet, so that the kinetic energy is dissipated (as heat rather than as potential energy) at a considerable distance from where it could otherwise be effective in producing a pressure change at the outlet.

To take into account the effect of kinetic energy on the time of flow t_f we can subtract Δp as given by Eq. (15) from p_1 in Eqs. (8) and (9), then invert and integrate as before. We obtain the integral appearing in Eq. (10) plus additional terms that are small. To simplify the latter, we assume that the kinetic-energy pressure drop Δp is small in comparison with p_1. We obtain

$$t_f = K\eta + \frac{\rho V}{8\pi\eta L} \ln \frac{p_1^o}{p_1^f} \tag{16}$$

where V is the gas volume below the capillary and K is the same apparatus constant as defined previously.

Slip Correction. It has been assumed throughout the above discussion that Eq. (1) is valid down to the smallest dimensions that are of any significant importance in the experiment. This assumption appears well founded for laminar *liquid* flow, but for gases it breaks down when the mean free path is not negligibly small in comparison with the apparatus dimensions. Thus, at very low pressures or at ordinary pressures in capillaries of very small diameter, gas viscosities appear to be lower than when measured under ordinary conditions. When the mean free path of a gas is small but not negligible in comparison with the radius of the capillary, the gas behaves as if it were "slipping" at the capillary walls, rather than having zero velocity at the walls as shown in Fig. 2b and given in Eq. (3). Indeed the mean velocity of the gas infinitesimally close to the wall is not zero, for although half the molecules in this region have suffered their most recent collision at the wall, the other half have suffered their most recent collision at some distance from the wall, of the order of a mean free path λ. Therefore the mass velocity, instead of being zero at the wall, is a fraction approximating $4\lambda/r$ of the average mass velocity $\bar{v}$. Thus, if Eq. (16) yields the "apparent viscosity" η_{app}, the true viscosity can be obtained from the equation

$$\eta = \eta_{app}\left(1 + \frac{4\lambda}{r}\right) \tag{17}$$

Dependence on Flow Velocity. With regard to the tacit assumption that the viscosity coefficient is independent of flow velocities and gradients of flow velocities, it must be remembered that the molecular velocities are essentially Maxwellian in a gas under shear only when variations in mass velocity over distances of the order of a mean free path are small in comparison with the molecular velocities themselves; if these differences are considerable, it is possible that the perturbed distribution of molecular velocities will manifest itself in an altered apparent viscosity coefficient. Without going into detail, it may be stated that if $\bar{v} \ll u$ (where u is the rms molecular velocity) and if $\lambda \ll r$, the error in η due to perturbation of the Maxwellian distribution will be negligible.

EXPERIMENTAL

Using the apparatus shown in Fig. 3, measure the flow times t_f for dry air, helium, argon, carbon dioxide (and/or any other gases provided) at the same temperature in the same apparatus. Moisture should be removed from the air by inserting a drying tube in the line between the compressed-air supply and the viscosity apparatus. This tube, containing a drying agent such as magnesium perchlorate, should be full but not too tightly packed. It should be removed from the line when other gases are studied. If CO_2 is studied, be sure it is at room temperature before starting the run, as it cools considerably on expansion at the outlet of the cylinder.

The pressure gauge used in this experiment should be a direct-reading gauge with a relatively small and constant internal volume. Reproducibility is more important than absolute accuracy since relative measurements are made on air (the standard gas used for calibration) and the other gases. *The critical feature is that the same initial p_1^o and final p_1^f inlet pressures are used in all runs.* Thus, one can use capacitance, reluctance, or strain-gauge manometers (see p. 596–597). The latter manometers are the least expensive and are adequate if models with the best resolution are chosen.

Admit one of the gases to the bulb B by slowly opening stopcock X while stopcock Y remains closed. Close stopcock X when the pressure gauge reads ~ 910 Torr. Carefully vent the bulb through the open outlet from the three-way stopcock Y to flush the system. Fill and flush a second time. Refill the bulb with gas, then open stopcock Y to the capillary and measure the time t_f for the pressure to drop from $p_1^o = 900$ Torr to $p_1^f = 800$ Torr. Four or more runs should be made if time permits.

Repeat the above procedure for each of the gases to be studied. Also record the temperature and the corrected barometric pressure p_2. Record at least approximate values for all the pertinent apparatus factors—L, r, V.

CALCULATIONS

For calibration purposes dry air was chosen as the standard, since it is easily available and has been carefully studied by many investigators. Data from many sources can be summarized in the empirical Sutherland expression:[4]

$$\eta = \frac{(145.8 \times 10^{-7})T^{3/2}}{T + 110.4} \tag{18}$$

which gives the viscosity of dry air in poise at 1 atm as a function of absolute temperature T. At 25°C, the value is 183.7 μP (0.01837 cP). For the temperature of your experiment, calculate η for dry air from Eq. (19).

For each gas studied, average the flow times for the individual runs to obtain an average t_f.

If Eq. (10) were valid, the ratio of viscosities for two gases would equal the corresponding ratio of flow times. Assuming Eq. (10) and using t_{air} and η_{air}, find approximate values of the viscosities of the other gases from their flow times.

For each gas (including air), calculate the maximum value of the Reynolds number from Eq. (11) and verify that it is below 1000. Use Eq. (12) to verify that L', the length of the transition region, is small compared with the length L of the capillary. Estimate the largest value of $\bar{v}$ from Eq. (5) and compare it with the molecular rms speed u and mean speed $\bar{c}$.

If the required conditions have been met, kinetic-energy and slip corrections may now be made in order to obtain more precise values of η. First calculate λ for each gas by combining Eqs. (V-2) and (V-12) and using the ideal-gas law:

$$\lambda = \frac{32\eta}{5\pi} \frac{1}{m} \frac{1}{\bar{N}} \left(\frac{\pi m}{8kT} \right)^{1/2} = \frac{16\eta}{5p} \left(\frac{RT}{2\pi M} \right)^{1/2} \tag{19}$$

With Eq. (17) find η_{app} for dry air and use this value in Eq. (16) to find the value of K. Now calculate η_{app} for the other gases from Eq. (16) (one can use the approximate value of η in evaluating the kinetic-energy correction term). Finally, apply the slip correction to obtain the true viscosity. The gas pressure p that appears explicitly in Eq. (19) and implicitly in Eq. (16), where $\rho = pM/RT$, can be replaced by the average inlet pressure $(p_1^o + p_1^f)/2$.

Report the final values obtained for η together with the temperature at which they were determined. Report also for each gas the calculated mean free path at 1 bar and the effective molecular diameter as calculated from the viscosity with Eq. (V-13).

SAFETY ISSUES

Gas cylinders must be chained securely to the wall or laboratory bench (see pp. 644–646 and Appendix C).

APPARATUS

Gas viscosity apparatus, complete with capillary tube and pressure gauge; gas supply outlet(s) with needle valve(s); drying tube; two lengths of gum-rubber tubing; 0 to 30°C thermometer; timer.

Cylinders of argon, helium, and carbon dioxide; source of compressed air; glass wool; drying agent (such as magnesium perchlorate); stopcock grease.

REFERENCES

1. J. F. Golubev, *Viscosity of Gases and Gas Mixtures,* chap. 2 [translated from Russian by Israeli Program for Scientific Translations, Jerusalem (1970)].

2. J. Reilly and W. N. Rae, *Physico-chemical Methods,* 5th ed., Vol. I. pp. 667, 690, and Vol. III, p. 247, Van Nostrand, Princeton, NJ (1953).

3. L. Prandtl, *Essentials of Fluid Dynamics,* Blackie, Glasgow (1952).

4. *Tables of Thermal Properties of Gases,* table 1-B, U.S. Natl. Bur. Stand. Circ. 564, U.S. Government Printing Office, Washington, DC (1955).

GENERAL READING

J. F. Johnson, J. R. Martin, and R. S. Porter, "Determination of Viscosity," in A. Weissberger and B. W. Rossiter (eds.), *Techniques of Chemistry: Vol. I. Physical Methods of Chemistry,* part VI, chap. 2, Wiley-Interscience, New York (1977).

S. Pai, *Viscous Flow Theory, Vol. I: Laminar Flow,* Van Nostrand-Reinhold, Princeton, NJ (1965).

EXPERIMENT 5

Diffusion of Gases

In this experiment the mutual diffusion coefficients for the $Ar-CO_2$ and $He-CO_2$ systems are to be measured using a modified Loschmidt apparatus. These transport coefficients are then compared with theoretical values calculated with hard-sphere collision diameters.

When a component of a continuous fluid phase (a liquid solution or a gas mixture) is present in nonuniform concentration, at uniform and constant temperature and pressure and in the absence of external fields, that component diffuses in such a way as to tend to render its concentration uniform. For simplicity, let the concentration of a given substance $\bar{n}_1$ be a function of only one coordinate x, which we shall take as the upward direction. The net "flux" z_1 of the substance passing upward past a given fixed point x_0 (i.e., amount per unit cross-sectional area per unit time) is under most conditions found to be proportional to the negative of the concentration gradient:

$$z_1 = -D\left(\frac{d\bar{n}_1}{dx}\right)_{x_0} \tag{1}$$

where D is the *diffusion constant,* which is characteristic of the diffusing substance and usually also of the other substances present. This is called *Fick's first law of diffusion.*

When only two substances are present in a closed gaseous system of fixed total volume at a uniform temperature, the distribution of one of the substances is completely determined when the distribution of the other is specified. The sum of the partial pressures of the two components is equal to the total pressure, and if the two components are perfect gases, the total pressure is constant and the sum of the two concentrations is also constant. Designating the two substances by the subscripts 1 and 2, we can write

$$z_1 = -D_{12}\frac{d\bar{n}_1}{dx} = D_{12}\frac{d\bar{n}_2}{dx} \tag{2}$$

$$z_2 = -z_1 = -D_{21}\frac{d\bar{n}_2}{dx} = D_{21}\frac{d\bar{n}_1}{dx} \tag{3}$$

from which it can be seen that D_{12}, the diffusion constant for the diffusion of component 1 into component 2, is equal to D_{21}, the diffusion constant for the diffusion of 2 into 1.

We can also define a diffusion constant for the diffusion of a substance into itself. Suppose that we have a single molecular species (substance 1 say) present in a vessel but that the molecules initially in a certain portion of the vessel can somehow be tagged or labeled (without changing their kinetic properties) so that later they can be distinguished from the others. After a period of time the distribution of the tagged molecules could then be determined experimentally and the *self-diffusion constant D_1* can be calculated. In actuality it is not possible to tag molecules without some effect on their kinetic properties. A close approach to such tagging can be obtained however with the use of different isotopes that may be distinguished with a mass spectrometer or a Geiger counter. Another close approach can be obtained in a few cases by the use of molecules that are very similar in size, shape, and mass but different in chemical constitution, for example, the pair N_2 and CO or the pair CO_2 and N_2O. For each of these pairs, the diffusion constant D_{12} should constitute a reasonably good estimate of the self-diffusion constants D_1 and D_2.

In the case where the two gases are distinctly different in their diffusing tendencies, the mutual diffusion constant D_{12} lies somewhere between the two self-diffusion constants D_1 and D_2 in value. It is also found that, when the two self-diffusion constants are very different, the mutual diffusion constant is largely determined by the more rapidly diffusing substance rather than by the more slowly diffusing one; thus there is a greater difference between the diffusion constants for the systems Kr–He and Kr–H_2 than for Kr–He and CO_2–He. The gas having the high self-diffusion constant (He or H_2) diffuses rapidly into the gas having the low self-diffusion constant. But there must be a *mass flow* of the resulting mixture in the opposite direction to maintain uniform pressure, so the gas with the lower diffusion constant gives the appearance of diffusing, although in reality it is mainly undergoing displacement.

THEORY

Let the concentration of gas 1 be $\bar{n}_1(x)$, a function of x only, in a vertical tube of uniform cross section A (x increasing upward). Let us calculate the rate of accumulation of gas 1 in the region from x to $x + dx$. The net flow of the gas into the region from below, per unit area per second, is given by Fick's first law,

$$z_1(x) = -D_{12}\left(\frac{\partial \bar{n}_1}{\partial x}\right)_x$$

The net flow of the gas out of the region in the upward direction, per unit area per second, is

$$z_1(x + dx) = -D_{12}\left(\frac{\partial \bar{n}_1}{\partial x}\right)_{x+dx}$$

$$= -D_{12}\left[\left(\frac{\partial \bar{n}_1}{\partial x}\right) + \left(\frac{\partial^2 \bar{n}_1}{\partial x^2}\right)dx\right]$$

The rate of increase of the concentration $\bar{n}_1(x)$ is given by

$$\frac{\partial \bar{n}_1(x)}{\partial t} = \frac{z_1(x) - z_1(x + dx)}{dx}$$

and therefore

$$\frac{\partial \bar{n}_1}{\partial t} = D_{12}\frac{\partial^2 \bar{n}_1}{\partial x^2} \tag{4}$$

This is the differential equation for diffusion in one dimension (called *Fick's second law*).

Let two perfect gases be placed in a *Loschmidt apparatus*[1]—a vertical tube of length $2L$, of uniform cross section, closed at both ends, and containing a removable septum in the center. The heavier of the two gases (gas 1 say) is initially confined to the lower half and the lighter (gas 2) to the upper half in order to avoid convective mixing. Both gases are at initial concentration $\bar{n}_0$. At time $t = 0$, the septum is removed, so that the gases are free to interdiffuse. Let us define a quantity $\Delta \bar{n}$ as follows:

$$\Delta \bar{n} = \bar{n}_1 - \bar{n}_2 \equiv 2\bar{n}_1 - \bar{n}_0 \tag{5}$$

The differential equation to be solved,

$$\frac{\partial \Delta \bar{n}}{\partial t} = D\frac{\partial^2 \Delta \bar{n}}{\partial x^2} \tag{6}$$

is the differential equation (4) in $\bar{n}_1$ minus a similar one in $\bar{n}_2$, and

$$D = D_{12} = D_{21}$$

Taking the origin at the middle of the tube, we write the boundary conditions

$$t = 0: \quad \Delta \bar{n} = \bar{n}_0 \quad -L \leq x < 0$$
$$\Delta \bar{n} = -\bar{n}_0 \quad 0 < x \leq L \tag{7}$$

$$t = \infty: \quad \Delta\bar{n} = 0 \quad -L \le x \le L \tag{8}$$

$$\text{All } t: \quad \frac{\partial \Delta\bar{n}}{\partial x} = 0 \quad x = -L$$

$$\frac{\partial \Delta\bar{n}}{\partial x} = 0 \quad x = L \tag{9}$$

The last boundary condition follows from the fact that there is no flow of gas through the two ends of the tube; see Eq. (1). Boundary condition (7) can be expressed in terms of a Fourier series expansion[2] of $\Delta\bar{n}$:

$$t = 0: \quad \Delta\bar{n} = -\frac{4}{\pi}\bar{n}_0\left(\sin\frac{\pi}{2L}x + \frac{1}{3}\sin\frac{3\pi}{2L}x + \frac{1}{5}\sin\frac{5\pi}{2L}x + \ldots\right) \tag{7'}$$

This is the Fourier series for a "square wave" of amplitude $\bar{n}_0$ and wavelength $2L$.

We shall omit the procedure for finding the solution of Eq. (6) and simply give the result:

$$\frac{\Delta\bar{n}}{\bar{n}_0} = \frac{\bar{n}_1 - \bar{n}_2}{\bar{n}_1 + \bar{n}_2} = -\frac{4}{\pi}\left(e^{-\pi^2 Dt/4L^2}\sin\frac{\pi}{2L}x + \frac{1}{3}e^{-9\pi^2 Dt/4L^2}\sin\frac{3\pi}{2L}x\right.$$
$$\left. + \frac{1}{5}e^{-25\pi^2 Dt/4L^2}\sin\frac{5\pi}{2L}x + \ldots\right) \tag{10}$$

The reader can easily verify that the solution satisfies the differential equation and the boundary conditions (7) or (7'), (8), and (9). The behavior of $\Delta\bar{n}$ with x and t is shown qualitatively in Fig. 1a.

After time t, let us replace the removable septum and analyze the two portions of the tube for the total quantity of one of the gases, say gas 1. The total amount of gas 1 in the upper part will be equal to the total amount of gas 2 in the lower part. Let W_1 and n_1 be the weight and number of moles of gas 1 in the lower part and W_1' and n_1' be the weight and the number of moles of gas 1 in the upper part, and let similar symbols with subscript 2 pertain to gas 2. Define a quantity f by

$$f \equiv \frac{n_1 - n_2}{n_1 + n_2} = \frac{(\bar{n}_1 - \bar{n}_2)_{av}}{\bar{n}_0} = \frac{1}{L}\int_{-L}^{0}\frac{\Delta\bar{n}}{\bar{n}_0}dx \tag{11}$$

Since $n_1' = n_2$, we have also

$$f = \frac{n_1 - n_1'}{n_1 + n_1'} = \frac{W_1 - W_1'}{W_1 + W_1'} \tag{12}$$

This expression permits the calculation of f from the results of analyzing separately the upper and lower parts of the tube for gas 1. From Eqs. (10) and (11), we obtain

$$f = \frac{8}{\pi^2}\left(e^{-\pi^2 Dt/4L^2} + \frac{1}{9}e^{-9\pi^2 Dt/4L^2} + \frac{1}{25}e^{-25\pi^2 Dt/4L^2} + \ldots\right) \tag{13}$$

At $t = 0$, the sum of the series in parentheses is equal to $\pi^2/8$, so the quantity f is initially unity. After a short time all terms except the first become negligible, and f therefore decays with time in a simple exponential manner.

At what time t should the run be stopped in order to obtain the most precise value of D, assuming that the only significant errors are those in determining f and that the uncertainty in f is constant (independent of the time)? Neglecting all but the first term of Eq. (13), we find on solving for D and differentiating that

$$\frac{\partial D}{\partial f} = -\frac{1}{t}\frac{4L^2}{\pi^2}\frac{1}{f}$$

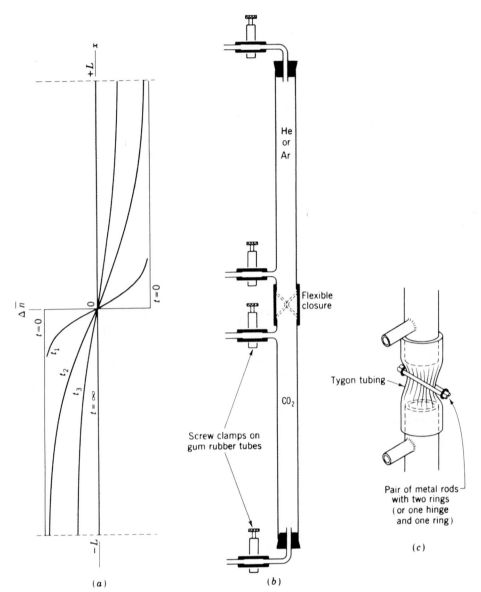

FIGURE 1
Loschmidt diffusion
apparatus: (*a*) curves showing
schematically the quantity
$\Delta \bar{n} = \bar{n}_1 - \bar{n}_2$; (*b*) cell;
(*c*) detail of flexible closure.

We wish to find at what value of t the derivative $\partial D/\partial f$ is at a minimum:

$$0 = \frac{\partial^2 D}{\partial t\, \partial f} = \frac{4L^2}{\pi^2}\left(\frac{1}{t}\frac{1}{f^2}\frac{\partial f}{\partial t} + \frac{1}{t^2}\frac{1}{f}\right) = \frac{4L^2}{\pi^2 ft}\left(-\frac{\pi^2 D}{4L^2} + \frac{1}{t}\right)$$

thus the optimum time is

$$t_{\text{opt}} = \frac{4L^2}{\pi^2 D} \tag{14}$$

At this optimum time, Eq. (13) gives

$$f = 0.81057(0.367879 + 0.000014 + \ldots) = 0.2982$$

Note that neglecting the second and higher terms of Eq. (13) proves to be justified if the experiment is run for any time in the neighborhood of the optimum time or for any greater time.

METHOD

In this experiment we shall measure the diffusion constants for the systems CO_2-He and CO_2-Ar at room temperature. In each run, CO_2, the heavier gas, will be initially in the lower portion of a vertical tube. The gas in the two portions at the end of the run will be analyzed for CO_2 by sweeping it through absorption tubes (U tubes) filled with Ascarite II (a nonfibrous silicate impregnated with NaOH, which reacts with CO_2) and with magnesium perchlorate (to absorb the water liberated in the first reaction); see Fig. 2. These tubes are weighed before and after sweeping the gas through them. The diffusion cell, a modified Loschmidt type, is shown in Fig. 1*b*. In place of a removable septum at the center, the two halves of this tube are joined by a flexible plastic (Tygon) sleeve of approximately the same inner diameter as the glass, which can be pinched tight to achieve a closure as shown in Fig. 1*c*.

A variant of the same basic method is shown later in Fig. 3. In this case, a vacuum system is used for handling the gases, a large-bore stopcock is used in place of the flexible closure, and a different method is used for the quantitative analysis of CO_2. Instead of absorbing the CO_2 on Ascarite II, the CO_2-Ar or CO_2-He mixture is slowly pumped through a copper coil immersed in liquid nitrogen. The CO_2 will be efficiently trapped, and after all the Ar or He has been removed, the trap is allowed to warm to room temperature and the CO_2 pressure is determined.

In round numbers, the length L in the apparatus to be employed is 50 cm and the two diffusion constants are of order of magnitude 0.6 $cm^2 \, s^{-1}$ for CO_2-He and 0.15 $cm^2 \, s^{-1}$ for CO_2-Ar. Using these figures the optimum diffusion times can be estimated from Eq. (14). The actual time should approximate this optimum time but in any case should not differ from it by more than 30 percent.

EXPERIMENTAL

PROCEDURE A

In carrying out the experiment with the apparatus shown in Figs. 1 and 2, you should follow a careful plan worked out with your partner, taking into account the following points:

1. Even a tiny leak is serious: all stoppers and clamps *must be tight*. Rubber and Tygon tubing must be in good condition.
2. When one gas is being flushed out by another, the gas being introduced should enter at the lower end of the vessel if it is the heavier and at the upper end if it is the lighter.

FIGURE 2
U tube for absorbing CO_2 gas.

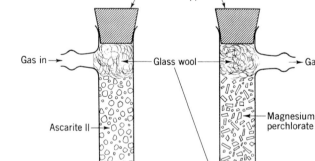

3. Considerable heat is evolved in the reaction of CO_2 with Ascarite II; a U tube that has absorbed a considerable amount of CO_2 should be allowed to cool for about 15 min before it is weighed.

4. Carbon dioxide gas emerges rather cool from the tank valve. Time should be allowed for it to warm to room temperature before the diffusion run is started. It is wise to vent the lower chamber, containing CO_2, momentarily at the side arm before the run to restore atmospheric pressure.

5. If the cylinder is not equipped with a regulator valve, turn off the gas supply at the cylinder before closing off the system. The inlet of the system should be clamped off before the outlet.

6. The CO_2-containing gas should be swept into the drying tubes no faster than five bubbles per second. Use a dibutyl-phthalate bubbler for metering gas flow. One bubble is approximately 0.1 cm^3. It is good to adjust the flow rate (with the outlet valve on the compressed air line) *before* attaching to the system. The sweeping air should go *first* through the bubbler and *then* through a purifying tube (to remove CO_2 and H_2O) before it enters the system. In case the flow rate is preadjusted, it is important to make sure that the system is open at both ends at the moment the air tube is attached.

7. Drying tubes should be closed off with rubber policemen when gas is not being passed through them. Take care that the drying tube always has the same two policemen.

8. Do not neglect the volumes of gas that are contained in any significant lengths of rubber tubing. Before using a rubber tube that may previously have contained CO_2, flush it out momentarily with air. Also, do not neglect the possibility of a heavy gas spilling out of the system if it is open at two places at different levels for more than a second or two.

Prepare as shown in Fig. 2 and weigh two U tubes, A and B. It is not necessary to correct weights to vacuum readings. A third U tube, which need not be weighed, should be prepared for purification of the air to be used for sweeping.

Flush out the lower half of the cell with CO_2, using a flushing rate of at least 10 to 15 bubbles per second through the bubbler. *Make sure the center partition is closed* before flushing. Pass at least six volumes of gas through to be sure of complete replacement.

Connect the lower half-cell to U tubes A and B in series, and sweep out the gas in the lower half-cell into the U tubes with three or four volumes of air (purified with the third U tube), at five bubbles per second. Without stopping the sweeping, *quickly* pull off tube A and reconnect tube B. Immediately close the ends of tube A. After one or two more additional volumes have been swept through tube B, weigh tube B first, then tube A. Tube B should not have gained more than a few milligrams. Add the two gains in weight; this quantity is $W_1 + W_1'$, the total weight of CO_2 in the cell.

Now flush out the lower half-cell with CO_2 and the upper one with the other gas (He or Ar) to be studied, as above. This should be done during the above weighings to save time.

Begin the diffusion by *carefully* (not too suddenly) opening the center partition and simultaneously starting a timer. After the estimated optimum time, close the partition.

Sweep the contents of the lower tube into U tubes A and B as before. After the weighings, repeat with the upper tube. The value of $W_1 + W_1'$ should agree with the value obtained above to within a very few milligrams; if greater disagreement is obtained, a leak must be suspected.

Repeat the procedure with the second gas combination. If time permits, repeat each of the above runs, with the diffusion cell inverted, and in each case take as your value of f the mean of the two values obtained.

Measure the length L as precisely as possible. Also record the barometric pressure and the ambient temperature.

PROCEDURE B

In carrying out the experiment with the apparatus shown in Fig. 3, the student should review the basic operation of a vacuum system (see Chapter XVIII).

The diffusion tube was designed and constructed such that the length L of the upper section is equal to the length of the lower section *plus* the length of the bore of stopcock A. Hence the stopcock bore and the lower section of the tube are both filled with CO_2.

The CO_2 and He (or Ar) gas tanks are connected to the diffusion tube as shown in Fig. 3. Stopcocks A through F should be opened and the line should be pumped on for about 20 min. Close stopcock F and check for an increase in pressure. The rate of increase should be no greater than 5 Torr h^{-1}. Then evacuate the entire apparatus again for a few minutes.

Close stopcock D and fill the system with CO_2 to a pressure between 750 and 770 Torr. Close stopcock B and read the CO_2 pressure exactly. After closing stopcock A, pump away the remaining CO_2. Fill the upper portion of the tube with He to a pressure *as close as possible* to the CO_2 pressure (the difference should be less than 1 Torr). Close stopcock C and evacuate the remaining portion of the apparatus. Begin the diffusion by *carefully* opening stopcock A and simultaneously starting a stopwatch or timer (be sure to align the stopcock bore with the vertical axis of the upper and lower tubes). After the estimated optimum time, close stopcock A and record the elapsed time.

Place a Dewar flask of liquid nitrogen around the copper coil. After the trap has been cooled (as indicated by an absence of violent boiling of the liquid nitrogen), *first* close stopcock D and then open C. After about 30 s, very slowly open stopcock D until the pressure (watch the pressure gauge) just begins to decrease. If the stopcock is opened too much, the trapping of CO_2 will not be complete. Adjust stopcock D to give a pressure decrease of about 20 Torr every 15 s. It should take not more than a few minutes to remove the helium. Then close both stopcocks D and C. To remove any He entrapped in the solid CO_2, remove the Dewar flask from the coil and allow the CO_2 to vaporize. However, do not wait for the coil to reach room temperature. Place the Dewar flask around the coil again and when the solidification is complete, fully open stopcock D for about a minute (*leave stopcock C closed*). If the above operations were performed properly, at least 99 percent of the CO_2 will have been trapped.

With stopcock D closed, remove the Dewar flask and allow the bulk of the CO_2 to vaporize. Then place a liter beaker containing water at room temperature around the coil (this will

FIGURE 3
Modified Loschmidt apparatus, to be used according to procedure B. Stopcock A should have a very large bore which is as close as possible to the diameter of the diffusion tube. Stopcocks B to E should be high-vacuum type. As a safety precaution, each gas lecture bottle should be connected to the system via a reducing valve or a pressure-relief valve.

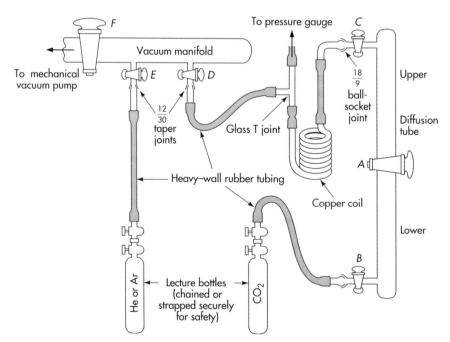

FIGURE 3
Modified Loschmidt apparatus, to be used according to procedure B. Stopcock A should have a very large bore which is as close as possible to the diameter of the diffusion tube. Stopcocks B to E should be high-vacuum type. As a safety precaution, each gas lecture bottle should be connected to the system via a reducing valve or a pressure-relief valve.

hasten the warming up of the coil). After about a minute, remove the beaker and dry the coil with a towel. Wait a few minutes for the coil to come to thermal equilibrium with the room and then measure the CO_2 pressure. This pressure in the upper half of the tube is denoted by p_1'.

Now evacuate all of the system above stopcock A. Then close stopcock C and open A. Repeat the procedure described above in order to determine the CO_2 pressure associated with the gas in the lower half of the tube.

The quantity f in Eq. (12) is defined in terms of pressures by

$$f = \frac{p_1 - p_1'}{p_1 + p_1'}$$

Finally, repeat the complete procedure with Ar as the second gas. In addition to determining the diffusion coefficients at atmospheric pressure, measure one of them at half an atmosphere.

Measure the length L as precisely as possible, and also record the ambient temperature.

CALCULATIONS

Your experimental data should be summarized in a table that includes for each run the initial pressure, the temperature, the length of the run, and the final CO_2 weights (or pressures) from the upper and lower parts of the cell.

By use of Eqs. (12) and (13), neglecting all but the first term in the series, calculate D for each gas combination. Then, with this value of D, calculate the second term in the series; if it is more than 1 percent of the first term, recalculate D. The resulting values of D are to be designated D_{12} for CO_2–He, D_{13} for CO_2–Ar.

Gas-viscosity measurements yield molecular diameters of 2.58 Å (angstroms) for helium, 3.42 Å for argon, and 4.00 Å for carbon dioxide.[3] With these values, calculate d_{12} and d_{13} from Eq. (V-36), and calculate D_{12} and D_{13} using Eq. (V-34). Compare with your experimental values.

The molecular diameters cited above were obtained from an analysis of viscosity data in terms of the Lennard–Jones "6-12 potential":[3]

$$U(r) = 4\varepsilon \left[\left(\frac{\sigma}{r} \right)^{12} - \left(\frac{\sigma}{r} \right)^{6} \right] \qquad (15)$$

where $U(r)$ is the potential energy of two molecules at a distance r. The diameters correspond to σ values (the distance at which the potential energy equals zero). Although these values are often called effective hard-sphere diameters (why?), one should remember that they do not result from the solution of the hard-sphere expression, Eq. (V-13). Alternatively, if one wishes to calculate an *accurate* diffusion coefficient, it is meaningless to use such σ values in conjunction with Eq. (V-34).

Optional: At 273 K, D_{23} (for He–Ar) is 0.653 cm^2 s^{-1}. From D_{12}, D_{13}, and D_{23} at *room temperature,* calculate d_{12}, d_{13}, and d_{23}. (To correct diffusion constants from one temperature to another, assume a $T^{3/2}$ dependence if the temperature change is small. This is only approximate, since the d's may vary in some degree with the temperature.) Then obtain d_1, d_2, and d_3 and use these to calculate D_1, D_2, and D_3 from Eq. (V-35). Determine the ratios of the self-diffusion constants to their respective viscosities. How do these compare with the theoretical ratios discussed in the introductory section of Chapter V?

DISCUSSION

The flexible closure or the large-bore stopcock, though simple and convenient, is admittedly less satisfactory than a removable diaphragm closure would be. Discuss its disadvantages from the points of view of (1) mixing produced in the opening and closing operations; (2) inexactness of the matching of its diameter with that of the tubes;

and (3) any other defects that occur to you. Try wherever possible to employ arguments that are not wholly qualitative; estimate orders of magnitude where you can. If possible, suggest an improved closure design. What additional techniques might be used with this experimental method to enable D_{23} (for He–Ar) to be measured directly? Using your value for D_{13}, estimate the time necessary for a diffusion run on CO_2-Ar in the present apparatus to achieve a state where $W_1 = 1.01W_1'$ (i.e., where the concentration of CO_2 is only 1 percent greater in the lower half than in the upper half).

SAFETY ISSUES

Gas cylinders must be chained securely to the wall or laboratory bench (see pp. 644–646). Procedure B: Work on a vacuum system requires preliminary review of procedures and careful execution in order to avoid damage to the apparatus and possible injury from broken glass; in addition, the liquid nitrogen used for cold traps must be handled properly (see Appendix C). Safety glasses must be worn.

APPARATUS

Procedure A. Diffusion cell (Loschmidt type) with sturdy support; clamping device for flexible closure; two one-hole rubber stoppers, with short bent tubes, for ends of cell; four 1-in. lengths of rubber tubing; four screw clamps; four short glass tubes; three glass U tubes with rubber stoppers and policemen; bubbler containing dibutyl phthalate; three long and two short lengths of rubber tubing; needle valve; two ring stands; four clamps with clamp holders; 0 to 30°C thermometer; stopwatch or timer; meter stick.

Cylinders of helium, argon, and carbon dioxide; source of compressed air; glass wool; Ascarite II; drying agent (such as magnesium perchlorate).

Procedure B. Vacuum system with mechanical pump and manifold having provision for two attachments via taper joints; Loschmidt diffusion cell with stopcocks attached at top and bottom; two standard taper joints to fit manifold; two ball–socket joints to fit diffusion cell; glass (or metal) T joint; five long lengths and one short length of heavy-wall rubber pressure tubing; tightly wound copper coil; large Dewar flask; pressure gauge; 1-L beaker; towel; stopwatch or electrical timer; meter stick; 0 to 30°C thermometer.

Liquid nitrogen; cylinders or lecture bottles of helium, argon, and carbon dioxide.

REFERENCES

1. P. J. Dunlop, K. R. Harris, and D. J. Young, "Experimental Methods for Studying Diffusion in Gases, Liquids, and Solids," in B. W. Rossiter and R. C. Baetzold (eds.), *Physical Methods of Chemistry*, 2d ed., Vol. VI, chap. 3, Wiley-Interscience, New York (1992).

2. Any text on advanced calculus, such as R. Courant, *Differential and Integral Calculus*, Wiley-Interscience, New York (1992); or F. B. Hildebrand, *Advanced Calculus for Applications*, 2d ed., Dover, Mineola, NY (1992).

3. J. O. Hirschfelder, C. F. Curtiss, and R. B. Bird, *Molecular Theory of Gases and Liquids*, pp. 1035–1044, 1110. Wiley, New York (1964).

GENERAL READING

T. I. Gombosi, *Gaskinetic Theory*, Cambridge Univ. Press, New York (1994).

R. D. Present, *Kinetic Theory of Gases*, McGraw-Hill, New York (1958).

P. P. Schram, *Kinetic Theory of Gases and Plasmas*, Kluwer, Norwell, MA (1991).

EXPERIMENTS

6. Heats of Combustion
7. Strain Energy of the Cyclopropane Ring
8. Heats of Ionic Reaction

PRINCIPLES OF CALORIMETRY

We are concerned here with the problem of determining experimentally the enthalpy change ΔH or the energy change ΔE accompanying a given isothermal change in state of a system, normally one in which a chemical reaction occurs. We can write the reaction schematically in the form

$$\underset{\text{initial state}}{A(T_0) + B(T_0)} = \underset{\text{final state}}{C(T_0) + D(T_0)} \tag{1}$$

If n is the number of moles of the limiting reagent, the molar quantities for the reaction are $\Delta \widetilde{H} = \Delta H/n$ and $\Delta \widetilde{E} = \Delta E/n$.

In practice we do not actually carry out the change in state isothermally; this is not necessary because ΔH and ΔE are independent of the path. In calorimetry we usually find it convenient to use a path composed of two steps:

Step I. A change in state is carried out *adiabatically* in the calorimeter vessel to yield the desired products but in general at another temperature:

$$A(T_0) + B(T_0) + S(T_0) = C(T_1) + D(T_1) + S(T_1) \tag{2}$$

where S represents those parts of the system (e.g., inside wall of the calorimeter vessel, stirrer, thermometer, solvent) that are always at the same temperature as the reactants or products because of the experimental arrangement; these parts, plus the reactants or products, constitute the system under discussion.

Step II. The products of step I are brought to the initial temperature T_0 by adding heat to (or taking it from) the system:

$$C(T_1) + D(T_1) + S(T_1) = C(T_0) + D(T_0) + S(T_0) \tag{3}$$

As we shall see, it is often unnecessary to carry out this step in actuality, since the associated change in energy or enthalpy can be calculated from the known temperature difference.

By adding Eqs. (2) and (3), we obtain Eq. (1) and verify that these two steps describe a complete path connecting the desired initial and final states. Accordingly, ΔH or ΔE for the change in state (1) is the sum of the values of this quantity pertaining to the two steps:

$$\Delta H = \Delta H_\mathrm{I} + \Delta H_\mathrm{II} \tag{4a}$$

$$\Delta E = \Delta E_\mathrm{I} + \Delta E_\mathrm{II} \tag{4b}$$

The particular convenience of the path described is that the heat q for step I is zero, while the heat q for step II can be either measured or calculated. It can be measured directly by carrying out step II (or its inverse) through the addition to the system of a measurable quantity of heat or electrical energy, or it can be calculated from the temperature change $(T_1 - T_0)$ resulting from adiabatic step I if the heat capacity of the product system is known. For step I,

$$\Delta H_\mathrm{I} = q_p = 0 \qquad \text{constant pressure} \tag{5a}$$

$$\Delta E_\mathrm{I} = q_v = 0 \qquad \text{constant volume} \tag{5b}$$

Thus, if *both* steps are carried out at constant pressure,

$$\Delta H = \Delta H_\mathrm{II} \tag{6a}$$

and if *both* are carried out at constant volume,

$$\Delta E = \Delta E_\mathrm{II} \tag{6b}$$

Whether the process is carried out at constant pressure or at constant volume is a matter of convenience. In nearly all cases it is most convenient to carry it out at constant pressure; the experiment on heats of ionic reaction is an example. An exception to the general rule is the determination of a heat of combustion, which is conveniently carried out at constant volume in a "bomb calorimeter." However, we can easily calculate ΔH from ΔE as determined from a constant-volume process (or ΔE from ΔH as determined from a constant-pressure process) by use of the equation

$$\Delta H = \Delta E + \Delta(pV) \tag{7}$$

When all reactants and products are condensed phases, the $\Delta(pV)$ term is negligible in comparison with ΔH or ΔE, and the distinction between these two quantities is unimportant. When gases are involved, as in the case of combustion, the $\Delta(pV)$ term is likely to be significant in magnitude, though still small in comparison with ΔH or ΔE. Since it is small, we can employ the perfect-gas law and rewrite Eq. (7) in the form

$$\Delta H = \Delta E + RT\Delta n_\mathrm{gas} \tag{8}$$

where Δn_gas is the *increase* in the number of moles of gas in the system.

We must now concern ourselves with procedures for determining ΔH or ΔE for step II. We might envisage step II being carried out by adding heat to the system or taking heat away from the system and measuring q for this process.† Usually, however, it is much easier to measure work than heat. In particular, electrical work done on the system by a heating coil (often referred to as *Joule heating*) can be used conveniently to carry out

---

†This could be done by placing the system in thermal contact with a heat reservoir (such as a large water bath) of known heat capacity until the desired change has been effected and calculating q from the measured change in the temperature of the reservoir.

either step II or its inverse, whichever is endothermic. Since the work is dissipated *inside* the system, the work is positive:

$$w_{el} = \int V_h \, dQ = \int V_h I \, dt \qquad (9a)$$

where V_h is the voltage drop across the heater, Q is the electric charge, and I is the electric current. For precise work, one should measure both V_h and I during the heating period. In many instances, however, it is possible to assume that the resistance of the heating coil R_h is constant and make only measurements of I by determining the potential drop V_s across a standard resistance R_s in series with the heating coil. In such a case, one can write

$$w_{el} = \int I^2 R_h \, dt = \frac{R_h}{R_s^2} \int V_s^2 \, dt \qquad (9b)$$

Note that the electrical work given by Eq. (9) is in joules when resistance is in ohms, potential (voltage) is in volts, and time is in seconds. If the electrical heating is done adiabatically,

$$\Delta H_{II} = w_{el} \qquad \text{constant pressure} \qquad (10a)$$

$$\Delta E_{II} = w_{el} \qquad \text{constant volume} \qquad (10b)$$

Our discussion so far has been limited to determining ΔH_{II} or ΔE_{II} by directly carrying out step II (or its inverse). However, it is often not necessary to carry out this step in actuality. If we know or can determine the heat capacity of the system, the temperature change $(T_1 - T_0)$ resulting from step I provides all the additional information we need:

$$\Delta H_{II} = \int_{T_1}^{T_0} C_p(\text{C} + \text{D} + \text{S}) \, dT \qquad (11a)$$

$$\Delta E_{II} = \int_{T_1}^{T_0} C_v(\text{C} + \text{D} + \text{S}) \, dT \qquad (11b)$$

The heat capacities ordinarily vary only slightly over the small temperature ranges involved: accordingly we can simplify Eqs. (11) and combine them with Eqs. (6) to obtain the familiar expressions

$$\Delta H = -C_p(\text{C} + \text{D} + \text{S})(T_1 - T_0) \qquad (12a)$$

$$\Delta E = -C_v(\text{C} + \text{D} + \text{S})(T_1 - T_0) \qquad (12b)$$

where C_p and C_v are average values over the temperature range.

The heat capacity must be determined if it is not known. A direct method, which depends on the assumption of the constancy of heat capacities over a small range of temperature, is to measure the adiabatic temperature rise $(T_2' - T_1')$ produced by the dissipation of a measured quantity of electrical energy. We then obtain

$$\left. \begin{aligned} C_p \\ C_v \end{aligned} \right\} = \frac{w_{el}}{T_2' - T_1'} \qquad \begin{aligned} (13a) \\ (13b) \end{aligned}$$

at constant pressure or at constant volume, respectively. This method is exemplified in the experiment on heats of ionic reaction.

An indirect method of determining the heat capacity is to carry out another reaction altogether, for which the heat of reaction is known, in the same calorimeter under the same conditions. This method depends on the fact that in most calorimetric measurements on chemical reactions the heat-capacity contributions of the actual product species (C and D) are very small, and often negligible, in comparison with the contribution due to the parts of the system denoted by the symbol S. In a bomb-calorimeter experiment the reactants or products amount to a gram or two, while the rest of the system is the thermal equivalent of about 2.5 kg of water. In calorimetry involving dilute aqueous solutions, the heat capacities of such solutions can in a first approximation be taken equal to that of equivalent weights or volumes of water. Thus we may write, in place of Eqs. (12a) and (12b),

$$\left.\begin{array}{l} \Delta H \\ \Delta E \end{array}\right\} = -C(S)(T_1 - T_0) \qquad\qquad \begin{array}{l} (14a) \\ (14b) \end{array}$$

for constant-pressure and constant-volume processes, respectively. In Eqs. (14a) and (14b) we have omitted any subscript from the heat capacity as being largely meaningless, since only solids and liquids, with volumes essentially independent of pressure, are involved. The value of $C(S)$ can be calculated from the heat of the known reaction and the adiabatic temperature change $(T_2' - T_1')$ produced by it, as follows:

$$C(S) = \begin{cases} \dfrac{-\Delta H_{known}}{T_2' - T_1'} & \text{constant pressure} & (15a) \\[2em] \dfrac{-\Delta E_{known}}{T_2' - T_1'} & \text{constant volume} & (15b) \end{cases}$$

This method is exemplified in the experiment on heats of combustion by bomb calorimetry.

Let us now consider step I, the adiabatic step, and the measurement of the temperature difference $(T_1 - T_0)$, which is the fundamental measurement of calorimetry. If this step could be carried out in an *ideal* adiabatic calorimeter, the temperature variation would be like that shown in Fig. 1a. In this case there would be no difficulty in determining the temperature change $\Delta T = T_1 - T_0$, since $(dT/dt) = 0$ before the time of mixing the reactants and after the products achieve thermal equilibrium. The only cause of temperature change here is the chemical reaction. However, it is an unrealistic idealization to assume that step I is truly adiabatic; as no thermal insulation is perfect, some heat will in general leak into or out of the system during the time required for the change in state to occur and for the thermometer to come into equilibrium with the product system.

In addition, we usually have a stirrer present in the calorimeter to aid in the mixing of reactants or to hasten thermal equilibration. The mechanical work done on the system by the stirrer results in the continuous addition of energy to the system at a small, approximately constant rate. During the time required for the change in state and thermal equilibration to occur, the amount of energy introduced can be significant. A typical temperature–time variation is shown in Fig. 1b, where a greatly expanded temperature

FIGURE 1

(a) Plot of temperature T versus time t for an ideal adiabatic calorimeter.
(b) Plot with interrupted temperature axis and greatly expanded temperature scale for a typical calorimeter run. Such a plot is designed to show clearly the temperature variation before and after the time t_i when the reaction is initiated.

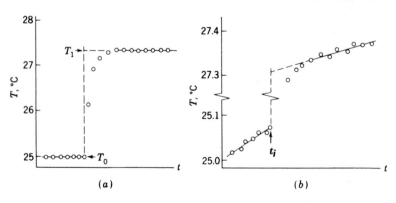

(a)   (b)

scale has been used with an interrupted temperature axis in order to show clearly the pre- and postreaction temperature changes. In this case, a more detailed analysis is needed in order to extract the adiabatic ΔT value from the observed data points.

Consider Fig. 2, which shows a schematic plot of $T(t)$ associated with step I as carried out in a typical calorimeter. This plot is based on the assumptions that the heat leakage through the calorimeter walls is small and the energy input due to the stirrer is moderate. These assumptions are appropriate for the thermochemistry experiments described in this book. The initial temperature T_i is conveniently taken to be the value at time t_i when the reaction is initiated following a period of essentially linear T, t readings that serve to establish the initial (prereaction) drift rate $(dT/dt)_i$. The temperature is followed after reaction for a period long enough to achieve a roughly linear variation and to establish the drift rate $(dT/dt)_f$ at an arbitrary point T_f, t_f.

The analysis of step I will be given for a constant-volume calorimeter run, but analogous results hold at constant pressure. The effect of heat leakage is evaluated using Newton's law of cooling,

$$\frac{dq}{dt} = -k(T - T_s) \tag{16}$$

where T_s is the temperature of the surroundings (air temperature in the laboratory) and k is a thermal rate constant that depends on the thermal conductivity of the calorimeter insulation. The mechanical power input to the system (dw/dt) will be denoted by P, which is assumed to be a constant independent of t. Thus

$$C_v \frac{dT}{dt} = \frac{dE}{dt} = P - k(T - T_s) \tag{17}$$

or

$$\left(\frac{dT}{dt}\right)_{\text{leak+stir}} = \frac{1}{C}[P - k(T - T_s)] \tag{18}$$

where C is C_v for constant-volume reactions and C_p for constant-pressure reactions. The temperature difference $T_f - T_i$ shown in Fig. 2 is given by

$$T_f - T_i = \Delta T + \int_{t_i}^{t_f} \left(\frac{dT}{dt}\right)_{\text{leak+stir}} dt \tag{19}$$

where $\Delta T = T_1 - T_0$ is the change due to an adiabatic chemical reaction and the integral is the net change due to heat leak and stirrer power input. Thus it follows that

$$T_1 - T_0 \equiv \Delta T = (T_f - T_i) - \frac{1}{C}\int_{t_i}^{t_f} \{P - k[T(t) - T_s]\}\, dt \tag{20}$$

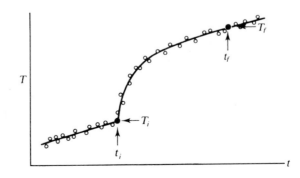

FIGURE 2

Schematic plot of temperature T versus time t observed in a typical calorimeter experiment. The temperature–time points (T_i, t_i) and (T_f, t_f) can be chosen somewhat arbitrarily as long as t_i is prior to or at t_{react} and t_f is a sufficiently long time after t_{react} (see text).

The following are the most important experimental approaches aimed at minimizing the effect of nonadiabatic conditions and the effect of the stirrer; they can be used separately or in combination.

1. The calorimeter may be built in such a way as to minimize heat conduction into or out of the system. A vessel with an evacuated jacket (Dewar flask), often with silvered surfaces to minimize the effect of heat radiation, may be used. This corresponds to making the thermal rate constant k very small.

2. One may interpose between system and surroundings an "adiabatic jacket," reasonably well insulated from both. This jacket is so constructed that its temperature can be adjusted at will from outside, as, for example, by supplying electrical energy to a heating circuit. In use, the temperature of the jacket is continuously adjusted so as to be as close as possible to that of the system, so that no significant quantity of heat will tend to flow between the system and the jacket. This corresponds to making $T - T_s$ very small.

3. The stirring system may be carefully designed to produce the least rate of work consistent with the requirement of thorough and reasonably rapid mixing. This corresponds to making P small.

4. We may assume that the rate of gain or loss of energy by the system resulting from heat leakage and stirrer work is reasonably constant with time at any given temperature. We may therefore assume that the temperature variation as a function of time should be initially and finally almost linear.

One simple possibility would be $k(T - T_s) = 0$ and $P = $ constant, in which case $(dT/dt) = P/C$ has the same value at all $t < t_i$ and $t > t_f$ and Eq. (19) gives

$$\Delta T = T_f - T_i - \frac{dT}{dt}(t_f - t_i) \qquad \text{if } k(T - T_s) = 0 \qquad (21)$$

As shown in Fig. 3a, ΔT in this case is just the vertical distance between two parallel $T-t$ lines. Naturally, this distance can be evaluated at any t. A more realistic possibility is that $k(T - T_s)$ is small but not negligible and $P = $ constant. In this case the $T-t$ plot looks like Fig. 3b, where the initial leak $+$ stir drift rate $(dT/dt)_i$ at t_i differs from the final rate $(dT/dt)_f$ at t_f. It can be shown[1] in this case that ΔT is given by

$$\Delta T = (T_f - T_i) - \left(\frac{dT}{dt}\right)_i (t_d - t_i) - \left(\frac{dT}{dt}\right)_f (t_f - t_d) \qquad (22)$$

FIGURE 3
Determination of the "adiabatic" temperature change ΔT from experimental $T(t)$ measurements made in nonideal calorimeters: (a) case of no heat leak but a constant stirring power input; (b) case of a small heat leak as well as stirrer energy input. The value of t_d is chosen so that the two shaded regions have equal areas; see text for details.

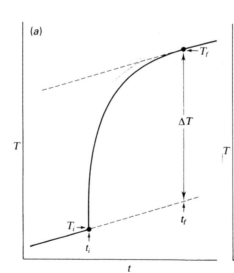
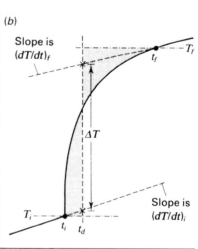

where t_d is chosen so that the two shaded regions in Fig. 3b are of equal area.

The derivation of Eq. (22) from Eq. (20) requires a bit of sleight-of-hand manipulation. Let us define a new variable $T_c \equiv T_s + (P/k)$, in terms of which Eq. (18) gives

$$\left(\frac{dT}{dt}\right)_i = -\frac{k}{C}(T_i - T_c) \quad \text{and} \quad \left(\frac{dT}{dt}\right)_f = -\frac{k}{C}(T_f - T_c) \tag{23}$$

and Eq. (20) becomes

$$\Delta T = (T_f - T_i) + \frac{k}{C}\int_{t_i}^{t_f}(T - T_c)dt \tag{24}$$

Note that

$$\int_{t_i}^{t_f}(T - T_c)\,dt = \int_{t_i}^{t_d}(T - T_i)\,dt + (T_i - T_c)(t_d - t_i)$$

$$+ \int_{t_d}^{t_f}(T - T_f)\,dt + (T_f - T_c)(t_f - t_d) \tag{25}$$

By the definition given above for t_d, the two integrals in Eq. (25) cancel out. Substituting Eq. (25) into Eq. (24) and making use of Eq. (23) leads directly to Eq. (22). It is clear from this derivation that Eq. (22) will be valid even if the T variation prior to t_i and after t_f is not linear. In such a case, one merely uses the tangent to the $T-t$ curves at T_i and t_f.

In summary, one should follow the temperature variation before and after reaction for a period long enough to allow a good evaluation of $(dT/dt)_i$ and $(dT/dt)_f$. Then make a plot like Fig. 3b, choose the point (T_f, t_f), determine the best value for t_d, and calculate ΔT from Eq. (22). In practice t_f should be chosen as close to t_i as is consistent with a well-characterized final drift rate. Note that t_d will lie much closer to t_i than to t_f for reactions that go to completion quickly. Whenever the time required for a change in state is long, as in the case of determining the heat capacity by electrical heating, the position of t_d will lie near the middle of the range t_i to t_f. As an approximate procedure for handling the analysis of electrical heating plots, one can choose t_i as the time the heater is turned on and t_f as the time it is turned off and choose t_d as the midpoint of this heating period.

In the experiments on heats of combustion, we make use of approaches 2 (optionally), 3, and 4. In the experiment on heats of ionic reaction, we make use of 1, 3, and 4. In all cases the small temperature changes can be measured with adequate precision with a relatively inexpensive mercury thermometer. Alternatively the measurements can be made using a sensitive thermistor (see Chapter XVII), which can be monitored repetitively by a computer. Calibration of the thermistor for improved linearity and accuracy is needed in this case, but this procedure itself can serve as a convenient introduction to interfacing a computer to a measurement device.

REFERENCE

1. H. C. Dickinson, *Bull. Natl. Bur. Stand.* **11**, 189 (1914); S. R. Gunn, *J. Chem. Thermodyn.* **3**, 19 (1971).

GENERAL READING

J. L. Oscarson and P. M. Izatt, "Calorimetry," in B. W. Rossiter and R. C. Baetzold (eds.), *Physical Methods of Chemistry,* 2d ed., Vol. VI, chap. 7, Wiley-Interscience, New York (1992).

EXPERIMENT 6

Heats of Combustion

In this experiment a bomb calorimeter is employed in the determination of the heat of combustion of an organic substance such as naphthalene, $C_{10}H_8$. The heat of combustion of naphthalene is $-\Delta\widetilde{H}$ for the reaction at constant temperature and pressure:

$$C_{10}H_8(s) + 12O_2(g) = 10CO_2(g) + 4H_2O(l) \qquad (1)$$

THEORY

A general discussion of calorimetric measurements is presented in the introductory section, Principles of Calorimetry. That material should be reviewed as background for this experiment. It should be noted that no specification of pressure is made for the reaction in Eq. (1) other than that it is constant. (Indeed, in this experiment, the reaction is not actually carried out at constant pressure, though the results are corrected to constant pressure in the calculations.) In fact energy and enthalpy changes attending physical changes are generally small in comparison with those attending chemical changes, and those involved in pressure changes on condensed phases (owing to their small molar volumes and low compressibilities) and even on gases (owing to their resemblance to perfect gases) at constant temperature are very small. Thus energy and enthalpy changes accompanying chemical changes can be considered as being independent of pressure for nearly all practical purposes.

EXPERIMENTAL

The calorimeter, as shown in Fig. 1, consists of a high-pressure cell (or "bomb"), which sits in a calorimeter pail containing 2 L of water. Projecting downward into this pail is a motor-driven stirrer and a precision thermometer. All these parts constitute what has been denoted by the letter S in the Principles of Calorimetry. The bomb contains a pair of electrical terminals to which are attached a short length of fine iron wire in contact with the specimen. Electrical ignition of this wire provides the means of initiating combustion of the sample, which burns in a small metal pan. The bomb is constructed in two parts, a cylindrical body and a headpiece, which are held together by a large threaded cap. A needle valve or check valve is provided at the top for filling the bomb with oxygen to about 380 psi.

The system shown in Fig. 1 is a Parr design with an adiabatic jacket. The purpose of this jacket is to match the temperature of the calorimeter pail (containing bomb and water) so that there will be no heat leak out of or into the pail via conduction or radiation, thus ensuring adiabatic conditions for the reaction.[1] The Parr calorimeter circulates water through a jacket surrounding the pail, and the temperature of this water is automatically controlled to match that of the calorimeter pail. Another type of bomb calorimeter, called the Emerson design, was often used in the past. In the Emerson apparatus, manual control of the adiabatic jacket is required. There are also differences in the bomb-closure design between Emerson and Parr calorimeters. Some Emerson bomb calorimeters are still in use, and operating instructions for their use can be found in the fifth edition of this book.

The use of an adiabatic jacket should in principle eliminate the need to plot and extrapolate temperature values. However, there are inevitable errors in adjusting the jacket temperature during a run, and the adiabatic jacket on the Parr is more for convenience in routine

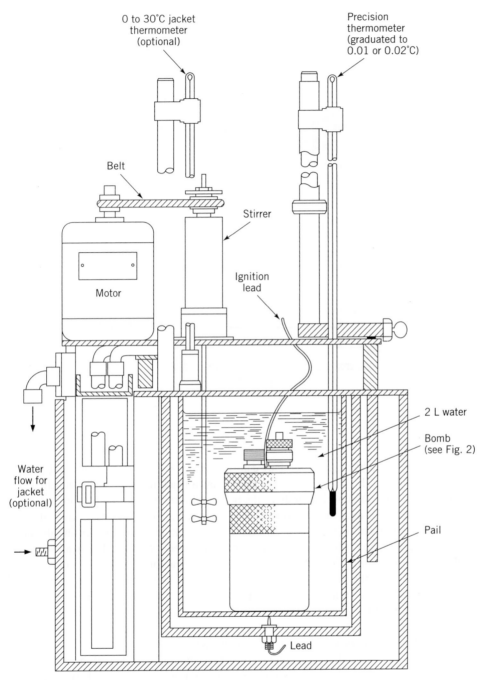

0 to 30°C jacket
thermometer
(optional)

Precision
thermometer
(graduated to
0.01 or 0.02°C)

Belt

Motor

Stirrer

Ignition
lead

Water
flow for
jacket
(optional)

2 L water

Bomb
(see Fig. 2)

Pail

Lead

FIGURE 1
Bomb calorimeter (Parr
design), shown with an
adiabatic jacket, which
may also be used empty
as an insulating air jacket.
The precision mercury
thermometer can be replaced
by a high-resolution
resistance thermometer or a
calibrated thermistor.

work than for high accuracy. It is also possible to operate the calorimeter without controlling the temperature of the adiabatic jacket. In this case, it is best to leave the jacket empty
to provide better thermal insulation. The heat leakage that does occur can be compensated
for very well by the extrapolation technique illustrated in Fig. VI-3. This insulating air-
jacket technique is strongly recommended, and a detailed procedure for use of the adiabatic
jacket will not be given here but is available elsewhere.[2] Indeed, some Parr calorimeters are
provided with insulating air jackets rather than adiabatic water jackets.

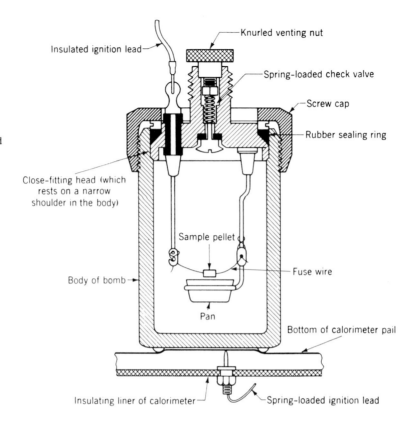

The principal advantage of the Parr calorimeter lies in the design of the self-sealing nickel-alloy bomb. This bomb, shown in Fig. 2, has a head that is precision-machined to fit into the top of the cylindrical body of the bomb. Closure is achieved by compressing a rubber ring that seals against the cylinder wall. The retaining cap is merely screwed down hand-tight; on pressurizing the bomb, a seal develops automatically. Releasing the pressure will break the seal and allow the cap to be unscrewed manually. The head shown in Fig. 2 has a single spring-loaded check valve that closes automatically when the bomb is pressurized. To vent this type of bomb, the knurled nut on the top of the valve is unscrewed one turn (to reduce the spring tension) and pressed down. A double-valve head, which has a check valve for filling and needle valve for venting the bomb, is also available. In the Parr bomb, one lead goes directly to an insulated terminal and the other lead makes contact automatically through the calorimeter pail that holds the bomb.

Following the procedure given below, two runs should be made with benzoic acid to determine the heat capacity of the calorimeter, and two runs should be made with naphthalene or another substance (perhaps an unknown).

Procedure.   The successful operation of this experiment requires close attention to detail, as there are possible sources of trouble that might prevent the experiment from working properly. The basic procedure is very similar for both Parr and Emerson calorimeters. Necessary changes in procedure will be provided by the instructor if an Emerson calorimeter is to be used. Although considerable detail is given below, further details on the operation of the Parr apparatus (including the adiabatic jacket) are available in a manual published by the manufacturer.[2]

The bomb is expensive and must be handled carefully. In particular, be very careful not to scratch or dent the closure surfaces or strip the threads on the screw cap. When the bomb is dismantled, its parts should be mounted on the assembly stand or placed gently on a clean, folded towel.

Condition of apparatus. The bomb must be clean and dry, with no bits of iron wire left on the terminals. Indeed, it is recommended that those electrical terminals be polished lightly with emery paper. Make sure that the jacket is completely empty and that all switches on the control box are off. Check to see that the inside of the calorimeter is dry.

Filling the bomb. Cut the iron wire, free from sharp bends or kinks, to the correct length and weigh it accurately. Press pellets of the substance concerned, of mass 0.5 ± 0.1 g for naphthalene and 0.8 ± 0.1 g for benzoic acid. Shave them to the desired mass with a knife or spatula if necessary. Fuse the wire into the pellet by heating the wire with current provided by a 1.5-V dry cell. Blow gently on the wire to prevent it from overheating. The pellet should be at the center of the wire. Weigh it accurately. The mass of the pellet alone is obtained by difference. Handle it very carefully after weighing.

Install the pellet and wire in the bomb. The pellet should be over the pan, and the wire should touch *only* the terminals. Carefully assemble the bomb and screw down the cap hand-tight; *never* use a wrench. Attach the bomb to the oxygen-filling apparatus. Line up the fitting carefully before tightening, as misalignment may result in damaged threads. *Carefully* open the main supply valve on the oxygen cylinder and *slowly* fill the bomb to 380 psi. **Do not exceed 450 psi.** Release the pressure to flush out most of the atmospheric nitrogen originally present in the bomb. Refill the bomb to 380 psi. If you encounter trouble in filling the bomb, check the condition of the gasket on the filling apparatus. Check the bomb for leaks by immersing it in water. If a leak does exist, vent the bomb, loosen and rotate the head slightly, and then retighten the screw cap. An occasional bubble—one every 5 or 10 s—is inconsequential. If trouble persists, check the condition of the O-ring and metal closure surfaces.

Assembly of calorimeter. Dry the bomb and set it in the dry pail. Then set the pail in the calorimeter, making sure that it is centered and does not touch the inner wall of the jacket. Make the electrical connection to the top of the bomb and be sure that it is tight.

Fill a 2-L volumetric flask with water at 25°C. A convenient way of doing this is to use both hot and cold water; add the two as required, swirling and checking with a thermometer until the flask is almost full, then make up to the mark. Pour the water from the flask carefully into the pail; avoid splashing. Allow the flask to drain for about 30 s.

Put the calorimeter lid in place and clamp the precision thermometer in place as low as it will go without obscuring the scale. The clamp should be as low as possible and *not too tight.* The jaws of the clamp should be rubber-covered to protect the thermometer. The temperature of the pail should be within half a degree of 25°C.

Using a test meter or an improvised tester consisting of a flashlight bulb and a dry cell, check the electrical continuity between the electrical lead and ground. With all switches off, plug the stirrer and ignition cords into the control box (see Fig. 3). Recheck the wiring and then plug the control box into a 110-V ac outlet. Turn on the stirrer and make sure that it runs smoothly.

Instead of the ignition circuit shown in the control box of Fig. 3, a low-voltage transformer can be used. Transformer ignition circuits are available commercially as accessories from calorimeter manufacturers or can be easily constructed. A suitable design is shown in Fig. 4.

Making the run. Begin time–temperature readings, reading the precision thermometer once every 30 s and recording both the time and the temperature. Estimate the temperature to thousandths of a degree if feasible. Tap the thermometer *gently* before

FIGURE 3

Control box containing electrical circuit for bomb calorimeter.

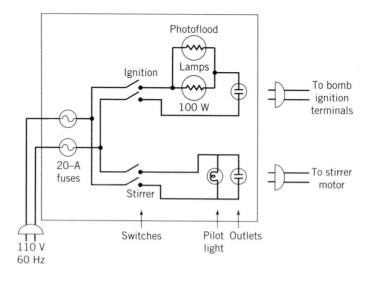

FIGURE 4

Transformer ignition circuit for bomb calorimeter. The switch is a momentary-contact push switch that is only on when it is held down. Release the switch if ignition does not occur after about 5 s (the pilot light will indicate when the fuse wire burns through). The choice of resistor R will depend on the type of calorimeter and the length of fuse wire to be used (a current of 3 to 4 A is desired).

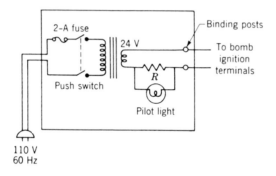

each reading. *Do not interrupt the time–temperature readings until the run is over.* The calorimeter pail temperature should change at a very slow linear rate (of the order of 0.001 K min^{-1}). After this steady rate has persisted for at least 5 min, the bomb may be ignited.

To ignite, turn the ignition switch on and then immediately off. Record the exact time. A dull brief flash may be observable in the lightbulbs, indicating the passage of current for the instant required to burn the wire through. In nearly all such cases, the burning wire will ignite the pellet, and after a period of 10 or 15 s the temperature will begin to rise. After a few minutes the pail temperature should again show a slow, steady rate of change. Do not discontinue readings; continue them until the time since ignition has been at least four times the period required for attainment of this steady rate. The purpose of this is to provide a valid basis for extrapolation. When readings are completed, turn all switches off.

Disassemble the apparatus, release the bomb pressure, and open the bomb. Remove and weigh any unburned iron wire; ignore "globules" unless attempts to crush them reveal that they are fused metal rather than oxide. Subtract the weight of unburned iron wire from the initial iron-wire weight to obtain the net weight of iron burned. If the inside of the bomb is found to be coated with soot, the amount of oxygen present at the time of ignition was presumably insufficient to give complete combustion and the run should be discarded. Wipe dry all bomb parts.

CALCULATIONS

For each run, plot temperature versus time using an expanded, interrupted temperature scale as shown in Fig. VI-1b and determine the initial and final drift rates $(dT/dt)_i$ and $(dT/dt)_f$. Then make an overall T versus t plot like Fig. VI-3b, choose t_f, and determine t_i and t_d. Finally, the "adiabatic" temperature change $\Delta T \equiv T_1 - T_0$ can be calculated from Eq. (VI-22). This determination of ΔT could be made directly from the overall T versus t plot as shown in Fig. VI-3b, but the procedure described above provides greater precision.

The heat capacity $C(\text{S})$ is found by determining the adiabatic temperature rise $(T_2' - T_1')$ obtained in the combustion of a known mass of benzoic acid (and of course a known mass of iron wire) and by making use of Eq. (VI-15b). For calculating the energy change produced, the specific (i.e., per gram) energies of combustion of benzoic acid (BA) and iron wire (Fe) given below can be used:[3]

$$\Delta \bar{E}_{\text{BA}} = -26.41 \text{ kJ g}^{-1}$$
$$\Delta \bar{E}_{\text{Fe}} = -6.68 \text{ kJ g}^{-1} \tag{2}$$

In this experiment ΔE for the combustion of a weighed specimen of naphthalene or other substance is determined from the rise in temperature $(T_1 - T_0)$ and the heat capacity $C(\text{S})$ by use of Eq. (VI-14b). The value of ΔE obtained includes a contribution for the combustion of the iron wire; this must be subtracted to yield the contribution from the naphthalene alone. The molar energy change $\Delta \widetilde{E}$ is then obtained by dividing by the number of moles of the reactant that is present in the least equivalent amount (in this case, naphthalene). The molar enthalpy change $\Delta \widetilde{H}$ can then be obtained by use of Eq. (VI-8).

Report the individual and average values of the heat capacity $C(\text{S})$ and the individual and average values of the molar enthalpy change $\Delta \widetilde{H}$ for the combustion of naphthalene, calculated using the average $C(\text{S})$.

DISCUSSION

How does the order of magnitude of the error introduced into the experimental result by the assumption of the perfect-gas law in Eq. (VI-8) compare with the uncertainties inherent in the measurements in this experiment? What is the magnitude of the uncertainty introduced by lack of knowledge of the specific heat of the sample? Does your ΔH value pertain to the initial or the final temperature?

SAFETY ISSUES

Gas cylinders must be chained securely to the wall or laboratory bench (see pp. 644–646 and Appendix C). There is some hazard of electrical shock or short circuit from exposed terminals. Check all electrical connections carefully before plugging the control box into the 110-V ac line. Do not pressurize the bomb above 450 psi.

APPARATUS

Bomb calorimeter complete with the bomb, pail, jacket, insulating table, cover, and stirring motor with power cord; electrical lead which connects to bomb; 110-V power source unit, equipped with fuse and bulbs or a commercial ignition unit; precision calorimeter thermometer covering range from 19 to 35°C with a magnifying thermometer reader (can be replaced by either a resistance thermometer or a calibrated thermistor with $\pm0.003°C$ resolution); 0 to 30°C thermometer; timer; 2-L volumetric flask; 1.5-V dry cell; 500-mL beaker.

Bomb-assembly stand; cylinder of oxygen with appropriate fittings; large pail for leak-testing the bomb; benzoic acid (5 g); naphthalene (5 g) or other solid to be studied; pellet press; spatula; 0.004-in.-diameter iron wire (50 cm); device for checking electrical continuity (optional).

REFERENCES

1. F. Daniels, *J. Amer. Chem. Soc.* **38,** 1473 (1916); T. W. Richards, *J. Amer. Chem. Soc.* **31,** 1275 (1909).

2. *Oxygen Bomb Calorimetry and Combustion Methods,* Tech. manual 130, Parr Instrument Co., Moline, IL.

3. R. D. Lide (ed.), *CRC Handbook of Chemistry and Physics,* 89th ed., CRC Press, Boca Raton, FL (2008/2009). Also available online for a fee at www.hbcpnetbase.com.

GENERAL READING

J. L. Oscarson and P. M. Izatt, "Calorimetry," in B. W. Rossiter and R. C. Baetzold (eds.), *Physical Methods of Chemistry,* 2d ed., Vol. VI, chap. 7, Wiley-Interscience, New York (1992).

EXPERIMENT 7
Strain Energy of the Cyclopropane Ring

The strain energy of the cyclopropane ring is determined from thermochemical measurements on two closely related compounds: the *n*-butyl ester of cyclopropanecarboxylic acid and the methyl ester of cyclohexanecarboxylic acid. Both compounds can be synthesized from the appropriate acid chlorides, and their heats of combustion measured with a bomb calorimeter.

THEORY

The molar enthalpy change $\Delta\widetilde{H}_d$ for dissociation of an organic compound into separated atoms in their gaseous state

$$C_mH_nO_pN_q \cdots = mC(g) + nH(g) + pO(g) + qN(g) + \ldots \tag{1}$$

is determined principally by the number and kinds of chemical bonds in the molecule. In many cases, the heat of dissociation $\Delta\widetilde{H}_d$ can be closely approximated by the sum of so-called bond energies (more properly bond enthalpies) B_i for all the bonds in the molecule;†

$$\Delta\widetilde{H}_d \cong \sum_i B_i \tag{2}$$

---

†Strictly speaking, a heat of dissociation calculated from bond energies applies to a compound in the vapor state, but the heat of vaporization from the liquid or solid state (typically 20 to 50 kJ mol⁻¹) is ordinarily smaller in magnitude than the errors inherent in the bond-energy treatment.

For all conventional chemical bonds, the B_i values are positive; $\Delta \widetilde{H}_d$ must have a positive sign for a stable molecule.

Certain molecules are made less stable by the presence of a *strain energy* **S** corresponding to the bending or stretching of bonds from their normal state as a result of geometric requirements, or are made more stable by the presence of a *resonance energy* **R** corresponding to aromatic character, conjugation, hyperconjugation, etc.:

$$\Delta \widetilde{H}_d = \sum_i B_i - \mathbf{S} + \mathbf{R} \tag{3}$$

The quantities **S** and **R** are defined so as to be positive in sign. We shall be concerned here with the strain energy of the cyclopropane ring, in which the C—C—C bond angles are constrained by geometry to be $60°$ rather than their normal values, which are in the neighborhood of the tetrahedral angle $109.5°$.

Consider the compounds *n*-butylcyclopropanecarboxylate (I) and methylcyclohexanecarboxylate (II), which have the same molecular composition $C_8H_{14}O_2$:

The quantity ΣB_i should have the same value for both compounds, since each has 7 C—C, 1 C=O, 2 C—O, and 14 C—H bonds. However, molecule I possesses appreciable strain energy while molecule II does not. (The cyclohexyl ring is staggered, so the angles are close to $109.5°$.) Neither molecule would be expected to have appreciable resonance energy; in any case resonance effects would be about the same for both molecules and would cancel out. Hence

$$\mathbf{S}(\text{I}) = \Delta \widetilde{H}_d\,(\text{II}) - \Delta \widetilde{H}_d\,(\text{I}) \tag{4}$$

The quantity $-\Delta \widetilde{H}_d$ is closely related to (but should be carefully distinguished from) the heat of formation of a compound, which is the molar enthalpy change resulting from the formation of the compound from the elements *in their standard states* at the given temperature and pressure (usually 298 K, 1 bar). The standard state of carbon at 298 K, 1 bar is graphite $C(s)$; those of hydrogen, oxygen, and nitrogen are the gases $H_2(g)$, $O_2(g)$, and $N_2(g)$. $\Delta \widetilde{H}_f$ for the change in state

$$m\mathrm{C}(s) + \tfrac{1}{2}n\mathrm{H}_2(g) + \tfrac{1}{2}p\mathrm{O}_2(g) + \tfrac{1}{2}q\mathrm{N}_2(g) \rightarrow \mathrm{C}_m\mathrm{H}_n\mathrm{O}_p\mathrm{N}_q \ (298 \text{ K, 1 bar}) \tag{5}$$

is related to $\Delta \widetilde{H}_d$ by

$$-\Delta \widetilde{H}_f = \Delta \widetilde{H}_d - m\,\Delta \widetilde{H}_d^0[\mathrm{C}(s)] - \tfrac{1}{2}n\,\Delta \widetilde{H}_d^0[\mathrm{H}_2(g)] - \tfrac{1}{2}p\,\Delta \widetilde{H}_d^0[\mathrm{O}_2(g)] - \tfrac{1}{2}q\,\Delta \widetilde{H}_d^0[\mathrm{N}_2(g)]$$

$$= \Delta \widetilde{H}_d - 716.68m - 217.97n - 249.17p - 472.70q \tag{6}$$

where the numerical values are given in kJ mol^{-1}. The quantities $\Delta \widetilde{H}_d^0$ are the heats of dissociation of the elements *in their standard states* into separated gaseous atoms; their values have been obtained from a variety of calorimetric and spectroscopic measurements.[1] The sign of $\Delta \widetilde{H}_f$ is usually but not always negative.

Since the change in state (5) will seldom take place as a single reaction in a laboratory calorimeter, the heat of formation is not measured directly. However, by Hess's law, the desired enthalpy change can be obtained by a suitable algebraic combination of the enthalpy changes for other changes in state corresponding to actual reactions. As an example of the kind of reaction that is useful for this purpose, we consider combustion with high-pressure oxygen gas:

$$C_mH_nO_p(s \text{ or } l) + (m + \tfrac{1}{4}n - \tfrac{1}{2}p)O_2(g) \rightarrow mCO_2(g) + \tfrac{1}{2}nH_2O(l)$$

$$(298 \text{ K}, 1 \text{ bar}) \quad (7)$$

We shall restrict ourselves to the combustion of solids and liquids; combustion of gases requires a very different experimental technique. Moreover we shall not consider compounds containing elements other than C, H, and O, since the combustion products of compounds containing other elements may be somewhat variable; nitrogen for example may appear in the product as $N_2(g)$, oxides of nitrogen, or nitric acid. Note that in Eq. (7) the products are gaseous CO_2 and liquid water. The heat of formation for the organic compound $C_mH_nO_p$ is related to the heat of combustion $\Delta\tilde{H}_c$, which is the enthalpy change for the change in state (7), by

$$-\Delta\tilde{H}_f = \Delta\tilde{H}_c + 395.51m + 142.92n \quad (8)$$

where, again, all quantities are in kJ mol^{-1} at 298 K, 1 bar. The sign of $\Delta\tilde{H}_c$ is, for all practical purposes, invariably negative. The second and third terms on the right arise from the heats of combustion of graphite and gaseous hydrogen. The heat of dissociation of the compound is then given by

$$\Delta\tilde{H}_d = \Delta\tilde{H}_c + 1112.19m + 360.89n + 249.17p \quad (9)$$

From Eqs. (4) and (9), we see that

$$\mathbf{S}(I) = \Delta\tilde{H}_c(II) - \Delta\tilde{H}_c(I) \quad (10)$$

all other terms having canceled out.

Equation (9) may be used to calculate $\Delta\tilde{H}_d$ from the experimentally determined heat of combustion of compound II for comparison with the sum of tabulated bond energies [see Eq. (2)]. A short list of bond energies[2] is given in Table 1. These bond energies are mostly averages of values derived from heats of formation of several compounds. The values for double and triple bonds are subject to greater variability than those for single bonds and should be used with some caution. Values of $\Delta\tilde{H}_d$ calculated with bond energies are typically in error by 1 or 2 percent in favorable cases and by as much as 5 percent (or even more) in bad cases. It should be no surprise if the value obtained from Eq. (2) for compound II (which is negligibly affected by strain) differs from that calculated with Eq. (9) by an amount that exceeds the small difference between the enthalpies of I and II

TABLE 1 Selected bond enthalpies at 298 K (in kJ mol^{-1})

Bond	B	Bond	B
C—C	344	C=C	615
C—H	415	C=O	725
C—O	350	C=N	615
C—N	292	O=O	498
C—Cl	328	N=N	418
C—Br	276		
C—I	240	C≡C	812
O—H	463	C≡N	890
N—H	391	N≡N	946
H—H	436		

due to strain. However, it may be expected that, since molecules I and II are very similar in all respects other than strain, the errors inherent in the application of Eq. (3) and Table 1 will be essentially the same for both molecules. Accordingly we may expect that Eq. (10) will give a fairly good measure of the strain energy **S** in molecule I.

It must be kept in mind that the strain energy **S** represents a small difference between two large quantities, $\Delta \widetilde{H}_c$ (I) and $\Delta \widetilde{H}_c$ (II), obtained from measurements on two different substances; the difference is only of the order of a few percent. Thus it is clear that all physical measurements involved in the calorimetry must be carried out with the highest possible precision (one or two parts per thousand) and also that the two different substances must be highly pure. The principal impurity to be feared in compound I is *n*-butanol, which boils only about 45°C lower than compound I at a pressure of 90 Torr. The replacement of a small percentage of compound I by an impurity such as butanol will produce a quite small percentage error in $\Delta \widetilde{H}_c$; the error will depend on the difference between the heats of combustion per gram of the two substances. Water, which has no heat of combustion, will cause a percentage error in $\Delta \widetilde{H}_c$ equal to its percentage as an impurity.

EXPERIMENTAL

Since the two esters I and II are not available commercially, they will be synthesized by a procedure that is basically that of Jeffery and Vogel.[3] The starting materials for the syntheses are the free acids, cyclopropanecarboxylic acid and cyclohexanecarboxylic acid. Each is first converted to the acid chloride with thionyl chloride, and the acid chloride is then allowed to react with the appropriate alcohol to form the ester. Alternatively the acid chlorides can be obtained commercially from Aldrich Chemical Co., Inc., Milwaukee, Wisconsin, and the first part of the syntheses can be dropped. The purity of the two esters is checked by refractometry, infrared spectroscopy, and vapor-phase chromatography; and the heats of combustion are determined with a bomb calorimeter. It is suggested that each student in the team synthesize and check the purity of one of the esters and that both work together on the calorimetric measurements.

Synthesis.   In following the directions in this section, one should adhere closely to the amounts of material used in each step. The yields given for the two esters are those obtained routinely and correspond to a competent execution of each preparation. Substantially lower yields may require repetition of the preparation in order to obtain sufficient material for the calorimetry work. Both HCl and SO_2 are evolved in parts of this experiment, and all operations in which either of these gases is evolved must be carried out in a hood.†

Preparation of acid chlorides.   Cyclopropanecarboxylic acid chloride is prepared according to the reaction

$$\text{S}\!\!\!\backslash\!\!-\!\!\overset{\displaystyle O}{\overset{\|}{C}}\!\!-\!\!OH + SOCl_2 \longrightarrow \text{S}\!\!\!\backslash\!\!-\!\!\overset{\displaystyle O}{\overset{\|}{C}}\!\!-\!\!Cl + HCl(g) + SO_2(g) \tag{11}$$

Assemble in a hood an apparatus consisting of a single-neck 50-mL round-bottom flask, a Claissen adapter, a vertical water-cooled condenser for reflux (joined to the main stem of the adapter), and a dropping funnel (joined at the side arm of the adapter). The flask should

†Reactions and distillations carried out in a hood are subject to strong drafts of air and should be protected. In particular distillation apparatus should be carefully wrapped in glass wool or aluminum foil or both, to prevent excessive and uneven cooling.

contain a small Teflon-covered stirring bar and should sit in a heating mantle on top of a magnetic stirring unit.

Allow 13.1 g (0.11 mol) of freshly distilled or reagent-grade thionyl chloride to run into the flask from the dropping funnel; close the stopcock as soon as the thionyl chloride has run into the flask. Start the magnetic stirrer and the flow of water through the condenser. Heat the flask until the thionyl chloride has begun to reflux. Add 6.25 g (0.073 mol) of cyclopropanecarboxylic acid dropwise from the dropping funnel to the flask. As soon as an evolution of a gas more vigorous than that due to the refluxing thionyl chloride becomes apparent, turn off the external source of heat. The rate of addition of acid should be adjusted so that constant evolution of gas is obtained—the addition usually requires about 15 min. After the addition of acid has been completed, close the stopcock and allow the system to stir at room temperature until the evolution of gas has slowed considerably. Reflux the mixture for 20 min. Then allow the system to cool and rearrange the apparatus to a conventional apparatus for simple distillation. In carrying out this rearrangement, close each entry to the reaction flask as each piece of equipment (dropping funnel, condenser) is removed from the flask, since the acid chloride decomposes slowly in contact with moist air. The small amount of residual thionyl chloride (bp 79°C/760 Torr) is then distilled from the acid chloride. At this point, one may continue and isolate the acid chloride by distillation (bp 119°C/760 Torr), or one may use the undistilled residue in the next step of the synthesis. The final yields of ester are comparable from both routes.

The procedure for preparing cyclohexanecarboxylic acid chloride is exactly the same as that described above for the preparation of cyclopropanecarboxylic acid chloride. In this case 11.9 g (0.10 mol) of freshly distilled or reagent-grade thionyl chloride and 7.66 g (0.06 mol) of cyclohexanecarboxylic acid are used. Cyclohexanecarboxylic acid is a solid that melts at 31°C, but it is easier to handle as a liquid. As an effective and safe (under normal precautions) method of melting the acid for weighing and addition to the thionyl chloride, apply steam to the exterior of the bottle or the dropping funnel. As with the previous synthesis, one may distill this acid chloride (bp 125°C/760 Torr), or one may use it without distillation in the next step.

Preparation of *n*-butyl cyclopropanecarboxylate.

$$\underset{\substack{\\ }}{\text{S}}\!\!>\!\!\overset{\displaystyle O}{\overset{\displaystyle \|}{\text{C}}}\!-\!\text{Cl} + n\text{-C}_4\text{H}_9\text{OH} \longrightarrow \underset{\substack{\\ }}{\text{S}}\!\!>\!\!\overset{\displaystyle O}{\overset{\displaystyle \|}{\text{C}}}\!-\!\text{OC}_4\text{H}_9 + \text{HCl}(g) \qquad (12)$$

Assemble in a hood the same apparatus as before, except that the condenser is replaced with a $CaCl_2$ drying tube and the heating mantle is omitted. Start the magnetic stirrer after 5.40 g (0.073 mol) of anhydrous *n*-butanol has been placed in the 50-mL flask. Pour the cyclopropanecarboxylic acid chloride into the dropping funnel, and stopper the dropping funnel with another $CaCl_2$ drying tube. [If commercially prepared acid chloride is used, add 7.63 g (0.073 mol).] Since the reaction is exothermic, before allowing the acid chloride to react with the alcohol, *prepare an ice bath* for use in case heat is evolved too rapidly. Add the acid chloride dropwise at a rate that maintains a vigorous evolution of HCl gas; the addition will usually take about 20 min. After the addition period, the ester reaction mixture is stirred for 40 min at room temperature.

After the solution is cool, add about 15 mL of ethyl ether, pour the mixture *carefully* into about 15 mL of water (**caution:** the reaction is exothermic), and mix well. Separate the aqueous layer, and extract it twice with 15-mL portions of ether. Wash the combined layer twice with 15-mL portions of saturated aqueous $NaHCO_3$ solution. Extract the washings twice with 15-mL portions of ether; in each case, use the ether layer from the extraction of the first portion of $NaHCO_3$ solution to extract the second. Dry the combined organic layers for 20 to 30 min with $MgSO_4$, swirling to hasten equilibrium. Filter into a 100-mL round-bottom flask, attach a Vigreux column, and drive off the ether with a steam bath (in a hood).

The ester should preferably be distilled at reduced pressure (bp 110–112°C/90 Torr) to minimize thermal decomposition. Simple distillation is likely to yield a product with a higher-than-acceptable contamination by butanol, so fractionation is recommended. The Vigreux column should be wrapped with glass wool and aluminum foil for thermal insulation and fitted with an Ace vacuum distilling head having Teflon greaseless stopcocks. This head has a "cold-finger" condenser, rotatable so that the reflux ratio can be adjusted, and has stopcocks to permit "fraction cutting" (change of receiving vials without releasing the vacuum). The stopcock through which the distillate flows is greaseless to avoid contamination of the product.

Before distilling the reduced liquid, transfer it (together with a 5-mL ether washing from the previous flask) to a 25-mL round-bottom flask. The distillation should be carried out with a reflux ratio of at least 4:1. Three fractions should be collected: a low-boiling one containing most of the ether and residual butanol, one boiling above the boiling point of butanol but below the stable boiling point of the ester, and the last containing the ester boiling at constant temperature. The usual yield in the last fraction is 6 to 6.5 g.

Take refractive indexes of the last two fractions (Jeffrey and Vogel[3] report $n_D = 1.42446$; at MIT, we have found $n_D^{25} = 1.4259$), and analyze both fractions by vapor-phase chromatography (VPC).† No impurities should show on the VPC trace when the peak height is about 90 percent full deflection. Decreased attenuation of the VPC may result in trace amounts of impurities being shown. To calibrate for the effect of butanol on the VPC spectrum, run also a sample consisting of 200 μL of the final ester fraction and 2 μL of *n*-butanol.

The infrared (IR) absorption spectrum of a 5 percent solution of the final fraction in carbon tetrachloride should be taken. Verification that the cyclopropyl group did not isomerize to propenyl-2 to any significant extent under the synthesis conditions should be found in the absence of any significant vinyl C—H bending band at 990 cm^{-1}. (Other such bands at 909 and 1320 cm^{-1} would be more or less obscured by other structural features of the spectrum. The vinyl C—H stretching bands at 3090 and 3030 cm^{-1} would be obscured by the cyclopropyl C—H stretching band at about 3060 cm^{-1}.) If there is any suspicion of such isomerization, a bromine test may be performed. Dilute about 0.5 g of the ester with 1 or 2 mL of CCl_4 and add one drop of a 5 percent solution of bromine in CCl_4; immediate fading of the bromine color would confirm that unsaturation is present. The presence of any significant quantity of butanol or free carboxylic acid would be indicated by absorption bands at 3625 cm^{-1} or 2700 to 2500 cm^{-1}, respectively.

Preparation of methyl cyclohexanecarboxylate.

$$\text{(13)}$$

The procedure for this synthesis is essentially the same as that for the *n*-butyl cyclopropanecarboxylate described above. In the present case 3.8 g (0.12 mol) of methanol is used. [If commercially prepared acid chloride is used, add 8.80 g (0.06 mol).] Since

---

†A silicone column (e.g., 10 percent 550 Dow–Corning silicone fluid on Chrom P) at 100°C and a flow rate of 60 mL s^{-1} is sufficient to show any significant impurities, although tailing of the ester peaks generally occurs. A Carbowax 20M column gives less tailing under the same conditions.

methanol boils substantially lower than butanol, simple distillation may be used instead of fractionation, but it is advisable to do it at reduced pressure. A conventional setup with a "pig"-type adapter for fraction cutting may be used. As with the other ester, the liquid to be distilled should be transferred, with an ether washing, to a 25-mL round-bottom flask. Distill the ester at about 35 Torr (bp 90–92°C/35 Torr) and collect three fractions: one low-boiling, one from 50 to 90°C at 35 Torr, and the last fraction after the boiling point of the ester is stable. The usual yield in the final fraction is 6.0 to 8.5 g of the ester.

Record the refractive indexes of the last two fractions ($n_D^{25} = 1.4410$), and analyze them by vapor-phase chromatography. Only minor impurity peaks should show on the VPC trace at low attenuation, with the ester peak off-scale. Obtain an infrared spectrum of a 5 percent solution of the final fraction in carbon tetrachloride. Isomerization of the ring to a terminally unsaturated straight chain is much less likely than with the cyclopropane ring. Check for absorption in the medium-strength 3085 cm^{-1} and 3030 cm^{-1} vinyl C—H stretching bands; vinyl C—H bending bands tend to be obscured by other features.

Calorimetry. The procedures for the operation of the bomb calorimeter are, with minor exceptions noted below, those described in detail in Exp. 6. That experiment and the general discussion presented in the section Principles of Calorimetry should be studied carefully.

The heat capacity of the calorimeter is to be determined with duplicate runs on solid benzoic acid, C_6H_5COOH. Two runs are then made on each liquid ester. Any run with unsatisfactory features in the time–temperature plot or indications of incomplete combustion (soot, etc.) must be rejected. If any pair of runs fails to give reasonably concordant results (agreement within 0.5 percent or better), additional runs on the same substance should be made.

For the present experiment, the adiabatic jacket is to be used empty as an air jacket. The procedure is experimentally simpler and capable of high accuracy, provided that the ambient temperature in the laboratory is reasonably steady.

Benzoic acid. Weigh a pellet of benzoic acid, 0.8 ± 0.1 g, accurately and place it in the sample pan in the bomb. Weigh also an 8-cm piece of iron ignition wire. Instead of fusing the wire into the pellet as described in Exp. 6, you may connect the wire to the terminals in such a way that it is held by light "spring action" against the surface of the pellet. In shaping the wire for this purpose, avoid kinks and sharp bends and make certain that the wire does not touch the sample pan itself.

Seal the bomb, flush and fill it with oxygen, test for leaks, and assemble the calorimeter as described in Exp. 6. Be sure to adjust the temperature of the water in the 2-L volumetric flask to 25 ± 1°C before introducing it into the calorimeter pail. When all is ready, turn on the stirring motor and begin time–temperature readings. Read the precision thermometer every 30 s, and record both time and temperature. Tap the thermometer gently before each reading, and estimate the temperature to thousandths of a degree. A temperature–time curve on an expanded scale (see Fig. VI-1*b*) should be plotted concurrently. The readings should fall on a straight line, with a small positive slope due to energy input from the stirrer. When the readings have shown straight-line behavior continuously for at least 5 min, ignite the bomb. Record the time of ignition and continue temperature readings until the total elapsed time following ignition is at least four times the period required to achieve a slow, steady rate of change in the upper temperature. Then turn off the control-box switches, release the bomb pressure, open the bomb, and remove and weigh any unburned iron wire. The presence of any soot indicates incomplete combustion, probably owing to insufficient oxygen; a leak should be suspected. In such a case, the result of the run must be rejected. A run should also be rejected if the time–temperature plot contains any unusual features that make extrapolation uncertain.

Esters. Introduce about 0.7 to 0.9 g of one of the esters into the sample pan from a 2-mL disposable hypodermic syringe. Fill the syringe nearly half full, weigh it on an analytical balance, discharge into the sample pan, and reweigh. Before each weighing, withdraw the plunger so as to pull a small quantity of air into the needle. This will prevent loss of liquid through dribbling from the needle tip. Naturally, a drop hanging from the tip should not be wiped off at any time between the two weighings. Although the 8-cm iron wire described above may be used, a double wire obtained by attaching the middle of a 16-cm length to one terminal and the two ends to the other terminal may provide better insurance against incomplete combustion. The two wires should be carefully shaped to dip into the liquid without touching each other or the metal of the pan. Care should be taken not to wet the walls of the pan above the meniscus any more than necessary, as liquid clinging to the walls may fail to burn. Handle the bomb very carefully; avoid bumping or tipping it. Perform the run as described above for benzoic acid.

CALCULATIONS

Use the plotting and calculation procedures described in Exp. 6 in order to determine the adiabatic temperature change associated with each combustion run. The same extrapolation procedure should be used for both esters and used in as consistent a manner as possible so that any systematic errors inherent in the procedure will cancel out in the calculation of the strain energy.

Determine $C(S)$, the heat capacity of the calorimeter system, from

$$C(S) = \frac{26.41m_{BA} + 6.68m_{Fe}}{\Delta T'} \quad kJ\ K^{-1} \tag{14}$$

where m_{BA} and m_{Fe} are the masses in grams of benzoic acid and iron wire *burned* (be sure to correct for any unburned wire), and $\Delta T'$ is the extrapolated temperature difference for the benzoic acid run. The value of $C(S)$ should be approximately 10.5 kJ K^{-1}. The molar energy of combustion of each ester is calculated with the equation

$$\Delta\widetilde{E}_c = -\frac{M}{m}[C(S)\Delta T - 6.68m_{Fe}] \quad kJ\ mol^{-1} \tag{15}$$

where $\Delta T = T_1 - T_0$ is the extrapolated temperature difference for the adiabatic ester combustion, m is the mass in grams of the ester, and M is the ester molar mass. It is the change in internal energy that is obtained directly in this experiment, since combustion in a bomb is a constant-volume process. The molar enthalpy change $\Delta\widetilde{H}_c$ for combustion at constant pressure can be calculated from $\Delta\widetilde{E}_c$ on the assumption that the gases O_2 and CO_2 obey the ideal-gas law reasonably well and that the enthalpies of all substances concerned are independent of pressure. Both of these assumptions are valid to an accuracy acceptable for our purposes. Thus, from Eq. (VI-8), we have

$$\Delta\widetilde{H}_c = \Delta\widetilde{E}_c + RT\ \Delta\tilde{n}_{gas} = \Delta\widetilde{E}_c - (\tfrac{1}{4}n - \tfrac{1}{2}p)RT \tag{16}$$

where $\Delta\tilde{n}_{gas}$ is the change in the number of moles of gas per mole of solid or liquid $C_mH_nO_p$ compound burned. Another necessary assumption, also valid to an accuracy sufficient for our purposes, is that the effect on $\Delta\widetilde{H}_c$ of the presence of a small fraction of the product water in the vapor phase rather than the liquid phase is negligible. This may be checked by a simple calculation based on the volume of the bomb interior (about 200 mL), the vapor pressure of water at room temperature (about 25 Torr), and the molar heat of vaporization of water (about 42 kJ mol^{-1}).

The strain energy **S** may be calculated from the heats of combustion for the two esters by use of Eq. (10).

Report the yields, boiling ranges, and refractive indexes of the two esters. If available, the VPC and IR curves should be attached to the report. The identification number of the calorimeter and bomb, the individual and average values of the calorimeter heat capacity $C(S)$ and the heats of combustion $\Delta \widetilde{H}_c$ for the two esters, and the apparent strain energy for the cyclopropane ring should be reported. An estimate of the uncertainties in $C(S)$, the two $\Delta \widetilde{H}_c$ values, and **S** should be given.

DISCUSSION

Calculate $\Delta \widetilde{H}_d$ for compound II with Eq. (9), and compare this value with the sum of bond energies obtained with Eq. (2) and Table 1. Comment on the agreement between these values. Also compare your **S** value with a literature estimate of the strain energy.[4]

SAFETY ISSUES

Gas cylinders must be chained securely to the wall or laboratory bench (see pp. 644–646 and Appendix C). There is some hazard of electrical shock or short circuit from exposed terminals. Check all electrical connections carefully before plugging into the 110-V ac line. Note also that CCl_4 is used in a minor way as a solvent in the preparatory work. This is a potentially hazardous chemical (see p. 197, CCl_4 hazards) and should be handled with care.

APPARATUS

Synthetic work.   One 25-, one 50-, and one 100-mL single-neck round-bottom flask; Claissen adapter; water-cooled condenser and rubber hoses; dropping funnel; Vigreux column; Ace vacuum distillation head; 0 to 150°C thermometer; magnetic stirring unit; heating mantle; two $CaCl_2$ drying tubes; ice bath; glass wool; aluminum foil.

Refractor; vapor-phase chromatograph; infrared spectrometer; fume hood; steam bath; pump for reduced pressure distillation; thionyl chloride (30 g); cyclopropanecarboxylic acid (7 g) and cyclohexanecarboxylic acid (8 g) *or* cyclopropanecarboxylic acid chloride (8 g) and cyclohexanecarboxylic acid chloride (9 g); *n*-butanol (6 g); methanol (4 g); ethyl ether (175 mL); saturated aqueous $NaHCO_3$ solution (40 mL); $MgSO_4$; CCl_4.

Calorimetry.   See the detailed list at the end of Exp. 6. An additional item required here is a 2-mL disposable hypodermic syringe.

REFERENCES

1. L. Pauling, *Nature of the Chemical Bond*, 3d ed., pp. 64–107, Cornell Univ. Press, Ithaca, NY (1960).

2. L. Pauling, *General Chemistry*, reprint ed., Dover, Mineola, NY (1988); I. N. Levine, *Physical Chemistry*, 6th ed., table 19.1, McGraw-Hill, New York (2009).

3. G. H. Jeffery and A. I. Vogel, *J. Chem. Soc.* **1948**, 1804 (1948).

4. R. A. Nelson and R. S. Jessup, *J. Res. Natl. Bur. Stand.* **48**, 206 (1952).

GENERAL READING

J. L. Oscarson and P. A. Izatt, "Calorimetry," in B. W. Rossiter and R. C. Baetzold (eds.), *Physical Methods of Chemistry,* 2d ed., Vol. VI, chap. 7, Wiley-Interscience, New York (1992).

L. Pauling, *Nature of the Chemical Bond,* 3d ed. Cornell Univ. Press, Ithaca, NY (1960).

EXPERIMENT 8
Heats of Ionic Reaction

It is desired in this experiment to determine the heat of ionization of water:

$$H_2O(l) = H^+(aq) + OH^-(aq) \qquad \Delta \tilde{H}_1 \tag{1}$$

and the heat of the second ionization of malonic acid (HOOC—CH$_2$—COOH, hereinafter designated H$_2$R):

$$HR^-(aq) = H^+(aq) + R^{2-}(aq) \qquad \Delta \tilde{H}_2 \tag{2}$$

These will be obtained by studying experimentally with a solution calorimeter the heat of reaction of an aqueous HCl solution with an aqueous NaOH solution, for which the ionic reaction is

$$H^+(aq) + OH^-(aq) = H_2O(l) \qquad \Delta \tilde{H}_3 \tag{3}$$

and that of the reaction of an aqueous NaHR solution with an aqueous NaOH solution,

$$HR^-(aq) + OH^-(aq) = R^{2-}(aq) + H_2O(l) \qquad \Delta \tilde{H}_4 \tag{4}$$

The last two equations can be written in the general form

$$A + B = \text{products} \tag{5}$$

where A represents the acid ion and B the basic ion.

THEORY

A general discussion of calorimetric measurements is presented in the section, Principles of Calorimetry, which should be reviewed in connection with this experiment. We shall not consider here the concentration dependence of these enthalpy changes. Such concentration dependence is generally a small effect, since the heats of dilution involved are usually much smaller than the heats of chemical reaction (indeed they are zero for perfect solutions). Since we are dealing here with solutions of moderate concentration, particularly in the case of the NaOH solution, it may be useful to make parallel determinations of heats of dilution of the solutions concerned by a procedure similar to that described here if time permits.

EXPERIMENTAL

In this experiment 500 mL of solution A, with a precisely known concentration in the neighborhood of 0.25 *M,* is reacted with 50 mL of solution B at a concentration sufficient to provide a slight excess over the amount required to react with solution A. The reaction is carried out in the solution calorimeter shown in Fig. 1. The calorimeter is a vacuum Dewar

FIGURE 1

Solution calorimeter.
The precision mercury
thermometer can be replaced
by a high-resolution
resistance thermometer or a
calibrated thermistor.

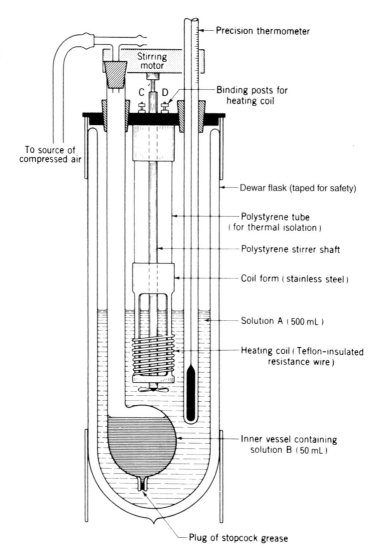

flask containing a motor-driven stirrer, a heating coil of precisely known electrical resistance, and a precision thermometer. Various methods of mixing the two solutions might be employed; here we use an inner vessel having an outlet hole plugged by stopcock grease, which can be blown out by applying air pressure at the top.

For determining the heat capacity, a dc electric current of about 1.5 A is passed through a heating coil of ~6Ω resistance during a known time interval. The magnitude of the current is measured as a function of time by determining the potential difference developed across a standard resistance in series with the coil. The electrical circuit is shown in Fig. 2.

It is recommended that two complete runs be made with HCl and NaOH and two with NaHR and NaOH.

Procedure. It is advisable to carry out a test of the electrical heating procedure in advance of the actual runs. This will also provide a chance to check on the heating-coil resistance R_h under actual operating conditions (i.e., when the coil is hot). Place roughly

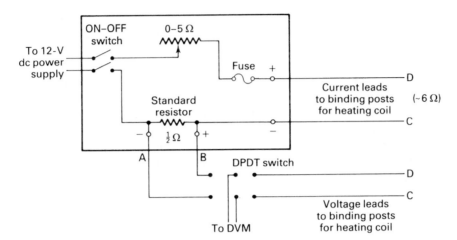

FIGURE 2
Electrical heating circuit for solution calorimeter. The standard resistor should be a wire-wound resistor with a low temperature coefficient and rated for 2 W. If a heating coil of higher resistance (say, ~60Ω is to be used, the current can be reduced to ~0.5 A and a 1-Ω standard resistor can be used.

550 mL of water in the calorimeter and introduce the stirrer–heater unit and thermometer. Make the electrical connections as shown in Fig. 2. Turn on the stirring motor and then turn on the current to the heater; **never** pass current through the heating coil when it is not immersed in a liquid, as it will overheat and may burn out. Measure the potential difference V_h across the heating coil by throwing the DPDT (double-pole, double-throw) switch to the right; i.e., measure the voltage drop between binding posts C and D on the stirrer–heater unit (see Fig. 1). Then measure V_s across the standard resistor in series with the heater by throwing the DPDT switch to the left; i.e., measure the voltage drop between A and B on the heater supply unit. The heater resistance can then be calculated from the known value of R_s and the measured ratio V_h/V_s. Also note the rate of temperature rise during heating. All of these observations will be useful during the actual runs.

Voltage measurements are best made with a good-quality digital voltmeter (DVM) having a high impedance. Such a meter, which can be read easily and rapidly, will facilitate measurements during the actual heating runs. Voltage measurements can also be made with a potentiometer; see Chapter XVI for a description of both digital multimeters and potentiometers. Either of two alternate procedures can be used to determine the electrical work dissipated inside the calorimeter during a heating run: (a) measure both V_h and V_s as a function of time and use Eq. (VI-9a); or (b) measure only V_h as a function of time, assume that R_h is constant with the value determined above, and use Eq. (VI-9b). If a DVM is used, procedure (a) is recommended. If a potentiometer is used, procedure (b) is preferable since potentiometer readings are somewhat time-consuming and, furthermore, a high-resistance potential divider would be needed in order to measure V_h. In this case, the value of R_h will be provided by the instructor.

Fill a 500-mL volumetric flask with solution A, the temperature of which should be within a few tenths of a degree of 25.0°C. (Adjust the temperature by swirling under running hot or cold water before the flask has been entirely filled, then make up to the mark.) Pour the solution into the clean and reasonably dry calorimeter and allow the flask to drain for a minute. Work a plug of stopcock grease into the capillary hole in the bottom of the inner vessel. The glass must be absolutely dry, or the grease will not stick. Pipette in 50 mL of solution B. Place the inner vessel in the calorimeter carefully.

Introduce the stirrer–heater unit and thermometer into the calorimeter, making sure that the thermometer bulb (or any other temperature sensor) is completely immersed (level with or just below the heating coil). The inner vessel should be held in a hole in the calorimeter cover by a split stopper so that it does not rest on the bottom of the Dewar flask.

Check the electrical connections to the heating coil. Connect the T tube to a compressed-air supply; turn on the air, adjust to a barely audible flow, and attach the T tube to the top of the inner vessel. Turn on the stirring motor.

Start temperature–time measurements. Read the thermometer every 30 s, estimating to thousandths of a degree if feasible. Tap the thermometer stem *gently* before each reading. After a slight but steady rate of temperature change due to stirring and heat leak has been observed for 5 min, initiate the reaction by blowing the contents of the inner vessel into the surrounding solution. This is done by placing a finger over the open end of the T tube. Release the pressure as soon as bubbling is heard. Record the time. After 15 s, blow out the inner vessel again to ensure thermal equilibrium throughout all the solution.

After a plateau with a slight, steady rate of temperature change has prevailed for 5 min, turn on the current to the heater and *record the exact time.* Immediately measure the potential difference across the standard resistor in series with the heater. Continue to make voltage measurements approximately every 30 s, alternating between readings of V_s and V_h (or readings of V_s at least once every minute if a potentiometer is being used, in which case the potentiometer should be standardized and set at the expected potential before the current is turned on). Note the time at which each voltage reading is taken.

When a temperature rise of about 1.5°C has been obtained by electrical heating, turn off the current, again noting the exact time. Simultaneously blow out the inner vessel again to mix the contents with the surrounding solution. After a final plateau with a slight, steady rate of temperature change has been achieved for about 7 min, the run may be terminated.

CALCULATIONS

For each run, plot temperature versus time using a greatly expanded temperature scale with an interrupted temperature axis (see Fig. VI-1*b*). Determine the drift rates $(dT/dt)_i$ and $(dT/dt)_f$ before and after the chemical reaction took place and the rate $(dT/dt)_h$ after electrical heating was completed. Carry out appropriate extrapolations to determine the temperature differences $(T_1 - T_0)$ and $(T_2' - T_1')$ shown in Fig. 3. In the case of $(T_1 - T_0)$ associated with the adiabatic reaction, use the method described in the section Principles of Calorimetry and illustrated in Fig. VI-3*b*. If $(dT/dt)_f$ and $(dT/dt)_h$ have reasonably

FIGURE 3

Schematic plot of temperature versus time: heat of ionic reaction.

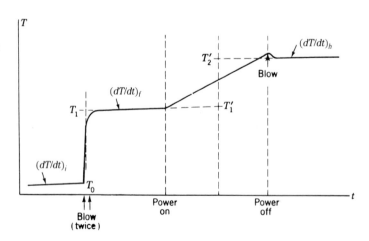

similar values, one can simplify the determination of $(T_2' - T_1')$ by choosing t_d as the midpoint of the electrical heating period as illustrated in Fig. 3.

The heat capacity is determined from Eqs. (VI-9) and (VI-13a). The integral in either Eq. (VI-9a) or (VI-9b) should be evaluated graphically unless the integrand is essentially constant during the heating period. The enthalpy change of the ionic reaction is calculated from Eq. (VI-12a). Calculate the number of moles of reaction from the number of moles of A (the limiting reactant) present and calculate the molar enthalpy change $\Delta\widetilde{H}_3$ or $\Delta\widetilde{H}_4$.

From the average values of $\Delta\widetilde{H}_3$ and $\Delta\widetilde{H}_4$ obtained above, calculate $\Delta\widetilde{H}_1$ and $\Delta\widetilde{H}_2$.

DISCUSSION

Using literature data for the appropriate heats of formation,[1] estimate the difference between the $\Delta\widetilde{H}_3$ value for the concentrations employed in this experiment and the value that would apply at infinite dilution. Compare this effect with a qualitative estimate of the experimental uncertainty in the measured $\Delta\widetilde{H}_3$. Indicate why there should be a large percentage error in $\Delta\widetilde{H}_4$ as determined here.

SAFETY ISSUES

Use a pipetting bulb; do not pipette by mouth.

APPARATUS

Solution calorimeter (1-L Dewar); stirrer–heater unit complete with motor and power cord; inner vessel; precision calorimeter thermometer covering range from 19 to 35°C with magnifying thermometer reader (can be replaced by either a resistance thermometer or a calibrated thermistor with a resolution of $\pm 0.003°C$); 0 to 30°C thermometer; two split rubber stoppers; glass T tube with stopper and rubber tubing; timer; dc power-supply unit for heater (see Fig. 2); double-pole, double-throw switch; digital multimeter or potentiometer setup (see Chapter XVI); 15 electrical leads with lugs attached; 250-mL beaker; 500-mL volumetric flask; 50-mL pipette; 100-mL beaker; rubber pipetting bulb.

Solutions.   NaOH, concentration slightly greater than 2.5 M (0.4 L); 0.25 M HCl (1.5 L); 0.25 M NaHC$_3$H$_2$O$_4$ (1.5 L; the sodium acid malonate solution can be prepared by neutralizing malonic acid with sodium hydroxide to a point just past the NaHC$_3$H$_2$O$_4$ end point); stopcock grease; source of compressed air.

REFERENCE

1. D. D. Wagman et al., "The NBS Tables of Chemical Thermodynamic Properties: Selected Values for Inorganic and C$_1$ and C$_2$ Organic Substances in SI Units," *J. Phys. Chem. Ref. Data* **11,** Suppl. 2 (1982).

VII
Solutions

EXPERIMENTS

9. Partial Molar Volume
10. Cryoscopic Determination of Molar Mass
11. Freezing-Point Depression of Strong and Weak Electrolytes
12. Chemical Equilibrium in Solution

EXPERIMENT 9
Partial Molar Volume

In this experiment the partial molar volumes of sodium chloride solutions will be calculated as a function of concentration from densities measured with a pycnometer.

THEORY

Most thermodynamic variables fall into two types. Those representing *extensive* properties of a phase are proportional to the amount of the phase under consideration; they are exemplified by the thermodynamic functions V, E, H, S, A, G. Those representing *intensive* properties are independent of the amount of the phase; they include p and T. Variables of both types may be regarded as examples of homogeneous functions of degree l: that is, functions having the property

$$f(kn_1, \ldots, kn_i, \ldots) = k^l f(n_1, \ldots, n_i, \ldots) \tag{1}$$

where n_i represents for our purposes the number of moles of component i in a phase. Extensive variables are functions of degree 1 and intensive variables are functions of degree 0.

Among intensive variables important in thermodynamics are *partial molar quantities,* defined by the equation

$$\bar{Q}_i = \left(\frac{\partial Q}{\partial n_i} \right)_{p,T,n_{j\neq i}} \tag{2}$$

where Q may be any of the extensive quantities already mentioned. For a phase of one component, partial molar quantities are identical with so-called molar quantities, $\widetilde{Q} = Q/n$. For an ideal gaseous or liquid solution, certain partial molar quantities $(\bar{V}_i, \bar{E}_i, \bar{H}_i)$ are equal to the respective molar quantities for the pure components while others $(\bar{S}_i, \bar{A}_i, \bar{G}_i)$ are not. For nonideal solutions all partial molar quantities differ in general from the corresponding molar quantities, and the differences are frequently of interest.

A property of great usefulness possessed by partial molar quantities derives from Euler's theorem for homogeneous functions, which states that, for a homogeneous function $f(n_1, \ldots, n_i, \ldots)$ of degree l,

$$n_1 \frac{\partial f}{\partial n_1} + n_2 \frac{\partial f}{\partial n_2} + \cdots + n_i \frac{\partial f}{\partial n_i} + \cdots = lf \tag{3}$$

Applied to an extensive thermodynamic variable Q, for which $l = 1$, we see that

$$n_1 \bar{Q}_1 + n_2 \bar{Q}_2 + \cdots + n_i \bar{Q}_i + \cdots = Q \tag{4}$$

Equation (4) leads to an important result. If we form the differential of Q in the usual way,

$$dQ = \frac{\partial Q}{\partial n_1} dn_1 + \cdots + \frac{\partial Q}{\partial n_i} dn_i + \cdots + \frac{\partial Q}{\partial p} dp + \frac{\partial Q}{\partial T} dT$$

and compare it with the differential derived from Eq. (4),

$$dQ = \bar{Q}_1 \, dn_1 + \cdots + \bar{Q}_i \, dn_i + \cdots + n_1 \, d\bar{Q}_1 + \cdots + n_i \, d\bar{Q}_i + \cdots$$

we obtain

$$n_1 \, d\bar{Q}_1 + \cdots + n_i \, d\bar{Q}_i + \cdots - \left(\frac{\partial Q}{\partial p} \right)_{n_i,T} dp - \left(\frac{\partial Q}{\partial T} \right)_{n_i,p} dT = 0 \tag{5}$$

For the important special case of constant pressure and temperature,

$$n_1 \, d\bar{Q}_1 + \cdots + n_i \, d\bar{Q}_i + \cdots = 0 \qquad (\text{const } p \text{ and } T) \tag{6}$$

This equation tells us that changes in partial molar quantities (resulting of necessity from changes in the n_i) are not all independent. For a binary solution we can write

$$\frac{d\bar{Q}_2}{d\bar{Q}_1} = -\frac{X_1}{X_2} \tag{7}$$

where the X_i are *mole fractions,* $X_i = n_i/\Sigma \, n_i$. In application to free energy, this equation is commonly known as the Gibbs–Duhem equation.

We are concerned in this experiment with the partial molar volume $\bar{V}_i$, which may be thought of as the increase in the volume of an infinite amount of solution (or an amount so large that insignificant concentration change will result) when 1 mole of component i is added. This is by no means necessarily equal to the volume of 1 mol of pure i.

Partial molar volumes are of interest in part through their thermodynamic connection with other partial molar quantities such as partial molar Gibbs free energy, known also as *chemical potential.* An important property of chemical potential is that for any given component it is equal for all phases that are in equilibrium with each other. Consider a system containing a pure solid substance (e.g., NaCl) in equilibrium with the saturated aqueous solution. The chemical potential of the solute is the same in the two phases. Imagine now that the pressure is changed isothermally. Will the solute tend to go from one phase to the other, reflecting a change in solubility? For an equilibrium change at constant temperature involving only expansion work, the change in the Gibbs free energy G is given by

$$dG = V\,dp \tag{8}$$

Differentiating with respect to n_2, the number of moles of solute, we obtain

$$d\bar{G}_2 = \bar{V}_2\,dp \tag{9}$$

where the partial molar free energy (chemical potential) and partial molar volume appear. For the change in state

$$\text{NaCl}(s) = \text{NaCl}(aq)$$

we can write

$$d(\Delta\bar{G}_2) = \Delta\bar{V}_2\,dp$$

or

$$\left[\frac{\partial(\Delta\bar{G}_2)}{\partial p}\right]_T = \Delta\bar{V}_2 \tag{10}$$

Thus, if the partial molar volume of solute in aqueous solution is greater than the molar volume of solid solute, an increase in pressure will increase the chemical potential of solute in solution relative to that in the solid phase; solute will then leave the solution phase until a lower, equilibrium solubility is attained. Conversely, if the partial molar volume in the solution is less than that in the solid, the solubility will increase with pressure.

Partial molar volumes, and in particular their deviations from the values expected for ideal solutions, are of considerable interest in connection with the theory of solutions, especially as applied to binary mixtures of liquid components, where they are related to heats of mixing and deviations from Raoult's law.

METHOD[1]

We see from Eq. (4) that the total volume V of an amount of solution containing 1 kg (55.51 mol) of water and m mol of solute is given by

$$V = n_1\bar{V}_1 + n_2\bar{V}_2 = 55.51\bar{V}_1 + m\bar{V}_2 \tag{11}$$

where the subscripts 1 and 2 refer to solvent and solute, respectively. Let $\tilde{V}_1^0$ be the molar volume of pure water ($= 18.016$ g mol^{-1}/0.997044 g cm$^{-3} = 18.069$ cm^3 mol^{-1} at 25.00°C). Then we define the *apparent molar volume* ϕ of the solute by the equation

$$V = n_1\tilde{V}_1^0 + n_2\phi = 55.51\tilde{V}_1^0 + m\phi \tag{12}$$

which can be rearranged to give

$$\phi = \frac{1}{n_2}(V - n_1\tilde{V}_1^0) = \frac{1}{m}(V - 55.51\tilde{V}_1^0) \tag{13}$$

Now

$$V = \frac{1000 + mM_2}{d}\,\text{cm}^3 \tag{14}$$

and

$$n_1 \tilde{V}_1^0 = \frac{1000}{d_0} \text{cm}^3 \tag{15}$$

where d is the density of the solution and d_0 is the density of pure solvent, both in units of g cm^{-3}, M_2 is the solute molar mass in grams, and 1000 g is the mass of water containing m mol of solute. Substituting Eqs. (14) and (15) into Eq. (13), we obtain

$$\phi = \frac{1}{d}\left(M_2 - \frac{1000}{m}\frac{d - d_0}{d_0}\right) \tag{16}$$

$$= \frac{1}{d}\left(M_2 - \frac{1000}{m}\frac{W - W_0}{W_0 - W_e}\right) \tag{17}$$

In Eq. (17), the directly measured weights of the pycnometer—W_e when empty, W_0 when filled to the mark with pure water, and W when filled to the mark with solution—are used. This equation is preferable to Eq. (16) for calculation of ϕ, as it avoids the necessity of computing the densities to the high precision that would otherwise be necessary in obtaining the small difference $d - d_0$.

Now, by the definition of partial molar volumes and by use of Eqs. (11) and (12),

$$\bar{V}_2 = \left(\frac{\partial V}{\partial n_2}\right)_{n_1, T, p} = \phi + n_2 \frac{\partial \phi}{\partial n_2} = \phi + m \frac{d\phi}{dm} \tag{18}$$

Also

$$\bar{V}_1 = \frac{1}{n_1}\left(n_1 \tilde{V}_1^0 - n_2^2 \frac{\partial \phi}{\partial n_2}\right) = \tilde{V}_1^0 - \frac{m^2}{55.51}\frac{d\phi}{dm} \tag{19}$$

We might proceed by plotting ϕ versus m, drawing a smooth curve through the points, and constructing tangents to the curve at the desired concentrations in order to measure the slopes. However, for solutions of simple electrolytes, it has been found that many apparent molar quantities such as ϕ vary linearly with $\sqrt{m}$, even up to moderate concentrations.[2] This behavior is in agreement with the prediction of the Debye–Hückel theory for dilute solutions.[3] Since

$$\frac{d\phi}{dm} = \frac{d\phi}{d\sqrt{m}}\frac{d\sqrt{m}}{dm} = \frac{1}{2\sqrt{m}}\frac{d\phi}{d\sqrt{m}} \tag{20}$$

we obtain from Eqs. (18) and (19),

$$\bar{V}_2 = \phi + \frac{m}{2\sqrt{m}}\frac{d\phi}{d\sqrt{m}} = \phi + \frac{\sqrt{m}}{2}\frac{d\phi}{d\sqrt{m}} = \phi^0 + \frac{3\sqrt{m}}{2}\frac{d\phi}{d\sqrt{m}} \tag{21}$$

$$\bar{V}_1 = \tilde{V}_1^0 - \frac{m}{55.51}\left(\frac{\sqrt{m}}{2}\frac{d\phi}{d\sqrt{m}}\right) \tag{22}$$

where ϕ^0 is the apparent molar volume extrapolated to zero concentration. Now one can plot ϕ versus $\sqrt{m}$ and determine the best *straight* line through the points. From the slope $d\phi/d\sqrt{m}$ and the value of ϕ^0, both $\bar{V}_1$ and $\bar{V}_2$ can be obtained.

EXPERIMENTAL

Make up 200 mL of approximately 3.2 m (3.0 M) NaCl in water. Weigh the salt accurately and use a volumetric flask; then pour the solution into a dry flask. If possible, prepare this solution in advance (since the salt dissolves slowly). Solutions of $\frac{1}{2}, \frac{1}{4}, \frac{1}{8}$, and $\frac{1}{16}$ of the initial molarity are to be prepared by successive volumetric dilutions; for each dilution pipette 100 mL of solution into a 200-mL volumetric flask and make up to the mark with distilled water.

Rinse the pycnometer with distilled water and dry it thoroughly before each use. Use an aspirator, and rinse and dry by suction; use a few rinses of acetone to expedite drying. The procedure given here is for the Ostwald–Sprengel type of pycnometer; the less accurate but more convenient stopper type can be used with a few obvious changes in procedure.† To fill, dip one arm of the pycnometer into the vessel containing the solution (preferably at a temperature *below* 25°C) and apply suction by mouth with a piece of rubber tubing attached to the other arm. Hang the pycnometer in the thermostat bath (25.0°C) with the main body below the surface but with the arms emerging well above. Allow at least 15 min for equilibration. While the pycnometer is still in the bath, adjust menisci to fiducial marks with the aid of a piece of filter paper. Remove the pycnometer from the bath and quickly but thoroughly dry the outside surface with a towel and filter paper. Weigh the pycnometer.

The pycnometer should be weighed empty and dry (W_e), and also with distilled water in it (W_0), as well as with each of the solutions in it (W). It is advisable to redetermine W_e and W_0 as a check, inasmuch as the results of all runs depend on them. All weighings are to be done on an analytical balance to the highest possible precision.

As an alternative and very convenient procedure, one can use a Cassia volumetric flask instead of a pycnometer. Although the precision of density measurements made with this flask is not as good as that obtainable with the Ostwald–Sprengel pycnometer, it is adequate for the present purposes. The Cassia flask, shown in Fig. 1c, is a special glass-stoppered volumetric flask with 0.1-mL graduations between 100 mL and 110 mL, calibrated to contain the indicated volume to within ± 0.08 mL. Since this flask is 26 cm high and will weigh over 120 g when full, it must be weighed on a top-loading balance. A high-quality balance of this type, capable of weighing to within ± 0.01 g, is required.

The Cassia flask should be weighed empty and dry (W_e) and twice with distilled water in it, once with the level near 100 mL (W_0) and once with the level near 105 mL (W_0'). It should then be weighed with each of the solutions in it (W) to a level just above 100 mL. In every case, record the liquid-volume reading to within ± 0.05 mL (V_0, V_0', V for each solution). As with the pycnometers, the Cassia flask must be rinsed well and dried prior to each filling. The filled flask is then thermally equilibrated in a constant-temperature bath (25°C) for at least 15 min. On removing the flask from the bath, dry the outside thoroughly before weighing. Since more solution is required to fill a Cassia flask than a pycnometer, you will need *at least* 225 mL of the 3 M stock solution. Using two 200-mL volumetric

---

†The filling of a Weld-type pycnometer must be carried out with great care. The temperature of the laboratory must be below the temperature at which the determination is to be made. Fill the pycnometer body with the liquid, and seat the capillary stopper firmly. Wipe off excess liquid around the tapered joint, cap the pycnometer, and immerse it in the thermostat bath to a level just below the cap. When temperature equilibrium has been reached, remove the cap, wipe off the excess liquid from the capillary tip, being careful not to draw liquid out of the capillary, and remove the pycnometer from the bath. As the pycnometer cools to room temperature, the liquid column in the capillary will descend; be careful not to heat the pycnometer with your hands, since this will force liquid out of the capillary. Carefully clean and wipe off the whole pycnometer, including the cap but not including the tip of the stopper. Cap the pycnometer, place it in the balance, and allow it to stand about 10 min before weighing.

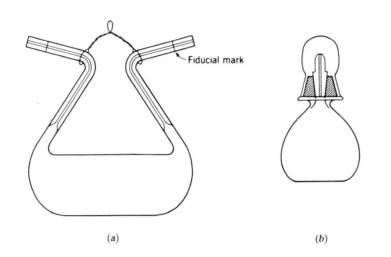

FIGURE 1
Pycnometers:
(*a*) Ostwald–Sprengel type;
(*b*) Weld (stopper) type;
(*c*) Cassia volumetric flask
(not drawn to scale).

Fiducial mark

(*a*)   (*b*)   (*c*)

flasks and two pipettes (one 50-mL and one 100-mL), you can prepare all the dilutions required.

CALCULATIONS

The success of this experiment depends greatly on the care with which the computations are carried out. For this reason, use of a spreadsheet program for analyzing the data is strongly recommended. If the work is done by two or more students, the partners are encouraged to work together, performing the same calculations independently and checking results after each step.

Calculate the density d of every solution to within an accuracy of at least one part per thousand:

$$d = \frac{W_{\text{soln}}}{V} = \frac{W - W_e}{V} \tag{23}$$

If a pycnometer was used, determine its volume from $W_0 - W_e$ and the density d_0 of pure water at 25°C (0.997044 g cm^{-3}). If a Cassia flask was used, carry out this volume calculation for both fillings with water and compare your results with the direct volume readings V_0 and V_0'. If necessary devise a calibration procedure that can be applied to correct the Cassia volume readings obtained on the solutions.

The molalities m (concentration in moles per kilogram of solvent) that are needed for the calculations can be obtained from the molarities M (concentration in moles per liter of solution) obtained from the volumetric procedures by using the equation

$$m = \frac{1}{1 - (M/d)(M_2/1000)}\frac{M}{d} = \frac{1}{(d/M) - (M_2/1000)} \tag{24}$$

where M_2 is the solute molar mass (58.45 g mol^{-1}) and d is the experimental density in g cm^{-3} units.

Calculate ϕ for each solution using Eq. (17) for pycnometer data or Eq. (16) for Cassia flask data. Plot ϕ versus $\sqrt{m}$. Determine the slope $d\phi/d\sqrt{m}$ and the intercept ϕ^0 at $m = 0$ from the best straight line through these data points. This should be done with a linear least-squares fitting procedure.

Calculate $\bar{V}_2$ and $\bar{V}_1$ for $m = 0$, 0.5, 1.0, 1.5, 2.0, and 2.5. Plot them against m and draw a smooth curve for each of the two quantities.

In your report, present the curves (ϕ versus $\sqrt{m}$, $\bar{V}_2$ and $\bar{V}_1$ versus m) mentioned above. Present also in tabular form the quantities d, M, m, $(1000/m)(W - W_0)/(W_0 - W_e)$, and ϕ for each solution studied. Give the values obtained for the pycnometer volume V_p and ϕ^0 and $d\phi/d\sqrt{m}$.

DISCUSSION

The density of NaCl(s) is 2.165 g cm^{-3} at 25°C. How will the solubility of NaCl in water be affected by an increase in pressure?

Discuss qualitatively whether the curves of $\bar{V}_1$ and $\bar{V}_2$ versus m behave in accord with Eq. (7).

SAFETY ISSUES

None.

APPARATUS

Pycnometer (approximately 70 mL) with wire loop for hanging in bath or Cassia flask; one or two 200-mL volumetric flasks; 100-mL pipette, and 50-mL pipette if a Cassia flask is used; pipetting bulb; 250-mL Erlenmeyer flask; one 250- and one 100-mL beaker; large weighing bottle; short-stem funnel; spatula; filter paper and gum-rubber tube (1 to 2 ft long) if an Ostwald–Sprengel pycnometer is used.

Constant-temperature bath set at 25°C; bath hardware for holding flasks and pycnometer; reagent-grade sodium chloride (35 g of solid or 200 mL of solution of an accurately known concentration, 50 g or 285 mL if a Cassia flask is used); acetone to be used for rinsing; cleaning solution.

REFERENCES

1. F. T. Gucker, Jr., *J. Phys. Chem.* **38,** 307 (1934).

2. D. O. Masson, *Phil. Mag.* **8,** 218 (1929).

3. O. Redlich and P. Rosenfeld, *Z. Phys. Chem,* **A155,** 65 (1931).

GENERAL READING

R. S. Davis and W. F. Koch, "Mass and Density Determinations," in B. W. Rossiter and R. C. Baetzold (eds.), *Physical Methods of Chemistry,* 2d ed., Vol. VI, chap. 1. Wiley-Interscience, New York (1992).

K. S. Pitzer, *Thermodynamics,* 3d ed., chap. 10, McGraw-Hill, New York (1995).

EXPERIMENT 10
Cryoscopic Determination of Molar Mass

When a substance is dissolved in a given liquid solvent, the freezing point is nearly always lowered. This phenomenon constitutes a so-called *colligative* property of the substance—a property with a magnitude that depends primarily on the number of moles of the substance that are present in relation to a given amount of solvent. In the present case, the amounts by which the freezing point is lowered, called the *freezing-point depression* ΔT_f, is approximately proportional to the number of moles of solute dissolved in a given amount of the solvent. Other colligative properties are vapor-pressure lowering, boiling-point elevation, and osmotic pressure.

THEORY

In Fig. 1, T_0 and p_0 are, respectively, the temperature and vapor pressure at which the pure solvent freezes. Strictly speaking, T_0 is the temperature at the "triple point," with a total pressure equal to p_0 rather than the freezing point at 1 atm. However, the pressure effect is small—of the order of a few thousandths of a degree—and moreover, as it is essentially the same for a dilute solution as for the pure solvent, it will cancel out in the calculation of ΔT_f. The curve labeled p_l is the vapor pressure of the pure liquid solvent as a function of temperature, and that labeled p_s is the vapor pressure ("sublimation pressure") of the pure solid solvent. The dashed curve labeled p_x is the vapor pressure of a solution containing solute at mole fraction X. We now introduce Assumption 1: *The solid in equilibrium with a solution at its freezing point is essentially pure solid solvent.* The validity of this assumption results primarily from the crystalline structure of the solid solvent, in which the molecules are packed in a regular manner closely dependent on their shape and size. The substitution of one or more solute molecules for solvent molecules, to yield a solid solution, is usually accompanied by a very large increase in free energy, for the solute molecule usually has a size and shape poorly adapted to suitably filling the cavity produced by removing a solvent molecule. Exceptions occur when solute and solvent molecules are very similar, as in the toluene–chlorobenzene, benzene–pyridine and other similar systems, and in some inorganic salt mixtures and many metal alloys.

By the second law of thermodynamics, when a solution is in equilibrium with solid solvent, the solvent vapor pressures of the two must be equal. Therefore, for a solution

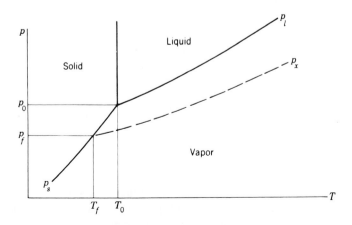

FIGURE 1

Pressure–temperature diagram for solvent and solution.

containing solute at mole fraction X, the freezing point T_f and the corresponding equilibrium vapor pressure p_f are determined by the intersection of the sublimation-pressure curve p_s with the solution vapor-pressure curve p_x. To determine the point of intersection, we write the equations for the two curves and solve them simultaneously for p and T.

The equation of the sublimation curve is conveniently obtained from the integrated Clausius–Clapeyron equation,

$$\ln \frac{p_s}{p_0} = \frac{\Delta \tilde{H}_s}{R} \left(\frac{1}{T_0} - \frac{1}{T} \right) \tag{1}$$

where $\Delta \tilde{H}_s$ is the molar heat of sublimation of the solid solvent. In order to use this equation, we must introduce Assumption 2: *All assumptions required in the derivation of the integrated Clausius–Clapeyron equation may be here assumed.* These are that (*a*) the vapor obeys the perfect-gas law, (*b*) the volume of a condensed phase is negligible in comparison with that of an equivalent amount of vapor, and (*c*) the enthalpy change accompanying vaporization or sublimation is independent of temperature. Part (*c*) of this assumption may be dispensed with, as we shall see later. The vapor-pressure curve p_x for the solution may be related to that for the liquid p_l, the equation for which we may take in accordance with the above assumption as

$$\ln \frac{p_l}{p_0} = \frac{\Delta \tilde{H}_v}{R} \left(\frac{1}{T_0} - \frac{1}{T} \right) \tag{2}$$

where $\Delta \tilde{H}_v$ is the molar heat of vaporization of the liquid solvent. To obtain the equation for the p_x curve we introduce Assumption 3: *Raoult's law applies.* Raoult's law is applicable to ideal solutions and, in the limit of infinite dilution, to real solutions. It is usually a fairly good approximation for moderately dilute real solutions. According to this law,

$$\frac{p_x}{p_l} = X_0 = 1 - X \tag{3}$$

where X_0 is the mole fraction of solvent and X is the mole fraction of solute in the solution. Combining Eqs. (2) and (3), we obtain, as the equation for p_x as a function of T,

$$\ln \frac{p_x}{p_0} = \ln \frac{p_x}{p_l} + \ln \frac{p_l}{p_0} = \ln(1 - X) + \frac{\Delta \tilde{H}_v}{R} \left(\frac{1}{T_0} - \frac{1}{T} \right) \tag{4}$$

Now we are ready to solve for the coordinates of the intersection point p_f, T_f. Setting p_x equal to p_s, we obtain from Eqs. (1) and (4),

$$\ln(1 - X) = \frac{\Delta \tilde{H}_s - \Delta \tilde{H}_v}{R} \left(\frac{1}{T_0} - \frac{1}{T_f} \right)$$

or

$$\ln(1 - X) \doteq -\frac{\Delta \tilde{H}_f}{R T_0 T_f} \Delta T_f \tag{5}$$

where

$$\Delta T_f \equiv T_0 - T_f \tag{6}$$

and $\Delta \tilde{H}_f$ is the molar heat of fusion of the liquid solvent.

It is worth noting here that Eq. (5) may be derived more elegantly on a somewhat different set of assumptions, in which those pertaining to properties of the solvent vapor do not appear. Those assumptions, together with Raoult's law, may be replaced by the single assumption that the *activity of the solvent is directly proportional to the mole fraction of the solvent in the solution.*

We wish to derive expressions for calculating the molality m of the solution (or the molar mass M of the solute) from a measured freezing-point depression ΔT_f. The molality of the solution is given by

$$ m = \frac{1000}{M_0} \frac{X}{1 - X} \tag{7} $$

where M_0 is the molar mass of the solvent in grams per mole. If g is the mass of solute and G is the mass of solvent,

$$ \frac{X}{X_0} = \frac{g/M}{G/M_0} $$

thus

$$ M = M_0 \frac{g}{G} \frac{X_0}{X} = M_0 \frac{g}{G} \frac{1 - X}{X} $$

Combining this with Eq. (7), we obtain

$$ M = \frac{g}{G} \frac{1000}{m} \tag{8} $$

The expressions commonly used for determination of molalities or molar mass from freezing-point depressions are derived with the following approximations:

$$ \ln(1 - X) \cong -X $$
$$ T_0 T_f \cong T_0^2 \tag{9} $$
$$ 1 - X = X_0 \cong 1 $$

With the aid of these approximations, Eq. (5) becomes

$$ X = \frac{\Delta \widetilde{H}_f}{RT_0^2} \Delta T_f \tag{10} $$

Combining Eq. (10) with Eqs. (7) and (8), we obtain

$$ m = \frac{1000 \, \Delta \widetilde{H}_f}{M_0 RT_0^2} \Delta T_f = \frac{\Delta T_f}{K_f} \tag{11} $$

and

$$ M = \frac{M_0 g RT_0^2}{G \Delta \widetilde{H}_f \, \Delta T_f} = 1000 \frac{g}{G} \frac{K_f}{\Delta T_f} \tag{12} $$

where

$$ K_f \equiv \frac{M_0 RT_0^2}{1000 \Delta \widetilde{H}_f} \tag{13} $$

K_f is called the *molal freezing-point depression constant,* since it is equal to the freezing-point depression predicted by Eq. (11) for a 1 molal solution.

It will often be the case that no further refinement of these expressions is justified. In some cases, however, the use of a higher order of approximation is worthwhile. In the treatment that follows, we shall retain terms representing no more than two orders in ΔT.

Let the right-hand side of Eq. (5) be called y,

$$y \equiv -\frac{\Delta \tilde{H}_f}{RT_0 T_f} \Delta T_f$$

and express e^y in a Maclaurin series, so that Eq. (5) becomes

$$1 - X = e^y = 1 + y + \frac{y^2}{2!} + \ldots$$

Since y is small in comparison with unity,

$$X \cong -y\left(1 + \frac{y}{2}\right) \cong -\frac{y}{1 - y/2} \tag{14}$$

The other approximations of Eq. (9) are replaced by

$$T_0 T_f = T_0^2 \frac{T_f}{T_0} = T_0^2\left(1 - \frac{\Delta T_f}{T_0}\right) \cong \frac{T_0^2}{1 + \Delta T_f/T_0} \tag{15}$$

and

$$1 - X = X_0 = 1 + y \cong \frac{1}{1 - y} \tag{16}$$

In addition, if it is not desired to retain Assumption 2(c) that $\Delta \tilde{H}_f$ is strictly constant, we can replace it in the expression for y by its mean value over the temperature range ΔT_f:

$$\overline{\Delta \tilde{H}_f} = \Delta \tilde{H}_f^0 - \Delta \tilde{C}_p \frac{\Delta T_f}{2} = \Delta \tilde{H}_f^0\left(1 - \frac{\Delta \tilde{C}_p}{2\Delta \tilde{H}_f^0}\Delta T_f\right) \tag{17}$$

where $\Delta \tilde{H}_f^0$ is the molar heat of fusion of the pure solvent at its freezing point.

In combining the above equations, we follow the usual rules for multiplying and dividing binominals and reject terms of second and higher order in ΔT_f in comparison with unity. We obtain

$$\frac{X}{1 - X} = \frac{\Delta \tilde{H}_f^0}{RT_0^2} \Delta T_f (1 + k_f \Delta T_f) \tag{18}$$

where

$$k_f = \frac{1}{T_0} + \frac{\Delta \tilde{H}_f^0}{2RT_0^2} - \frac{\Delta \tilde{C}_p}{2\Delta \tilde{H}_f^0} \tag{19}$$

We further obtain, by use of Eqs. (7) and (8),

$$m = \frac{\Delta T_f}{K_f}(1 + k_f \Delta T_f) \tag{20}$$

$$M = 1000\frac{g}{G}\frac{K_f}{\Delta T_f}(1 - k_f \Delta T_f) \tag{21}$$

In Table 1 are given values of the pertinent constants for two common solvents: cyclohexane and water. It will be seen that, for freezing-point depressions of about 2 K, omission of the correction term $k_f \Delta T_f$ leads to errors of the order of 1 percent in m or M.

Where Raoult's law fails, we may expect in the case of a nondissociating and nonassociating solute that the experimental conditions will at least lie in a concentration range

TABLE 1

		Cyclohexane	Water
Molar mass (g)	M_0	84.16	18.02
Celsius freezing point (°C)	t_0	6.68	0.00
Absolute freezing point (K)	T_0	279.83	273.15
Molar heat of fusion at T_0 (J mol^{-1})	$\Delta \widetilde{H}_f^0$	2678	6008
Molar heat-capacity change on fusion (J K^{-1} mol^{-1})	$\Delta \widetilde{C}_p$	15.1	38.1
Molal freezing-point depression constant (K molal^{-1})	K_f	20.4	1.855
Correction constant (K^{-1})	k_f	0.003	0.005

over which the deviation in vapor pressure can be expressed fairly well by a quadratic term in the solute mole fraction X. Within this range the main effect on the above equations will be to change k_f by a small constant amount. Thus we may expect that a plot of M [calculated either with Eq. (12) or with Eq. (21)] against ΔT_f, with a number of experimental points obtained at different concentrations, should yield an approximately straight line that, on extrapolation to $\Delta T_f = 0$, should give a good value for M. In the event of failure of the assumption of no solid solution however, the limiting value of M itself should be expected to be in error.†

METHOD

In this experiment the freezing point of a solution containing a known weight of an "unknown" solute in a known weight of cyclohexane is determined from cooling curves. From the result at each of two concentrations, the molar mass of the unknown is determined.

The apparatus is shown in Fig. 2. The inner test tube, containing the solution, stirrer, and thermometer, is partially insulated from a surrounding ice–salt cooling bath through being suspended in a larger test tube with an air space between them. To provide additional insulation, the space between may be filled by a hollow plug of expanded polystyrene foam. The thermometer is either a special cryoscopic mercury thermometer of appropriate range, with graduations every 0.01 or 0.02°C, a digital resistance thermometer with a resolution of ±0.01°C, or a calibrated thermistor.

Under the conditions of the experiment, heat flows from the inner system, at temperature T, to the ice–salt bath, at temperature T_b, at a rate which is approximately proportional to the temperature difference:

$$-\frac{dH}{dt} = A(T - T_b) \tag{22}$$

where H is the enthalpy of the inner system and A is a constant incorporating shape factors and thermal-conductivity coefficients. If $(T - T_b)$ is sufficiently large in comparison with the temperature range covered in the experiment, we can write

$$-\frac{dH}{dt} \cong \text{const} \tag{23}$$

†This can easily be shown by a treatment parallel to the derivation here given, taking account of the facts that the equilibrium concentration of solute in the solid phase increases (and that of solvent decreases) with increasing solute concentration in the liquid phase and that the solvent vapor pressure of the solid decreases as the solvent concentration in the solid decreases. Thus for the solid a new vapor-pressure curve should be drawn below p_s in Fig. 1, and its intersection point with p_x is to the right of that shown.

FIGURE 2
Apparatus for cryoscopic
determination of molar mass.
The cryoscopic mercury
thermometer can be replaced
by a resistance thermometer
or a calibrated thermistor.

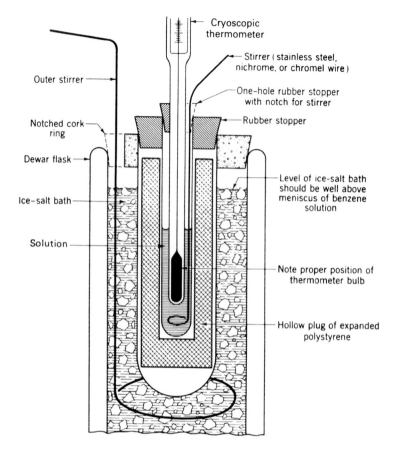

In the absence of a phase change, the rate of change of the temperature is given by

$$-\frac{dT}{dt} = \frac{1}{C}\left(-\frac{dH}{dt}\right) \tag{24}$$

where C is the heat capacity of the inner system. When a pure liquid freezes, dT/dt vanishes as long as two phases are present and we have a "thermal arrest." When pure solid solvent separates from a liquid solution on freezing, the temperature does not remain constant because the solution becomes continually more and more concentrated and the freezing point T_f correspondingly decreases. It can be shown that, in this case,

$$-\frac{dT}{dt} = \frac{1}{(N_0\,\Delta\tilde{H}_f\,/\,\Delta T_f) + C}\left(-\frac{dH}{dt}\right) \tag{25}$$

where N_0 is the number of moles of solvent present in the liquid phase. Thus, when the solid solvent begins to freeze out of solution on cooling, the slope changes discontinuously from that given by Eq. (24) to the much smaller slope given by Eq. (25), and we have what is called a "break" in the cooling curve.

Figure 3 shows schematically the types of cooling curves that are expected as a result of these considerations. It will be noted that the solutions may "supercool" before solidification of solvent takes place. In the present experiment, the extent of supercooling rarely exceeds about 2 K and is best kept below 1 K by seeding—introducing a small crystal of frozen solvent.

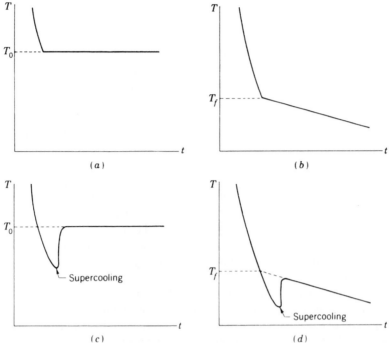

FIGURE 3
Schematic cooling curves:
(a) and (c) show the cooling curves for pure solvent;
(b) and (d) show the cooling curves for a solution.

When supercooling occurs, the recommended procedure for estimating the true freezing point of the solution (the temperature at which freezing would have started in the absence of supercooling) is to extrapolate back that part of the curve that corresponds to freezing out of the solvent until the extrapolate intersects the cooling curve of the liquid solution. To a good approximation, the extrapolation may be taken as linear.†

This procedure is valid only when the extent of supercooling is small in comparison with the difference in temperature between the system and the ice–salt cooling bath. When this is not the case, Eq. (22) must be used in place of Eq. (23), and it is then seen that the rate of decrease of enthalpy is dependent in significant degree on whether or not supercooling takes place. Therefore the amount of supercooling must be kept small, and the bath must be kept as cold as possible.

EXPERIMENTAL

The success of this experiment depends primarily on a careful experimental technique. The solvent to be used is cyclohexane, which must be of reagent-grade purity and scrupulously dry. Pour about 25 mL of cyclohexane into a clean, dry, glass-stoppered flask

---

†To justify this extrapolation procedure, consider two experiments starting at the same temperature and the same time under conditions identical in all respects except that in one case supercooling is allowed to take place and in the other it is somehow prevented. By Eq. (23) the enthalpies of the two systems remain identical throughout the experiment. Except during the interval of supercooling, both systems are in equilibrium, and therefore, when their enthalpies are equal, they are identical in all respects, including temperature. Thus, except for the interval of supercooling, the two curves when superimposed are congruent throughout their entire length. The dashed line representing the extrapolation in Fig. 3c or 3d is therefore part of the curve for the hypothetical experiment in which supercooling was prevented.

for your own use. During the experiment, take care to keep the inner part of the apparatus dry and to expose the inner test tube to the air as little as possible in order to avoid condensation of moisture inside.

The ice–salt mixture should be made up freshly for each run. Mix about one part by volume of coarse rock salt with about four parts by volume of finely crushed ice in a beaker or battery jar. Allow it to stand a few minutes to become slushy, and then pour the desired amount into the Dewar flask. (This ice–salt mixture can be replaced by a mixture of 10 parts ice with 1 part denatured alcohol.)

Fill a small beaker with crushed ice and a little water, and place in it a test tube containing 1 or 2 mL of cyclohexane and a very thin glass rod. Stopper the tube with cotton or glass wool to keep out moist air from the room. This frozen cyclohexane is to be used for seeding. When it is desired to seed the system during a run, remove the glass rod, making sure it carries a small amount of frozen cyclohexane, and carefully insert it into the solution with the least possible disruption of the experiment.

Cyclohexane should be introduced into the inner test tube from a weighing bottle (weighed before and after delivery) or from a pipette. *Do not pipette cyclohexane by mouth;* use a rubber bulb. The quantity normally required is 15 mL. If a pipette is used, record the ambient temperature.

The thermometer and stirrer should be inserted with the thermometer carefully mounted in such a way that the sensing element (bulb of a mercury thermometer, resistance coil of a platinum resistance thermometer, or calibrated thermistor) is about halfway between the bottom of the test tube and the upper surface of the liquid and concentric with the tube so that the stirrer can easily pass around it. If the thermometer is too close to the bottom, a bridge of frozen solvent can easily form, which will conduct heat away from the thermometer and result in low readings.

Much depends on the technique of stirring the solution. The motion of the stirrer should carry it from the bottom of the tube up to near the surface of the liquid; before assembling the apparatus, it is well to observe carefully how high the stirrer can be raised without too frequent splashing. The stirring should be continuous throughout the run and should be at the rate of about one stroke per second. The outer bath should be stirred a few times per minute.

The test tube containing cyclohexane, stopper, thermometer, and stirrer is held in the beaker of ice water, and the cyclohexane is stirred until it visibly starts to freeze. The outside of the tube is wiped dry, and the tube is allowed to warm up at least 1°C above the freezing point. Then the tube is placed in the assembled apparatus and temperature readings are taken every 30 s. If a mercury cryoscopic thermometer is being used, tap the thermometer gently before each reading, and estimate the temperature readings to tenths of the smallest scale division. If the temperature falls 0.5°C below the normal freezing point without evidence of freezing, seed the liquid. Once freezing has occurred, continue temperature readings for about 5 min. If the cyclohexane and apparatus are suitably dry, the temperature should remain constant or fall by no more than about 0.02°C during that time.

The inner test tube is removed and warmed (with stirring) in a beaker of water until the cyclohexane has completely melted. The stopper is lifted, and an *accurately weighed* pellet of the unknown compound, of about 0.6 g, is introduced. The pellet is dissolved by stirring, the inner tube is cooled with stirring in an ice bath to about the freezing temperature of pure cyclohexane and then dried on the outside, and the apparatus is reassembled. Temperature readings are taken every 30 s throughout the run. After a depression of about 2°C has been reached, it is advisable to seed the system at intervals of about 0.5°C until freezing starts. After freezing has begun, temperature readings should be continued for a period at least four times as long as the estimated period of supercooling to provide data for an adequate extrapolation.

The inner tube is removed, the cyclohexane is melted as before, and a second weighed pellet of the unknown is added and stirred into solution. Before this pellet is added, it is well to make a rough calculation, based on the results of the first run, to make sure that the temperatures in the second run will remain on scale and, if necessary, to modify the weight of the pellet accordingly. The second run is carried out as before, after cooling in an ice bath to about the temperature at which freezing was obtained in the first run.

If the quantities $g/\Delta T_f$ from these two runs are not in agreement and time allows, a repeat of both runs should be made using somewhat different pellet weights.

CALCULATIONS

If the cyclohexane was introduced with a pipette, its weight G can be calculated using the density:

$$\rho(\text{cyclohexane}) = 0.779 - 9.4 \times 10^{-4}(t - 20) \text{ g cm}^{-3}$$

where t is the Celsius temperature. Plot the cooling-curve data for pure cyclohexane and for each solution studied. If supercooling took place, perform the extrapolations as shown in Fig. 3. Report the weight g of solute and the freezing-point depression ΔT_f for each run.

Calculate the molar mass from both Eqs. (12) and (21) using the appropriate constants in Table 1. An extrapolated value of M can be obtained from the calculated values from either equation by plotting the calculated molar mass against the depression ΔT_f. If the elementary analysis or empirical formula of the unknown is given, deduce the molecular formula and the exact molar mass.

SAFETY ISSUES

Use a pipetting bulb; do not pipette by mouth. Dispose of waste chemicals as instructed.

APPARATUS

Dewar flask; notched cork ring to fit top of Dewar; large test tube with polystyrenefoam insert; large ring stirrer; inner test tube that fits into polystyrene insert; large rubber stopper with hole for inner test tube; medium stopper with hole for thermometer and a notch for small ring stirrer; 1-qt battery jar or 1000-mL beaker; 250-mL beaker; 125-mL glass-stoppered flask for storing cyclohexane; small test tube; glass rod (3 mm diameter and 20 cm long); 15-mL pipette; small rubber pipetting bulb; precision cryoscopic thermometer and magnifying thermometer reader (can be replaced by either a digital resistance thermometer with $\pm0.01°C$ resolution or a calibrated thermistor); stopwatch or timer.

Dry reagent-grade cyclohexane (50 mL); glass wool; naphthalene or other solid; pellet press; acetone for rinsing; ice (3 lb); ice grinder or shaver; coarse rock salt (2 lb).

GENERAL READING

J. B. Ott and J. P. Goates, "Temperature Measurement with Application to Phase Equilibria Studies," in B. W. Rossiter and R. C. Baetzold (eds.), *Physical Methods of Chemistry*, 2d ed., Vol. VI, chap. 7, Wiley-Interscience, New York (1992).

EXPERIMENT 11

Freezing-Point Depression of Strong and Weak Electrolytes

In this experiment, the freezing-point depression of aqueous solutions is used to determine the degree of dissociation of a weak electrolyte and to study the deviation from ideal behavior that occurs with a strong electrolyte.

THEORY

Use will be made of the theory developed in Exp. 10 for the freezing-point depression ΔT_f of a given solvent containing a known amount of an ideal solute; this material should be reviewed.

In the case of a dissociating (or associating) solute, the molality given by Eq. (10-11) or (10-20) is ideally the *total* effective molality—the number of moles of all solute species present, whether ionic or molecular, per 1 kg of solvent. As we shall see, ionic solute species at moderate concentrations do not form ideal solutions and, therefore, do not obey these equations. However, for a weak electrolyte, the ionic concentration is often sufficiently low to permit treatment of the solution as ideal.

Weak Electrolytes.   As an example, let us discuss a weak acid HA with nominal molality m. Owing to the dissociation,

$$HA = H^+ + A^-$$

the equilibrium concentrations of HA, H^+, and A^- will be $m(1 - \alpha)$, $m\alpha$, and $m\alpha$, respectively, where α is the fraction dissociated. The total molality m' of all solute species is

$$m' = m(1 + \alpha) \tag{1}$$

This molality m' can be calculated from the observed ΔT_f using Eq. (10-20). Thus freezing-point measurements on weak electrolyte solutions of known molality m enable the determination of α.

The equilibrium constant in terms of concentrations can be calculated from

$$K_c = \frac{(H^+)(A^-)}{(HA)} = m\frac{\alpha^2}{(1 - \alpha)} \tag{2}$$

In this experiment α and K_c are to be determined at two different molalities. Since these will be obtained at two different temperatures, the values of K_c should be expected to differ slightly.

Strong Electrolytes.   Solutes of this type, such as HCl, are completely dissociated in ordinary dilute solutions. However, their colligative properties when interpreted in terms of ideal solutions appear to indicate that the dissociation is a little less than complete. This fact led Arrhenius to postulate that the dissociation of strong electrolytes is indeed incomplete. Subsequently this deviation in colligative behavior has been demonstrated to be an expected consequence of interionic attractions.

For a nonideal solution, Eq. (10-5) is replaced by

$$\ln a_0 = -\frac{\Delta \widetilde{H}_f}{RT_0 T_f}\Delta T_f \cong -\frac{\Delta \widetilde{H}_f}{RT_0^2}\Delta T_f \tag{3}$$

where a_0 is the *activity* of the solvent and is related to the mole fraction of solvent X_0 by

$$a_0 = \gamma_0 X_0 \tag{4}$$

The quantity γ_0 is the activity coefficient for the solvent and in electrolytic solutions differs from unity even at moderately low concentrations. Let us write $\ln a_0$ as $(\ln \gamma_0 + \ln X_0)$ in Eq. (3) and divide both sides by $\ln X_0$ to obtain

$$-\frac{\Delta \tilde{H}_f}{RT_0^2} \frac{\Delta T_f}{\ln X_0} = 1 + \frac{\ln \gamma_0}{\ln X_0} \equiv g_1 \tag{5}$$

where g_1 is called the *osmotic coefficient* of the solvent.[1,2] Now

$$X_0 = \frac{n_0}{n_0 + \nu n_1} \tag{6}$$

where n_0 is the number of moles of solvent and νn_1 is the total number of moles of ions formed from n_1 moles of solute (for example, $\nu = 2$ for HCl). Thus

$$-\ln X_0 \equiv \ln\left(1 + \frac{\nu n_1}{n_0}\right) \cong \frac{\nu n_1}{n_0} \tag{7}$$

for dilute solutions. Substituting this expression for $\ln X_0$ into Eq. (5), we have

$$\frac{\Delta \tilde{H}_f}{RT_0^2} \frac{n_0}{\nu n_1} \Delta T_f = g_1 \tag{8}$$

For a solution of molality m in a solvent with molar mass M_0 in g mol^{-1}, we can replace $(n_0/\nu n_1)$ by $(1000/M_0 \nu m)$. Equation (8) can then be written as

$$\left(\frac{\Delta \tilde{H}_f}{RT_0^2} \frac{1000}{M_0}\right) \frac{\Delta T_f}{\nu m} = \frac{\Delta T_f}{\nu m K_f} = g_1 \tag{9}$$

where K_f is the molal freezing-point depression constant defined by Eq. (10-13).

For an ideal solution, $\gamma_0 = 1$ and g_1 is unity. Then Eq. (9) is consistent with Eq. (10-11), since the total molality of all solute species is νm for a completely dissociated solute of molality m. For ionic solutions, the Debye–Hückel theory predicts a value of γ_0 different from unity and therefore a deviation of g_1 from unity. A treatment of this aspect of the Debye–Hückel theory is beyond the scope of this book, and we shall merely state the result. The osmotic coefficient g_1 at 0°C for dilute solutions of a single strong electrolyte in water is given[1] by

$$g_1 = 1 - 0.376\sigma \left|z_+ z_-\right| I^{1/2} \tag{10}$$

where z_+ is the valence of the positive ion, z_- is the valence of the negative ion, and I is the ionic strength:

$$I \equiv \tfrac{1}{2} \sum_i m_i z_i^2 \tag{11}$$

the sum being taken over all ionic species present. The quantity σ is a function of κa where, for aqueous solutions at 0°C, κa is given by MacDougall[1] as

$$\kappa a = 0.324 \times 10^8 \, a I^{1/2}$$

when the effective ionic diameter a is expressed in centimeters. Values of σ are given for several values of κa in Table 1.

TABLE 1 σ **values from Ref. 1**

κa	0.20	0.25	0.30	0.35	0.40	0.45	0.50	0.55
σ	0.7588	0.7129	0.6712	0.6325	0.5988	0.5673	0.5376	0.5108

For a small ion, a is approximately 3×10^{-8} cm and we can take $\kappa a \cong I^{1/2}$. For a uni-univalent electrolyte such as HCl, this becomes simply

$$\kappa a \cong m^{1/2} \qquad (12)$$

Therefore Eq. (10) can be written as

$$g_1 \cong 1 - 0.38\sigma m^{1/2} \qquad (13)$$

For a known value of m, κa for a uni-univalent electrolyte is given by Eq. (12) and one can find the appropriate value of σ by interpolating in Table 1. Use of this value in Eq. (13) allows one to calculate g_1. It should be emphasized that Eq. (13) is an approximation based on the Debye–Hückel theory and thus valid only in dilute solutions.[2]

EXPERIMENTAL

The apparatus used in this experiment is shown in Fig. 1. The thermometer is either a special cryoscopic mercury thermometer of appropriate range, with graduations every 0.01 or 0.02°C, a resistance thermometer with a resolution of ± 0.01°C, or a calibrated thermistor. In this experiment an aqueous solution of a weak or strong acid is mixed with crushed ice until equilibrium is attained. The temperature is recorded, and two or more aliquots of the liquid phase are withdrawn for titration to determine the equilibrium nominal concentration m_0. The ice to be used should preferably be distilled-water ice.

The most difficult part of the experimental technique is the achievement of thorough mixing. This difficulty is aggravated by the fact that water has a maximum density near 4°C, and solution at that temperature tends to settle to the bottom of the Dewar flask while colder solution tends to float near the top with the ice. The stirring must therefore be *vigorous and prolonged*. A good technique is to work the stirrer frequently above the mass of ice and with a vigorous downward thrust propel the ice all the way to the bottom of the flask. *It must not be assumed that equilibrium has been obtained until the temperature shown by the thermometer has become quite stationary and does not change when the stirring is stopped or when the manner or vigor of stirring is changed.*

Procedure. The solutions to be studied are the strong electrolyte HCl and weak electrolyte monochloroacetic acid, each at two concentrations—roughly 0.25 and 0.125m. About 150 mL of each solution will be needed, and the 0.125m solutions required should be prepared by diluting the 0.25m stock solutions with distilled water. Place each solution in a clean, glass-stoppered flask packed in crushed ice. A flask of distilled water should also be packed in ice.

Wash about 300 mL of crushed, distilled-water ice with several small amounts of the chilled distilled water, then fill the Dewar flask about one-third full with this washed ice. Add about 100 mL of the chilled distilled water to the Dewar, assemble the apparatus, and stir the mixture well to achieve equilibrium. Record the temperature T_0. If a cryoscopic mercury thermometer is being used, tap it gently before reading and estimate to tenths of the smallest division. It should be realized that a cryoscopic thermometer may not read

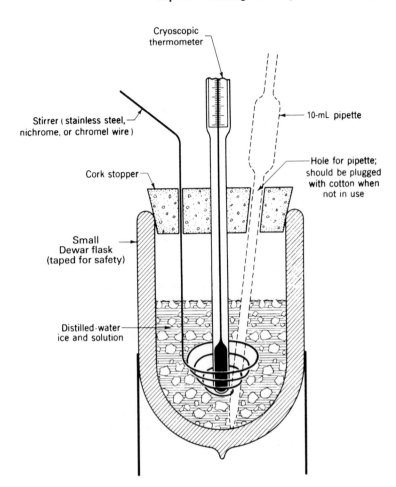

Cryoscopic thermometer

Stirrer (stainless steel, nichrome, or chromel wire)

Cork stopper

10-mL pipette

Hole for pipette; should be plugged with cotton when not in use

Small Dewar flask (taped for safety)

Distilled-water ice and solution

FIGURE 1

Apparatus for measuring the freezing point of aqueous solutions. The cryoscopic mercury thermometer can be replaced by a resistance thermometer or a calibrated thermistor.

exactly 0°C in this ice–water bath and may deviate by as much as a ±0.1°C without coming under suspicion of being defective. The important function of a cryoscopic thermometer is to measure temperature *differences*. For this experiment, absolute accuracy in temperature values is not important, but high resolution (i.e., precision in ΔT values) is important.

Pour off water and replace it by 100 mL of chilled 0.25m HCl solution. After stirring to achieve equilibrium, as described previously, read the temperature and withdraw a 10-mL aliquot with a pipette. Introduce the pipette quickly, while blowing a gentle stream of air through it to prevent solution from entering it until the tip touches the bottom of the flask. This will prevent small particles of ice from being drawn into the pipette. Alternatively, attach to the tip of the pipette a filter consisting of a short length of rubber tubing containing a wad of cotton or glass wool. Discharge this aliquot into a clean weighing bottle, warm to room temperature, and weigh accurately. Then transfer it quantitatively to a flask and titrate with 0.1 M NaOH to a methyl red or phenolphthalein end point.†
Resume stirring vigorously for about 5 min, then take another temperature reading and

†This NaOH solution should be prepared from carbonate-free NaOH as described in Chapter XX. The solution is standardized by titrating, with phenolphthalein as an indicator, a weighed quantity of dry potassium acid phthalate, $KH(C_8H_4O_4)$.

withdraw a second aliquot. If these results are not consistent with each other, take a third aliquot after further stirring.†

Pour off the solution and repeat the experiment with 0.125m HCl. Be sure that an adequate amount of ice is present. If time permits, a run should be made with 0.0625m HCl also.

Carry out similar runs with the solutions of monochloroacetic acid, preferably repeating the measurement of T_0.

CALCULATIONS

Calculate the equilibrium molality m (in mol/kg of water) for each aliquot. If the results for the two aliquots from a given run are consistent, the average values of m and ΔT_f may be used in further calculations. For monochloroacetic acid, a weak electrolyte, calculate the effective total molality m' from Eq. (10-20) using the appropriate constants in Table 10-1. Then calculate α and K_c for each of the two concentrations studied.

For hydrochloric acid, a strong electrolyte, calculate an experimental value of g_1 with Eq. (9) for each of the concentrations studied. In addition, use Eq. (13) to obtain a value of the osmotic coefficient g_1 based on the Debye–Hückel theory for each concentration. Compare these experimental and theoretical values.

DISCUSSION

The colligative behavior of strong electrolytes is sometimes expressed in terms of the Van't Hoff factor $i \equiv m_{app}/m$, where m_{app} is the "apparent" total molality as deduced from any colligative property when the solution is treated as ideal. Using the expressions for freezing-point depression, show the relation between i and g_1.

What additional data would be required in order to compare the values of K_c obtained for monochloroacetic acid at two different temperatures? Can you predict the direction of the change in K_c with T—that is, the sign of dK_c/dT?

SAFETY ISSUES

Use a pipetting bulb; do not pipette by mouth.

APPARATUS

Dewar flask (short, wide mouth, 1 pt); cork stopper with three holes; coil-type stirrer; stopwatch; two 10-mL weighing bottles; 50-mL burette; burette clamp and stand; two 100-mL volumetric flasks; one 10-, one 25-, and one 50-mL pipette; pipetting bulb; battery jar; wash bottle; precision cryoscopic thermometer and magnifying thermometer reader, or a digital resistance thermometer with ±0.01°C resolution, or a calibrated thermistor.

Distilled- or deionized-water ice (500 g); 0.1 M sodium hydroxide solution (500 mL); 0.25m HCl solution (400 mL) and 0.25m monochloroacetic acid solution (400 mL); phenolphthalein indicator; stopcock grease.

†The temperatures and the concentrations of the two aliquots may differ slightly, owing to some melting of ice, but the differences should be consistent. If they are not consistent, at least one of the aliquots was presumably not withdrawn at equilibrium.

REFERENCES

1. F. H. MacDougall, *Thermodynamics and Chemistry,* 3d ed., pp. 296–302, Wiley, New York (1939).

2. P. W. Atkins and J. de Paula, *Physical Chemistry,* 8th ed., pp. 163–165, Freeman, New York (2006).

GENERAL READING

J. B. Ott and J. P. Goates, "Temperature Measurement with Application to Phase Equilibria Studies," in B. W. Rossiter and R. C. Baetzold (eds.), *Physical Methods of Chemistry,* 2d ed., Vol. VI, chap. 7, Wiley-Interscience, New York (1992).

EXPERIMENT 12

Chemical Equilibrium in Solution

In this experiment a typical homogeneous equilibrium in aqueous solution is investigated and the validity of the law of mass action is demonstrated. The equilibrium to be studied is that which arises when iodine is dissolved in aqueous KI solutions:

$$I_2 + I^- = I_3^- \tag{1}$$

In determining the concentration of the I_2 species present at equilibrium in the aqueous solution, use is made of a heterogeneous equilibrium: the distribution of molecular iodine I_2 between two immiscible solvents, water and carbon tetrachloride.

THEORY

Homogeneous Equilibrium. For the change in state in aqueous solution given in (1), the thermodynamic equilibrium constant K_a, which must be expressed in terms of the activities of the chemical species involved, is given to a good approximation by the analogous expression involving concentrations:

$$K_a = \frac{a_{I_3^-}}{a_{I_2} a_{I^-}} = \frac{(I_3^-)}{(I_2)(I^-)} \frac{\gamma_{I_3^-}}{\gamma_{I_2}\gamma_{I^-}} \cong \frac{(I_3^-)}{(I_2)(I^-)} = K_c \tag{2}$$

where the a's are activities, the γ's are activity coefficients, and K_c is the equilibrium constant in terms of concentration. The parentheses denote dimensionless quantities equal in magnitude to the concentrations of the species in mol L^{-1}; i.e., $(X) = c_X/c_0$, where c_X is the molar concentration of species X and $c_0 = 1\ M$. The approximation $K_a = K_c$ follows from the fact that $\gamma_{I^-} = \gamma_{I_3^-}$ and $\gamma_{I_2} = 1$, as we show below.

A commonly used approximate form of the Debye–Hückel theory for the activity coefficients of ionic species at 25°C is[1]

$$\log \gamma_i = \frac{-0.509 z_i^2 \sqrt{I}}{1 + \sqrt{I}} \tag{3}$$

Since both I^- and I_3^- have the same charge z and are influenced by the same ionic strength I, the activity coefficients γ_{I^-} and $\gamma_{I_3^-}$ are equal within the accuracy of Eq. (3). It should be pointed out however that Eq. (3) does not take account of variations in the size and shape of ions. For the I_2 species, it will be observed that the I_2 concentration in aqueous solution is very small. Thus γ_{I_2} is approximately unity, since the activity coefficient of a dilute neutral solute species does not deviate greatly from its limiting value at infinite dilution. Therefore the approximation given in Eq. (2) should be good to within 1 percent or better in moderately dilute solutions, and K_c at any specified temperature should be a constant independent of the concentrations of the individual species.

If, then, one starts with a solution of KI whose concentration C is known accurately and dissolves iodine in it, the iodine will be present at equilibrium partly as neutral I_2 molecules and partly as I_3^- ions (formed when I_2 reacts with some of the I^- ions initially present). The total iodine concentration, $T = (I_2) + (I_3^-)$, can be found by titration with a thiosulfate solution. If there is an independent way to measure (I_2) at equilibrium, then (I^-) and (I_3^-) can be obtained from

$$(I_3^-) = T - (I_2) \tag{4}$$

and

$$(I^-) = C - (I_3^-) \tag{5}$$

We can measure (I_2) at equilibrium by taking advantage of the fact that CCl_4, which is immiscible with aqueous solutions, dissolves molecular I_2 but not any of the ionic species involved.

Heterogeneous Equilibrium. We are concerned here with the distribution of molecular iodine I_2 as the solute between two immiscible liquid phases, aqueous solution and CCl_4. At equilibrium the concentrations of I_2 in the two phases, $(I_2)_w$ and $(I_2)_{CCl_4}$, are related by a distribution constant k:

$$k = \frac{(I_2)_w}{(I_2)_{CCl_4}} \tag{6}$$

This distribution law applies only to the distribution of a definite chemical species, as does Henry's law. The distribution constant k is not a true thermodynamic equilibrium constant, since it involves concentrations rather than activities. Thus it may vary slightly with the concentration of the solute (particularly because of the relatively high concentration of I_2 in the CCl_4 phase); it is therefore advantageous to determine k at a number of concentrations. It can be determined directly by titration of both phases with standard thiosulfate solution when I_2 is distributed between CCl_4 and pure water. Once k is known, (I_2) in an aqueous phase containing I_3^- can be obtained by means of a titration of the I_2 in a CCl_4 layer that has been equilibrated with this phase. The use of a distribution constant in this manner depends upon the assumption that its value is unaffected by the presence of ions in the aqueous phase.

EXPERIMENTAL

Distribution Ratio. Measure the distribution constant $k = (I_2)_w/(I_2)_{CCl_4}$, using CCl_4 and pure H_2O as solvents and I_2 as solute. Distilled water (200 mL) and solutions of I_2 in CCl_4 (50 mL) should be put into 500-mL glass-stoppered Erlenmeyer flasks and equilibrated at 25°C. The quantities to use are given in Table 1 (runs 1 to 3); three sets of conditions are used in order to check the variation of the distribution constant with concentration. The flasks containing the solutions should be shaken vigorously for 5 min and clamped in a thermostat bath. After 10 min of thermal equilibrium, remove the flasks one at a time, wrap them in a dry towel, shake them vigorously for 3 to 5 min, and then return them to the thermostat bath. Repeat this procedure for at least an hour. Before removing

TABLE 1[a]

Run No.	CCl$_4$ (50 mL), Molarity I$_2$	Aqueous Layer (200 mL), Molarity KI	CCl$_4$ layer (use 20-mL pipette)		Aqueous layer (use 50-mL pipette)	
			Burette Size, mL	Molarity S$_2$O$_3^{2-}$	Burette Size, mL	Molarity S$_2$O$_3^{2-}$
1	0.080	0.0	50	0.1	10	0.01
2	0.040	0.0	50	0.1	10	0.01
3	0.020	0.0	10	0.1	10	0.01
4	0.080	0.15	10	0.1	50	0.1
5	0.040	0.15	10	0.1	10	0.1
6	0.080	0.03	50	0.1	10	0.1

[a] The table shows the nominal *initial* concentrations and volumes of the aqueous KI solutions and of the solutions of I$_2$ in CCl$_4$ to be used. Also given are the nominal concentrations of the thiosulfate solutions to be used for titrating and the sizes of burettes and pipettes to be used. The *actual*, precise concentrations of the solutions used should be read from the labels on the respective bottles.

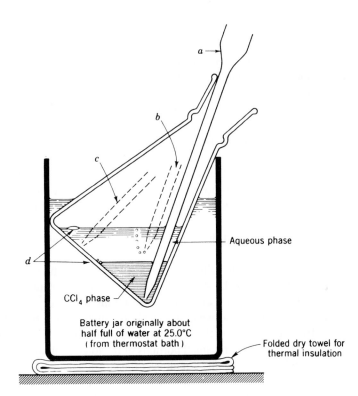

CCl$_4$ phase

Aqueous phase

Battery jar originally about half full of water at 25.0°C (from thermostat bath)

Folded dry towel for thermal insulation

FIGURE 1
Arrangement for withdrawing samples. Use a rubber bulb for drawing up liquid in the pipette.
(*a*) Position of pipette for withdrawing sample of CCl$_4$ phase. Tip should touch bottom at deepest point and should not be moved around. (*b*) On inserting pipette, blow a *small* stream of air to keep aqueous phase from entering tip. (*c*) Position of pipette for withdrawing aqueous sample. (*d*) On inserting pipette, avoid contact with drops of CCl$_4$ phase (spheres on bottom or lenses floating on top).

samples for analysis, let the flasks remain in the bath for 10 min after the last shaking to allow the liquid layers to separate completely. After equilibration is complete, remove one flask at a time from the thermostat bath and place in a battery jar containing water at 25°C as shown in Fig. 1. A sample of the aqueous layer and a sample of the CCl$_4$ layer are removed with pipettes, **using a pipetting bulb** (see Appendix C for safety precautions for potentially hazardous chemicals). Stopper the flask and return it to the bath for an additional 30 min of equilibration with shaking as described above, then take a second sample of each phase for titration. Additional information is given in Table 1.

The purpose of taking a second sample after further equilibration is to verify that equilibrium has been achieved. If the two titrations are in substantial agreement, the average value can be used. If the results of the titrations indicate that equilibrium had not been reached when the first samples were taken, the results of the second titrations should be used although proof of equilibrium is in this case lacking.

In removing the samples for titration, it is important to avoid contamination of the sample by drops of the other phase, especially when the other phase is very much more concentrated, as is the CCl_4 layer in this case. When pipetting samples of the CCl_4 solutions, blow a slow stream of air through the pipette while it is being introduced into the solution in order to minimize contamination by the aqueous layer. It is necessary to use a rubber bulb with the pipette for these solutions; **do not pipette by mouth.** Between samples, rinse the pipette well with acetone and dry it before taking the next sample.

Transfer each sample from the pipette to a 250-mL Erlenmeyer flask containing 10 mL of 0.1 M KI and titrate with the appropriate thiosulfate solution (see Table 1). The iodide is added to reduce loss of I_2 from the aqueous solution by evaporation during the titration, by forming the nonvolatile I_3^-. The reaction taking place during the titration is

$$2S_2O_3^{2-} + I_3^- \rightarrow S_4O_6^{2-} + 3I^-$$

Near the end point, as indicated by a very light yellow color of aqueous iodine solution, add Thyodene† or 1 mL of soluble starch solution; this acts as an indicator by adsorbing iodine and giving a deep blue color. At the end point, this blue color disappears sharply. If the starch is added too soon, the color may redevelop after an apparent end point owing to diffusion of iodine from the interior of the colloidal starch particles; the titration should be continued until no blue color reappears. However, the blue color may reappear on prolonged standing because of air oxidation of the iodide ion, and this effect should be disregarded. In the two-phase titration, shake the flask vigorously after adding each portion of thiosulfate solution.

If desired, an excellent end point for the two-phase titration can be taken as the disappearance of the reddish-violet iodine color from the CCl_4 layer. This may even be done in the titration of the aqueous sample by adding 1 or 2 mL of pure CCl_4. In this case, do not add the starch indicator.

It is recommended that a practice titration be performed with one of the stock solutions of I_2 in CCl_4.

Since a thiosulfate solution is susceptible to attack by sulfur-metabolizing bacteria, it may be wise to check the standardization of the stock solution with a standard solution of KIO_3.[2] Place 1.3 to 1.4 g of KIO_3 in a weighing bottle, dry it for several hours at 110°C, cool in a desiccator, and weigh it exactly on an analytical balance. Transfer this salt to a clean 100-mL volumetric flask and make up to the mark with distilled water. Rinse a clean 5-mL pipette with several small portions of the iodate solution and then carefully deliver a 5-mL sample into a 250-mL Erlenmeyer flask. Add about 20 mL of distilled water, 5 mL of a 0.5 g/ml KI solution, and 10 mL of 1 M HCl. Titrate at once with the thiosulfate solution until the reddish color turns orange and then yellow and becomes pale. At this point add about 0.5 g of Thyodene indicator, mix well, and titrate until the blue color disappears.

Equilibrium Constant. To determine the concentrations of I_2 and I_3^- in equilibrium in aqueous solutions we equilibrate aqueous KI solutions (200 mL) with solutions of

---

†Thyodene is a proprietary starch derivative that gives very sharp and reliable end points. It may be added directly as a powder and is therefore much more convenient for occasional use than the traditional "soluble starch" suspension. Should Thyodene be unavailable, a starch indicator solution may be prepared as follows. Mull 1 g of soluble starch powder in a mortar with several milliliters of boiling water. Pour the paste into 200 mL of boiling water, boil for 2 to 3 min, and allow to cool. When cool, the indicator is ready to use (about 2 mL/titration).

I_2 in CCl_4 (50 mL) as shown in Table 1 (runs 4 to 6). The equilibration and titration procedures are identical with those described above; additional information is given in Table 1. With the distribution constant determined above and the initial I^- concentration, the equilibrium concentrations of I_2, I_3^-, and I^- can be calculated.

Adequate time for equilibration is of great importance; the flasks should be placed in the thermostat bath as soon as possible. The temperature of the thermostat bath should be checked several times throughout the equilibration.

CALCULATIONS

Calculate the distribution constant k from the results of runs 1 to 3. Plot k versus the iodine concentration in the CCl_4 solution, $(I_2)_{CCl_4}$. If k is not constant, discuss its variation with concentration.

Using Eqs. (4), (5), and (6) and the results of runs 4 to 6, calculate the equilibrium concentrations of I_2, I_3^-, and I^- in each of the aqueous solutions. In each case use the appropriate value of k as read from the smooth curve of k versus $(I_2)_{CCl_4}$. Calculate the equilibrium constant K_c for each run. If any variation of K_c with concentration is found, do you regard it as experimentally significant?

SAFETY ISSUES

Solid iodine is corrosive to the skin; handle it with care if it is necessary to make up your own stock solutions. Iodine solutions can cause bad stains, but if handled properly, they are not a significant health hazard. The important safety precautions are: (1) *Do not pipette by mouth,* as already stressed in the procedure above; (2) *store CCl_4 liquid and solutions of I_2 in CCl_4 in a fume hood* and carry out all transfers from the stock bottle to a stoppered flask in this hood; and (3) *dispose of waste materials properly,* i.e., place used CCl_4 in a storage bottle kept in the hood.

Like many other chlorinated hydrocarbons, CCl_4 is a toxic substance that can cause liver and/or kidney damage if ingested or inhaled.[3] In many cases of CCl_4 poisoning, the victims were chronic alcoholics or heavy drinkers, so there appears to be a synergistic effect of alcohol and CCl_4. The prescribed short-term exposure limit is 20 ppm CCl_4 vapor in the air (~125 mg m^{-3}).[3] In order to reach this level in a typical laboratory room of 750 m^3 volume, 100 g of liquid CCl_4 would have to evaporate into the atmosphere. Since CCl_4 liquid has a moderately low vapor pressure and as used in the present procedure lies below an aqueous layer in which it is not soluble, there should be no serious hazard in carrying out this experiment.

If desired, one could replace CCl_4 with hexane (a mixture of several isomers), since hexane is not very toxic and has an exposure limit of 100 ppm vapor in the air. However, hexane is more volatile, and the distribution constant for I_2 is different in different hexane isomers,[4] leading to problems with the data analysis. Perhaps a better alternative is p-xylene.[5] Although less toxic this solvent is however more flammable than CCl_4, with a flash point of 17°C. Finally, we note that it is also possible to determine I_2 concentrations spectrophotometrically, which eliminates the titrations.

APPARATUS

Three to six 500-mL glass-stoppered Erlenmeyer flasks; 1-qt battery jar or 1500-mL beaker; three pipettes—20, 50, 100 (or 200) mL; one 10- and one 50-mL burette; six

250-mL Erlenmeyer flasks; 10-mL graduated cylinder; three 250-mL beakers; burette clamp and stand; large rubber bulb for pipetting; wash bottle.

Constant-temperature bath set at 25°C; bath clamps for holding 500-mL Erlenmeyer flasks; pure CCl_4; acetone for rinsing; large carboy for waste liquids. Solutions: 0.08 M solution of I_2 in CCl_4 (300 mL); 0.04 M I_2 in CCl_4 (250 mL); 0.02 M I_2 in CCl_4 (100 mL); 0.15 M KI solution (800 mL); 0.03 M KI solution (400 mL); 0.1 M $Na_2S_2O_3$ solution (500 mL); 0.01 M $Na_2S_2O_3$ solution (100 mL); approximately 0.1 M KI solution (500 mL); Thyodene or 0.2 percent soluble starch solution, containing a trace of HgI_2 as preservative (50 mL). See the *warning* given above about the storage and handling of CCl_4.

REFERENCES

1. I. N. Levine, *Physical Chemistry,* 6th ed., sec. 10.6, McGraw-Hill, New York (2009).

2. D. A. Skoog, D. M. West, and F. J. Holler, *Fundamentals of Analytical Chemistry,* 7th ed., Harcourt College, Fort Worth, TX (1996).

3. *Documentations of the Threshold Limit Values,* Amer. Conf. of Governmental Industrial Hygienists, Cincinnati, OH (issued annually).

4. M. N. Ackermann, *J. Chem. Educ.* **55,** 795 (1978).

5. S. C. Petrouic and G. M. Bodner, *J. Chem. Educ.* **68,** 509 (1991).

GENERAL READING

K. S. Pitzer, *Thermodynamics,* 3d ed., McGraw-Hill, New York (1995).

VIII

Phase Behavior

EXPERIMENTS

13. Vapor Pressure of a Pure Liquid
14. Binary Liquid–Vapor Phase Diagram
15. Ordering in Nematic Liquid Crystals
16. Liquid–Vapor Coexistence Curve and the Critical Point

EXPERIMENT 13

Vapor Pressure of a Pure Liquid

When a pure liquid is placed in an evacuated bulb, molecules will leave the liquid phase and enter the gas phase until the pressure of the vapor in the bulb reaches a definite value, which is determined by the nature of the liquid and its temperature. This pressure is called the vapor pressure of the liquid at a given temperature. The equilibrium vapor pressure is independent of the quantity of liquid and vapor present, as long as both phases exist in equilibrium with each other at the specified temperature. As the temperature is increased, the vapor pressure also increases up to the critical point, at which the two-phase system becomes a homogeneous, one-phase fluid.

If the pressure above the liquid is maintained at a fixed value (say by having the bulb containing the liquid open to the atmosphere), then the liquid may be heated up to a temperature at which the vapor pressure is equal to the external pressure. At this point vaporization will occur by the formation of bubbles in the interior of the liquid as well as at the surface; this is the boiling point of the liquid at the specified external pressure. Clearly the temperature of the boiling point is a function of the external pressure; in fact, about a given T, p point, the variation of the boiling point with external pressure is the inverse of the variation of the vapor pressure with temperature.

In this experiment the variation of vapor pressure with temperature will be measured and used to determine the molar heat of vaporization.

THEORY

We are concerned here with the equilibrium between a pure liquid and its vapor:

$$X(l) = X(g) \qquad (p, T) \tag{1}$$

It can be shown thermodynamically[1] that a definite relationship exists between the values of p and T at equilibrium, as given by

$$\frac{dp}{dT} = \frac{\Delta S}{\Delta V} \tag{2}$$

In Eq. (2) dp and dT refer to infinitesimal changes in p and T for an equilibrium system composed of a pure substance with both phases always present; ΔS and ΔV refer to the change in S and V when one phase transforms to the other at constant p and T. Since the change in state (1) is isothermal and ΔG is zero, ΔS may be replaced by $\Delta H/T$. The result is

$$\frac{dp}{dT} = \frac{\Delta H}{T\Delta V} \tag{3}$$

Equation (2) or (3) is known as the *Clapeyron equation.* It is an exact expression that may be applied to phase equilibria of all kinds, although it has been presented here in terms of the one-component liquid–vapor case. Since the heat of vaporization ΔH_v is positive and ΔV is positive for vaporization, it is seen immediately that the vapor pressure must increase with increasing temperature.

For the case of vapor–liquid equilibria in the range of vapor pressures less than 1 atm, one may assume that the molar volume of the liquid $\tilde{V}_l$ is negligible in comparison with that of the gas $\tilde{V}_g$, so that $\Delta \tilde{V} = \tilde{V}_g$. This assumption is very good in the low-pressure region, since $\tilde{V}_l$ is usually only a few tenths of a percent of $\tilde{V}_g$. Thus we obtain

$$\frac{dp}{dT} = \frac{\Delta \tilde{H}_v}{T\tilde{V}_g} \tag{4}$$

Since $d \ln p = dp/p$ and $d(1/T) = -dT/T^2$, we can rewrite Eq. (4) in the form

$$\frac{d \ln p}{d(1/T)} = -\frac{\Delta \tilde{H}_v}{R} \frac{RT}{p\tilde{V}_g} = -\frac{\Delta \tilde{H}_v}{RZ} \tag{5}$$

where we have introduced the *compressibility factor* Z for the vapor:

$$Z = \frac{p\tilde{V}_g}{RT} \tag{6}$$

Equation (5) is a convenient form of the Clapeyron equation. We can see that, *if* the vapor were a perfect gas ($Z \equiv 1$) and $\Delta \tilde{H}_v$ were independent of temperature, then a plot of $\ln p$ versus $1/T$ would be a straight line, the slope of which would determine $\Delta \tilde{H}_v$. Indeed, for many liquids, $\ln p$ is almost a linear function of $1/T$, which implies at least that $\Delta \tilde{H}_v/Z$ is almost constant.

Let us now consider the question of gas imperfections, i.e., the behavior of Z as a function of temperature for the saturated vapor. It is difficult to carry out p–V–T measurements on gases close to condensation and such data are scarce, but data are available for water[2] and theoretical extrapolations[3] have been made for the vapor of "normal" liquids based on data obtained at higher temperatures. Figure 1 shows the variation of the compressibility factor Z for a saturated vapor as a function of temperature in the

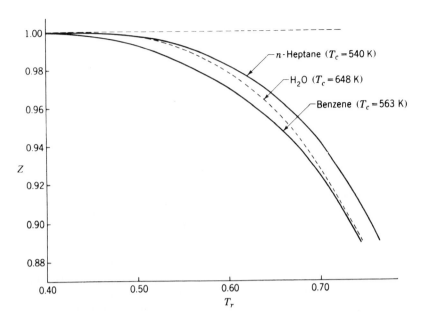

FIGURE 1
The compressibility factor
Z of saturated vapor as
a function of reduced
temperature T_r for water,
benzene, and n-heptane.

case of water and two normal liquids, benzene and n-heptane. For the temperature axis, a "reduced" temperature T_r is used; $T_r = T/T_c$, where T_c is the critical temperature. This has the effect of almost superimposing the curves of many different substances; indeed, by the law of corresponding states, such curves would be exactly superimposed. In general it is clear that Z decreases as the temperature increases. Water, owing to its high critical temperature, is a reasonably ideal gas even at 100°C, where Z equals 0.986. But n-heptane at its 1-atm boiling point of 98°C has a value of Z equal to 0.95 and is relatively nonideal. For many substances, sizable gas imperfections are present even at pressures below 1 atm.

Next we must consider the variation of $\Delta \widetilde{H}_v$ with temperature. For a change in state such as Eq. (1),

$$\Delta H_{T_2} = \Delta H_{T_1} + \int_{T_1}^{T_2} \Delta C_p \, dT + \int_{p_1}^{p_2} \left(\frac{\partial \Delta H}{\partial p} \right)_T dp \tag{7}$$

Since $(\partial H_g/\partial p)_T$ is zero for a perfect gas and very small for most real gases and $(\partial H_l/\partial p)_T$ is always very small for liquids, it is possible to neglect the final term and approximate Eq. (7) by

$$\Delta H_{T_2} \simeq \Delta H_{T_1} + \overline{\Delta C_p}(T_2 - T_1) \tag{8}$$

where $\overline{\Delta C_p}$ is the average value over the temperature interval. For $\Delta \widetilde{H}_v$ to be independent of temperature, the average value of $\overline{\Delta C_p}$ must be very close to zero, which is generally not true. Heat capacities for water, benzene, and n-heptane are given in Table 1 as typical examples.[4,5] For n-heptane, the specific heat of both gas and liquid changes rapidly with temperature; use of average values will give only an order-of-magnitude result. In general the value of $\Delta \widetilde{H}_v$ will decrease as the temperature increases.

Since both $\Delta \widetilde{H}_v$ and Z decrease with increasing temperature, it is possible to see why $\Delta \widetilde{H}_v/Z$ might be almost constant, yielding a nearly linear plot of $\ln p$ versus $1/T$.

TABLE 1

Compound	Temp. Range, °C	Average values, J K^{-1} mol^{-1}		
		$\widetilde{C}_p(g)$	$\widetilde{C}_p(l)$	$\Delta\widetilde{C}_p$
Water	25–100	33.5	75	−41.5
Benzene	25–80	92	146	−54
n-Heptane	25–100	~197	~243	−46

METHODS

There are several experimental methods of measuring the vapor pressure as a function of temperature.[6] In the *gas-saturation method,* a known volume of an inert gas is bubbled slowly through the liquid, which is kept at a constant temperature in a thermostat. The vapor pressure is calculated from a determination of the amount of vapor contained in the outcoming gas or from the loss in weight of the liquid. A common *static method* makes use of an *isoteniscope,* a bulb with a short U tube attached. The liquid is placed in the bulb, and some liquid is placed in the U tube. When the liquid is boiled under reduced pressure, all air is swept out of the bulb. The isoteniscope is then placed in a thermostat. At a given temperature the external pressure is adjusted so that both arms of the U tube are at the same height. At this setting, the external pressure, which is equal to the pressure of the vapor in the isoteniscope, is measured with a pressure gauge. A common *dynamic method* is one in which the variation of the boiling point with external applied pressure is measured. The total pressure above the liquid can be varied and maintained at a given value by use of a large-volume ballast bulb; this pressure is then measured with a pressure gauge. The liquid to be studied is heated until boiling occurs, and the temperature of the refluxing vapor is measured in order to avoid any effects of superheating. The experimental procedure for both the isoteniscope method and a boiling-point method is given below.

EXPERIMENTAL

1 BOILING-POINT METHOD

The apparatus should be assembled as shown in Fig. 2. Use pressure tubing to connect the condenser and the pressure gauge to the ballast bulb. Fill the flask about one-third full with the liquid to be studied. A few carborundum boiling chips should be added to reduce "bumping." If a mercury thermometer is used, be sure that the thermometer scale is visible over a range of at least 50°C below the boiling point at 1 atm. Also, make sure that the thermometer bulb (or any other thermometer sensor) is positioned carefully so that the temperature of coexisting vapor and liquid is measured. Heating should be accomplished with an electrical heating mantle. Turn on the circulating water to the condenser before heating the liquid, and turn it off at the end of the experiment.

To make a reading, adjust the heating so as to attain steady boiling of the liquid, but avoid heating too strongly. When conditions appear to be steady (i.e., when pressure gauge and thermometer readings appear to be as constant as they can be maintained), record p and T values as nearly simultaneously as possible. The thermometer should be read to the nearest 0.1°C, and vapor should be condensing on and dripping from the thermometer to ensure that the equilibrium temperature is obtained. In reading the pressure gauge, estimate the pressure to the nearest 0.1 Torr. Record the ambient air temperature at the manometer several times during the run.

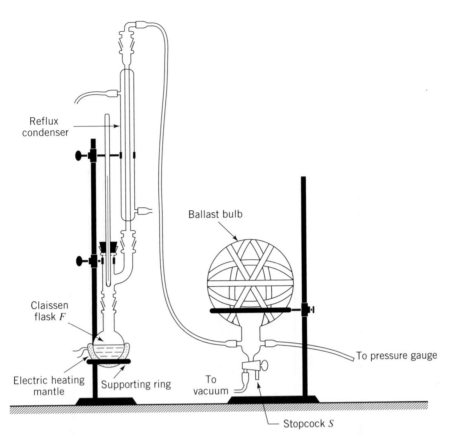

To change the pressure in the system between measurements, first remove the heating mantle. After a short time, admit some air or remove some air by opening stopcock S for a few seconds. Be especially careful in removing air from the ballast bulb to avoid strong bumping of the hot liquid. Check the pressure. Repeat as often as necessary to attain the desired pressure, and then restore the heating mantle to its position under the flask.

Take readings at approximately the following pressures:

Pressure descending: 760, 600, 450, 350, 260, 200, 160, 130, 100, 80 Torr

Pressure ascending: 90, 110, 140, 180, 230, 300, 400, 520, 670 Torr

2   ISOTENISCOPE METHOD

In this technique much of the equipment shown in Fig. 2 is used, but the distilling flask and reflux condenser are replaced by an isoteniscope mounted in a glass thermostat as shown in Fig. 3. The water in the bath should be stirred vigorously (or circulated) to ensure thermal equilibrium. Place the liquid to be studied in the isoteniscope so that the bulb is about two-thirds full and there is no liquid in the U tube. Then place the isoteniscope in the thermostat (which should be at room temperature) and connect it to the ballast bulb and pressure gauge (which are assembled and connected as in Fig. 2).

Sweep the air out of the bulb by **cautiously** and slowly reducing the pressure in the ballast bulb until the liquid boils very gently. Continue pumping for about 4 min, but be careful to avoid evaporating too much of the liquid. Then tilt the isoteniscope so that some liquid from the bulb is transferred into the U tube. Carefully admit air to the ballast bulb through stopcock S until the levels of the liquid to both arms of the U tube are the same.

FIGURE 3

Isoteniscope. (*a*) Schematic diagram of the apparatus, which must be connected to the ballast bulb and manometer shown in Fig. 2. A temperature controller is also needed, or a commercial circulating-type temperature control bath can be used. (*b*) An alternate design for the isoteniscope. (An even more elaborate type of isoteniscope is described by Arm, Daeniker, and Schaller.[7])

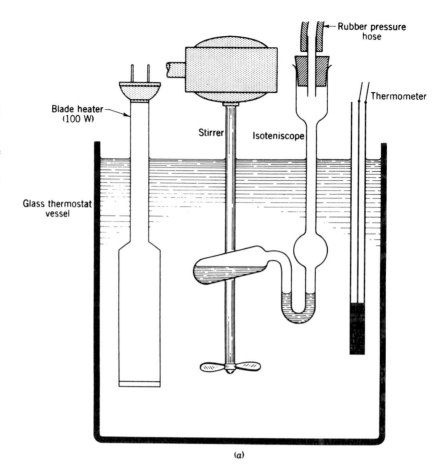

(a)

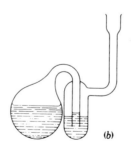

(b)

Read and record the temperature and the pressure. To establish that all the air has been removed from the isoteniscope, **cautiously** reduce the pressure in the ballast bulb until a few bubbles are observed passing through the U-tube liquid. Then determine the equilibrium pressure reading. Repeat the above procedure, if necessary, until successive vapor-pressure readings are in good agreement.

Once the air is removed and a good pressure reading at room temperature is obtained, heat the thermostat bath to a new temperature about 5°C above room temperature. Keep the liquid levels in the U tube approximately equal at all times. When the bath temperature is steady at its new value, adjust the pressure in the ballast bulb until the levels in the U tube are equal and record both temperature and pressure.

Take readings at approximately 5°C intervals until the bath is at about 75°C, and then take readings at decreasing temperatures that are between the values obtained on heating. The thermometer shown in Fig. 3 can be a thermocouple or a digital resistance thermometer and must have a resolution of ±0.1°C or better.

CALCULATIONS

Correct all pressure readings to absolute values in Torr if calibration corrections are needed.

Convert all Celsius temperature readings to absolute temperatures T and plot $\ln p$ versus $1/T$. If there is no systematic curvature, draw the best straight line through the points. If there is noticeable curvature, draw a smooth curve through the points and also draw a straight line tangent to the curve at about the midpoint. Determine the slope of the straight line or tangent. From Eq. (5) it follows that this slope is $-\Delta \widetilde{H}_v/RZ$.

In addition to this graphical analysis, carry out a least-squares fit to your p-versus-T data. If visual inspection of your $\ln p$-versus-$1/T$ plot does not indicate any systematic curvature, make a direct linear fit of $\ln p$ as a function of $1/T$. If systematic curvature is observed, make a least-squares fit of $\ln p$ with the empirical power-series form $a + b/T + c/T^2$. Then differentiate the resulting expression to obtain $\Delta \widetilde{H}_v/RZ$.

Report both the graphical and the least-squares values of $\Delta \widetilde{H}_v/Z$. Estimate the value of Z for the saturated vapor at the appropriate temperature from Fig. 1 and calculate $\Delta \widetilde{H}_v$ in J mol^{-1}. Report the value of the heat of vaporization and the vapor pressure for the liquid at the applicable temperature (corresponding to the midpoint of the range studied).

DISCUSSION

Using Fig. 1 and Eq. (8), estimate the variation in $\Delta \widetilde{H}_v/Z$ expected over the range of temperatures studied. Indicate clearly whether $\Delta \widetilde{H}_v/Z$ should increase or decrease with increasing temperature. Does your $\ln p$-versus-$1/T$ plot show a curvature of correct sign?

Evaluate a quantitative uncertainty in your value of $\Delta \widetilde{H}_v/Z$ by the method of "limiting slopes" and compare this with the standard deviation obtained from the least-squares fit to your data. Comment on this uncertainty in relation to the variation with temperature calculated above. Discuss possible sources of systematic errors.

If water or some other compound with a simple molecular structure has been studied, it is possible to combine the entropy of vaporization, $\Delta \widetilde{S}_v = \Delta \widetilde{H}_v/T$, with the third-law calorimetric entropy of the liquid to obtain a thermodynamic value for the entropy of the vapor. The statistical mechanical value of $\widetilde{S}_g$ can be calculated using the known molar mass and the spectroscopic parameters for the rotation and vibration of the gas-phase molecule. A comparison of $\widetilde{S}_g$ (thermodynamic) with $\widetilde{S}_g$ (spectroscopic) provides a test of the validity of the third law of thermodynamics. The case of H_2O is particularly interesting, since ice has a nonzero residual entropy at 0 K due to frozen-in disorder in the proton positions.[8]

The pertinent thermodynamic data needed to carry out the calculation of $\widetilde{S}_g$ (thermodynamic) for water vapor are as follows: $\widetilde{S}_l$(1 bar, 298.15 K) = 66.69 J K^{-1} mol^{-1}, which is the calorimetric value obtained by assuming (erroneously) that the third law is valid for ice;[9] $\widetilde{C}_p(l)$, which is essentially independent of temperature over the range 298–355 K, has the average value 75.36 J K^{-1} mol^{-1}.[5] You may neglect the effect of pressure on the value of the molar entropy of the liquid, an approximation that has very little effect on the calculated thermodynamic value of $\widetilde{S}_g(p_m, T_m)$ for $H_2O(g)$ at the vapor pressure

p_m associated with the temperature T_m at which $\Delta \widetilde{H}_v$ has been determined. To facilitate comparison with the ideal gas statistical entropy discussed below, one can also make the simplifying assumption that corrections for the nonideality of water vapor at p_m, T_m have negligible effect on the value of $\widetilde{S}_g(p_m, T_m)$.

The experimental data needed to calculate the statistical entropy $\widetilde{S}_g$ are the rotational temperatures $\Theta_r = hcB/k$ and the vibrational temperatures $\Theta_v = h\nu/k = hc\widetilde{\nu}/k$, where $B \,(= h/8\pi^2 Ic)$ and $\widetilde{\nu}$ are frequencies in wavenumber units (cm^{-1}), and c is the speed of light in cm s^{-1}. Since the nonlinear H_2O molecule is an asymmetric top, the three principal moments of inertia are all different. The three rotational temperatures are $\Theta_A = 40.1$ K, $\Theta_B = 20.9$ K, and $\Theta_C = 13.4$ K.[10] There are three nondegenerate normal modes of vibration (one bend, one symmetric stretch, and one antisymmetric stretch), and the vibrational temperatures are $\Theta_{v1} = 2290$ K, $\Theta_{v2} = 5160$ K, and $\Theta_{v3} = 5360$ K.[10] The statistical-mechanical entropy of an ideal gas of *nonlinear* molecules is a sum of a translational contribution (the so-called Sackur–Tetrode term), a rotational contribution for three degrees of rotational freedom, and a vibrational contribution for $3N - 6$ vibrational degrees of freedom, where N is the number of atoms in the molecule. These three contributions are given by[10]

$$\widetilde{S}(\text{trans}) = R[1.5 \ln M + 2.5 \ln T - \ln p - 1.15171] \tag{9}$$

where M is the molar mass in grams and p is the pressure in bar,

$$\widetilde{S}(\text{rot}) = R\left[1.5 + \ln \frac{\pi^{1/2}T^{3/2}}{\sigma(\Theta_A\Theta_B\Theta_C)^{1/2}}\right] \tag{10}$$

where σ is the symmetry number[1,10] ($\sigma = 2$ for the case of H_2O), and

$$\widetilde{S}(\text{vib}) = R\left\{\sum_i \left[\frac{\Theta_{vi}}{T}\frac{1}{(e^{\Theta_{vi}/T} - 1)} - \ln(1 - e^{\Theta_{vi}/T})\right]\right\} \tag{11}$$

with $i = 1$ to 3 in the case of H_2O.

Calculate a statistical/spectroscopic value of $\widetilde{S}_g(p_m, T_m)$, compare this with your thermodynamic value, and report the discrepancy. For most substances, these two $\widetilde{S}_g$ values agree and the third law is valid for the solid as T approaches 0 K. However, you should find that $\widetilde{S}_g(\text{spectroscopic}) > \widetilde{S}_g(\text{thermodynamic})$ for H_2O, which means that ice does not have the perfect order at 0 K required by the third law. The explanation for this in terms of the disorder in the proton configurations in ice was first given by Pauling[11] and is well described by Davidson.[8]

SAFETY ISSUES

Method 1—The ballast bulb must be taped to prevent flying glass in the unlikely event of breakage. As always, safety glasses must be worn in the laboratory.

Method 2—None.

APPARATUS

Ballast bulb (5 L); pressure gauge; three long pieces of heavy-wall rubber pressure tubing. If the boiling-point method is used: distillation flask; appropriate thermometer to cover the desired range, with one-hole rubber stopper to allow adjustment of thermometer position; reflux condenser; electrical heating mantle; two long pieces of rubber tubing for circulating water through condenser; two condenser clamps and clamp holders; ring stand; iron ring (and clamp holder if necessary).

If an isoteniscope is used: a glass thermostat (e.g., large battery jar) with mechanical stirrer, electrical blade heater, and temperature controller (alternatively a commercial water bath circulator/heater); thermometer; isoteniscope.

Liquid, such as water, *n*-heptane, cyclohexane, or 2-butanone.

REFERENCES

1. Any standard text, such as R. J. Silbey, R. A. Alberty, and M. G. Bawendi, *Physical Chemistry,* 4th ed., pp. 181, 586, Wiley, New York (2005).

2. J. H. Keenan, *Steam Tables,* 2d ed., Krieger, Melbourne, FL (1992).

3. K. S. Pitzer, D. Z. Lippmann, R. F. Curl, Jr., C. M. Huggins, and D. E. Petersen, *J. Amer. Chem. Soc.* **77,** 3433 (1955).

4. *Selected Values of Properties of Hydrocarbons,* Natl. Bur. Std. Circ. C461, U.S. Government Printing Office, Washington, DC (1947).

5. R. D. Lide (ed.), *CRC Handbook of Chemistry and Physics,* 82nd ed., CRC Press, Boca Raton, FL (2001/2002).

6. G. W. Thomson and D. R. Douslin, "Determination of Pressure and Volume," in A. Weissberger and B. W. Rossiter (eds.), *Techniques of Chemistry: Vol. I, Physical Methods of Chemistry,* part V, chap. 2, Wiley-Interscience, New York (1971).

7. H. Arm, H. Daeniker, and R. Schaller, *Helv. Chim. Acta* **48,** 1772 (1966).

8. N. Davidson, *Statistical Mechanics,* p. 371*ff.* and chap. 11, McGraw-Hill, New York (1962); L. Pauling, *The Nature of the Chemical Bond,* 3d ed., pp. 464–469, Cornell Univ. Press, Ithaca, NY (1960).

9. W. F. Giauque and J. W. Stout, *J. Amer. Chem. Soc.* **58,** 1144 (1936).

10. D. A. McQuarrie, *Statistical Thermodynamics,* reprint ed., chap. 8, University Science Books, Sausalito, CA (1985).

11. L. Pauling, *J. Amer. Chem. Soc.* **57,** 2680 (1935).

GENERAL READING

G. W. Thomson and D. R. Douslin, "Determination of Pressure and Volume," in A. Weissberger and B. W. Rossiter (eds.), *Techniques of Chemistry: Vol. 1, Physical Methods of Chemistry,* part V, chap. 2, Wiley-Interscience, New York (1971).

EXPERIMENT 14
Binary Liquid–Vapor Phase Diagram

This experiment is concerned with the heterogeneous equilibrium between two phases in a system of two components. The particular system to be studied is cyclohexanone–tetrachloroethane at 1 atm pressure. This system exhibits a strong negative deviation from Raoult's law, resulting in the existence of a maximum boiling point.

THEORY

For a system of two components (A and B), we have from the phase rule,[1]

$$F = C - P + 2 = 4 - P \tag{1}$$

where C is the number of *components* (minimum number of chemical constituents necessary to define the composition of every phase in the system at equilibrium), P is the number of *phases* (number of physically differentiable parts of the system at equilibrium), and F is the *variance* or number of *degrees of freedom* (number of intensive variables pertaining to the system that can be independently varied at equilibrium without altering the number or kinds of phases present).

When a single phase is present, the pressure p, the temperature T, and the composition X_B (mole fraction of component B) of that phase can be varied independently; thus a single-phase two-component system at equilibrium is defined, except for its size,† by a point in a three-dimensional plot in which the coordinates are the intensive variables p, T, and X_B (see Fig. 1). When two phases, e.g., liquid L and vapor V, are present at equilibrium, there are four variables, but only two of them can be independently varied. Thus, if p and T are specified, X_{BL} and X_{BV} (the mole fractions of B in L and V) are fixed at their *limiting* values (X_{BL}^0 and X_{BV}^0) for the respective phases at this p and T. The loci of points X_{BL}^0 (p, T) and X_{BV}^0 (p, T) constitute two surfaces, shown in Fig. 1. The shaded region between them may be interpreted as representing the coexistence of two phases L and V if in this region X_B is interpreted as a mole fraction of B for the system as a whole. Within the two-phase region, X_B is *not* to be regarded as one of the intensive variables constituting the variance (although in a single-phase region it is indeed one of these variables). In a two-phase region the value of X_B determines the relative proportions of the two phases in the system; as X_B varies from X_{BL}^0 to X_{BV}^0, the molar proportion x_V of vapor phase varies from zero to unity:

$$x_V = 1 - x_L = \frac{X_B - X_{BL}^0}{X_{BV}^0 - X_{BL}^0} \tag{2}$$

Figure 1 is drawn for the special case of two components that form a complete range of *ideal solutions*, i.e., solutions that obey Raoult's law with respect to both components at all compositions. According to this law, the vapor pressure (or partial pressure in the vapor) of a component at a given temperature T_1 is proportional to its mole fraction in the liquid. Thus in Fig. 1 the light dashed lines representing the partial pressures p_A and p_B and the total vapor-pressure line (L, joining p_A^0 and p_B^0) are straight lines when plotted against the liquid composition. However, the total vapor pressure as plotted against the *vapor* composition is not linear. The curved line V, joining p_A^0 and p_B^0, is convex downward, lying *below* the straight line on the constant-temperature section; its slope has everywhere the same sign as the slope of L.

The vapor pressures p_A^0 and p_B^0 of the pure liquids increase with temperature (in accord with the Clapeyron equation) as indicated by the curves joining p_A^0 with $p_A'^0$ and p_B^0 with $p_B'^0$. At a constant pressure, say 1 atm, the boiling points of the pure liquids are indicated as T_A^0 and T_B^0. The boiling point of the solution, as a function of X_{BL} or X_{BV}, is represented by the curve L^* or V^* joining these two points. Neither curve is in general a straight line. If Raoult's law is obeyed, both are convex upward in temperature, the vapor curve lying above the liquid curve in temperature and being the more convex.

In most binary liquid–vapor systems, Raoult's law is a good approximation for a component only when its mole fraction is close to unity. Large deviations from this law are com-

---

†The complete definition of the system would of course include also its shape, description of surfaces, specification of fields, etc.; ordinarily these have negligible effects as far as our present discussion is concerned.

FIGURE 1
Schematic three-dimensional
vapor–liquid equilibrium
diagram for a two-component
system obeying Raoult's law.

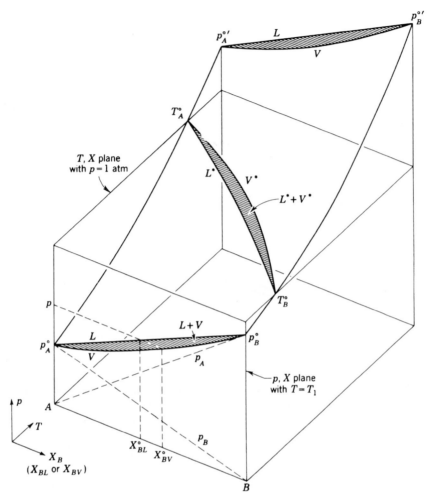

monplace for the dilute component or for both components when the mole fraction of neither
is close to unity. If at a given temperature the vapor pressure of a solution is higher than that
predicted by Raoult's law, the system is said to show a *positive deviation* from that law. For
such a system, the boiling-point curve L^* at constant pressure is usually convex downward
in temperature. If at a given temperature the vapor pressure of the solution is lower than that
predicted by Raoult's law, the system is said to show a *negative deviation;* in this case the
curve L^* is more convex upward. These deviations from Raoult's law are often ascribed to
differences between "heterogeneous" molecular attractions (A – – – B) and "homogeneous"
attractions (A – – – A and B – – – B). Thus the existence of a positive deviation implies that
homogeneous attractions are stronger than heterogeneous attractions, and a negative deviation implies the reverse. This interpretation is consistent with the fact that positive deviations
are usually associated with positive heats of mixing and volume expansions on mixing, while
negative deviations are usually associated with negative heats and volume contractions.

In many cases the deviations are large enough to result in maxima or minima in the
vapor-pressure and boiling-point curves, as shown in Fig. 2. Systems for which the boiling-point
curves have a maximum include acetone–chloroform and hydrogen chloride–water; systems
with a minimum include methanol–chloroform, water–ethanol, and benzene–ethanol. At
a maximum or a minimum, the compositions of the liquid and of the vapor are the same;
accordingly, there is a *point of tangency* of the curves L and V and of the curves L^* and V^*

FIGURE 2

Schematic vapor-pressure
and boiling-point diagrams
for systems showing (*a*) a
strong positive deviation
and (*b*) a strong negative
deviation from Raoult's law.

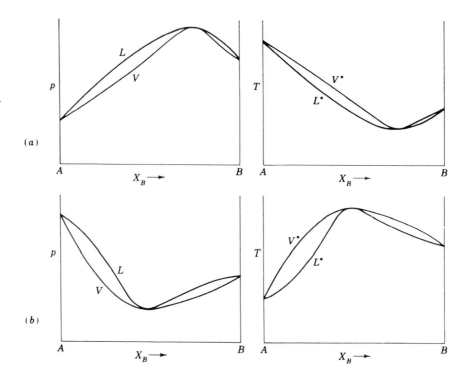

at the maximum or minimum. At every value of X_B, the slope of V (or V^*) has the same sign as the slope of L (or L^*); one is zero where and only where the other is zero, at the point of tangency. (A common error in curves of this kind, found even in some textbooks, is to draw a cusp—a point of discontinuity of slope—in one or both curves at the point of tangency; both curves are in fact smooth and have continuous derivatives.)

Liquid–vapor phase diagrams, and boiling-point diagrams in particular, are of importance in connection with *distillation,* which usually has as its object the partial or complete separation of a liquid solution into its components. Distillation consists basically of boiling the solution and condensing the vapor into a separate receiver. A simple "one-plate" distillation of a binary system having no maximum or minimum in its boiling-point curve can be understood by reference to Fig. 3. Let the mole fraction of B in the initial solution be represented by X_{BL1}. When this is boiled and a small portion of the vapor is condensed, a drop of distillate is obtained, with mole fraction X_{BV1}. Since this is richer in A than is the residue in the flask, the residue becomes slightly richer in B, as represented by X_{BL2}. The next drop of distillate X_{BV2} is richer in B than was the first drop. If the distillation is continued until all the residue has boiled away, the last drop to condense will be virtually pure B. To obtain a substantially complete separation of the solution into pure A and B by distillations of this kind, it is necessary to separate the distillate into portions by changing the receiver during the distillation, then subsequently to distill the separate portions in the same way, and so on, a very large number of successive distillations being required. The same result can be achieved in a single distillation by use of a fractionating column containing a large number of "plates"; discussion of the operation of such a column is beyond the scope of this book. If there is a maximum in the boiling-point curve (Fig. 2*b*), the compositions of vapor and residue do not approach pure A or pure B but rather the composition corresponding to the maximum. A mixture with this composition will distill without change in composition and is known as a *constant-boiling mixture* or *azeotrope*. These terms are also applied to a mixture with a minimum boiling point. Azeotropes are important in chemical technology. Occasionally they are useful (as in constant-boiling aqueous hydrochloric acid, used as an analytical standard); often they are nuisances

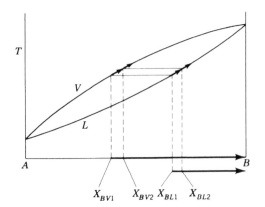

FIGURE 3
Variation of liquid and
vapor compositions during
distillation of a system with
small positive deviations
from Raoult's law.

(as in the case of the azeotrope of 95 percent ethanol with 5 percent water, the existence of which prevents preparation of absolute ethanol by direct distillation of dilute solutions of ethanol in water). Extensive lists of azeotropes have been compiled.[2]

METHOD

A boiling-point curve can be constructed from data obtained in actual distillations in an ordinary "one-plate" distilling apparatus. Small samples of the distillate are taken directly from the condenser, after which small samples of the residue are withdrawn with a pipette. The samples of distillate and residue are analyzed, and their compositions are plotted on a boiling-point diagram against the temperatures at which they were taken. In the case of the distillate, the temperature to be plotted for each sample should be an average of the initial and final values during the taking of the sample. In the case of the residue, the temperature to be plotted should be that recorded at the point where the distillation is stopped to take the sample of residue.

For analysis of the samples, a physical method is often preferable to chemical methods. Chemical analysis usually is appropriate only when a simple titration of each sample is involved, as in the case of the system $HCl–H_2O$. If a physical property is chosen as the basis for an analytical method, it should be one that changes significantly and smoothly over the entire composition range to be studied. The refractive index is one property that can be used for the cyclohexanone–tetrachloroethane system. The values of n_D^{20} are 1.4507 for cyclohexanone and 1.4942 for 1,1',2,2'-tetrachloroethane, and log n_D^{20} is almost a linear function of the weight percent of cyclohexanone. Thus one can interpolate linearly between the values listed in Table 1, and then convert weight percentages into mole fractions.

TABLE 1 Logarithm of refractive index for cyclohexanone–tetrachloroethane mixtures

Log n_D^{20}	W% $C_6H_{10}O$	Log n_D^{20}	W% $C_6H_{10}O$	Log n_D^{20}	W% $C_6H_{10}O$
0.17441	0	0.16864	40	0.16360	80
0.17298	10	0.16719	50	0.16256	90
0.17155	20	0.16582	60	0.16158	100
0.17010	30	0.16473	70		

Another property that could also be used is the density, which varies in a nonlinear way between 0.9478 g/mL for cyclohexanone and 1.600 g/mL for tetrachloroethane at 20°C. In this case, a calibration curve should be constructed from known solutions or provided by the instructor. The density of each distillation sample can be measured by pipetting 1 mL into a small, previously weighed vial and then weighing again (to the nearest 0.1 mg).

Warning: These solutions will decompose slowly at room temperature (~1 day) and more rapidly during distillation at high temperatures. Impure solutions become yellowish, which can interfere with the refractive index measurements. Use well-purified starting materials and do not prolong the distillations unnecessarily. It should also be noted that many chlorinated hydrocarbons, including tetrachloroethane, are toxic chemicals. *Chronic* exposure to tetrachloroethane can cause liver damage. The toxic oral dose is high (~0.5 g per kilogram of body weight, or 27 g for someone weighing 120 lb), but smaller doses can cause short-term medical problems.[3] **Do not pipette by mouth.** The vapor is irritating to eyes and mucous membranes, and the toxic level is 4.5 g m^{-3}. **Do not inhale the vapor while taking samples.**

EXPERIMENTAL

A simple distilling apparatus that can be used for this experiment is shown in Fig. 4. The mercury thermometer bulb (or any other thermometer sensor) should be about level with the side arm to the condenser so that the temperature of coexisting vapor and liquid is measured. Except when samples of distillate are being taken for analysis, an adequate receiving flask should be placed at the lower end of the condenser.

Before beginning the distillations, prepare twenty 5-mL vials for taking samples and label with the designations 1*L*, 1*V*, 2*L*, . . . , 10*V* (*L* = *liquid* residue; *V* = condensed

FIGURE 4

Distillation apparatus. The mercury thermometer can be replaced by a resistance thermometer or other direct-reading thermometric device.

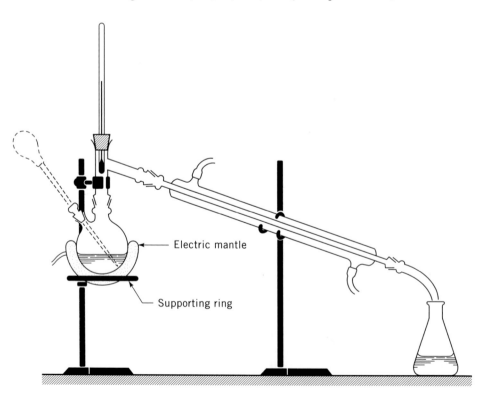

Electric mantle

Supporting ring

vapor or distillate). Take samples of about 2 mL in size. Turn on the circulating water to the condenser before heating the liquid, and turn it off at the end of the experiment.

When the distillation is proceeding at a normal (not excessive) rate at about the desired temperature, quickly replace the receiver with a vial and read the thermometer. After about 2 mL has been collected, read the thermometer again, replace the receiver, and seal the vial tightly. Turn off and lower the heating mantle to halt the distillation. When the temperature just begins to fall, record another thermometer reading. After the flask has cooled about 15°C, remove the glass plug on the side arm of the flask and insert a 2-mL pipette equipped with a rubber bulb. Fill the pipette, discharge it into the appropriate vial, and stopper the vial.

The following procedure is recommended for economical use of materials in carrying out this experiment. The paragraph numbers correspond to sample numbers. A graduated cylinder is adequate for measuring liquids. The temperatures recommended are those appropriate for 760 Torr; at ambient pressure differing markedly from this value, the temperatures should be adjusted accordingly. For example at Denver (altitude 1609 m) the average atmospheric pressure is 836 mbar = 627 Torr. Since the enthalpies of vaporization of cyclohexane and tetrachloroethane are both near 40 kJ mol^{-1}, this pressure change will reduce the boiling points of the pure materials by $\sim$7°C compared with the values at 760 Torr. Comparable changes are expected for the azeotrope and for intermediate compositions.

1. Pure tetrachloroethane: Introduce 125 mL ($\sim$200 g) of 1,1′,2,2′-tetrachloroethane into the flask. Distill enough to give a constant temperature (should be near 146°C at 760 Torr). Collect samples (1V and 1L) for analysis.
2. 149°C (tetrachloroethane-rich side of azeotrope): Cool the distilling flask, and return the excess distillate of paragraph 1 to the flask. Add 38 mL ($\sim$36 g) of cyclohexanone. Begin distillation. When the temperature reaches 149°C, collect about 2 mL of distillate (2V) and 2 mL of residue (2L).
3. 151°C: Resume the distillation. Distill until the temperature reaches 151° C (this may take some time) and collect samples (3V, 3L).
4. 154°C: Resume the distillation. When the temperature reaches 154° C, collect samples (4V, 4L).
5. 157°C: Cool the flask somewhat, and add 35 mL of tetrachloroethane and 25 mL of cyclohexanone. Resume the distillation. When the temperatures reaches approximately 157°C, collect samples (5V, 5L).
6. Azeotrope: Cool the flask somewhat and add 36 mL of tetrachloroethane and 54 mL of cyclohexanone. Resume the distillation until the boiling point ceases to change significantly, and take samples (6V, 6L). (If the boiling point does not become sufficiently constant, analyze the remaining residue, and make up 100 mL of solution to the composition found. Distill to constant temperature and take samples.)
7. Pure cyclohexanone: Introduce 105 mL of cyclohexanone into the clean flask and determine the boiling point as in paragraph 1. (The temperature should be near 155°C at 760 Torr). Collect samples (7V and 7L).
8. 156.5°C (cyclohexanone-rich side of azeotrope): Cool the distilling flask, return the excess distillate of paragraph 7, and add 20 mL of tetrachloroethane. Resume the distillation, and take samples (8V, 8L) at about 156.5°C.
9. 157°C: Cool the flask somewhat, and add 50 mL cyclohexanone and 17 mL tetrachloroethane. Resume the distillation and collect samples (9V, 9L) at about 157°C.
10. Azeotrope: Resume the distillation, continue to constant boiling temperature, and take samples (10V, 10L).

The indexes of refraction should be measured and recorded as soon as possible (the samples decompose on standing). The refractometer and the procedure for its use are described in Chapter XIX. (If the experiment is being done by several teams using the same refractometer, it is wise to take samples to the refractometer as soon as six or eight samples are ready, or fewer if the instrument happens to be free.) If careful attention is given to the proper technique of using the refractometer, it should be possible to take readings at the rate of one sample per minute.

At the end of the experiment, all cyclohexanone–tetrachloroethane mixtures should be poured into a designated waste vessel.

At some time during the laboratory period, the barometer should be read. The ambient temperature should be recorded for the purpose of making thermometer stem corrections if these are necessary; see pp. 563.

CALCULATION

Determine the weight-percent compositions by interpolation in Table 1 and then convert the weight percentages to mole fractions. Plot the temperatures (after making any necessary stem corrections; see Chapter XVII) against the mole fractions. Draw one smooth curve through the distillate points V and another through the residue points L. Label all fields of the diagram to indicate what phases are present. Report the azeotropic composition and temperature, together with the atmospheric pressure (i.e., the properly corrected barometer reading).

SAFETY ISSUES

Tetrachloroethane has a known toxicity, as described in the warning in the Method section. Do not pipette this or any other chemical by mouth, and avoid inhaling the vapor while withdrawing samples. Dispose of waste chemicals as instructed.

APPARATUS

Distilling flask; mercury thermometer, graduated to 0.1°C or a digital resistance thermometer with a resolution of ±0.1°C; one-hole stopper for thermometer to fit flask and allow adjustment in thermometer position; straight-tube condenser; two lengths of rubber hose for condenser cooling water; distilling adapter to fit flask and condenser; two clamps and clamp holders; two ring stands; one iron ring; electrical heating mantle (or steam bath); 20 small vials; 100-mL graduated cylinder; two wide-mouth 250-mL flasks; 2-mL pipette; pipetting bulb; two 500-mL glass-stoppered Erlenmeyer flasks.

Refractometer, thermostated at 25°C; sodium-vapor lamp (optional); eye droppers; *clean* cotton wool; acetone wash bottles; pure 1,1′,2,2′-tetrachloroethane (300 mL) and pure cyclohexanone (350 mL); acetone for rinsing; large bottle for disposal of waste solutions.

REFERENCES

1. P. W. Atkins and J. de Paula, *Physical Chemistry,* 8th ed., chap. 6, Freeman, New York (2006).

2. L. H. Horsley, *Azeotropic Data* (Advances in Chemistry, no. 116), American Chemical Society, Washington, DC (1973).

3. *Documentations of the Threshold Limit Values,* Amer. Conf. of Governmental Industrial Hygienists, Cincinnati, OH (issued annually).

GENERAL READING

R. J. Silbey, R. A. Alberty, and M. G. Bawendi, *Physical Chemistry,* 4th ed., Wiley, New York (2005).

M. Hillert, *Phase Equilibria, Phase Diagrams and Phase Transformations,* Cambridge Univ. Press, New York (1998).

EXPERIMENT 15
Ordering in Nematic Liquid Crystals

Liquid crystals are interesting materials with properties intermediate between those of normal liquids and crystalline solids, and they have many important practical uses, particularly for LCD displays. One important class of liquid crystal materials consists of uniaxial molecules in the form of long semirigid rods with aromatic cores and aliphatic chain tails, like those shown in Fig. 1. Their distinguishing characteristic is some sort of structural order in a fluid phase.[1–4] Depending on the temperature, such liquid crystals can exist in several distinct phases, the simplest of which are *isotropic, nematic,* and *smectic-A* (see Fig. 2). The possible types of structural long-range order for liquid crystals are orientational order and positional order. The former designates order in the spatial orientation for the long axes of the molecules, while the latter designates order involving the position of the molecular centers of mass. The high-temperature isotropic (I) phase is like an ordinary liquid, with no order, whereas the nematic (N) phase obtained on cooling has orientational but not positional order. The smectic-A (SmA) phase observed at still lower temperatures in some liquid crystals retains orientational order but also has one degree of positional order corresponding to a layerlike structure with a periodic array of the z-components of the centers of mass but no x or y periodicity (i.e., SmA is a 1-D crystal of stacked 2-D liquid layers that have nematic orientational order).

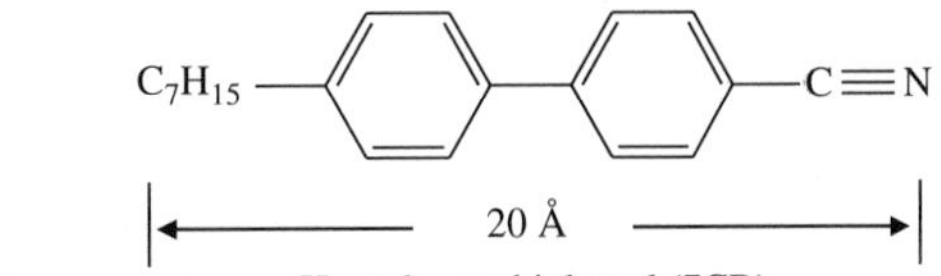

FIGURE 1

Structural formulae for typical liquid crystal molecules with their shorthand names given in parentheses. The lateral thickness perpendicular to the long axis is ~4.5Å in each case.

C_7H_{15}—⟨⟩—⟨⟩—$C\equiv N$

20 Å

Heptylcyanobiphenyl (7CB)

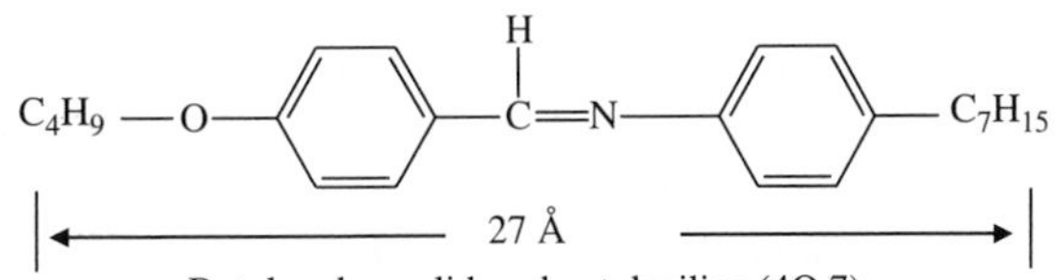

C_4H_9—O—⟨⟩—$\overset{H}{\underset{}{C}}$=N—⟨⟩—$C_7H_{15}$

27 Å

Butyloxybenzylidene heptylaniline (4O.7)

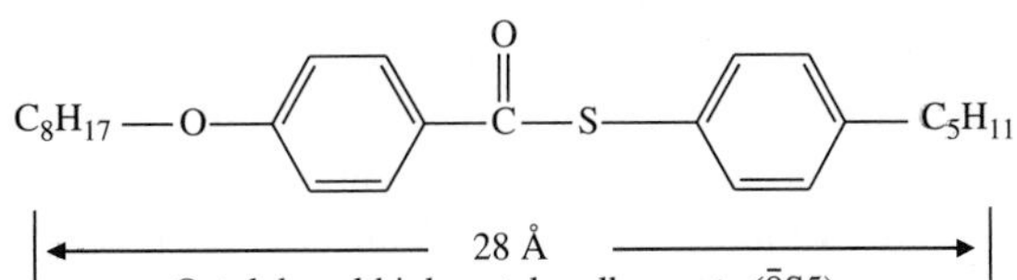

C_8H_{17}—O—⟨⟩—$\overset{O}{\underset{}{C}}$—S—⟨⟩—$C_5H_{11}$

28 Å

Octylphenylthiol pentyloxylbenzoate ($\bar{8}$S5)

FIGURE 2

(a) Simplified sketches of the isotropic (I), nematic (N), and smectic-A (SmA) phases of rodlike liquid crystal molecules. (b) The orientational ordering of uniaxial liquid crystal molecules in a given nematic domain is shown in a somewhat more realistic exploded view using elongated ellipsoids to represent the molecules. The direction of the local director **n** is taken to be the z axis, and θ is the angle between the long axis of any given molecule and the director. Note that there is no orientational order with respect to the x or y axis in a uniaxial nematic phase.

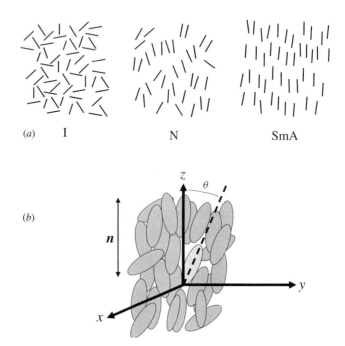

The orientational phase transition of a liquid crystal as it passes from the nematic to isotropic phase is interesting because it involves first a significant gradual loss of orientational order and then an abrupt change as all such order is lost. This is in contrast to the ordinary melting transition from a 3-D positionally ordered crystal to a disordered isotropic liquid, where only a discontinuous change in order occurs. The present experiment will use optical methods to examine the change in orientational nematic order in the temperature region below the N-I phase transition point. It will be shown that nematic ordering can be understood in terms of a simple free energy expression and that the temperature dependence of nematic order exhibits a power-law form.

THEORY

Sketches of the isotropic, nematic, and smectic-A phases for uniaxial rodlike molecules are given in Fig. 2. The *director* ***n*** shown in this figure defines the local z direction of the preferred orientation of the long axes of the molecules in any arbitrary region, and θ denotes the angle between ***n*** and the long axis of any given molecule. In the nematic phase, the ***n*** directions for all regions (domains) can be aligned by an electric or magnetic field or, as in this experiment, by a series of parallel microgrooves on the window surface of the cell containing the liquid crystal.

The orientation of a molecule with respect to ***n*** can be given by a Legendre polynomial $P(\cos\theta)$ and is typically represented by a P_1 dipole term and/or a P_2 quadrupolar term.[1-4] In the case of liquid crystals like those shown in Fig. 1, the electric dipole orientation forces are very small relative to the van der Waals forces of quadrupolar symmetry produced by the long side-to-side contact of the molecules. Even if the molecule has a dipole oriented along the rod axis (as for the CN group in 7CB), it is equally probable for the dipole to point "up" or "down". As a result, $\langle P_1 \rangle = \langle \cos\theta \rangle$, the statistical average over many molecules, is zero. Thus, the simplest measure of the long-range orientational order of a liquid crystal is given by the scalar order parameter S, defined by averaging the orientational function of quadrupolar symmetry, the second Legendre polynomial $P_2(\cos\theta) = (3\cos^2\theta - 1)/2$; i.e.,

$$S = <P_2> = \frac{1}{2}<3\cos^2\theta - 1> \tag{1}$$

This is equivalent to saying that the order formed in the nematic phase is not governed by electric dipole interactions but rather by the molecular shape and the van der Waals forces between molecules. Note that, although entropy is also related to the order of a system, the symbol S used here does not refer to entropy.

In the isotropic I phase, molecular orientations are random and thus $S = 0$. If all the molecules were perfectly aligned (i.e., all $\theta = 0°$ or $180°$), $S = 1$. In the nematic N phase, S has intermediate positive values that are strongly temperature dependent. Note that negative S values imply that the molecular long axes lie preferentially at angles between $55°$ and $125°$ with respect to $\textbf{\textit{n}}$; the most negative possible value is $S = -0.5$, corresponding to all $\theta = 90°$. Such structures with negative S values have a higher free energy than that of the I phase and are not stable.

The variation of $S(T)$ near the N-I phase transition will be measured in this experiment and will be compared with the behavior predicted by Landau theory,[1,4–6] which is a variant of the "mean-field" theory[6] first introduced for magnetic order-disorder systems. In this theory, local variations in the environment of each molecule are ignored and interactions with neighbors are represented by an average. This type of theory for order-disorder phase transitions is a very useful approximate treatment that retains the essential features of the transition behavior. Its simplicity arises from the suppression of many complex details that make the statistical mechanical solution of 3-D order-disorder problems impossible to solve exactly.

In any discussion of order-disorder phase transitions, it is important to distinguish between first-order and second-order transitions. For first-order phase transitions, there are discontinuities in the first derivatives of the free energy G with respect to temperature or pressure; the latent heat of melting is an example. For second-order transitions, the first derivatives of G are continuous at the transition but the second derivatives (like the heat capacity or the thermal expansion coefficient) exhibit infinite singularities at a "critical" transition temperature T_c. Such second-order transitions are called critical because the behavior at the transition is qualitatively the same as that at the familiar liquid-gas critical point. The order parameter undergoes a continuous variation with temperature for second-order transitions, whereas the order parameter for first-order transitions shows a discontinuous jump at the transition temperature. For a "strongly" first-order transition like the freezing of a simple liquid, the positional order parameter jumps at the transition temperature from the value zero in the liquid to a temperature-independent maximum value in the solid. The N-I transition in liquid crystals is called a "weakly" first-order transition since there is a discontinuous jump in S at the transition temperature but there is also a significant further increase in S as the sample is cooled below that temperature. Landau theory assumes that the temperature variation in the free energy can be represented by a power series in S in the region near the phase transition. It has proven to be a very useful phenomenological model for describing the essential features of second-order and weakly first-order transitions.

Before considering the Landau theory for an N-I transition, it is very helpful to examine first the simple case of a uniaxial magnetic solid that undergoes a second-order paramagnetic-ferromagnetic transition.[6] Such a solid can be viewed as a lattice array of atoms with spins, and hence magnetic moments, pointing either up or down. In the high-temperature paramagnetic phase, the individual moments are randomly up or down and no net spontaneous magnetization M is observed. On cooling below a critical temperature T_c, a magnetization

M is observed to develop as clusters of spins tend to line up in the same direction. For this paramagnetic-ferromagnetic transition, the quantity $\varphi = M / |M_{max}|$ is the magnetic order parameter, where $|M_{max}|$ is the magnitude of the maximum magnetization observed when all spins point in the same direction. Thus $\varphi = 0$ for $T > T_c$ and $1 \geq |\varphi| > 0$ for $T < T_c$, where T_c is the paramagnetic-ferromagnetic critical transition temperature. As shown below, it is important to use different symbols for the magnetic order parameter φ and the liquid crystal orientational order parameter S since the dependence of the free energy G on the order parameter differs in these two cases.

For the magnetic system in zero external field, the Landau expression for the free energy can be written as an expansion in φ about the G value for $\varphi = 0$:[6]

$$G = G_o + at\,\varphi^2 + b\,\varphi^4 + c\,\varphi^6 \qquad (2)$$

with coefficients $a > 0$, b, $c \geq 0$ that are independent of p and T. The free energy G_o for the disordered paramagnetic phase is a slowly varying function of p and T, and $t = (T - T_c)/T_c$ is a dimensionless quantity called the *reduced temperature*. Note that only even powers of the order parameter φ appear in Eq. (2). This is required for ferromagnets in zero external field since a system with positive φ (spins up) has the same free energy as one with negative φ (spins down). An essential feature of the Landau expression is that the coefficient of the φ^2 term must change sign at T_c, and the quantity at is the simplest form that satisfies that requirement. The coefficient a must be positive in order to achieve the desired transition at $t = 0$ (i.e., at T_c). The curvature at $\varphi = 0$, $\partial^2 G/\partial \varphi^2 = 2at$, will then be positive for $t > 0$, corresponding to a minimum in G at $\varphi = 0$, while the curvature becomes negative for $t < 0$, corresponding to a local maximum in G; see Fig. 3a. Also, as will be seen from the results given below, $c \geq 0$ is needed to avoid nonphysical negative values for φ^2 (and thus imaginary values for the order parameter φ) when $t < 0$.

We consider here the cases where coefficient $b \geq 0$, and Fig. 3a shows a schematic plot of the resulting Landau free energy as a function of φ for various choices of t. In order to find the equilibrium value of φ as a function of t, one uses the equilibrium condition $\partial G/\partial \varphi = 0$, which yields

$$\partial G/\partial \varphi = 2at\,\varphi + 4b\,\varphi^3 + 6c\,\varphi^5 = 0 \qquad (3)$$

For $T > T_c$ or t positive, Eq. (3) can be satisfied only for $\varphi = 0$, which corresponds to the disordered phase with random up/down spin orientations and no net magnetization (i.e., the paramagnetic phase).

For $T < T_c$ or t negative, there are two stable free energy minima with $G < G_o$ and nonzero φ values, as shown in Fig. 3a. These correspond to ferromagnets with net positive or negative magnetization states, which are equivalent in the absence of an external field. For t negative, we consider three cases for the evolution of φ with t according to Eq. (3):

(a) coefficient $c = 0$ — states with $\varphi = \pm(a/2b)^{0.5}|t|^{0.5}$ are the stable solutions to Eq. (3).

(b) coefficient $b = 0$ — states with $\varphi = \pm(a/3c)^{0.25}|t|^{0.25}$ are the stable solutions to Eq. (3). Since $b < 0$ leads to a first-order transition† while $b > 0$ is second order, the case $b = 0$ is a special one and this transition is called *tricritical*.

(c) general case with b, $c > 0$ — the stable state has an order parameter given by

$$\varphi = \pm\left\{\frac{b}{3c}\left[\left(1 + \frac{3ac|t|}{b^2}\right)^{1/2} - 1\right]\right\}^{1/2}$$

---

†When $b < 0$, a first-order transition occurs at $t_1 = b^2/4ac$, where $G(\varphi_1) = G_o$ and $\varphi_1^2 = -b/2c > 0$.

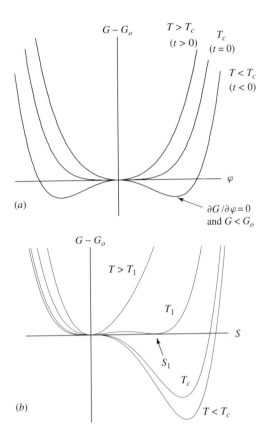

$G - G_o$

$T > T_c$
$(t > 0)$

T_c
$(t = 0)$

$T < T_c$
$(t < 0)$

φ

(a)

$\partial G/\partial \varphi = 0$
and $G < G_o$

$G - G_o$

$T > T_1$

T_1

S

S_1

T_c

(b)

$T < T_c$

FIGURE 3

(a) $G - G_o$ as a function of the order parameter φ for a ferromagnetic solid according to Eq. (2). For $t > 0$, the minimum G corresponds to $\varphi = 0$ (disordered paramagnet). At $t = 0$ (i.e., $T = T_c$), the minimum is still at $\varphi = 0$ but the curvature $\partial^2 G/\partial \varphi^2 = 0$ there. For $t < 0$, the minimum G occurs at two equivalent positive and negative nonzero φ values (ordered ferromagnet). Note that for $t < 0$, the curvature is *negative* at $\varphi = 0$, which means that the disordered phase is unstable. (b) $G - G_o$ vs S for the weakly first-order N-I transition according to Eq. (5). The transition occurs at T_1, where $G(0) = G(S_1)$. At the lower temperature T_c, the curvature $\partial^2 G/\partial S^2 = 0$ at $S = 0$ and this would have been the transition temperature if there were no S^3 term in Eq. (5).

It can easily be shown that this general solution yields a type (a) solution $\varphi \sim |t|^{0.5}$ when c is small and yields a type (b) solution $\varphi \sim |t|^{0.25}$ when b is small. For both the special cases (a) and (b) above, the temperature dependence of φ in the ordered phase obeys a *power law*

$$\varphi \sim |t|^{\beta} \qquad (4)$$

where $\beta = 0.5$ is the second-order Landau (mean-field) value of the exponent beta in case (a) and $\beta = 0.25$ is the tricritical value of the exponent in case (b). If Eq. (4) is assumed to apply approximately for intermediate values of c/b, physically acceptable values of β must fall in the range 0.25 to 0.5.

Unlike ferromagnets, where $G(\varphi) = G(-\varphi)$, the free energy of nematic liquid crystals exhibits an asymmetry $G(S) \neq G(-S)$, as discussed previously and shown in Fig. 3b. To reproduce this behavior, a term that is an odd power of the order parameter S is needed in the Landau expression for the free energy of a nematic liquid crystal.[4-6] A term linear in S is not allowed since the equilibrium condition $\partial G/\partial S = 0$ could not then be satisfied in the disordered isotropic phase where $S = 0$. The presence of a cubic term will lead to the desired asymmetry in G as a function of S and to the emergence on cooling of a second minimum in G at a finite S value.

Thus, one has for the Landau free energy of a nematic liquid crystal

$$G = G_o + atS^2 - dS^3 + bS^4 + cS^6 \qquad (5)$$

where all four coefficients a, b, c, d are taken to be positive. Note that a negative coefficient $(-d)$ for the S^3 term is necessary to insure that the second minimum occurs at positive S values. The presence of this cubic term means that the N-I transition must be first order. That is, there is a temperature $T_1 > T_c$ where $G(S_1) = G_o$. For all $T > T_1$, the lowest free energy corresponds to $S = 0$. For all $T < T_1$, the lowest free energy corresponds to a nonzero $S > 0$ value; see Fig. 3b. At the first-order transition temperature T_1, there is a discontinuous jump in S from $S = 0$ to $S = S_1$.

Unfortunately, the presence of the S^3 term in Eq. (5) makes it much more difficult to extract the equilibrium $S(T)$ behavior than it was to find $\varphi(T)$ for an Ising magnet. However, the general approach is the same—find the S value at which $\partial G/\partial S = 0$ and G is a global minimum. The temperature dependence of S arising from Eq. (5) for a nematic liquid crystal turns out to be quite complicated since there is no analytic solution to the quartic equation arising from $\partial G/\partial S = 0$. However, the behavior of $S(T)$ for $T < T_1$ can still be very well approximated by a power-law expression[5,7]

$$S(T) = S^{**} + A\,|\tau|^\beta \tag{6}$$

where $\tau \equiv (T - T^{**})/T^{**}$ is a new reduced temperature relative to an effective phase transition temperature T^{**} for N phase behavior. The value of T^{**} is slightly larger than the first-order transition temperature T_1 and $S^{**} < S_1$. Note that three characteristic temperatures $T^{**} > T_1 > T_c$ play a role for the N-I phase transition. When $T = T_c$, the coefficient of the S^2 term in Eq. (5) is zero, and T_c represents the metastability limit on cooling the I phase. The quantity T^{**} represents the metastability limit on warming the N phase, and S^{**} is the order parameter corresponding to T^{**}. Although T^{**} and S^{**} are not physically accessible, they are useful variables that allow the complicated $S(T)$ variation in the N phase to be written in a simple power law form involving the critical exponent β.

Equation (6) predicts that S values at low temperatures will decrease as T increases, heading toward the value S^{**} at T^{**}. But when T reaches T_1 (a temperature below T^{**}), S will undergo a first-order jump from S_1 to 0 since the isotropic phase is the stable phase for all $T > T_1$. The observed variation of $S(T)$ for a typical thermotropic nematic liquid crystal is shown in Fig. 4. For the simplified case where the sixth-order coefficient c in Eq. (5) is set equal to zero, it can be shown[6] that

$$T^{**} - T_c = 9d^2 T_c /32ab, \quad T_1 - T_c = d^2 T_c /4ab$$

and $S_1 = d/2b$. Thus, if the cubic term were to be "turned off" by letting $d \to 0$, one finds that $T^{**} \to T_c$, $T_1 \to T_c$, $S_1 \to 0$, $S^{**} \to 0$ and Eq. (6) would become a $|t|^\beta$ second-order power law like that in Eq. (4). This conclusion is valid for any $c > 0$ value, but explicit expressions can no longer be written for $T^{**} - T_c$, $T_1 - T_c$, and S_1.

FIGURE 4

Temperature variation of the nematic order parameter S for hexylcyanobiphenyl (6CB) in the immediate vicinity of $T_1 \approx 28.1°C$. All of the $S(T)$ data observed for $T < T_1$ are well represented by Eq. (8) with $S^{**} = 0.121$, $\beta = 0.25$, and $T^{**} = T_1 + 0.15$ K.[7]

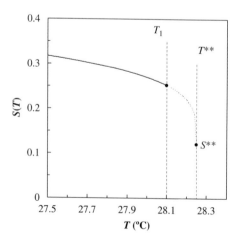

At $T = 0$ K, the nematic order would become perfect; i.e., $S(0) = 1$. Thus we find

$$S(0) = S^{**} + A = 1, \quad A = 1 - S^{**} \tag{7}$$

This allows us to rewrite Eq. (6) as

$$S(T) = S^{**} + (1 - S^{**}) |\tau|^{\beta} \tag{8}$$

which involves three free parameters—S^{**}, T^{**}, and β.[7] When the coefficient d of the S^3 term in Eq. (5) is small, as it is in all practical cases, one finds $\beta = 0.5$ if the coefficient b is large and $\beta = 0.25$ if b is very small. These are the limiting values of β for physically realistic fits to $S(T)$ data. A power-law form like that for φ in Eq. (4) or $S - S^{**}$ in Eqs. (6) and (8) is a general feature of all phase transitions where pretransitional effects are substantial. The critical exponent β has a universal value for a given type of transition; e.g., mean-field second-order transitions for all kinds of ordering in all kinds of systems have $\beta = 0.5$. Thus, the value obtained for β from the best least-squares fit of experimental $S(T)$ data with Eq. (8) is of interest since it tells us whether the critical behavior underlying the N-I transition is close to mean-field second-order ($\beta = 0.5$) or tricritical ($\beta = 0.25$).

METHOD

To study the N-I phase transition, it is necessary to find some physical property proportional to $S(T)$ that can be measured accurately as a function of temperature. There are many possibilities, which include depolarization of Raman spectra,[4,8] diamagnetic susceptibility anisotropy,[4,9] NMR deuterium quadrupole splitting,[4,10,11] dielectric constant anisotropy,[4,12,13] and optical birefringence.[4,7,14] The last of these is perhaps the simplest and is the one employed in this experiment.

The use of birefringence to determine the behavior of $S(T)$ is a natural choice since the principal characteristic of the nematic phase is optical birefringence; i.e., the refractive index differs for light polarized parallel ($n_{\parallel}$) or perpendicular ($n_{\perp}$) to the axis of molecular alignment. For a nematic liquid crystal, the director $\boldsymbol{n}$ specifies this optical z axis and $n_{\parallel} = n_e$ and $n_{\perp} = n_o$ are called the extraordinary and ordinary refractive indices, respectively. In general, $n_e > n_o$ and the difference is the refractive index anisotropy (birefringence)

$$\Delta n = n_e - n_o = n_{\parallel} - n_{\perp} \tag{9}$$

In order to link Δn to the orientational order parameter S, one needs to develop a molecular model to connect n, a macroscopic property, to the related molecular property, the polarizability α. One widely used model is that described by Vuks and others,[4,7,15] which leads to a quantity K_V that is directly proportional to $S(T)$:

$$K_V \equiv \frac{n_e^2 - n_o^2}{\langle n^2 \rangle - 1} = \frac{\Delta \alpha}{\langle \alpha \rangle} S(T) \tag{10}$$

where $\langle n^2 \rangle = (n_e^2 + 2n_o^2)/3$. The molecular polarizability anisotropy is $\Delta \alpha = \alpha_l - \alpha_t$, and $\langle \alpha \rangle = (\alpha_l + 2\alpha_t)/3$. The quantities α_l and α_t are the polarizabilities longitudinal (parallel) and transverse (perpendicular) to the long axis of the liquid crystal molecule and these quantities are independent of T while n and $S(T)$ are not.

Noting that $n_e^2 - n_o^2 = 2\bar{n}\Delta n$, where $\Delta n = n_e - n_o$ and $\bar{n} = (n_e + n_o)/2$, and taking $\bar{n} \approx n_I$ and $\langle n^2 \rangle \approx \bar{n}^2 \approx n_I^2$, where n_I is the n value in the isotropic phase just above T_I, the macroscopic quantity K_V can be obtained as

$$K_V = \frac{2n_I \Delta n}{n_I^2 - 1} \tag{11}$$

This approximation is excellent (typical error $< 0.7\%$ over a 10 K range) due to partial cancellation of changes caused by small systematic variations in $\bar{n}$ with T.

Although one could measure both n_e and n_o independently,[7] accurate values of the quantity $\Delta n(T)$ can be determined directly with an interference method.[3,14] In such a scheme, a thin birefringent slab of a planar aligned nematic liquid crystal is illuminated by polarized light with the plane of polarization making an angle of 45° to the optic axis (director). The transmitted beam is then passed through a second polarizer, oriented for minimum signal when the sample is absent or is warmed above T_1 so that the liquid crystal is in the isotropic phase. At temperatures below T_1, changes in the polarization angle $\Delta\Phi$ can be determined by measuring the intensity of light transmitted through the second polarizer. (The detailed procedure for this is described below in the experimental section.) The value of Δn is related to $\Delta\Phi$ by the relation

$$\Delta n = \frac{\lambda}{2\pi D}\Delta\Phi \tag{12}$$

where λ is the wavelength of the light and D is the sample thickness.

The experimental value of K_V can then be related to the order parameter by combining Eqs. (8) and (10) to obtain

$$K_V = \frac{\Delta\alpha}{<\alpha>}\left[S^{**} + (1 - S^{**})|\tau|^{\beta}\right] \tag{13}$$

where $\Delta\alpha/<\alpha>$, S^{**}, T^{**}, and β are fit parameters.[3,4]

EXPERIMENTAL

Cell Construction. The optical interference technique to be used here is a modification of that described by Gramsbergen and de Jeu.[14] This method uses a thin parallel-plate cell containing a planar aligned nematic liquid crystal sample. The sample cell consists of two parallel 15 × 50 mm microscope slides separated by a gap D in the range 50–100 μm, which is determined by spacer shims. A description of its construction is given below.

To achieve uniform films, the glass slides should be cleaned in an ultrasonic bath with an alkaline detergent such as Contrad 70, rinsed with methanol, dried, and then coated on one side with a film formed by adding drops of a 5% polyvinyl alcohol solution in water. The drops can be spread uniformly on the slides by using a commercial spin coating apparatus. If such an apparatus is not available, it can be improvised from a computer case cooling fan run at about 2000 rpm, with the slide attached symmetrically to the central rotor using double-sided Scotch tape. (For safety, the fan should be enclosed in an open-topped box.) After coating, the slides are dried for 10 min in an oven at 110°C or overnight at room temperature.

A uniform planar orientation for the liquid crystal with the director $\boldsymbol{n}$ parallel to the surface of the glass plates can be achieved by gently rubbing the coated surface of each cell window with a velvet cloth for about 10 min.[2,3,16] The cloth can be held in a small sanding block and a holding template can be formed using extra slides attached with double-sided Scotch tape to a flat surface. A guide for the sanding block will ensure that the rubbing direction is maintained parallel to the long axis of each window. The shallow microgrooves created in this way will align the liquid crystal molecules at the surfaces of the cell plates, and on cooling below T_1, this will promote the formation of a single aligned N domain (in order to minimize the elastic energy).[3,4]

The glass slides and spacers should be sandwiched together and the two long lateral sides sealed using an epoxy cement such as Torr Seal or Kwik. To ensure a uniform spacing, the slides should be evenly clamped between two flat 0.5"-wide bars while the epoxy dries. The edges of the slides should first be scraped with a razor blade to remove any polyvinyl alcohol film so the epoxy will bind directly to the glass. The epoxy should have the consistency of toothpaste to avoid being drawn into the cell by capillary action. After the epoxy has dried, the cell thickness D can be determined from a measurement of weak interference fringes seen by passing a visible or infrared beam through the middle of the empty cell. Microscope slides transmit in the visible and also in the infrared in the region above 2000 cm^{-1}. By using as a reference two similarly coated slides (which need not be rubbed, but their glass surfaces should not be parallel), weak fringes should be visible in the ratioed transmittance spectrum. Successive maxima (or minima) ν_p, expressed in cm^{-1}, are measured and indexed with an integer p, starting with $p = 1$ for the lowest ν_p. Using a least-squares fit of ν_p versus p, obtain $d\nu_p/dp$; the thickness $D(cm) = (2\ d\nu_p/dp)^{-1}$. This value of D will typically be close to the nominal spacer thickness and is usually larger.

The cell is next filled with the liquid crystal. Pentylcyanobiphenyl (5CB) is recommended as the liquid crystal to be studied and the temperature values cited below pertain to 5CB, but other possible liquid crystal choices are listed in the apparatus section below. The tip of a small dropper containing the liquid crystal is placed on one of the unsealed ends of the cell, and liquid will be drawn into the cell by capillary action. This filling procedure works best if the liquid crystal is in the I phase but will also succeed if it is in the N phase close to the N-I transition temperature. Excess liquid crystal is wiped off, and the top and bottom edges of the cell are sealed with Torr Seal or some other epoxy cement to ensure a water-tight seal. The uniformity of the cell spacing can be judged by viewing the cell through crossed polarizers; the intensity pattern of the nematic liquid in the middle region should vary uniformly as the cell is rotated.

The nCB compounds are especially stable and, if stored in a cool dry place, such a filled cell should be useable for several years. If laboratory time is limited, it is suggested that the instructor provide the students with a filled and sealed cell with a specified thickness D. Construction and filling of a cell and measurement of D can also be carried out as a special project if desired.

Optical Measurements.   The apparatus for the optical rotation measurement and a detailed view of the cell are shown in Fig. 5. The sample cell can be suspended from above

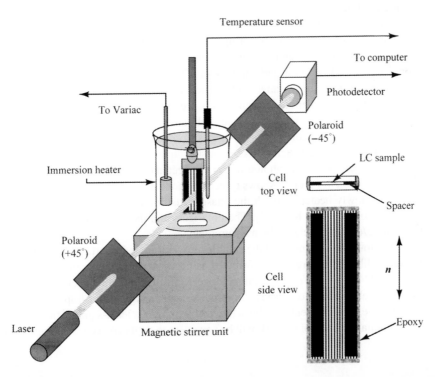

FIGURE 5

Schematic diagrams of the cell and apparatus for measuring the temperature dependence of the optical rotation of an aligned nematic liquid crystal cell. The birefringence cell is formed from microscope slides and the lines in the cell denote the rubbing direction for the glass plates and thus the direction for the nematic director *n*.

using a clamp-rod assembly and submerged in 900 mL of water contained in a 1000-mL beaker with a magnetic stir bar, all positioned above a magnetic stirrer unit. The temperature sensor should be positioned as close to the cell as possible, and the water is then heated by an immersion heater controlled with a Variac. The water can be heated fairly rapidly to about 30°C, but the Variac should then be set to a predetermined value that will produce a heating rate of about 0.05 K per minute at 31°C. Do not change the Variac setting while data are being collected from 31°C to a few tenths of a degree above the N-I phase transition temperature. The heater can then be switched off and data collection continued as the sample cools to about 31°C.

The probe laser should be 1 to 5 mW in power; convenient sources are a He-Ne laser (633 nm) or a red (650 nm) or green (532 nm) laser diode module. The beam is passed successively through a polarizer (accurately oriented at +45° to the vertical), the beaker of water, and then through a second polarizer (and diffuser/attenuator as needed) to a photodetector whose signal can be digitized and stored on a computer. After the second polarizer is adjusted for minimum detector signal (−45°), the LC cell is then lowered into the beam, with the long axis of the cell accurately oriented to be vertical. Note that the same minimum detector signal would be expected if the liquid crystal were in the isotropic phase but, since the cell is at room temperature, the birefringence of the nematic liquid will produce a room-temperature polarization rotation Φ_{RT} such that the signal will in general be larger than the minimum.

When the temperature of the liquid crystal sample increases from room temperature, the polarization angle of the beam decreases as a consequence of a decrease in the birefringence of the sample; see Eq. (12). This causes a rotation of Φ in the relative phase of the two E_x, E_y field components of the light and produces a rotation of $\Phi/2$ in the resultant field polarization. If the emergent beam is initially linearly polarized at say +45° (i.e., $\Phi_{RT} = 2\pi k$ where $k = 0, 1, 2, \ldots$), heating causes the polarization angle to decrease and the polarization becomes successively elliptical, circular ($\Phi = \Phi_{RT} - \pi/2$), elliptical, linear at −45° ($\Phi = \Phi_{RT} - \pi$), etc. Since I is proportional to E^2, the change in intensity of the light passing through the second polarizer is thus given by $I(\Phi) \sim \sin^2(\Phi/2)$ and the separation between intensity maxima (or minima) corresponds to a Φ rotation by 2π. On varying the temperature of a 5CB sample from about 4 K below to about 0.5 K above the N-I transition temperature, several cycles of rotation will occur and, from an analysis of such data, the behavior of the birefringence Δn in the nematic phase can be deduced.

The experimental setup should be shielded from ambient light, or the experiment should be carried out in a darkened room. The photomultiplier or photodiode output is adjusted to be in its linear range and, if needed, amplified to match the input range of an analog-to-digital converter. Measurements are conveniently made using a computer and data collection system such as a Vernier LabPro, which is a small inexpensive module that does analog-to-digital conversion of signals from various sensors, displays the results, and passes them to software such as Excel for subsequent analysis. Signals from the calibrated temperature sensor and the photodiode are collected every second and displayed as the cell is heated or cooled. Figure 6 shows typical photodetector voltage changes that occur as the sample is slowly heated through the phase transition and then, when the heater is turned off, cools to about 4 K below T_1. The polarizers and the cell should not be moved during the heating/cooling cycle of the measurements.

Accurate measurement of the sample temperature is important, and the sensor should be well calibrated and readings corrected as necessary. The temperature digitization steps (about 0.024 K for a Vernier LabPro logger) limit the absolute accuracy for T but, since the temperature variation varies smoothly in time, a significant improvement in relative temperature measurements can be achieved by making a least-squares fit of temperature to a polynomial function of time. (It will be necessary to do this piecewise if the residuals of the

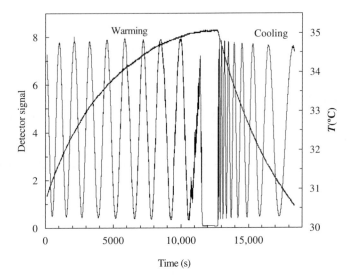

FIGURE 6

Typical data obtained on 5CB in the nematic-isotropic transition region. The sample is first heated from the nematic into the isotropic phase and then cooled back into the nematic. Note that the detector voltage is independent of temperature and is near zero when T is above the N-I transition temperature. The slightly larger values observed at the minima for the nematic phase result from some residual ellipticity of the beam if the first polarizer and the $\boldsymbol{n}$ director are not oriented at exactly 45 degrees.

fit exceed the digitization step value.) The resultant function can then be used to produce a more precise plot of photodetector voltage versus temperature, as shown for the transition region in Fig 7. The transition temperature T_1 is taken as the temperature at which the photodiode signal undergoes an abrupt change from I_{min} in the isotropic phase to some larger I value. For low heating/cooling rates and a well-stirred water bath, T_1 values obtained on heating and cooling should differ by less than 0.05 K. Report the average of $T_1(\text{heat})$ and $T_1(\text{cool})$ as the best T_1 value for your sample.

When the liquid crystal is in the isotropic phase, Δn and $\Delta\Phi$ equal zero, but these values change discontinuously when the nematic phase forms at T_1. The measurements do not give the actual polarization rotation $\Delta\Phi$ for the nematic liquid crystal. So it is convenient to express $\Delta\Phi$ at temperature $T < T_1$ as

$$\Delta\Phi(T) = \Phi_{obs} + \Phi_o = 2\pi D\Delta n/\lambda \tag{14}$$

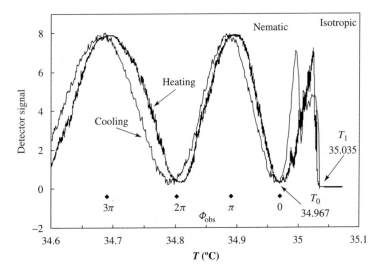

FIGURE 7

Plot of detector signal from Fig. 6 as a function of temperature near T_1. T_1 is taken as the point at which the signal becomes constant (sample is completely in the isotropic phase) and is the average for heating and cooling runs. The ◆ symbols show the average temperature for each minimum and maximum in the detector signal. T_0 is the average temperature of the first minimum (or maximum) observed near T_1. The relative optical rotation Φ_{obs} increases from 0 at T_0 by π radians for each maximum/minimum.

where Φ_0 is a constant representing the rotation at the temperature corresponding to the first well-defined minimum or maximum in the photodetector signal obtained just below T_1. Φ_{obs} thus measures the rotation change relative to this first extremum. It has, by definition, the value zero at the first extremum, and it changes by exactly π radians between adjacent minima and maxima in the photodetector signal. The temperatures of these maxima/minima should be determined for both heating and cooling curves, smoothing the voltage data if necessary for an accurate determination. Tabulate for each observed maximum/minimum the Φ_{obs} value and the associated T(heat) and T(cool) values. Due to the use of fairly rapid scan rates, the maxima/minima observed on heating will occur at slightly higher temperature values than those observed on cooling. As a reasonable estimate of the temperature that one would obtain under truly equilibrium conditions, the average of each pair of heat/cool values should be used, and such averages should be listed in the table.

CALCULATIONS

By combining Eqs. (11)–(14), one gets

$$K_V = \frac{2n_1}{n_1^2 - 1} \frac{\lambda}{2\pi D} (\Phi_{obs} + \Phi_0) = \frac{\Delta\alpha}{<\alpha>}\left[S^{**} + (1 - S^{**})|\tau|^\beta\right] \tag{15}$$

which can be arranged to the form

$$\Phi_{obs} = C_1 + C_2 |\tau|^\beta \tag{16}$$

where

$$C_1 = \frac{n_1^2 - 1}{2n_1} \frac{2\pi D}{\lambda} \frac{\Delta\alpha}{<\alpha>} S^{**} - \Phi_0 \quad \text{and} \quad C_2 = \frac{n_1^2 - 1}{2n_1} \frac{2\pi D}{\lambda} \frac{\Delta\alpha}{<\alpha>}(1 - S^{**}) \tag{17}$$

Note that values for the four quantities λ, n_1, D, and Φ_0 are not actually needed to fit the Φ_{obs} data with Eq. (16) in order to test the power-law form and to get T^{**} and β values.

Since Eq. (16) is nonlinear, one must use a nonlinear least-squares fitting program or, as described here, make use of the Solver option available as an add-in tool in Excel. An example of the use of Solver is given in Chapter III. In the present application, initial estimated values for the four fitting parameters (C_1, C_2, T^{**}, and β) are entered into four worksheet cells. For each of the N data points, these cells are used to calculate τ and then to obtain a theoretical Φ_{obs} value from Eq. (16). The difference between the experimental and theoretical Φ_{obs} value (residual) is squared and the sum of these squares (essentially proportional to $N\chi_v^2$) is placed in a test location. Solver is then run iteratively to adjust the fitting parameters so as to minimize this sum of residuals squared.

Because there is some correlation between the β value and the values of some of the other fitting parameters, it may be difficult to obtain a unique fit of even very good data if all the C_1, C_2, T^{**}, and β parameters are allowed to vary freely.[7] In the present experiment, you should first test the compatibility of the data with the two limiting theoretical values of β by fixing β at (a) the mean-field second-order value of 0.50 and (b) the tricritical value of 0.25. Carry out two such fits with three (C_1, C_2, T^{**}) free parameters and compare the two sums of residuals squared to determine which limiting β provides a better fit to the data. Then explore the region of the better fit by varying β in steps over a range of 0.1 (i.e., 0.2–0.3 or 0.45–0.55). Plot the resulting C_1, C_2, and ($T^{**} - T_1$) values as well as the sum of residuals squared versus β. Choosing initial parameter values close to those obtained for the best of these three-parameter fits, allow Solver to vary all four parameters to see if it will converge to an even better minimum.

Since λ is known and D is either given or measured in the experiment, it is possible to extract $\Delta\alpha/<\alpha>$ and S^{**} from the parameters C_1 and C_2, provided n_I and Φ_0 values are also known. Typical values of n_I for the nCB liquid crystals are 1.55–1.65 and, luckily, these are slowly varying functions of T and wavelength in the isotropic phase.[2,3] For 5CB an n_I value of 1.587 can be used with an error estimated at less than 0.5%.

In principle, Φ_0 can be deduced by accurately measuring Φ_{obs} values for two cells with different (known) D spacings. Alternatively, a literature value of Δn near the phase transition temperature can be used as a calibration point to deduce Φ_0. For 5CB, Chirtoc et al.[7] give $T_1 = 35.32°C = 308.47$ K as the N-I transition temperature but this value can vary by a few tenths of a degree, depending upon sample purity and, of course, calibration of the temperature sensor. Thus, it is best to define a reference temperature relative to the T_1 value measured for a given sample. Accordingly, we choose $T_{ref} = T_1 - 0.40$ K as a calibration point, at which we estimate from literature data[7,17] that Δn is 0.1206 (532 nm), 0.1145 (633 nm), and 0.1134 (650 nm). Substitute the appropriate Δn value for your laser wavelength into Eq. (14) to obtain $\Delta\Phi(T_{ref})$. Use your T_1 value to calculate T_{ref} and then determine by interpolation of your data a value for $\Phi_{obs}(T_{ref})$ and hence $\Phi_0 = \Delta\Phi(T_{ref}) - \Phi_{obs}(T_{ref})$. Use this Φ_0 value with your best values for C_1 and C_2 to calculate S^{**} and $\Delta\alpha/<\alpha>$ from Eq. (17).

DISCUSSION

What does your best value for the exponent β indicate about the nature of the critical behavior underlying the N-I transition (mean-field second-order or tricritical)? Note that β and T^{**} describe the temperature variation of the nematic order parameter S, which is a basic characteristic of the liquid crystal studied. Thus, the same β and T^{**} values could have been obtained from measurements of several other physical properties, such as those mentioned in the methods section.

As shown by Eq. (10), $\Delta\alpha/<\alpha>$ is a multiplicative ("normalization") factor that allows one to relate the experimental quantity K_V to the order parameter S. For each experimental T value listed in your table, calculate $\Delta n(T)$ from Eq. (12) and $K_V(T)$ from Eq. (11). Then use your best value for $\Delta\alpha/<\alpha>$ to calculate a value of S at each experimental T value and plot these $S(T)$ points versus T. The theoretical power-law curve for $S(T)$ given by Eq. (8) should also be shown as a line on this plot. Draw vertical lines at T_1 and at T^{**} and show how the theoretical curve for $S(T)$ extrapolates in this region to the value S^{**} at T^{**}. What value of S_1 do you get from this curve? How do your values of β, $\Delta\alpha/<\alpha>$, $T^{**} - T_1$, and S^{**} compare with those reported for 5CB in Ref. 7? How does the $S(T)$ curve for a liquid crystal differ from the positional order parameter variation for an ordinary solid-liquid phase transition?

Recalling that $\Delta\alpha = \alpha_l - \alpha_t$ and $<\alpha> = (\alpha_l + 2\alpha_t)/3$, calculate the ratio α_l/α_t. Is this value reasonable? As an optional exercise, α_l and α_t could be calculated using ab initio programs such as Gaussian or Hyperchem.

SAFETY ISSUES

A laser power of 5 mW or less is sufficient for this experiment. At this power level, such sources are relatively harmless but of course the beam should never be viewed directly. Further discussion of laser safety practices is given in Appendix C. A safety shield should be used in any spin coating operations.

APPARATUS

A suitable sample cell as described in the experimental section, constructed from microscope slides with spacers (cut from shim stock Teflon sheets from Fluoro-Plastics, Inc., or plastic sheets from Artus Corp.) and sealed with an epoxy cement such as Torr Seal (Varian) or Kwik (J. B. Weld); liquid crystal to be studied. Pentylcyanobiphenyl = 5CB (CAS Number 40817-08-1) with $T_1 = 35.3°C$ is the recommended choice; other possible materials with suitably large birefringence values are heptylcyanobiphenyl (7CB CAS Number 41122-71-8) with $T_1 = 42.4°C$, pentyloxycyanobiphenyl (5OCB CAS Number 52364-71-3) with $T_1 = 68.0°C$, butyloxybenzylidene octylaniline (4O.8 CAS Number 39777-26-9) with $T_1 = 79.0°C$. Frinton Laboratories in Vineland, New Jersey (www.frinton.com), and Aldrich Chemicals (www.sigmaaldrich.com) are good sources of liquid crystals.

A 100-W immersion heater (such as VWR Scientific Inc. #33897-140; a 200-W to 300-W tea/coffee immersion heater can also be used), controlled by a Variac; magnetic stirrer and stir bar; 1000-mL beaker; thermometer with a resolution of $±0.02$ K (platinum resistance thermometer or calibrated thermistor such as Vernier model TMP-BTA); 1- to 5-mW He-Ne laser (633 nm) or a red (545 nm) or green (532 nm) laser — a battery-operated laser pointer is suitable, but a module with a separate 3-volt power supply, available from Z-Bolt and other sources, will give better power stability for long runs; two Polaroid sheets in rotation holders; photodetector such as photomultiplier tube or Thorlabs model 201/579-7227 silicon photodiode; analog-to-digital data collection system such as Vernier LabPro.

REFERENCES

1. P. J. Collings and M. Hird, *Introduction to Liquid Crystals: Chemistry and Physics,* esp. pp. 23–27, 223–33, 245–50, 271–76, Taylor & Francis, London (1997). See also http://en.wikipedia.org/wiki/liquid_crystal and http://plc.cwru.edu/.

2. D. Dunmur, A. Fukuda, and G. Luckhurst (eds.), *Physical Properties of Liquid Crystals: Nematics,* esp. p. 268, INSPEC: Institution of Electr. Engs., London (2001).

3. W. H. de Jeu, *Physical Properties of Liquid Crystalline Materials,* esp. pp. 8–9, 34–71, Gordon & Breach, New York (1980).

4. G. Vertogen and W. H. de Jeu, *Thermotropic Liquid Crystals: Fundamentals,* esp. pp. 70–74, 167–201, 221–28, Springer-Verlag, New York (1988).

5. M. A. Anisimov, *Critical Phenomena in Liquids and Liquid Crystals,* pp. 306–11, 337, Gordon & Breach, New York (1991).

6. P. M. Chaikin and T. C. Lubensky, *Principles of Condensed Matter Physics,* secs. 4.2–4.6 and pp. 230–34, Cambridge University Press, Cambridge (1995).

7. I. Chirtoc, M. Chirtoc, C. Glorieux, and J. Thoen, *Liq. Cryst.* **31,** 229 (2004).

8. S. Jen, N. A. Clark, P. S. Pershan, and E. B. Priestley, *J. Chem. Phys.* **66,** 4635 (1977).

9. A. Buka and W. H. de Jeu, *J. Phys. (Paris)* **43,** 361 (1982).

10. J. W. Emsley, G. R. Luckhurst, P. J. Parsons, and B. A. Timimi, *Molec. Phys.* **56,** 767 (1985).

11. G. S. Iannacchione, S. Qian, D. Finotello, and F. M. Aliev, *Phys. Rev. E* **56,** 554 (1997); X. Shen and R. Y. Dong, *J. Chem. Phys.* **108,** 9177 (1998).

12. J. Thoen and G. Menu, *Mol. Cryst. Liq. Cryst.* **97,** 163 (1983).

13. G. Heppke, S. Pfeiffer, C. Nagabhushan, et al., *Mol. Cryst. Liq. Cryst.* **170,** 89 (1989); S. J. Rzoska et al., *Phys. Rev. E* **64,** 052701 (2001).

14. E. F. Gramsbergen and W. H. de Jeu, *J. Chem. Soc. Faraday Trans. 2* **84,** 1015 (1988); see also K. C. Lim and J. T. Ho, *Mol. Cryst. Liq. Cryst.* **47,** 173 (1978).

15. M. F. Vuks, *Opt. Spectrosc.* **20,** 361 (1966); S. Chandrasekhar and N. V. Madhusudana, *J. Physique Colloq.* **30,** C4-24 (1969).

16. E. R. Waclawik, M. J. Ford, P. S. Hale, J. G. Shapter, and N. H. Voelcker, *J. Chem. Educ.* **81,** 854 (2004).

17. H. J. Coles in *The Optics of Thermotropic Liquid Crystals,* S. Elston and R. Sambles (eds.), pp. 57–84, esp. Table 4.2, Taylor & Francis, London (1998).

GENERAL READING

P. J. Collings and M. Hird, *Introduction to Liquid Crystals: Chemistry and Physics,* Taylor & Francis, London (1997).

P. Oswald and P. Pieranski, *Nematic and Cholesteric Liquid Crystals: Concepts and Physical Properties Illustrated by Experiments,* Taylor & Francis, London (2005).

S. Singh, *Liquid Crystals: Fundamentals,* World Science Publ., Singapore (2002).

EXPERIMENT 16
Liquid–Vapor Coexistence Curve and the Critical Point

The object of this experiment is to measure the densities of coexisting carbon dioxide liquid and vapor near the critical temperature. The critical density of carbon dioxide and a value of the critical temperature will also be obtained from a Cailletet–Mathias curve constructed from the data.

THEORY

At its critical point, the liquid and vapor densities of a fluid become equal. Since this point is very difficult to determine directly, an extrapolation procedure is often used. This procedure is based on the *law of rectilinear diameters,*[1] which states that

$$\rho_{av} \equiv \tfrac{1}{2}(\rho_l + \rho_v) = \rho_0 - cT \tag{1}$$

where ρ_{av} is the average of the densities of the liquid and vapor that are in coexistence with each other at temperature T, and ρ_0 and c are constants that are characteristic of the fluid. The critical point is taken to be that point at which the straight line drawn through ρ_{av} as a function of T intersects the coexistence curve (locus of the liquid and vapor density values as a function of temperature). Such a plot is known as a *Cailletet–Mathias* curve.

METHOD

Six to eight heavy-wall glass capillary tubes of uniform bore and known cross-sectional area are filled with different, known amounts of CO_2. The diameters of the capillary tubes can be obtained prior to filling by weighing a thread of mercury of measured length contained in them. The capillaries are then filled on a vacuum system by freezing into them

CO_2 gas contained in a known volume at a known pressure and sealing them off.[2] Residue corrections for the amount of CO_2 left behind in the vacuum system after sealing off the tubes should be made if necessary. Heavy-walled capillary tubing with a nominal bore of 1 mm is convenient for use in this experiment. Lengths of 10 to 30 cm are suitable; perhaps 20 cm is the best length to use. (It may be desirable to prepare a few tubes of different lengths, which are filled to the same pressure as a way of testing the reproducibility of the method.) The amounts of CO_2 to be loaded into the tubes should be chosen to yield a range of pressures between 18 and 30 bar at 25°C. *The data on the calibration and loading of the capillaries should be made available at the beginning of the laboratory period.*

The filled capillary tubes are mounted in a well-stirred, temperature-regulated water bath that can be warmed or cooled and whose temperature can be measured accurately. If a tube contains less than a critical amount of CO_2, the meniscus between the liquid and vapor will fall toward the bottom of the tube on warming. At a certain temperature, say T_0, the meniscus in a given tube will reach the bottom; thus the filling density in that tube will equal the density of the saturated vapor at the temperature T_0. This density is the total mass of CO_2 in the tube (which is known) divided by the internal volume of the tube (= cross-sectional area × length of capillary).

At temperature T_0, all of the other tubes *that still have both liquid and vapor present* will contain vapor whose density is the same as that just calculated. The volume of vapor in each of these tubes can be determined (= cross-sectional area × length of tube occupied by vapor). Therefore, the mass of CO_2 present as vapor in each of these tubes can be found (= density of vapor × volume of vapor). The mass of liquid CO_2 in each tube can then be determined (= total CO_2 − vapor). Since the volume of the liquid in each tube can also be determined (= cross-sectional area × length of tube occupied by liquid), the density of the liquid phase can be found at temperature T_0.

EXPERIMENTAL

A glass-walled constant-temperature bath with an adjustable thermoregulator and an efficient stirrer is required. This bath should also be provided with a precision thermometer and a light to illuminate the capillary tubes. The heights of the menisci in the capillaries, which are completely immersed in the thermostat bath, should be measured with a *cathetometer*. This instrument is used for the accurate measurement of vertical distances, such as the height of the menisci in a wide-bore manometer. It consists of a heavy steel bar (or rod) mounted on a sturdy tripod stand. This steel bar is graduated in millimeters and supports a traveling telescope, which can be moved vertically through about 100 cm and can also be rotated in a horizontal plane. Leveling screws in the base of the stand are used to obtain accurate vertical alignment of this steel scale, and the telescope mounting is equipped with a fine-adjustment screw and a spirit level to ensure accurate horizontal positioning of the telescope. The meniscus to be measured is brought into focus and aligned with respect to a cross-hair in the eyepiece; the position of the telescope on the scale is then read to the nearest 0.1 or 0.05 mm with a vernier scale. A cathetometer is especially convenient for reading levels on an apparatus that must be immersed in a constant-temperature bath (e.g., the CO_2 capillaries in the present experiment).

Procedure. If the cathetometer is not level, it should be adjusted using the spirit levels on the base and on the telescope. Focus the telescope on one of the capillaries in the bath. The vertical cross-hair should be lined up parallel to the side of the capillary. This may be achieved if necessary by rotating the eyepiece. When reading the height of an object, such as a meniscus, always use the same position on the horizontal cross-hair—usually near the center of the field. Note that the image is inverted by the telescope.

Now check the temperature of the bath. If it is above 24°C, add a little ice to cool it to that temperature; while the bath is coming to equilibrium, you can begin to measure the dimensions of the tubes (see below). If the temperature of the bath is constant at 24°C or below, record the temperature and measure the positions of the menisci in all the tubes. The readings of the cathetometer should be recorded to the nearest 0.005 cm, and you should measure one of the tubes several times to establish the reproducibility of these readings. During the period in which these cathetometer measurements of the menisci levels are being made, the bath temperature should be determined several times and recorded to the nearest 0.01°C. The temperature should not vary over a range of more than 0.04°C during the measurements, and the average of these temperature readings should be used in the subsequent calculations.

It will be convenient to number the tubes in sequence from left to right, in order to simplify the tabulation of data. Once the first set of data are recorded, begin heating the bath to the next desired temperature by making a suitable adjustment to the temperature controller (see Chapter XVII). Recommended bath temperatures are ~24, 26.0, 28.0, 29.2, 29.7, 30.1, 30.5°C, and higher by increments of 0.3°C (or less) until the critical temperature is reached. The bath should regulate at each temperature for at least 10 min (preferably more near T_c) before the meniscus levels are measured.

Once you have begun making readings, be very careful not to jar the cathetometer or to make any more leveling adjustments. Throughout the course of the entire experiment, the only permissible movements of the cathetometer are the raising, lowering, and traversing of the telescope.

While the bath is heating up from one temperature setting to the next, carry out measurements of the dimensions of the capillary tubes. Since the top and bottom ends of these tubes are tapered or rounded, it is necessary to make two cathetometer readings at each end. As shown in Fig. 1, readings are made at the extreme tips of the inside space (a and d) and at levels where the tapering begins (b and c). The center section (bc) is a cylinder of uniform cross-sectional area; the small ends (ab and cd) can be described by cones or hemispheres.

If possible, try to observe the *critical opalescence* in carbon dioxide near the critical temperature. This milky opalescence is caused by large spatial fluctuations in the density and can only be seen if the filling density is close to the critical density of CO_2. In order to look for the opalescence, darken the room (or at least shield the bath from bright lights) and shine a flashlight on the capillary tube whose meniscus is closest to the center of the tube. Look for the scattered light at roughly 90° from the direction of the flashlight beam.

Do not be surprised if, during the experiment, you find that some menisci move toward the top of the tube with increasing temperature. Some of the menisci will become difficult to see as the critical temperature is approached, but do not give up looking too soon because only a few will disappear until just before the critical temperature.

Precaution: Do not raise the temperature of the thermostat bath appreciably above the critical temperature, as the pressure increase in the capillaries could possibly cause them to explode (the critical pressure of CO_2 is 73.8 bar = 72.9 atm). Although the possibility of explosion is remote, the wearing of *safety glasses,* which should be mandatory practice in all laboratory work, should be particularly emphasized in this experiment.

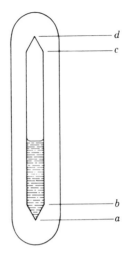

FIGURE 1

Sketch of a capillary tube with conical ends. The internal volume is given by $V = \pi r^2 (h_{bc} + \frac{1}{3}h_{ab} + \frac{1}{3}h_{cd})$, where $2r$ is the inner diameter of the tube and h_{ij} represents the vertical distance between levels i and j. If the ends look more hemispherical than conical, one should find $h_{ab} \simeq h_{cd} \simeq r$; in this case, $V = \pi r^2(h_{bc} + \frac{4}{3}r)$.

CALCULATIONS

First calculate the total internal volume V^i of each of the $i = 1, 2, \ldots, N$ capillaries. The drawn-out ends can be assumed to be conical (or hemispherical) in shape. Calculate the volume of liquid V_l^i in each capillary at each temperature, and obtain the volume of the coexisting vapor V_v^i from

$$V_v^i = V^i - V_l^i \qquad (2)$$

The liquid and vapor densities can be determined either by a graphical method, or, preferably, by a numerical least-squares analysis as described in Chapter XXI. Both of these methods are described below.

Graphical Method.   Plot V_l versus temperature for each capillary in which the meniscus has reached the bottom, and extrapolate to zero volume. At this temperature, call it T_0, you know the density of the vapor in each of the capillaries still containing liquid and vapor. Obtain by interpolation the volume of vapor and liquid, V_v and V_l, for each of these capillaries at T_0. From the known total mass of CO_2 in each tube, several values of the liquid density at T_0 will be obtained. Not all of these values are equally reliable. You should take this into account in arriving at the "best" value for the liquid density at T_0. If the meniscus reaches the top of any of the capillaries, a similar extrapolation of vapor volume V_v versus temperature to zero volume gives a temperature at which the liquid density in all capillaries containing both liquid and vapor is known. Several values for the vapor density can be obtained from capillaries containing liquid and vapor at this temperature, and a "best" value of the vapor density can be chosen.

Least-Squares Analysis.   In this experiment there are two phases, and only one component is present in the tubes. Thus the phase rule tells us that there is only one degree of freedom,

$$F = C - P + 2 = 1$$

and by fixing the temperature, the vapor pressure and the densities of the liquid and vapor are also fixed. At a given temperature T, there will be N equations of the form

$$m_i = \rho_l V_l^i + \rho_v V_v^i \tag{3}$$

where m_i is the total mass of CO_2 in the ith tube. V_l^i and V_v^i are the volumes occupied by liquid and vapor in the ith tube at T. In this experiment the number of tubes N is six or more, while only two are needed to find the two unknowns $\rho_l(T)$ and $\rho_v(T)$ in Eq. (3). Owing to experimental errors, no pair of density values will satisfy all N equations exactly. Thus one wants to make a least-squares fit, and Eq. (3) has a form suitable for a multiple linear regression using a spreadsheet.

For an insight into this type of fitting procedure, let us rewrite Eq. (3) in terms of the residual r_i for the ith tube:

$$r_i \equiv \rho_l V_l^i + \rho_v V_v^i - m_i \tag{4}$$

The method of least squares tells us that the best possible fit to all the data is obtained when the sum of the squares of the residuals is a minimum. Thus, for equal weight fitting (see Chapter XXI),

$$X^2 = \sum r_i^2 = \sum_{i=1}^{N} (\rho_l V_l^i + \rho_v V_v^i - m_i)^2 \tag{5}$$

should be a minimum. This quantity can be minimized with respect to the values of the two parameters ρ_l and ρ_v by using the equations

$$\frac{\partial X^2}{\partial \rho_l} = 0 = 2 \sum (\rho_l V_l^i + \rho_v V_v^i - m_i) V_l^i \tag{6a}$$

$$\frac{\partial X^2}{\partial \rho_v} = 0 = 2 \sum (\rho_l V_l^i + \rho_v V_v^i - m_i) V_v^i \tag{6b}$$

These equations reduce to two equations in two unknowns:

$$\rho_l[\Sigma \ (V_l^i)^2] + \rho_v \ [\Sigma \ (V_v^i V_l^i)] \ = \ \Sigma \ m_i V_l^i \tag{7a}$$

$$\rho_l[\Sigma \ (V_l^i V_v^i)] + \rho_v \ [\Sigma \ (V_v^i)^2] \ = \ \Sigma \ m_i V_v^i \tag{7b}$$

These equations can then be solved for the best values of ρ_l and ρ_v at each temperature.

Using the resulting pairs of density values, construct a Cailletet–Mathias plot (liquid and vapor densities versus temperature). From this plot, obtain the critical temperature T_c, test the "law" of rectilinear diameters [see Eq. (1)], and obtain the critical density ρ_c.

DISCUSSION

The critical temperature and molar volume can be related to the constants of the van der Waals equation of state:

$$\left(p + \frac{n^2 a}{V^2}\right)(V - nb) = nRT \tag{8}$$

The resulting expressions are[3]

$$\tilde{V}_c \ = 3b \qquad T_c \ = \ \frac{8a}{27bR} \tag{9}$$

Using these equations and your values of T_c and $\tilde{V}_c$, calculate a and b. How well do your values agree with values tabulated in the literature? For a van der Waals fluid, the critical pressure is given by

$$p_c \ = \ \frac{a}{27b^2} = \ \frac{3RT_c}{8\tilde{V}_c} \tag{10}$$

Calculate p_c from Eq. (10) and your T_c and $\tilde{V}_c$ values, and compare it with the accepted literature value. From the empirical corresponding-states behavior of many fluids, one would expect that $p_c \tilde{V}_c / RT_c \ \simeq 0.28$. Will this yield a better p_c value than Eq. (10)?

Explain why the meniscus in some tubes moved down and disappeared at the bottom, while the meniscus in other tubes may have moved up and disappeared at the top. Is it possible for the meniscus to remain stationary as the temperature is increased?

In recent years, it has been established experimentally and theoretically that the variation of many thermodynamic properties near a critical point can be described by the use of *critical exponents*.[4] In the case of fluids, the density difference $\rho_l(T) - \rho_v(T)$ is well represented by

$$\frac{\rho_l(T) - \rho_v \ (T)}{\rho_c} \ = \ B\left(\frac{T_c - T}{T_c}\right)^\beta \tag{11}$$

where the exponent β has a value of 0.325. Thus a log–log plot of $(\rho_l - \rho_v)$ versus $(T_c - T)$ should yield a straight line of slope β. Make such a plot and determine the value of β either graphically or from a least-squares fit of your $\Delta\rho(\Delta T)$ data with Eq. (11).

Binary-Liquid Option.   As an alternative to this study of critical behavior in a pure fluid, one can use quite a similar technique to investigate the coexistence curve and critical point in a binary-liquid mixture.[5] Many mixtures of organic liquids (call them A and B) exhibit an upper critical point, which is also called a *consolute point*. In this case, the system exists as a homogeneous one-phase solution for all compositions if T is greater than

T_c. For temperatures below T_c, the system separates into two coexisting liquid phases—one with $X_A < X_c$ and one with $X_A > X_c$, where X_c denotes the mole fraction of component A at the critical point. Since this binary-liquid phase separation can be studied at a constant pressure of 1 atm, the cells are much simpler and easier to fill than those required for a pure fluid.

SAFETY ISSUES

The prospect that a sealed capillary tube will explode due to excessively high internal pressure is quite remote, but it is prudent not to heat the capillaries above 32°C. Even if one of the tubes were to explode, no damage is likely since they are immersed in a thermostat water bath. Nevertheless, the importance of wearing safety glasses for all laboratory work should be stressed.

APPARATUS

Thermostat bath with a heater, adjustable thermoregulator, and stirrer; thermometer with a resolution of at least ± 0.01 K (a platinum resistance thermometer, thermocouple or quartz thermometer could be used; see Chapter XVII); illuminating light and cathetometer for reading the vertical position of the menisci; number of sealed capillaries of uniform bore containing varying amounts of carbon dioxide, prepared in advance.

REFERENCES

1. J. S. Rowlinson and F. L. Swinton, *Liquids and Liquid Mixtures,* 2d ed., pp. 90–94, Plenum Trade/Perseus, Cambridge, MA (1969).

2. M. S. Banna and R. D. Mathews, *J. Chem. Educ.* **56,** 838 (1979).

3. I. N. Levine, *Physical Chemistry,* 6th ed., sec. 8.4, McGraw-Hill, New York (2009); M. W. Zemansky and R. Dittman, *Heat and Thermodynamics,* 6th ed., McGraw-Hill, New York (1981).

4. H. E. Stanley, *Introduction to Phase Transitions and Critical Phenomena,* pp. 9–12, 39–49, Oxford Univ. Press, New York (1987); M. R. Moldover, *J. Chem. Phys.* **61,** 1766 (1974).

5. S. B. Ngubane and D. T. Jacobs, *Am J. Phys.* **54,** 542 (1986).

GENERAL READING

J. S. Rowlinson and F. L. Swinton, *Liquids and Liquid Mixtures,* 2d ed., Plenum Trade/Perseus, Cambridge, MA (1969).

H. E. Stanley, *Introduction to Phase Transitions and Critical Phenomena,* Oxford Univ. Press, New York (1987).

M. A. Anisimov, *Critical Phenomena in Liquids and Liquid Crystals,* Gordon and Breach, Newark, NJ (1991).

EXPERIMENTS

17. Conductance of Solutions
18. Temperature Dependence of emf
19. Activity Coefficients from Cell Measurements

EXPERIMENT 17
Conductance of Solutions

In this experiment we shall be concerned with electrical conduction through aqueous solutions. Although water is itself a very poor conductor of electricity, the presence of ionic species in solution increases the conductance considerably. The conductance of such electrolytic solutions depends on the concentration of the ions and also on the nature of the ions present (through their charges and mobilities), and conductance behavior as a function of concentration is different for strong and weak electrolytes. Both strong and weak electrolytes will be studied at a number of dilute concentrations, and the ionization constant for a weak electrolyte can be calculated from the data obtained.

THEORY

Electrolyte solutions obey Ohm's law just as metallic conductors do. Thus the current I passing through a given body of solution is proportional to the applied potential difference V. The resistance R of the body of solution in ohms (Ω) is given by $R = V/I$, where the potential difference is expressed in volts and the current in amperes. The *conductance*, defined as the reciprocal of the resistance, of a homogeneous body of uniform cross section is proportional to the cross-sectional area A and inversely proportional to the length l:

$$\frac{1}{R} = \frac{\kappa A}{l} \qquad (1)$$

or

$$\kappa = \frac{1}{R}\frac{l}{A} = \frac{k}{R} \tag{2}$$

where κ is the *conductivity* with units $\Omega^{-1}\,m^{-1}$. (By international agreement, the reciprocal ohm Ω^{-1} is now called a *siemen*, $1\,S \equiv 1\,\Omega^{-1}$, but this nomenclature will not be used here.) The conductivity of a given solution in a cell of arbitrary design and dimensions can be obtained by first determining the cell constant k (the "effective" value of l/A) by measuring the resistance of a solution of known conductivity. The standard solution used for this purpose will be 0.02000 *M* potassium chloride. Once the cell constant k has been determined, conductivities can be calculated from experimental resistances by using Eq. (2).

The conductivity κ depends on the equivalent concentrations and the mobilities of the ions present. Consider a single electrolyte $A_{\nu_+}B_{\nu_-}$ giving ions A^{z_+} and B^{z_-} and having fractional ionization α at a solute concentration c in *moles per liter*. The concentration in equivalents per liter for the positive ions (and also for the negative ions) is $\nu\alpha c$, where $\nu = \nu_+ z_+ = \nu_- |z_-|$ denotes the number of equivalents of positive or negative ions per mole of the electrolyte. The conductivity is given by[1]

$$\kappa = 1000\alpha c \mathcal{F}(\nu_+ z_+ U_+ + \nu_- |z_-| U_-)$$
$$= 1000\alpha c \nu \mathcal{F}(U_+ + U_-) \tag{3}$$

The quantity $1000\alpha c$ is the concentration of A^+ ions (and also of B^- ions) in mol m^{-3}, $\mathcal{F} = N_0 e$ is the Faraday constant (96485.31 C equiv^{-1}), and U_i is the ionic mobility of charged species i. Note that the mobility is defined as the migration speed of an ion under the influence of unit potential gradient and hence has the units $m^2\,s^{-1}\,V^{-1}$. It is now convenient to define a new quantity, the *equivalent conductance* Λ, by

$$\Lambda = \frac{\kappa}{1000\nu c} = \alpha \mathcal{F}(U_+ + U_-) \tag{4}$$

Note that Λ has units of $\Omega^{-1}\,m^2\,equiv^{-1}$. In this experiment, we are concerned with simple one-one electrolytes A^+B^-. In this case, where $\nu = 1$ equiv mol^{-1}, there is no distinction between equivalents and moles and the equivalent conductance is the same as the *molar conductance* Λ_m.[1]

For a strong electrolyte, the fraction ionized is unity at all concentrations; thus Λ is roughly constant, varying to some extent owing to changes in mobilities with concentration but approaching a finite value Λ_0 at infinite dilution. The effect of ionic attraction reduces the mobilities, and Onsager[2] showed theoretically for strong electrolytes in dilute solution that

$$\Lambda = \Lambda_0 (1 - \beta\sqrt{c}) \tag{5}$$

Using this relation, Λ_0 for strong electrolytes can be obtained experimentally from measurements of conductance as a function of concentration. At infinite dilution the ions act completely independently, and it is then possible to express Λ_0 as the sum of the limiting conductances of the separate ions:

$$\Lambda_0 = \lambda_0^+ + \lambda_0^- \tag{6}$$

for a one-one electrolyte A^+B^-, where $\lambda_0^+ = \mathcal{F}U_+^0$ and $\lambda_0^- = \mathcal{F}U_-^0$.

For a weakly ionized substance, Λ varies much more markedly with concentration because the degree of ionization α varies strongly with concentration. The equivalent conductance, however, must approach a constant finite value at infinite dilution, Λ_0, which again corresponds to the sum of the limiting ionic conductances. It is usually impractical to determine this limiting value from extrapolation of Λ values obtained with the weak electrolyte itself, since to obtain an approach to complete ionization the concentration must be made too small for effective measurement of conductance. However, Λ_0 for a weak electrolyte can be deduced from Λ_0 values obtained for *strong electrolytes* by the use of Eq. (6). As an example, let us consider acetic acid (CH_3COOH, denoted as HAc) as a typical weak electrolyte. It follows from Eq. (6) that

$$\Lambda_0(HAc) = \Lambda_0(HX) + \Lambda_0(MAc) - \Lambda_0(MX) \tag{7}$$

where M^+ is any convenient univalent positive ion such as K^+ or Na^+ and X^- is a univalent negative ion such as Cl^- or Br^-. The only restriction on M and X is the requirement that HX, MAc, and MX must all be strong electrolytes so that their Λ_0 values can be obtained by extrapolation using Eq. (5).

For sufficiently weak electrolytes, the ionic concentration is small and the effect of ion attraction on the mobilities is slight; thus we may assume the mobilities to be independent of concentration and obtain the approximate expression

$$\alpha \simeq \frac{\Lambda}{\Lambda_0} \tag{8}$$

If one measures Λ for a weak electrolyte at a concentration c and calculates Λ_0 from the conductivity data for strong electrolytes as described above, it is possible to obtain the degree of ionization of the weak electrolyte at concentration c.

Equilibrium Constant for Weak Electrolyte. Knowing the concentration c of the weak electrolyte, say HAc, and its degree of ionization α at that concentration, the concentrations of H^+ and Ac^- ions and of un-ionized HAc can be calculated. Then the equilibrium constant in terms of concentrations K_c can be calculated from

$$K_c = \frac{(H^+)(Ac^-)}{(HAc)} = c\frac{\alpha^2}{1 - \alpha} \tag{9}$$

The equilibrium constant K_c given by Eq. (9) using α values obtained from Eq. (8) differs from K_a, the true equilibrium constant in terms of activities, owing to the omission of activity coefficients ($\gamma_\pm^2$) from the numerator of Eq. (9) and the approximations inherent in Eq. (8). At the very low ionic concentrations encountered in the dissociation of a weak electrolyte, a simple extrapolation procedure can be developed to obtain K_a from the values of K_c. Since $K_a = K_c \gamma_\pm^2$ is an excellent approximation, it follows that

$$\log K_c = \log K_a - 2 \log \gamma_\pm \tag{10}$$

According to Debye–Hückel theory,[3] the mean activity coefficient at low and moderate ionic concentrations is given by

$$\log \gamma_\pm = -|z_+ z_-| \frac{A\sqrt{I}}{1 + B\sqrt{I}} \tag{11}$$

The ionic strength I is defined by the expression

$$I = \tfrac{1}{2} \sum_i m_i z_i^2 \tag{12}$$

where m_i is the concentration of the ith ionic species in moles per kilogram of solvent (i.e., the molality), z_i is the charge on ion i in units of the electron charge, and the sum is taken over *all* ionic species present in the solution. For dilute aqueous solutions at 25°C, the quantity A in Eq. (11) has the value $0.509 \text{ kg}^{1/2} \text{ mol}^{-1/2}$ and the quantity B is found to be close to unity for many salts.[4] Furthermore, the difference between molarity and molality is sufficiently small in dilute aqueous solutions that one can approximate m_i by the molar concentration c_i in mol L^{-1}.

For dilute solutions of HAc, $I = \alpha c$ has a very small value and Eq. (10) is well approximated by

$$\log K_c = \log K_a + 2(0.509)\sqrt{\alpha c} \tag{13}$$

Thus, if K_c has been determined from conductance measurements at a number of low HAc concentrations, one can plot $\log K_c$ against $\sqrt{\alpha c}$ and make a linear extrapolation to $c = 0$ to obtain K_a.

METHOD

The use of dc circuitry is impractical for determining ionic conductance from measurements of the resistance of the solution in a conductivity cell, since the electrodes quickly become *polarized;* that is, electrode reactions take place that set up an emf (electromotive force) opposing the applied emf, leading to a spuriously high apparent cell resistance. Polarization can be prevented by (1) using a high (audio)-frequency *alternating current,* so that the quantity of electricity carried during one half-cycle is insufficient to produce any measurable polarization, and at the same time by (2) employing platinum electrodes covered with a colloidal deposit of "platinum black," having an extremely large surface area, to facilitate the adsorption of the tiny quantities of electrode reaction products produced in one half-cycle so that no measurable chemical emf is produced.[5]

Although conductance/resistance meters are available commercially (e.g., YSI, Yellow Springs Instruments, and others), the simple ac Wheatstone bridge of Fig. 1 is adequate and better illustrates some of the principles for accurate conductivity measurements. Here the ionic solution is placed in the conductivity cell, which is part of a bridge subjected to a small ac voltage from a 1-kHz oscillator. The condition of balance for the bridge is detected with an oscilloscope and requires that the alternating potential at points B and D be of equal amplitude and exactly in phase. This corresponds to a balance condition

$$\frac{Z_1}{Z_2} = \frac{Z_3}{Z_4} \tag{14}$$

FIGURE 1

Conductance bridge. The slide-wire reading X is on a scale from 0 to 1000. The oscillator should be isolated from the bridge circuit by a good-quality transformer, and another transformer should be used to isolate the bridge from the oscilloscope. The oscillator and cables should be well shielded.

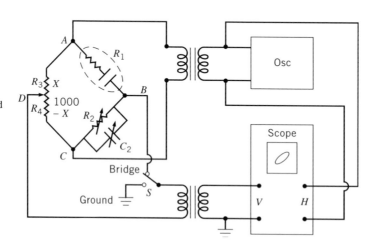

where Z is the impedance. Since the cell has both resistance and capacitance, it represents a complex impedance Z_1; therefore at least one of the other arms must be complex in order to satisfy Eq. (14). In other words it is necessary to include some variable reactance in order to balance both the real (resistive) and imaginary (reactive) components and achieve a null signal through the detector. In the bridge circuit shown in Fig. 1, arm 2 is complex owing to the presence of an adjustable parallel capacitor C_2. Arms 3 and 4 are pure resistance ratio arms (i.e., $Z_3 = R_3$ and $Z_4 = R_4$).

Once a balance condition has been achieved, there is still the problem of relating the cell resistance R_1 to the known resistance R_2. The principal difficulty lies in the need to adopt an equivalent electrical circuit for the cell. This question is discussed in detail elsewhere,[5,6] but the simplest realistic model is

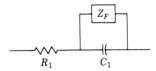

where the *series* capacitance C_1 is due to the electrical double layer at the interface between the electrode and the electrolyte and the frequency-dependent faradaic impedance Z_F is associated with the transfer of charge through the interface. Note that capacitances in parallel with R_1 have been neglected in this model. For "ideally polarized" electrodes, one can neglect the faradaic leakage (set $Z_F = \infty$). The resulting approximation for the impedance of the cell is a series combination of a resistance R_1 and a capacitance C_1, yielding

$$Z_1 = R_1 + iX_1 = R_1 + \frac{1}{i\omega C_1} = R_1 - i\frac{1}{\omega C_1} \tag{15}$$

where $i = \sqrt{-1}$, X is the capacitive reactance, and $\omega = 2\pi f$. The impedance of the parallel combination of resistance R_2 and capacitance C_2 used in arm 2 is given by

$$\frac{1}{Z_2} = \frac{1}{R_2} + \frac{1}{iX_2} = \frac{1}{R_2} + i\omega C_2 \tag{16}$$

Thus Eq. (14) becomes

$$\left(\frac{R_1}{R_2} + \frac{C_2}{C_1}\right) + i\left(\omega R_1 C_2 - \frac{1}{\omega R_2 C_1}\right) = \frac{R_3}{R_4} \tag{17}$$

where $\omega R_1 C_2 - (1/\omega R_2 C_1)$ must equal zero since R_3/R_4 is real. Using this condition, one finds from the real part

$$R_2 \frac{R_3}{R_4} = R_1\left(1 + \frac{1}{\omega^2 R_1^2 C_1^2}\right) \tag{18}$$

Typical $R_1 C_1$ values are so large that the second term above can be neglected,† and one should find at all frequencies that

$$R_1 = R_2 \frac{R_3}{R_4} \tag{19}$$

†Since the double layer is very thin ($\sim$1 Å), large capacitances are possible. A platinized electrode may have an effective surface area 10 to 100 times the area of a bright metal electrode of the same size. Thus the capacitance of the double layer for a 1-cm² platinized electrode can be 100 μF or greater. Taking as typical conditions $R_1 > 10^3\ \Omega, f = 10^3$ Hz and $C_1 > 100\ \mu$F, the quantity $1/\omega^2 R_1^2 C_1^2 < 2.5 \times 10^{-6}$ which is negligible compared to 1 in Eq. (18).

This is, fortunately, the behavior observed for cells having platinum electrodes with a heavy deposit of platinum black such as those to be used in this experiment. The expression that is valid for bright platinum electrodes is considerably more complicated.[6]

The use of a parallel combination of R_2 and C_2, as in Fig. 1, is attractive because the value of C_2 needed to compensate for the capacitance of the cell is quite low, and small capacitors are cheaper, more accurate, and less frequency dependent than large ones.[6] By adjusting C_2, and thus compensating for the phase shift in the cell, one can improve the balance of the bridge. The ideal detector is an oscilloscope since this permits separate observation of both the capacitive and the resistive balance[7] and can also reveal any waveform distortion that might occur if there is serious polarization of the electrodes.

As described in Chapter XIX, a 1:1 Lissajous figure can be generated on an oscilloscope screen when the imbalance signal between B and D is applied to the vertical input and a sine-wave signal from the audio oscillator is applied to the horizontal input. This configuration is shown in Fig. 1. If all the bridge arms were pure resistance elements, bridge imbalance would be indicated by an oscilloscope trace that is a tilted straight line whose vertical projection is proportional to the voltage difference between B and D. Thus a horizontal trace would indicate bridge balance. If the unknown arm is a conductivity cell, the oscilloscope trace will in general be an ellipse, since the cell is not a pure resistance and the potentials at B and D are generally not in phase. A capacitive balance of the bridge will close the ellipse to produce a tilted line; resistive balance is then indicated by a horizontal line. In practice the balance procedure is more complex than this, and the bridge is considered balanced when the minor axis of the ellipse is as short as possible and the major axis is as nearly horizontal as possible. Under the conditions encountered in this experiment, very good null patterns should be achieved for all but the most dilute solutions studied.

EXPERIMENTAL

Set up the circuit shown in Fig. 1, making sure that the leads or lugs on all hook-up wires are clean before making connections; if necessary use fine emery paper to remove any dirt or oxide coating. All connections must be *tight* to avoid any unwanted contact resistances. A 10-turn Helipot variable-resistance slide wire can be used for the ratio arms 3 and 4. Alternatively the slide wire in a potentiometer can also be used; see Chapter XVI for more details. A high-quality four- or five-decade resistance box with a range from 0 Ω to 100 kΩ is used as R_2, while a variable or decade capacitor serves as C_2. Note that it is not necessary to know the value of C_2, but it must be variable over a sufficient range (0 to 0.5 μF) to allow compensation of the phase shift due to the cell.

The bridge should be operated with as low a power level and for as short a time as possible to prevent polarization effects.† If necessary, a preamplifier can be used to increase the imbalance signal before it is applied to the vertical input terminals of the oscilloscope. Be careful that the output leads of the preamplifier are properly connected to the input terminals of the scope, one of which is usually grounded. In order to optimize the accuracy of the measurements, choose R_2 values whenever possible so that the variable terminal D is near the center of the slide wire, i.e., the reading X is near 500 and $R_3/R_4 \simeq 1$).

Two methods of using the oscilloscope as a null detector are described below.

Sine wave. Set the selector switch on the scope to "internal sweep." As the null point is approached, the amplitude of the sinusoidal imbalance signal will diminish. At the maximum gain setting, the scope pattern near the balance point will be a 60-Hz leakage

---

†In the unlikely event that significant polarization occurs, very poor balance patterns will be observed because the severe distortion of the waveform cannot be compensated by varying capacitor C_2.

signal with a small 1-kHz bridge signal superimposed. Do not attempt to balance out the 60-Hz signal; merely balance for minimum 1-kHz ripple.

Lissajous figure. Set the selector switch on the scope to "horizontal input." Now both the horizontal and the vertical inputs come from the oscillator. Generally an ellipsoidal pattern will be seen on the scope. The width (minor axis) of the ellipse can be reduced by balancing the reactance. The null point occurs when the major axis of the ellipse is as nearly horizontal as possible.

Calibration. There may be small systematic errors in the slide wire due to nonlinearity of the wire or failure of the indicator to correspond precisely to the position of the movable contact. In this case the reading X on a scale from 0 to 1000 may yield a value of $X/(1000 - X)$ that does not equal the true value of R_3/R_4. This can be checked, and corrected if necessary, by carrying out calibration measurements using a high-precision resistance box in place of the cell in arm 1.

Set both the resistances R_1 and R_2 at 3000 Ω and balance the bridge. If there are no errors, the X reading should be exactly 500. To obtain a correction figure, make four readings of the balance approaching the null twice from each side. Then reverse the positions of the resistance boxes and take four more readings. The average of all eight readings will be called X^*. A correction quantity $(500 - X^*)$ should be added to all future X readings that are close to the center of the dial.

Conductivity cell. A satisfactory conductivity cell design is shown in Fig. 2. This cell is fragile and should be handled with care. The leads should be arranged to avoid placing any strain on the cell while it is mounted in the thermostat bath. The electrodes are sensitive to poisoning if very concentrated electrolytes are placed in the cell and will deteriorate if allowed to become dry. (Details concerning the preparation of "platinized" platinum electrodes are given in Chapter XX.) After each emptying, the cell must *immediately* be rinsed or filled with water or solution. While the cell is in the bath, the filling arms must be stoppered by placing rubber policemen over the ends. If the cell is allowed to stand for any long time between runs, leave it filled with conductivity water rather than solution.

Draw liquid into the cell through the tapered tip as you would with a pipette. Empty the cell by allowing it to drain out the other end. The procedure for filling the cell with a new solution is to empty the cell, rinse it at least *four times* with aliquots of the desired

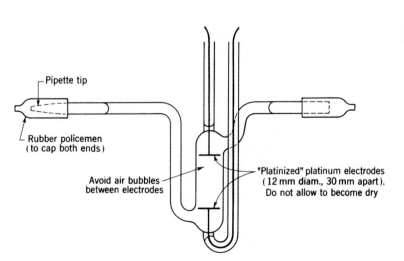

FIGURE 2
Conductivity cell.

Pipette tip

Rubber policemen
(to cap both ends)

Avoid air bubbles
between electrodes

"Platinized" platinum electrodes
(12 mm diam., 30 mm apart).
Do not allow to become dry

solution, using about enough to fill the cell one-fourth full, shake well on each rinsing, then fill the cell, being careful to avoid bubbles on the electrodes.

Conductivity water. Ordinary distilled water is not suitable for use in this experiment, since its conductance is too high. Much of this conductivity results from dissolved CO_2 gas from the air, which can be removed by boiling distilled water and capping a full bottle while it is still hot. Water of better conductivity can be prepared by special distillation[5] or by passing distilled water through an ion-exchange resin to "deionize" it. Store conductivity water in screw-cap polyethylene bottles to avoid contamination by ionic substances leached from glass.

Rinse thoroughly and fill the cell with conductivity water. Place the cell in arm 1 of the bridge and determine its resistance. High precision is not required, but the resistance should be high enough to yield a water conductivity κ_w less than $2 \times 10^{-4}\,\Omega^{-1}\,m^{-1}$. For a cell like that shown in Fig. 2, which will have a cell constant k of roughly 250 m^{-1}, this means a resistance $R_w > 1.3 \times 10^6\,\Omega$. If the cell to be used has significantly different dimensions, estimate k from roughly measured values of the spacing l and the area A of the electrodes and determine the minimum resistance corresponding to the upper bound $2 \times 10^{-4}\,\Omega^{-1}\,m^{-1}$ on κ_w. If the cell resistance is too low, rinse the cell a few more times and repeat the measurement. Record the value of R_w obtained.

Standardization. Rinse and fill the cell with the 0.02000 M KCl solution.† Then mount the cell in a 25°C thermostat bath and connect it to the bridge circuit. Balance the bridge, adjusting the value of R_2 so that the slide-wire reading is roughly in the middle of its range (i.e., X near 500). If necessary, improve the null point by adjusting the value of the capacitor C_2. If excessive capacitance C_2 appears to be required, polarization is probably taking place and an instructor should be consulted.

Once the best possible null point has been achieved, do not disturb any of the settings except the slide-wire contact. Final measurements on the solution consist of a set of four readings, the balance being approached twice from each side. The value of X on the slide-wire scale is recorded for each balance. There should be no appreciable drift in these values; if drift is observed, the cell is probably not in thermal equilibrium with the bath. Wait a few minutes and repeat the measurements. Be sure to record R_2 for the resistance box. The measurement on this KCl solution should be made with great care, since it determines the cell constant k and will affect all subsequent calculations.

Procedure. Determine the resistance R of the cell containing acetic acid solutions at concentrations $c_a = 0.05\,M$, $c_a/4$, $c_a/16$, and $c_a/64$. Also determine R for solutions of the strong electrolyte KAc at $c_b = 0.02\,M$, $c_b/4$, $c_b/16$, and $c_b/64$.

If time permits, determine R for KCl and/or HCl solutions at c_b, $c_b/4$, $c_b/16$, and $c_b/64$ as well.

The procedure for each solution is the same as that described above for the standardization measurement on 0.02 M KCl. For several of the HAc solutions and for the most dilute strong electrolytes, the resistance may be so high that it is not possible to balance the bridge with X near 500. One can either use two decade resistance boxes in series for arm 2 or merely take slide-wire readings at larger X values. In either case be sure that your bridge calibration procedure will provide any necessary correction factors.

For dilute solutions with high resistance, the bridge balance is more difficult to achieve. Iterative adjustments of R_2 and C_2 will be required to achieve a good null point.

---

†It is recommended that a *precisely* 0.02000 M KCl solution be used for standardization. If this is not available, record the actual concentration used and make the appropriate correction as described in the Calculations section.

Dilutions. The quality of your results depends on accurate dilutions and scrupulous care in avoiding contamination of the dilute solutions by stray electrolytes. Follow the dilution procedure described below.

Clean thoroughly a 100-mL volumetric flask, a 125-mL wide-mouth flask, and a 25-mL pipette. Rinse these with conductivity water. Rinse the 125-mL flask with two or three *small* aliquots of the stock solution, and then take not more than 100 mL of the solution, noting the concentrations given on the stock bottle.

Rinse the pipette with two or three small aliquots from the flask, and then pipette exactly 25 mL of the solution from the flask into the volumetric flask. Make up to the mark with conductivity water and mix thoroughly. Set this aside for the next measurement and dilution. Use the solution remaining in the 125-mL flask for rinsing and filling the conductivity cell. When the measurement of resistance of the cell containing that solution has been completed, discard the remaining solution from the 125-mL flask and rinse with two or three aliquots of the new solution from the volumetric flask. Pour the contents of the volumetric flask into the 125-mL flask, and rinse the volumetric flask with conductivity water.

Repeat the process described above as many times as necessary to obtain the dilutions needed. Care is important in this work, since dilution errors are cumulative.

Make sure that the thermostat bath is regulating properly at all times; record the bath temperature several times during the experiment. It is a good idea to carry out some rough calculations while you are taking the data, in order to verify that everything is working properly. For example, the product of the cell resistance times the dilution factor $(1, \frac{1}{4}, \frac{1}{16}, \frac{1}{64})$ should be almost constant for strong electrolytes but should decrease for HAc as the solution becomes more dilute.

At the end of each laboratory session, rinse the cell well, fill it with conductivity water, and store it with the filling arms stopped.

CALCULATIONS

The cell resistance R is obtained from Eq. (19). If the calibration procedure showed that significant slide-wire corrections are needed, apply these corrections during the process of converting X into the desired ratio (R_3/R_4).

From the result of the 0.02000 M KCl measurement, calculate the cell constant k from Eq. (2) using $\kappa = 0.27653\ \Omega^{-1}\ m^{-1}$ at 25°C.[8] If the KCl solution was not precisely 0.02000 M, an appropriate κ value can be calculated from the following empirical equation, valid at 25°C over the range $c = 10^{-4}\ M$ to 0.04 M:[8]

$$\kappa(\Omega^{-1}\ m^{-1}) = 14.984c - 9.484c^{3/2} + 5.861c^2 \log c + 22.89c^2 - 26.42c^{5/2} \quad (20)$$

Calculate the conductivity κ for each solution studied. If the conductivity κ_w of the water used in the dilutions is significant compared to the measured conductivity κ_{obs} of any solution, a correction should be made:

$$\kappa = \kappa_{obs} - \kappa_w \quad (21)$$

Using Eq. (4), calculate the equivalent conductance Λ for each solution. The values of R, κ, and Λ should be tabulated for all solutions studied. Cite the units for each quantity, and also give the value of the cell constant k (with units) in the table caption.

For the KAc solutions and any other strong electrolytes studied as a function of concentration, plot Λ versus $\sqrt{c}$ and extrapolate linearly to $c = 0$ in order to obtain Λ_0. In making these extrapolations, beware of increasingly large experimental uncertainties at the lowest concentrations and also of systematic errors due to conducting impurities or dilution errors at low concentrations. If you are making a least-squares fit with Eq. (5) rather

TABLE 1

Solution	Λ_0 at 25°C	$d\Lambda_0/dT$
HCl	0.04262	6.4×10^{-4}
KCl	0.014986	2.8×10^{-4}
KAc	0.01144	2.1×10^{-4}

than a graphical extrapolation, be sure to estimate the relative weights associated with each data point. If runs have not been made on one or more of the strong electrolytes, the data in Table 1 may be used,[5,9] where Λ_0 is given in units of Ω^{-1} equiv^{-1} m^2.

Combine your extrapolated Λ_0 values and/or the data cited above to obtain Λ_0 for acetic acid. If the temperature of any of your runs was different from 25°C, the $d\Lambda_0/dT$ data above will allow a correction to be made.

For each dilution of acetic acid, calculate α from Eq. (8) and then K_c from Eq. (9). Present in tabular form Λ, α, c, and K_c for each dilution of acetic acid. Plot log K_c against $\sqrt{\alpha c}$. If an extrapolation to zero concentration is feasible with a slope of correct sign and magnitude [see Eq. (13)], make it and obtain a value for K_a. If the data do not appear to be sufficiently good to warrant an extrapolation, report an average or best value for K_c. Report the temperature at which the HAc measurements were made.

DISCUSSION

How does the cell constant k compare with the geometric value l/A obtained from an approximate measurement of the dimensions of your cell? Why is the equivalent conductance Λ_0 so large for an HCl solution? How do the slopes of your Λ versus $\sqrt{c}$ plots for strong electrolytes compare with literature values and the values expected from Onsager's theory?[2] Find a literature (or textbook) value for the equilibrium constant K_a for HAc ionization. Using this value and Eq. (13), draw a dashed literature/theory line on your plot of log K_c versus $\sqrt{\alpha c}$. Are the deviations of your data points from this line reasonable in view of the experimental errors expected in this work? What is the limiting factor in the accuracy of your K_c measurements?

SAFETY ISSUES

None.

APPARATUS

AC Wheatstone-bridge circuit; oscillator (1 kHz, with power supply); transformers; preamplifier; oscilloscope; precision resistance decade box or plug box; decade or adjustable capacitor; conductivity cell, filled with conductivity water and capped off with clean rubber policemen; holder for mounting cell in bath; two leads to connect cell to bridge; electrical leads with lugs or soldered tips; electrical switch; 100-mL volumetric flask; 25-mL pipette; pipetting bulb; two 100- or 250-mL beakers; two 125-mL Erlenmeyer flasks; 500-mL glass-stoppered flask for storing conductivity water.

Constant-temperature bath set at 25°C; emery paper; conductivity water (1500 mL); precisely 0.02000 M KCl solution (300 mL); 0.02 M HCl solution (200 mL); 0.02 M potassium acetate solution (200 mL); 0.05 M acetic acid (250 mL).

REFERENCES

1. S. I. Smedley, *The Interpretation of Ionic Conductivity in Liquids,* reprint ed., Books on Demand, Ann Arbor, MI (1980).

2. L. Onsager, *Phys. Z.* **28,** 277 (1927); P. W. Atkins and J. de Paula, *Physical Chemistry,* 8th ed., p. 762, Freeman, New York (2006).

3. P. W. Atkins and J. de Paula, *op. cit.,* pp. 163–165.

4. G. Scatchard, *Chem. Rev.* **19,** 309 (1936).

5. M. Spiro, "Conductance and Transference Determinations," in B. W. Rossiter and J. F. Hamilton (eds.), *Physical Methods of Chemistry,* 2d ed., Vol. II, chap. 8, Wiley-Interscience, New York (1986).

6. R. A. Robinson and R. H. Stokes, *Electrolyte Solutions,* 2d ed., pp. 88–95, Butterworth, London (1959); J. Braunstein and G. D. Robbins, *J. Chem. Educ.* **48,** 52 (1971).

7. D. Edelson and R. M. Fuoso, *J. Chem. Educ.* **27,** 610 (1950).

8. M. Spiro, *op. cit.,* pp. 718–719; E. Juhasz and K. N. Marsh, *Pure Appl. Chem.* **53,** 1841 (1981).

9. H. S. Harned and B. B. Owen, *The Physical Chemistry of Electrolytic Solutions,* 2d ed., app. A, Reinhold, New York (1950).

GENERAL READING

S. I. Smedley, *The Interpretation of Ionic Conductivity in Liquids,* reprint ed., Books on Demand. Ann Arbor, MI (1980).

M. Spiro, "Conductance and Transference Determinations," in B. W. Rossiter and J. F. Hamilton (eds.), *Physical Methods of Chemistry,* 2d ed., Vol. II, chap. 8, Wiley-Interscience, New York (1986).

EXPERIMENT 18

Temperature Dependence of emf

In this experiment the following electrochemical cell is studied:

$$Cd(s)|Cd^{2+}SO_4^{2-}(aq,\ c)|Cd(Hg,\ X_2) \qquad 1\ bar,\ T \qquad (1)$$

The change in state accompanying the passage of 2 faradays of *positive* electricity from left to right through the cell is given by

$$
\begin{array}{ll}
\text{Anode:} & Cd(s) = Cd^{2+}(aq,\ c) + 2e^- \\
\text{Cathode:} & \underline{2e^- + Cd^{2+}(aq,\ c) = Cd(Hg,\ X_2)} \\
\text{Net:} & Cd(s) = Cd(Hg, X_2) \qquad 1\ bar,\ T
\end{array}
\qquad (2)
$$

In the above, (s) refers to the pure crystalline solid, $(aq,\ c)$ refers to an aqueous solution of concentration $c,$ and $(Hg,\ X_2)$ represents a liquid (single-phase) cadmium amalgam in which the mole fraction of Cd is X_2.

From the emf (electromotive force) of this cell and its temperature coefficient, the changes in free energy, entropy, and enthalpy for the above change in state are to be determined.

THEORY

When the cell operates reversibly at constant pressure and temperature, with no work being done except electrical work and expansion work,

$$\Delta G = -n\mathscr{E}\mathscr{F} \tag{3}$$

where ΔG is the increase in free energy of the system attending the change in state produced by the passage of n faradays of electricity through the cell, $\mathscr{F}$ is the Faraday constant, and $\mathscr{E}$ is the emf of the cell.

From the Gibbs–Helmholtz equation we have, for a change in state at constant pressure and temperature,

$$\left(\frac{\partial \Delta G}{\partial T}\right)_p = -\Delta S \tag{4}$$

Combining Eqs. (3) and (4), we find that

$$\Delta S = n\mathscr{F}\left(\frac{\partial \mathscr{E}}{\partial T}\right)_p \tag{5}$$

Knowing both ΔG and ΔS, one can find ΔH by using

$$\Delta G = \Delta H - T\Delta S \tag{6}$$

METHOD

The emf of a cell is best determined by measurements with a potentiometer, since this method gives a very close approach to reversible operation of the cell. The cell potential is opposed by a potential drop across the slide wire of the potentiometer, and at balance only very small currents are drawn from the cell (depending on the sensitivity of the null detector used and the fineness of control possible in adjusting the slide wire). Alternatively a high-resolution digital voltmeter with a large internal impedance can be used. Potentiometers and digital voltmeters are described and compared in Chapter XVI.

EXPERIMENTAL

The cell consists of a small beaker with a top that will accommodate two electrodes as shown in Fig. 1. The *cadmium electrode* is made by plating cadmium onto a platinum wire that is sealed through the bottom of a small glass tube. The *amalgam electrode* is made by placing a small quantity of the cadmium amalgam in the cup of a special J-shaped glass tube with a platinum wire sealed into it. Electrical contacts are made by copper wires spot-welded to the platinum lead wires.

Procedure. A cadmium–mercury amalgam containing 2 percent Cd by weight will be used. This amalgam can be prepared as follows. Remove the oxide coating from a thin rod of pure Cd with dilute acid, rinse the rod thoroughly in distilled water, dry, and weigh. Then dissolve the rod in the proper (weighed) amount of triple-distilled mercury. Using a medicine dropper, fill the cup of the J electrode with this amalgam. Carefully insert this electrode into the cell, which previously has been rinsed and filled with a 0.1 M $CdSO_4$ solution. The cell should be filled only half full to prevent contact of the solution with the cover.

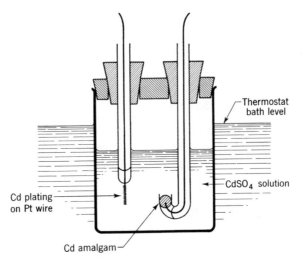

FIGURE 1
Cadmium amalgam cell.

The platinum wire of the other electrode is plated with a cadmium deposit by immersing the electrode in a beaker containing a 0.1 M solution of $CdSO_4$ and a pure cadmium rod. The electrode is connected to the negative terminal of a 1.5-V dry cell, and the cadmium rod is connected to the positive terminal. Current is passed through this plating bath until a heavy deposit of cadmium is visible on the electrode. This cadmium electrode is then transferred to the cell. Handle the electrode with care to avoid flaking off any of the cadmium deposit. It is advisable to keep the electrode wet at all times.

Mount the assembled cell in a thermostat bath. Either a potentiometer circuit or a high-resolution digital voltmeter can be used to measure the cell voltage (see Chapter XVI). In connecting the cell, consider the sign of ΔG for the change in state to decide on the proper connection. Have an instructor check your circuit.

After the cell has been in the thermostat bath for at least 10 min, measure the emf and repeat this measurement at least three times at 5-min intervals to verify that there is no systematic drift in the emf. The potentiometer circuit should be checked against the standard cell immediately before each reading. The emf should be determined as a function of temperature at four values in the range from 0 to 40°C, and an ice bath can be used for the 0°C measurement.

CALCULATIONS

Plot the emf of the cell $\mathcal{E}$ versus the absolute temperature T. Report the value of $\mathcal{E}$ and the slope $(\partial \mathcal{E}/\partial T)_p$ at 25.0°C (298.15 K) from the best smooth curve through your experimental points. Using Eqs. (3), (5), and (6), calculate ΔG, ΔS, and ΔH in J mol^{-1}, and give the change in state to which they apply.

From the known weight percentage of Cd in the amalgam, compute the mole fraction X_2 of cadmium. Assuming for this concentration an activity coefficient of unity (so that the activity a_2 is equal to X_2), compute the standard free-energy change $\Delta G°$ for the change in state

$$Cd(s) = Cd(Hg)$$

by means of the equation

$$\Delta G = \Delta G° + RT \ln a \tag{7}$$

DISCUSSION

Explain why $\Delta G°$ is not equal to zero, as would be expected if the standard state were an "amalgam" with $X_2 = 1$.

SAFETY ISSUES

Any spilled mercury is a concern, since mercury vapor is toxic and a hazardous concentration can build up in the atmosphere over time if spills are not cleaned up promptly and carefully (see Appendix C). Dispose of waste chemicals as instructed.

APPARATUS

Complete potentiometer setup or high-resolution digital voltmeter (see Chapter XVI); cell (50-mL beaker or weighing bottle); platinum electrode; J electrode for holding amalgam; two leads for connections to cell; two beakers; battery jar; large ring stirrer; ring stand; large clamp and clamp holder.

Provision for electroplating cadmium from 0.1 M $CdSO_4$ onto platinum electrode; dilute (2 percent) cadmium amalgam (1 mL); eye dropper; constant-temperature baths set at 15, 25, and 35°C; ice bath; approximately 0.1 M $CdSO_4$ solution (150 mL); ice (1 kg).

GENERAL READING

P. Reiger, *Electrochemistry,* Prentice-Hall, Upper Saddle River, NJ (1987).

V. K. LaMer and W. G. Parks, *J. Amer. Chem. Soc.* **56,** 90 (1934).

R. J. Silbey, R. A. Alberty, and M. G. Bawendi, *Physical Chemistry,* 4th ed., sec. 7.6, Wiley, New York (2005).

EXPERIMENT 19
Activity Coefficients from Cell Measurements

We shall be concerned with the electrochemical cell

$$Pt|H_2(g, p \text{ bar})|H^+Cl^-(aq, c)|AgCl(s)|Ag(s) \tag{1}$$

Measurements of emf (electromotive force) are to be made with this cell under reversible conditions at a number of concentrations c of HCl. From these measurements relative values of activity coefficients at different concentrations can be derived. To obtain the activity coefficients on such a scale that the activity coefficient is unity for the reference state of zero concentration, an extrapolation procedure based on the Debye–Hückel limiting law is used. By this means, the standard electrode emf of the silver–silver chloride electrode is determined, and activity coefficients are determined for all concentrations studied.

THEORY

When 1 faraday of positive electricity passes reversibly through cell (1) from left to right, the overall change in state is

$$AgCl(s) + \tfrac{1}{2}H_2(g) = H^+Cl^-(aq) + Ag(s) \tag{2}$$

Cell (1) may conveniently be considered as a combination of two half-cells or "electrodes," which are conventionally written as undergoing reduction reactions.

Hydrogen Electrode.   This can be written as

$$H^+(aq, c)|H_2(g, p \text{ bar})|Pt \tag{3}$$

With the reversible passage of 1 faraday of positive electricity from left to right through this electrode, the change in state is

$$H^+(aq, c) + e^- = \tfrac{1}{2}H_2(g) \tag{4}$$

and the electrode emf is accordingly

$$\mathcal{E}_3 = \mathcal{E}_3^0 - \frac{RT}{\mathcal{F}} \ln \frac{f_{H_2}^{1/2}}{a_{H^+}} \tag{5}$$

where f_{H_2} is the fugacity of the hydrogen gas, a_{H^+} is the activity of the aqueous hydrogen ion, and $\mathcal{F}$ is the Faraday constant. By convention, the standard potential for the hydrogen electrode is taken to be zero. Thus

$$\mathcal{E}_3^0 = \mathcal{E}_{H^+/H_2}^0 = 0 \tag{6}$$

Silver–Silver Chloride Electrode.   This can be written as

$$Cl^-(aq, c)|AgCl(s)|Ag(s) \tag{7}$$

When 1 faraday of positive electricity passes reversibly from left to right through this electrode, the change in state is

$$AgCl(s) + e^- = Ag(s) + Cl^-(aq, c) \tag{8}$$

and the electrode emf is

$$\mathcal{E}_7 = \mathcal{E}_7^0 - \frac{RT}{\mathcal{F}} \ln a_{Cl^-} \tag{9}$$

where $\mathcal{E}_7^0 = \mathcal{E}_{AgCl/Ag^+Cl^-}^0$ is the standard electrode potential for the silver–silver chloride electrode (a reduction potential) and a_{Cl^-} is the activity of the chloride ion in the aqueous solution.

The Cell.   Noting that cell (1) corresponds to half-cell (7) minus half-cell (3) and of course that change in state (2) is Eq. (8) minus Eq. (4), the emf of the cell is given by

$$\mathcal{E} = \mathcal{E}_7 - \mathcal{E}_3 = \mathcal{E}^0 - \frac{RT}{\mathcal{F}} \ln \frac{a_{H^+} a_{Cl^-}}{f_{H_2}^{1/2}} \tag{10}$$

where

$$\mathcal{E}^0 = \mathcal{E}_7^0 - \mathcal{E}_3^0 = \mathcal{E}_{AgCl/Ag^+Cl^-}^0 \tag{11}$$

Activity Coefficients.   Let us write

$$f_{H_2} = \gamma' p \qquad a_{H^+} a_{Cl^-} = \gamma_\pm^2 c^2 \tag{12}$$

where γ' is the activity coefficient for $H_2(g)$ and $\gamma_\pm$ is the mean activity coefficient for $H^+Cl^-(aq)$. Equation (10) can now be written

$$\mathcal{E} = \mathcal{E}^0 - \frac{2.303RT}{\mathcal{F}} \log \frac{c^2}{p^{1/2}} - \frac{2.303RT}{\mathcal{F}} \log \frac{\gamma_\pm^2}{\gamma'^{1/2}} \tag{13}$$

where p is in bar and c is in mol L^{-1}.† From this equation it is clear that, if emf measurements are made on two or more cells that differ only in the concentrations c of HCl, the ratios of the corresponding activity coefficients $\gamma_\pm$ can be determined from the differences in emf. For the determination of the individual values of these activity coefficients, it is necessary to know the values of $\mathscr{E}^0$ and γ'. At 25°C and 1 bar, the activity coefficient γ' is 1.0006, which for the purpose of this experiment may be taken as unity. To determine $\mathscr{E}^0$ however requires a procedure equivalent to determining the emf with the solute in its "reference state," at which the activity coefficient $\gamma_\pm$ is unity. However, the reference state for a solute in solution is the limiting state of zero concentration, which is inaccessible to direct experiment. However, the Debye–Hückel theory predicts the limiting behavior of $\gamma_\pm$ as the concentration approaches zero, and we can make use of this predicted behavior in devising an extrapolation procedure for the determination of $\mathscr{E}^0$. According to the Debye–Hückel limiting law,[1]

$$\log \gamma_\pm \cong -A\sqrt{I} \tag{14}$$

In the present case, the ionic strength I can be set equal to the HCl molar concentration c. The value of the constant A given by the Debye–Hückel theory for a uni-univalent electrolyte in aqueous solution at 25°C is 0.509. While knowledge of this value may be helpful to the extrapolation, it is not necessary.

Let us rearrange Eq. (13) and set γ' equal to unity:

$$\mathscr{E}^0 = \mathscr{E} - \frac{2.303RT}{\mathscr{F}} \log \frac{p^{1/2}}{c^2} + \frac{2.303RT}{\mathscr{F}} \log \gamma_\pm^2 \tag{15}$$

Now define $\mathscr{E}^{0\prime}$ by

$$\mathscr{E}^{0\prime} \equiv \mathscr{E} - \frac{2.303RT}{\mathscr{F}} \log \frac{p^{1/2}}{c^2} - 2\frac{2.303RT}{\mathscr{F}}(0.509)\sqrt{c} \tag{16}$$

The value of $\mathscr{E}^{0\prime}$ should be close to that of $\mathscr{E}^0$ [depending on the validity of Eq. (14) as an approximation for $\log \gamma_\pm$]; in any case $\mathscr{E}^{0\prime}$ will approach $\mathscr{E}^0$ as the concentration approaches zero. A plot of $\mathscr{E}^{0\prime}$ versus c should be approximately linear and have only a small slope, thus permitting a good extrapolation to zero concentration.[2]

Alternatively, one could define a quantity $\mathscr{E}^{0\prime\prime}$ by

$$\mathscr{E}^{0\prime\prime} \equiv \mathscr{E} - \frac{2.303RT}{\mathscr{F}} \log \frac{p^{1/2}}{c^2} \tag{17}$$

which would also equal $\mathscr{E}^0$ at infinite dilution. If this quantity $\mathscr{E}^{0\prime\prime}$ is plotted against $\sqrt{c}$, we expect to obtain a curve that will give a limiting straight-line extrapolation to zero concentration.

In either case the extrapolated intercept is the desired value of $\mathscr{E}^0$. Often it is best to make both extrapolations: with only a few points, use of $\mathscr{E}^{0\prime}$ is recommended.

Once the value of $\mathscr{E}^0$ is determined, the activity coefficients for the various concentrations studied can be determined individually by use of Eq. (13). These experimental $\gamma_\pm$

†In the treatment given here, the standard-state pressure has been taken to be 1 bar $\equiv 10^5$ Pa, as recommended by the IUPAC. Prior to 1983 the standard-state pressure was 1 atm. Fortunately the effect on standard free energies and therefore on standard electrode potentials is very small, since 1 atm = 101325 Pa is close to 1 bar. For an electrode involving only condensed phases, such as the $AgCl/Ag^+Cl^-$ electrode, one finds that $\mathscr{E}^0(atm) - \mathscr{E}^0$ (bar) = +0.169 mV, which corresponds to a free-energy difference of 16.4 J mol^{-1}. This difference arises from the new choice of $\mathscr{E}^0_{H^+/H_2} = 0$ when hydrogen gas is in its standard state of 1 bar rather than the old choice of $\mathscr{E}^0_{H^+/H_2} = 0$ when hydrogen was at a pressure of 1 atm. For the standard potential of some arbitrary half-cell, $\mathscr{E}^0(atm) - \mathscr{E}^0(bar) = -0.338(\Delta n - 0.5)$mV, where Δn is the number of moles of gas produced in the half-cell change in state.

values can then be compared with those predicted by a more complete form of the Debye–Hückel equation, such as

$$\log \gamma_\pm = -0.509 \, |z_+ z_-| \, \frac{\sqrt{I}}{1 + B\sqrt{I}} \tag{18}$$

where B is approximately equal to unity.[1]

EXPERIMENTAL

The cell vessel consists of a small beaker with a stopper or cover with holes through which the hydrogen electrode assembly and the silver–silver chloride electrode assembly may be introduced. The assembled cell is shown in Fig. 1.

The hydrogen electrode consists of a mounted platinum gauze square contained within a glass sleeve having large side holes at about the level of the gauze and a side arm for admission of hydrogen near the top. The platinum gauze is "platinized," that is, coated with a deposit of platinum black by electrolytic deposition from a solution containing platinic chloride and a trace of lead acetate. This deposit should be removed with warm aqua regia and renewed if the electrode has been allowed to dry out or if there is evidence that the deposit has been "poisoned."

The silver–silver chloride electrode consists of a mounted platinum screen that has been heavily plated with silver from a cyanide bath, rinsed, aged in an acidified silver nitrate solution, rinsed, coated with a thin layer of silver chloride by anodizing in a dilute HCl solution (preferably no more than a few days before use), and kept in dilute HCl pending use. This is mounted in a glass sleeve with a small hole in the bottom to admit the cell

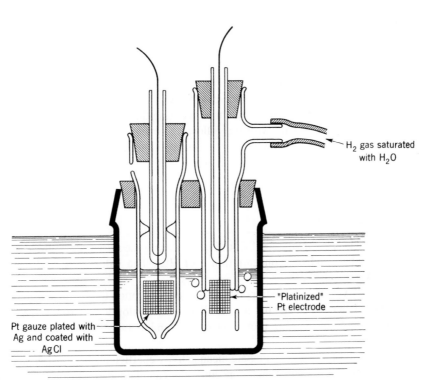

FIGURE 1

Electrochemical cell with H_2 and Ag|AgCl electrodes.

solution and a small side hole near the top for passage of air; this sleeve protects the electrode from mechanical damage and also prevents the attainment of any significant concentration of dissolved H_2 in the solution in contact with the electrode. (Further details concerning the preparation of Pt and of Ag|AgCl electrodes can be found in Chapter XX.)

Both electrode assemblies should be furnished ready for use. Care should be exercised to prevent the electrodes from becoming dry. The use of hydrogen in any significant quantity (as might occur when many students were carrying out this experiment simultaneously) is attended by explosion hazard. Hydrogen usage must be kept at the minimum necessary for the experiment, and wastage and leakage must be guarded against.

The hydrogen to be used should be oxygen free. Oxygen is most conveniently removed by passing tank hydrogen through a commercial catalytic purifier. The hydrogen should also be saturated with water vapor at room temperature (about 25°C) by passing it through a fritted-glass bubbler filled with water.

The partial pressure p of the hydrogen at the electrode should be determined from the atmospheric pressure in the laboratory by subtracting the partial pressure of water vapor at 25°C and adding the mercury equivalent of the "water head" (the average difference in liquid levels inside and outside the hydrogen electrode shell). Read the barometer to determine the atmospheric pressure.

Procedure. Fill the cell about half full with the appropriate HCl solution. Rinse the two electrodes in separate small portions of the same solution, and insert them. When the liquid level in the silver–silver chloride electrode shell has reached its equilibrium level, pass hydrogen through the cell at a moderate rate to sweep out the air and saturate the solution. After about 10 min, slow down the rate to a few bubbles per second and begin to make emf readings. Make at least four readings at 5-min intervals. These should show no significant drift.

Readings of emf are obtained with a potentiometer circuit. If a null detector reading cannot be obtained with any setting of the potentiometer dial, reverse the leads and try again. The sign of the emf will depend on whether the null reading is obtained with the right-hand terminal of the cell (Ag + AgCl) connected to the positive or to the negative terminal of the potentiometer. Alternatively a high-resolution digital voltmeter with a large internal impedance can also be used very successfully. See Chapter XVI for a description and comparison of potentiometers and digital multimeters.

Care should be exercised to prevent the cell from becoming polarized by accidental shorting or by passage of excessive currents during the emf measurements.

Runs should be made with the following concentrations of HCl: 0.1, 0.05, 0.025, 0.0125, and 0.00625 M, obtained by successive volumetric dilution of 0.1 M HCl stock solution. Make up 200 mL of each solution.

CALCULATIONS

For each run, calculate $\mathscr{E}^{0\prime}$ from Eq. (16) using the measured $\mathscr{E}$, the partial pressure p of H_2 in bar, and the concentration c of HCl. The value of $2.303RT/\mathscr{F}$ at 25°C is 0.05916 V.

Plot $\mathscr{E}^{0\prime}$ versus c and extrapolate to zero concentration to obtain $\mathscr{E}^0$ (which is equal to the standard electrode potential for the silver–silver chloride electrode). If you wish, also calculate $\mathscr{E}^{0\prime\prime}$ from Eq. (17), plot it versus $\sqrt{c}$, and extrapolate to obtain another value of $\mathscr{E}^0$ as a check.

With this value of $\mathscr{E}^0$, calculate mean activity coefficients for all concentrations studied by use of Eq. (13). For each concentration, calculate also a theoretical mean activity coefficient by use of the Debye–Hückel equation in the form given in Eq. (18). Present your experimental and theoretical activity coefficients in tabular form.

SAFETY ISSUES

Gas cylinders must be chained securely to the wall or laboratory bench (see pp. 644–646 and Appendix C). The general rule against smoking in the laboratory should be stressed, due to the hazard of a hydrogen explosion if the H_2 partial pressure is too high and any open flames are present. Good laboratory ventilation is important, particularly if several students are carrying out this experiment simultaneously.

APPARATUS

Complete potentiometer setup or high-resolution digital voltmeter (see Chapter XVI); 50-mL weighing bottle as cell; special three-hole stopper to fit cell and hold electrodes; two leads for connections to cell; hydrogen electrode, and Ag–AgCl electrode; 200-mL volumetric flask; 100-mL pipette; pipetting bulb; 250-mL flasks; two 250-mL beakers; small gas bubbler; two pieces of gum-rubber tubing; large clamp and clamp holder.

Cylinder of hydrogen gas, regulator fitted with a Deoxo purifier to remove any oxygen and a flow reducer to limit flow to 5 ft^3 h^{-1}; large fritted-glass bubbler to saturate hydrogen with water vapor; constant-temperature bath set at 25°C; 0.1 M HCl solution (350 mL).

REFERENCES

1. I. N. Levine, *Physical Chemistry,* 6th ed., sec. 10.6, McGraw-Hill, New York (2009).

2. H. S. Harned and R. W. Ehlers, *J. Amer. Chem. Soc.* **55,** 2179 (1933).

GENERAL READING

R. J. Silbey, R. A. Alberty, and M. G. Bawendi, *Physical Chemistry,* 4th ed., sec. 7.5, Wiley, New York (2005).

A. Hammett, *Electrochemistry,* Wiley, New York (1998).

X

Chemical Kinetics

EXPERIMENTS

20. Method of Initial Rates: Iodine Clock
21. NMR Study of a Reversible Hydrolysis Reaction
22. Enzyme Kinetics: Inversion of Sucrose
23. Kinetics of the Decomposition of Benzenediazonium Ion
24. Gas-Phase Kinetics

EXPERIMENT 20
Method of Initial Rates: Iodine Clock

The homogeneous reaction in aqueous solution

$$IO_3^- + 8I^- + 6H^+ \rightarrow 3I_3^- + 3H_2O \tag{1}$$

like virtually all reactions involving more than two or three reactant molecules, takes place not in a single molecular step but in several steps. The detailed system of steps is called the *reaction mechanism.* It is one of the principal aims of chemical kinetics to obtain information to aid in the elucidation of reaction mechanisms, which are fundamental to our understanding of chemistry.

THEORY

The several steps in a reaction are usually consecutive and tend to proceed at different speeds. Usually, when the overall rate is slow enough to measure at all, it is because one of the steps tends to proceed so much more slowly than all the others that it effectively controls the overall reaction rate and can be designated the *rate-controlling* step. A steady state is quickly reached in which the concentrations of the reaction intermediates are controlled by the intrinsic speeds of the reaction steps by which they are formed and consumed. A study of the rate of the overall reaction yields information of a certain kind regarding the nature of the rate-controlling step and closely associated steps. Usually, however, only part of the information needed to formulate uniquely and completely the correct reaction mechanism is supplied by rate studies.

When the mechanism is such that steady state is attained quickly, the rate law for Eq. (1) can be written in the form

$$-\frac{d(\mathrm{IO_3^-})}{dt} = f\,[(\mathrm{IO_3^-}),(\mathrm{I^-}),(\mathrm{H^+}),(\mathrm{I_3^-}),(\mathrm{H_2O}),\ldots] \tag{2}$$

where parenthesized quantities are concentrations. In the general case, the brackets might also contain concentrations of additional substances, referred to as *catalysts,* whose presence influences the reaction rate but that are not produced or consumed in the overall reaction. The determination of the rate law requires that the rate be determined at a sufficiently large number of different combinations of the concentrations of the various species present to enable an expression to be formulated that accounts for the observations and gives good promise of predicting the rate reliably over the concentration ranges of interest. The rate law can be written to correspond in form to that predicted by a theory based on a particular type of mechanism, but basically it is an empirical expression.

The most frequently encountered type of rate law is of the form [again using the reaction of Eq. (1) as an example]

$$-\frac{d(\mathrm{IO_3^-})}{dt} = k(\mathrm{IO_3^-})^m(\mathrm{I^-})^n(\mathrm{H^+})^p\cdots \tag{3}$$

where the exponents $m, n, p, \ldots$ are determined by experiment. Each exponent in Eq. (3) is the *order* of the reaction with respect to the corresponding species; thus the reaction is said to be mth order with respect to $\mathrm{IO_3^-}$ etc. The algebraic sum of the exponents, $m + n + p$ in this example, is the *overall order* (or commonly simply the *order*) of the reaction. Reaction orders are usually but not always positive integers within experimental error.

The order of a reaction is determined by the reaction mechanism. It is related to and is often (but not always) equal to the number of reactant molecules in the rate-controlling step—the "molecularity" of the reaction. Consider the following proposed mechanism for the hypothetical reaction $3\mathrm{A} + 2\mathrm{B} = $ products:

> *a.* $\mathrm{A} + 2\mathrm{B} = 2\mathrm{C}$ (fast, to equilibrium, K_a)
> *b.* $\mathrm{A} + \mathrm{C} = $ products   (slow, rate controlling, k_b)

The rate law predicted by this mechanism is

$$-\frac{d(\mathrm{A})}{dt} = \tfrac{3}{2}k_b(\mathrm{A})(\mathrm{C}) = \tfrac{3}{2}k_b(\mathrm{A})\mathrm{K}_a^{1/2}(\mathrm{A})^{1/2}(\mathrm{B}) = k(\mathrm{A})^{3/2}(\mathrm{B})$$

The overall reaction involves five reactant molecules, but it is by no means necessarily of fifth order. Indeed the rate-controlling step in this proposed mechanism is bimolecular, and the overall reaction order predicted by the mechanism is $\frac{5}{2}$. It is also important to note that this mechanism is not the only one that would predict the above $\frac{5}{2}$-order rate law for the given overall reaction; thus experimental verification of the predicted rate law would by no means constitute proof of the validity of the above proposed mechanism.

It occasionally happens that the observed exponents deviate from integers or simple rational fractions by more than experimental error. A possible explanation is that two or more simultaneous mechanisms are in competition, in which case the observed order should lie between the extremes predicted by the individual mechanisms. A possible alternative explanation is that no single reaction step is effectively rate controlling.

We now turn our attention to the experimental problem of determining the exponents in the rate law. Except in first- and second-order reactions, it is usually inconvenient to determine the exponents merely by determining the time behavior of a reaction system in which many or all reactant concentrations are allowed to change simultaneously and comparing the observed behavior with integrated rate expressions. A procedure is desirable that permits the dependencies of the rate on the concentrations of the different reactants to be isolated from one another and determined one at a time. In one such procedure, all the species but the one to be studied are present at such high initial concentrations relative to that of the reactant to be studied that their concentrations may be assumed to remain approximately constant during the reaction; the apparent reaction order with respect to the species of interest is then obtained by comparing the progress of the reaction with that predicted by rate laws for first order, second order, and so on. This procedure would often have the disadvantage of placing the system outside the concentration range of interest and thus possibly complicating the reaction mechanism.

In another procedure, which we shall call the *initial-rate method,* the reaction is run for a time small in comparison with the "half-life" of the reaction but large in comparison with the time required to attain a steady state, so that the actual value of the initial rate [the initial value of the derivative on the left side of Eq. (3)] can be estimated approximately. Enough different combinations of initial concentrations of the several reactants are employed to enable the exponents to be determined separately. For example the exponent m is determined from two experiments which differ only in the IO_3^- concentration.

In the present experiment the rate law for the reaction shown in Eq. (1) will be studied by the initial rate method, at 25°C and a pH of about 5. The initial concentrations of iodate ion, iodide ion, and hydrogen ion will be varied independently in separate experiments, and the time required for the consumption of a definite small amount of the iodate will be measured.

METHOD

The time required for a definite small amount of iodate to be consumed will be measured by determining the time required for the iodine produced by the reaction (as I_3^-) to oxidize a definite amount of a reducing agent, arsenious acid, added at the beginning of the experiment. Under the conditions of the experiment, arsenious acid does not react directly with iodate at a significant rate but reacts with iodine as quickly as it is formed. When the arsenious acid has been completely consumed, free iodine is liberated, which produces a blue color with a small amount of soluble starch that is present. Since the blue color appears rather suddenly after a reproducible period of time, this series of reactions is commonly known as the *iodine clock reaction.*

The reaction involving arsenious acid may be written, at a pH of about 5, as

$$H_3AsO_3 + I_3^- + H_2O \rightarrow HAsO_4^{2-} + 3I^- + 4H^+ \qquad (4)$$

The overall reaction, up to the time of the starch end point, can be written, from reactions (1) and (4), as

$$IO_3^- + 3H_3AsO_3 \rightarrow I^- + 3HAsO_4^{2-} + 6H^+ \qquad (5)$$

Since with ordinary concentrations of the other reactants hydrogen ions are evidently produced in quantities large in comparison to those corresponding to pH 5, it is evident that buffers must be used to maintain constant hydrogen-ion concentration. As is apparent from the method used, the rate law will be determined under conditions of essentially zero concentration of I_3^-; the dependence of the rate on triiodide, which in fact has been shown to be very small,[1] will not be measured. Under these conditions, Eq. (3) is an appropriate expression for the rate.

A constant initial concentration of H_3AsO_3 is used in a series of reacting mixtures having varying initial concentrations of IO_3^-, I^-, and H^+. Since the amount of arsenious acid is the same in each run, the amount of iodate consumed up to the color change is constant and related to the amount of arsenious acid by the stoichiometry of Eq. (5). The initial reaction rate in mol L^{-1} s^{-1} is thus approximately the amount consumed (per liter) divided by the time required for the blue end point to appear. From the initial rates of two reactions in which the initial concentration of only one reactant is varied and all other concentrations are kept the same, it is possible to infer the exponent in the rate expression associated with the reactant that is varied. This is most conveniently done by taking logarithms of both sides of Eq. (3) and subtracting the expressions for the two runs.

EXPERIMENTAL

Solutions. Two acetate buffers, with hydrogen-ion concentrations differing by a factor of 2, will be made up by the student from stock solutions. Use will be made of the fact that at a given ionic strength the hydrogen-ion concentration is proportional to the ratio of acetic-acid concentration to acetate-ion concentration,

$$(H^+) = K \frac{(HAc)}{(Ac^-)} \frac{1}{\gamma_\pm^2} \tag{6}$$

where at 25°C, $K = 1.753 \times 10^{-5}$ when concentrations are expressed in units of mol L^{-1}. The experiments will all be carried out at about the same ionic strength (0.16 ± 0.01), and accordingly the activity coefficient is approximately the same in all experiments, by the Debye–Hückel theory. It will also be seen that within wide limits the amount of buffer solution employed in a given total volume is inconsequential, provided the ionic strength of the resultant solution is always kept about the same. The solutions required are as follows:

Buffer A: Pipette 100 mL of 0.75 *M* NaAc solution, 100 mL of 0.22 *M* HAc solution, and about 20 mL of 0.2 percent soluble starch solution into a 500-mL volumetric flask, and make up to the mark with distilled water [yields $(H^+) \cong 10^{-5}$ *M*].

Buffer B: Pipette 50 mL of 0.75 *M* NaAc solution, 100 mL of 0.22 *M* HAc solution, and about 10 mL of 0.2 percent soluble starch solution into a 250-mL volumetric flask, and make up to the mark with distilled water [yields $(H^+) \cong 2 \times 10^{-5}$ *M*].

H_3AsO_3, 0.03 M: Should be made up from $NaAsO_2$ and brought to a pH of about 5 by addition of HAc.

KIO_3: 0.1 M; KI: 0.2 M.

Suggested sets of initial volumes of reactant solutions, based on a final volume of 100 mL, are given in Table 1.

TABLE 1

		Initial volumes of solutions, mL			
Solution	Pipette Sizes (mL)	1	2	3	4
H_3AsO_3	5	5	5	5	5
IO_3^-	5	5	10	5	5
Buffer A	20, 25	65	60	40	
Buffer B	20, 25				65
I^-	25	25	25	50	25

Two or three runs should be made on each of the four sets. Two or more runs should also be made on a set with proportions chosen by the student, in which the initial compositions of *two* reacting species differ from those in set 1. In each case, the amount of buffer required is that needed to obtain a final volume of 100 mL.

It is convenient to use each pipette only for a single solution, if possible, to minimize time spent in rinsing. The pipettes should be marked to avoid mistakes.

The buffer solutions and the iodide solution should be equilibrated to 25°C by clamping flasks containing them in a thermostat bath set at that temperature. Two vessels of convenient size (ca. 250 mL) and shape (beakers or Erlenmeyer flasks), rinsed and drained essentially dry, should also be clamped in the bath. One of them should have a white-painted bottom surface (or have a piece of white cloth taped under the bottom) to aid in observing the blue end point unless other means are available to obtain a light background.

To make a run, pipette all the solutions *except* KI into one of the vessels and the KI solution into the other. Remove both vessels from the bath, and begin the reaction by pouring the iodide rapidly but quantitatively into the other solution, simultaneously starting the timer. Pour the solution back and forth once or twice to complete the mixing, and place the vessel containing the final solution back into the bath. Stop the timer at the appearance of the first faint but definite blue color.

CALCULATIONS

The student should construct a table giving the actual initial concentrations of the reactants IO_3^-, I^-, and H^+. The H^+ concentrations should be calculated from the actual concentrations of NaAc and HAc in the stock solutions employed, with an activity coefficient calculated by use of the Debye–Hückel theory for the ionic strength ($I = 0.16$) of the reacting mixtures.

Using the known initial concentration of H_3AsO_3, calculate the initial rate for each run. From appropriate combinations of sets 1, 2, 3, and 4, calculate the exponents in the rate expression (3). Also calculate a value of the rate constant k from each run, and obtain an average value of k from all runs.

Write the rate expression, with the numerical values of the rate constant k and the experimentally obtained values of the exponents. Beside it write the temperature and ionic strength at which this expression was obtained.

Write another rate expression, in which those exponents that appear to be reasonably close (within experimental error) to integers are replaced by the integral values. Use this expression to calculate values for the initial rates of all sets studied, and compare them with the observed initial rates.

DISCUSSION

The kinetics of this reaction have been the subject of much study, and the mechanism is not yet completely elucidated with certainty. The following is an incomplete list of the mechanisms that have been proposed.

1. $IO_3^- + 2I^- + 2H^+ \longrightarrow 2HIO + IO^-$ (slow)

Followed by fast reactions (Dushman[1])

2. $IO_3^- + H^+ \overset{K}{\rightleftharpoons} HIO_3$ (fast, to equil.)

   $I^- + H^+ \overset{K'}{\rightleftharpoons} HI$ (fast, to equil.)

   $HIO_3 + HI \overset{k}{\longrightarrow} HIO + HIO_2$ (slow)

   Followed by fast reactions (at low iodide concentrations; Abel and Hilferding[2])

3. $IO_3^- + I^- + 2H^+ \overset{K}{\rightleftharpoons} H_2I_2O_3$ (fast, to equil.)

   $H_2I_2O_3 \overset{K'}{\rightleftharpoons} I_2O_2 + H_2O$ (fast, to equil.)

   $I_2O_2 + I^- \overset{k}{\longrightarrow} I_3O_2^-$ (slow)

   Followed by fast reactions (Bray[3])

4. $IO_3^- + I^- + 2H^+ \overset{K}{\rightleftharpoons} H_2I_2O_3$ (fast, to equil.)

   $H_2I_2O_3 \overset{k}{\longrightarrow} HIO + HIO_2$ (slow)

   Followed by fast reactions (at low iodide concentrations; Bray[3])

5. $IO_3^- + H^+ \overset{K}{\rightleftharpoons} IO_2^+ + OH^-$ (fast, to equil.)

   $H^+ + OH^- \overset{1/K_w}{\rightleftharpoons} H_2O$ (fast, to equil.)

   $IO_2^+ + I^- \overset{K'}{\rightleftharpoons} IOIO$ (fast, to equil.)

   $IOIO + I^- \overset{k}{\longrightarrow} I^+ + 2IO^-$ (slow)

   Followed by fast reactions (Morgan, Peard, and Cullis[4])

6. $IO_3^- + H^+ \overset{K}{\rightleftharpoons} IO_2^+ + OH^-$ (fast, to equil.)

   $H^+ + OH^- \overset{1/K_w}{\rightleftharpoons} H_2O$ (fast, to equil.)

   $IO_2^+ + I^- \overset{K'}{\rightleftharpoons} IOIO$ (fast, to equil.)

   $IOIO \overset{k}{\longrightarrow} IO^+ + IO^-$ (slow)

   Followed by fast reactions (at low iodide concentrations; Morgan, Peard, and Cullis[4])

The student should discuss the above mechanisms in connection with his or her experimentally determined rate law.

The oxidation of iodide ion by chlorate ion ClO_3^- has also been studied.[5,6] Although the reaction appears to be attended by complications that make it difficult to study, under certain conditions it can be carried out as an iodine clock experiment. (For the interested student, suggested concentration ranges for 20 to 25°C are ClO_3^-, 0.05 to 0.10 M; I^-, 0.025 to 0.10 M; H^+, 0.02 to 0.04 M. A sulfate–bisulfate buffer may be used.) The resulting rate law is not identical with that for the reaction with iodate but appears to be compatible with mechanisms analogous to several of those given above.

A different type of "clock" reaction, suitable for student investigation, is the reaction of formaldehyde with bisulfite ion.[7,8] The reaction involves a single slow step,

$$HCHO + HSO_3^- \longrightarrow HOCH_2SO_3^- \qquad (7)$$

followed by a rapid buffer equilibrium,

$$HSO_3^- \overset{K_a}{\rightleftharpoons} H^+ + SO_3^{2-} \qquad (8)$$

As bisulfite ion is used up in reaction (7), the hydrogen-ion concentration adjusts itself according to the buffer equation (8). If an indicator such as phenolphthalein is added to the mixture, it will undergo a sudden color change when the pH of the solution reaches the pK_i of the indicator. The time τ required for the color change is related to the rate constant for reaction (7) by the equation[7]

$$\frac{1}{F_0 - B_0}\left[\log_{10}\frac{B_0(F_0 - B_0)K_a}{F_0 S_0 K_i'}\right] \cong 0.43 k_7 \tau \qquad (9)$$

where B_0 is the initial bisulfite concentration, S_0 is the initial sulfite concentration, F_0 is the initial formaldehyde concentration, K_a is the bisulfite equilibrium constant (6.7×10^{-8} at 25°C when using mol L^{-1} concentration units), $K_i' = rK_i$, where K_i is the indicator equilibrium constant and r is the indicator color ratio at the end point for the indicator concentration used.

"Flowing Clock" Modification of the Experiment. In conventional "clock" reactions, it is difficult to obtain reliable results for reaction times of less than 20 or 30 s. Since much of modern chemical kinetics is concerned with reactions occurring on much faster time scales, it is worthwhile to introduce a modification of such experiments that makes use of a simple flow system to explore reaction times of the order of a few seconds or less.[9] Both the iodide oxidation and formaldehyde–bisulfite reactions can be investigated in this way.

The simple apparatus is shown in Fig. 1. The reactants are contained in two thermostated 250-mL graduated cylinders, each with a siphon tube connected to the T-shaped mixing chamber A, which is a 1-mm-bore three-way stopcock. The flow tube is a 2-mm-bore Pyrex capillary, which lies between the mixing chamber and an in-line stopcock B, used to start and stop the flow. A water aspirator or a house vacuum line protected with a 1-L trap is connected to stopcock B. During tests of the apparatus carried out using water, stopcock A should be carefully adjusted so that equal volumes of fluid are removed from each of the two cylinders in a given time interval. Stopcock A is then left at that setting throughout all the runs.

FIGURE 1
Schematic diagram of the "flowing clock" apparatus.

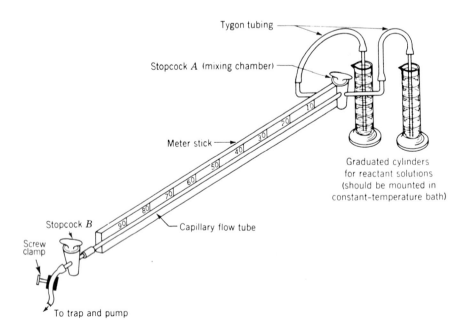

Tygon tubing

Stopcock A (mixing chamber)

Meter stick

Graduated cylinders for reactant solutions (should be mounted in constant-temperature bath)

Stopcock B

Screw clamp

Capillary flow tube

To trap and pump

TABLE 2

| Run No. | Reservoir 1, vol. in mL | | | Reservoir 2, vol. in mL | | | |
	4 M H_2O_2	2 M HCl	Distilled Water	0.05 M KI	0.01 M $Na_2S_2O_3$	Distilled Water	Starch Solution
1	125	125	0	100	100	0	50
2	62.5	125	62.5	100	100	0	50
3	125	62.5	62.5	100	100	0	50
4	125	125	0	50	100	50	50

A volume flow rate f in the range 1 to 10 mL s^{-1} should be suitable for carrying out the experiments described below, and this rate can be fine-tuned with an adjustable screw clamp on the tube between stopcock B and the trap. The best rate of flow for a given experiment will depend on the reaction being studied and the concentrations being used for the reactants. It is assumed that this value, determined from preliminary trial runs, has been provided by the instructor. The student should measure and adjust the flow rate using water in the graduated cylinders. Apply suction and open stopcock B *briefly* in order to fill the tubing and capillary with water. Then read the initial level of the water in each cylinder. Fully open stopcock B for a measured time interval τ_0, close B, and read the final water levels. The flow rate is given by $f = V/\tau_0$, where V is the total volume of water removed from the cylinders.

The cylinders should now be emptied of water, rinsed with the solutions to be studied, and then filled with the appropriate solutions. Apply suction and fully open stopcock B to carry out a run.

The reaction time τ for a given experiment is determined by measuring the distance x along the flow tube from stopcock A at which the appropriate color change occurs, and using the relationship

$$\tau = \frac{xA}{f} \tag{10}$$

where A is the cross-sectional area of the flow tube and f is the previously determined volume flow rate.

The iodine clock experiment can be carried out using peroxide in place of iodate and sodium thiosulfate in place of arsenious acid. The rate law, by analogy to Eq. (3), is

$$-\frac{d(H_2O_2)}{dt} = k(H_2O_2)^m (I^-)^n (H^+)^p \tag{11}$$

A set of experimental conditions suitable for determining the exponents in Eq. (11) is given in Table 2.

For the formaldehyde–bisulfite reaction, a set of runs is given in Table 3 where the buffer is 0.3 M in sulfite ion and 0.05 M in bisulfite ion.

A plot of $(F_0 - B_0)^{-1}$ versus τ will test Eq. (9), since the quantity in brackets is constant. Assuming that $K_a/K_i' = 79,$[7] one can estimate k_7.

SAFETY ISSUES

Use a pipetting bulb; do not pipette by mouth. Dispose of waste chemicals properly as instructed.

TABLE 3

Run No.	Reservoir 1, vol. in mL			Reservoir 2, vol. in mL	
	Buffer Solution	Phenol-phthalein	Distilled Water	1 M Formaldehyde Solution	Distilled Water
1	120	2	128	120	130
2	60	2	188	60	190
3	30	2	218	30	220
4	15	2	233	15	235

APPARATUS

Three 200-mL beakers (bottom painted white); two 250-mL and one 100-mL beakers; two 5-, one 20-, two 25-, and one 50-mL pipettes; pipetting bulb; one 250- and one 500-mL volumetric flask; four 250-mL Erlenmeyer flasks with four corks to fit; one 10-mL graduated cylinder; glass-marking pencil; stopwatch or timer.

Constant-temperature bath (set at 25°C), with provision for mounting beakers and flasks; 0.75 M NaAc solution (300 mL); 0.22 M HAc solution (300 mL); 0.03 M H_3AsO_3 solution (150 mL); 0.1 M KIO_3 solution (150 mL); 0.2 M KI solution (500 mL); 0.2 percent soluble starch solution, with trace of HgI_2 as preservative (75 mL).

For the flowing clock modification: two 250-mL graduated cylinders; three-way stopcock; in-line stopcock; 1 m of 2-mm-bore Pyrex capillary tubing; meter stick; connecting tubing; water aspirator or other rough-vacuum source; 1-L trap; assorted clamps and ring stands; stopwatch; thermostat bath; appropriate solutions.

REFERENCES

1. S. Dushman, *J. Phys. Chem.* **8,** 453 (1904).

2. E. Abel and K. Hilferding, *Z. Phys. Chem.* **136A,** 186 (1928).

3. W. C. Bray, *J. Amer. Chem. Soc.* **52,** 3580 (1930).

4. K. J. Morgan, M. G. Peard, and C. F. Cullis, *J. Chem. Soc.* **1951,** 1865.

5. W. C. Bray, *J. Phys. Chem.* **7,** 92 (1903).

6. A. Skrabal and H. Schreiner, *Monatsh. Chem.* **65,** 213 (1934).

7. P. Jones and K. B. Oldham, *J. Chem. Educ.* **40,** 366 (1963).

8. R. L. Barrett, *J. Chem Educ.* **32,** 78 (1955).

9. M. L. Hoggett, P. Jones, and K. B. Oldham, *J. Chem. Educ.* **40,** 367 (1963).

GENERAL READING

P. W. Atkins and J. de Paula, *Physical Chemistry,* 8th ed., pp. 797–798, Freeman, New York (2006).

K. L. Laidler and K. J. Laidler, *Chemical Kinetics,* 3d ed., Addison-Wesley, Reading, MA (1997).

I. N. Levine, *Physical Chemistry,* 6th ed., chap. 16, McGraw-Hill, New York (2009).

EXPERIMENT 21
NMR Study of a Reversible Hydrolysis Reaction

The hydrolysis of pyruvic acid to 2,2-dihydroxypropanoic acid is a reversible reaction giving rise to the following equilibrium in aqueous solution:

$$CH_3COCOOH + H_2O \underset{k^r}{\overset{k^f}{\rightleftharpoons}} CH_3C(OH)_2COOH \tag{1}$$

In this experiment, nuclear magnetic resonance (NMR) techniques will be used to determine the specific rate constants k^f and k^r for the forward and reverse reactions as well as the value for the equilibrium constant K. Like the hydrolysis of many other organic compounds, this reaction can be acid catalyzed and the effect of hydrogen-ion concentration on the kinetics can be studied. Furthermore, the dependence of k^f, k^r, and K on temperature will be measured and used to evaluate activation energies.

THEORY

A first-order reaction is ordinarily understood to be one in which the rate of disappearance of a single species is proportional to the concentration of that species in the reaction mixture. This terminology is often used even when strictly speaking the total reaction order is different from unity, owing to the participation of additional species in the reaction that are themselves not consumed (i.e., catalysts—H^+ in the present instance) or the participation of substances that are present in such large amounts that their concentrations do not undergo significant percentage change during the reaction (H_2O in the present instance). It would be more correct in such cases to state that the reaction is first-order with respect to a given disappearing species.

The reversible pyruvic acid hydrolysis proceeds kinetically along both an uncatalyzed and an acid-catalyzed path:

Path I

$$\underset{\text{O}}{\overset{\text{O}}{\underset{\|}{CH_3C}}}-COOH + H_2O \underset{k_0^r}{\overset{k_0^f}{\rightleftharpoons}} CH_3\underset{\overset{|}{OH}}{\overset{\overset{OH}{|}}{C}}-COOH \tag{2}$$

Path II

$$CH_3\overset{\overset{O}{\|}}{C}-COOH + H^+ \rightleftharpoons CH_3\underset{\overset{|}{OH}}{C^+}-COOH \qquad \text{very rapid to equil. } K_1 \tag{3}$$

$$CH_3\underset{\overset{|}{OH}}{C^+}-COOH + H_2O \underset{k_1^r}{\overset{k_1^f}{\rightleftharpoons}} CH_3\underset{\overset{|}{OH}}{\overset{\overset{OH}{|}}{C}}-COOH + H^+ \tag{4}$$

Introducing an abbreviated notation for convenience such that $A = CH_3COCOOH$, $B = CH_3C(OH)_2COOH$, and $AH^+ = CH_3C^+(OH)COOH$, one finds

$$+\frac{d(B)}{dt} = k_0^f(A) - k_0^r(B) \qquad \text{for path I} \qquad (5)$$

$$+\frac{d(B)}{dt} = k_1^f(AH^+) - k_1^r(H^+)(B) \qquad (6a)$$

$$= k_1^f K_1(H^+)(A) - k_1^r(H^+)(B) \qquad (6b)$$

$$= k_H^f(H^+)(A) - k_H^r(H^+)(B) \qquad \text{for path II} \qquad (6c)$$

where (X) denotes the concentration of species X. In Eqs. (5) and (6a), the quantity (H_2O), which is constant, has been incorporated into k_0^f and k_1^f, respectively. Equation (6b) has made use of the equilibrium approximation $K_1 = (AH^+)/(A)(H^+)$ for reaction step (3). The overall rate of change in (B) is given by the pseudo-first-order expression

$$+\frac{d(B)}{dt} = k^f(A) - k^r(B) \qquad (7)$$

where

$$k^f = k_0^f + k_H^f(H^+) \qquad (8a)$$

$$k^r = k_0^r + k_H^r(H^+) \qquad (8b)$$

When reaction (1) achieves equilibrium, the thermodynamic equilibrium constant K is given by

$$K = \frac{(B)_{eq}}{(A)_{eq}} \qquad (9)$$

and the principle of detailed balancing[1] then yields

$$\frac{d(B)}{dt} = k_0^f(A)_{eq} - k_0^r(B)_{eq} = 0 \qquad \text{for path I}$$

$$K = \frac{k_0^f}{k_0^r} \qquad (10a)$$

and

$$\frac{d(B)}{dt} = k_H^f(H^+)(A)_{eq} - k_H^r(H^+)(B)_{eq} = 0 \qquad \text{for path II}$$

$$K = \frac{k_H^f}{k_H^r} \qquad (10b)$$

Dependence on Temperature.   The effect of temperature on specific rate constants is well described by the Arrhenius expression

$$k = A \exp\left(-\frac{E_a}{RT}\right) \qquad (11)$$

where A is a temperature-independent *frequency factor* and E_a is an activation energy, interpretable as the energy required to reach the transition state or *activated complex*.

Arrhenius plots of $\ln k$ versus $1/T$ should yield a straight line with a negative slope, and the activation energies for the forward and reverse reactions are given by

$$E_a^f = -R\frac{d \ln k^f}{d(1/T)} \qquad E_a^r = -R\frac{d \ln k^r}{d(1/T)} \tag{12}$$

In addition, the effect of temperature on the equilibrium constant K is governed by the Gibbs–Helmholtz equation,

$$\frac{d \ln K}{d(1/T)} = -\frac{\Delta \widetilde{H}^0}{R} \tag{13}$$

where $\Delta \widetilde{H}^0$ is the molar standard enthalpy change for the reaction. It follows from Eqs. (10) to (13) that

$$\Delta \widetilde{H}^0 = E_a^f - E_a^r \tag{14}$$

METHOD

The use of NMR techniques to investigate the present system is based on an experiment first described in Ref. 2. Two quantitative aspects of NMR spectra are involved. The first is the familiar fact that the integrated band area is directly proportional to the concentration of nuclei that give rise to that band. Thus the equilibrium constant K can be obtained from the ratio of integrated band areas for $CH_3C(OH)_2COOH$ (species B) and $CH_3COCOOH$ (species A). The second aspect, concerning the dynamic information that can be obtained from band-shape analysis, is less familiar and will be summarized below.

For an idealized case (perfectly homogeneous magnet, low transverse fields, steady-state conditions, no saturation effects), an NMR absorption line in the frequency domain or the Fourier transform of a free induction decay has a Lorentzian profile[3]

$$I(\nu) = \frac{I_0}{1 + 4\pi^2(\nu - \nu_0)^2 T_2^2} \tag{15}$$

When $\nu = \nu_0$, which corresponds to the resonance condition where the frequency of the field matches the Larmor frequency ν_0, the intensity achieves its maximum value $I_0 = CT_2$.[4] The constant C depends on several nuclear and instrumental parameters that do not vary in a given experiment. An ideal Lorentzian line shape is characterized by a full-width-at-half-maximum (FWHM) W given by

$$W(\mathrm{Hz}) = \frac{1}{\pi T_2} \ (\text{ideal}) \tag{16}$$

since $I(\nu) = I_0/2$ at $\nu = \nu_0 \pm (2\pi T_2)^{-1}$. The quantity T_2 is called the *spin-spin relaxation time* or the *transverse relaxation time,* since it governs the decay of the transverse magnetization $M_y(t) \sim \exp(-t/T_2)$.

Typical proton NMR T_2 values are of the order of 1 s, so $W \approx 0.3$ Hz would be expected if there were no chemical exchange and no instrumental contributions to W. However, there are always small magnetic field inhomogeneities that spread the Larmor frequencies slightly. If one assumes that the line shape remains Lorentzian, Eq. (15) can easily be modified by replacing T_2 with an effective value T_2^*, where

$$W(\mathrm{Hz}) = \frac{1}{\pi T_2^*} = \left(\frac{1}{\pi T_2}\right) + W_{\mathrm{inst}} \tag{17}$$

and T_2^* reflects the instrumental contribution as well as the natural line width.[3,5]

FIGURE 1

Changes in the NMR spectrum for a two-site exchange system as a function of $k^f = k_{A \to B}$. Model parameters used: $K = (B)/(A) = 2$, $\nu_A - \nu_B = 20$ Hz, and $T_{2A}^* = T_{2B}^* = 0.5$ s.

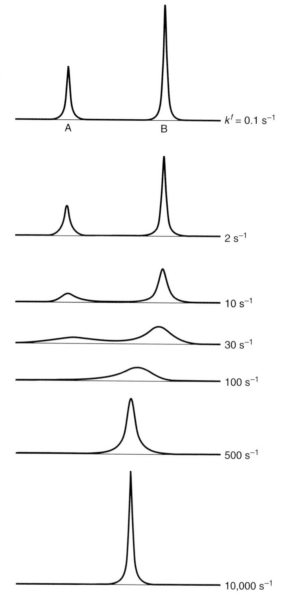

$k^f = 0.1$ s^{-1}

2 s^{-1}

10 s^{-1}

30 s^{-1}

100 s^{-1}

500 s^{-1}

$10{,}000$ s^{-1}

Chemical Exchange. If the chemical system under investigation contains magnetic nuclei that can transfer rapidly between two different environments, this chemical exchange can have dramatic effects on the NMR spectrum. In our case, the methyl protons occupy site A in pyruvic acid and site B in dihydroxypropanoic acid, and they switch back and forth due to the reversible hydrolysis reaction. The general equations for two-site exchange are fairly complex,[4,6] but the qualitative changes in the spectrum due to $A \rightleftharpoons B$ exchange are shown in Fig. 1. At slow exchange rates (small values of k^f and k^r), two non-overlapping lines are observed with FWHM values W_A and W_B given by

$$\pi W_A = \frac{1}{T_{2A}^*} + k^f \qquad \pi W_B = \frac{1}{T_{2B}^*} + k^r \tag{18}$$

Thus, for very small rate constants, the widths are dominated by the effective T_2^* values that would pertain in the absence of exchange. In this limit the mean lifetimes $\tau_A = 1/k^f$ for a nucleus in environment A and the analogous $\tau_B = 1/k^r$ for environment B are long compared to the time required for a magnetic transition, and both lines are "sharp." As k^f (and $k^r = k^f/K$) increases due to changes in (H^+) or temperature, the lines grow wider. If k^f grows sufficiently large, the lines become very broad and overlap, then merge into a single line, and this line becomes progressively sharper as k^f continues to increase, as shown in Fig. 1. The initial broadening of the lines described by Eq. (18) is an example of lifetime broadening[2,5] that follows from the Heisenberg uncertainty principle in the form $\Delta E \cdot \Delta t \simeq h/2\pi$. In our case, since $\Delta E = h\Delta\nu$ and $\Delta t = \tau$, this condition is $\Delta\nu \cdot \tau \simeq 1/2\pi$ and therefore $W = 2\Delta\nu \simeq 1/\pi\tau$.

EXPERIMENTAL

Descriptions of the general features of continuous-wave NMR (CW–NMR) and Fourier-transform NMR (FT–NMR) spectrometers are given in Exps. 32 and 43, respectively. The instructor will provide specific operating procedures for the particular instrument to be used in this experiment and will suggest choices of operational parameters such as sweep rates (CW spectrometer) or acquisition times (FT spectrometer). The shimming of the NMR probe to achieve the narrowest possible line width (smallest W_{inst}) is important and should be practiced on some convenient reference sample, perhaps pure pyruvic acid. It is also essential to obtain expanded-scale spectra, and care should be taken that enough digitized data points are acquired across a peak to allow an accurate determination of integrated band areas and widths. It should be stressed that high-quality spectra with optimal resolution are required, and the radio-frequency (B_1) field should be well below saturation levels.[4]

Pyruvic acid is hygroscopic and may polymerize or decompose somewhat on standing at room temperature; impure samples are yellowish. A supply of pyruvic acid should be purified in advance by vacuum distillation to remove water, decomposition products, and any other impurities. With the use of a laboratory pump, this distillation can be carried out anywhere between 20 and 50°C depending on the pressure achieved. Be sure to add boiling chips or use a magnetic stirrer (better at lower pressures) to prevent bumping. The distilled material should be a clear, colorless liquid. Store the purified pyruvic acid in an airtight container in a refrigerator. If solutions are made up in advance, they should also be stored in small sealed tubes kept in a refrigerator. Under such conditions, solutions should remain sufficiently stable for 4 or 5 days.

Prepare the solutions listed in Table 1. The HCl is reagent-grade concentrated hydrochloric acid containing ~38 percent HCl by weight (about 12 M). Note that the solvent consists of a mixture of H_2O and D_2O, the latter added to provide a NMR lock.

TABLE 1

Soln.	Pyruvic Acid, mL	HCl, mL	H_2O, mL	D_2O, mL
1	1.0	0	1.10	0.10
2	1.0	0.10	1.00	0.10
3	1.0	0.20	0.90	0.10
4	1.0	0.30	0.80	0.10
5	1.0	0.40	0.70	0.10
6	1.0	0.50	0.60	0.10

The concentration of D_2O is fixed at the same value for all the solutions in order to hold constant any possible isotope effect on the rates. **Caution:** Since pyruvic acid is irritating to the skin and concentrated HCl can cause severe burns, solution handling should be carried out wearing gloves.

Each solution should be prepared in a small test tube or vial that can be sealed airtight for refrigerated storage prior to transfer to an NMR sample tube. In order to calculate the H^+ concentration in each solution, the molarity of the concentrated HCl is needed. If this is not given by the instructor, it can be determined either by titration or by a density measurement (using tables in the *CRC Handbook of Chemistry and Physics*).

The NMR spectrum of pure pyruvic acid consists of two resonances—one due to the carboxyl proton at 8.67 ppm and another due to methyl protons at 2.55 ppm (relative to tetramethylsilane as the reference).[7] The separation between these two peaks is dependent on the purity of the sample—the greater the purity, the greater is the separation. Taking the spectrum of pyruvic acid will allow a qualitative check on sample purity and will also provide practice with shimming a sample for optimal resolution (narrow lines).

The spectrum of an aqueous solution of pyruvic acid consists of a total of three bands for the species involved in the equilibrium given by Eq. (1). The methyl protons of pyruvic acid give a band A with a chemical shift $\delta_A = 2.6$ ppm; and the methyl proton band for dihydroxypropanoic acid, denoted band B, occurs at $\delta_B = 1.75$ ppm. A third band at a larger δ value represents the resonances of the carboxyl, hydroxyl, and water protons. This band is a singlet since the proton exchange rate between —COOH, —OH, and H_2O environments is very rapid. The positions of all three bands will vary somewhat with the composition of the solution.

Dependence on Hydrogen-Ion Concentration. Solutions 1 through 6 will be studied at a constant temperature of 25°C in order to determine the acid-catalyzed rate constants k_H^f and k_H^r. If the NMR sample tubes have been filled with cold solutions, bring these tubes up to room temperature before inserting them into the spectrometer and allow about 5 min for the tubes to come to thermal equilibrium with the probe. You should then lock on and shim each sample tube to achieve the narrowest possible line widths.

The two methyl peaks should be recorded on an expanded scale in order to achieve good width measurements. As the H^+ concentration changes, the peak positions will shift appreciably and it will probably be necessary to change the screen scan parameters. Integrate the band area for each peak and determine the widths W_A (for pyruvic acid) and W_B (for dihydroxypropanoic acid). As these data are being obtained, make preliminary plots of line widths versus the hydrogen-ion concentration. These plots should be linear, as indicated by Eq. (19) in the Calculations section. If the points for any of the samples deviate substantially from the linear trend of the other data points, this is probably due to improper shimming and you should reshim that sample and repeat the measurements.

Temperature Dependence of Uncatalyzed Rate. The line widths and band areas for solution 1 (containing no HCl) will be measured as a function of temperature over the range 45 to 65°C. This will allow a determination of the T dependence of the equilibrium constant K and an approximate evaluation of the activation energies for the forward and reverse reactions.

Reinstall sample tube 1 in the spectrometer and adjust the probe temperature controller to 45°C. After the sample has reached 45°C (~10 min), relock and reshim the sample. This shim setting will be maintained for all the temperature runs. Once the methyl spectrum has been recorded and plotted on an expanded scale and the determination of integrated band areas and FWHM line widths has been completed at 45°C, raise the temperature by 5°C intervals and repeat the process until you reach 65°C. As you increase the sample temperature, the peaks will shift laterally across the screen. This drift in the peak positions will stop when the sample comes to thermal equilibrium at each temperature setting, and this fact can be used as an indication of how long you should wait before taking data at each temperature.

As before, it is useful to make preliminary plots as the data are being obtained; in this case, plot log W versus $1/T$ (K^{-1}). Any data points that deviate seriously from linearity should be repeated if time allows.

CALCULATIONS

For both the isothermal series of runs on samples 1 through 6 and the temperature-dependence runs on sample 1, make a table of your results. For each run, specify the H^+ concentration and/or temperature and list for both methyl band A (pyruvic acid) and band B (dihydroxypropanoic acid) the band position, the FWHM width W (Hz), and the integrated area. If possible, obtain an ASCII file of $I(\nu)$ intensity data, which will allow a least-squares fit to the peaks with a Lorentzian form with baseline corrections; see Exp. 43. This permits a check on the actual line shape and an independent determination of area and width values.

For the constant-temperature data at 25°C, plot πW versus the molar hydrogen-ion concentration. It follows from Eqs. (8) and (18) that

$$\pi W_A = \frac{1}{T_{2A}^*} + k_0^f + k_H^f(H^+) = C_A + k_H^f(H^+) \tag{19a}$$

$$\pi W_B = \frac{1}{T_{2B}^*} + k_0^r + k_H^r(H^+) = C_B + k_H^r(H^+) \tag{19b}$$

where C_A and C_B depend on T but not on (H^+). Thus the slopes of the best linear fits to these plots will yield values for k_H^f and k_H^r and the intercepts $C_{A,B}$ will give estimates of $(1/T_2^*) + k_0$.

Since the integrated area of each methyl band is proportional to the concentration of that solute species, it follows from Eq. (9) that the equilibrium constant $K = (\text{area})_B/(\text{area})_A$. Calculate K values at 25°C from the areas measured on samples 1 through 6. Report a best average value for K at 25°C, weighing the results for various samples differently if that seems appropriate.

Compare the K value obtained above from integrated area data with the value of K calculated from Eq. (10b) using your rate constants k_H^f and k_H^r. One can also use Eq. (10a) to obtain a rough estimate of the magnitude of the effective T_2 values T_{2A}^* and T_{2B}^* and thus get an idea of the values for k_0^f and k_0^r at 25°C. If one *assumes* $1/T_{2A}^* = 1/T_{2B}^* = C_0$ at 25°C, Eqs. (10a), (19a), and (19b) yield

$$K = \frac{C_A - C_0}{C_B - C_0} \tag{20}$$

Calculate a value for C_0 using the known values of K and C_A, C_B (either the intercepts from your plots or the πW values for sample 1), and then obtain approximate values of k_0^f and k_0^r.

For the high-temperature runs on sample 1, calculate K at each temperature from the integrated area values and plot $\ln K$ versus the reciprocal of the absolute temperature. Use Eq. (13) to determine the standard enthalpy change $\Delta \widetilde{H}^0$. With the assumption that $\Delta \widetilde{H}^0$ is independent of T down to 25°C, calculate an extrapolated value for $K(25°C)$ and compare this to your experimental K value determined at 25°C.

An effort should be made to evaluate the uncatalyzed activation energies E_a^f and E_a^r from the W_A and W_B width data obtained at high temperatures on sample 1. In this case $k_0^f = \pi W_A - 1/T_{2A}^*$ and $k_0^r = \pi W_B - 1/T_{2B}^*$, and your data analysis will depend on what values are assigned to $1/T_{2A}^*$ and $1/T_{2B}^*$. One possibility is to set $1/T_{2A}^*$ and $1/T_{2B}^*$ equal to zero, based on the assumption that their values are negligibly small compared to k_0^f and k_0^r at high temperatures. For this limiting case, make plots of $\ln(\pi W_A)$ and $\ln(\pi W_B)$ versus $1/T$ and use Eq. (12) to determine E_a^f and E_a^r. These values will be lower bounds on the correct activation energies. A second possibility is to assume that $1/T_{2A}^*$ and $1/T_{2B}^* = C_0$ (the value estimated at 25°C). For this case make plots of $\ln(\pi W_A - C_0)$ and $\ln(\pi W_B - C_0)$ versus $1/T$ and determine new values for E_a^f and E_a^r. Compare the difference $(E_a^f - E_a^r)$ obtained both ways with the value of $\Delta \widetilde{H}^0$.

DISCUSSION

List the chemical shifts observed in the spectrum of pyruvic acid and comment (qualitatively) on the purity of the acid as inferred from a comparison with literature values.[7] Also note changes in the chemical shifts of the two methyl peaks as a function of H^+ concentration and temperature and suggest a reason for such changes.

Give the standard errors associated with the values of area and width that are determined from the line profile. Is the scatter in the data (K and πW_A, πW_B) as a function of (H^+) or T consistent with those error estimates? Discuss other sources of error beyond linefit uncertainties. Comment on the agreement between $(area)_B/(area)_A$ and k_H^f/k_H^r as values of K at 25°C. If the temperature dependence of the area ratio had been measured for sample 6 rather than sample 1, would the $\Delta \widetilde{H}^0$ value for the acid-catalyzed reaction be the same as $\Delta \widetilde{H}^0$ for the uncatalyzed reaction? What difficulty would complicate matters if one tried to determine activation energies from the temperature dependence of the line widths of acid solutions (in addition to the problem of assessing the values of $1/T_{2A}^*$ and $1/T_{2B}^*$)?

On the basis of general concepts from statistical thermodynamics, what would you predict to be the *sign* of ΔS^0 for reaction (1)? Use your data to estimate a value for ΔS^0 at 55°C.

An interesting variation of this experiment involves using the same technique of NMR line-width measurements to study the cis–trans exchange rate in N,N-dimethylacetamide. By determining this rate as a function of temperature over the range 300 to 500 K, one can evaluate the rotational barrier for this cis–trans isomerization.[8]

SAFETY ISSUES

Solution handling should be done wearing gloves, since both pyruvic acid and concentrated HCl can cause skin damage. Use a pipetting bulb; do not pipette by mouth. Dispose of waste chemicals properly.

APPARATUS

NMR spectrometer with peak-integrating and line-width measurement capability; several stoppered bottles for pyruvic acid, HCl, and D_2O; small beaker; six small stoppered tubes or vials for solutions; precision 1-mL graduated pipette; small pipetting bulb; NMR tubes; distillation apparatus for purifying pyruvic acid; refrigerated storage space; pyruvic acid (10 mL); reagent-grade concentrated HCl (3 mL); D_2O (2 mL); fume hood (optional).

REFERENCES

1. R. J. Silbey, R. A. Alberty, and M. G. Bawendi, *Physical Chemistry,* 4th ed., sec. 18.5, Wiley, New York (2005).

2. E. F. H. Brittain, W. O. George, and C. H. J. Wells, *Introduction to Molecular Spectroscopy: Theory and Experiment,* pp. 244–247, 291–293, Academic Press, New York (1970).

3. R. J. Abraham, J. Fisher, and P. Lofthus, *Introduction to NMR Spectroscopy,* Wiley, New York (1992).

4. J. Sandstrom, *Dynamic NMR Spectroscopy,* pp. 6–18, 65–76, Academic Press, New York (1982).

5. P. W. Atkins and J. de Paula, *Physical Chemistry,* 8th ed., p. 538, Freeman, New York (2006).

6. G. Binsch, "Band-Shape Analysis," in L. M. Jackman and F. A. Cotton (eds.), *Dynamic Nuclear Magnetic Resonance Spectroscopy,* Academic Press, New York (1975).

7. C. J. Pouchert and J. R. Campbell, *The Aldrich Library of NMR Spectra,* Vol. II, Aldrich Chemical Co., Milwaukee, WI (1974).

8. F. P. Gasparro and N. H. Kolodny, *J. Chem. Educ.* **54,** 258 (1977).

GENERAL READING

L. M. Jackman and F. A. Cotton (eds.), *Dynamic Nuclear Magnetic Resonance Spectroscopy,* Academic Press, New York (1975).

J. Sandstrom, *Dynamic NMR Spectroscopy,* Academic Press, New York (1982).

EXPERIMENT 22
Enzyme Kinetics: Inversion of Sucrose

A very important aspect of chemical kinetics is that dealing with the rates of enzyme-catalyzed reactions. Enzymes are a class of proteins that catalyze virtually all biochemical reactions. In this experiment we shall study the inversion of sucrose, as catalyzed by the enzyme invertase (β-fructofuranidase) derived from yeast. The rate of the enzyme-catalyzed reaction will then be compared to that of the same reaction catalyzed by hydrogen ions.

THEORY

The basic mechanism for enzyme-catalyzed reactions was first proposed by Michaelis and Menten and confirmed by a study of the kinetics of the sucrose inversion. A simple reaction mechanism by which an enzyme converts a reactant S, usually called a substrate, into products P is

$$E + S \underset{k_{-1}}{\overset{k_1}{\rightleftharpoons}} ES \tag{1}$$

$$ES \underset{k_{-2}}{\overset{k_2}{\rightleftharpoons}} E + P \tag{2}$$

The initial steps involving the enzyme–substrate complex ES are rapid (k_1 and k_{-1} large) and the decomposition of the complex to form products is relatively slow. The back-reaction in which E and P recombine to form complex will be ignored here for two reasons.

The rate constant k_{-2} is generally small, and the concentration of P is also small since we will be concerned with the initial stages of product formation. In some enzyme-catalyzed reactions, more complicated mechanisms involving several different complexes may be required.[1] In the present experiment, however, the simple mechanism given in Eqs. (1) and (2) is adequate to describe the kinetics.

The rate of formation of products when one ignores the back-reaction in Eq. (2) is

$$r = +\frac{d(P)}{dt} = k_2(ES) \tag{3}$$

where parentheses indicate the concentration of the specified species in mol L^{-1}. In order to proceed further, we need an expression for (ES) in terms of enzyme and substrate concentrations. The key innovation made by Michaelis and Menten was to use the steady-state approximation for the complex ES. That is, it is assumed that $d(ES)/dt$, the net rate of change of ES, is very small compared to the rates of formation and destruction of ES. Thus we have

$$\frac{d(ES)}{dt} = k_1(E)(S) - k_{-1}(ES) - k_2(ES) \cong 0 \tag{4}$$

which yields

$$(ES) = \frac{k_1(E)(S)}{k_{-1} + k_2} = \frac{(E)(S)}{K_m} \tag{5}$$

where the Michaelis–Menten constant K_m is defined by $K_m = (k_{-1} + k_2)/k_1$. Note that if $k_2 \ll k_{-1}$, the quantity K_m has a simple interpretation as the equilibrium constant for the dissociation of the ES complex. The instantaneous concentration of enzyme (E) is not known as a function of time but will not be needed. Since the enzyme is conserved, its total concentration in the form of free enzyme and enzyme–substrate complex must be constant and equal to the initial enzyme concentration $(E)_0$:

$$(E)_0 = (E) + (ES) \tag{6}$$

Combining Eqs. (3), (5), and (6), we obtain

$$r = \frac{k_2(E)_0(S)}{K_m + (S)} \tag{7}$$

A schematic plot of r versus (S) is shown in Fig. 1. When the substrate concentration is sufficiently low, $(S) \ll K_m$, the kinetics are first order with respect to S: $r = [k_2(E)_0/K_m](S)$.

FIGURE 1

Reaction velocity versus substrate concentration.

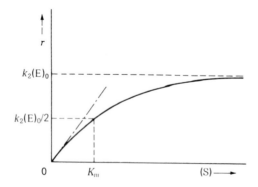

When (S) is very high, $(S) \gg K_m$, the reaction becomes zero order with respect to S and the rate approaches the limiting value $r = k_2(E)_0$. The rate will be equal to half of this maximum value when $(S) = K_m$. Thus a determination of the dependence of r on the substrate concentration (S) will give both $k_2(E)_0$ and K_m values. If the molar concentration $(E)_0$ is known, the rate constant k_2 can also be determined. This quantity, often referred to as the *turnover number*, is of special interest because of its simple interpretation. The value of k_2 corresponds to $r_{max}/(E)_0$; thus the turnover number is the number of sucrose molecules hydrolyzed per second per molecule of enzyme when the enzyme is saturated with substrate (i.e., virtually all the enzyme is in the form of the complex ES).

In this experiment, you will measure the initial rate r_0 as a function of $(E)_0$ and $(S)_0$. The most practical way of analyzing such data is to plot the results in a manner that should yield a straight-line relationship. Equation (7) can be rewritten in two useful forms:

$$\text{L–B:} \qquad \frac{1}{r_0} = \frac{1}{k_2(E)_0} + \frac{K_m}{k_2(E)_0} \frac{1}{(S)_0} \qquad (8)$$

$$\text{E–H:} \qquad \frac{r_0}{(S)_0} = \frac{k_2(E)_0}{K_m} - \frac{r_0}{K_m} \qquad (9)$$

For a plot of $1/r_0$ versus $1/(S)_0$, called a Lineweaver–Burk plot, the intercept determines $k_2(E)_0$ and K_m can then be determined from the slope. For a plot of $r_0/(S)_0$ versus r_0, called an Eadie–Hofstee plot, the slope determines K_m and $k_2(E)_0$ can then be determined from the intercept. The classic plot is the Lineweaver–Burk plot, which has the advantage of displaying the variables separately on different axes. However, the Eadie–Hofstee plot has some practical advantages as demonstrated by a statistical comparison of several different methods of data analysis.[2]

Acid-Catalyzed Reaction. A number of enzyme-catalyzed reactions, including the sucrose inversion to be studied in this experiment, can also be carried out under nonphysiological conditions by using H^+ ions as a less efficient catalyst. In the present case, the acid-catalyzed reaction rate has the form

$$r = +\frac{d(P)}{dt} = -\frac{d(S)}{dt} = k_H(H^+)(S) \qquad (10)$$

This rate law will be studied in the absence of the enzyme and at H^+ concentrations in the range 0.1 to 0.5 *M*, which are found to give suitable values for the rate. On the other hand, the enzyme-catalyzed rate will be studied in a buffer solution that maintains (H^+) at $\sim 10^{-5}$ *M*, which effectively eliminates the acid-catalyzed path during those runs.

Activation Energy. Rate constants are frequently represented by an Arrhenius equation,

$$k = A\, e^{-E_a/RT} \qquad (11)$$

where the activation energy E_a represents an "energy threshold" that must be overcome before the reaction can take place. A conceptual model for the action of an enzyme, or any other catalyst, is that it lowers the activation barrier, thus permitting the reaction to take place more rapidly. One of the objectives of this experiment is to determine the activation energies of both the enzyme-catalyzed and the H^+-catalyzed reactions from measurements of the temperature dependence of the rates. This will allow you to decide whether the higher efficiency of the enzyme catalyst can be attributed simply to a lowering of the energy barrier for the reaction.

METHOD

The enzyme yeast invertase catalyzes the conversion of sucrose to fructose and glucose according to the hydrolysis reaction

$$\text{Sucrose} + H_2O \longrightarrow \text{Glucose} + \text{Fructose} \tag{12}$$

For this example of the simple Michaelis–Menten mechanism, we hypothesize that an intermediate complex ES is formed between one molecule of sucrose and one molecule of invertase. This complex then reacts with a molecule of water to yield the products P_1 and P_2 and to regenerate a molecule of the free enzyme. Thus the rate of product formation is given by

$$r = \frac{d(P_1)}{dt} = \frac{d(P_2)}{dt} = k_2'(H_2O)(ES) \tag{13}$$

where k_2' is the rate constant of the rate-determining step. For reactions occurring in dilute aqueous solutions, the water concentration does not change significantly during the reaction. Thus $k_2'(H_2O)$ can be replaced by a new constant k_2, and Eq. (13) becomes identical to Eq. (3).†

One of the most striking consequences of the sucrose hydrolysis is a large change in the optical-rotatory power of the solution as the sucrose is progressively converted to glucose and fructose. This conversion actually produces a change in the direction in which the plane of polarization of light is rotated by the solution, and it is from this "inversion" that the enzyme derives its name. The inversion of sucrose in the presence of yeast was noted as early as 1832 by Persoz; and Michaelis and Menten used enzyme extracted from yeast in their classic study of the enzyme–substrate complex, in which the course of the reaction was monitored by observation of the change in optical-rotatory power accompanying it.

In the present experiment, we shall use a rather simpler, although equally accurate, way of following the progress of the reaction. This method hinges on the fact that glucose and fructose, like all other monosaccharides, are active reducing agents, while the sucrose from which they are formed is not a reducing sugar. Thus any measure of the reducing capacity of a reaction mixture becomes, in effect, a measurement of the extent to which the conversion of sucrose to glucose and fructose has proceeded in that mixture. In this experiment we shall use a 3,5-dinitrosalicylic acid anion reagent in assaying the reducing capacity of various reaction mixtures.[3] Since the oxidized and reduced form of this reagent

---

†It should be kept in mind that there are other enzymatic reactions with the overall stoichiometry $S + B \rightarrow P$ in which the concentration (B) of the second reactant does vary. In such cases, the simple mechanism given above must be modified.

possess significantly different visible absorption spectra, measurement of the absorbance at a suitable wavelength (540 nm) can be used to measure the extent of hydrolysis that has occurred.

The procedure we will use is first to prepare a particular reaction system containing appropriate amounts of sucrose, enzyme, and buffer solution. This mixture is hydrolyzed for some specified time, and then "quenched" by addition of sodium hydroxide (contained in the 3,5-dinitrosalicylate reagent added at this point). The solution is heated to "develop" the color produced by the reagent in the presence of reducing activity in the solution, and the intensity of that color is determined with a spectrophotometer. After a calibration has been carried out, the spectrophotometric reading directly establishes the number of moles of fructose and glucose present in the solution. Then, knowing the time during which the reaction proceeded, we can readily establish the average rate at which sucrose was converted to glucose and fructose.

EXPERIMENTAL

There are a large number of runs described in this section, and a given team may not be required to carry out all of them. In any event, it is recommended that students work in teams of two or three, with the work divided equitably among the partners. To avoid any needless expenditure of time, each partner should carry out several runs concurrently. Do not, however, undertake so many that you are unable to stop each run at precisely the correct time. In making several runs of equal duration, a staggered schedule is recommended. For a given set of runs, set up a series of labeled assay tubes, each containing all but the last solution—the one that will initiate the reaction. Add the prescribed amount of this last solution to the first tube and note the time; exactly 1 min later, add the prescribed amount of the last solution to the second tube; and so on. You will then be able to quench these reactions at 1-min intervals with all the times precisely controlled.

Most of the runs are carried out at room temperature without any constant-temperature control. Note the ambient air temperature near your laboratory workbench at the beginning of each set of runs.

Procedure.   The standard assay procedure to be used (with variations as indicated elsewhere) is as follows.

1. Pipette into a clean test tube—*in the order given*—the appropriate amounts of enzyme solution, distilled water, buffer solution, and sucrose solution. Swirl the assay tube vigorously before and *immediately* after the addition of the sucrose.
2. Begin timing when the sucrose solution is added and the tube swirled to mix the reagents. Let the reaction proceed for *exactly* 5.0 min, or whatever other period of time is specified in a particular run.
3. At the end of the allotted period, quench the reaction by adding 2.0 mL of dinitro-salicylate reagent (containing NaOH as the active quenching agent), and once again swirl the tube vigorously to mix the reagents.
4. Immerse the test tube in boiling water for 5 min; cover *loosely* with Parafilm to prevent the escape of water vapor.
5. Cool the reaction tube by holding it in a slanted position and running cold tap water down the outside of the tube.
6. Dilute the solution with 15.0 mL of distilled water from a graduated pipette. Cover *tightly* and shake the tube to ensure thorough mixing.

7. Using a suitable spectrophotometer, determine the absorbance of the solution at 540 nm with distilled water as the reference. Read carefully the operating instructions provided for the instrument to be used. A general discussion of spectrophotometers and absorbance measurements is given in Chapter XIX.

Enzyme Activity. Carry through the standard assay procedure described above with 1.0 mL of enzyme solution, 0.5 mL of distilled water, 0.5 mL of buffer, and 1.0 mL of 0.3 M sucrose solution. The observed absorbance should lie in the range 0.6 to 0.9. If it is greater than 0.9, dilute the enzyme solution with sterile, previously boiled distilled water to reduce the activity to an appropriate level. If the absorbance is less than 0.6, the enzyme solution is too weak and a new solution must be made up.

A. Blank Runs. Our first concern is to establish that the dinitrosalicylate reagent gives no appreciable test in the absence of reducing sugars. To establish this point, one should systematically examine the results of the action of the reagent on each of the other substances present in the reaction mixture. The solutions to be tested are given below. In run A0, add the enzyme, water, and buffer; then add 2.0 mL of the dinitrosalicylate reagent *before adding* the 1.0 mL of sucrose solution. Pick up the standard assay procedure at step 4. In the following table and in subsequent tables, an asterisk (*) is placed before the volume of sucrose to remind you to add dinitrosalicylate reagent before adding the sucrose. Run A0 constitutes a so-called zero-time blank—since the sucrose is added only after the reagent that should completely halt the enzymatic reaction—and constitutes a particularly sensitive blank run. In runs A1 to A3, mix the indicated reagents, wait 5.0 min, and then pick up the standard assay at step 3. None of these blank runs should develop an absorbance greater than about 0.05 at 540 nm.

Run	Enzyme, mL	Water, mL	Buffer, mL	0.3 M Sucrose, mL
A0	1.0	0.5	0.5	*1.0
A1	0.0	2.5	0.5	0.0
A2	0.0	1.5	0.5	1.0
A3	1.0	1.5	0.5	0.0

B. Standardization Runs. We must establish the absorbance at 540 nm produced by the action of our reagent in the presence of known amounts of glucose and fructose. The reagent offers an exceedingly delicate test, and to standardize such a reagent we use a very dilute aqueous glucose–fructose solution (hereafter denoted by GFS) containing only 0.9 g of glucose and 0.9 g of fructose per liter. Five standardization runs are recommended.

In each case make up the indicated solution and then pick up the standard assay procedure at step 3. The results of blank A2 should have shown that, in the absence of glucose and fructose, the mixture of reagents used in the standardization runs will yield no significant color. The absorbances obtained in runs B1 to B5 can thus be attributed to the action of the reagent on the GFS. These data will be used to construct a calibration curve for the spectrophotometric assay.

Run	Enzyme, mL	Water, mL	Buffer, mL	0.3 M Sucrose, mL	GFS, mL
B1	0.0	1.1	0.5	1.0	0.4
B2	0.0	0.8	0.5	1.0	0.7
B3	0.0	0.5	0.5	1.0	1.0
B4	0.0	0.3	0.5	1.0	1.2
B5	0.0	0.0	0.5	1.0	1.5

C–D. Progress of the Reaction with Time.   For the second of the following two sets of runs, prepare some 0.03 M sucrose solution by diluting 1 mL of the 0.3 M stock solution with 9 mL of distilled water. Runs in series C use the concentrated sucrose solution, and those in series D use the dilute solution.

Run	Enzyme, mL	Water, mL	Buffer, mL	Sucrose, mL	Time, min
C0	0.2	1.3	0.5	*1.0 (0.3 M)	0
C1	0.2	1.3	0.5	1.0 (0.3 M)	1
C3	0.2	1.3	0.5	1.0 (0.3 M)	3
C5	0.2	1.3	0.5	1.0 (0.3 M)	5
C10	0.2	1.3	0.5	1.0 (0.3 M)	10
C20	0.2	1.3	0.5	1.0 (0.3 M)	20
D0	1.0	0.5	0.5	*1.0 (0.03 M)	0
D1	1.0	0.5	0.5	1.0 (0.03 M)	1
D3	1.0	0.5	0.5	1.0 (0.03 M)	3
D5	1.0	0.5	0.5	1.0 (0.03 M)	5
D10	1.0	0.5	0.5	1.0 (0.03 M)	10
D20	1.0	0.5	0.5	1.0 (0.03 M)	20

Runs C0 and D0 are zero-time blanks in which the sucrose addition is made only *after* inactivation of the enzyme by the addition of 2.0 mL of dinitrosalicylate reagent (denoted by * in the tables). For these two runs, pick up the standard assay procedure at step 4. In all other runs, begin timing when the sucrose solution is added, and let the reaction proceed for *precisely* the time indicated in each case. Then pick up the standard assay procedure at step 3.

E. Dependence of Initial Rate on Substrate Concentration.   Prepare from the 0.3 M stock solution of sucrose, another 10 mL of 0.03 M sucrose solution. Taking care to use the sucrose solution called for in each case, proceed to the following runs—all of which are allowed to proceed for the standard time of 5.0 min except run E0, which is a zero-time blank. This set of runs will be analyzed to obtain k_2 and K.

Run	Enzyme, mL	Water, mL	Buffer, mL	Sucrose, mL
E0	1.0	0.5	0.5	*1.0 (0.3 M)
E1	1.0	0.5	0.5	1.0 (0.3 M)
E2	1.0	1.3	0.5	0.2 (0.3 M)
E3	1.0	0.5	0.5	1.0 (0.03 M)
E4	1.0	1.0	0.5	0.5 (0.03 M)
E5	1.0	1.3	0.5	0.2 (0.03 M)
E6	1.0	1.4	0.5	0.1 (0.03 M)

F. Nonenzymatic Hydrolysis of Sucrose.[3]   The basic procedure for carrying out hydrolysis runs with H^+ ions as the catalyst is as follows. Pipette the indicated volume of 0.3 M sucrose solution into an assay tube. Add sufficient distilled water so that the final volume of the reaction mixture will be 4 mL (*not 3 mL, as in all preceding runs*). Add the indicated volume of 1 M HCl solution to initiate the catalytic hydrolysis, starting your timing as the acid is added. After exactly 5.0 min, stop the reaction by adding 5 mL of 1 M NaOH, which immediately neutralizes all of the catalytically active H^+ present. Add 2.0 mL of dinitrosalicylate reagent, and heat the reaction tube in a boiling-water bath for

5 min. Cool the reaction tube, and dilute by adding 9 mL of distilled water (*not 15 mL as in earlier runs*). Thus the final volume is as before 20 mL. Measure the absorbance of the solution at 540 nm.

A set of suggested concentrations at which this procedure can be carried out is given below. Run F0 is a blank, which will allow you to see how much (if any) hydrolysis takes place in the absence of any catalyst.

Run	0.3 *M* Sucrose, mL	Water, mL	1 *M* HCl, mL
F0	2.0	2.0	0.0
F1	1.0	2.0	1.0
F2	1.0	1.0	2.0
F3	2.0	1.0	1.0
F4	2.0	0.0	2.0

G–H. Dependence of the Rate on Temperature. In order to determine the temperature dependence of the enzyme-catalyzed rate, the reaction is carried out at several different temperatures, which are held constant to within $\pm 0.3°C$ or better. Choose from runs E a set of initial concentrations that gives an absorbance of ~0.5 at room temperature after the standard assay. Convenient temperatures for these runs, denoted as runs G, are 0°C (ice bath), 12, 25, 35, and 45°C. In each case, prepare the enzyme–water–buffer mixture in the assay tube, put about 2 mL of the sucrose stock solution in another test tube, and immerse both tubes in the thermostat bath for a few minutes to achieve temperature equilibrium. Next, pipette 1.0 mL of the equilibrated sucrose solution into the assay tube and read the bath temperature. Leave the assay tube in the constant-temperature bath until, 5.0 min after the addition of the sucrose solution, the reaction is terminated by the addition of 2.0 mL of the dinitrosalicylate reagent. Then complete the assay as usual.

Finally, choose from runs F a set of initial concentrations that gives a moderate rate of reaction. Then carry out a study of the temperature dependence of the acid-catalyzed rate over the range 0 to 50°C. It is convenient to do these measurements, denoted as runs H, at the same temperatures and at the same times as runs G are carried out.

Precautions. In order to obtain successful results from this experiment, careful volumetric technique and judicious handling of reagents are required. The following points should be noted.

1. The pipettes, especially those to be used for dispensing the enzyme, should be cleaned carefully and rinsed thoroughly with boiled distilled water. When first using any pipette, rinse it with a small portion of the solution to be used and make sure that the pipette drains well. If possible, always use the same pipette for a given solution; *it is imperative that the pipettes used to dispense enzyme never be used for any other solution.* The pipettes needed are:

    Two 1-mL transfer pipettes, used to dispense the "standard" amounts of sucrose and enzyme solution

    Two 2-mL graduated pipettes, used for dispensing "nonstandard" amounts of sucrose and enzyme solution

    One 2-mL transfer pipette, used exclusively for dispensing the dinitrosalicylate reagent

    One 5-mL transfer pipette, used exclusively for dispensing the NaOH solution required in parts F and H of the experiment

One 1-mL graduated pipette, used for dispensing buffer solution and distilled water in step 1 of the assay procedure

One 25-mL graduated pipette, used for dilution in step 6

2. The enzyme, a protein, is highly susceptible to attack by airborne microorganisms. All glassware that comes in contact with the enzyme must be cleaned and rinsed very carefully. Place about 50 mL of the enzyme solution in a glass-stoppered Erlenmeyer flask, and keep this flask chilled with ice throughout the experiment. Any dilution of the enzyme stock solution must be done with chilled water that has been previously sterilized by boiling.

3. Sucrose solutions are also susceptible to attack by microorganisms. In this instance, it is sufficient to use a freshly prepared solution each day.

Solutions. A variety of comments will be given here about the preparation of the required solutions. As noted above, a new sucrose solution must be prepared each day by the student team doing the experiment. The enzyme solution must be prepared and tested in advance, presumably by the teaching staff. Other solutions may be made available to the students or be prepared by them, depending upon the policy of the instructor. A complete list of all solutions is given in the Apparatus section.

Enzyme solution. Powdered invertase obtained from yeast is available from many biochemical supply houses, such as Sigma Chemical Co., P.O. Box 14508, St. Louis, MO 63178. A sterile solution of invertase must be prepared with great care. All apparatus used to make up this solution must be sterilized prior to use. More than 1 L of distilled water is boiled for 10 min, covered with aluminum foil, and chilled in an ice bath. Then add a precisely weighed amount of invertase (approx. 5–10 mg) to water in a 1-L volumetric flask and make up to the mark. *The enzyme solution should be kept tightly stoppered and chilled at all times.* The correct invertase concentration is very sensitive to the specific activity of the enzyme as purchased; it is necessary to carry out the standard assay and adjust the solution to an appropriate final concentration.

Reducing-sugar reagent. This reagent is made up by mixing two solutions: a solution of about 25.0 g of 3,5-dinitrosalicylic acid in 1000 mL of warm 1 M NaOH and one of 750 g of sodium potassium tartrate, $NaKC_4H_4O_4 \cdot 4H_2O$, in 750 mL of warm distilled water. These solutions are mixed slowly and diluted to a total volume of 2500 mL with warm distilled water.

Glucose-fructose standard solution (GFS). Each individual team will need 100 mL of this solution. Weigh out exactly 0.090 g of glucose and 0.090 g of fructose, dissolve both solids in distilled water contained in a 100-mL volumetric flask, and make up to the mark. Note that fructose is somewhat hygroscopic; it may be necessary to handle it in a dry-box if the humidity is unusually high.

Acetate buffer. A suitable buffer with pH 4.8 to 5.0 can be prepared by dissolving 4.10 g of sodium acetate in 1000 mL of distilled water and then adding 2.65 mL of glacial acetic acid.

Optional Investigations. A number of additional investigations can be carried out on this system. Three suggestions are given below.

1. Confirm the first-order dependence of the rate on the initial enzyme concentration $(E)_0$. A set of recommended runs based on the use of the 0.3 M stock sucrose solution is given below. Each run in series I is allowed to proceed for the standard time of 5.0 min except run I0, which is a zero-time blank.

Run	Enzyme, mL	Water, mL	Buffer, mL	0.3 M Sucrose, mL
I0	1.0	0.5	0.5	*1.0
I1	1.5	0.0	0.0	1.0
I2	1.0	0.5	0.5	1.0
I3	0.5	1.0	0.5	1.0
I4	0.2	1.3	0.5	1.0
I5	0.1	1.4	0.5	1.0

2. By a series of runs at different pH values, determine the pH at which invertase functions most effectively as a catalyst. Is the reaction rate highly dependent on pH? Why should pH have anything to do with the catalytic activity of invertase?

3. Investigate the inhibition of the enzymatic reaction by p-mercuribenzoate at concentrations of about 10^{-5} M. Show how an investigation of inhibition or "poisoning" of the catalyst might be used to establish the number of catalytically active sites on an invertase molecule.

DATA ANALYSIS

A spreadsheet program will be useful for the plots and linear regressions to be made in the following analysis. In order to convert the observed absorbance A into the desired product concentration (P) in the reaction mixture, a calibration must be carried out using the results of runs A2 and B1 to B5. According to the Beer–Lambert law, commonly called *Beer's law*,[4]

$$A = \log\left(\frac{I_0}{I}\right) = \varepsilon cd \tag{14}$$

where I/I_0 is the fraction of the light transmitted through an absorbing solution of path length d and c is the concentration of the absorbing species (in this case, the reduced form of the dinitrosalicylate reagent in the final 20 mL solution). The absorption coefficient ε is a constant at any specified wavelength, 540 nm in this case; and d will also be a constant if matched cuvettes with the same path length are used.

The value of c is *proportional* to $n_1 + n_2$, where n_1 and n_2 are the numbers of moles of glucose and of fructose formed in the reaction up to the time it was quenched. It is obvious from the stoichiometry of Eq. (12) that $n_1 = n_2 = x$, where x is the number of moles of sucrose hydrolyzed. Thus c is proportional to x, and we can rewrite Eq. (14) as

$$A = \alpha x \tag{15}$$

where α is a calibration constant. The product concentration (P) = (P_1) = (P_2) is given by

$$(P) = \frac{x}{V} = \frac{A}{\alpha V} \tag{16}$$

where V is the volume of the reaction mixture (3 mL for enzyme runs and 4 mL for acid runs).

For a determination of α, plot the absorbance measured for runs A2 and B1 to B5 versus the value of $(n'_1 + n'_2)/2$, where n'_1 and n'_2 are the numbers of moles of glucose and fructose added to the assay tube in the standardization runs. Since the GFS solution is made up by weighing, n_1' may not be precisely equal to n_2' and it is appropriate to use the average value. The molar mass of both glucose and fructose is 180.16 g mol^{-1}. Make a graphical or least-squares evaluation of the slope and give its units.

Using the value of α determined above, the results of the standard assay made initially to check the enzyme activity, the assay in part C, and the given concentration of the enzyme stock solution in g L^{-1}, calculate the *specific activity* of the enzyme—that is, the number of micromoles of sucrose hydrolyzed per minute per gram of enzyme present. (The specific activity of an enzyme preparation is of course a function of the purity of the enzyme. As inactive protein is removed from the preparation, the specific activity will rise. When the specific activity can no longer be increased by any purification method, a homogeneous enzyme preparation may have been achieved; but proof of this depends on other criteria.) The exact chemical composition of invertase is still unknown, but its molar mass has been estimated at 100,000 g mol^{-1}. Combining this datum with your calculated specific activity, estimate the turnover number for the enzyme.

From the data of runs C1 to C20 and D1 to D20, calculate x, the number of moles of sucrose hydrolyzed in each time interval. If the reaction were zero order in sucrose, then we would expect that $(x/0.003) = k_0 t$, where $x/0.003$ is the concentration of either of the product species in mol L^{-1} units. Prepare a graph of the results obtained in these two series of runs, plotting x versus t, and indicate whether the data are consistent with the hypothesis that the reaction is zero order in sucrose. Note that, even if a reaction starts out being zero order in sucrose, this cannot continue indefinitely. Indeed, we expect the inversion reaction to become first order in sucrose when (S) becomes sufficiently small.

If the reaction were first order in sucrose, then letting a stand for the number of moles of sucrose originally present, we should find that $\ln[a/(a - x)] = k_1 t$. Prepare a graph of the results obtained in the above two series of runs, plotting log $[a/(a - x)]$ versus t, and indicate whether your data are consistent with the hypothesis that the reaction is first order in sucrose.

Show that the average rate measured over the first 5 min of reaction provides an acceptable approximation to the true initial rate r_0 by estimating the percent error associated with this approximation.

If the optional runs I1 to I5 were carried out, determine the dependence of the initial rate on the initial concentration of the enzyme $(E)_0$.

Using the data from runs E, prepare a Lineweaver–Burk plot of $1/r_0$ versus $1/(S)_0$ and an Eadie–Hofstee plot of $r_0/(S)_0$ versus r_0. Determine the values at room temperature of $k_2(E)_0$ and K_m from both of these plots. Using the nominal molar mass of 100,000 g mol^{-1} for invertase, calculate $(E)_0$ in mol L^{-1} units and obtain the value of k_2 in clearly stated units.

From the data of part F, estimate the rate constant k_H for the acid-catalyzed reaction. Also calculate a turnover number for the acid-catalyzed reaction, i.e., the number of molecules of sucrose hydrolyzed per second per hydrogen ion present. Note that Eq. (10) predicts that this turnover number depends on (S). Do your data confirm this?

From the data of part G, determine the initial rate r_0 for each temperature and plot log r_0 versus $1/T$. From the slope of the line, determine the activation energy E_a from

$$\frac{d \log r_0}{d(1/T)} = -\frac{E_a}{2.303R} \tag{17}$$

Finally, use the data from runs H for the temperature dependence of r_0 to determine E_a for the acid-catalyzed reaction.

DISCUSSION

A number of questions might be addressed in the discussion of the results. How reproducible are the initial rate measurements? (Note that runs D5 and E3 are duplicates; also, runs E1 and the standard assay for enzyme activity have identical initial concentrations.) Are the enzyme-catalyzed data compatible with the Michaelis–Menten mechanism? Do the data from both runs C and D follow apparent zero-order kinetics, and how does this agree with expectations based on comparing (S) with K_m? Which of the two types of analysis, Lineweaver–Burk or Eadie–Hofstee, seems to give the better results and why? How does k_2 agree with the estimate of the turnover number based on the specific activity? Are the acid-catalyzed data consistent with the rate law given in Eq. (10)?

Compare the enzyme turnover number with a typical hydrogen-ion turnover number. What does this tell you about the efficiency of enzymes? Compare the activation energy for the enzyme-catalyzed reaction with that for the acid-catalyzed reaction. Does the difference in E_a values account completely for the ratio of turnover numbers for the enzyme and H^+?

SAFETY ISSUES

Use a pipette bulb; do not pipette by mouth.

APPARATUS

Sterile preparation of yeast invertase (50 mL per team); pH 4.8 to 5 acetate buffer; 1 M sodium hydroxide; 3,5-dinitrosalicylate reagent solution; 1 M HCl; glucose-fructose standard solution; sucrose.

Two 1-mL transfer pipettes; one 2-mL transfer pipette; one 5-mL transfer pipette; one 1-mL graduated pipette; two 2-mL graduated pipettes; one 25-mL graduated cylinder or pipette; pipetting bulb; glass-stoppered flask; volumetric flasks, beakers, test tubes.

Boiling-water bath; stopwatch or other timer; spectrophotometer (Bausch & Lomb Spectronic 20, Sequoia-Turner model 340, or other suitable type); 0 to 50°C constant-temperature baths.

REFERENCES

1. F. J. Kezdy and M. L. Bender, *Biochem.* **1**, 1097 (1962); M. L. Bender, F. J. Kezdy, and F. C. Wedler, *J. Chem. Educ.* **44**, 84 (1967).

2. H. B. Dunford, *J. Chem. Educ.* **61**, 129 (1984).

3. J. G. Dawber, D. R. Brown, and R. A. Reed, *J. Chem. Educ.* **43**, 34 (1966).

4. I. N. Levine, *Physical Chemistry*, 6th ed., sec. 20.2, McGraw-Hill, New York (2009); P. W. Atkins and J. de Paula, *Physical Chemistry*, 8th ed., p. 432, Freeman, New York (2006).

GENERAL READING

G. G. Hammes, *Thermodynamics and Kinetics for the Biological Sciences*, Wiley, New York (2000).

J. I. Steinfeld, J. S. Francisco, and W. L. Hase, *Chemical Kinetics and Dynamics*, 2d ed., Prentice-Hall, Upper Saddle River, NJ (1999).

I. Tinoco, Jr., K. Sauer, and J. C. Wang, *Physical Chemistry: Principles and Applications in Biological Sciences,* 2d ed., chap. 8, Prentice-Hall, Upper Saddle River, NJ (1985).

C. T. Walsh, *Enzymatic Reaction Mechanisms,* Freeman, San Francisco (1979).

EXPERIMENT 23

Kinetics of the Decomposition of Benzenediazonium Ion

In an acidic aqueous solution, benzenediazonium ion $(C_6H_5N_2^+)$ will decompose[1] to form nitrogen and phenol:

$$C_6H_5N_2^+ + H_2O \longrightarrow C_6H_5OH + N_2(g) + H^+ \tag{1}$$

In this experiment you will follow the course of the above reaction by a spectrophotometric measurement of the unreacted diazonium ion.[2] The reaction order with respect to $C_6H_5N_2^+$, the rate constant k, and the activation energy for the reaction will be determined.

THEORY

The rate of reaction (1), $-d(C_6H_5N_2^+)/dt$, can be written in terms of the concentrations of the reacting species as

$$-\frac{d(C_6H_5N_2^+)}{dt} = k'(C_6H_5N_2^+)^n(H_2O)^m \tag{2}$$

As written here, the reaction is nth-order with respect to $C_6H_5N_2^+$, mth-order with respect to water, and has an overall order of $n + m$. The reaction is not acid catalyzed, although very high acid concentrations (say, 12 M HCl) seem to increase the rate slightly.[1] Thus the effect of changes in the H^+ concentration due to reaction (1) can be completely neglected. Since the present experiment will be performed in a dilute aqueous solution, the concentration of water (H_2O) will be very nearly constant throughout the reaction. Thus the factor $(H_2O)^m$ can be absorbed into the rate constant and Eq. (2) can be written as

$$-\frac{dc}{dt} = kc^n \tag{3}$$

where c represents the instantaneous concentration $(C_6H_5N_2^+)$.

Equation (3) can readily be integrated to give

$$c = c_0 e^{-kt} \qquad \text{for } n = 1 \tag{4a}$$

$$c^{1-n} - c_0^{1-n} = (n-1)kt \qquad \text{for } n \neq 1 \tag{4b}$$

where c_0 is the concentration at $t = 0$. In other words, if the reaction is first order, a plot of log c versus t should give a straight line of slope $-k/2.303$ and intercept log c_0. If the reaction is nth order where n (which need not be an integer) is not equal to unity, a plot of c^{1-n} versus t should give a straight line of slope $(n-1)k$ and intercept c_0^{1-n}. By making such plots for various trial values of n and determining which gives the best straight-line dependence over a wide range of concentration, one can obtain the order n of the reaction and then determine the appropriate rate constant k_n.

If the value of a rate constant is measured at several different temperatures, it is almost always found that the temperature dependence can be represented by

$$k = \mathscr{A}e^{-E_a/RT} \tag{5}$$

where the factor $\mathscr{A}$ is independent of temperature. The Arrhenius activation energy E_a can be easily determined by plotting $\log k$ versus $1/T$. This should give a straight line of slope $-E_a/2.303R$.

METHOD

Any physical variable giving an accurate measure of the extent to which a reaction has gone toward completion can be used to obtain rate data. In this experiment we shall monitor the concentration of unreacted diazonium ion by measuring the absorption of ultraviolet light by the solution.

According to the Beer–Lambert law,[3] the intensity of light I transmitted by an absorbing medium is given by

$$I = I_0 e^{-c\varepsilon'd} \tag{6}$$

where c is the concentration of absorbing molecules, d is the path length, I_0 is the intensity of incident light, and ε' is an "extinction coefficient." The *absorbance A,* defined as $\log(I_0/I)$, thus gives a direct measure of the concentration:

$$A \equiv \log\frac{I_0}{I} = \frac{c\varepsilon'd}{2.303} = c\varepsilon d \tag{7}$$

where ε is called the *molar absorption coefficient* when the concentration is expressed in $\mathrm{mol\ L^{-1}}$ units. When the sample is a solute in solution, I is the intensity of light transmitted by a cell filled with the solution and I_0 is the intensity transmitted by the cell filled with pure solvent.

For first-order reactions, the quantity εd will cancel out of Eq. (4a) and k can be obtained directly from a plot of $\log A$ versus t. However, for orders different from first order, εd does not cancel out of Eq. (4b). In such cases εd must be evaluated by measuring the absorbance of a solution of known concentration in order to permit k to be calculated in concentration units.

A description of spectrophotometers and their use in determining the absorbance is given in Chapter XIX.

EXPERIMENTAL

Kinetic runs are to be made at 25, 30, and 40°C. Well-regulated thermostat baths operating at about these temperatures will be required. The actual temperatures of each of these baths should be measured with a good thermometer and recorded.

The concentration of benzenediazonium ion can be determined by the absorbance at wavelengths between 295 and 325 nm. Below 295 nm, products of the reaction produce interfering absorption; and above 325 nm, the molar absorption coefficient is too small to permit effective measurement of changes in the benzenediazonium ion concentration.[4] The absorbance will be measured using a suitable spectrophotometer such as a Beckman model DU-20 UV; detailed instructions for operating the instrument will be provided in the laboratory. A wavelength of 305 nm should be used, with a slit width of 0.3 mm. Make two absorbance readings on each sample.

The diazonium salt that should be used in this experiment is benzenediazonium fluoborate ($C_6H_5N_2BF_4$, $M = 191.9$). The great majority of diazonium salts are notoriously unstable solids and can decompose with explosive violence. The fluoborates are by far the safest to use and are not known to explode; however, reasonable caution should be used in preparing the compound. Since even benzenediazonium fluoborate will decompose slowly, it should not be prepared too far in advance, and it must be stored in a refrigerator. A simple high-yield procedure for its preparation has been given by Dunker, Starkey, and Jenkins.[5] Recrystallization of the product from 5 percent fluoboric acid yields white needlelike crystals, which can be dried by vacuum pumping at 1 Torr for several hours.†

The diazonium salt should be available at the beginning of the experiment. Remove a *small* quantity from the refrigerator and warm it rapidly to room temperature. Prepare approximately 10^{-3} M solutions by placing an accurately weighed 15- to 20-mg sample of the salt into each of three 100-mL volumetric flasks, and then make up to the mark with a 0.2 M HCl solution. Label the flasks and be sure to record the exact weight of salt placed in each flask. Return the unused diazonium salt to the refrigerator at once. Be careful not to waste it.

Using a hypodermic syringe, withdraw a sample of about 5 mL of each solution *as soon as possible* after it is made up. Chill these samples by placing them in test tubes set in crushed ice, then put them aside for absorbance measurements (to be made as soon as conveniently possible). These measurements will provide a value for εd in case it is needed later on.

Suspend one of the volumetric flasks in each of the three constant-temperature baths, and record the time. Allow about 20 min for the solutions to achieve thermal equilibrium. After thermal equilibrium has been attained, you may begin taking samples for spectrophotometric analysis.

When you take a sample, use the following procedure. Withdraw about 1 mL of the solution with the hypodermic syringe and use this to rinse out the syringe. Then quickly withdraw a 5-mL sample and place it in a labeled test tube set in crushed ice for chilling. Record the time of discharge into the test tube. After the sample is chilled (3 to 5 min), use about 1 mL of it to rinse out the spectrophotometer cell. Fill the cell with the remaining sample and measure its absorbance, using the 0.2 M HCl solution as a blank. These cells are fragile and expensive; handle them with care.

The reason for chilling the sample is to slow down the reaction rate so that a negligible amount of reactant will decompose between the time you chill the sample and the time you make the absorbance measurement. Since it is impossible to eliminate completely errors due to reaction after withdrawal and cooling, it is important to try to perform the sample removal and chilling procedure in as reproducible a fashion as possible. This will lead to a partial cancellation of such errors.‡

---

†After storage at 0°C in a vacuum desiccator for several weeks, these crystals tend to stick together. After six months there is clear evidence of decomposition; fortunately, the main impurity is phenol (2 to 3 percent), which does not interfere much with rate studies in aqueous solution. By a very careful purification, it is possible to obtain a benzenediazonium fluoborate sample that can be stored at 0°C for several months without any signs of decomposition. The sample is dissolved in acetone, and then chloroform is added until a few crystals are formed. When the solution is then chilled to −20°C for 30 min, the compound will crystallize out in the form of tiny white needles. After three such recrystallizations, the sample can again be dried by pumping and can be stored in a vacuum desiccator.

‡If a spectrophotometer with a cell compartment that can be temperature controlled is available, the procedure can be greatly improved and simplified. Adjust the bath temperature to 30°C, turn on the circulating pump, and wait until the cell compartment (with cell holder in place) has become stable. Then prepare a solution as described above and fill two spectrophotometer cells as soon as possible. (A third cell should already have been filled with 0.2 M HCl solution, for use as a blank.) Place the cells in the cell holder, and record the time at which the cell holder is returned to the cell compartment. Obtain absorbance readings on both samples as soon as possible; these initial readings will provide values of εd. Allow about 20 min for the solutions to achieve thermal equilibrium, and then begin measuring the absorbance every 15 min. Since the runs are made sequentially rather than simultaneously, it is advisable to replace the slow run at 25°C with a run at 45°C (use 5-min intervals). Fresh solutions are needed for each temperature studied.

Take as many measurements as you reasonably can without rushing. Since the runs at the higher temperatures will proceed more rapidly, it will be necessary to take samples from them at more frequent intervals. A suggested interval might be once every 25 min for the 25°C bath, every 15 min for the 30°C bath, and every 10 min for the 40°C bath.

At the end of the experiment, wash the cells thoroughly with distilled water and dry them very carefully. Store them in a safe place.

CALCULATIONS

Using all the data points from any single run, prepare plots of the following: (1) $A^{1/2}$ versus t (order $\frac{1}{2}$), (2) log A versus t (order 1), (3) A^{-1} versus t (order 2), and (4) any additional plots you feel to be necessary in order to establish the order of the reaction.

Having determined the order of the reaction, make the appropriate plot for each of the runs and determine the value of k (in concentration units) for each temperature. It is convenient to make these plots using A values rather than c values. The value of the slope can then be corrected, if necessary, to obtain k in concentration units by utilizing the value of εd. You should also carry out a least-squares analysis of your data using the appropriate form of Eq. (4). Report both sets of k values, together with the standard deviations for the least-squares values. Finally, plot log k versus $1/T$ and determine E_a.

Report your result for the order of the reaction and list the best values obtained for $k(T_1)$, $k(T_2)$, and $k(T_3)$. Report also the activation energy E_a. Give the correct units for each quantity, using s as the unit of time, mol L^{-1} for concentration, and kJ for energy.

DISCUSSION

From your results, calculate the extent of reaction that would occur in a sample of the initial solution after 30 min at 0°C. Does this indicate that varying lengths of time between the chilling of a sample and its absorbance measurement would introduce serious or negligible errors in your data?

SAFETY ISSUES

None.

APPARATUS

Spectrophotometer, such as one of the Beckman model DU series; two or more quartz sample cells; constant-temperature baths set at 25, 30, and 40°C; lens tissue; three 100-mL volumetric flasks; clock, stopwatch, or other timer; hypodermic syringe; about 20 test tubes; plastic pail or large battery jar; beaker.

Benzenediazonium fluoborate (75 mg), stored in a refrigerator; 0.2 M hydrochloric acid (400 mL); crushed ice.

REFERENCES

1. E. A. Moelwyn-Hughes and P. Johnson, *Trans. Faraday Soc.* **36,** 948 (1940); M. L. Crossley, R. H. Kienle, and C. H. Benbrook, *J. Amer. Chem. Soc.* **62,** 1400 (1940).

2. J. E. Sheats, Ph.D. thesis, Department of Chemistry, Massachusetts Institute of Technology, Cambridge, MA (1965).

3. P. W. Atkins and J. de Paula, *Physical Chemistry,* 8th ed., p. 432, Freeman, New York (2006).

4. A. Wohl, *Bull. Soc. Chim. Fr.* **6,** 1319 (1939).

5. M. F. W. Dunker, E. B. Starkey, and G. L. Jenkins, *J. Amer. Chem. Soc.* **58,** 2308 (1936).

GENERAL READING

J. F. Bunnett, "Kinetics in Solution," in E. S. Lewis (ed.), *Techniques of Chemistry: Vol. VI, Investigation of Rates and Mechanisms of Reactions,* 3d ed., part I, chap. 4, Wiley-Interscience, New York (1974).

L. G. Hargis and J. A. Howell, "Visible and Ultraviolet Spectroscopy," in B. W. Rossiter and R. C. Baetzold, *Physical Methods of Chemistry,* 2d ed., Vol. VIII, chap. 1, Wiley-Interscience, New York (1993).

EXPERIMENT 24
Gas-Phase Kinetics

Although a vast majority of important chemical reactions occur primarily in liquid solution, the study of simple gas-phase reactions is very important in developing a theoretical understanding of chemical kinetics. A detailed molecular explanation of rate processes in liquid solution is extremely difficult. At the present time reaction mechanisms are much better understood for gas-phase reactions; even so this problem is by no means simple. This experiment will deal with the unimolecular decomposition of an organic compound in the vapor state. The compound suggested for study is cyclopentene or di-*t*-butyl peroxide, but several other compounds are also suitable; see, for example, Table XI.4 of Ref. 1.

THEORY

A discussion of bimolecular reactions (which usually give rise to second-order kinetics) can be based quite naturally on collision theory or on transition-state theory.[1,2] We shall assume a general knowledge of these treatments as background for the discussion of unimolecular reactions (which usually give rise to first-order kinetics). The crucial question is: how does a molecule acquire the necessary energy to undergo a spontaneous unimolecular decomposition? If the "activation" occurs via a collision, one could normally expect second-order kinetics. However, Lindemann pointed out that this would not be true if the time between collisions is short compared with the average lifetime of an activated molecule.[3]

Let us consider a simple unimolecular decomposition† in the gas phase:

$$A \rightarrow B + C$$

The proposed reaction mechanism will consist of three steps: (1) collisional activation, (2) collisional deactivation, and (3) spontaneous decomposition of the activated molecule. Thus

$$A + A \rightarrow A + A^* \quad k_1 \tag{1}$$

$$A^* + A \rightarrow A + A \quad k_2 \tag{2}$$

$$A^* \rightarrow B + C \quad k_A \tag{3}$$

---

†In the following treatment, several details and difficulties will be overlooked; more advanced developments are given by Benson and by Trotman-Dickenson.[1]

where A* is an activated A molecule with a definite internal energy ε (above its ground state) that is greater than a certain critical value ε^* necessary for reaction to occur, k_1 and k_2 are second-order rate constants, and k_A is a first-order rate constant. The overall rate of reaction is given by

$$+\frac{d(B)}{dt} = +\frac{d(C)}{dt} = -\frac{d(A)}{dt} = k_A(A^*) \tag{4}$$

where parentheses indicate concentrations of the species. As discussed below, the value of k_A will depend on the particular value of ε involved. If the concentration of A* is small, one can make the steady-state approximation, $d(A^*)/dt = 0$, and obtain

$$+\frac{d(A^*)}{dt} = k_1(A)^2 - k_2(A^*)(A) - k_A(A^*) = 0 \tag{5}$$

Solving Eq. (5) for (A*) and substituting the resulting expression into Eq. (4) gives

$$-\frac{d(A)}{dt} = \frac{k_1 k_A (A)^2}{k_A + k_2(A)} \tag{6}$$

High-Pressure Limit. At high concentrations deactivation is much more probable than decomposition, since the collision frequency is high, and deactivation should occur at the first collision suffered by A*. Thus at high pressure, $k_2(A) \gg k_A$ and the reaction becomes first order:

$$-\frac{1}{(A)}\frac{d(A)}{dt} = \frac{k_1}{k_2}k_A \tag{7}$$

Also

$$\frac{k_1}{k_2} = \frac{(A^*)}{(A)} \cong f_A \tag{8}$$

where f_A is the fraction of A molecules in an activated state with some internal energy ε. Obviously f_A depends on the value of ε, as does k_A. Thus

$$-\frac{1}{(A)}\frac{d(A)}{dt} = \int_{\varepsilon^*}^{\infty} k_A(\varepsilon) f_A(\varepsilon) d\varepsilon = k_\infty \tag{9}$$

where k_∞, the value of the integral, corresponds to the experimental first-order rate constant k_{exp} determined at high pressures.

Rice and Ramsperger[4] and Kassel[5] have given an expression for k_A as a function of ε:

$$k_A(\varepsilon) = \begin{cases} 0 & \text{for } \varepsilon < \varepsilon^* \\ \nu\left(\dfrac{\varepsilon - \varepsilon^*}{\varepsilon}\right)^{n-1} & \text{for } \varepsilon \geq \varepsilon^* \end{cases} \tag{10}$$

Equation (10) is based on a model (called the RRK model, after the authors) in which the molecule contains a total internal energy ε distributed among n weakly coupled harmonic oscillators. One of these oscillators is localized in a weak bond, which will break when energy ε^* is concentrated in it; the other $n - 1$ oscillators are assumed to act as a reservoir of freely available energy. The probability that an energy at least as large as ε^* is localized in one oscillator is given by $(1 - \varepsilon^*/\varepsilon)^{n-1}$, and ν is the frequency of energy transfer among the various oscillators in the molecule. For such a model, one would expect ν to be of about the same order of magnitude as molecular vibrational frequencies—namely, 10^{13} s^{-1}. [A more realistic and sophisticated molecular model,[1] proposed by N. B. Slater, gives results very similar to those based on Eq. (10) and also predicts a value for ν of about 10^{13} s^{-1}.]

To get a rough approximation for $f_A(\varepsilon)$, we can use a classical model in which the normal modes of vibration of the molecule A are represented by n classical harmonic oscillators. With this drastic assumption, one can use Boltzmann statistics to obtain[6]

$$f_A(\varepsilon) = \frac{1}{(n-1)!kT}\left(\frac{\varepsilon}{kT}\right)^{n-1} e^{-\varepsilon/kT} \tag{11}$$

When Eqs. (10) and (11) are substituted into Eq. (9), we find that

$$k_\infty = \frac{\nu}{(n-1)!} e^{-\varepsilon^*/kT} \int_0^\infty x^{n-1} e^{-x}\, dx \tag{12}$$

where $x = (\varepsilon - \varepsilon^*)/kT$. The definite integral is the gamma function $\Gamma(n) = (n-1)!$. Thus

$$k_\infty = \nu e^{-\varepsilon^*/kT} = \nu e^{-E^*/RT} \tag{13}$$

where $E^* = N_0 \varepsilon^*$. Equation (13) can be compared directly with the empirical Arrhenius equation, which is used to obtain the experimental activation energy:

$$k = \mathscr{A} e^{-E_a/RT} \tag{14}$$

The temperature-independent factor $\mathscr{A}$ has now been given a physical significance and an estimated magnitude.

The theory presented above can be improved by using a quantum mechanical model for the oscillations of the activated complex; this will permit a consideration of the effect of possible structural changes in the activated complex. The final result from this more sophisticated model[7] is

$$k_\infty = \nu^* e^{\Delta S^*/R} e^{-\Delta H^*/RT} \tag{15}$$

where ν^* is an average frequency of the vibrational modes in the activated complex and ΔS^* and ΔH^* are the entropy and enthalpy changes for forming the activated complex (transition state) from a normal species. Comparing Eqs. (14) and (15), one can identify $\mathscr{A}$ with $\nu^* e^{\Delta S^*/R}$, but ν^* is not well known, since it depends on the properties of an activated molecule (which are themselves not well known). Marcus[8] has proposed a modification of the RRK model, termed the RRKM treatment, in which ν^* is calculated by quantum statistical mechanical methods. Vibrational frequencies of reactive species and activated complexes are considered explicitly, along with the effects of rotation and zero-point energies. On the whole this theory has been quite successful in describing unimolecular reactions, but the details (given in the General Reading section) are beyond the scope of the present discussion. For simplicity we will assign ν^* a value of 10^{13} s^{-1} (the value of $\mathscr{A}$ in many reactions). Then, when the experimental value of $\mathscr{A}$ is less than 10^{12} or greater than 10^{14}, we can look for a possible explanation in terms of a value of ΔS^* that differs appreciably from zero.†

Low-Pressure Limit.   We now wish to comment on the behavior of Eq. (6) at low concentrations, where the collision frequency is so low that deactivation is very slow

---

†Transition-state theory gives the same result as Eq. (15) except that ν^* is replaced by a universal factor $kT/h(= 6 \times 10^{12}$ s^{-1} at 300 K). Although this would appear to resolve the uncertainty as to the value of ν^* (and thus allow better calculation of ΔS^* from $\mathscr{A}$ values), the theory involves several assumptions that are open to question on physical grounds.[7]

compared with decomposition. At low pressures $k_2(A) \ll k_A$ and we find second-order kinetics. It is convenient to write the rate law in a pseudo-first-order form:

$$-\frac{1}{(A)}\frac{d(A)}{dt} = k_{exp} = k_1(A) \tag{16}$$

A detailed treatment of the change with pressure of the apparent first-order rate constant k_{exp} is quite complicated,[1,2] but we can easily see that

$$k_{exp}(\text{low } p) = k_1(A) \ll \frac{k_1 k_A}{k_2} \cong k_\infty = k_{exp}(\text{high } p) \tag{17}$$

Thus, at low initial pressures, if one analyzes the data according to a first-order rate law, the "apparent first-order rate constant" k_{exp} should decrease as the pressure is decreased (i.e., the half-time for the reaction at constant temperature is independent of pressure at high pressure but rises as the pressure approaches zero). For reactions where $\mathcal{A} \gg 10^{13}$ s^{-1}, this change has been observed to occur, but normally (i.e., where $\mathcal{A} \sim 10^{13}$ s^{-1}) the change will occur at very low pressure and is not observed experimentally.

Chain Mechanisms.[9] The fact that first-order kinetics are observed for a gas-phase reaction does not prove that the unimolecular mechanism described above must be involved. Indeed very many organic decompositions that experimentally are first order have complicated free-radical chain mechanisms.

As an example let us consider a simple *hypothetical* chain mechanism for the dehydrogenation of a hydrocarbon:

$$
\begin{aligned}
M_a &\rightarrow H + R & k_1 \\
H + M_a &\rightarrow H_2 + R & k_2 \\
R &\rightarrow H + M_b & k_3 \\
H + R &\rightarrow M_a & k_4
\end{aligned}
\tag{18}
$$

where the M's are stable molecules and R is a free radical. The chain is initiated by the first step, propagated by many repetitions of the second and third steps (to form the stable products H_2 and M_b), and terminated by the last step. By writing the equations for $d(H)/dt$ and $d(R)/dt$ and making the steady-state assumption that (H) and (R) are constant, one can obtain

$$(H) = \left(\frac{k_1 k_3}{k_2 k_4}\right)^{1/2} \qquad (R) = \left(\frac{k_1 k_2}{k_3 k_4}\right)^{1/2}(M_a) \tag{19}$$

Now, the rate of disappearance of M_a is given by

$$
\begin{aligned}
-\frac{d(M_a)}{dt} &= k_1(M_a) + k_2(H)(M_a) \\
&= \left(\frac{k_1 k_2 k_3}{k_4}\right)^{1/2}(M_a)
\end{aligned}
\tag{20}
$$

and we see that the overall rate is first order.

The presence of a chain reaction can usually be detected by adding small quantities of an inhibitor such as nitric oxide or propylene, which will markedly reduce the rate by reacting with the radicals and greatly shortening the chain length.[10] Another technique is to add small amounts of a compound that is known to provide relatively large concentrations of free radicals and see if the overall rate is increased. A detailed treatment of chain mechanisms will not be given here, but it should be pointed out that the overall rate law is often very complex and the apparent order of the reaction often changes considerably with pressure. To obtain a complete picture of the kinetics may require measurements over the pressure range from a few Torr to 10 bar or more.

Wall Effects. In the above discussion, we have assumed that the reaction is homogeneous (i.e., no catalytic reaction at the walls of the reaction bulb). The fact that the data give first-order kinetics is not a proof that wall effects are absent. This point can be checked by packing a reaction bulb with glass spheres or thin-walled tubes and repeating the measurements under conditions where the surface-to-volume ratio is increased by a factor of 10 to 100. This will not be done in this experiment, but the system chosen for study must be free from serious wall effects or it may not be possible to discuss the experimental results in terms of the theory of unimolecular reactions.

METHOD[11]

The three basic experimental features of gas-phase kinetic studies are temperature control, time measurement, and the determination of concentrations. Of these, the principal problem is that of following the composition changes in the system. Perhaps the most generally applicable technique is the chemical analysis of aliquots; however, continuous methods are much more convenient. By far the easiest method is to follow the change in total pressure. This technique will be used in the present experiment. Obviously the pressure method is possible only for a reaction that is accompanied by a change in the number of moles of gas. Also the stoichiometry of the reaction should be straightforward and well understood, so that pressure changes can be related directly to extent of reaction.

For the simple apparatus designs shown in Figs. 1 and 2, both the reactant and the products must be sufficiently volatile to avoid condensation in the manometer, which is much cooler than the reaction flask. (Heating wire may be wound around the connecting tubing between the furnace and the manometer to permit warming this section to about 50°C if necessary.) Also there should be no chemical reaction between the vapors and glass or mercury. If conventional stopcocks are used, the vapors must not react with or dissolve in the stopcock grease; this concern can be eliminated by using greaseless vacuum stopcocks. The furnace in Fig. 1 should be capable of operating at temperatures up to 500°C, and the temperature should remain constant to within less than 1°C during a run. A large, well-insulated furnace has the disadvantage of heating up very slowly, but it will maintain a reasonably constant temperature without regulation. The Pyrex reaction bulb should be mounted in the center of the furnace. A large bulb (about 1000 mL) with a well to hold the thermocouple junction is preferable. However, a small bulb (as small as 100 mL) can be used, and the thermocouple can be wound around the outside of the bulb and then covered with aluminum foil. A two-junction Chromel–Alumel thermocouple, with the reference junction immersed in an ice bath, is used to measure the high temperatures achieved with this apparatus (see Chapter XVII for more information on thermocouples). The oil bath shown in Fig. 2 should be capable of operating in the range 140 to 160°C with a stability of ±0.5°C during a run. In the case of an oil bath, there is no problem in achieving thermal contact between the reaction vessel and the thermometer. The connecting tubing between the reaction bulb and the mercury in the manometer should be short and of as small bore as feasible. The rate of the reaction should be slow enough that the time for filling the

FIGURE 1
Apparatus for high-
temperature gas-phase
kinetics experiment.

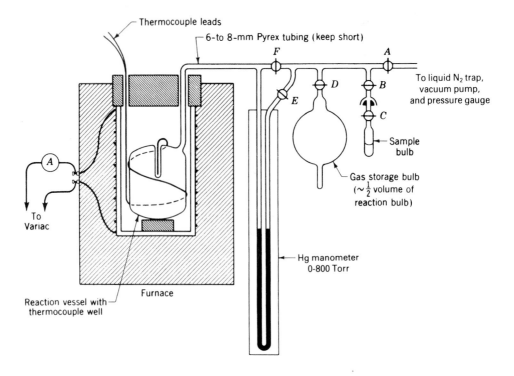

FIGURE 1
Apparatus for high-
temperature gas-phase
kinetics experiment.

FIGURE 2
Apparatus for moderate-
temperature gas-phase
kinetics experiment.

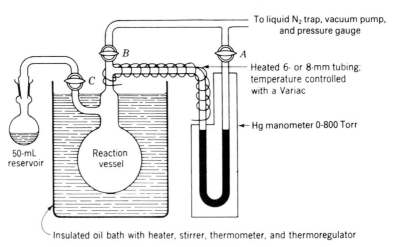

reaction bulb and for the gas to warm up to the furnace temperature is small compared with the length of the run. Since this initial period is about 30 to 60 s, the temperature should be chosen to give a half-life of at least 30 min.

We must finally derive an expression for the rate law in terms of the total pressure of the system. As an example, consider the gas-phase decomposition $A \rightarrow B + C$, and assume that the ideal-gas law holds for all species. Let the initial total pressure be p_A^0; this will also be the partial pressure of species A at zero time (p_A^0). At some arbitrary time t, $p_B = p_C = p_A^0 - p_A$, and since $p = p_A + p_B + p_C$,

$$p_A = 2p^0 - p \tag{21}$$

For a first-order reaction, $\ln[(A)_0/(A)] = kt$; therefore

$$\ln p^0 - \ln(2p^0 - p) = \ln\frac{p^0}{2p^0 - p} = kt \tag{22}$$

Thus a plot of $\ln(2p^0 - p)$ versus t should be a straight line. For a reaction $A \rightarrow B + C + D$ or $A \rightarrow B + 2C$ in which three product species are created, it can easily be shown that $\ln p^0 - \ln\frac{1}{2}(3p^0 - p) = kt$. Thus a plot of $\ln(3p^0 - p)$ versus t should be a straight line in this case.

For a moderately fast reaction, it is often difficult to make a direct measurement of p^0. It is then best to back-extrapolate the early pressure data to zero time to obtain a value of p^0. Another possibility is to allow the reaction to go to completion (assuming there is no back-reaction) and measure p^∞, which is equal to $2p^0$ or $3p^0$ for the particular reactions under consideration. This method is usually not so reliable, since side reactions that have little effect on the early stages of the reaction may influence the final pressure. (The experimental value of p^∞ is often found to be slightly less than the expected value.)

Another approach is to use the Guggenheim method,[12] which eliminates the need for either a p^0 or a p^∞ value. This method requires only that one know the time dependence of some quantity that is a linear function of the reactant concentration. In the present case,

$$p = a(A) + b \tag{23}$$

where $a = -RT$, $b = 2p^0$ for $A \rightarrow B + C$, as follows from Eq. (21), and $a = -2RT$, $b = 3p^0$ for $A \rightarrow B + 2C$. For first-order kinetics,

$$p(t) = a(A)_0 e^{-kt} + b \tag{24}$$

and

$$\Delta p \equiv p(t + \Delta t) - p(t) = a(A)_0(e^{-k\Delta t} - 1)e^{-kt} \tag{25}$$

Thus, for *constant* Δt,

$$\ln \Delta p = -kt + \text{const} \tag{26}$$

and a plot of $\ln \Delta p$ versus t will yield a straight line of slope $-k$. The Guggenheim treatment and its results are in principle independent of the particular constant value chosen for Δt. The choice of a value of Δt depends on practical considerations: if it is too small, the scatter of points on the logarithmic plot will be too great, and if it is too large, the number of points that can be plotted becomes too small. For a clear demonstration of first-order kinetics and a good determination of the rate constant, the points plotted should cover at least one natural logarithmic cycle (one power of e). A good procedure would be to run the experiment over a total time sufficient to obtain about $1\frac{1}{2}$ cycles (i.e., until increments between successive readings have decreased to about one-fifth of the initial value) and to take about one-third of this total time for Δt.

In deriving Eq. (22), the dead-space volume V_D in the manometer and connecting tube has not been considered. Since V_D changes during the reaction owing to the change of the mercury level in the manometer, $p^0/(2p^0 - p)$ in Eq. (22) must be multiplied by a correction factor for very precise work.[13] With $V_D \sim 20$ mL and $V_B \sim 1000$ mL, this correction factor varies linearly with pressure from 1.00 to $\sim$0.98 during the entire run and can be neglected.

Once the order of the reaction is established by a plot of $\ln(np^0 - p)$ versus t, where n is the number of product molecules per A molecule dissociated, there is a rapid method of evaluating specific rate constants which utilizes the time for the reaction to proceed to a given fraction reacted. The half-time and third-time for a first-order reaction are given by

$$kt_{1/2} = \ln 2 = 0.691$$
$$kt_{1/3} = \ln 1.5 = 0.406 \tag{27}$$

For the reaction $A \rightarrow B + C$, $t_{1/2}$ is the time required for the total pressure p to reach $3p^0/2$ and $t_{1/3}$ is the time required to reach $4p^0/3$. For the reaction $A \rightarrow B + 2C$, the total pressure will equal $2p^0$ at $t_{1/2}$ and $5p^0/3$ at $t_{1/3}$. Either (or both) times can be used to obtain k; it is also possible to check that the pressure data are consistent with first-order kinetics by calculating $t_{1/2}/t_{1/3}$. This ratio should be 1.70 for a first-order reaction; it is 1.50 and 2.00 for zero-order and second-order reactions, respectively.

EXPERIMENTAL

1 HIGH-TEMPERATURE REACTION

The reaction suggested for study is the decomposition of cyclopentene to cyclopentadiene and hydrogen:

$$\tag{28}$$

This reaction is quite clean-cut, and at least 95 percent of the products are accounted for by Eq. (28). Vanas and Walters[14] have studied the reaction in the gas phase and find that p^∞/p^0 is about 1.9, which indicates the occurrence of some side reactions. However, they carried out a careful chemical analysis of the products and showed that the partial pressures of cyclopentadiene and hydrogen are equal to each other and to the increase in total pressure $p - p^0$ over at least the first half of the reaction. With the exception of the last stages of the decomposition, this reaction is very well suited to the pressure method. In addition it has been shown that the reaction is homogeneous and does not involve a chain mechanism.

The rate of reaction should be studied at a single temperature for several different initial pressures in the range 50 to 200 Torr. A furnace temperature of about 510°C will give a convenient half-life (~30 min). Ordinarily it is necessary to follow the reaction only until p exceeds $1.5p^0$. However, data should be taken on one of the runs until p is at least $1.75p^0$. (If possible let this reaction mixture stand overnight and obtain a value of p^∞.) If NO gas is available, add about 2 to 3 percent of NO to the cyclopentene for one of the runs and check for any indication of inhibition; consult the instructor regarding any necessary changes in the procedure. If more than one day is available for experimental work, raise the temperature of the furnace by about 10°C and make several more runs.

Procedure. The furnace should be turned on the day before measurements are to be made so that it will achieve a steady temperature. Set the power supply to a predetermined voltage appropriate for the desired reaction temperature. The heating current should be measured with a series ammeter and recorded periodically. At the start of the experiment, place the reference thermocouple junction in a Dewar flask filled with ice and distilled water (see Chapter XVII) and connect the thermocouple to a potentiometer or digital voltmeter (see Chapter XVI). Measure the thermocouple emf and check to see if the furnace is at the proper temperature. If the temperature shows appreciable long-term drift (greater than $\pm 0.5 \text{ K h}^{-1}$), adjustment of the heater current will be necessary. On making any change in voltage setting, be careful to note the time response of the furnace as an aid in making later adjustments.

While the temperature stability of the furnace is being checked, carefully open all of the stopcocks in Fig. 1 *except C*, which isolates the sample bulb containing liquid cyclopentene. Evacuate the system until the pressure is 10^{-2} Torr or lower. Then close stopcocks E and F and check for leaks in the reaction bulb by monitoring the manometer reading over a period of ~ 10 min. Next close stopcock D, slowly freeze the cyclopentene with liquid nitrogen, and then open stopcock C in order to pump off any air. Close C, remove the liquid nitrogen, and allow the solid to melt so that it will liberate any dissolved gases. Repeat this degassing process at least once more, and then allow the cyclopentene to warm to room temperature. Close stopcock A and open D and then C to permit the storage bulb to be filled with vapor. The vapor pressure at 25°C is about 350 Torr.

To start a run, close stopcock C and then **slowly** open stopcock F and watch the manometer. When the pressure has reached about 95 percent of the desired value for p^0, close F and start a timer. Record pressure and time readings every minute during the early stages of the reaction and every 2 min thereafter until the end of the run.

Analysis of Products. It is possible to design an apparatus that will permit the efficient removal of the entire reaction mixture for analysis. With the apparatus shown in Fig. 1, a qualitative analysis of the products can be made if desired. Analytical details are given by Vanas and Walters.[14]

2 MODERATE-TEMPERATURE REACTION

The reaction suggested for study is the decomposition of *tert*-butyl peroxide (*t*BP).†
Batt and Benson[15] have shown that the stoichiometry of this reaction is three (i.e., 3 mol of product for each mol of reactant) with 90 percent of the reaction accounted for by

$$(CH_3)_3 COOC(CH_3)_3 \rightarrow C_2H_6 + 2(CH_3)_2 CO \tag{29}$$

The mechanism of this reaction is given by the following scheme:

$$tBP \rightarrow 2tBO\cdot \tag{30}$$

$$tBO\cdot \rightarrow Me\cdot + (Me)_2 CO \tag{31}$$

$$Me\cdot + Me\cdot \rightarrow C_2H_6 \tag{32}$$

where the cleavage of the O—O bond in the first step is rate determining. The other 10 percent of the *t*BP decomposes according to

$$tBP \rightarrow CH_4 + (Me)_2 CO + MeCOEt$$
$$tBP \rightarrow 2CH_4 + \text{biacetonyl} \tag{33}$$

both of which involve intermediate $MeCOCH_2\cdot$ radicals. The reaction is homogeneous, and chain contributions are less than 2 percent.[15] Thus this reaction is quite well suited to the pressure method if two minor precautions are observed. (1) Fresh stopcock grease may absorb *t*BP, leading to a p^∞/p^0 ratio of ~ 2.9. Greaseless stopcocks are therefore recommended; if these are not available, use the minimum amount of grease necessary and

---

†No special handling procedures are required for *t*BP, since it is one of the most stable organic peroxides known. It can be distilled at 1 atm without decomposition.

if possible do not regrease stopcocks just before a run. (2) The side arm to the manometer must be heated to avoid condensation of acetone.

The apparatus shown in Fig. 2 has a 500-mL reaction vessel, which is connected to a 50-mL reservoir for liquid *t*BP via a stopcock and a standard-taper joint.† The connecting tubing between the reaction vessel and the mercury in the manometer must be wrapped with commercial heating tape or wound carefully with heating wire. The temperature of this section, which should be as short and of as small bore as feasible, is controlled with a variable-voltage power supply. It is not necessary to control this temperature accurately, but it should be close to the bath temperature (say, within 20°C) and *must* be above the boiling point of acetone (56°C).

The rate of the reaction should be studied at a single temperature for several initial pressures in the range 30 to 100 Torr. A reaction temperature of about 160°C will give a convenient half-life. At least one run should be made at a lower temperature (say 150°C). In general it is necessary to follow the reaction only until p exceeds $2p^0$, but data should be taken on one run until p is at least $2.5p^0$.

Procedure. If necessary, vacuum distill the *t*BP. Place about 10 mL of *t*BP in the side reservoir and close stopcock C. Open stopcocks A and B and evacuate the system to a pressure of 10^{-2} Torr or lower while the oil bath is being heated to the desired temperature. Note that the pumping system should include a liquid nitrogen trap to prevent condensation of *t*BP or acetone in the pump. Check for leaks by closing stopcocks A and B; then open them again and continue pumping.

Heating of the oil bath should be started well in advance of the measurements so that the temperature will have stabilized at about 160°C. The temperature, which can be monitored with a thermocouple as described in the procedure for the high-temperature reaction, should not drift by more than ±0.5°C/h.

When the reaction vessel is at constant temperature, close stopcock A and **slowly** open stopcock C. Pump down the reservoir as low as possible, but do not waste *t*BP by prolonged pumping. Next close stopcock C and place a beaker of hot water (60 to 65°C) around the reservoir. Stopcock B is then closed and C is opened until the desired value of p^0 is achieved. Finally, close stopcock C and start a stopwatch or timer. Record the pressure and time readings every minute during the early stages of the reaction and every 2 min thereafter.

CALCULATIONS

Plot the pressure reading for each run versus time and extrapolate to zero time to obtain a value for p^0. For each run determine $t_{1/3}$ and $t_{1/2}$, and check their ratio with that expected for first-order kinetics. Use Eqs. (27) to calculate k, in units of s^{-1}, from each of these times. Analyze the data from your longest run by carrying out a linear least-squares fit of $\ln(np^0 - p)$ as a function of t, where $n = 2$ or 3 depending on the stoichiometry and p^0 is taken to be the value determined previously by extrapolation. The early data points should be weighted more heavily than the late ones. Plot $\ln(np^0 - p)$ versus t and add the best straight-line fit to the graph. Report the value of the resulting specific rate constant k. If possible, carry out a *nonlinear* least-squares fit with Eq. (22), or its analog for the case of three product species, taking both p^0 and k as freely adjustable fitting parameters. Do the

---

†This apparatus is very similar in design to an apparatus described by Ellison;[16] see also the design of Trotman-Dickenson.[17]

resulting values of the parameters differ much from the previous values? If so, use the new p^0 value to make another plot of $\ln(np^0 - p)$ versus t. Alternatively, carry out an analysis using the Guggenheim method.

Tabulate T, p^0, $t_{1/3}$, and k for each run, and list your best overall value of k for each temperature studied. If data were obtained at two different temperatures, use Eq. (14) to calculate a value of the activation energy. Compare your result with the appropriate literature value. For cyclopentene, E_a is reported to be 246 kJ mol^{-1} on the basis of a linear plot of $\ln k$ versus $1/T$ for data over the range 485 to 545°C.[14] For *tert*-butyl peroxide, Batt and Benson[15] report 156.5 kJ mol^{-1} for the temperature range 130 to 160°C.

DISCUSSION

Calculate the frequency factor ν and discuss its value in terms of the theory of unimolecular decompositions.

In the case of *t*BP, there is a complication owing to the fact that reaction (29) is exothermic enough to cause temperature gradients in a spherical reaction vessel. Batt and Benson have shown that such gradients increase with increasing temperature, pressure, and reaction volume. At 160°C there is a gradient of roughly 0.5°C during the first half of the reaction.[15] Estimate the effect that such a gradient will have on the rate constant determined in this experiment.

SAFETY ISSUES

Work on a vacuum system requires preliminary review of procedures and careful execution in order to avoid damage to the apparatus and possible injury from broken glass; in addition, the liquid nitrogen used for cold traps must be handled properly (see Appendix C). Safety glasses must be worn. In the event of breakage, any mercury spill must be cleaned up promptly and carefully as described in Appendix C. Avoid contact with the hot furnace or hot bath oil.

APPARATUS

High-Temperature Reaction. Gas-kinetics apparatus as shown in Fig. 1, including vacuum line with greaseless stopcocks, sample bulb, gas storage bulb, manometer, Pyrex reaction bulb; high-temperature furnace; variable-voltage power supply; ammeter for measuring heater current; two-junction Chromel–Alumel thermocouple; two 1-qt Dewar flasks; millivolt-range potentiometer setup or digital voltmeter (see Chapter XVI); thermocouple calibration table; stopwatch or other timer.

Cyclopentene; ice (3 lb); liquid nitrogen (2 L); cylinder of NO gas, with needle valve and pressure tubing (optional).

Moderate-Temperature Reaction. Gas-kinetics apparatus as shown in Fig. 2, including vacuum line with greaseless stopcocks, sample reservoir bulb, reaction vessel, heated mercury manometer; regulated oil bath; thermometer; 500-mL beaker; 1-qt Dewar; stopwatch or other timer.

tert-Butyl peroxide (15 mL); liquid nitrogen (2 L).

REFERENCES

1. S. W. Benson, *The Foundations of Chemical Kinetics,* reprint ed., chaps. X and XI, Krieger, Melbourne, FL (1982); A. F. Trotman-Dickenson, *Gas Kinetics,* secs. 2.3, 2.4, 3.2, Butterworth, London (1955).

2. Any standard physical chemistry text, such as P. W. Atkins and J. de Paula, *Physical Chemistry,* 8th ed., chap. 24, Freeman, New York (2006); or I. N. Levine, *Physical Chemistry,* 6th ed., chap. 22, McGraw-Hill, New York (2009).

3. F. A. Lindemann, *Trans Faraday Soc.* **17,** 598 (1922).

4. H. C. Ramsperger, *Chem. Rev.* **10,** 27 (1932).

5. L. S. Kassel, *Kinetics of Homogenous Gas Reactions,* chap. 5, Reinhold (ACS Monograph), New York (1932).

6. S. W. Benson, *op cit.,* pp. 222–223.

7. *Ibid.,* pp. 250–252.

8. R. A. Marcus, *J. Chem. Phys.* **20,** 359 (1952).

9. S. R. Logan, *Fundamentals of Chemical Kinetics,* Addison-Wesley Longman, Harlow, England, (1996).

10. A. F. Trotman-Dickenson, *op. cit.,* pp. 153–160.

11. H. W. Melville and B. G. Gowenlock, *Experimental Methods in Gas Reactions,* Macmillan, London (1963).

12. E. A. Guggenheim, *Phil. Mag.* **2,** 538 (1926).

13. A. O. Allen, *J. Amer. Chem. Soc.* **56,** 2053 (1934).

14. D. W. Vanas and W. D. Walters, *J. Amer. Chem. Soc.* **70,** 4035 (1948).

15. L. Batt and S. W. Benson, *J. Chem. Phys.* **36,** 895 (1962).

16. H. R. Ellison, *J. Chem. Educ.* **48,** 205 (1971).

17. A. F. Trotman-Dickenson, *J. Chem. Educ.* **46,** 396 (1969).

GENERAL READING

S. W. Benson, *The Foundations of Chemical Kinetics,* reprint ed., Krieger, Melbourne, FL (1982).

K. L. Laidler and K. J. Laidler, *Chemical Kinetics,* 3d ed., especially sec. 5.3, Addison-Wesley, Reading, MA (1997).

J. W. Moore and R. G. Pearson, *Kinetics and Mechanism,* 3d ed., especially pp. 121–129, Wiley-Interscience, New York (1981).

K. A. Holbrook et al., *Unimolecular Reactions,* 2d ed., Wiley-Interscience, New York (1996).

XI

Surface Phenomena

EXPERIMENTS

25. Surface Tension of Solutions
26. Physical Adsorption of Gases

EXPERIMENT 25
Surface Tension of Solutions

The capillary-rise method is used to study the change in surface tension as a function of concentration for aqueous solutions of n-butanol and sodium chloride. The data are interpreted in terms of the surface concentration using the Gibbs isotherm.

THEORY

If a body of material is homogeneous, the value of any *extensive* property is directly proportional to the quantity of matter contained in the body:

$$Q = \overline{Q}_V V = \overline{Q}_m m = \overline{Q}_n n \tag{1}$$

where Q is the extensive quantity; V, m, and n are, respectively, the volume, mass, and number of moles of the substance involved. $\overline{Q}_V$, $\overline{Q}_m$, and $\overline{Q}_n$ are, respectively, the specific values of Q per unit volume, per unit mass, and per mole and have *intensive* magnitudes.

It is known however that the values of extensive properties of bodies often thought of as homogeneous are not always independent of the surface area. In fact liquid bodies with surfaces are in general not entirely homogeneous, for the value of a given intensive quantity (say Q_V) in the region of the surface may at equilibrium deviate from the value that this quantity has in the bulk of the solution ($\overline{Q}_V$). We can write

$$Q = \int_V Q_V \, dV = \overline{Q}_V V + \int_\tau (Q_V - \overline{Q}_V) \, dV \tag{2}$$

299

where the second integral needs to be taken only over a region τ in the neighborhood of the surface, within which Q_V is different from $\overline{Q}_V$, the *bulk* value that prevails through the body except near the surface. We can replace the volume element dV in the second integral by $dA\,dx$, where dA is an element of surface area and dx is an element of distance *inward* from the surface of the body. The integration over dA extends over the surface of the body, and that over dx extends to a distance τ inside the body beyond which Q_V is substantially equal to $\overline{Q}_V$. If $(Q_V - \overline{Q}_V)$ is independent of position on the surface of the body, the integration over dA can be carried out at once, and we have

$$Q = \overline{Q}_V V + \overline{Q}_A A \tag{3}$$

where we have introduced a new quantity, called a *specific surface quantity*,

$$\overline{Q}_A \equiv \int_\tau (Q_V - \overline{Q}_V)\,dx \tag{4}$$

This quantity represents the *excess*, per unit area of surface, of the quantity Q over what Q would be for a perfectly homogeneous body of the same magnitude with $Q_V = \overline{Q}_V$ throughout; see Fig. 1. In some cases $\overline{Q}_A$ may be negative, corresponding to a deficiency rather than an excess.

If the body has surfaces of several different kinds (such as a crystal with different kinds of crystal faces or a liquid with part of its surface in contact with the air and the remainder of its surface in contact with solid or other liquid phases), the parenthesized quantity in Eq. (2) is independent of surface positional coordinates only within the boundaries of each kind of surface. Each kind of surface will have in general a different $\overline{Q}_A$, and we write in such a case

$$Q = \overline{Q}_V V + \sum_i \overline{Q}_{A_i} A_i \tag{5}$$

Surface Concentration. When the extensive quantity concerned is the number n of moles of solute, the corresponding specific bulk property is $n_V \equiv c$, the bulk concentration (concentration in the interior of the solution). The corresponding specific surface quantity $n_A \equiv \Gamma$ is called *surface concentration* and represents excess of solute per unit area of the surface over what would be present if the internal (bulk) concentration prevailed all the way to the surface. It may be expressed in moles per square centimeter.

FIGURE 1

Schematic plot of Q_V versus x. The shaded area is equal to $\overline{Q}_A$.

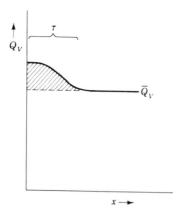

Surface Free Energy or Surface Tension. Under conditions of constant temperature and pressure, the equilibrium state sought by any system is that of lowest free energy G, and the maximum work (other than expansion work) done on the system in any change of state under these conditions is equal to the free-energy increase ΔG. A body of liquid with a free surface will tend to assume the shape that gives it the lowest possible free energy at the given temperature and pressure prevailing.

The free energy of a system containing variable surface areas can be written, from Eq. (5), as follows:

$$G = G_0 + \sum_i \gamma_i A_i \tag{6}$$

where $\overline{G}_{A_i} \equiv \gamma_i$ is the specific surface free energy or *surface tension* of surface i, which may be a free surface (exposed to air or vapor or vacuum) or an interface with another liquid or solid. In the event that the surface is an interface, this quantity is called *interfacial tension*. For a stable free liquid surface, it clearly must be positive. The surface tension γ_i is equal to the expenditure of work required to increase the net area of surface i by one unit of area. If the increase in surface area is accomplished by moving a line segment of unit length in a direction perpendicular to itself, γ_i is equal to the force (or "tension") opposing this motion. Accordingly, the SI units for surface tension are N m^{-1} or J m^{-2}; cgs units, which are still widely used in many tabulations, are dyn cm^{-1} or erg cm^{-2} (1 N m^{-1} = 1000 dyn cm^{-1}).

The variation of surface tension with temperature will not be discussed here, except to remark that surface tension decreases with temperature and that the rate of decrease is large enough to require that the temperature of measurement of surface tension be kept constant, within the order of 0.1°C, by means of a thermostat.

The Gibbs Isotherm. It is found that the surface tensions of solutions are in general different from those of the corresponding pure solvents. It has also been found that solutes whose addition results in a decrease in surface tension tend to concentrate slightly in the neighborhood of the surface (positive surface concentration); those whose addition results in an increase in surface tension tend to become less concentrated in the neighborhood of the surface (negative surface concentration). The migration of solute either toward or away from the surface is always such as to make the surface tension of the *solution* (and thus the free energy of the system) lower than it would be if the concentration of solute were uniform throughout (surface concentration equal to zero). Equilibrium is reached when the tendency for free-energy decrease due to lowering surface tension is balanced by an opposing tendency for free-energy increase due to increasing nonuniformity of solute concentration near the surface.

Let us imagine a body of solution of volume V, surface area A, bulk concentration c, and bulk osmotic pressure Π at constant temperature T and external pressure p. For arbitrary changes dA and dV in the area and volume† of the solution, the free-energy change can be written

$$dG = \gamma\, dA - \Pi\, dV \tag{7}$$

This is an exact differential; from the well-known reciprocity relation,[1] we find that

$$-\left(\frac{\partial \gamma}{\partial V}\right)_A = \left(\frac{\partial \Pi}{\partial A}\right)_V \tag{8}$$

---

†The volume change may be thought of as resulting from the motion of a piston, containing a semipermeable membrane, against the osmotic pressure Π.

FIGURE 3

Variation of ionic solute concentration near a free surface: (*a*) schematic diagram of the distribution of ions, showing negative surface adsorption; (*b*) concentration versus distance for two different bulk concentrations (the shaded area is equal to $-\Gamma$ for a solution of bulk concentration c_0); (*c*) the "equivalent empty-layer thickness" x_0, chosen such that the two shaded areas will be equal.

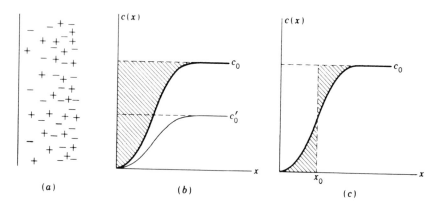

(*a*)   (*b*)   (*c*)

METHOD

There are many experimental techniques for measuring the surface tension of liquids. In the *ring method,* one determines the force necessary to pull a metal ring free from the surface of a liquid. The Du Nouy tensiometer, in which the ring is hung from the beam of a torsion balance, is often used, but it cannot easily be thermostated. In calculating the surface tension, it is necessary to apply a correction factor that takes into account the shape of the liquid held up by the ring. In the *drop-weight method,* one determines the weight of a drop that falls from a tube of known radius. Again a correction factor is needed, because the drop that actually falls does not represent all the liquid that was supported by surface tension. In the *bubble-pressure method,* one determines the maximum gas pressure obtained in forming a gas bubble at the end of a tube of known radius immersed in the liquid. Detailed descriptions of these and other methods may be found elsewhere.[3]

Capillary Rise. In the absence of external forces, a body of liquid tends to assume a shape of minimum area. It is normally prevented from assuming spherical shape by the force of gravity, as well as by contact with other objects. When a liquid is in contact with a solid surface, there exists a specific surface free energy for the interface, or interfacial tension γ_{12}. A solid surface itself has a surface tension γ_2, which is often large in comparison with the surface tensions of liquids. Let a liquid with surface tension γ_1 be in contact with a solid with surface tension γ_2, with which it has an interfacial tension γ_{12}. Under what circumstances will a liquid film spread freely over the solid surface and "wet" it? This will happen if, in creating a liquid–solid interface and an equal area of liquid surface at the expense of an equal area of solid surface, the free energy of the entire system decreases:

$$\gamma_1 + \gamma_{12} - \gamma_2 < 0 \tag{16}$$

If we have a vertical capillary tube that dips into a liquid, a film of the liquid will tend to run up the capillary wall if condition (16) is obeyed. Then, in order to reduce the surface of the liquid, the meniscus will tend to rise in the tube. It will rise until the force of gravity on the liquid in the capillary above the outside surface, $\pi r^2 (h + r/3)\rho g$, exactly counterbalances the tension at the circumference, which is $2\pi r \gamma_1$. In these expressions ρ is the density of the liquid, g is the acceleration of gravity, h is the height of the liquid above the outside surface, r is the radius of the cylindrical capillary, and $r/3$ is a correction

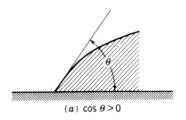

FIGURE 4
Capillary rise h in a tube of radius r.

for the amount of liquid above the bottom of the meniscus, assuming it to be hemispherical (see Fig. 4). Thus we obtain

$$\gamma_1 = \frac{1}{2}\left(h + \frac{r}{3}\right)r\rho g \qquad (17)$$

If Eq. (16) is not obeyed, but if instead

$$\gamma_1 \cos \theta + \gamma_{12} - \gamma_2 = 0 \qquad (18)$$

for some value of θ, the liquid will not tend to spread indefinitely on the solid surface but will tend instead to give a *contact angle* θ (see Fig. 5). This may be the case with aqueous solutions or water on glass surfaces that are not entirely clean. We shall then have, instead of Eq. (17),

$$\gamma_1 \cos \theta = \frac{1}{2}\left(h + \frac{r}{3}\right)r\rho g \qquad (19)$$

However, in practice, there is usually some "hysteresis": that is, the contact angle finally attained is somewhat variable, depending on whether the liquid has been advancing over the solid surface or receding from it. Thus two different capillary rise heights are to be

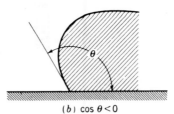

(a) $\cos \theta > 0$

(b) $\cos \theta < 0$

FIGURE 5
Contact angle between a liquid and solid surface: (a) $\cos \theta > 0$ (e.g., water on a glass surface that is not completely clean); (b) $\cos \theta < 0$ (e.g., Hg on glass or water on paraffin).

expected. If the *same* height is obtained regardless of whether the liquid was allowed to rise from below or fall from above in the capillary, it may be assumed that Eqs. (16) and (17) are almost certainly valid. This is nearly always true of aqueous solutions in carefully cleaned glass capillary tubes.

EXPERIMENTAL

If the capillary tube has not been cleaned recently, it should be soaked in hot nitric acid for several minutes and rinsed copiously with distilled water. A clean capillary is essential to obtaining good results. When it is not in use, the capillary should be stored by immersing it in a tall flask of distilled water. Assemble the apparatus as shown in Fig. 6. Adjust the capillary tube upward or downward until the outside liquid level is at or slightly above the zero position on the scale.

Determine the height h of capillary rise for pure water at 25°C. Take at least four readings, alternately allowing the meniscus to approach its final position from above and below. Also be sure to read the position of the outside level. If there is not good agreement among these readings, reclean the capillary and repeat the measurements.

FIGURE 6

Apparatus for measuring surface tension by the method of capillary rise.

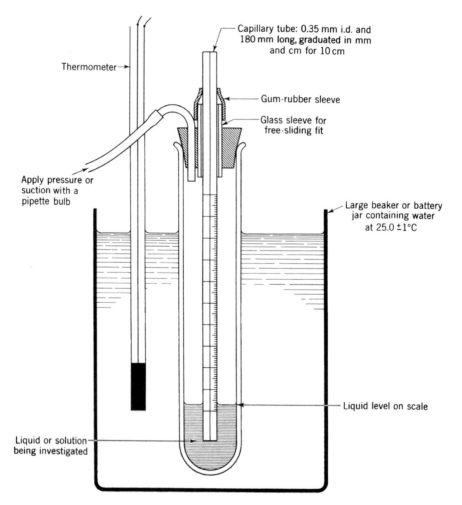

Repeat the above procedure with 0.8 M n-butanol solution, dilute to precisely three-quarters the concentration and repeat, and so on until eight concentrations have been used (the last being 0.11 M). Rinse the apparatus and capillary with one or two small aliquots of fresh solution at each concentration change.

If time permits repeat the above procedure with NaCl solutions. Use solutions of concentrations approximately 4, 3, 2, and 1 M.

CALCULATIONS

From the data obtained for pure water, calculate the capillary radius r and use it in calculating the surface tension of each solution studied. Use Eq. (17) for these calculations. For the butanol solutions one may assume that the density is equal to that of pure water; however, for NaCl solutions it is necessary to use values of ρ obtained by interpolation in Table 1. At 25.0°C for pure water, the surface tension is 72.0 dyn cm^{-1} (0.072 N m^{-1}) and the density is 0.9971 g cm^{-3}. Tabulate all your γ values and give the units in which they are expressed.

For the runs on n-butanol solutions, plot the surface tension of the solution γ versus the logarithm of the bulk concentration c and determine the slope. The surface concentration (in units of mol cm^{-2} or mol m^{-2}) can be calculated from Eq. (14). Express this surface concentration in molecules per square Angstrom, and obtain the "effective cross-sectional area" per molecule of adsorbed butanol in Å^2.

For the runs on NaCl solutions, plot γ versus the bulk concentration c expressed in mol cm^{-3}. As seen by Eq. (13) the ratio Γ/c can be determined from the slope of the straight line obtained. Calculate the "effective empty-layer thickness" x_0 in Å for NaCl. Since the laws of ideal solutions do not hold at the NaCl concentrations studied, the results obtained with NaCl have only qualitative significance and the x_0 obtained represents only a rough order of magnitude value.

TABLE 1   Density of NaCl solutions at 25°C[a]

Percentage by wt.	NaCl, g L^{-1}	mol L^{-1}	Density, g cm^{-3}
0	0	0	0.9971
1	10.05	0.1719	1.0045
2	20.22	0.3460	1.0108
4	41.00	0.7015	1.0250
6	62.36	1.067	1.0394
8	84.31	1.443	1.0539
10	106.9	1.829	1.0686
12	130.0	2.224	1.0836
14	153.8	2.632	1.0986
16	178.2	3.049	1.1139
18	203.3	3.479	1.1295
20	229.1	3.920	1.1453
22	255.5	4.372	1.1614
24	282.6	4.836	1.1777
26	310.5	5.313	1.1944

[a] Calculated from data given in Landolt-Bornstein, New Series, Group IV, vol. Ib, p. 80, Springer-Verlag, Berlin/New York (1977).

SAFETY ISSUES

If the capillary tube is cleaned with hot nitric acid, handle the acid with care and wear safety goggles and gloves (to avoid any chemical skin burns).

APPARATUS

One (or two) graduated capillary tubes (cleaned with concentrated HNO_3, rinsed thoroughly with distilled water, and stored in distilled water); ring stand; battery jar; test tube, with two-hole stopper assembly, to hold capillary; thermometer; large clamp and clamp holder; 20-in. length of gum-rubber tubing; 200-mL volumetric flask; 50-mL pipette; pipetting bulb; 250-mL beakers.

Water aspirator with attached clean gum-rubber tube; 0.8 M aqueous solution of n-butanol (250 mL); 4 M solution of NaCl (400 mL).

REFERENCES

1. P. W. Atkins and J. de Paula, *Physical Chemistry,* 8th ed., pp. 103–104, 968, Freeman, New York (2006).

2. W. D. Harkins, and R. W. Wampler, *J. Amer. Chem. Soc.* **53,** 850 (1931).

3. A. Couper, "Surface Tension and Its Measurement," in B. W. Rossiter and R. C. Baetzold (eds.), *Physical Methods of Chemistry,* 2d ed., vol. IXA, chap. 1, Wiley-Interscience, New York (1993).

GENERAL READING

A. W. Adamson, *Physical Chemistry of Surfaces,* 6th ed., Wiley, New York (1997).

R. Defay and I. Prigogine, *Surface Tension and Adsorption,* reprint ed., Books on Demand, Ann Arbor, MI (1966).

EXPERIMENT 26
Physical Adsorption of Gases

This experiment is concerned with the multilayer physical adsorption of a gas (the *adsorbate*) on a high-area solid (the *adsorbent*). Since such adsorption is caused by forces very similar to those that cause the condensation of a gas to a bulk liquid, appreciable adsorption occurs only at temperatures near the boiling point of the adsorbate. The adsorption of N_2 gas on a high-area solid will be studied at 77.4 K (the boiling point of liquid nitrogen), and the surface area of the solid will be obtained.

THEORY

We shall be concerned with the adsorption isotherm, i.e., the amount adsorbed as a function of the equilibrium gas pressure at a constant temperature.[1] For physical adsorption (sometimes referred to as van der Waals adsorption), five distinct types of isotherms have

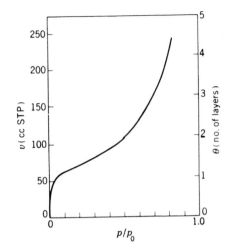

FIGURE 1

A typical isotherm for the physical adsorption of nitrogen on a high-area solid at 77 K.

been observed; we shall discuss the theory for the most common type, a typical example of which is shown in Fig. 1. It has unfortunately become standard practice to express the amount adsorbed by v, the volume of gas in cubic centimeters at STP (standard pressure and temperature: $p = 1$ atm, $T = 273.15$ K) rather than in moles. If the adsorbed volume required to cover the entire surface with a complete *monolayer* is denoted by v_m, one can describe the isotherm in terms of the coverage θ:

$$\theta = \frac{v}{v_m} \tag{1}$$

As shown in Fig. 1, adsorption increases rapidly at high pressures, and several layers of adsorbate are present even at a relative pressure $p/p_0 = 0.8$. The pressure p_0 is the saturation pressure of the gas (i.e., the vapor pressure of the liquid at that temperature). When $p/p_0 = 1.0$, bulk condensation will occur to form a liquid film on the surface ($\theta \rightarrow \infty$). One can also see from Fig. 1 that there is no clear, sharp indication of the formation of a first layer; indeed, the second and higher layers usually begin to form before the first layer is complete. Clearly, we cannot obtain a good value of v_m from the adsorption isotherm without the aid of a theory that will explain the shape of the isotherm.

Brunauer, Emmett, and Teller[2] were the first to propose a theory for multilayer adsorption (BET theory). Since the behavior of adsorbed molecules is even more difficult to describe in detail than that of molecules in the liquid state, the BET theory contains some rather drastic assumptions. In spite of this, it is still a generally useful theory of physical adsorption. The BET theory gives a correct semiquantitative description of the shape of the isotherm and provides a good means of evaluating v_m (which is then used to estimate the surface area of the solid).

The usual form of the BET isotherm is derived for the case of a free (exposed) surface, where there is no limit on the number of adsorbed layers that may form. Such an assumption is clearly not good for a very porous adsorbent, such as one with deep cracks having a width of a few monolayer thicknesses, since the surface of these cracks can hold only a few layers even when the cracks are filled. Fortunately, the BET theory is most reliable at low relative pressures (0.05 to 0.3), at which only a few complete layers have formed, and it can be applied successfully to the calculation of v_m even for porous solids.

The original derivation of the BET equation[1,2] was an extension and generalization of Langmuir's treatment of monolayer adsorption. This derivation is based on kinetic considerations—in particular on the fact that at equilibrium the rate of condensation of gas molecules to form each adsorbed layer is equal to the rate of evaporation of molecules from that layer. In order to obtain an expression for θ as a function of p (the isotherm), it is necessary to make several simplifying assumptions. The physical nature of these assumptions in the BET theory can be seen most clearly from a different derivation based on statistical mechanics.[3] Neither derivation will be presented here, but we shall discuss the physical model on which the BET equation is based. A detailed review of this model has been given by Hill;[4] the important assumptions are as follows:

1. The surface of the solid adsorbent is uniform; i.e., all "sites" for adsorption of a gas molecule in the first layer are equivalent.
2. Adsorbed molecules in the first layer are localized; i.e., they are confined to sites and cannot move freely over the surface.
3. Each adsorbed molecule in the first layer provides a site for adsorption of a gas molecule in a second layer, each one in the second layer provides a site for adsorption in a third, with no limitation on the number of layers.
4. There is no interaction between molecules in a given layer. Thus the adsorbed gas is viewed as many independent stacks of molecules built up on the surface sites.
5. All molecules in the second and higher layers are assumed to be like those in the bulk liquid. In particular the energy of these molecules is taken to be the same as the energy of a molecule in the liquid. Molecules in the first layer have a different energy owing to the direct interaction with the surface.

In brief the statistical derivation is based on equilibrium considerations—in particular on finding that distribution of the heights of the stacks that will make the free energy a minimum.

The BET isotherm obtained from either derivation is

$$\theta = \frac{N}{S} = \frac{cx}{(1 - x)[1 + (c - 1)x]} \tag{2}$$

where S is the number of sites, N is the number of adsorbed molecules, x is the relative pressure (p/p_0), and c is a dimensionless constant greater than unity and dependent only on the temperature. Note that θ equals zero when $x = 0$ and approaches infinity as x approaches unity, in agreement with Fig. 1. Making use of Eq. (1), we can rearrange Eq. (2) to the usual form of the BET equation:

$$\frac{x}{v(1 - x)} = \frac{1}{v_m c} + \frac{(c - 1)x}{v_m c} \tag{3}$$

Thus a plot of $[x/v(1 - x)]$ versus x should be a straight line; in practice, deviations from a linear plot are often observed below $x = 0.05$ or above $x = 0.3$. From the slope s and the intercept I, both v_m and c can be evaluated:

$$v_m = \frac{1}{s + I} \qquad c = 1 + \frac{s}{I} \tag{4}$$

The volume adsorbed (in cm^3 at STP) is related to n, the number of moles adsorbed, by

$$v = nRT_0 \tag{5}$$

where $T_0 = 273.15$ K and $R = 82.06$ cm^3 atm. The total area of the solid is

$$A = N_0 n_m \sigma \tag{6}$$

where N_0 is Avogadro's number and σ is the cross-sectional area of an adsorbed molecule.

Although we shall not be concerned experimentally with measuring heats of adsorption, it is appropriate to comment that ΔH for the physical adsorption of a gas is always negative, since the process of adsorption results in a decrease in entropy. The *isosteric heat of adsorption* (the heat of adsorption at constant coverage θ) can be obtained by application of the Clausius–Clapeyron equation if isotherms are determined at several different temperatures; the thermodynamics of adsorption have been fully discussed by Hill.[5]

METHOD

Adsorption from the gas phase can be measured by either gravimetric or volumetric techniques. In the gravimetric method, the weight of adsorbed gas is measured by observing the stretching of a helical spring from which the adsorbent is hung (see Fig. 2).† Alternatively,

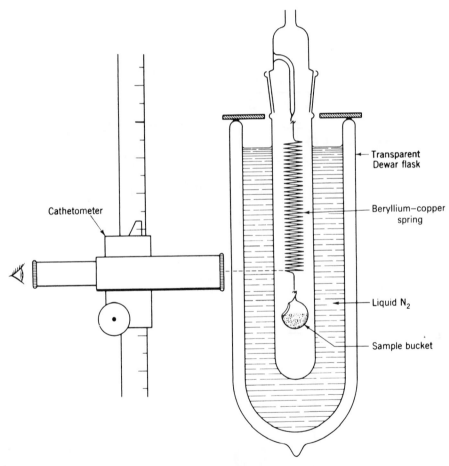

FIGURE 2
Gravimetric adsorption apparatus.

Transparent Dewar flask

Beryllium–copper spring

Cathetometer

Liquid N_2

Sample bucket

†A suitable spring can be made by winding 0.010-in. beryllium–copper wire on a 8- to 10-mm carbon rod for a length of 4 to 6 in., heating it on the rod in a furnace to about 450°C, and then slowly cooling to room temperature. The total load on the spring (bucket plus contents) should not exceed about 5 g, for which a total deflection of about 10 to 15 cm should be obtained; the length of the spring should be adjusted accordingly, and the apparatus shown in Fig. 2 must be dimensioned to accommodate the maximum deflection. The spring should be tested for the presence of any hysteresis; if any is found, the maximum load must be decreased enough to eliminate it. The spring must be protected from mercury vapor if a mercury manometer, McLeod gauge, or mercury diffusion pump is used. Adequate protection is obtained by placing a wad of fine copper wool in the vacuum line ahead of the spring balance; this should be replaced each time the experiment is performed. (Alternatively, springs made of fused silica may be used; their performance is excellent and they are chemically inert, but they are very fragile.) The sample bucket can be made from a thin blown glass or silica bubble, or from aluminum foil.

an electrostatic balance may be used. In the volumetric method, the amount of adsorption is inferred from pV measurements that are made on the gas before and after adsorption takes place. An excellent review of the many apparatus designs in common use has been given by Joy;[6] we shall consider only a conventional volumetric apparatus similar to one that has been very completely described by Barr and Anhorn.[7] This apparatus is shown in Fig. 3. The central feature in this design is a gas burette connected to a manometer that can be adjusted to maintain a constant volume in the arm containing the gas.

The gas burette shown in Fig. 3 is constructed from two 50-mL bulbs, large-bore capillary tubing, and a stopcock (F). Fiducial marks are engraved above the top bulb, between the bulbs, and below the bottom bulb. Before the burette is attached to the vacuum system,

FIGURE 3

Apparatus for measuring adsorption isotherms by the volumetric method.

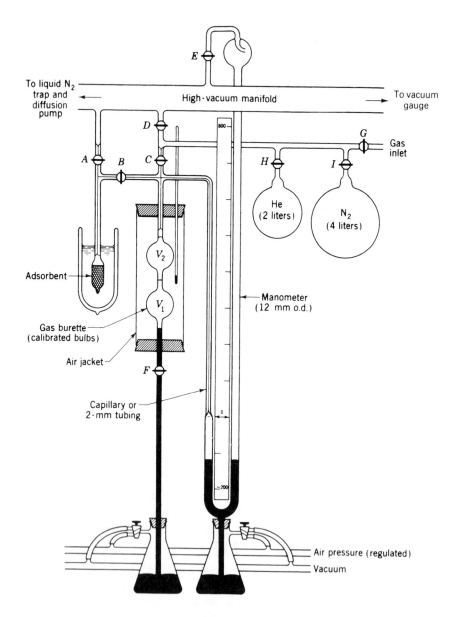

the volume between fiducial marks (V_1 and V_2) can be measured accurately by weighing the mercury required to fill each bulb. More elegant designs[6] achieve greater flexibility in operation by using five or six bulbs, but two are sufficient for the purpose of evaluating V_3, the volume of the tubing between the top of the gas burette and the zero level in the manometer. This is accomplished by measuring the pressure p of gas with both bulbs in use, then filling the lower bulb with mercury and determining the new pressure value p'. From the perfect-gas law,

$$V_3 = \frac{pV_1 - (p' - p)V_2}{p' - p} \tag{7}$$

When an isotherm is being determined, the temperature of the gas in the burette system must be known and should remain constant. Although a water jacket is often used, an air jacket is sufficient for the present experiment.

Adsorbent. For the apparatus described in Fig. 3, a sample of a high-area solid adsorbent that has a total surface area of 150 to 250 m^2 should be used. Of the adsorbents most commonly used (charcoal, silica, alumina), silica is recommended as an excellent choice. The sample bulb should be small enough that it is almost completely filled with the powder. Figure 4 shows a type of bulb that can easily be filled; a loose plug of Pyrex wool in the capillary will prevent loss of powder during filling and later during degassing. The exact weight of the sample should be determined before the bulb is attached to the system. After it is attached to the vacuum line, the sample must be degassed (i.e., pumped on to remove any physically adsorbed substances, principally water). This can be done by mounting a small electrical tube furnace around the sample bulb and heating to 200 to 250°C for several hours. Before heating, stopcock A should be opened to pump out all the air (stopcock B remains closed); this must be done cautiously to avoid forming a tight plug of powder in the capillary tube (gentle tapping will help also). After this initial degassing, a shorter degassing period will suffice between adsorption runs.

Adsorbate Gas. For physical adsorption, any pure gas that does not react with the solid adsorbent can be used. However, for quantitative area measurements, a molecule of known adsorption area is required. Nitrogen gas is a very common choice and will be used in this experiment.

Liquid nitrogen is the most convenient constant-temperature bath for isotherm measurements with N_2 gas. For high-precision work, the temperature of this bath should be measured with a thermocouple so that an accurate value of p_0 can be obtained, or p_0 should be measured directly with a separate nitrogen vapor-pressure manometer.[6,7] For this experiment it is adequate to assume that p_0 is equal to the atmospheric pressure in the room.

We can now develop the necessary equations for calculating the amount adsorbed from the pV data obtained during a run. Let us denote the known volume of the gas burette by V_B ($= V_1 + V_2 + V_3$ or $V_2 + V_3$ as the case may be) and its temperature by T_B. Since the sample bulb is not completely filled by the solid adsorbent, we must also consider the so-called dead-space volume occupied by gas. Let us denote this volume by V_s and consider it to be at a single temperature T_s. If the burette is filled with gas at an initial pressure p_1^0 and then stopcock B is opened, the pressure will drop to a new equilibrium value p_1. The number of moles adsorbed is

$$n_1 = \frac{p_1^0 V_B}{RT_B} - \left(\frac{p_1 V_B}{RT_B} + \frac{p_1 V_s}{RT_s} \right) \tag{8}$$

FIGURE 4

Detail drawing of the sample bulb, showing method of filling.

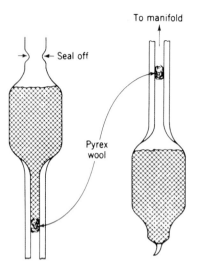

where typically the V's are expressed in cm^3, the p's in atm, and $R = 82.06 \ cm^3$ atm. Equation (8) can be rewritten in terms of v_1 (the amount adsorbed in cm^3 at STP) by using Eq. (5):

$$v_1 = p_1^0 V_B^* - p_1(V_B^* + V_s^*) \tag{9}$$

where the p's are now in Torr and

$$V_B^* = \frac{273.15 V_B}{760 T_B} \qquad V_s^* = \frac{273.15 V_s}{760 T_s} \tag{10}$$

The value of V_B^* can be calculated from V_B and T_B, which are known. V_s^* is constant but unknown; however, it can be determined if a run is made using helium gas. Since He is not significantly adsorbed at liquid-nitrogen temperature, $v_1 = 0$ and V_s^* can be calculated from Eq. (9).

If N_2 gas is admitted to the bulb, adsorption does occur and Eq. (9) yields a value of v_1, the "volume" adsorbed at the equilibrium pressure p_1. Now stopcock B is closed and more N_2 gas is added to the burette to achieve a new initial pressure p_2^0. When B is opened, more adsorption will occur and the pressure will change to an equilibrium value p_2. The total volume adsorbed at p_2 is

$$v_2 = [p_1^0 V_B^* + (p_2^0 - p_1)V_B^*] - p_2(V_B^* + V_s^*) \tag{11}$$

The generalization of Eq. (11) for v_j, the volume adsorbed at p_j after j additions of gas to the burette, is obvious; for example

$$v_3 = [p_1^0 V_B^* + (p_2^0 - p_1)V_B^* + (p_3^0 - p_2)V_B^*] - p_3(V_B^* + V_s^*)$$

Note that the brackets contain an expression for the *total* amount of gas added to the system and the second term represents the amount remaining in the gas phase at equilibrium. Since the effect of errors is cumulative in this method, care should be exercised in measuring all pressures.

Normally the entire gas burette is used ($V_B = V_1 + V_2 + V_3$), but on occasion it may be convenient to obtain two points on the isotherm from a single filling of the burette; this can be accomplished after the first point is obtained by raising the mercury level so as to fill bulb V_1. Equation (11) is still applicable, but the value of the last term is changed, since V_B now equals $V_2 + V_3$ (in that term only) and a new equilibrium pressure must be used.

EXPERIMENTAL

Details of the operation of the vacuum system (Fig. 3) should be reviewed carefully in consultation with an instructor before beginning the experiment. In the following it is assumed that both the helium and nitrogen storage bulbs have already been filled to a pressure slightly over 1 atm, the diffusion pump is operating, and the sample has been degassed.

Stopcocks B, C, D, E, and of course G should be closed and the pressure determined with the vacuum gauge; it should be 10^{-5} Torr or less. Now set the mercury level exactly at the lower mark in the gas burette and close stopcock F. Also raise the mercury in the manometer to approximately the zero level. After closing stopcock A and opening C, D, and E, read the vacuum gauge again and also verify that the mercury level is the same in both arms of the manometer.

Slowly raise a narrow-mouth Dewar flask filled with liquid nitrogen into place, and clamp it so that the sample bulb is completely immersed. Then plug the mouth of the Dewar loosely with glass wool or a piece of clean towel to retard condensation of water and oxygen from the air. (Liquid O_2 dissolved in the liquid N_2 would raise the temperature of the bath.)

Close stopcocks D and E, and **very slowly** turn the stopcock on the helium storage bulb while watching the manometer. As soon as the mercury levels begin to move, cease turning the stopcock and let the gas slowly fill the gas burette to a pressure of about 300 Torr. When this pressure is reached, immediately close the stopcock on the helium storage bulb and then close C. Adjust the mercury level in the manometer so that the left arm is exactly at the zero level. Wait a few minutes for equilibrium to be achieved, readjust to the zero level if necessary, and record the pressure. Now carefully raise the mercury level in the gas burette to the center mark (i.e., fill bulb V_1 with Hg) and again adjust the manometer to the zero level. After a short wait, record the new pressure. Using the known values of V_1 and V_2 (given by the instructor), one can now calculate V_3 from Eq. (7).

Next open stopcock B slowly and allow He to enter the sample bulb. Wait for at least 3 min, then reset the manometer to the zero level and read the pressure. After several minutes more, readjust this level (if necessary) and read the pressure again. When the pressure is constant, record its value and also record T_B, the ambient temperature at the gas burette. The quantity V_s^* can be calculated from these data using Eq. (9) with $v_1 = 0$. Note that, in calculating V_B^* from Eq. (10), the appropriate value of V_B is $V_2 + V_3$.

Slowly open stopcock A and pump off the He gas. Lower the Dewar flask and allow the sample to warm up to room temperature. Open stopcocks C and D and continue pumping for at least 15 min. During this time, reset the Hg level in the gas burette to the lower mark and lower the mercury in the manometer to approximately the zero level. (Do the two arms still read the same? Open stopcock E and see if there is any change.) Now close stopcocks A and B and replace the liquid-nitrogen Dewar around the sample bulb. If time is limited and values of V_3 and V_s^* have been provided, this helium run may be omitted and the following procedure is then carried out directly after first cooling the sample.

Procedure for Measuring the N$_2$ Isotherm. Close stopcocks D and E, and fill the gas burette with N_2 gas to a pressure of about 300 Torr. Follow the procedure given above in filling the burette and measuring the pressure. Record this pressure p_1^0. *Slowly* open stopcock B and allow N_2 to enter the sample bulb. Again follow the procedure given previously for obtaining the new equilibrium pressure p_1. In the case of N_2, a longer period may be necessary to achieve equilibrium, since adsorption is now taking place. The volume adsorbed v_1 can be calculated from Eq. (9). [In this case V_B in Eq. (10) is $V_1 + V_2 + V_3$.]

Now close stopcock B and add enough N_2 gas to the burette to bring the pressure up to about 100 Torr. Record this pressure p_2^0; then open B and obtain the next equilibrium pressure p_2. The volume v_2 is calculated from Eq. (11). Continue to repeat this process so as to obtain about seven points in the pressure range from 0 to about 250 Torr. Each time start with an initial pressure p_j^0 somewhat greater (by say 15 to 25 percent) than the desired equilibrium pressure p_j. If time permits obtain additional points somewhat more widely spaced over the range 250 to 650 Torr. At all times be sure that the liquid-nitrogen level completely covers the sample bulb.

Several times during the period, record the barometric pressure and T_B, the temperature at the gas burette. Record the mass of the adsorbent sample given by the instructor.

CALCULATIONS

From your helium data and the known values of V_1 and V_2, calculate V_3 from Eq. (7). If the temperature at the gas burette has been fairly constant throughout the experiment, use the average value as T_B in Eq. (10) and calculate values of V_B^* when $V_B = V_1 + V_2 + V_3$ and when $V_B = V_2 + V_3$. These values can then be used in all further calculations. If T_B has varied by more than $\pm0.5°C$, appropriate changes in V_B^* should be made where necessary. Now use Eq. (9) to calculate V_s^*.

For each equilibrium point on the isotherm, calculate a value of v, the volume adsorbed at pressure p. If the barometric pressure has been almost constant, its average value may be taken as p_0, the vapor pressure of nitrogen at the bath temperature. For each isotherm point, calculate $x = p/p_0$.

Plot the isotherm (v versus x) at 77 K and compare it qualitatively with the one shown in Fig. 1. Finally, calculate $x/v(1 - x)$ for each point and plot that quantity versus x; see Eq. (3). Draw the best straight line through the points between $x = 0.05$ and 0.3, and determine the slope and intercept of this line. This evaluation can be made graphically or by a linear least-squares fit of the data with Eq. (3) (see Chapter XXI). From Eq. (4), calculate v_m and c for the sample studied.

Using Eqs. (5) and (6) and the known mass m of the adsorbent sample, calculate the *specific area* $\bar{A}$ in square meters per gram of solid. The cross-sectional area for an adsorbed N_2 molecule may be taken as 15.8 $Å^2$.

DISCUSSION

What factors can you think of that would tend to make the BET theory less reliable above $\theta = 0.3$ or below $\theta = 0.05$?

SAFETY ISSUES

Work on a vacuum system requires preliminary review of procedures and careful execution in order to avoid damage to the apparatus and possible injury from broken glass; in addition, the liquid nitrogen used for cold traps must be handled properly (see Appendix C). Safety glasses must be worn. In the event of breakage, any mercury spill must be cleaned up promptly and carefully as described in Appendix C.

APPARATUS

Adsorption apparatus (Fig. 3), containing high-area sample, attached to high-vacuum line equipped with a vacuum gauge, diffusion pump, and liquid-nitrogen trap (see Chapter XVIII); high-purity helium and nitrogen gas; small electrical tube furnace; narrow-mouth taped Dewar flask and clamp for mounting; glass wool or clean towel; 0 to 30°C thermometer; stopcock grease. Liquid nitrogen.

REFERENCES

1. A. W. Adamson, *Physical Chemistry of Surfaces,* 6th ed., Wiley, New York (1997).

2. S. Brunauer, P. H. Emmett, and E. Teller, *J. Amer. Chem. Soc.* **60,** 309 (1938).

3. T. L. Hill, *J. Chem. Phys.* **14,** 263 (1946) and **17,** 772 (1949).

4. T. L. Hill, "Theory of Physical Adsorption," in *Advances in Catalysis,* Vol. IV, pp. 225–242, Academic Press, New York (1952).

5. *Ibid.,* pp. 242–255.

6. A. S. Joy, *Vacuum* **3,** 254 (1953).

7. W. E. Barr and V. J. Anhorn, *Instruments* **20,** 454, 542 (1947).

GENERAL READING

A. W. Adamson, *Physical Chemistry of Surfaces,* 6th ed., Wiley, New York (1997).

S. Brunauer, *Physical Adsorption,* chaps. I–IV, Princeton Univ. Press, Princeton, NJ (1945).

XII
Macromolecules

EXPERIMENTS

27. Intrinsic Viscosity: Chain Linkage in Polyvinyl Alcohol
28. Helix–Coil Transition in Polypeptides

EXPERIMENT 27
Intrinsic Viscosity: Chain Linkage in Polyvinyl Alcohol

While the basic chemical structure of a synthetic polymer is usually well understood, many physical properties depend on such characteristics as chain length, degree of chain branching, and molar mass, which are not easy to specify exactly in terms of a molecular formula. Moreover the macromolecules in a given sample are seldom uniform in chain length or molar mass (which for a linear polymer is proportional to chain length); thus the nature of the distribution of molar masses is another important characteristic.

A polymer whose molecules are all of the same molar mass is said to be *monodisperse;* a polymer in which the molar masses vary from molecule to molecule is said to be *polydisperse.* Specimens that are approximately monodisperse can be prepared in some cases by fractionating a polydisperse polymer; this fractionation is frequently done on the basis of solubility in various solvent mixtures.

This experiment is concerned with the linear polymer polyvinyl alcohol (PVOH), $-(CH_2-CHOH)_n$, which is prepared by methanolysis of the polyvinyl acetate (PVAc) obtained from the direct polymerization of the monomer vinyl acetate, $CH_2=CH-OOCCH_3$. As ordinarily prepared, polyvinyl alcohol shows a negligible amount of branching of the chains. It is somewhat unusual among synthetic polymers in that it is soluble in water. This makes polyvinyl alcohol commercially important as a thickener, as a component in gums, and as a foaming agent in detergents.

A characteristic of interest in connection with PVOH and PVAc is the consistency of orientation of monomer units along the chain. In the formula given above, it is assumed that all monomer units go together "head to tail." However, occasionally a monomer unit will join onto the chain in a "head-to-head" fashion, yielding a chain of the form

$$+CH_2-CHX \rangle_n CH_2-CHX-CHX-CH_2-CH_2-CHX+CH_2-CHX \rangle_n \cdots$$

head-to-head linkage   reversed monomer unit

where X is Ac or OH. The frequency of head-to-head linkage depends on the relative rates of the normal growth-step reaction α,

$$R \cdot + M \xrightarrow{k_\alpha} R-M \cdot \qquad (1a)$$

and the abnormal reaction β,

$$R \cdot + M \xrightarrow{k_\beta} R-M \cdot \qquad (1b)$$

(where $R \cdot$ is the growing polymer radical, the arrow representing the predominant monometer orientation). The rates in turn must depend on the activation energies:

$$\frac{k_\beta}{k_\alpha} = \frac{A_\beta e^{-E_{a\beta}/RT}}{A_\alpha e^{-E_{a\alpha}/RT}} = S e^{-\Delta E_a/RT} \qquad (2)$$

where $\Delta E_a = E_{a\beta} - E_{a\alpha}$ is the additional thermal activation energy needed to produce abnormal addition and S is the *steric factor* representing the ratio of the probabilities that the abnormally and normally approaching monomer will not be prevented by steric or geometric obstruction from being in a position to form a bond. Presumably the activation energy for the normal reaction is the lower one, in accord with the finding of Flory and Leutner[1,2] that the frequency of head-to-head linkages in PVAc increases with increasing polymerization temperature. The quantities S and ΔE_a can be determined from data obtained at two or more polymerization temperatures by a group of students working together. It should be of interest to see whether steric effects or thermal activation effects are the more important in determining the orientation of monomer units as they add to the polymer chain.

In this experiment, the method of Flory and Leutner will be used to determine the fraction of head-to-head attachments in a single sample of PVOH. The method depends on the fact that in PVOH a head-to-head linkage is a 1,2-glycol structure, and 1,2-glycols can be specifically and quantitatively cleaved by periodic acid or periodate ion. Treatment of PVOH with periodate should therefore break the chain into a number of fragments, bringing about a corresponding decrease in the effective molar mass. All that is required is a measurement of the molar mass of a specimen of PVOH before and after treatment with periodate.

METHOD[3]

For the determination of very high molar masses, freezing-point depressions, boiling-point elevations, and vapor-pressure lowerings are too small for accurate measurement. Osmotic pressures are of a convenient order of magnitude, but measurements are time-consuming. The technique to be used in this experiment depends on the determination of the intrinsic viscosity of the polymer. However, molar-mass determinations from osmotic pressures are valuable in calibrating the viscosity method.

Two other methods, light scattering and gel permeation chromatography (GPC), could also be used as alternatives or in addition to the viscosity technique. Light scattering is a classical method for determining molar masses, and if scattering as a function of angle is measured, it is also possible to learn something of the polymer shape, i.e., to distinguish rodlike from spherical forms. In the case of gel permeation chromatography, molecules of different sizes are carried by a liquid solvent over a substrate gel network of uniformly sized pores, which can be chosen in a range from 10 to 100,000 nm. Smaller solute molecules diffuse into and out of the pores and hence exhibit a longer retention time than larger solute molecules, which are not retained and travel through the column with the solvent. This method is simple, but to obtain accurate molar masses of a given polymer type from the retention times it is necessary to reproduce the flow rates carefully and to calibrate the column with polymers of known molar mass and similar chemical composition.

The coefficient of viscosity η of a fluid is defined in Exp. 4. It is conveniently measured, in the case of liquids, by determination of the time of flow of a given volume V of the liquid through a vertical capillary tube under the influence of gravity. For a virtually incompressible fluid such as a liquid, this flow is governed by Poiseuille's law in the form

$$\frac{dV}{dt} = \frac{\pi r^4 (p_1 - p_2)}{8\eta L}$$

where dV/dt is the rate of liquid flow through a cylindrical tube of radius r and length L and $(p_1 - p_2)$ is the difference in pressure between the two ends of the tube. In practice a

FIGURE 1

Viscosimeter arrangement to be used if a glass-walled thermostat bath is not available.

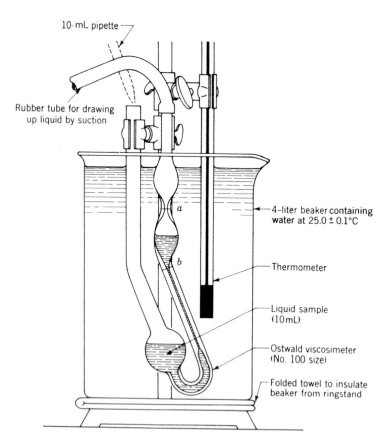

10·mL pipette

Rubber tube for drawing up liquid by suction

a

b

4–liter beaker containing water at 25.0 ± 0.1°C

Thermometer

Liquid sample (10mL)

Ostwald viscosimeter (No. 100 size)

Folded towel to insulate beaker from ringstand

viscosimeter of a type similar to that shown in Fig. 1 is used. Since $(p_1 - p_2)$ is proportional to the density ρ, it can be shown that, for a given total volume of liquid,

$$\frac{\eta}{\rho} = Bt \tag{3}$$

where t is the time required for the upper meniscus to fall from the upper to the lower fiducial mark (a to b) and B is an apparatus constant that must be determined through calibration with a liquid of known viscosity (e.g., water). The derivation of this equation is similar to that given for Eq. (4-10). For high-precision work, it may be necessary to consider a kinetic-energy correction to Eq. (3). As in Eq. (4-16), we can write

$$\frac{\eta}{\rho} = Bt - \frac{V}{8\pi Lt}$$

but the correction term is usually less than 1 percent of Bt and may be neglected in this experiment.

THEORY

Einstein showed that the viscosity η of a fluid in which small rigid spheres are present in dilute and uniform suspension is related to the viscosity η_0 of the pure fluid (solvent) by the expression[4]

$$\frac{\eta}{\eta_0} - 1 = \frac{5}{2}\frac{v}{V}$$

where v is the volume occupied by all the spheres and V is the total volume. The quantity $(\eta/\eta_0) - 1$ is called the *specific viscosity* η_{sp}. For nonspherical particles the numerical coefficient of v/V is greater than $\frac{5}{2}$ but should be a constant for any given shape provided the rates of shear are sufficiently low to avoid preferential orientation of the particles.

The *intrinsic viscosity*, denoted by $[\eta]$, is defined as the ratio of the specific viscosity to the weight concentration of solute, in the limit of zero concentration:

$$[\eta] \equiv \lim_{c \to 0} \frac{\eta_{sp}}{c} = \lim_{c \to 0}\left(\frac{1}{c}\ln\frac{\eta}{\eta_0}\right) \tag{4}$$

where c is conventionally defined as the concentration in grams of solute per 100 mL of solution. In this case, the units for $[\eta]$ are 10^2 cm^3 g^{-1}. Both η_{sp}/c and $(1/c)(\ln \eta/\eta_0)$ show a reasonably linear concentration dependence at low concentrations. A plot of $(1/c)(\ln \eta/\eta_0)$ versus c usually has a small negative slope, while a plot of η_{sp}/c versus c has a positive and larger slope.[3] Careful work demands that either or both of these quantities be extrapolated to zero concentration, although $(1/c)(\ln \eta/\eta_0)$ for a single dilute solution will give a fair approximation to $[\eta]$.

If the internal density of the spherical particles (polymer molecules) is independent of their size (i.e., the volume of the molecule is proportional to its molar mass), the intrinsic viscosity should be independent of the size of the particles and hence of no value in indicating the molar mass. This, however, is not the case; to see why we must look into the nature of a polymer macromolecule as it exists in solution.

Statistically Coiled Molecules. A polymer such as PVOH contains many single bonds, around which rotation is possible. If the configurations around successive carbon atoms are independent and unrelated, it will be seen that two parts of the polymer chain more than a few carbon atoms apart are essentially uncorrelated in regard to direction in space. The molecule is then "statistically coiled" and resembles a loose tangle of yarn:

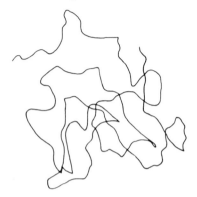

Simple statistical treatments[2] show that the mean distance between the two ends of the chain, and indeed also the effective mean diameter d of the coiled molecule regarded as a rough sphere, should be proportional to the square root of the chain length and thus to the square root of the molar mass:

$$d \propto M^{1/2}$$

The volume v_m occupied by a molecule should then vary as $M^{3/2}$. The number of molecules in a given weight of polymer varies inversely with the molar mass; hence the total volume of the spheres is

$$v \propto \frac{cV}{M} M^{3/2} = cVM^{1/2}$$

Therefore

$$[\eta] = KM^{1/2} \tag{5}$$

where K is a constant. This treatment is much simplified; it ignores among other things the problem of "excluded volume"; that is, the fact that the chain cannot coil altogether randomly because it is subject to the restriction that no two parts of the chain may be at the same point in space at the same time. This restriction becomes more and more important the higher the molar mass. Even more serious is the effect of solvent; the above treatment tacitly assumes a "poor solvent," which would barely get the polymer into solution. A "good solvent," by solvating the polymer, makes the size of the statistical coil increase faster with chain length than it otherwise would, owing to enhancement of the excluded volume effect. Accordingly, instead of Eq. (5) we might write

$$[\eta] = KM^a \tag{6}$$

where K and a are empirical parameters characteristic both of the polymer itself and of the solvent. The exponent a varies from about 0.5, for well-coiled polymer molecules in a poor solvent, to as much as 1.7 for a rigidly extended "rodlike" polymer molecule.

Flory and Leutner,[1] working with monodisperse specimens of PVOH differing from one another in molar mass over a wide range (obtained by fractionating polydisperse commercial PVOH), established a correlation between the molar mass, as determined from osmotic pressure measurements, and the intrinsic viscosity. They found that for PVOH in aqueous solution at 25°C,

$$[\eta] = 2.0 \times 10^{-4} M^{0.76}$$
$$M = 7.6 \times 10^4 [\eta]^{1.32} \tag{7}$$

Equation (7) also holds for a polydisperse sample of PVOH, but the molar mass in this case is $\bar{M}_v$, the viscosity-average molar mass defined below.

Number-Average and Viscosity-Average Molar Mass.   For a polydisperse polymer, any determination of molar mass must yield an average of some sort. When a colligative property such as osmotic pressure is used, the average is a number average:

$$\bar{M}_n = \frac{\int_0^\infty MP(M)\, dM}{\int_0^\infty P(M)\, dM} \tag{8}$$

where $P(M)$ is the molar-mass distribution function; that is, $P(M)\, dM$ is proportional to the number of molecules with molar masses between M and $M + dM$. However, the average obtained from intrinsic viscosity is not the same kind of average. It is called the viscosity average and is given by[2]

$$(\bar{M}_v)^a = \frac{\int_0^\infty M^{1+a}P(M)\, dM}{\int_0^\infty MP(M)\, dM} \tag{9}$$

For a nonodisperse polymer, $\bar{M}_v = \bar{M}_n = M$; for a polydisperse polymer the two kinds of averages are not equal but are related by a constant factor that depends on the distribution function $P(M)$ and the parameter a. A commonly encountered distribution function, and one that is likely to be valid for PVOH, is one that arises if the probability of a chain-termination reaction during the polymerization is constant with time and independent of the chain length already achieved. It is also the most likely function for the product resulting from cleavage with periodate if the head-to-head structures can be assumed to be randomly distributed along the PVOH chains. This distribution function is

$$P(M) = \frac{1}{\bar{M}_n} e^{-M/\bar{M}_n} \tag{10}$$

When this function is used as the weighting function in evaluating averages, it can be shown[3,5] that

$$\frac{\bar{M}_v}{\bar{M}_n} = [(1 + a)\Gamma(1 + a)]^{1/a}$$

where Γ is the gamma function. For $a = 0.76$ (the value given above for polyvinyl alcohol),

$$\frac{\bar{M}_v}{\bar{M}_n} = 1.89 \tag{11}$$

If a much exceeds 0.5, $\bar{M}_v$ is much closer to a mass-average molar mass.

$$\bar{M}_m = \frac{\int_0^\infty M^2 P(M)\, dM}{\int_0^\infty MP(M)\, dM} \tag{12}$$

than it is to the number average $\bar{M}_n$. In fact, when $a = 1$, $\bar{M}_v$ and $\bar{M}_m$ are identical, and with the distribution assumed in Eq. (10), their ratio to $\bar{M}_n$ is 2.

Determination of Frequency of Head-to-Head Occurrences. We wish to calculate the fraction of linkages that are head to head (that is, the ratio of "backward" monomer units to total monomer units), on the assumption that degradation arises exclusively from cleavage of 1,2-glycol structures and that all such structures are cleaved. Let us denote this ratio by Δ. It is equal to the *increase* in the number of molecules present in the system, divided by the total number of monomer units represented by all molecules in the system. Since these numbers are in inverse proportion to the respective molar masses,

$$\Delta = \frac{1/\bar{M}'_n - 1/\bar{M}_n}{1/M_0} \tag{13}$$

where $\bar{M}_n$ and $\bar{M}'_n$ are number-average molar masses before and after degradation, respectively, and M_0 is the monomer molar mass, equal to 44 g mol^{-1}. Thus

$$\Delta = 44 \left(\frac{1}{\bar{M}'_n} - \frac{1}{\bar{M}_n} \right) \tag{14}$$

Making use of Eq. (11), we can write

$$\Delta = 83 \left(\frac{1}{\bar{M}'_v} - \frac{1}{\bar{M}_v} \right) \tag{15}$$

which permits viscosity averages to be used directly.

EXPERIMENTAL

Clean the viscosimeter thoroughly with cleaning solution,† rinse copiously with distilled water (use a water aspirator to draw large amounts of distilled water through the capillary), and dry with acetone and air. Immerse in a 25°C thermostat bath to equilibrate. Place a small flask of distilled water in a 25°C bath to equilibrate. Equilibration of water or solutions to bath temperature, in the amounts used here, should be complete in about 10 min.

If a stock solution of the polymer is not available, it should be prepared as follows. Weigh out accurately in a weighing bottle or on a watch glass 4.0 to 4.5 g of the dry polymer. Add it slowly, with stirring, to about 200 mL of hot distilled water in a beaker. To the greatest extent possible, "sift" the powder onto the surface and stir gently so as not to entrain bubbles or produce foam. When all of the polymer has dissolved, let the solution cool and transfer it carefully and quantitatively into a 250-mL volumetric flask. Avoid foaming as much as possible by letting the solution run down the side of the flask. Make the solution up to the mark with distilled water and mix by *slowly* inverting a few times. If the solution appears contaminated with insoluble material that would possibly interfere with the viscosity measurements, filter it through Pyrex wool. (Since making up this solution may take considerable time, it is suggested that the calibration of the viscosimeter with water be carried out concurrently.)

In all parts of this experiment, avoid foaming as much as possible. This requires careful pouring when a solution is to be transferred from one vessel to another. It is also important to be meticulous about rinsing glassware that has been in contact with polymer

---

†The cleaning of glassware is discussed in Chapter XX. An excellent cleaning solution is chromic acid, but this must be handled with care (see p. 642). Alternative choices are hot 50–50 nitric + sulfuric acid (caution in handling also required), alcoholic KOH, or a commercial cleaning solution such as CONTRAD 70.

solution promptly and *very thoroughly,* for once the polymer has dried on the glass surface it is quite difficult to remove. Solutions of the polymer should not be allowed to stand for many days before being used, as they may become culture media for airborne bacteria.

Pipette 50 mL of the stock solution into a 100-mL volumetric flask and make up to the mark with distilled water, observing the above precautions to prevent foaming. Mix the liquids and place the flask in the bath to equilibrate. In this and other dilutions, rinse the pipette *very thoroughly* with water and dry with acetone and air.

To cleave the polymer, pipette 50 mL of the stock solution into a 250-mL flask and add up to 25 mL of distilled water and 0.25 g of solid KIO_4. Warm the flask to about 70°C, and stir until all the salt is dissolved. Then clamp the flask in a thermostat bath and stir until the solution is at 25°C. Transfer quantitatively to a 100-mL volumetric flask and make up to the mark with distilled water. Mix carefully and place in the bath to equilibrate. (This operation can be carried out while viscosity measurements are being made on the uncleaved polymer.)

At this point two "initial" solutions have been prepared: 100 mL of each of two aqueous polymer solutions of the same concentration (~0.9 g/100 mL), one cleaved with periodate and one uncleaved. To obtain a second concentration of each material, pipette 50 mL of the "initial" solution into a 100-mL volumetric flask and make up to the mark with distilled water. Place all solutions in the thermostat bath to equilibrate. The viscosity of distilled water and of both solutions of each polymer (cleaved and uncleaved) should be determined. If time permits a third concentration (equal to one-quarter of the initial concentration) should also be prepared and measured.

The recommended procedure for measuring the viscosity is as follows:

1. The viscosimeter should be mounted vertically in a constant-temperature bath so that both fiducial marks are visible and below the water level. If a glass-walled thermostat bath that will allow readings to be made with the viscosimeter in place is not available, fill a large beaker or battery jar with water from the bath and set it on a dry towel for thermal insulation. The temperature should be maintained within ±0.1°C of 25°C during a run. This will require periodic small additions of hot water.
2. Pipette the required quantity of solution (or water) into the viscosimeter. Immediately rinse the pipette copiously with water and dry it with acetone and air before using it again.
3. By suction with a pipette bulb through a rubber tube, draw the solution up to a point well above the upper fiducial mark. Release the suction and measure the flow time between the upper and lower marks with a stopwatch or timer. Obtain two or more additional runs with the same filling of the viscosimeter. Three runs agreeing within about 1 percent should suffice.
4. Each time the viscosimeter is emptied, rinse it *very thoroughly* with distilled water, then dry with acetone and air. Be sure to remove *all* polymer with water before introducing acetone.

If time permits, the densities of the solutions should be measured with a Westphal balance.[6] Otherwise, the densities may be taken equal to that of the pure solvent without introducing appreciable error.

CALCULATIONS

At 25°C, the density of water is 0.99708 g cm^{-3} and the coefficient of viscosity is 8.909×10^{-3} g cm^{-1} s^{-1} = 0.8909 cP, where the *poise* (P) is a convenient cgs unit of viscosity equivalent to 10^{-1} N m^{-2} s = 10^{-1} Pa s. In case a temperature correction is needed for your calibration run, the density and viscosity of water are 0.99757 g cm^{-3} and 0.9317 cP at 23°C and 0.99654 g cm^{-3} and 0.8525 cP at 27°C. Using your time of flow for pure water, determine the apparatus constant B in Eq. (3).

For each of the polymer solutions studied, calculate the viscosity η and the concentration c in grams of polymer per 100 mL of solution. Then calculate η_{sp}/c and $(1/c)\,(\ln \eta/\eta_0)$. Plot η_{sp}/c and $(1/c)(\ln \eta/\eta_0)$ versus c and extrapolate linearly to $c = 0$ to obtain $[\eta]$ for the original and for the degraded polymer.

Calculate $\bar{M}_v$ and $\bar{M}_n$ for both the original polymer and the degraded polymer, then obtain a value for Δ. Report these figures together with the polymerization temperature for the sample studied. Discuss the relationship between Δ and the rate constants k_α and k_β.

If a group of students has studied a variety of samples polymerized at different temperatures, it will be possible to obtain estimates of the difference in thermal activation energies ΔE_a and the steric factor S [see Eq. (2)]. It may be difficult to obtain a quantitative interpretation unless the details of the polymerization (amount of initiator, polydispersity, etc.) are well known. However, you should comment qualitatively on your results in the light of what is known about the theory of chain polymerization.[2,7]

DISCUSSION

Equations (14) and (15) were derived on the assumption that a 1,2-glycol structure will result from every abnormal monomer addition. What is the result of two successive abnormal additions? Can you derive a modified expression on the assumption that the probability of an abnormal addition is independent of all previous additions? Would such an assumption be reasonable? Are your experimental results (or indeed those of Flory and Leutner) precise enough to make such a modification significant in experimental terms? What polymerization factors besides temperature might influence the frequency of abnormal head-to-head linkages?

SAFETY ISSUES

Chromic acid or nitric + sulfuric acid cleaning solutions are corrosive and need to be handled with care; see p. 642. Wear safety goggles and gloves while cleaning the viscosimeter.

APPARATUS

Ostwald viscosimeter; two 100- and two 250-mL volumetric flasks; a 10- and a 50-mL pipette; pipetting bulb; 250-mL glass-stoppered flask; one 100- and two 250-mL beakers; stirring rod; hot plate; thermometer; stopwatch or timer; length of gum-rubber tubing. If polymer stock solution is to be prepared by the student: weighing bottle; funnel; Pyrex wool.

Glass-walled thermostat bath at 25°C or substitute; polyvinyl alcohol (M ~60,000 to 80,000); 5 g of solid or 200 mL of solution containing 18 g L^{-1}; KIO$_4$ (1 g); chromic acid or other cleaning solution (50 mL); Westphal balance, if needed. The polyvinyl alcohol can be obtained from Polysciences Inc., 400 Valley Road, Warrington, PA 18976, and many other suppliers.

REFERENCES

1. P. J. Flory and F. S. Leutner, *J. Polym. Sci.* **3**, 880 (1948); **5**, 267 (1950).

2. P. J. Flory, *Principles of Polymer Chemistry,* Cornell Univ. Press, Ithaca, NY (1953).

3. P. J. Dunlop, K. R. Harris, and D. J. Young, "Experimental Methods for Studying Diffusion in Gases, Liquids, and Solids," in B. W. Rossiter and R. C. Baetzold (eds.), *Physical Methods of Chemistry,* 2d ed., vol. VI, chap. 3, Wiley-Interscience, New York (1992).

4. A. Einstein, *Investigations on the Theory of Brownian Movement,* chap. III, Dover, New York (1956); R. J. Silbey, R. A. Alberty, and M. G. Bawendi, *Physical Chemistry,* 4th ed., p. 776, Wiley, New York (2005).

5. J. R. Schaefgen and P. J. Flory, *J. Amer. Chem. Soc.* **70**, 2709 (1948).

6. N. Bauer and S. Z. Lewin, "Determination of Density," in A. Weissberger and B. W. Rossiter (eds.), *Techniques of Chemistry: Vol. I. Physical Methods of Chemistry,* part IV, chap. 2, Wiley-Interscience, New York (1972).

7. R. W. Lenz, *Organic Chemistry of Synthetic High Polymers,* pp. 261–271, 305–369, Wiley-Interscience, New York (1967).

GENERAL READING

F. W. Billmeyer, Jr., *Textbook of Polymer Science,* 3d ed., Wiley-Interscience, New York (1984).

P. J. Flory, *Principles of Polymer Chemistry,* Cornell Univ. Press, Ithaca, NY (1953).

EXPERIMENT 28
Helix–Coil Transition in Polypeptides

Polymer molecules in solution can be found in many different geometric conformations, and there exist a variety of experimental methods (e.g., viscosity, light scattering, optical rotation) for obtaining information about these conformations.[1] In this experiment, the measurement of optical rotation will be used to study a special type of conformational change that occurs in many polypeptides.

The two important kinds of conformation of a polypeptide chain are the helix and the random coil. In the helical form, the amide hydrogen of each "amide group,"

$$-N-C-C-$$

is internally hydrogen bonded to the carbonyl oxygen of the third following amide group along the chain. Thus this form involves a quite rigid, rodlike structure (see Fig. 1). Under different conditions, the polypeptide molecule may be in the form of a statistically random coil (see Exp. 27). The stable form of the polypeptide will depend on several factors—the nature of the peptide groups, the solvent, and the temperature. For example, poly-γ-benzyl-*l*-glutamate† (PBG),

†Note that the usual chemical description of polypeptides is in terms of amino acid residues, as in the PBG formula shown, rather than in terms of amide groups, which are more convenient for the present discussion.

$$\text{--}NH\text{--}CH\text{--}\overset{\overset{\displaystyle O}{\|}}{C}\text{--}_{\overline{N}}$$

has a helical conformation when dissolved in ethylene dichloride at 25°C, but it is in the random-coil form when dissolved in dichloroacetic acid at the same temperature. This difference is quite reasonable, since a hydrogen-bonding solvent such as dichloroacetic acid can form strong hydrogen bonds with the amide groups and thus disrupt the internal hydrogen bonds that are necessary for the helical form. For a mixed solvent of dichloroacetic acid and ethylene dichloride, PBG can be made to transform from the random coil to the helix by raising the temperature over a fairly narrow range. This rapid reversible transition will be investigated here.

FIGURE 1

The Pauling–Corey alpha helix. In addition to the right-handed helix shown, a left-handed one is also possible (with the same *l*-amino acids). In proteins, the helix always seems to be right-handed. [*Reprinted by permission from L. Pauling,* The Nature of the Chemical Bond, *p. 500, Cornell Univ. Press, Ithaca, NY (1960).*]

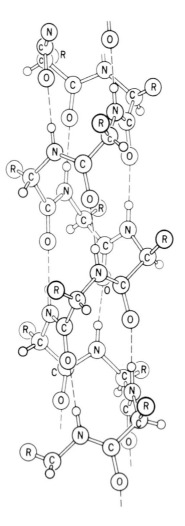

As we shall see below, the solvent plays a crucial role in determining which form is stable at low temperatures. When the helix is stable at low temperatures, the transition to the random coil at high temperatures is called a "normal" transition. For the case in which the random coil is the more stable form at low temperatures, the transition is called an "inverted" transition.

THEORY

We wish to present here a very simplified and approximate statistical-mechanical theory of the helix–coil transition. The treatment is closely related to that given by Davidson,[2] which is based on a model by Zimm et al.[3] Let us consider the change

$$\text{Coil} \cdot NS = \text{helix} + NS \tag{1}$$

where the polymer molecules each consist of N segments (monomer units) and are dissolved in a solvent S that can hydrogen bond to the amide groups when the chain is in the random-coil form. We shall (for convenience) artificially simplify the physical model of internal hydrogen bonding in the helix by assuming that the hydrogen bond formed by each amide group is with the *next* amide group along the chain (see Fig. 2) rather than the third following amide group as in the actual helix. We can then assume that the chain segments are independent of each other. Let q_1 be the molecular partition function of a *segment in the random-coil form* with a solvent molecule S hydrogen bonded to it, and q_2 be the product of the partition function of a segment in the *helical form* times the partition function of a "free" solvent molecule S. We will then define a parameter s by

$$q_2 = sq_1 \tag{2}$$

Note that s is the ratio of partition functions and many of the contributions to q_1 and q_2 (e.g., vibrational terms) will cancel out. The principal contributions to s will involve differences between the helix and random-coil form, and we may guess that s can be represented in the general form[2]

$$s = s_0 e^{-\varepsilon/kT} \tag{3}$$

where $\varepsilon = \varepsilon_h - \varepsilon_c$ is the energy change per segment and s_0 is related to the entropy change per segment for the change in state of Eq. (1).

First-Order Transition Model. To simplify the model even further, let us assume for the time being that a given polymer molecule is either completely in the helical form

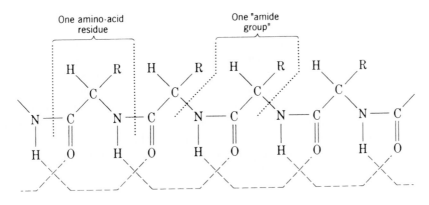

FIGURE 2

Hypothetical model of an internally hydrogen-bonded chain, with the simplification that *adjacent* "amide groups" are connected by hydrogen bonds. Note the distinction between the *amino acid residue* and the "amide group"; the latter is the more convenient unit of structure for the present discussion.

or completely in the random-coil form. This assumption will lead to a first-order transition between these two forms. The partition function for a random-coil molecule q_c is then given by

$$q_c = q_1^N \tag{4}$$

while that for a helical molecule q_h will be

$$q_h = q_2^N = s^N q_1^N \tag{5}$$

if one neglects end effects (e.g., the fact that at the beginning of the chain the first carbonyl oxygen is not involved in an internal hydrogen bond).

At constant pressure, the two forms will be in equilibrium with each other at some temperature T^* at which $\Delta G = 0$ for the change in state (1). Since ΔV will be quite small for change (1), we can take $\Delta A \cong 0$ and $\widetilde{A}_c = \widetilde{A}_h$ at T^*, where $\widetilde{A}$ is the Helmholtz free energy per mole of polymer. For independent polymer molecules, we have[4]

$$\widetilde{A} = -kT \ln Q = -kT \ln q^{N_0} = -RT \ln q \tag{6}$$

where N_0 is Avogadro's number, q is the partition function for a single polymer molecule, and Q is the (canonical) partition function for a mole of molecules. Thus one finds that $q_h = q_c$ (or $s = 1$) at T^*. With the use of Eq. (3), the value of T^* can be related to the parameters ε and s_0:

$$\frac{\varepsilon}{kT^*} = \ln s_0 \tag{7}$$

For an inert (non-hydrogen-bonding) solvent, one observes experimentally a "normal" transition: the helix, which is stable at low temperatures, is transformed at higher temperatures into the random coil. This case is represented by our model when $\varepsilon < 0$ and $s_0 < 1$. Since the solvent is inert, random-coil segments are essentially unbonded, whereas helical segments have internal hydrogen bonds and thus a lower energy ($\varepsilon_h - \varepsilon_c < 0$). In the random-coil polymer molecule, considerable rotation can occur about the single bonds in the chain skeleton, but the helical form has a rather rigid structure. Therefore there is a decrease in the entropy per segment on changing from the relatively free rotational configurations of the flexible random coil to the more rotationally restricted helix (and thus $s_0 < 1$). From Eqs. (3) to (6), it follows that $s > 1$ at low temperatures ($T < T^*$) and thus $\widetilde{A}_h < \widetilde{A}_c$ and the helix is the more stable form. At high temperatures ($T > T^*$), $s \cong s_0 < 1$ and the random coil is the more stable form.

For an active (hydrogen-bonding) solvent, one observes experimentally an "inverted" transition: the random coil is stable at low temperatures and changes into the helix when the solution is heated. This case is represented by our model when $\varepsilon > 0$ and $s_0 > 1$. Here the solvent plays a dominant role, since a solvent molecule is strongly hydrogen bonded to each random-coil segment. In terms of energy, there is little difference between polymers with random-coil segments and those with helical segments, since there is comparable hydrogen bonding in both forms. It is now almost impossible to predict the sign of ε, but it is certainly quite reasonable to find that the energy change associated with (1) may be positive ($\varepsilon_h - \varepsilon_c > 0$). Although a helical segment will still have a lower entropy than a random-coil segment owing to the rigidity of the helix, the solvated random coil is now less flexible than previously. More important, the free solvent molecules will have a higher entropy than solvent molecules that are bonded to the random coil. Thus the entropy change associated with change in state (1) is positive because of the release of S molecules (and therefore $s_0 > 1$). From Eqs. (3) to (6), we see that at low temperatures $s < 1$ and the random coil will be stable in this inverted case.

Recalling that ΔV is small so that $\Delta G \cong \Delta A$, we can make use of Eqs. (3) to (6) to obtain

$$\Delta \widetilde{S} = -\left(\frac{\partial \Delta \widetilde{G}}{\partial T}\right)_p \cong R\frac{\partial}{\partial T}\left(T \ln \frac{q_h}{q_c}\right) = NR \ln s_0 \tag{8}$$

The entropy change *per mole of monomer*, ΔS_m, is then given by

$$\Delta S_m \equiv \frac{\Delta \widetilde{S}}{N} = R \ln s_0 \tag{9}$$

Since $\Delta G = 0$ for the first-order transition at p and T^*, we have for the enthalpy change *per mole of monomer*,

$$\Delta H_m = T^* \Delta S_m = RT^* \ln s_0 = N_0\varepsilon \tag{10}$$

The first-order theory presented above is related to the treatment of Baur and Nosanow,[5] who used a similar but perhaps physically less realistic model to derive the same results [that is, Eqs. (7), (9), and (10)].†

Cooperative Transition Model. It is an experimental fact that the helix–coil transition is *not* a first-order transition. In order to account for this, one must adopt a more realistic model by eliminating the assumption that an entire polymer molecule is all in a given form. Indeed, a polymer molecule can have some sections that are helical and others that are randomly coiled. The formation of a helical region will be a cooperative process. Since segment (i.e., amide group) n is hydrogen bonded to segment $n + 3$, $n + 1$ to $n + 4$, $n + 2$ to $n + 5$, etc., it is difficult to initiate this ordered internal bonding; but once a single hydrogen-bond link is made, the next ones along the chain are much easier to achieve (a sort of "zipper" effect).

No attempt will be made here to develop the details of such a cooperative model. Readers with a sufficient background in statistical mechanics should refer to Davidson's presentation[2] or to the original papers in the literature.[3] The crucial idea is to introduce a "nucleation" parameter σ that is independent of temperature and very small in magnitude. It is then assumed that the partition function q_2 is changed to the value σq_2 for the *initial* segment of a helical section (i.e., the internally bonded segment directly adjacent to a solvent-bonded segment). Physically this means that it is difficult to initiate a new helical section. For such a model, it can be shown[2,3] that X_h, the mole fraction of the segments in helical sections, will undergo a very rapid but *continuous* change from $X_h \cong 0$ when $s < 1$ to $X_h \cong 1$ when $s > 1$. For $\sigma \lesssim 10^{-2}$, X_h is very well represented by the approximation

$$X_h \begin{cases} \cong sF(s) & \text{for } s < 1 \\ = \frac{1}{2} & \text{for } s = 1 \\ \cong F(s) & \text{for } s > 1 \end{cases} \tag{11}$$

where the function $F(s)$ is given by

$$F(s) \equiv \frac{1}{2}\left\{1 + \frac{(s - 1) + 2\sigma}{[(s - 1)^2 + 4\sigma s]^{1/2}}\right\} \tag{12}$$

---

†In addition, Baur and Nosanow showed for the inverted case that the helical form will be stable only between T^* and another, much higher transition temperature T', where the helix reverts to the random-coil form. (For PBG, it is estimated[5] that $T' \cong 700$ K.) This is physically reasonable in terms of our model: at very high temperatures, all hydrogen bonds (internal or with solvent) will be broken and the random coil will predominate owing to its higher entropy.

FIGURE 3
FIGURE 3

Percentage of polypeptide in helix form, as a function of the parameter s (on a log scale), for $\sigma = 10^{-4}$ and $\sigma = 10^{-2}$. For the case of an inverted transition, $s > 1$ corresponds to $T > T^*$.

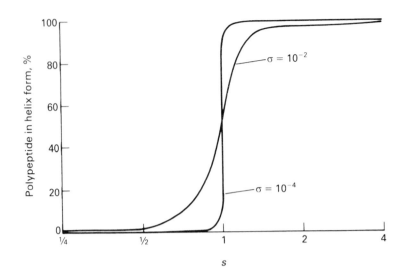

The resulting variation of X_h with the parameter s is shown in Fig. 3, which implies that a rapid cooperative transition occurs in the vicinity of a critical temperature T^* (the value of T for which $s = 1$). It is the fact that σ is very small that ensures a sharp transition. If σ were taken to be 1 (which corresponds to ignoring the nucleation problem), this model would yield $X_h = s/(1 + s)$, which corresponds to a very gradual transition. For high-molar-mass PBG, a value of $\sigma \simeq 2 \times 10^{-4}$ seems to be appropriate to obtain a fairly good quantitative fit to the experimental data.[2] It should be noted that the transition is sharpest, and the model is most successful, when polypeptides with $N \gtrsim 1000$ are used.

METHOD

Since the helical form of a polypeptide is rigid and rodlike, its solutions are more viscous than corresponding solutions of the random-coil form, and the transition can be detected by measuring the intrinsic viscosity (see Exp. 27). However, the highly ordered structure of the helical form causes it to have an optical-rotatory power markedly different from that of a random coil. Indeed, use of optical rotation for following the transition gives better results than does intrinsic viscosity, since the optical rotation is related directly to the fraction X_h of segments in the helical form, whereas the hydrodynamic properties of the polymer molecule change substantially at the first appearance of random-coil regions in the helix (in effect, causing the otherwise straight rod to bend randomly in one or several places).

The polarimeter and its operation are described in Chapter XIX. Although not stressed there, it is obviously necessary to use a polarimeter in which the temperature of the sample tube can be well controlled. The *specific rotation* $[\alpha]^t$ as determined with the sodium D doublet at $t°C$ is defined by Eq. (XIX-9), and we note here that for a solution containing two solutes (helix and random coil) one can show that

$$[\alpha]^t = X_h[\alpha]_h^t + X_c[\alpha]_c^t \tag{13}$$

where $[\alpha]_h^t$ and $[\alpha]_c^t$ are the specific rotations of helix and random-coil form. In a narrow range of temperature near T^*, $[\alpha]_h^t$ and $[\alpha]_c^t$ will be essentially constant. Thus measurements of $[\alpha]^t$ will determine the mole fraction X_h as a function of T.

EXPERIMENTAL

A solution of poly-γ-benzyl-l-glutamate (PBG) in a mixed solvent of dichloroacetic acid (DCA) and ethylene dichloride will be used. The dichloroacetic acid should be purified by vacuum distillation; the ethylene dichloride may be purified by conventional distillation in air. Mix 38 mL of DCA with 12 mL of ethylene dichloride to make about 50 mL of the desired solvent. In order to prepare approximately 25 mL of a 2.5 weight percent polymer solution, weigh accurately a sample of about 1 g of PBG and dissolve it in about 38 g of solvent (also weighed accurately). The resulting solution is very viscous; it must be mixed well and then allowed to stand for two or three days until any small undissolved particles have settled and the solution is clear. During this period, the solution should be placed in a polyethylene bottle and stored in a refrigerator. The remaining solvent should also be stored for future density determinations.

Since the solution is difficult to prepare, it is recommended that it be made available to the student. **Warning: The solvent is extremely corrosive.**

Detailed instructions for using the polarimeter and the temperature-regulating bath will be provided in the laboratory. After determining the path lengths of the cells to be used, fill a cell with pure solvent and zero-adjust the instrument. Next, rinse and fill a cell with the polymer solution.†

After mounting the solution cell in the polarimeter, it should not be necessary to move this cell during the rest of the experiment. Provisions must be made for measuring the temperature of the sample without obscuring the light path. Begin measurements with the temperature at about 10°C and read the optical rotation α at reasonably spaced temperature intervals until about 50°C is reached. The exact temperature intervals to be used are at your discretion, but should be close enough to follow the essential features of the variation of optical rotation; small intervals (perhaps 1 or 2°C) should be used when the optical rotation is changing rapidly with temperature (near 30°C), and larger intervals may be used when the changes with temperature are small. Be sure to allow sufficient time for thermal equilibration between readings.

In order to calculate $[\alpha]^t$, the density of the solutions must be known at each temperature. The density of the solution is close enough to that of the solvent that they can be assumed to be identical. The density of the solvent should be determined with a pycnometer for at least three different temperatures in the range 10 to 50°C. The volume of the pycnometer should first be determined by weighing it empty and also filled with distilled water at a known temperature. Since the density of water as a function of temperature is well known, the volume of the pycnometer can be determined. This procedure (see Exp. 9) is then repeated using the solvent, and since the volume of the pycnometer is now known, the density can be calculated. The solvent density should be plotted against temperature and the best straight line drawn through the points obtained.

Record the composition of the solution and the molecular weight (or N value) of the polymer used.

---

†A difficulty in filling these cells is the problem of avoiding any air bubbles that might obscure part of the light path. Since the solution is viscous, this can be a tedious job and may require some time. If possible it is wise to fill the cells about an hour before use and allow them to stand.

CALCULATIONS

For each data point, calculate the specific rotation at temperature t from the expression

$$[\alpha]^t = 100\frac{\alpha}{Lp\rho} \tag{14}$$

where α is the observed rotation in degrees, L is the path length of the cell in decimeters, p is the weight percent of solute in the solution, and ρ is the density of the solution in g cm^{-3}. These density values are interpolated from the ρ-versus-T plot for the solvent.

Construct a plot of $[\alpha]^t$ versus temperature, and determine the limiting (asymptotic) values of the rotation: $[\alpha]_h$ from the high-temperature plateau, $[\alpha]_c$ from the low-temperature plateau. If it is assumed that these represent *temperature-independent* specific rotations for the helix and random-coil forms, respectively, Eq. (13) gives for X_h:

$$X_h = \frac{[\alpha]^t - [\alpha]_c}{[\alpha]_h - [\alpha]_c} \tag{15}$$

Using this equation, you can add to your plot a scale of X_h values so that the plot will also represent X_h versus T. Determine and report the "transition" temperature T^* at which $X_h = 0.5$.

DISCUSSION

If one *arbitrarily* defined an equilibrium constant by $K = X_h/X_c$ and used the data in this experiment together with the van't Hoff equation $d \ln K/d(1/T) = \Delta \tilde{H}^0/R$ to determine the value of $\Delta \tilde{H}^0$, to what (if anything) would such an enthalpy change refer?

SAFETY ISSUES

Many chlorinated hydrocarbons are toxic to some degree. Thus the mixed solvent of dichloroacetic acid and ethylene dichloride should be handled with care. This solvent is also very corrosive. If it is necessary to prepare the PBG solutions, gloves should be worn. Use a pipette bulb; do not pipette by mouth. Dispose of waste chemicals as instructed.

APPARATUS

Polarimeter, with one or more optical cells (complete with windows, caps, and Teflon or neoprene washers); constant-temperature control for range 10 to 50°C; pycnometer; 25-mL pipette for filling the pycnometer and a small pipette for filling the cell; rubber bulb for pipetting; 0 to 50°C thermometer; 250-mL beaker; gum-rubber tubing; lint-free tissues for wiping cell windows; 50-mL polyethylene bottles for PBG solution and excess solvent.

Poly-γ-benzyl-*l*-glutamate, M = 300,000 (1 g of solid or 25 mL of 2.5 weight percent solution); ethylene dichloride–dichloroacetic acid solvent (76 volume percent DCA) stored in polyethylene bottle; wash acetone. The polypeptide can be obtained from Sigma Chemical Co., P.O. Box 14508, St. Louis, MO 63178. Check the current *Chem. Sources* for other suppliers.

REFERENCES

1. M. A. Stahmann (ed.), *Polyamino Acids, Polypeptides and Proteins,* reprint ed., Books on Demand, Ann Arbor, MI (1962).

2. N. Davidson, *Statistical Mechanics,* pp. 385–393, McGraw-Hill, New York (1962).

3. B. H. Zimm and J. K. Bragg, *J. Chem. Phys.* **31,** 526 (1959); B. H. Zimm, P. Doty, and K. Iso, *Proc. Natl. Acad. Sci.* **45,** 160A (1959).

4. I. N. Levine, *Physical Chemistry,* 6th ed., sec. 21.3, McGraw-Hill, New York (2009).

5. M. E. Baur and L. H. Nosanow, *J. Chem. Phys.* **38,** 578 (1963).

GENERAL READING

P. Doty, J. H. Bradbury, and A. M. Holtzer, *J. Amer. Chem. Soc.* **78,** 947 (1956); P. Doty and J. T. Yang, *J. Amer. Chem. Soc.* **78,** 498 (1956).

J. H. Gibbs and E. A. DiMarzio, *J. Chem. Phys.* **30,** 271 (1959).

XIII

Electric, Magnetic, and Optical Properties

EXPERIMENTS

29. Dipole Moment of Polar Molecules in Solution
30. Dipole Moment of HCl Molecules in the Gas Phase
31. Magnetic Susceptibility
32. NMR Determination of Paramagnetic Susceptibility
33. Dynamic Light Scattering

EXPERIMENT 29
Dipole Moment of Polar Molecules in Solution

When a substance is placed in an electric field, such as exists between the plates of a charged capacitor, it becomes to some extent electrically polarized. The polarization results at least in part from a displacement of electron clouds relative to atomic nuclei; polarization resulting from this cause is termed *electronic polarization*. For molecular substances, atomic polarization may also be present, owing to a distortion of the molecular skeleton. Taken together, these two kinds of polarization are called *distortion polarization*. Finally, when molecules possessing permanent dipoles are present in a liquid or gas, application of an electric field produces a small preferential orientation of the dipoles in the field direction, leading to *orientation polarization*.

The permanent dipole moment μ of a polar molecule, as a solute molecule in a liquid solution in a nonpolar solvent or as a molecule in a gas, can be determined experimentally from measurements of the *dielectric constant* κ. This quantity is the ratio of the electric permittivity ε of the solution or gas to the electric permittivity ε_0 of a vacuum $(8.854 \times 10^{-12} \text{ F m}^{-1})$:

$$\kappa = \frac{\varepsilon}{\varepsilon_0} \tag{1}$$

For that reason the dielectric constant is also known as the *relative permittivity* (with symbol ε_r). The dielectric constant is determined with a capacitance cell, incorporating a capacitor of fixed dimensions in a suitable containment vessel. If C is the capacitance of the cell when the medium between the capacitor plates is the solution or gas of interest and C_0 is the capacitance when the medium is a vacuum, then[1]

$$\kappa = \frac{C}{C_0} \tag{2}$$

The present experiment deals with polar molecules in solution in a nonpolar solvent. Experiment 30 deals with polar molecules in a gas. The basic theoretical framework is the same for both.

Either of two systems can be studied in this experiment: (1) *o*- and *m*-dichlorobenzene, or (2) succinonitrile and propionitrile. In the first case, the results can be compared with the vector sum of C—Cl bond moments obtained from the known dipole moment of monochlorobenzene. In the second case, one obtains direct information about internal rotation in a simple 1,2-disubstituted ethane

The angle ϕ is defined so that $\phi = 180°$ when the two C≡N groups eclipse each other (on viewing the molecule along the C—C axis). Thus $\phi = 0$ corresponds to the most stable trans configuration, and $\phi = \pm 120°$ corresponds to two equivalent gauche configurations. Hindered rotation about the C—C sigma bond can be taken into account by defining temperature-dependent mole fractions in the trans, gauche plus, and gauche minus states. The dipole moment of each of these forms can be predicted from the C—C≡N bond moment measured in propionitrile (ethyl cyanide).

THEORY

An electric dipole[2] consists of two point charges, $-Q$ and $+Q$, with a separation represented by a vector **r**, the positive sense of which is from $-Q$ to $+Q$. The electric dipole moment is defined by

$$\mathbf{m} = Q\mathbf{r} \tag{3}$$

A molecule possesses a dipole moment whenever the center of gravity of negative charge does not coincide with the center of gravity of positive charge. In an electric field, all molecules have an induced dipole moment (which is aligned approximately parallel to the field direction) owing to distortion polarization. In addition *polar molecules* have a *permanent dipole moment* (i.e., a dipole moment that exists independent of any applied field) of constant magnitude μ and a direction that is fixed relative to the molecular skeleton.

The resultant (vector sum) electric moment of the medium per unit volume is known as the *polarization* **P**.[3] For an isotropic medium, **P** is parallel to the electric field intensity **E**, and to a first approximation it is proportional to **E** in magnitude. For a pure substance, **P** is given by

$$\mathbf{P} = \bar{\mathbf{m}}\frac{N_0}{\tilde{V}} = \bar{\mathbf{m}}\frac{N_0\rho}{M} \tag{4}$$

where $\bar{\mathbf{m}}$ is the average dipole moment of each molecule, N_0 is Avogadro's number, $\tilde{V}$ is the molar volume, M is the molar mass, and ρ is the density.

The moment of a polarized dielectric is equivalent to a moment that would result from electric charges of opposite sign on opposite surfaces of the dielectric. In a capacitor these "polarization charges" induce equal and opposite charges in the metal plates that are in contact with them. These induced charges are in addition to the charges that would be present at the same applied potential for the capacitor with a vacuum between the plates. Accordingly, the capacitance is increased by the presence of a polarizable medium. Thus the dielectric constant κ is greater than unity. By electrostatic theory it can be shown[4] that

$$\mathbf{D} = \varepsilon\mathbf{E} = \varepsilon_0\mathbf{E} + \mathbf{P} \tag{5a}$$

or

$$\kappa\mathbf{E} = \mathbf{E} + \frac{1}{\varepsilon_0}\mathbf{P} \tag{5b}$$

where $\mathbf{D}$ is the electric displacement and $\mathbf{E}$ is the electric field strength.

The average dipole moment $\bar{\mathbf{m}}$ for an atom or molecule in the medium is given by

$$\bar{\mathbf{m}} = \alpha\mathbf{F} \tag{6}$$

where we denote by $\mathbf{F}$ the *local* electric field and α is the polarizability (which is independent of field intensity if the field is not so intense that saturation is incipient). To a good approximation, $\mathbf{F}$ may be taken as the electric field intensity at the center of a spherical cavity in the dielectric within which the atom or molecule is contained. Thus $\mathbf{F}$ is the resultant of $\mathbf{E}$ and an additional contribution due to the polarization charges on the surface of the spherical cavity; electrostatic theory[4] gives

$$\mathbf{F} = \mathbf{E} + \frac{1}{3\varepsilon_0}\mathbf{P} \tag{7}$$

Combining this with Eq. (5b) we obtain

$$\mathbf{F} = \frac{\kappa + 2}{\kappa - 1}\frac{1}{3\varepsilon_0}\mathbf{P} \tag{8}$$

and using Eqs. (4) and (6) we obtain, for a pure substance,

$$\frac{\kappa - 1}{\kappa + 2}\frac{M}{\rho}\frac{1}{3\varepsilon_0}N_0\alpha \equiv P_M \tag{9}$$

This is the *Clausius–Mossotti* equation. The quantity P_M is called the *molar polarization* and has the dimensions of volume per mole.

If the molecules have no permanent dipole moment, only distortion polarization takes place. The corresponding polarizability is denoted by α_0. If each molecule has a permanent dipole moment of magnitude μ, there is a tendency for the moment to become oriented parallel to the field direction, but this tendency is almost completely counteracted by thermal motion, which tends to make the orientation random. The component of the moment in the field direction is $\mu \cos \theta$, where θ is the angle between the dipole orientation and the field direction. The potential energy U of the dipole in a local field of intensity $\mathbf{F}$ is $-(\mu \cos \theta)F$, which is small in comparison with kT under ordinary experimental conditions. By use of the Boltzmann distribution, the average component of the permanent moment in the field direction is found to be

$$\bar{m}_\mu = [\mu \cos \theta \, e^{-U/kT}]_{av} = [\mu \cos \theta \, e^{\mu F \cos \theta/kT}]_{av}$$

$$\cong \mu\left[\cos \theta\left(1 + \frac{\mu F \cos \theta}{kT}\right)\right]_{av}$$

where the average is taken over all orientations in space. The average of $\cos\theta$ vanishes, but the average of its square is $\frac{1}{3}$; accordingly, as found by Debye,

$$\bar{m}_\mu = \frac{\mu^2}{3kT}F \qquad (10)$$

Thus the total polarizability is given by

$$\alpha = \alpha_0 + \frac{\mu^2}{3kT} \qquad (11)$$

and the molar polarization can be written in the form

$$P_M = \frac{\kappa-1}{\kappa+2}\frac{M}{\rho} = \frac{1}{3\varepsilon_0}N_0\left(\alpha_0 + \frac{\mu^2}{3kT}\right)$$
$$= P_d + P_\mu \qquad (12)$$

where P_d and P_μ are, respectively, the distortion and orientation contributions to the molar polarization:

$$P_d = \frac{1}{3\varepsilon_0}N_0\alpha_0 \qquad \text{and} \qquad P_\mu = \frac{1}{3\varepsilon_0}N_0\frac{\mu^2}{3kT} \qquad (13)$$

Equation (12) is known as the *Debye–Langevin equation.*

Immediately it is clear that a plot of P_M versus $1/T$ utilizing measurements of κ as a function of temperature will yield both α_0 and μ. This technique is readily applicable to gases. In principle it is applicable to liquids and solutions also, but it is seldom convenient owing largely to the small temperature range accessible between the melting point and boiling point.

In the foregoing derivation, a static (dc) electric field was assumed. The equations apply also to alternating (ac) fields, provided the frequency is low enough to enable the molecules possessing permanent dipoles to orient themselves in response to the changing electric field. Above some frequency in the upper radio-frequency or far-infrared range, the permanent dipoles can no longer follow the field and the orientation term in Eq. (12) disappears. At infrared and visible frequencies, the dielectric constant cannot be measured by the use of a capacitor. However, it is known from electromagnetic theory that, in the absence of high magnetic polarizability (which does not exist at these frequencies for any ordinary materials),

$$\kappa = n^2$$

where n is the *index of refraction.* We then obtain from Eq. (12) the relation of Lorentz and Lorenz:

$$R_M \equiv \frac{n^2-1}{n^2+2}\frac{M}{\rho} = \frac{1}{3\varepsilon_0}N_0\alpha_0 = P_d \qquad (14)$$

where R_M is known as the *molar refraction.* Thus the distortion polarizability α_0 can in principle be obtained from a measurement of the refractive index at some wavelength in the far infrared where the distortion polarization is virtually complete. However, it is not experimentally convenient to measure the index of refraction in the infrared range. The index of refraction n_D measured with the visible sodium D line can usually be used instead. Although the atomic contribution to the distortion polarization is absent in the visible, and the electronic contribution is not in all cases at its dc value, these variations in the distortion polarization are small and usually negligible in comparison with the orientation polarization. The measurement of n_D can be made conveniently in an Abbe or other type of refractometer (see Chapter XIX). Thus P_d can be measured independently from P_M, and P_μ can be determined by difference. By this means μ can be determined from measurements made at a single temperature.

Measurements in Solution. We are concerned here with a dilute solution containing a polar solute 2 in a nonpolar solvent 1. The molar polarization can be written

$$P_M = X_1 P_{1M} + X_2 P_{2M} = \frac{\kappa - 1}{\kappa + 2} \frac{(M_1 X_1 + M_2 X_2)}{\rho} \tag{15}$$

where the X's are mole fractions, M's are molar masses, and κ and ρ (without subscripts) pertain to the solution. Since a nonpolar solvent has only distortion polarization, which is not greatly affected by interactions between molecules, we can take P_{1M} to have the same value in solution as in the pure solvent:

$$P_{1M} = \frac{\kappa_1 - 1}{\kappa_1 + 2} \frac{M_1}{\rho_1} \tag{16}$$

We can then get P_{2M} from

$$P_{2M} = \frac{1}{X_2} (P_M - X_1 P_{1M}) \tag{17}$$

obtained by rearrangement of Eq. (15).

Values of P_{2M} calculated using Eq. (17) are found to vary with X_2, generally increasing as X_2 decreases. This effect arises from strong solute–solute interactions due to the permanent dipoles. This difficulty can be eliminated by extrapolating P_{2M} to infinite dilution ($X_2 = 0$) to obtain P_{2M}^0. Although this could be done by plotting P_{2M} for each of a series of solutions against X_2, it is easier and more accurate to follow the procedure of Hedestrand,[5] which is given below.

Let us assume a linear dependence of κ and ρ on the mole fraction X_2:

$$\kappa = \kappa_1 + aX_2 \tag{18}$$

$$\rho = \rho_1 + bX_2 \tag{19}$$

On writing out Eq. (17) explicitly and using Eqs. (16), (18), and (19), it is possible to rearrange terms and obtain the *limiting* expression

$$P_{2M}^0 = \frac{3M_1 a}{(\kappa_1 + 2)^2 \rho_1} + \frac{\kappa_1 - 1}{(\kappa_1 + 2)\rho_1} \left(M_2 - \frac{M_1 b}{\rho_1} \right) \tag{20}$$

Thus measurements on solutions of the *slope* of κ versus X_2 and the *slope* of ρ versus X_2 enable us to calculate the limiting molar polarization of the solute in solution. If we assume that the molar distortion polarization in an infinitely dilute solution is equal to that in the pure solute, then

$$P_{2d}^0 = R_{2M} = \frac{n_2^2 - 1}{n_2^2 + 2} \frac{M_2}{\rho_2} \tag{21}$$

where n_2 is the index of refraction and ρ_2 is the density measured for the solute in the pure state. From these two expressions, we can obtain the molar orientation polarization of the solute at infinite dilution:

$$P_{2\mu}^0 = P_{2M}^0 - P_{2d}^0 = \frac{1}{3\varepsilon_0} N_0 \frac{\mu^2}{3kT} \tag{22}$$

On substituting the numerical values of the physical constants, we obtain

$$\mu = 42.7(P^0_{2\mu}T)^{1/2} \times 10^{-30} \text{ C m} \tag{23a}$$

$$= 12.8(P^0_{2\mu}T)^{1/2} \text{ debye} \tag{23b}$$

where $P^0_{2\mu}$ is given in $m^3 \text{ mol}^{-1}$ units and T is in kelvin. The unit of molecular dipole moment universally used by chemists is the debye (D), historically defined as 10^{-18} esu cm, where esu is the electrostatic unit of charge having the value 3.33564×10^{-10} C. Thus $1 \text{ D} = 3.33564 \times 10^{-30}$ C m.

An alternative expression for $P^0_{2\mu}$ has been given by Smith,[6] who improved a method of calculation first suggested by Guggenheim.[7] This expression presupposes knowledge of the index of refraction n of the *solution*, and assumes that

$$n^2 = n_1^2 + cX_2 \tag{24}$$

where c, like a and b, is a constant determined by experiment. It follows[6] that

$$P^0_{2\mu} = \frac{3M_1}{\rho_1} \left[\frac{a}{(\kappa_1 + 2)^2} - \frac{c}{(n_1^2 + 2)^2} \right] \tag{25}$$

This expression is an approximate one that holds when $\kappa_1 - n_1^2$ is small (as in the case of benzene, where it is 0.03) and M_2 and b are not too large. It is a useful form when the index of refraction n_2 of the pure solute is inconvenient to determine; this is often the case when the pure solute is solid. Note also that Eq. (25) does not require a knowledge of the densities of the solutions.

Solvent Effects.   The values of P^0_{2M} obtained from dilute solution measurements differ somewhat from P_M values obtained from the pure solute in the form of a gas. This effect is due to solvent–solute interactions in which polar solute molecules induce a local polarization in the nonpolar solvent. As a result μ as determined in solution is often smaller than μ for the same substance in the form of a gas, although it may be larger in other cases. Usually the two values agree within about 10 percent. A discussion of solvent effects is given in Ref. 4.

METHOD

A capacitance cell suitable for work with liquids or solutions is shown in Fig. 1; it is made with a small variable-air capacitor of the type formerly in common use in radios and electronic circuits. It should have a maximum capacitance of 50 to 200 pF. This device is more convenient than a fixed-plate capacitor, since with the latter device it is necessary to measure separately the stray capacitance due to electrical leads, etc. In the cell shown, the variable capacitor is used in two positions: fully closed (maximum capacitance) and fully open (minimum capacitance); these positions are defined by mechanical stops for the pointer on the knob that rotates the capacitor shaft.† The difference ΔC between the closed (b) and open (a) positions is independent of the stray capacitance. Thus the dielectric constant of the liquid or solution is given by

$$\kappa = \frac{\Delta C(\text{liq})}{\Delta C(\text{air})} \tag{26}$$

---

†For high-precision work with very stable electronics, a fixed-plate capacitor cell is preferable, since it eliminates any error due to lack of reproducibility in setting a variable capacitor. However, in this case, one must determine the capacitance of the leads from measurements on a material of known dielectric constant.

FIGURE 1

Variable-capacitance
dielectric cell for solutions.

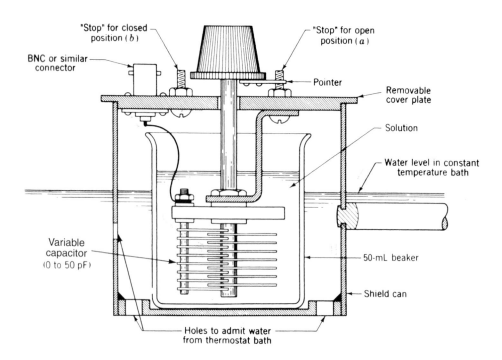

where $\Delta C = C_b - C_a$. For liquids and solutions it is possible to use air instead of vacuum, since the dielectric constant of air is very close to unity.

We now turn our attention to methods of measuring the capacitance. A very precise method, and one that is particularly useful when the liquid or solution has a high conductance, involves the use of a capacitance bridge.[8] For the present system, in which the conductance is small, less expensive instruments can be used. Among these are digital capacitance meters made by a variety of manufacturers (see listings given by Tucker Electronics Co., Garland, TX) with capacitance ranges down to 1 to 200 pF, readability of ±0.1 pF, and accuracy of 1 pF or better. These are very simple to operate, give capacitance readings directly, and are suitable for use in the present experiment, but their use gives little insight into the physical principles involved in capacitance measurement.

Several traditional methods are based on use of an electrical oscillator incorporating an inductance L and a capacitance C in parallel; this combination is known as a "tank" circuit. The frequency of such an oscillator is given by

$$f = \frac{1}{2\pi\sqrt{LC}} \tag{27}$$

The inductance L is fixed, while the capacitance C is the sum of several separate capacitances: that of the conductance cell, C_X; that of a separate variable tank capacitor incorporated in the oscillator, C_T (if present); and stray capacitances due to leads, etc., C_S. Also present in some methods is a precision air capacitor C_P graduated directly in capacitance units (generally picofarads), used when the equipment is operated in a null mode. In this mode the precision air capacitor is used to bring the frequency f to the same value in each measurement, and the change in reading of the precision air capacitor is then equal in magnitude to the change in capacitance of the cell.

An important and precise traditional method of measuring capacitance for dipole moment determinations is the *heterodyne beat* method, a particular form of the null method mentioned above. The output of an LC oscillator incorporating the capacitance cell and a precision variable air capacitor is mixed with the output of a fixed-frequency oscillator to produce a difference frequency in the audio range. This is typically applied to the vertical plates of an oscilloscope while the fixed-frequency output of a stable audio oscillator is applied to the horizontal plates. In each determination the precision air capacitor is adjusted so that the two audio frequencies are equal, as indicated by the appropriate Lissajous figure (circle, ellipse, or slant line) on the oscilloscope screen; see Fig. XIX-6. An instrument that has been widely used for this purpose, incorporating the fixed oscillator, the variable oscillator, and the mixer, is the WTW Dipolmeter, model DM01. This instrument unfortunately is no longer produced; however, such instruments are present in many physical chemistry laboratories. This instrument requires the minor modification of removing an existing capacitor from the tank circuit and introducing connections for leads for external capacitors, namely, the capacitance cell and the precision air capacitor (such as GenRad 1304B).

With the advent of inexpensive, fast frequency counters, which count the individual cycles over a precisely fixed period (usually 1 s) and display the frequency digitally, it is more convenient to connect the radio-frequency output of the variable-frequency oscillator directly to the frequency counter and determine the total capacitance with the aid of Eq. (27). This technique is highly suitable for the present experiment if a WTW Dipolmeter or another LC oscillator is available or can be constructed. (With the Dipolmeter only the variable-frequency oscillator is used.) A simple LC oscillator circuit that can be constructed from inexpensive components has been described by Bonilla and Vassos:[9] this circuit, with a small modification to provide for one side of the tank to be grounded, is shown in Fig. 2. In this circuit, as in the WTW Dipolmeter circuit, all tank capacitances are in *parallel*. [This is *not* true of the circuit described in Ref. 4 of Exp. 30, as that circuit incorporates some *series* capacitance. If that circuit is employed, Eqs. (28) to (30) are not valid and Eqs. (30-3) to (30-5) must be used instead, unless the null mode is employed.]

It follows from Eq. (27) that

$$C = C_X + C_T + C_S = \frac{1}{4\varepsilon\pi^2 Lf^2} \tag{28}$$

The capacitance of the cell can be determined from measurements of the frequency at the closed (*b*) and open (*a*) positions using the following equation, in which the capacitances C_T and C_S cancel out since they are held constant:

$$C_X = C(b) - C(a) = \frac{1}{4\pi^2 L}\left(\frac{1}{f_b^2} - \frac{1}{f_a^2}\right) \tag{29}$$

To eliminate the apparatus constant $1/4\pi^2 L$ and determine the dielectric constant, we combine this with Eq. (26) and obtain

$$\kappa = \left(\frac{1}{f_b^2} - \frac{1}{f_a^2}\right)_{sample} \bigg/ \left(\frac{1}{f_b^2} - \frac{1}{f_a^2}\right)_{air} \tag{30}$$

Another type of electronic circuit, the relaxation oscillator, can be used to measure capacitance; a simple apparatus for this purpose has been described by Kurtz, Anderson, and Willeford.[10]

EXPERIMENTAL

Although several methods for making the capacitance measurements have been suggested, we will limit our discussion to the use of the LC oscillator and frequency counter shown in Fig. 2.

FIGURE 2

A simple oscillator circuit for capacitance measurements, adapted from Bonilla and Vassos.[9] The transistor is 1N3904 or a similar type. Suggested parameters: radio-frequency transformer T, 0.6 mH; R_1 and R_2, 12 kΩ; C, 300 pF; C_T, 0 to 100 pF. The circuit should be housed in a metal box for shielding, and connections with the capacitance cell and the frequency counter should be made with shielded coaxial cable with BNC or similar connectors.

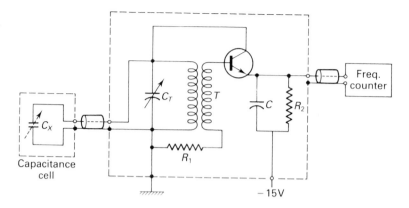

With the cell capacitor and cell beaker clean and dry, assemble the cell and mount it in a constant-temperature bath set at 25°C. Set the cell capacitor at the open position (a). Turn on the oscillator and the frequency counter, and adjust the tank capacitance C_T to yield a frequency in the range of 1.3 to 1.5 MHz. Wait a while to make sure that the apparatus is operating stably and not drifting in frequency. Determine the frequencies f_a and f_b in alternation at intervals of 30 s by alternating the position of the pointer knob between (a) and (b) until four to six frequency values have been recorded at each position. Make certain that you do not alter C_T or move the leads so as to affect C_S during either set of measurements.

Rinse the cell beaker and variable capacitor with the pure nonpolar solvent, fill the cell beaker to a level that will completely immerse the capacitor plates, and reassemble the cell. Make measurements in the open and closed positions as described above. Repeat this procedure with all the solutions of the polar solute in the nonpolar solvent. Note the temperature of the bath.

Handle the capacitor with care; damage to the plates will affect the frequency values.

Average the frequency readings for each determination, and calculate the 95 percent confidence limit for each mean (see Chapter IIB).

Dichlorobenzene.   Measurements should be made on pure benzene and on dilute solutions of o- and m-dichlorobenzene. Make up 50 or 100 mL of each solution as follows. Weigh a dry, clean volumetric flask; add an appropriate amount of solute using a Mohr pipette; and weigh again. Now carefully make up to the mark with benzene and reweigh. Suggested concentrations are 1, 2, 3, and 4 mole percent of solute. The densities can be calculated from these weighings. If desired, the refractive index of each solution can also be measured with an Abbe refractometer so that Guggenheim's method of calculation can be used.

Succinonitrile.   Measurements should be made on pure benzene and on dilute solutions of succinonitrile and propionitrile (or acetonitrile). Make up 50 or 100 mL of each solution in the same manner as described above. Densities can be calculated and/or refractive indexes can be measured.

CALCULATIONS

For each solution studied, calculate κ from Eq. (30) as well as ρ and X_2. Plot κ and ρ versus the mole fraction of solute X_2, and draw the best straight lines through your points; see Eqs. (18) and (19). Obtain the slopes a and b; the intercepts should agree with the pure solvent results. Using Eq. (20), calculate P_{2M}^0, the molar polarization at infinite dilution. Estimate P_{2d}^0 from Eq. (21) using the literature value of the refractive index of the solute n_2, and obtain $P_{2\mu}^0$.

Alternatively, Guggenheim's method can be used if the indexes of refraction were measured for each solution. Plot κ and n^2 versus X_2 and draw the best straight lines through your points; see Eqs. (18) and (24). Using Eq. (25) and the appropriate slopes a and c, calculate $P_{2\mu}^0$. It would be instructive to follow both of these procedures and see what difference (if any) the method of extrapolation to infinite dilution has on the $P_{2\mu}^0$ values.

Finally, calculate the dipole moment from Eq. (23) in units of 10^{-30} C m, and report it also in debye units (D).

DISCUSSION

Dichlorobenzene.   Compare your experimental results with the values computed from a vector addition of carbon–chlorine bond moments obtained from the dipole moment of monochlorobenzene (5.17×10^{-30} C m, or 1.55 D). If they do not agree, suggest possible physical reasons for the disagreement.

Succinonitrile.   An excellent discussion of this system has been presented by Braun, Stockmayer, and Orwoll,[11] who were the first to propose studying the dipole moment of succinonitrile. They show that the average square of the dipole moment is given by

$$\langle \mu^2 \rangle = \tfrac{8}{3}(1 - X_t)\mu_1^2 \tag{31}$$

where X_t is the mole fraction in the trans form and μ_1 is the C—C≡N bond moment as determined from the dipole moment of either propionitrile or acetonitrile. Thus a measurement of μ^2 for succinonitrile will determine the distribution of molecules among the three "rotational isomeric" states at the given temperature. [Note that $X_+ = X_- = \tfrac{1}{2}(1 - X_t)$, since the gauche plus and gauche minus states are equivalent.] It can also be shown from the appropriate Boltzmann populations that

$$X_t = (1 + 2e^{-\Delta E/RT})^{-1} \tag{32}$$

where $\Delta E = E(\text{gauche}) - E(\text{trans})$. Thus the value of X_t obtained from Eq. (31) will determine the value of ΔE and one can predict the temperature dependence of μ. Calculate a value of X_t and ΔE from your results and predict the value of μ at 280 and 350 K.

Theoretical Calculations.   As a project exercise for the class, dipole moments of the molecules studied might be calculated using an ab initio program such as Gaussian (see discussion of this program in Chapter III). Calculations at the Hartree–Fock level using the STO-3G basis set will be reasonably fast but will generally yield low values for the dipole moments of substituted benzenes.[12] Use of a 6-31G* or higher basis set would give better results, and this might be explored if computer time permits. Do the calculation also for chlorobenzene and use the calculated dipole moments of the three compounds to test the idea of bond dipole addition. A similar test could be done for succinonitrile using the dipole calculated for propionitrile or acetonitrile. It would also be interesting to calculate the energies of the trans and gauche forms of succinonitrile for comparison of the difference with your experimental ΔE value.

SAFETY ISSUES

Benzene can have both chronic and acute toxic effects. The risk of acute effects is low, since acute symptoms occur only at 1000 ppm or higher. *Chronic* vapor inhalation at the level of 25 to 50 ppm can cause changes in blood chemistry, and *continual* exposure at 100 ppm can cause severe blood disorders. The OSHA exposure limits for benzene vapor are 1 ppm as an 8-hour time-weighted average and a ceiling of 50 ppm for no more than 10 min. In order to reach the level of 10 ppm in a laboratory of 750 m^3 volume, 23 g of liquid benzene would have to evaporate into a closed atmosphere. Thus the hazards associated with the infrequent use of liquid benzene in a well-ventilated laboratory are very low.

Chlorobenzenes are also toxic substances that must be handled with care. Both benzene and solutions containing benzene or chlorobenzenes must be disposed of properly in a designated waste container.

APPARATUS

Capacitance cell as shown in Fig. 1; oscillator as described in the text and Fig. 2; frequency counter (range at least 0.5 to 5 MHz); shielded coaxial cables with connectors; five 50- or 100-mL volumetric flasks; a 5-mL Mohr pipette; acetone wash bottle; rubber pipette bulb.

Benzene (analytical grade, 500 to 1000 mL; o- and p-dichlorobenzene (10 to 20 mL each), or succinonitrile and propionitrile (10 to 20 mL each); constant-temperature bath set at 25°C.

REFERENCES

1. P. W. Atkins and J. de Paula, *Physical Chemistry,* 8th ed., p. 628, Freeman, New York (2006).

2. *Ibid.,* p. 620.

3. *Ibid.,* p. 623.

4. P. W. Atkins and R. S. Friedman, *Molecular Quantum Mechanics,* 3d ed., pp. 393–395, Oxford Univ. Press, Oxford (1997).

5. G. Hedestrand, *Z. Phys. Chem.* **B2,** 428 (1929).

6. J. W. Smith, *Trans. Faraday Soc.* **46,** 394 (1950).

7. E. A. Guggenheim, *Trans. Faraday Soc.* **45,** 714 (1949).

8. J. M. Pochan, J. J. Fitzgerald, and G. Williams, "Experimental Methods for Chemists: A Simplified Approach. Dielectric Properties of Polymers and Other Materials," in B. W. Rossiter and R. C. Baetzold (eds.), *Physical Methods of Chemistry,* 2d ed., vol. VIII, chap. 6, Wiley-Interscience, New York (1993).

9. A. Bonilla and B. Vassos, *J. Chem. Educ.* **54,** 130 (1977).

10. S. R. Kurtz, O. T. Anderson, and B. R. Willeford Jr., *J. Chem. Educ.* **54,** 181 (1977).

11. C. L. Braun, W. H. Stockmayer, and R. A. Orwoll, *J. Chem. Educ.* **47,** 287 (1970).

12. W. J. Hehre, L. Radom, P. v.R. Schleyer, and J. A. Pople, *Ab Initio Molecular Orbital Theory,* p. 334, Wiley. New York (1986). Out of print but available from Gaussian, Inc.; see http://www. gaussian.com/allbooks.htm.

GENERAL READING

A. Chelkowski, *Dielectric Physics,* Elsevier, Amsterdam (1980).

C. J. F. Böttcher, O. C. Van Belle, P. Bordewijk, and A. Rip, *Theory of Electric Polarization,* Elsevier, Amsterdam (1978).

EXPERIMENT 30
Dipole Moment of HCl Molecules in the Gas Phase

The permanent dipole moment μ of a polar molecule is determined in Exp. 29 from measurements of the dielectric constant of a solution containing such molecules as solute. In the present experiment, the permanent dipole moment of a gas molecule is determined. The orientation polarization can be separated from the distortion polarization by means of measurements at more than one temperature, making use of the fact that the former is temperature dependent while the latter is not. An alternative method, which is recommended for this experiment, is to obtain the orientation polarization by subtracting from the molar polarization the distortion polarization as determined separately from the refractive index of the gas, which is determined by means of a laser interferometer.[1] Thus the molar polarization needs to be determined at only one temperature.

The gas recommended is hydrogen chloride, HCl, because it has a large permanent dipole moment and because it has a boiling point low enough ($-83.7°C$) to permit measurements of dielectric constant to be made down to Dry Ice temperature ($-78.5°C$) if the option of measuring the dielectric constant at more than one temperature is chosen.

Thus the experiment consists of two parts: the measurement of the dielectric constant of the gas by a method similar in principle to that of Exp. 29, and the measurement of the refractive index with an interferometer. The measurement of a dielectric constant at more than one temperature constitutes a complete stand-alone experiment, independent of the refractive index measurement; the measurement of refractive index can be used as a stand-alone experiment if a value for the low-frequency dielectric constant is supplied to permit the calculation of the dipole moment.

THEORY

The concepts of permanent dipole moment, induced dipole moment, and molar polarization are discussed in Exp. 29; this material should be reviewed. We assume further that deviations from the perfect-gas law are small in comparison with the experimental uncertainties.

FIGURE 1

Schematic representation of the frequency dependence of the molar polarization of a gas of permanent dipoles. Dashed lines indicate ranges of complex behavior.

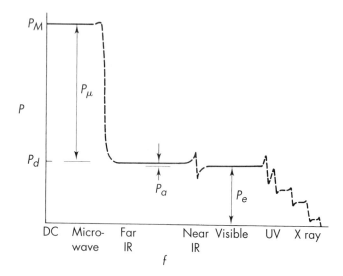

Since the dielectric constant $\kappa \equiv \varepsilon/\varepsilon_0$ is only very slightly greater than unity and perfect-gas behavior is being assumed, we may write Eq. (29-12) in the form

$$\frac{(\kappa - 1)}{3} \frac{RT}{p} = \frac{N_0}{3\varepsilon_0}\left[\frac{\mu^2}{3kT} + \alpha_0\right] \equiv P_M = P_\mu + P_d \tag{1}$$

where N_0 is Avogadro's number and ε_0 is the dielectric permittivity of a vacuum. A plot of P_M versus $1/T$ will yield $N_0\mu^2/9\varepsilon_0 k$ as the slope and $N_0\alpha_0/3\varepsilon_0$ as the intercept.

Properly speaking the distortion polarization P_d is the sum of two parts, the atomic polarization P_a and the electronic polarization P_e. The atomic polarization results from distortion of the nuclear framework of the molecule in response to the electric field, while the electronic polarization results from distortion of the electron cloud on a time scale short compared to that in which the nuclei are able to move.

A separation of the molar polarization P_M into its three parts, P_μ, P_a, and P_e can in principle be accomplished by measurements of the dielectric constant at three different frequencies, as shown in Fig. 1. At zero-frequency (dc) and low-frequency alternating electric field, all three parts are present: The molecules have time to reorient, the nuclei have time to move with respect to each other, and the electron cloud has time to distort. At frequencies corresponding to wavenumbers in the far-infrared range of about 20 to 300 cm^{-1}, the molecules can absorb energy and change their rotational states, since their rotations are in or close to resonance with the alternating field. At frequencies well above this range, the molecules can no longer reorient in response to the field. In principle $P_d \equiv P_a + P_e$ can be measured directly with radiation in the infrared frequency range above 300 cm^{-1}. In practice this is not done, since this frequency range is too high for dielectric constant measurements and is too low for refractive index measurements with any convenient means (from the points of view of radiation sources and detectors and the very long path length required). At about 3000 cm^{-1} in the infrared range, the molecule can absorb vibrational energy from the alternating field and become promoted to the first excited vibrational state. Well above this frequency, the nuclei can no longer respond to the alternating field; thus P_e can be measured in the visible range (with index of refraction measurements). In the ultraviolet and x-ray ranges electronic excitations can take place; with very hard X rays even the electrons are unable to respond to the rapidly alternating field, and all polarization becomes negligible.

The dielectric constant becomes complex at frequencies near those at which the above-mentioned resonances and excitations take place. Measurements should be taken at frequencies that differ from such special frequencies by at least one order of magnitude.

If P_μ and P_d are separated through the temperature dependence of P_M, and P_e is determined at the frequency of visible light, it is possible in principle to determine P_a as $P_d - P_e$. Smyth[2] gives a value of $P_a = 0.5$ cm^3 for HCl gas, determined in this way. However, analysis of literature data[1,3] indicates that this value, which is a small difference between large quantities, carries an uncertainty possibly as large as the value itself. Moreover, theoretical considerations (see Discussion) indicate that P_a should be very much smaller than this value. We will for this experiment assume that P_a is negligible in comparison to the experimental uncertainties in P_d and P_e, and reserve further consideration of P_a for the Discussion. Since the magnetic permeability of HCl gas is the same as that of a vacuum, $\kappa = n^2$ and we may use Eq. (29-14). Since the refractive index n of a gas at ordinary pressures is only very slightly greater than unity and since perfect-gas behavior is being assumed, we may rewrite that equation in the form

$$P_d \simeq P_e = R_M = \frac{2(n-1)}{3} \frac{RT}{p} = \frac{N_0}{3\varepsilon_0} \alpha_0 \qquad (2)$$

The interferometric determination of the index of refraction consists of determining the increase in the number of wavelengths in the light path of a Michelson interferometer as the gas is admitted to a previously evacuated cell that is placed in one of the two mutually perpendicular beam arms of the interferometer.

EXPERIMENTAL

Dielectric Constant. The measurement of the dielectric constant of a gas is more difficult than the measurement of that of a solution. Because of the low density of a gas at ordinary pressures, its dielectric constant differs from that of a vacuum by a very small amount, of the order of a part per thousand. This requires a much higher order of experimental precision in the instrumentation used and in the procedure followed. To measure $(\kappa - 1)$ to 1 percent accuracy with a capacitance cell requires that the capacitance be measurable to a precision and reproducibility of one part in 10^5 or (preferably) better. This rules out the use of inexpensive capacitance meters for determining the small capacitance difference between the evacuated capacitance cell and the same cell containing a gas. The use of a high-resolution capacitance bridge (e.g., GenRad 161JA or Wayne-Kerr 4210) is probably ruled out for this experiment unless one is lucky enough to be able to borrow one from a nearby research laboratory, as they are extremely expensive. For the experiment as described here, a dimensionally stable and well-shielded capacitance cell, a very stable LC oscillator, and an accurate frequency counter are required.

The stable LC oscillator contained in the WTW Dipolmeter mentioned in Exp. 29 can be used if that equipment is available. The oscillator circuit shown in Fig. 29-2 is probably not sufficiently stable; it lacks an amplifier stage to isolate the oscillator stage from the output load. A very stable solid-state LC oscillator for operation at about 1.5 to 2 MHz, constructed inexpensively from a published circuit diagram,[4] has been found to be satisfactory. This should be built in a metal box (which will serve as an electrical shield), with two BNC connectors—one for connecting with the cell and one for the output connection to the frequency counter. All such connections must be made with shielded cables. The capacitance cell replaces the capacitances shown in the *LC* tank circuit of the published diagram.† For best performance the temperature inside the oscillator box should be controlled with a small heating element, a temperature sensor, and a proportionating circuit (see Chapter XVII).

Suitable gas capacitance cells are not available commercially; it is necessary to construct one. A design for a capacitance cell comprising a cylindrical capacitor contained in a Pyrex glass jacket was given in the second edition of this book. A capacitance cell of superior stability, comprising a cylindrical capacitor in a metal jacket, is shown in Fig. 2. A satisfactory material for the cell is stainless steel. Although that material is not immune to corrosion by hydrogen chloride gas, our experience with a stainless steel cell has been good. Monel or Inconel, which are superior in corrosion resistance, are much more expensive materials. The cell is demountable for cleaning (in case of corrosion) and for changing the number of plates if necessary to obtain the optimum capacitance (200 to 250 pF). The vacuum seal uses a gasket made from a wire-form hard solder such as Eutectoid 157; the wire solder is bent into a circle with the ends overlapping; and when the screws are uniformly tightened, the ends of the wire cold-weld together.

The cell and oscillator circuit should oscillate stably in the range 1 to 3 MHz and should have a frequency drift of less than 1 ppm per minute. The cell temperature should be held constant to 0.1°C by submerging it in a bath of water (or acetone and Dry Ice for low temperatures) contained in a Dewar flask. If the frequency is sensitive to placement of nearby objects (including the experimenter's hands), look for defects in the electrostatic shielding; all visible parts of the apparatus should be at ground potential.

This cell has high thermal inertia. After transient changes in the internal temperature resulting from filling or evacuating the cell, thermal equilibrium is restored very slowly. This is particularly notable when a gas at room temperature is admitted to the cell at Dry Ice temperature; in this case, effective restoration of equilibrium may require as much as 10 or 15 min. If the oscillator is stable with a small *but constant* drift rate, the approach to thermal equilibrium may be monitored by following the frequency. When the drift in oscillator frequency again becomes constant, thermal equilibrium may be considered to have been reached. Thus the frequency should be plotted as a function of time. When it achieves a constant slope, it should be extrapolated to a time that is the same for the determinations being directly compared (i.e., gas and vacuum). *It is important to take measurements for a long enough time to permit a valid extrapolation.*

The recommended procedure for determining the dielectric constant of an unknown gas with the cell described here depends on the relationship of changes in the capacitance of the cell to changes produced in the effective capacitance in the tank circuit. These are not the same because other capacitances are present besides that of the capacitor contained in the cell, C_{cell}. These always include the capacitance of the shielded cable connecting the cell to the oscillator and stray capacitances in the oscillator tank; these are *in parallel* with the cell, and lumped together may be called C_{par}. In addition there is in the circuit described

---

†The tank inductor shown in that diagram may have to be replaced by one having a value or range of values of the inductance L that, with the tank capacitance C_{eff}, gives stable oscillatory behavior at a convenient frequency (preferably about 1.5 to 2 MHz) in accordance with Eq. (29-27). Suitable inductors are available from the J. W. Miller Division of Bell Industries, Compton, CA. Inductors of the 4400 and 4500 series from that source have powdered iron cores; the user-adjustable position of the core determines the inductance L. (The Miller 4508 or 4509 inductor is suitable with the capacitance cell described here.) The magnetic permeability of the core has a negative temperature coefficient, while the thermal expansion of the copper wire in the coils tends by itself to yield a positive temperature coefficient of inductance. Thus the magnitude and sign of the temperature coefficient of inductance are dependent on the core position.

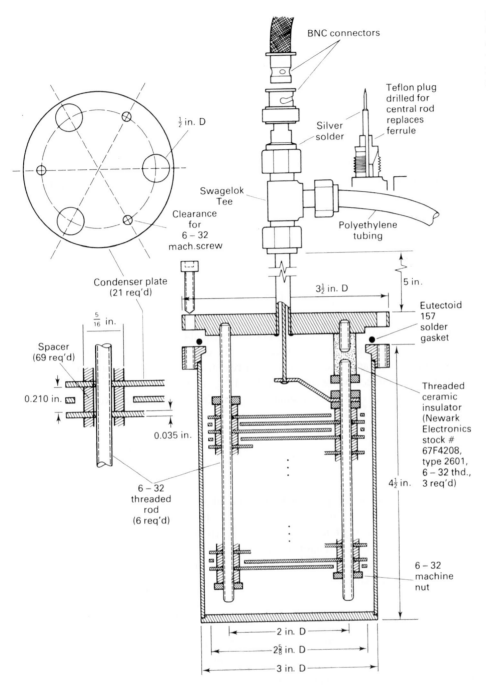

FIGURE 2
Suggested design for gas capacitance cell with a vacuum capacitance of about 250 pF. The recommended material is stainless steel.

in Ref. 4 additional capacitance C_{ser} *in series* with the capacitances already mentioned. On following the rules for combining series and parallel capacitances, we obtain for the effective capacitance in the tank circuit

$$C_{eff} = \frac{(C_{cell} + C_{par})C_{ser}}{C_{cell} + C_{par} + C_{ser}}$$

Only C_{cell} varies with the dielectric constant κ of the medium in the cell. On taking logarithms of both sides of this equation and differentiating with respect to C_{cell}, we obtain

$$\frac{1}{C_{eff}} \frac{dC_{eff}}{dC_{cell}} = \frac{1}{C_{cell} + C_{par}} - \frac{1}{C_{cell} + C_{par} + C_{ser}}$$

$$= \frac{C_{ser}}{(C_{cell} + C_{par})(C_{cell} + C_{par} + C_{ser})} \tag{3}$$

The dependence of the oscillator frequency f on the dielectric constant κ of the gaseous medium in the cell is obtained by differentiating Eq. (29-27) and combining the result with Eq. (3):

$$\frac{df}{d\kappa} = -\frac{f}{2C_{eff}} \frac{dC_{eff}}{dC_{cell}} \frac{dC_{cell}}{d\kappa}$$

$$= -\frac{f}{2}\left[\frac{C_{ser}C_{cell}}{(C_{cell} + C_{par})(C_{cell} + C_{par} + C_{ser})}\right] \equiv -\frac{f}{2}K \tag{4}$$

where K is the quantity in brackets (with C_{cell} set equal to its vacuum value). K is a dimensionless apparatus constant that, once carefully determined, need not be redetermined unless and until the apparatus is changed or readjusted. When C_{ser} approaches ∞ (equivalent to that capacitance being shorted out) and C_{par} approaches zero, K approaches unity.

On integrating over the very small change in dielectric constant, we obtain (reversing signs for convenience)

$$f_{vac} - f_{gas} = \frac{f_{vac}}{2} K(\kappa_{gas} - 1) \tag{5}$$

While in principle it is possible to measure all of the capacitances required for the calculation of K with Eq. (4), it is very difficult to do this in practice with sufficient accuracy. For our purposes it is very much preferable to determine K by measuring the frequency shift produced on introducing a reference gas of known dielectric constant into the previously evacuated cell. (Where students' time is limited, this may be done in advance by the teaching staff.) Dielectric constants of some convenient reference gases are given in Table 1.[5]

TABLE 1   Dielectric constants κ of some reference gases[a]

Gas	$(\kappa - 1) \times 10^6$ at 20°C, 1 atm		
	Radio Frequency	**Microwave**	**Optical**
Ar	514.7	517.7	517.1
Air (dry, CO_2-free)	537.0	536.6	536.3[b]
N_2	547.2	547.8	548.3
CO_2	921.5	921.5	—

[a] From Ref. 5. Microwave and optical values are averages of two or more independent determinations. All values are stated to be uncertain in the final digit.

[b] Averaged from nine independent determinations; 95 percent confidence limits for the average are ±0.3.

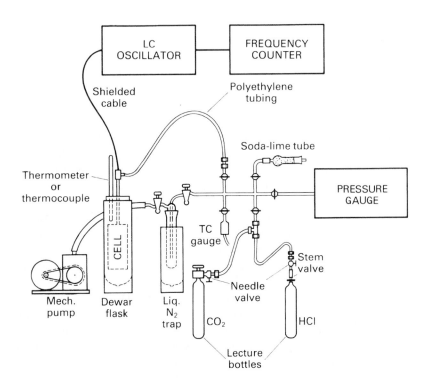

FIGURE 3
Gas-handling system for gas dielectric constant determination. Many connections are made with Swagelok and/or Ultra-Torr fittings. The gas lecture bottles must be securely fastened to the laboratory bench.

The molar polarization of these gases may be regarded as independent of temperature, and $\kappa - 1$ may be considered to be proportional to the density of the gas. Thus, at temperatures and pressures other than 293.2 K and 1 atm, assuming the perfect-gas law,

$$(\kappa - 1)_{T,p} = (\kappa - 1)_{293.2,1} \frac{293.2}{T(\text{K})} p \text{ (atm)} \tag{6}$$

This equation may be considered as valid for the reference gases and also for HCl gas, within the precision of this experiment.

Given a value of κ_{gas} for the chosen reference gas from Table 1, adjusted if necessary to the experimental temperature and pressure with Eq. (6), and given the value of the frequency shift obtained with this gas, Eq. (5) may be used to obtain the apparatus constant K. Once K is known, Eq. (5) may be used to determine the dielectric constant of the subject gas, say HCl, from the frequency shift obtained with that gas.

The overall experimental arrangement, including the gas-handling system, is shown in Fig. 3. A 0- to 1000-Torr pressure gauge is needed for measuring the pressure of the gas in the cell. A corrosion-resistant capacitance manometer (e.g., MKS—highly accurate but expensive) or variable-reluctance manometer (e.g., Validyne—less accurate but also less expensive) should be used. The thermocouple (TC) pressure gauge is handy for leak testing but should not be exposed to HCl gas; it must be isolated with a stopcock. The HCl gas is provided from a lecture bottle fitted with a stem valve and a high-pressure needle valve. Connections between this and polyethylene tubing leading to the system are made with Swagelok fittings; Cajon Ultra-Torr fittings are preferable for all other connections made with polyethylene tubing. Since HCl is corrosive, the needle valve will deteriorate if left in contact with HCl gas for a long period. After the experiment is completed, close the stem valve and either pump on the open needle valve or remove this valve and blow it out with compressed air or dry nitrogen gas.

Several hours before the experiment, the oscillator and the frequency counter should be turned on, and the capacitance cell, which should be in a Dewar flask containing water and a thermometer (or a thermocouple, taped to the cell), should be evacuated. Before any runs are made, the vacuum should be tested by closing off the cell from the vacuum pump and monitoring the pressure for at least 10 min. The frequency of the oscillator should then be recorded, and if the oscillator is drifting it should be recorded at intervals of 30 s. When the drift is steady, HCl gas should be slowly and carefully admitted to a pressure of approximately 1 atm. When equilibrium has been attained as indicated by restoration of a steady drift rate, the frequency should be recorded periodically over a long enough period to permit any needed linear extrapolation, and the pressure and the temperature should be recorded. The cell should then be evacuated, and the frequency difference determined in the same manner. This procedure should be repeated until at least four fillings have been made with HCl gas, yielding eight frequency differences. *During the experiment, great care must be taken to avoid bumping the cell, cable, or oscillator.*

If the dielectric constant is also to be measured at Dry Ice temperature ($-78.5°C$), the evacuated cell should be immersed in a large Dewar flask containing barely enough acetone to submerge the main body of the cell, and crushed Dry Ice should be added slowly until further addition of Dry Ice causes no increase in gas evolution; enough more is added to allow for further vaporization over the time of the experiment. The procedure described for determining the oscillator frequencies should be repeated, with allowance for the additional time that may be required for the attainment of equilibrium at the lower temperature. It is possible also to make measurements above room temperature if means are available to obtain satisfactory temperature control for the duration of the measurements.

Since the capacitance of the cell, C_{cell}, varies somewhat with temperature owing to thermal expansion and contraction of the metal parts, it is advisable to determine the apparatus constant K for each temperature at which the cell is to be used. If this is to be done with CO_2 as the reference gas when the cell is cooled with Dry Ice, care must be taken to limit the filling pressure to significantly less than 1 atm to avoid condensation of CO_2 in the cell.

Refractive Index. The Michelson interferometer divides an incoming beam of light into two beams that are exactly perpendicular to each other. This is done by a beam splitter, a half-silvered plane mirror inclined at 45° to the incoming beam. The two light beams, which are reflected back along the same light paths by adjustable plane mirrors at the ends of the two beam arms, recombine at the beam splitter, where they undergo mutual interference. Depending on the relative phases of the two beams as they meet at the beam splitter, they produce an outgoing beam perpendicular to the incoming beam with an amplitude that is the sum of those of the two beams, or the difference, or some amplitude between these two values. If the transmission of the beam splitter is exactly 50 percent, the minimum amplitude is zero. The beam splitter and mirrors are mounted on a sturdy base for high dimensional stability. One (or both) of the mirrors has adjustment screws for making very small changes in the path length.

Figure 4 shows the interferometer setup. The interferometer is of the type used in undergraduate physics demonstrations and laboratory courses. A small diverging lens (with a focal length of -20 to -50 mm) between the 0.5- to 2-mW He–Ne laser ($\lambda = 632.8$ nm) and the beam splitter of the interferometer diverges the beam so that a large area of the

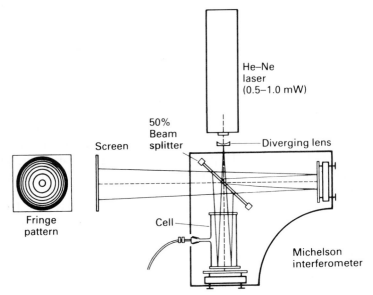

FIGURE 4

Laser interferometer setup
for gas refractive index
determination.

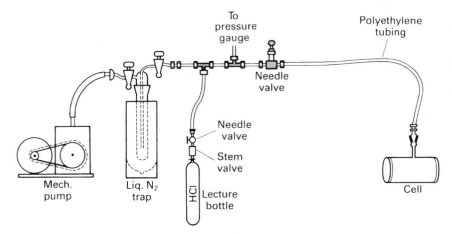

FIGURE 5

Gas-handling system
for gas refractive index
determination. Many
connections are made with
Swagelok and/or Ultra-Torr
fittings. The HCl lecture
bottle must be securely
fastened to the laboratory
bench.

screen is illuminated, and (unless by accident the two beam path lengths are exactly equal)
a pattern of concentric interference fringes appears on the screen. The fringe pattern may
be optimized by adjusting one of the mirrors. (Preliminary adjustment of the mirrors *without* the diverging lens may be necessary to correct the alignment of the reflected beams
if fringes are not apparent.) As the index of refraction changes, the fringes move outward
or inward, appearing or disappearing at the center of the pattern. The direction of fringe
movement with increasing index of refraction depends on the relative effective lengths of
the two beam paths.

The gas-handling system is shown in Fig. 5. The gas cell is similar to gas cells used
in infrared spectroscopy (see Exp. 37), but ordinary Pyrex glass or quartz is more suitable for windows. Suitable cells with sealed-on windows are available commercially, or
cells can be made by cementing windows to the ends of a cylindrical tube that will serve

as the body of the cell with epoxy cement or Apiezon wax. In any case, the windows must be of accurately uniform thickness and must be accurately plane and parallel to each other, and of accurately known distance apart, preferably 10 cm. A suitable diameter is in the range of 2.5 to 5 cm. The commercially available cell normally has a taper joint; this can be attached by a Cajon Ultra-Torr fitting to a polyethylene tube through which the HCl gas is admitted to the cell. A thermocouple junction should be taped or waxed to the cell.

As gas is *slowly* admitted to the evacuated cell through a needle valve, the fringes are counted as they appear or disappear at the center of the screen or as they move past some designated point. (It is also possible to mount a photodiode on the screen and connect its amplified output to a strip-chart or X–Y recorder; this should give an oscillatory output, the cycles of which can be counted to a fraction of a fringe.) After the pressure in the cell has built up to approximately 1 atm and the gas supply has been shut off, counting of the fringes should continue until movement stops, indicating the attainment of equilibrium. At this point the pressure and the temperature should be recorded. The count of fringes made on filling the cell should now be checked by slowly evacuating the cell and counting the fringes as they move in the opposite direction. If time permits, the entire procedure should be executed at least four times.

CALCULATIONS

Orientation Polarization. If oscillator drift is negligible, the frequencies obtained under equilibrium conditions can be used directly in Eq. (5); otherwise the vacuum and HCl frequencies should be plotted against time and the readings at steady drift should be extrapolated back to a common time as mentioned earlier. The difference between the extrapolated values is $f_{vac} - f_{gas}$. The dielectric constant κ of the HCl gas at pressure p and temperature T is then calculated with Eq. (5), and corrected to 20°C and 1 atm with Eq. (6).

P_M is calculated with Eq. (1) from the measured values of $\kappa_{HCl} - 1$, p, and T for each run. Report the values of P_M together with their average in m^3 and in the conventional units of cm^3.

If values of P_M have been obtained at two (or more) temperatures, plot them against $1/T$. Calculate the permanent dipole moment μ from the slope; present it in units of 10^{-30} C m and also in debye. (1 D = 3.33564×10^{-30} C m, see Exp. 29.) From the intercept determine P_d in m^3 and cm^3, and calculate α_0 in units of $C^2\ m^2\ J^{-1}$.

If P_M was measured at a single temperature, it is necessary to use data on the index of refraction n and Eq. (2) to estimate P_d. The orientation polarization P_μ, and the value of μ itself, can then be obtained with Eq. (1).

Distortion Polarization. In the interferometer the light traverses the cell twice because of reflection in a mirror. If the cell is of length d, the number of wavelengths when the cell is evacuated is

$$N_{vac} = \frac{2d}{\lambda_0} = \frac{2d\nu}{c} \tag{7}$$

When the cell is filled with the gas to the desired pressure p at temperature T, the number of wavelengths is

$$N_{gas} = \frac{2d}{\lambda} = \frac{2d\nu}{v} \tag{8}$$

Here λ_0 and λ are the wavelengths of the light in the vacuum and in the gas, respectively, c and v are the speeds of light in the vacuum and in the gas, respectively, and ν is the frequency of the light:

$$\nu = \frac{c}{\lambda_0} = \frac{v}{\lambda} \tag{9}$$

The index of refraction is then

$$n \equiv \frac{c}{v} = \frac{\lambda_0}{\lambda} = \frac{N_{gas}}{N_{vac}} = 1 + \frac{\Delta N}{N_{vac}} \tag{10}$$

where $\Delta N = N_{gas} - N_{vac}$.

From the measured values of n, p, and T, use Eq. (2) to calculate the distortion polarization P_d and the distortion polarizability α_0 in units of $C^2\,m^2\,J^{-1}$.

If the low-frequency dielectric constant has been determined at more than one temperature *and* the refractive index has *also* been measured with visible light, calculate the atomic polarization from

$$P_a = P_M - P_\mu - P_e = P_d - P_e \tag{11}$$

Since the result is a small difference between two numbers having substantial uncertainties, it is particularly important to estimate the *uncertainty* in P_a relative to the value of P_a itself, in order to establish whether the determined value of P_a is significantly different from zero.

DISCUSSION

It is instructive to compare the magnitude of the permanent dipole moment as determined in this experiment with the value that would result if net charges of $+e$ and $-e$ were centered at the H and Cl nuclei, respectively. For this purpose we require the value of the internuclear distance,[6]

$$r = 1.275 \times 10^{-10}\ m$$

as well as that of the magnitude of the electronic charge,

$$e = 1.602 \times 10^{-19}\ C$$

Discuss this comparison in terms of mixed ionic and covalent character of the H—Cl bond. The dipole moment can also be discussed in terms of the difference in electronegativity between the H and Cl atoms.[7]

Let us now consider whether neglecting the atomic polarization P_a is justified. A simple classical treatment of atomic polarization of a gas molecule has been given by Van Vleck,[8] Coop and Sutton,[9] and Smyth.[10] It is based on the approximation that the vibrations of the molecule are harmonic. Van Vleck argues that the result of this classical treatment is (to the same level of approximation) valid also under quantum mechanics.

We consider a diatomic molecule the axis of which is at an angle θ with respect to the electric field direction. The electric field F has two effects: a turning moment on the molecule, which we ignore here because it is presumably taken account of in the orientation polarization, and a stretching or compression of the H—Cl bond due to interaction of the field with the charge distribution in the molecule. As the internuclear distance r changes from the equilibrium value r_e that it has in the absence of an applied electric field, the permanent dipole moment μ changes continuously from its equilibrium value μ_e. The energy of the molecule in the field may be written

$$U = -\mu F \cos \theta + \tfrac{1}{2} k (r - r_e)^2 \tag{12}$$

where k is the bond force constant. The molecule comes to a new equilibrium with a value of r for which U is a minimum:

$$\frac{dU}{dr} = -\frac{d\mu}{dr} F \cos \theta + k(r - r_e) = 0 \tag{13}$$

$$r - r_e = \frac{d\mu}{dr} \frac{F}{k} \cos \theta \tag{14}$$

The change in internuclear distance produces a change in permanent dipole moment, thereby producing a change in the component of polarization moment in the field direction (parallel to the z axis):

$$m_z = \frac{d\mu}{dr}(r - r_e) \cos \theta = \left(\frac{d\mu}{dr}\right)^2 \frac{F}{k} \cos^2 \theta \tag{15}$$

When we average over all molecular orientations, $\cos^2 \theta$ averages to $\tfrac{1}{3}$, and we obtain an *atomic polarizability*

$$\alpha_a = \frac{\langle m_z \rangle_{av}}{F} = \frac{1}{3k}\left(\frac{d\mu}{dr}\right)^2 \tag{16}$$

if we express the force constant in terms of the vibrational frequency ν with the harmonic oscillator expression

$$k = 4\pi^2 \mu_{mm} \nu^2 \tag{17}$$

where in terms of the atomic masses m_H and m_{Cl}

$$\mu_{mm} = \frac{m_H m_{Cl}}{m_H + m_{Cl}} \tag{18}$$

is the reduced mass, then we have

$$\alpha_a = \frac{1}{12\pi^2 \mu_{mm} \nu^2}\left(\frac{d\mu}{dr}\right)^2 \tag{19}$$

The resulting expression for the atomic polarization is

$$P_a = \frac{N_0 \alpha_a}{3\varepsilon_0} = \frac{N_0}{36\pi^2 \varepsilon_0 \mu_{mm} \nu^2}\left(\frac{d\mu}{dr}\right)^2 \tag{20}$$

For evaluating this expression, we obtain from spectroscopic data[6]

$$\tilde{\nu} = \frac{\nu}{c} = 2991 \text{ cm}^{-1}$$

(where c is the velocity of light). The magnitude of the dipole derivative has been calculated theoretically by several quantum mechanical models[11] and determined experimentally from the transition moment from the ground state to the first excited vibrational state;[12] the values are

$$\left(\frac{d\mu}{dr}\right)_{\text{theor}} = 1.205 \text{ D Å}^{-1}, \ 1.260 \text{ D Å}^{-1}$$

$$\left(\frac{d\mu}{dr}\right)_{\text{exp}} = 1.206 \text{ D Å}^{-1}$$

The student is encouraged to calculate α_a and P_a, and to compare their magnitudes with the experimental quantities α_0 and P_e, respectively. Was the assumption that P_a is negligible justified?

The student should discuss the relative merits of the two methods described here for determining the permanent dipole moment of the HCl molecule.

Theoretical Calculations.   A class project to calculate the physical properties of HCl using an ab initio program such as Gaussian is described in Chapter III. The dipole moment and the polarizability obtained for different basis set choices and levels of calculation can be compared with values measured in the present experiment. It should be noted that Gaussian computes all the elements of the polarizability tensor and the average of the diagonal elements $(\alpha_{xx} + \alpha_{yy} + \alpha_{zz})/3$, should be computed for comparison with the α_0 determined from the measured R_M value. Some units conversion may be needed for this comparison, since the α values given by Gaussian and other programs may be in atomic volume units of a_0^3, where $a_0 = 0.529$ Å is the Bohr radius. Finally it would be interesting to calculate and plot the dipole moment as a function of bond length and to compare the slope $d\mu/dr$ at the equilibrium position with the experimental value of 1.206 D Å^{-1} mentioned above.

SAFETY ISSUES

Lecture bottles of HCl and CO_2 gas must be strapped or chained securely to the laboratory bench. Work on a vacuum system requires preliminary review of procedures and careful execution in order to avoid damage to the apparatus and possible injury from broken glass; in addition, the liquid nitrogen used for cold traps must be handled properly (see Appendix C). Safety glasses must be worn. Acetone is volatile at room temperature and should be handled carefully; clean up any spills and avoid breathing the vapor.

APPARATUS

Gas Dielectric Constant.   Gas dielectric cell (shown in Fig. 2); WTW Dipolmeter (modified as in Exp. 29) or very stable variable-frequency LC oscillator with regulated 12-V power supply (see text); frequency counter (range at least 0.5 to 5 MHz); shielded coaxial cables with BNC connectors; gas-handling system (shown in Fig. 3) incorporating mechanical vacuum pump, liquid nitrogen trap, corrosion-resistant pressure gauge (0 to 1000 Torr), thermocouple pressure gauge and associated electronics for vacuum testing, high-pressure stem valve for HCl lecture bottle, high-pressure needle valve(s), Swagelok and Cajon Ultra-Torr fittings, and polyethylene tubing; thermometer or thermocouple and readout meter; Dewar flask for liquid-nitrogen trap; large Dewar flask for constant-temperature bath for capacitance cell.

Lecture bottle of HCl gas; cylinder or lecture bottle of Ar gas, N_2 gas, or CO_2 gas (if used); soda-lime tube for drying air (if used); acetone and Dry Ice for a constant-temperature bath if low-temperature measurements are to be made; liquid nitrogen for trap.

Gas Refractive Index. Michelson interferometer [sources for such instruments are Sargent-Welch Scientific Company (e.g., Cat. No. 3559), Central Scientific Company (e.g., Cat. No. 30666), and Ealing Scientific Company (e.g., Cat. No. 25-9069)]; 0.5 to 2 mW He–Ne laser and power supply; diverging lens (−20 to −50 mm focal length); small projection screen (a sheet of stiff white cardboard in a sturdy mount will do); gas cell (see text and Fig. 4); gas-handling system (see Fig. 5) incorporating mechanical vacuum pump, liquid-nitrogen trap, corrosion-resistant pressure gauge (0 to 1000 Torr), high-pressure stem and needle valves for lecture bottle, Swagelok and Cajon Ultra-Torr fittings, and polyethylene tubing; thermometer or thermocouple and readout meter; Dewar flask for liquid-nitrogen trap.

Lecture bottle of HCl gas; liquid nitrogen for trap.

REFERENCES

1. D. A. Coe and J. W. Nibler, *J. Chem. Educ.* **50,** 82 (1973).

2. C. P. Smyth, *Dielectric Behavior and Structure,* p. 420, McGraw-Hill, New York (1955).

3. R. D. Nelson, Jr., D. R. Lide, Jr., and A. A. Maryott, *Selected Values of Electric Dipole Moments for Molecules in the Gas Phase,* National Bureau of Standards NSRDS-NBS 10, p. 10, U.S. Government Printing Office, Washington, DC (1967).

4. *The ARRL Handbook for the Radio Amateur,* chap. 10 (especially Fig. 14), American Radio Relay League, Newington, CT (1987) [the same circuit also appears in some earlier editions].

5. R. D. Lide (ed.), *CRC Handbook of Chemistry and Physics,* 89th ed., CRC Press, Boca Raton, FL (2008/2009); A. A. Maryott and F. Buckley, *Table of Dielectric Constants and Electric Dipole Moments of Substances in the Gaseous State,* National Bureau of Standards Circular 537, U.S. Government Printing Office, Washington, DC (1953).

6. K. P. Huber and G. Herzberg, *Molecular Spectra and Molecular Structure IV: Constants of Diatomic Molecules,* p. 286, Van Nostrand Reinhold, New York (1979); although this book is now out of print, its contents are conveniently available as part of the *NIST Chemistry WebBook* (http:// webbook.nist.gov).

7. L. Pauling, *The Nature of the Chemical Bond,* 3d ed., pp. 98–100, Cornell Univ. Press, Ithaca, NY (1960).

8. J. H. Van Vleck, *The Theory of Electric and Magnetic Susceptibilities,* pp. 45–47, Oxford Univ. Press, Oxford (1965).

9. I. E. Coop and L. E. Sutton, *J. Chem. Soc.,* 1269 (1932).

10. C. P. Smyth, *op cit.,* pp. 416–422.

11. M. Kobayashi and I. Suzuki, *J. Mol. Spectrosc.* **116,** 422 (1986).

12. J. F. Ogilvie, W. R. Rodwell, and R. H. Tipping, *J. Chem. Phys.* **73,** 5221 (1980).

GENERAL READING

J. M. Pochan, J. J. Fitzgerald, and G. Williams, "Experimental Methods for Chemists: A Simplified Approach. Dielectric Properties of Polymers and Other Materials," in B. W. Rossiter and R. C. Baetzold (eds.), *Physical Methods of Chemistry*, 2d ed., vol. VIII, chap. 6, Wiley-Interscience, New York (1993).

C. P. Smyth, *Dielectric Behavior and Structure*, McGraw-Hill, New York (1955).

J. H. Van Vleck, *The Theory of Electric and Magnetic Susceptibilities*, Oxford Univ. Press, Oxford (1965).

N. E. Hill, W. E. Vaughan, A. H. Price, and M. Davies, *Dielectric Properties and Molecular Behavior*, Van Nostrand Reinhold, London (1969).

EXPERIMENT 31
Magnetic Susceptibility

When an object is placed in a magnetic field, in general a magnetic moment is induced in it. This phenomenon is analogous to the induction of an electric moment in an object by an electric field (see Exps. 29 and 30) but differs from it in that an induced magnetic moment may have either direction in relation to the applied field. If the induced moment is parallel to the external field (as in the electric case), the material is called *paramagnetic* or *ferromagnetic*, depending on whether the field due to the induced moment is small or large in comparison with the external field. If the moment is antiparallel to the external field, the material is called *diamagnetic*; the moment in this case is always small. This experiment will deal only with paramagnetic and diamagnetic substances in solution.

THEORY

If **M** is the magnetization (magnetic dipole moment per unit volume, analogous to the dielectric polarization **P**) induced by the field **H,** the volume magnetic susceptibility χ is defined by the equation

$$\mathbf{M} = \chi\mathbf{H} \tag{1}$$

For a paramagnetic substance, χ is positive; for a diamagnetic substance, it is negative. It is a dimensionless number, ordinarily very small in comparison with unity (except in the case of ferromagnetism) and essentially independent of **H** for fields readily available in the laboratory.

Magnetic susceptibilities are usually given in the literature on a mass or molar basis. Thus, while the volume susceptibility χ is induced moment per unit volume per unit applied field and is dimensionless, the mass susceptibility

$$\chi_{\text{mass}} = \frac{\chi}{\rho} \tag{2}$$

(where ρ is the density) is induced moment per unit mass per unit applied field and has SI units of $m^3\,kg^{-1}$. The molar susceptibility

$$\chi_M = M\chi_{\text{mass}} = \frac{M}{\rho}\chi = \tilde{V}\chi \tag{3}$$

(where M is the molar mass and $\tilde{V}$ is the molar volume) is induced moment per mole per unit applied field and has SI units of $m^3 \, mol^{-1}$.†

Diamagnetism. Nearly all known substances are diamagnetic. Diamagnetism results from the precession of the electronic orbits in atoms that occurs when a magnetic field is present. Volume diamagnetic susceptibilities are generally very small in magnitude compared with volume paramagnetic susceptibilities for pure substances. In paramagnetic substances the observed susceptibility is the net result of a paramagnetic contribution and a very much smaller diamagnetic contribution. In an estimation of the paramagnetism this diamagnetic contribution is often neglected, but in the case of aqueous solutions a correction should be made for the diamagnetic susceptibility of the water owing to the relatively large amount of it present.

Paramagnetism. The most important source of paramagnetism is the magnetic moment associated with the spin of the electron. The electron has two spin states, having spin magnetic quantum numbers $-\frac{1}{2}$ and $+\frac{1}{2}$, with the principal component of magnetic moment, respectively, parallel and antiparallel to the magnetic field direction. Spin paramagnetism (or in some cases ferromagnetism) exists in a substance if the atoms, molecules, or ions in it contain unequal numbers of electrons in the two possible spin states. This condition obviously exists when the atom, molecule, or ion contains an odd number of electrons [as in Fe^{3+}, Cu^{2+}, $(C_6H_5)_3C\cdot$ and other free radicals, etc.]. It may also exist when the number of electrons is even, if a degenerate electronic level (such as a d or f atomic subshell) is only partially filled. For example, the free ferrous ion Fe^{2+} has six electrons outside the argon shell. The available orbitals of lowest energy are the five degenerate (i.e., equal-energy) $3d$ orbitals. Each of these may contain two electrons with their spins opposed (one with spin $+\frac{1}{2}$, the other with spin $-\frac{1}{2}$, in accord with the Pauli exclusion principle) or a single electron with either spin. No spin paramagnetism would occur if the six outer electrons of Fe^{2+} occupied three of the five $3d$ orbitals in pairs so that all spin magnetic moments cancelled. However, this would be contrary to a principle known as *Hund's first rule,* which states that when several electronic orbitals of equal or very nearly equal energy are incompletely filled, the electrons tend to occupy as many as possible of the orbitals singly rather than in pairs, the electrons in singly occupied orbitals all having the same spin. In Fe^{2+} this rule predicts that one $3d$ orbital will contain a pair of electrons with spins opposed and the other four orbitals will each contain a single electron, the four spins being the same. Thus the free ferrous ion is paramagnetic. In molecular oxygen O_2, two molecular orbitals of equal energy each contain a single electron in accordance with Hund's first rule; consequently oxygen gas is paramagnetic.

An atom, molecule, or ion containing one or more unpaired electrons with the same spin has a permanent magnetic dipole moment μ. In the absence of orbital contributions to the moment (see below), the magnitude μ of this moment is completely determined by the number of unpaired electrons n:[1]

$$\mu(\text{spin only}) = g_e \mu_B \sqrt{S(S+1)} = \sqrt{n(n+2)}\, \mu_B \qquad (4)$$

---

†The use of $cm^3 \, mol^{-1}$ units for χ_M was standard practice in the past (cgs system) and still persists. Such values are almost always cgs values for $\chi_M(\text{ir})$, the irrational molar susceptibility given by $\chi_M(\text{ir, cgs}) = \chi_M(\text{cgs})/4\pi = 10^6 \chi_M(\text{SI})/4\pi$.

where S is the spin quantum number (equal to the sum of the individual electron spin quantum numbers s_i), the electron g-factor g_e has been taken as 2 for simplicity rather than 2.0023, and μ_B is the *Bohr magneton,* given by

$$\mu_B = \frac{eh}{4\pi m_e} = 9.274 \times 10^{-24} \text{ J T}^{-1} \tag{5}$$

where T denotes tesla (1 T = 10^4 gauss). Thus, for example, the "spin-only" magnetic moment of Fe^{2+} is

$$\mu = \sqrt{4 \times 6}\mu_B = 4.90 \text{ Bohr magneton}$$

Orbital magnetic moments may also contribute to paramagnetism. An electron in an orbital with one or more units of angular momentum behaves like an electric current in a circular loop of wire and produces a magnetic moment. When all orbitals in a subshell (e.g., all five $3d$ orbitals) are equally filled (with one electron each, as in Fe^{3+}, or two electrons each, as in Cu^+), the orbital moments cancel one another and there is no orbital contribution to the observed moment. In other cases (e.g., Fe^{2+}) an orbital contribution may arise, although usually it is "quenched" to a large extent by interactions with neighboring molecules or ions and does not contribute more than a few tenths of a Bohr magneton to the total moment. (Important exceptions are certain rare-earth ions, since quenching occurs to a much smaller extent for $4f$ orbitals than for $3d$ orbitals.) Atomic nuclei often possess spin magnetic moments, but these nuclear moments are so small as to have a negligible effect on magnetic susceptibility. They are important, however, in NMR spectroscopy.

In the absence of an applied field, the atomic moments in a paramagnetic substance orient themselves essentially at random owing to thermal motion and there is no net observable moment. In the presence of a magnetic field, the atomic moments tend to line up with the field, but the degree of net alignment is slight because of the disorienting effect of thermal motion. It is possible to show[1] that the paramagnetic contribution to the molar susceptibility is $N_0\mu_0\mu^2/3kT$, where N_0 is Avogadro's number, μ_0 is the vacuum permeability, and k is the Boltzmann constant. The molar susceptibility can be written

$$\chi_M = N_0\mu_0\xi + \frac{N_0\mu_0\mu^2}{3kT} \tag{6}$$

where ξ is the magnetizability and $\mu_0\xi$ is the small (negative) diamagnetism per molecule. We can write this equation in the form

$$\chi_M = N_0\mu_0\xi + \frac{C}{T} \tag{7}$$

where C is called the *Curie constant* for the substance concerned. If C is determined by experiment, the magnetic dipole moment of the atom, molecule, or ion is obtained from it with the equation

$$\mu = \left(\frac{3kC}{N_0\mu_0}\right)^{1/2} \tag{8}$$

Expressing this result in Bohr magneton units, we obtain

$$\mu = 797.8\sqrt{C} \text{ Bohr magneton} \tag{9}$$

where C is in SI units ($m^3 \text{ mol}^{-1} \text{ K}$).

Transition-Metal Complexes. The present experiment is largely concerned with complex ions of transition-group metals, such as hexahydrated or ammoniated ferrous or ferric ions and ferro- or ferricyanides. Here each metal ion is surrounded by a number of negative or neutral groups called *ligands*. This number is six in the cases cited and in other common cases may be four or eight.

Two distinct groups of complexes can be distinguished on the basis of experimental paramagnetic susceptibilities. *High-spin complexes* are those for which the effective magnetic moment is very close to the spin-only value for the free (gaseous) transition ion. *Low-spin complexes* have much lower moments than would be predicted for the free metal ion and can even be diamagnetic.

The early theory of transition-metal complexes, due largely to Pauling,[2] involved an explanation based on distinguishing between essentially ionic and essentially covalent bonding between the metal ion and its ligands. Where the number of unpaired electrons in the complex as deduced from the measured susceptibility is the same as that expected for the free metal ion (high-spin case), the bonding with the ligands was considered to be ionic (i.e., due to Coulomb attraction as in $[Fe(III)F_6]^{3-}$ or due to electrostatic polarization of neutral ligands by the central ion as in $[Co(III)(H_2O)_6]^{3+}$). Where the number of unpaired electrons found in the complex is considerably less than the free-metal-ion value (low-spin case, as in most complexes with cyanides, ammonia, carbon monoxide, etc.), the bonding was considered to be covalent. It is assumed that the electrons are paired owing to the necessity of accommodating, in the atomic orbitals, some additional electrons donated by the ligands for forming electron-pair bonds. Thus in $[Co(III)(NH_3)_6]^{3+}$ the six electrons outside the argon shell of Co^{3+} are augmented by 6 electron pairs from the ligands, giving 18 electrons. Of these 12 electrons (or 6 pairs) are shared with the ligands to form 6 octahedral covalent bonds, using two $3d$, one $4s$, and three $4p$ orbitals of cobalt. The other 6 electrons are paired in the remaining three $3d$ orbitals. Since all electrons are paired with spins opposed, salts of this complex are diamagnetic.

The "crystal-field" or "ligand-field" theory of transition-metal complexes, first proposed by Van Vleck[3] and subsequently developed extensively,[4] has proven to be of great value in the interpretation of a wide range of properties. This theory explains transition-metal complexes in terms of the splitting of the five-fold-degenerate d level into two or more levels of different energy by perturbations due to the ligands. In the simplest version, this splitting is due purely to the electrostatic crystal field of the ligands. However, it is often necessary to consider also the metal–ligand orbital overlap (so-called adjusted crystal-field theory[4]). In any case, ligand-field theory is an adequate description as long as the d orbitals of the metal ion are well defined. If there is strong mixing between metal ion and ligand orbitals (as in metal carbonyls), a molecular-orbital theory is required.†

We shall discuss ligand-field theory in the one-electron approximation, in which a one-electron d state is split by the ligand field and the individual states are then filled by the d electrons of the central ion (taking account of interactions). As an example, consider a d^6 complex with octahedral symmetry; Fig. 1 shows the situation for Co(III) complexes. For weak ligand fields, the splitting Δ is small and the electrons distribute themselves according to Hund's first rule. Thus a complex such as $[Co(III)F_6]^{3-}$ is a high-spin complex with four unpaired electrons. The energy of this configuration can be written as $(-\frac{2}{5}\Delta + P)$, where P is the average energy required to form an electron pair. When the ligand field is strong enough (Δ sufficiently large), the energy difference between the levels can no longer be overcome by the electron-unpairing tendency of Hund's rule. Thus a complex such as $[Co(III)(NH_3)_6]^{3+}$ is a low-spin complex: the six d electrons of the cobalt

---

†As shown by Van Vleck,[3] Pauling's theory is a special case of the more general MO theory.

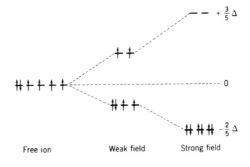

FIGURE 1
Weak field, e.g.,
$[Co(III)F_6]^{3-}$, and strong
field, e.g., $[Co(III)(NH_3)_6]^{3+}$,
splitting of the $3d$ level of
Co(III) due to ligands with
O_h symmetry. The splitting
between the low-lying triplet
and the upper doublet is
defined as Δ. In the weak-
field case, there are four
unpaired spins; whereas all
the spins are paired in the
strong-field case.

ion fill the lower triplet state with spins paired and the complex is diamagnetic. The energy of this configuration is $(-\frac{12}{5}\Delta + 3P)$. Analogous arguments can be applied to d^4, d^5, and d^7 complexes.

Thus ligand-field theory allows one to understand both high-spin and low-spin complexes. In particular low-spin complexes can be explained without assuming a covalent electron-pair bond between the metal ion and the ligand. Indeed there is no clear-cut distinction made between ionic and weak covalent character. According to the ligand-field theory, a low-spin complex simply means that the ligand field strength (and thus the splitting Δ) is greater than some critical value. It should be stressed that this critical value will be different for different complexes and one cannot equate increasing ligand field strength with increasing "covalency" in any simple way.

METHOD

Most methods for the determination of a magnetic susceptibility depend upon measuring the force resulting from the interaction between a magnetic field gradient and the magnetic moment induced in the sample by the magnetic field.[5] Assuming for simplicity that H varies only as a function of x, the x component of this force, per unit volume of the sample, is

$$f_x = \mu_0 M \frac{dH}{dx} = \chi\mu_0 H \frac{dH}{dx} \qquad (10)$$

where M is the magnetization and μ_0 is the vacuum permeability, equal to $4\pi \times 10^{-7}$ N A^{-2} (N = newton, A = ampere). This force is such as to tend to draw the sample into the strongest part of the field if the sample is paramagnetic (χ positive) or repel it into the weakest part if the sample is diamagnetic (χ negative). The work done on the system when a volume dV of the sample is carried from a point of field strength H_1 to a point of field strength H_2 is

$$dw = dV \int f_x dx = dV\chi\mu_0 \int H \frac{dH}{dx} dx = dV\chi\mu_0 \int_{H_1}^{H_2} H \, dH$$

$$= \tfrac{1}{2}\chi\mu_0 (H_2^2 - H_1^2)dV \qquad (11)$$

The Gouy Balance.[5] In the Gouy balance (Fig. 2), a long tube, divided into two regions by a septum, is suspended from one side of an analytical balance so as to hang vertically

FIGURE 2

Gouy balance.

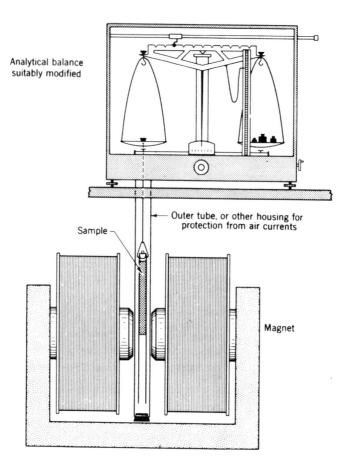

Analytical balance
suitably modified

Sample

Outer tube, or other housing for
protection from air currents

Magnet

in a magnetic field. The septum is in the strongest part of the field and the two ends of the tube are in regions of essentially zero field strength. The part of the tube above the septum is filled with the sample to be investigated, and the part below is empty.

In addition to the downward gravitational force acting on the tube, a force is exerted by the magnetic field. Let us calculate the work done in *lowering* the tube, with cross-sectional area A, by an amount $|\Delta x|$. This is equivalent to bringing a volume $A|\Delta x|$ of the specimen from a region of zero field strength to a region of field strength H. Thus, ignoring gravitational work,

$$w = \tfrac{1}{2}\chi\mu_0 H^2 A |\Delta x| = f|\Delta x|$$

Therefore the *downward force f* on the tube due to its interaction with the magnetic field is†

$$f = \tfrac{1}{2}\chi\mu_0 H^2 A \tag{12}$$

†In case the top of the tube is not at zero field strength, we should write

$$f = \tfrac{1}{2}\chi\mu_0 (H_{max}^2 - H_{min}^2) A$$

Note that this force f as defined is a positive quantity for a paramagnetic sample, whereas f_x is negative when the positive x direction is taken to be upward.

In making a measurement, the "apparent weight" W is determined in the absence of a field and then with a field present, and the difference is equated to f:

$$(W_{\text{field}} - W_{\text{no field}}) = f = \tfrac{1}{2}\chi\mu_0 H^2 A \qquad (13)$$

The weight W in each case is a *force* (in newtons) obtained by multiplying the mass of the weights used (in kg) by the acceleration due to gravity (in m s^{-2}).

The magnet should provide a field of at least 0.4 T ($= 4$ kgauss) and preferably 0.6 T or more. The field should be reasonably homogeneous over a region considerably larger than the diameter of the tube. For a sample tube up to 15 mm in diameter, a magnet with a gap of 1 in. and a pole diameter of at least 3 in. is convenient.

To obtain the weights in the presence and absence of a field, it is most convenient to have an electromagnet, the field of which can easily be turned on and off. Such a magnet, with a regulated power supply, has the disadvantage of being rather expensive. A permanent magnet is usually less expensive, but special arrangements for making the no-field measurements are required. If the magnet is mounted on rails or on a pivot, it can be rolled or swung in and out of its normal position. If the magnet is stationary, the Gouy tube can be hung at two different levels (Fig. 3). If the latter method is to be used, the septum should be in the strongest part of the field in the lower position and in essentially zero field in the upper position. The sample tube should be long enough that in either position the bottom end is in an essentially zero-field region *below* the strong part of the field.

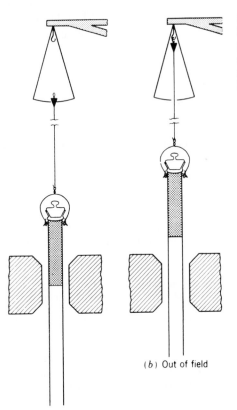

FIGURE 3
Positioning of Gouy tube for (*a*) in-field and (*b*) out-of-field weighings when a stationary permanent magnet is used.

(*b*) Out of field

(*a*) In field

A suitable analytical balance is mounted on a sturdy table over the magnet as shown in Fig. 2. Displayed is a two-pan balance but an electronic single-pan balance such as the Mettler model AE 100 can also be used and is simpler to operate. The Gouy tube is supported by a nonmagnetic wire or nylon thread that passes through a hole drilled in the base of the balance. If an electromagnet is used, the left-hand pan may be dispensed with and the wire attached directly to the stirrup. If, instead, a permanent magnet is used with the scheme shown in Fig. 3, a hole must be drilled in the pan so that the wire can pass through it to a hook that is attached to the stirrup for the no-field weighing or allowed to rest on the pan for the in-field weighing. The Gouy tube should be protected against air currents. Provision should also be made for mounting a thermometer near the Gouy tube.

EXPERIMENTAL

The procedure to be used in operating the Gouy balance will necessarily depend on the details of construction and cannot be given here with any completeness; a set of instructions should be compiled by the instructor and posted near the apparatus. The balance should be operated in the normal manner; the beam must be off the knife edges when the Gouy tube is being mounted, demounted, or changed in position. It is important to **remove your watch** when working near the magnet and to keep steel and iron tools or instruments out of the way.

In order to determine the "apparatus constant" $\mu_0 H^2 A/2$ that appears in Eq. (13), measurements are made on a material of known susceptibility. For this purpose, a common standard is an aqueous solution of nickel chloride, about 30 percent $NiCl_2$ by weight. Prepare 100 mL of such a solution with an accurately known weight fraction of $NiCl_2$ in air-free distilled water. When this and other solutions are weighed on the Gouy balance, record the ambient temperature.

Solutions of the following salts† in air-free distilled water should be studied. In each case the concentration ($\sim$0.5 molar) must be known precisely. If possible, prepare the solutions in advance, since some of these salts dissolve quite slowly.

KMn(VII)O_4, potassium permanganate (*Note:* This is soluble only to $\sim$0.4 molar)

Mn(II)SO_4, manganous sulfate

$[Fe(II)(H_2O)_6](NH_4)_2(SO_4)_2$, ferrous ammonium sulfate (*Note:* This is subject to slow air oxidation)

$K_4[Fe(II)(CN)_6]$, potassium ferrocyanide

$K_3[Fe(III)(CN)_6]$, potassium ferricyanide

It is important to know the densities of all solutions. A satisfactory procedure is to weigh the Gouy tube empty and then filled with water to a fiducial mark near the top. From

---

†Magnetic measurements may be made on powdered crystalline salts rather than on aqueous solutions. The volume susceptibilities are very much larger (an attractive feature if a strong magnet is not available), but the molar susceptibilities are not so easily interpretable in terms of atomic moments, owing to interaction effects in the crystalline state. In place of Eq. (7), we write (neglecting the diamagnetic term)

$$\chi_M = \frac{C}{T - \Delta}$$

This is called the *Curie-Weiss law.* The constant Δ may be positive or negative and for most compounds is less than 75 K in magnitude. To obtain a value of the Curie constant from which an atomic moment can be calculated, it is necessary to measure χ_M at more than one temperature. When the reciprocal of χ_M is plotted against absolute temperature, the slope is the reciprocal of the Curie constant C. The density ρ required in the calculations is not the crystal density but an effective powder density determined in the manner described for solutions.

the known density of water at this temperature, calculate the volume. Then fill the tube with the solution to be studied up to the same fiducial mark and weigh it. (All these weighings are in the absence of a magnetic field.) Alternatively the density of each solution can be determined with a Westphal balance[6] or a good hydrometer.

CALCULATIONS

Calibration.   The mass susceptibility of an aqueous nickel chloride solution is given by[7]

$$\chi_{mass} = \left[\frac{10\,030\,p}{T} - 0.720(1 - p) \right] \times 4\pi \times 10^{-9} \text{ m}^3 \text{ kg}^{-1} \tag{14}$$

where p is the mass fraction of $NiCl_2$ and T is the absolute temperature. The second term in the brackets is the correction for the diamagnetism of the water used as solvent. This expression assumes that the solution is free of dissolved atmospheric oxygen.

From the χ_{mass} given by this expression and the density of the $NiCl_2$ solution, the volume susceptibility χ can be calculated from Eq. (2). With this and the measured weight difference, determine and report the apparatus constant $\mu_0 H^2 A/2$ appearing in Eq. (13).

Measurements on Unknown Solutions.   The mass susceptibility of a solution in which the solute has molar mass M is related to the molar susceptibility by

$$\chi_{mass} = \frac{\chi_M(\text{solute})}{M} p - 0.720 \times 4\pi \times 10^{-9}(1 - p) \text{m}^3 \text{ kg}^{-1} \tag{15}$$

where p is again the mass fraction of solute. If the concentration c of the solute in units of mol L^{-1} is known, we can write the volume susceptibility directly in the form

$$\chi = 1000 c \chi_M - 0.720 \times 4\pi \times 10^{-9}(\rho - 1000 cM) \tag{16}$$

where χ_M, ρ, and M are all expressed in SI units (m^3 mol^{-1}, kg m^{-3}, kg mol^{-1}, respectively). From the experimental weight difference, χ is determined by use of Eq. (13) and the known apparatus constant $\mu_0 H^2 A/2$. The solute molar susceptibility χ_M is obtained from Eq. (16).

If the material is paramagnetic (χ_M positive), the small negative diamagnetic term $N_0 \mu_0 \xi$ can be neglected in Eq. (7) and the constant C can be determined. The atomic moment μ can then be calculated from Eq. (9). With neglect of any orbital contribution, the number of unpaired electrons can be found approximately with Eq. (4). Calculate the number of unpaired electrons for each paramagnetic substance studied.

DISCUSSION

In potassium permanganate Mn(VII) has no unpaired electrons and thus no permanent magnetic moment. However, the magnetic field induces a small, temperature-independent paramagnetism because the field couples the ground state to paramagnetic excited states.[4] The other four compounds illustrate high- and low-spin cases of octahedral d^5 and d^6 complexes. (Neutral solutions of manganese sulfate are very pale pink and contain $[Mn(II)(H_2O)_6]^{2+}$ ions.) In the case of Fe(III) and the isoelectronic Mn(II), the ^6S ground state of the free ion has no orbital angular momentum and no effective coupling to excited states. Thus the magnetic moment of their high-spin complexes should be very close to the spin-only value. Comment on the spin type and ligand-field strength of each complex studied.

SAFETY ISSUES

None.

APPARATUS

Gouy balance (comprising a magnet, power supply if needed, suitably modified analytical balance); glass-stoppered Gouy tube; several 100-mL volumetric flasks and glass-stoppered 200-mL flasks; thermometer; Westphal balance or hydrometer (optional).

$NiCl_2$ (40 g); $MnSO_4$ (10 g); $KMnO_4$ (10 g); $K_4Fe(CN)_6$ (25 g); $K_3Fe(CN)_6$ (20 g); $Fe(NH_4)_2(SO_4)_2 \cdot 6H_2O$ (20 g); or a solution of each salt of an accurately known concentration (100 mL).

REFERENCES

1. P. W. Atkins and J. de Paula, *Physical Chemistry,* 8th ed., p. 735, Freeman, New York (2006).

2. L. Pauling, *The Nature of the Chemical Bond,* 3d ed., pp. 161ff., Cornell Univ. Press, Ithaca, NY (1960).

3. J. H. Van Vleck, *J. Chem. Phys.* **3,** 807 (1935).

4. F. A. Cotton, G. Wilkinson, C. A. Murillo, and M. Bochmann, *Advanced Inorganic Chemistry,* 6th ed., sec. 17, Wiley-Interscience, New York (1999).

5. L. N. Mulay and I. L. Mulay, "Static Magnetic Techniques and Applications," in B. W. Rossiter and J. F. Hamilton (eds.), *Methods of Physical Chemistry,* 2d ed., vol. IIIB, chap. III, Wiley-Interscience, New York (1989).

6. N. Bauer and S. Z. Lewin, "Determination of Density," in A. Weissberger and B. W. Rossiter (eds.), *Techniques of Chemistry: Vol. 1. Physical Methods of Chemistry,* part IV, chap. 2, Wiley-Interscience, New York (1972).

7. P. W. Selwood, *Magnetochemistry,* 2d ed., p. 26, Interscience, New York (1956).

GENERAL READING

R. L. Carlin, *Magnetochemistry,* Springer-Verlag, Berlin/New York (1986).

R. S. Drago, *Physical Methods for Chemists,* 2d ed., chaps. 10 and 11, Saunders, New York (1992).

E. M. Purcell, *Electricity and Magnetism,* 2d ed., McGraw-Hill, New York (1985).

R. M. White, *Quantum Theory of Magnetism,* Springer-Verlag, Berlin/New York (1983).

J. H. Van Vleck, *Electric and Magnetic Susceptibilities,* Oxford University Press, New York (1932).

S. Vulfson, *Molecular Magnetochemistry,* Gordon and Breach, Newark, NJ (1998).

EXPERIMENT 32

NMR Determination of Paramagnetic Susceptibility

The energy levels of a nucleus with a magnetic moment are changed in the presence of a magnetic field. Transitions between these levels can be induced by electromagnetic radiation in the radio-frequency region, and this resonance is useful in characterizing the chemical environment of the nucleus. For this reason, nuclear magnetic resonance (NMR) spectroscopy has developed as one of the most powerful experimental methods used in chemistry; for other applications, see Exps. 21, 42, and 43. In a slightly different application, in this experiment we will examine the effect of paramagnetic "impurities" on NMR solvent resonances and will use the predicted resonance shifts to deduce the paramagnetic susceptibility and election spin of several transition-metal complexes.

THEORY

The magnetic moment of a nucleus with nuclear spin quantum number I is

$$\mu = g_N \mu_N \sqrt{I(I + 1)} \tag{1}$$

where g_N is the nuclear g factor (5.5856 for a proton) and $\mu_N = eh/4\pi m_p$ is the nuclear magneton. Substitution of the charge e and mass m_p of a proton gives a value of 5.051×10^{-27} J T^{-1} for μ_N. The symbol μ_N is the unit of nuclear magnetic moment and is smaller than the electronic Bohr magneton μ_B (defined in Exp. 31) by the electron-to-proton mass ratio.

The nuclear moment will interact with a local *magnetic induction* (flux density)† B_{loc} to cause an energy change (Zeeman effect)

$$E_N = -g_N \mu_N M_I B_{\text{loc}} \tag{2}$$

Here M_I is the quantum number measuring the component of nuclear spin angular momentum (and magnetic moment) along the field direction, and it can have values $-I$, $-I + 1, \ldots, +I$. The effect of the field is thus to break the $2I + 1$ degeneracy and to produce energy levels whose spacing increases linearly with B_{loc} (or B) (Fig. 1). Transitions among these levels can be produced by electromagnetic radiation provided that the selection rule $\Delta M_I = \pm 1$ is satisfied. In this case, the resonant frequency is given by

$$\nu = \frac{\Delta E_N}{h} = \frac{g_N \mu_N}{h} B_{\text{loc}} \tag{3}$$

For protons ν (Hz) $= (4.26 \times 10^7)B_{\text{loc}}$ (tesla), so that, for typical fields of 1 to 5 T, ν falls in the radio-frequency region. In practice ν is usually fixed at some convenient value (e.g., 60, 100, or 220 MHz) and the induction B is varied until resonance is achieved.

---

†The vector quantity **B** is called either the *magnetic induction* or the *magnetic flux density,* although "magnetic field strength" would be a more appropriate name. Unfortunately, the name magnetic field strength was given to **H** at the time when **H** was considered to be the fundamental magnetic-field vector. It is now known that **B** is the fundamental vector (analogous to **E**, the fundamental electric-field vector). To add to the confusion, **B** = **H** in a vacuum when gaussian units are used. This book uses SI units, for which **B** = μ_0**H** in a vacuum and $\mu_0 \neq 1$. In SI units **B** is expressed in tesla (1 T = 1 weber m^{-2} = 10^4 gauss) and **H** is expressed in A m^{-1}. In many texts **B** is loosely called the *magnetic field.*

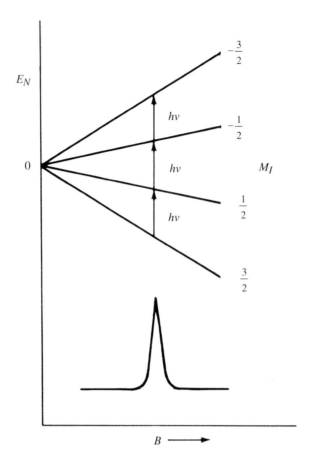

In general, the local induction B_{loc} at the nucleus will differ from the externally applied induction B because of the magnetization M that is induced by B:

$$B_{\text{loc}} = B + \mu_0 M = (1 + \chi)B \tag{4}$$

where μ_0 is the vacuum permeability, defined as $4\pi \times 10^{-7}$ N A^{-2} = 1.2566 $\times$ 10^{-6} N A^{-2} (N = newton; A = ampere). The volume susceptibility χ (a dimensionless quantity) is the magnetic analog of the dielectric polarizability (Exp. 29), and it can be converted to a mass susceptibility χ_{mass} by dividing by the density ρ. Multiplication by the molar mass M gives the molar susceptibility

$$\chi_M = M\chi_{\text{mass}} = \frac{M\chi}{\rho} = \tilde{V}\chi \tag{5}$$

where $\tilde{V}$ is the molar volume. See Exp. 31 for a discussion of units. As noted in Exp. 31, χ_M has the form

$$\chi_M = N_0\mu_0\xi + \frac{N_0\mu_0\mu^2}{3kT} = N_0\mu_0\xi + \chi_M^{\text{para}} \tag{6}$$

where N_0 is Avogadro's number. $N_0\mu_0\xi$ is a diamagnetic contribution while the second term is a temperature-dependent paramagnetic contribution that is dominant if the magnetic moment μ is not zero. This will be true only if the electron spin or electronic orbital angular momentum is not zero, a case that is fairly common for transition-metal compounds.

Most organic compounds are not paramagnetic, and so it is the diamagnetic susceptibility that is important in determining the resonance condition. The diamagnetic contribution arises because the orbital motion of the electrons is altered by the presence of B so
that there is a net orbiting of electrons about the field lines. This circulating charge in turn
generates a magnetic induction B_d, which is proportional to the applied field and *opposed*
to B. Thus $B_{loc} = B + B_d$, with

$$B_d = \chi B = \frac{\rho}{M} N_0 \mu_0 \xi B \equiv -\sigma B \tag{7}$$

where M is the molar mass and σ is a positive constant. The resonant frequency is

$$\nu = \frac{g_N \mu_N}{h} B(1 - \sigma) = \nu_0(1 - \sigma) \tag{8}$$

where ν_0 is the resonance for a bare proton. The diamagnetic shielding constant σ is
usually quite small ($\sim 10^{-5}$) and increases as the electron density about the nucleus is
increased. Changes in the local induction $B_{loc} = B(1 - \sigma)$, and thus in ν, of a few parts
per million (ppm) are typical when the chemical environment about a nucleus is changed.
These chemical shifts relative to a convenient standard, such as tetramethylsilane, are easily measured with modern NMR instruments and hence serve to characterize the chemical
bonding about a given nucleus. In addition, the relative intensities of for example proton resonances give a measure of the relative number of protons with different chemical
environments (e.g., —CH_3 versus —CH_2— groups). Finally, the coupling of magnetic
moments of nearby nuclei can produce spin–spin splitting patterns that are quite useful in
identifying the functional groups present in the molecule. Some further discussion of these
applications is presented in Exp. 42.

If a proton of a diamagnetic molecule is present in a solution containing a paramagnetic solute, the induction ("local field") B at the nucleus will be increased because of the
alignment of the solute magnetic moments in the applied field. Evans[1] has shown that this
increase in the local field is given by

$$\Delta B = \frac{1}{6}(\chi_s - \chi_0)B \tag{9}$$

where χ_s and χ_0 are the susceptibilities of the solution with and without the paramagnetic
solute, respectively. According to Eqs. (3) and (9), the proton resonance will shift by an
amount

$$\frac{\Delta \nu}{\nu} = \frac{\Delta B}{B_{loc}} \approx \frac{\Delta B}{B} = \frac{1}{6}(\chi_s - \chi_0) \tag{10}$$

with the approximation $B_{loc} \approx B$ leading to negligible error.

To obtain the mass susceptibility χ_{mass} of the pure paramagnetic material, we assume
that the volume susceptibility χ_s of the solution can be written as a sum of parts,

$$\chi_s = \chi_{mass\,s}\rho_s = \chi_{mass}m + \chi_{mass\,0}(\rho_s - m) \tag{11}$$

Here ρ_s is the density of the solution containing m kg of paramagnetic solute per cubic
meter and $\chi_{mass\,0}$ is the mass susceptibility of the solution without the paramagnetic material. The density of the latter is ρ_0 so that $\chi_0 = \chi_{mass\,0}\rho_0$ and we obtain from Eqs. (10) and
(11) the expression

$$\chi_{mass} = \frac{6}{m}\frac{\Delta \nu}{\nu} + \chi_{mass\,0} + \chi_{mass\,0}\frac{\rho_0 - \rho_s}{m} \tag{12}$$

The third term is a small correction that is unimportant for highly paramagnetic materials and is often neglected.[1] The value of χ_{mass} for a paramagnetic material can be determined therefore by measuring the difference in chemical shift of a proton in the solvent and in a solution containing a known weight of the paramagnetic solute. The value of $\chi_{mass\ 0}$ can be obtained from tables such as those contained in Ref. 2 or by summing atomic susceptibilities χ_i according to Pascal's empirical relation:

$$\chi_{mass\ 0} = \frac{\chi_{M_0}}{M_0} = \frac{1}{M_0}\left[\sum_i t_i\chi_i + \sum_j \lambda_j\right] \tag{13}$$

where M_0 is the solvent molar mass, t_i is the number of atoms of type i in the molecule, and the constitutive correction constants λ_j depend on the nature of the multiple bonds. See Table 1 for some typical values of χ_i and λ_j.

The molar susceptibility χ_M for the solute is obtained by multiplying χ_{mass} by the molar mass of the paramagnetic complex. As can be seen from Eq. (6), the paramagnetic contribution χ_M^{para}, and hence the paramagnetic moment μ, can be extracted by a temperature-dependence study (see also Exp. 31). For accurate results, a correction should be made to m to account for solvent-density changes with temperature.[3] Alternatively a measurement of χ_M at a single temperature can usually be combined with an adequate estimate of $N_0\mu_0\xi$ from Pascal's constants[2] to allow one to deduce the paramagnetic contribution. If χ_M^{para} is obtained in this manner, the paramagnetic moment is given by

$$\mu = \sqrt{\frac{3kT\chi_M^{para}}{N_0\mu_0}}$$

$$= 797.8\sqrt{T\chi_M^{para}}\ \text{Bohr magneton} \tag{14}$$

where the Bohr magneton (μ_B) is 9.274×10^{-24} J T^{-1}. In the absence of any orbital contributions, the value of μ is related to the number n of unpaired electrons in the d shell of the transition metal by

$$\mu(\text{spin only}) = \sqrt{n(n+2)}\ \text{Bohr magneton} \tag{15}$$

TABLE 1   Pascal's constants χ_i for diamagnetic susceptibility[a] (units of 10^{-11} m^3 mol^{-1})

Co^{2+}	-15	Br^-	-45	B	-8.8	N Open chain	-7.00
Co^{3+}	-13	CN^-	-23	Br	-38.5	Ring	-5.79
Cr^{2+}	-19	Cl^-	-33	C	-7.5	Monamides	-1.94
Cr^{3+}	-14	F^-	-14	Cl	-25.3	Diamides, imides	-2.65
Cu^{2+}	-14	NO_3^-	-25	F	-7.9	O Alcohol, ether	-5.79
Fe^{2+}	-16	OH^-	-15	H	-3.68	Aldehyde, ketone	$+2.17$
Fe^{3+}	-13	SO_4^{2-}	-50	I	-56.0	Carboxylic $=$ O	-4.22
K^+	-16			P	-33.0		
Ni^{2+}	-15			S	-18.8		

λ_j Corrections for bonds			
C=C	$+6.9$	C=N	$+10.3$
C≡C	$+1.0$	C≡N	$+1.0$
C=C—C=C	$+13.3$	C in aromatic ring	-0.30

[a] Adapted from Ref. 2.

As noted in Exp. 31, magnetic susceptibility measurements are thus quite useful in determining the electron configuration of a paramagnetic ion in a particular ligand field.

METHOD

The basic elements of an NMR spectrometer are outlined in Fig. 2. The principal magnetic field is provided by a permanent magnet (~ 1.5 T), an electromagnet (2.5 to 5.0 T), or a superconducting electromagnet (5.0 to 7.5 T). Since shifts of a few ppm are to be measured, the field must be quite stable and uniform. This is achieved by adding small adjustment coils to the magnet and by spinning the sample tube to average out residual inhomogeneities. An additional set of Helmholtz coils is wound around the pole faces to permit small linear variation in the field in recording spectra.

The sample tube is inserted into a probe region containing two radio-frequency (RF) coils wound at 90° to each other and to the magnetic-field coils. Radiation of some fixed frequency, usually 60 MHz or higher for protons, is sent through the sample by the transmitter coil. If the "local field" induction is such that Eq. (3) is satisfied, sample absorption and emission will occur. This reradiated signal is detected by the receiver coil, which, being oriented at right angles to the transmitter, senses no signal in the absence of sample coupling. The signal from the receiver coil is amplified and displayed on a chart recorder to yield the NMR spectrum as a function of field. Since the absolute value of the magnetic field is not easily determined with high precision, field shifts are measured relative to some reference compound such as tetramethylsilane and are expressed as chemical shifts in ppm,

$$\delta_i = \frac{B_r - B_i}{B_r} \times 10^6 \qquad (16)$$

Here, B_r and B_i are induction values for resonance by the reference nucleus and by nucleus i, respectively, for a fixed spectrometer frequency ν. If B is fixed and ν is varied, δ_i is given by

$$\delta_i \equiv \frac{\nu_i - \nu_r}{\nu_r} \times 10^6 \qquad (17)$$

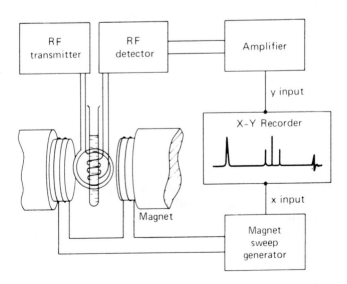

FIGURE 2
Schematic drawing of a nuclear magnetic resonance spectrometer.

The latter is the conventional definition of δ_i, but, to the accuracy of chemical shift measurements, Eq. (16) gives equivalent results. Many NMR chart displays are calibrated both in δ units and in frequency shift units appropriate to the fixed frequency source ν_r (60 MHz for most small proton NMR instruments).

EXPERIMENTAL

Operation of the NMR instrument will be described by the instructor. Particular care should be taken in inserting and removing the NMR sample tubes to prevent damage to the probe. All tubes should be wiped clean prior to insertion to avoid contamination of the probe.

To provide an internal reference, the solvent is sealed in a capillary, which is placed at the bottom of the NMR tube used for the sample solution. For this purpose, a melting-point capillary is closed at one end and a syringe is used to add the reference solution. For aqueous studies, a 2% solution of t-butyl alcohol in water can be used as reference and as solvent for the paramagnetic solute. The shift with respect to the methyl resonance of the t-butyl group is then monitored. With organic ligands such as acetylacetonate (acac) groups, complexes such as $Cr(acac)_3$, $Fe(acac)_3$, and $Co(acac)_2$ are soluble in benzene and the proton resonance of the solvent is a convenient reference. The capillary is filled one-third full, the lower end of the capillary is cooled in ice, and the upper part is sealed off with a small hot flame. Alternatively, the reference solution can be placed in one compartment of a coaxial pair of cylindrical NMR tubes, which are available commercially. In either case the spectral display should be expanded to permit an accurate measurement of the frequency shift.

The following complexes are suitable for study at room temperature. Either toluene or benzene can be used as the solvent, but it should be noted that toluene has *two* proton resonances.

d^3 $Cr(acac)_3$, chromium acetylacetonate ($M = 349.3$ g mol^{-1})

d^5 $Fe(acac)_3$, ferric acetylacetonate ($M = 353.2$ g mol^{-1})

d^8 $Co(acac)_2$, cobaltous acetylacetonate ($M = 257.2$ g mol^{-1})

The concentrations are not critical (0.02 to 0.05 molar), but they must be known precisely. These salts are available commercially, or they can be readily synthesized by methods described in Ref. 4.

If the NMR spectrometer is equipped with a variable-temperature probe, the $Fe(acac)_3$ shifts should be measured at four or five temperatures. A 0.025 molar solution in toluene can be employed over the liquid range from $-95°C$ to $+110°C$. Care should be taken to seal and test the NMR tube if the high-temperature range is to be studied (i.e., heat the tube to $\sim135°C$ in a fume hood before mounting it in the spectrometer). Since the temperature variation of toluene density is appreciable,[3] the solute concentration m in kg per cubic meter should be corrected at each temperature T:

$$m_t \cong m_{rt}\left(\frac{\rho_t}{\rho_{rt}}\right) \tag{18}$$

where ρ_t is the density of *pure* solvent at temperature t and ρ_{rt} is that at room temperature. The density of toluene as a function of temperature is given to a fairly good approximation by $\rho_t = 0.8845 - 0.92 \times 10^{-3}\ t$, where t is in degrees Celsius; a more detailed temperature dependence from $-95°C$ to $+99°C$ is given in Ref. 5. Approximately 15 min should be allowed for equilibrium to be reached after changing the temperature. An accurate measurement of the probe temperature should be made in the manner described in the spectrometer manual. This usually involves a measurement of the frequency separation between OH and CH resonances in methanol or ethylene glycol, since this separation changes by about 0.5 Hz K^{-1} for a 60-MHz instrument.

Appropriate salts for alternative aqueous measurements are

d^5 $Fe(NO_3)_3 \cdot 9H_2O$ (0.015 M), ferric nitrate† (M = 404.0 g mol^{-1})

d^5 $K_3Fe(CN)_6$ (0.06 M), potassium ferricyanide (M = 329.3 g mol^{-1})

d^6 $FeSO_4 \cdot 7H_2O$ (0.02 M), ferrous sulfate (M = 278.0 g mol^{-1})

d^8 $NiCl_2$ (0.08 M), nickel chloride (M = 129.6 g mol^{-1})

d^9 $CuSO_4$ (0.08 M), cupric sulfate (M = 159.6 g mol^{-1})

Three or more of these solutions should be made up in 10- or 25-mL volumetric flasks using as reference solvent ~2% (by volume) t-butyl alcohol in deoxygenated water. The Fe^{2+} and Fe^{3+} solutions illustrate the effect of oxidation state on the electron configuration of a single element. The Fe^{3+} and $K_3Fe(CN)_6$ solutions demonstrate the role of weak and strong ligand-field splitting on the electron configuration (see discussion in Exp. 31).

The $Fe(NO_3)_3$ or $NiCl_2$ solutions can be studied from 0 to 100°C in the same manner as described for the acetylacetonate complexes. In correcting for changes in m with temperature, the density variations of pure H_2O may be used.

CALCULATIONS

Neglecting the third term of Eq. (12), the mass susceptibility of a paramagnetic solute is readily determined from m and $\Delta\nu$. The diamagnetic mass susceptibilities $\chi_{mass\ 0}$ of the solvents, in SI units of m^3 kg^{-1}, are -8.8×10^{-9} (benzene), -9.0×10^{-9} (toluene), and -9.0×10^{-9} (t-butyl alcohol–water solvent). For temperature-dependence studies, correct m according to Eq. (18), calculate χ_M and plot χ_M versus $1/T$. The slope of the best straight line through these points is the Curie constant C (see Exp. 31), and the magnetic moment μ is given by

$$\mu = \left(\frac{3kC}{N_0\mu_0}\right)^{1/2} = 797.8\sqrt{C} \text{ Bohr magneton} \qquad (19)$$

The value of $N_0\mu_0\xi$ is obtained from the intercept of the plot. Compare this value of the diamagnetic susceptibility of the solute with that estimated from Pascal's constants in Table 1. For single-temperature measurements, estimates of the latter type should be used to obtain χ_M^{para} for use in Eq. (14).

Report your values of μ in Bohr magneton and calculate the number of unpaired electrons using the spin-only formula, Eq. (15).

DISCUSSION

In an aqueous solvent, the ions Fe^{2+}, Fe^{3+}, Ni^{2+}, and Cu^{2+} are all complexed to six H_2O molecules in an octahedral arrangement. The magnetic moments can be understood in terms of the weak-field splitting of the five d orbitals as described in Exp. 31. In $Fe(CN)_6^{3-}$ the CN^- ligands are strongly bound to the Fe^{3+} ion and the ligand-field splitting is larger.

---

†This compound is hygroscopic, so the weighing should be done rapidly. It is also necessary to make the solution ~0.3 M in HNO_3 to prevent precipitation of $Fe(OH)_3$. The reference solvent used with this solution should have the same acid concentration.

The structures of $Cr(acac)_3$ and $Fe(acac)_3$ are also pseudo-octahedral, with the two oxygen atoms of each $acac^-$ ion occupying adjacent metal ligand positions. For $Co(acac)_2$ a square planar or tetrahedral arrangement of the ligands might be expected, but the energetics favoring octahedral coordination are such that there is intermolecular association to fill the two vacant ligand sites. The result is a tetramer in which a pseudo-octahedral arrangement is achieved by each cobalt atom.[6]

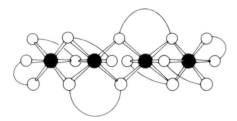

Discuss your observed magnetic moment values in terms of the number of unpaired electrons predicted for each transition-metal complex assuming octahedral structures. Comment on any differences from the predicted spin-only moments.

SAFETY ISSUES

Benzene or toluene is used as a solvent in this experiment. These potentially hazardous chemicals should be handled with care and disposed of properly. See p. 346 for further details about benzene.

APPARATUS

NMR spectrometer, with variable temperature capability if possible; several 10- or 25-mL volumetric flasks; NMR tubes; melting-point capillaries; syringe; torch for sealing capillaries; ethylene glycol and methanol in NMR tubes for temperature calibrations.

$Cr(acac)_3$, $Fe(acac)_3$, $Co(acac)_2$, where acac = acetylacetonate ($\sim$1 g each); benzene or toluene (250 mL); $Fe(NO_3)_3 \cdot 9H_2O$, $K_3Fe(CN)_6$, $FeSO_4 \cdot 7H_2O$, $NiCl_2$, $CuSO_4$ ($\sim$1 g each); 2% by volume *t*-butyl alcohol in deoxygenated water (250 mL); 3 *M* HNO_3 for acidifying $Fe(NO_3)_3$ and corresponding reference solution.

REFERENCES

1. D. F. Evans, *J. Chem. Soc.* 2003 (1959).

2. L. N. Mulay and E. A. Boudreaux, *Theory and Applications of Molecular Diamagnetism,* p. 303, Wiley-Interscience, New York (1976); R. L. Carlin, *Magnetochemistry,* p. 3, Springer-Verlag, New York (1986).

3. D. Ostfield and I. A. Cohen, *J. Chem. Educ.* **49,** 829 (1972).

4. T. H. Crawford and J. Swanson, *J. Chem. Educ.* **48,** 382 (1971).

5. E. W. Washburn (ed.), *International Critical Tables,* vol. III, p. 28, McGraw-Hill, New York (1929).

6. F. A. Cotton, G. Wilkinson, C. A. Murillo, and M. Bochmann, *Advanced Inorganic Chemistry,* 6th ed., Wiley, New York (1999).

GENERAL READING

F. A. Cotton et al., *Advanced Inorganic Chemistry,* 6th ed., Wiley, New York (1999).

R. S. Drago, *Physical Methods for Chemists,* 2d ed., chaps. 10–12, Saunders, New York (1992).

D. F. Shriver and P. W. Atkins, *Inorganic Chemistry,* 3d ed., Freeman, New York (1999).

EXPERIMENT 33
Dynamic Light Scattering

Light scattering will be used in this experiment to study the diffusive motion of submicron-sized polystyrene spheres suspended in a fluid.[1,2] Both static and dynamic light scattering are powerful techniques for the investigation of systems with spatial inhomogeneities with sizes ranging from a few nanometers (much smaller than the wavelength λ of the incident light) to a few micrometers. Such inhomogeneities cause local variations in the refractive index that result in the scattering of a light beam. Examples are polymer solutions, micelles, gels, colloids, and binary liquid mixtures near a critical point. Perhaps the most familiar example is the Tyndall effect, which is light scattered by colloidal particles—for instance, automobile headlights in a fog (due to water droplets) or a beam of sunlight in a smoky room (due to dust particles).

One of the practical applications of dynamic light scattering involves the determination of particle sizes in media dispersed as dilute suspensions in a liquid phase.[3,4] This aspect of dynamic light scattering is the focus here. Analysis of the scattering data will yield the translational diffusion constant D for a dilute aqueous suspension of polystyrene spheres, and this is directly related to the radius of the spheres. In addition, scattering will be studied from dilute skim milk, which reveals that a distribution of particle sizes exists for this system.

THEORY

Polystyrene (PS) spheres suspended in a liquid will undergo Brownian motion due to the impacts of the solvent molecules.[5] The impact forces on these PS spheres are random in magnitude, direction, and time. Since the positions of the particles are continually changing due to the Brownian motion, when a PS dispersion is exposed to an incident laser beam, the scattered electric field (which is a function of the particle positions) is also constantly changing. Thus, the intensity of light scattered at any given angle will fluctuate with time due to the diffusive motion of the scattering particles. Such time-dependent light scattering is called dynamic light scattering (DLS). The scattered photons are also subject to tiny frequency shifts (roughly a few Hz to a few kHz) from the input laser frequency due to the motion of the scattering particles (quasi-elastic scattering). If this is analyzed in the frequency domain with a spectrum analyzer, DLS is a form of extremely high resolution translational spectroscopy. In this experiment, we will be concerned with DLS data obtained in the time domain, but it should be noted that time-domain and frequency-domain measurements are related by a simple Fourier transform operation.

As an introduction to the discussion of dynamic light scattering, a brief description of the general concept of autocorrelation functions will be useful.[2,6] Consider a macroscopic (bulk) property $A(t)$ whose value depends on the positions and momenta of all the particles in the system. A simple example would be the pressure of a gas on the walls of the container. The equilibrium value $\langle A \rangle$ is a long-time average given by

$$\langle A \rangle = \lim_{T \to \infty} \frac{1}{T} \int_0^T A(t)dt \tag{1}$$

and this is independent of time. However, the quantity $A(t)$ will vary about $\langle A \rangle$ over short time scales due to fluctuations and $A(t + \Delta t) \neq A(t)$ in general, as shown in Fig. 1a.

For equilibrium or steady-state systems, an autocorrelation function $\langle A(0)A(\Delta t) \rangle$ can be defined by

$$\langle A(0)A(\Delta t) \rangle = \lim_{T \to \infty} \frac{1}{T} \int_0^T A(t)A(t + \Delta t)dt \tag{2}$$

and this quantity depends only on the time difference Δt and not on the choice of the initial time. The correlation function $\langle A(0)A(\Delta t) \rangle$ is a measure of the "phase relation" between two instantaneous values $A(t)$ and $A(t + \Delta t)$ of a "signal" that oscillates erratically (quasi-randomly) about the average value $\langle A \rangle$. When Δt is close to zero, the signals are in phase (i.e., they are correlated) and $\langle A(0)A(\Delta t) \rangle \approx \langle A(0)A(0) \rangle = \langle A^2 \rangle$. When Δt is very large, the phases are random (i.e., the signals are uncorrelated) and $\langle A(0)A(\Delta t) \rangle = \langle A(0) \rangle \langle A(\Delta t) \rangle = \langle A \rangle^2$, which is smaller than $\langle A^2 \rangle$. In many cases, the autocorrelation function decays like a single exponential, as shown in Fig. 1b:

$$\langle A(0)A(\Delta t) \rangle = \langle A \rangle^2 + [\langle A^2 \rangle - \langle A \rangle^2]\exp(-\Delta t/\tau) \tag{3}$$

where the quantity τ is the relaxation time (also called the correlation time).

The discussion here of dynamic light scattering will be limited to the simple case of Brownian diffusion of scattering spheres dispersed as a dilute solute in a liquid solvent.[1] As indicated in Fig. 2, there is a collimated incident light beam with wave vector $\mathbf{k_0}$ and the scattered light leaves the sample in all possible directions, each corresponding to some scattering angle θ and associated exit wave vector $\mathbf{k}$. The scattering wave vector $\mathbf{q}$ is given by $\mathbf{q} = \mathbf{k_0} - \mathbf{k}$. Since $\mathbf{k} \approx \mathbf{k_0} = 2\pi/\lambda \gg \mathbf{q}$, the magnitude of $\mathbf{q}$ is given by

$$q = |\mathbf{k_0} - \mathbf{k}| = (4\pi/\lambda)\sin(\theta/2) = (4\pi n/\lambda_{vac})\sin(\theta/2) \tag{4}$$

FIGURE 1

(a) Variation of a macroscopic property $A(t)$, which fluctuates in time. $\langle A \rangle$ represents the long-time average. (b) The autocorrelation function $\langle A(0)A(\Delta t) \rangle$ for a simple system that undergoes exponential decay; τ is the relaxation time (correlation time).

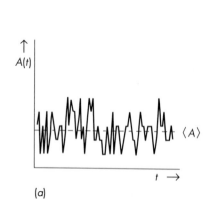

(a)

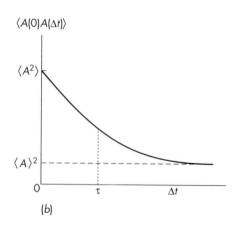

(b)

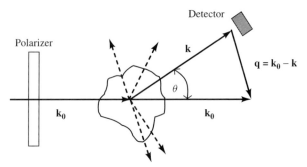

FIGURE 2
Incident light with wave vector $\mathbf{k_0}$ is scattered in all directions. The scattered light of wave vector $\mathbf{k}$ that is observed depends on the position of the detector. The scattering wave vector $\mathbf{q} = \mathbf{k_0} - \mathbf{k}$ for this detected light has a magnitude $q = (4\pi/\lambda)\sin(\theta/2)$, where λ is the laser wavelength in the sample.

where λ is the wavelength of light in the solution and n is the refractive index of the solution at λ_{vac}, the laser wavelength in vacuum (or air).

The amplitude of the electric field at the detector due to scattering by all N particles in the illuminated volume is

$$E(q,t) = E_0(q) \sum_{j=1}^{N} \exp[i\mathbf{q} \cdot \mathbf{r}_j(t)] \tag{5}$$

where $E_0(q)$ is the amplitude of the scattered field from a single particle and $\mathbf{r}_j(t)$ is the position of the jth particle. The exponential term in Eq. (5) is of the form $\exp(i\varphi_j)$, where φ_j is the phase shift in a wave scattered by a particle at $\mathbf{r}_j$ relative to that scattered by a particle at $\mathbf{r} = 0$. Since the scattering particles undergo random Brownian motion, the intensity of the scattered light $I = |E|^2$ will fluctuate with time at a rate dictated by the particle motion. Thus, the scattered radiation consists of a random "speckle" pattern of bright and dim regions that vary in intensity irregularly with time for any given $\mathbf{q}$. If the scattered light intensity is sufficiently large, this speckle pattern can be observed directly with the eye by placing a screen near the scatterer. The photodetector used to record the time-dependent scattering intensity at a given angle θ should collect light through an aperture of area comparable to the speckle size (corresponding to a diameter of about 0.1 to 0.3 mm for the imaging used in this experiment).

Two kinds of time autocorrelation functions can be defined for light scattering: $G_I(q, \Delta t) = \langle I(q,0)I(q,\Delta t) \rangle$ for the *intensity* and $G_E(q,\Delta t) = \langle E(q,0)E(q,\Delta t) \rangle$ for the *field*. It is convenient to consider normalized versions of these:

$$g_I \equiv G_I / \langle I(q) \rangle^2 = \langle I(q,0)I(q,\Delta t) \rangle / \langle I(q) \rangle^2 \tag{6a}$$

$$g_E \equiv G_E / \langle |E^2(q)| \rangle = \langle E(q,0)E(q,\Delta t) \rangle / \langle |E^2(q)| \rangle = \langle E(q,0)E(q,\Delta t) \rangle / \langle I(q) \rangle \tag{6b}$$

For scattering by Brownian particles, $E(q,t)$ is a Gaussian variable (varies with time in a Gaussian fashion) when the number of particles in the observed scattering volume is large.[1] The two autocorrelation functions are then related by

$$g_I(q,\Delta t) = 1 + B\, g_E^2(q,\Delta t) \tag{7}$$

where B is an experimental constant factor less than 1 that depends on the sample, the beam profile, and the detector collection optics. For the simple diffusion of rigid noninteracting (dilute) monodisperse spherical scattering particles with translational diffusion constant D, the normalized *field* autocorrelation function g_E is given by

$$g_E(q,\Delta t) = \exp(-\Gamma_q \Delta t) \tag{8}$$

where $\Gamma \equiv 1/\tau$ is the correlation decay rate. One also finds that $\Gamma_q = Dq^2$ for a given scattering wave vector q.[4,7] Thus, the intensity autocorrelation function $g_I(q,\Delta t)$ is given by

$$g_I(q,\Delta t) = 1 + B \exp(-2Dq^2\Delta t) \tag{9}$$

which is a normalized form of what is known as the Siegert relation.[1]

The Siegert relationship is rarely violated unless the number of scattering particles in the scattering volume is very small or very large or the viscosity of the solvent is extremely high. In the case of very low particle concentrations, the scattering is weak since it comes from photons scattered by only a few particles. Then number fluctuations dominate rather than diffusive fluctuations, and one gets some kind of "swimming speed" distribution. In the case of very high particle concentrations, particle-particle interactions and extensive multiple scattering complicate the analysis and interpretation of data.[8] In the case of highly viscous solvents, the system becomes nonergodic† with very long nonexponential time scales for relaxation. Much interesting and challenging "glassy" behavior is seen in this regime, but special DLS experimental techniques are required.[9]

It should be noted that most physical properties of interest are directly related to the field autocorrelation function [Eq. (8) is an example], whereas DLS measures the intensity autocorrelation function. The Siegert relationship, Eq. (7) or (9), is a bridge between these two types of autocorrelation function. As stated above, the validity of Eq. (7), and thus of Eq. (9), depends on the assumption that the particle positions are statistically independent and that there is a reasonably high particle density. There will then be a sufficient number, say 30 or more, of scatterers in the volume illuminated and detected (about 10^{-6} cm$^3 = 1$ nL) so that the scattered electric field is a statistical Gaussian random variable. If the particle concentration is too low, number fluctuations begin to play a role, causing an additional slowly decaying relaxation term in the autocorrelation function. Typical concentrations of scattering particles should lie in the range 10^7–10^9 cm^{-3}. In the case of 100-nm diameter particles, a number density of 10^9 cm^{-3} means a volume fraction of 5×10^{-7}, and for a 0.1-mm diameter beam that means ~ 500 particles in the scattering volume.

The fluctuating motion of suspended PS spheres can be related to the random-walk problem. Einstein was the first to show the connection between this random-walk Brownian motion and Fick's laws of diffusion. The resulting Stokes-Einstein relation is[10]

$$D = kT/6\pi\eta R \tag{10}$$

where η is the solvent viscosity and R is the radius of a spherical scattering particle.‡ Thus, DLS provides a method for evaluating the size of small particles suspended in a fluid by measurement of $g_I(q,\Delta t)$ for a well-defined scattering angle θ.

METHOD

The traditional method of using DLS to determine the intensity autocorrelation function is to detect the scattered light at a fixed angle with a photomultiplier tube and then process this signal with a hard-wired electronic correlator.[4,11] Commercial instruments with hardware and software optimized for this purpose are available from manufacturers such as Brookhaven Instruments, Malvern Instruments, and others. Here we will utilize

---

†For ergodic systems, the statistical ensemble average equals the long-time time average for a single system. For nonergodic systems like glasses and gels, this is no longer true and very slow nonexponential dynamics are observed.

‡If the scattering particle is irregularly shaped, the value of R obtained from Eq. (10) is an effective average called the hydrodynamic radius.

a simpler and less-expensive apparatus and data collection procedure that is adequate for this experiment and that more directly illustrates the principles of DLS. Figure 3 shows the experimental setup, which consists of the elements described below.

Laser. Almost any laser source of power greater than 1 mW can be used for this experiment, but since the scattered intensity increases with power, helium-neon or red diode laser sources of 10 to 100 mW are commonly employed in commercial instruments. Inexpensive green 532-nm modules providing 10 to 50 mW or red laser diodes at 650 to 660 nm are recommended here but, if available, argon or krypton ion lasers can also be used. **As in any laser experiment, it is important to follow all safety guidelines (Appendix C) and to avoid any eye exposure to direct or reflected laser beams. Laser safety goggles should be worn when setting up the experiment and when inserting or removing samples.**

An important characteristic of the laser is its coherence length, the distance over which a phase measurement of the light wave can vary by π radians. This length is related to the spectral width $\Delta\lambda$ of the laser and is given by $L_{coh} \sim \lambda^2/\Delta\lambda$. Helium-neon lasers typically have a coherence length of 20 cm, while lengths for narrow-band semiconductor lasers can exceed 100 m. Inexpensive 532-nm sources are multimode sources with much shorter coherence lengths (1 to 3 mm) but even these have proven adequate for this experiment since the length viewed in the sample is only 0.1 to 0.3 mm. Whatever the source, the incident beam should be polarized with the electric vector vertical to the horizontal scattering plane, and a polarizer sheet should be added and oriented to ensure a polarization ratio in excess of 500:1.

Imaging. The sample cell should be a 1-cm square cross-section cuvette mounted in a suitable holder so that samples can be reproducibly positioned. The laser beam is focused at the center of the cuvette with a lens of focal length f of about 75 mm, and the beam then passes through a viewing sheet with a hole in it to a blackened beam stop to minimize

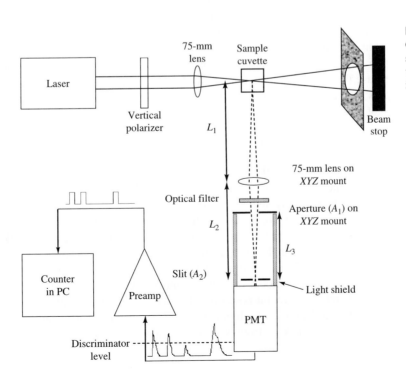

FIGURE 3

Optical layout for a 90° scattering experiment. The vertical polarization direction is normal to the page.

background light. (This arrangement allows one to visually observe the forward scattered speckle pattern of a sample.) For the measurements at a 90° scattering angle made in this experiment, the scattering volume in the region of the focus can be considered to be that of a horizontal cylinder of diameter $d_s \approx 2.44 \lambda f/d_l$, where d_l is the diameter of the laser beam. For a 532-nm beam with $d_l = 1$ mm imaged by a 75-mm focal length lens, d_s is about 0.1 mm and, if a length of 0.1 mm along the cylinder is imaged to the detector, the scattering volume is about 10^{-3} mm^3 or 1 nL.

The light scattered in the horizontal plane at 90° is imaged with a 75-mm focal length lens positioned such that L_1 and L_2 in Fig. 3 are each 150 mm (1:1 imaging, no magnification). This lens should be mounted on an *XYZ* stage to allow precise positioning of the image of a highly scattering test sample of polystyrene onto an aperture (A_1) of 0.1- to 0.3-mm diameter or, better, onto an adjustable slit attached to the front flange of the photomultiplier tube (PMT) housing. The slit A_2 is positioned for 90° scattering and is oriented vertically so that its width determines the length of the scattering image viewed by the photomultiplier. To limit the angular spread collected by the lens, an adjustable iris A_1 on an *XYZ* stage is placed behind the lens at a distance $L_3 \sim 100$ mm from the slit. Detection of coherently scattered light is optimal when the aperture diameter d_1 and slit width d_2 pass one speckle area, which corresponds to the condition $d_1 d_2 = \lambda L_3$.[3] For weakly scattering samples, the $d_1 d_2$ product can be increased up to $\sim 5\lambda L_3$ in order to have reasonable signal levels.

The alignment of the collection optics should be optimized using a test PS sample for which the strongly scattering laser track can be clearly seen in the cuvette. With the PMT off and the laboratory room darkened, aperture A_1 can be opened fully to permit visual centering of the scattering image onto the slit A_2. With the room lights still extinguished and the slit A_2 initially closed, the PMT is then turned on and the photon count rate is measured as A_2 is slowly opened. During the image optimization, the slit A_2 should be adjusted to keep the photon count rate from the PMT at less than about 10^6 cps (counts per second). With aperture A_1 wide open, the *xyz* lens position is iteratively adjusted to give a maximum count rate. After this imaging is optimized, aperture A_1 is closed down to its minimum of about 0.5 mm and its *xyz* position is varied to give maximum count rate. For the samples examined in this experiment, count rates in the range 10^5 to 10^6 cps are required for good counting statistics. When the laser is blocked and all light is extinguished, the background count level of the detector should be of the order of 200 cps or less. For convenience in operating under ambient lighting conditions, a light-tight enclosure can be constructed to encase the region from aperture A_1 to the PMT. Although not necessary, a narrow band-pass filter at the laser wavelength placed just in front of A_1 will further restrict background light from reaching the detector.

Detection System. A good photon counting module with an amplifier-discriminator is important for this experiment. Suitable photomultiplier tubes with low dark counts are the RCA 31034A and the Hamamatsu R649 tubes. Hamamatsu also offers the photon-counting modules HC124 and HC125 that provide the advantage that they contain a built-in high-voltage supply and amplifier/discriminator. A comparable module PMT120-OP is available from Correlator.com. In each case, the high-voltage level is optimized for photon counting and the discriminator level is adjusted for highest signal-to-background ratio at the operating voltage. The output for each photon should be a TTL pulse of 2- to 5-volt height and 10- to 50-ns width so that it is easily detected by an inexpensive digital counter board such as the National Instruments model M622x. This board fits into a PC expansion slot and is general purpose since it also does analog-to-digital conversions. As shown below, data collection and subsequent processing can be conveniently done using the Lab-VIEW software package from National Instruments.

EXPERIMENTAL

Sample Preparation. The polystyrene spheres to be used should be monodisperse with a particle radius R of about 50 nm, although any size in the range $R = 30$ to 100 nm is suitable. Such nanospheres are available commercially as aqueous latex suspensions with 1% to 10% PS by weight. A small amount of this latex suspension should be diluted 100- to 1000-fold. Using a microsyringe, take 0.1 mL from the PS stock, deliver this into a rinsed dilution bottle, and then add 10 mL of a filtered 10-mM solution of NaCl or other 1:1 electrolyte. The purpose of this electrolyte is to partially suppress coulombic interactions (electrostatic double-layer repulsion) that can influence the diffusion constant and lead to R values that are artificially high by $\sim 10\%$. The electrolyte solution should be prepared from distilled water and stored at room temperature. Before use, it must be filtered through a suitable membrane (0.1-μm pore size) to remove dust particles. Avoidance of dust is crucial, and capped dilution bottles should be used.

Further dilution can be made as needed to achieve the desired final concentration of PS in about 10 mL of saline solution. Rinse all dilution bottles with filtered salt solution before use. In carrying out the dilution, samples should be mixed by gently inverting the dilution bottles many times, with care being taken to avoid the formation of air bubbles. *Do not shake samples.*

Rinse a clean cuvette three times with filtered salt solution and then *pour* 2–3 mL of diluted latex suspension into the cuvette; pouring avoids the introduction of new dust particles from the use of syringes or pipettes. It may be useful to subject the cuvette to an ultrasonic bath for about 30 seconds, to reduce the effect of any aggregation of particles and to eliminate bubbles. The surfaces of the sample cuvette should be scratch free and wiped to eliminate any fingerprints. The final PS concentration should be such that the attenuation of the light beam on passing through the cell is ~ 5 percent or less (about 20 to 40 μg/mL for particles of about 50-nm radius). At such a concentration, the solution will appear slightly turbid and the focused beam in the sample will look like a crisp rather than a fuzzy line.

Because the viscosity of water changes with temperature, the samples should ideally be in a thermostatted cell at 25°C but, for convenience, the experiment can be done at room temperature, which should be recorded. The sample cuvette can be glass or plastic, and care should be taken to avoid scattering by scratches, smudges, or fingerprints on the cell walls. *Again, it is important to eliminate all traces of dust particles and any air bubbles in the sample* since scattering from these can overwhelm that from the polystyrene spheres.

Measurements will also be made on a sample of *skim* milk (i.e., nonfat milk that is free from all fat globules), in which the small spherical casein micelles dispersed in water scatter all wavelengths of light to produce the characteristic white color. Dilution of about 500:1 with the filtered 10-mM 1:1 electrolyte solution will give a suitable nonfuzzy laser track through the sample. The sample should be fresh, made up from a small sample of cold milk, and diluted by about a factor of 500 with room temperature saline solution.

Data Collection and Processing. A PC counter board with timers will be used to count the number of scattered photons in each of N_c successive time channels, each of width δt, where δt is chosen to be about one-twentieth the expected correlation decay time τ. For particles of 50-nm radius, a value of $\delta t = 20$ μs is reasonable; for significantly larger or smaller size particles, δt can be increased or decreased proportionately. For test runs, a total data collection time $N_c \delta t$ of 0.2 second ($N_c = 10{,}000$) is convenient. The resultant array of N_i photon counts for the ith channel of width δt is the raw data to be saved and processed. An expression for the sum that approximates the integration given in Eq. (2) for the autocorrelation function is, for the jth element of g_I at time $\Delta t_j = j\delta t$,

$$g_I(\Delta t_j) \equiv \frac{\langle I(q,0)I(q,\Delta t_j)\rangle}{\langle I(q)\rangle^2} = \frac{\left(\sum_{i=1}^{N_c-j} N_i N_{i+j}\right)/(N_c - j)}{\left(\sum_{i=1}^{N_c} N_i/N_c\right)^2} \qquad (j = 1, 2, ... M). \quad (11)$$

The value of $g_I(0)$ at time 0, corresponding to $j = 0$, could be calculated but it must be excluded in fitting any data since it produces a discontinuous spike even for totally random data. The maximum value of j is $N_c - 1$, corresponding to the total time of the measurement, but there is little point in calculating g_I much beyond a time $\Delta t = 5\tau/2$ where the exponential in Eq (9) has decreased to $e^{-5} = 0.0067$. After that period, correlation has been effectively lost and the g_I values are essentially reflecting fluctuations in the background level. For this reason, the j range could be limited to an upper value $M =$ about 50 when the channel width δt is about $\tau/20$, as recommended for this experiment. Because an initial estimate of τ may not be available, and in order to obtain a good measure of the background level and its noise, it is recommended that the j range be extended somewhat to $M = 200$ points.

For the final runs, the scan length $N_c\delta t$ should be set at 60 seconds, corresponding to $N_c = 3,000,000$, and the computer is then used to collect all the data and to calculate the g_I values, yielding a dramatic reduction of noise in the g_I calculated curves. This is most conveniently done using programming methods such as Visual Basic or, as employed here, LabVIEW. The latter requires some learning effort but is particularly useful for this experiment, and two LabVIEW Virtual Instruments (VIs) created for this experiment are shown in Figs. 4 and 5.† Although written for a specific counter board (National Instruments M series 622x), the details of these VIs will vary only slightly for other boards.

Figure 4 shows a simple VI that serves as a frequency counter for use in monitoring the PMT count levels during the initial optical imaging of the scattered light when settings are being optimized. It uses internal clocks on the board for a time reference and a counter (ctr0) to sense the rising edge of each TTL pulse from the PMT amplifier/discriminator, which is introduced to the appropriate pin (PFI8) on the board input connector. The result is shown every 0.5 seconds on the Front Panel display on the screen until the mouse is clicked on the stop icon. For an actual data run, the count rate should ideally be a few hundred thousand; thus, some adjustment of slit widths and sample concentration may be necessary.

Figure 5 shows a VI that can be used for data collection and for performing the auto-correlation calculation. It consists of three basic parts. The first (A) uses a 10-MHz clock and a counter (ctr1) on the board to generate a square wave at a specified frequency f, say 50,000 Hz, such that the period (1/f) is 20 μs, the desired δt channel time over which scattered photons are to be counted. The second part (B) uses a counter (ctr0) to sense the rising edge of each TTL pulse from the PMT/amplifier-discriminator and to keep a running total. The rising edge of the square wave from A is sent internally on the board from the output of ctr1 (PFI3) to ctr0. This causes part B to store the counts (a running total) in a buffer every 20 μs and this process is repeated for N_c precise time intervals during a run.

---

†It is not practical to give the complete details of each of the VI icons here, but these can be seen when running LabVIEW by placing the mouse cursor on each element and reading the extensive help files provided. Further information and helpful examples are provided on the National Instruments LabVIEW site http://www.ni.com/labview/.

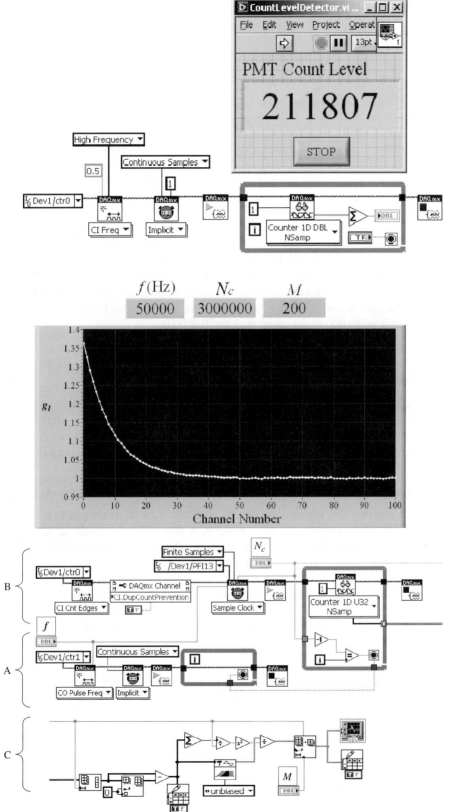

FIGURE 4

LabVIEW Front Panel display (top) of photomultiplier count rate in counts per second (cps) and block diagram (bottom) for a Virtual Instrument (VI) that measures this. The VI is used to count scattered photons using a counter (ctr0) on a National Instruments M622x PC board and is useful in optimizing the imaging optics and for ensuring adequate count levels for the experiment.

FIGURE 5

LabVIEW Front Panel display of g_I function (top) and block diagram (bottom) for a Virtual Instrument (VI) for dynamic light scattering measurements. The VI is used to count scattered photons using a National Instruments M622x PC board, and the basic functions of parts A to C are described in the text. The detailed meaning of each icon, along with descriptions of the controlling inputs and outputs, can be seen in LabVIEW by positioning the mouse cursor on the icon. See also the discussion in Chapter III of the use of LabVIEW in interfacing situations.

In part C the data in the counter board buffer are transferred into the memory of the computer and the differences between successive count values are calculated to obtain the counts for each δt interval. These differences (N_j values) are stored as an array of N_c values that can be saved as an output file, if desired. If N_c is chosen to be, say, 10,000, this output file can be used for a direct Excel calculation of the autocorrelation, as described as an exercise at the beginning of the next section. For the final run measurements, where N_c is set at 3,000,000, there are too many points to handle in a spreadsheet and it is much more efficient to calculate these with a program or using autocorrelation features of programs such as IGOR PRO, Origin, and others. In the present VI, we use an autocorrelation operation available as an option for the LabVIEW program.† This is done in the middle portion of part C, and the resultant $g_I(\Delta t_j)$ values for $j = 1$ to 200 are computed and stored as a single column in an output file that can be read by Excel for subsequent analysis. The g_I values are also displayed versus the channel number ($j = 1$ to 100) in the Front Panel, as shown at the top of Fig. 5.

The experiment consists of preparing a PS sample, optimizing the imaging of the scattered light to obtain a count level of about 10^5 or more, and then collecting data. To initiate a measurement, f, N_c and M values are entered into the boxes on the Front Panel and the VI is activated. Runs should be made with $N_c = 10,000$ and with $N_c = 3,000,000$ and the output files saved for both of these. For the skim milk measurements, only a run with $N_c = 3,000,000$ is needed but the correlation time is expected to be longer and f should be decreased from 50,000 to the range 10,000–20,000 Hz.

CALCULATIONS

Polystyrene Spheres. As an initial exercise, Excel should be used to calculate the autocorrelation function $g_I(q,\Delta t)$ from PS data obtained using $N_c = 10,000$. In Excel, a calculation equivalent to that of Eq. (11) can be done by placing the raw data values in cells A1 to A10000. With nothing (or 0) in cells A10001 to A20000, one can calculate g_I in column B by entering in cell B1 the formula

SUMPRODUCT($A$1:$A$10000,A1:A10000)/ROWS(A1:$A$10000)/(AVERAGE($A$1:$A$10000)^2

and then copying this formula to cells B2 to B200. Cell B1 corresponds to $g_I(0)$ and is excluded in fitting the data to get a correlation time. Plot the resulting $g_I(q,\Delta t)$ values (B2 to B200) versus the delay time (a column C2 to C200 containing Δt, $2\Delta t$, $3\Delta t$. . .). Compare this result obtained using an Excel calculation of $g_I(\Delta t)$ with that given by the LabVIEW program.

You should observe a dramatic improvement in the noise in $g_I(q,\Delta t)$ when a data set is taken with $N_c = 3,000,000$. The LabVIEW $g_I(q,\Delta t)$ values should be used to determine the diffusion constant D and thus the particle radius R, using Eq. (9) in the form

$$g_I(q,\Delta t) - 1 = A + B \exp(-2Dq^2\Delta t) \tag{12a}$$

$$\ln[g_I(q,\Delta t) - 1] = \ln[A + B \exp(-2Dq^2\Delta t)] \approx \ln B - 2Dq^2\Delta t \tag{12b}$$

If the evaluation of the long-time baseline value $\langle I(q)\rangle^2$ for each run were perfect, then the quantity A would be 0, as seen from Eq. (9), but it is better as a practical matter to let A be an adjustable fitting parameter.[12] Its deviation from zero serves as a test that the

---

†The "unbiased" form of the LabVIEW autocorrelation operation should be used. This generates a symmetric two-sided function of 20,000 points from 10,000 data, with the time 0 data point in the middle. The first half of this array is redundant so only the desired region of the second half is stored in part C.

equipment and analysis are working properly; fit values for the parameter A of less than about ± 0.02 are reasonable.

Since the function in Eq. (12) is nonlinear, fitting should be done with a nonlinear least-squares program, which can be found in LabVIEW and in other software packages such as IGOR PRO and Origin. A useful website that allows one to easily paste data and enter a function for on-line nonlinear fitting is http://statpages.org/nonlin.html. One can also use the Solver feature of Excel for nonlinear fitting, as described in Chapter III. It is recommended that only data up to $j = 50$ be fitted, using Eq. (12a) rather than (12b), since the latter can give problems with negative arguments of the natural log function.

Report the values of the fit parameters A, B, Dq^2 and the corresponding correlation time τ. Using the q value obtained from Eq. (4) for $90°$ scattering, determine the diffusion constant D and from that the radius R of the polystyrene particles. For dilute PS solutions in the 20–25°C range, the water refractive index value of 1.334(2) at 25°C can be used in these calculations since n varies only slightly with temperature and wavelength in the visible region. The solvent viscosity η that appears in Eq. (10) is 0.891 centipoise $= 0.891 \times 10^{-3}$ N m^{-2} s for water at 25°C. Since η changes by about 2% per K, its value should be calculated at the temperature of the measurements. The Arrhenius expression $\ln \eta(cP) = -7.045 + 2066/T(K)$ can be used as a reasonable approximation for the 290–300 K range.[13]

In order to test the near-linearity of a semilog plot expected from Eq. (12b) and to show any experimental scatter in the data points or systematic deviations from exponential decay, plot your PS data in the form $\ln[g_I(q,\Delta t) - 1 - A]$ versus Δt and add the best fit line $\ln B - 2Dq^2 \Delta t$. Plot also the difference between these and examine these residuals for any evidence of nonrandom or systematic errors. Report the standard errors in the fitting parameters provided by the fitting program and calculate the error in the PS radius. (Solver does not give the standard error in nonlinear fitting parameters like other nonlinear least-squares programs do, but it is possible to estimate these using procedures described in Ref. 14.)

Dilute Milk Sample. For the $g_I(\Delta t)$ data acquired on the dilute milk sample, carry out an initial least-squares fit with Eq. (12a) and plot $\ln[g_I(q,\Delta t) - 1 - A]$ versus Δt for that system, adding the best fit line $\ln B - 2Dq^2 \Delta t$. A linear variation for the experimental points is unlikely since the casein micelles in milk are not monodisperse, that is, they exhibit a distribution about some mean size. Although this distribution is fairly narrow, its presence should be obvious from the curvature of the semilog plot. Thus, we need to address the issue of the analysis of polydisperse samples.

Since DLS does not detect single particles, the interpretation of data on polydisperse samples is quite a bit more complex than that for monodisperse samples. In essence, the expression for the field autocorrelation function g_E given in Eq. (8) must be generalized to

$$g_E(\Delta t) = \int_0^{\infty} f(\Gamma) \exp(-\Gamma \Delta t) \, d\Gamma \tag{13}$$

where $f(\Gamma)$ is the distribution function for the decay rate of different-sized particles. The quantity $f(\Gamma) \, d\Gamma$ is the fraction of particles with decay rates between Γ and $\Gamma + d\Gamma$. This is a normalized value; i.e., $\int_0^{\infty} f(\Gamma) \, d\Gamma = 1$. Although it is easy to calculate $g_E(\Delta t)$ from a known distribution $f(\Gamma)$, the reverse transformation to extract $f(\Gamma)$ from an experimentally known $g_E(\Delta t)$ is extremely difficult.[4]

The simplest data analysis technique for dealing with polydisperse systems is the *method of cumulants*.[3,4] In this method, Eq. (13) is expanded about an average decay rate $\overline{\Gamma}$ to yield

$$g_E(\Delta t) = \exp\left\{(-\overline{\Gamma} \Delta t)\left[1 + \frac{\mu_2}{2!}(\Delta t)^2 - \frac{\mu_3}{3!}(\Delta t)^3 + \ldots\right]\right\} \tag{14}$$

where

$$\overline{\Gamma} = \int_0^\infty \Gamma f(\Gamma) d\Gamma = \overline{D}q^2 \tag{15}$$

is the first moment and the nth moments for $n \geq 2$ are given by

$$\mu_n = \int_0^\infty (\Gamma - \overline{\Gamma})^n f(\Gamma) d\Gamma \tag{16}$$

As a result, Eqs. (12) have added terms and now take the form

$$g_I(\Delta t) - 1 = A + B \exp\left[-2\overline{\Gamma}\Delta t + \mu_2(\Delta t)^2 - \frac{1}{3}\mu_3(\Delta t)^3 + \ldots\right] \tag{17a}$$

$$\ln[g_I(\Delta t) - 1] = \ln\left(A + B \exp\left[-2\overline{\Gamma}\Delta t + \mu_2(\Delta t)^2 - \frac{1}{3}\mu_3(\Delta t)^3 + \ldots\right]\right) \tag{17b}$$

where the coefficients of $(\Delta t)^n$ are known as cumulants. In practice, only the first two moments can be evaluated with reasonable certainty; so the terms $(\Delta t)^3$ and higher will be neglected here.

For simple center-of-mass diffusive motion, the mean decay rate $\overline{\Gamma}$ can be shown[1] to be $\overline{D} q^2$, where $\overline{D}$ is the intensity-weighted average $\langle D \rangle$. The use of $\overline{D}$ in the Stokes-Einstein equation, Eq. (10), yields an effective radius $\overline{R}$. The second moment μ_2 is given by

$$\mu_2 = \langle D^2 \rangle - \langle D \rangle^2 \tag{18}$$

and a polydispersity index Q can be defined as

$$Q \equiv \mu_2/\overline{\Gamma}^2 = [\langle D^2 \rangle - \langle D \rangle^2]/\langle D \rangle^2 \tag{19}$$

This index is a measure of the width of the decay rate distribution $f(\Gamma)$ and thus the range of diffusion constants (and hence sizes) present in the sample. When $Q = 0$, the sample is monodisperse. A nonzero value of Q will, however, not tell one whether the distribution is a broad symmetric distribution, a skewed distribution, or possibly a bimodal distribution since Q models only the width and not the shape of the distribution.

Since the PS reference sample is almost monodisperse, a cumulant analysis of that material would yield a very small Q, say $Q < 0.03$. That is, all the correction terms are negligible and Eqs. (17) collapse to Eqs. (12). But cumulant analysis is a useful way to handle practical samples such as pigments, inks, microemulsions, swollen micelles, globular proteins, and spherical virus particles, where there is a size distribution but one that is not very broad (say $Q < 0.3$). This analysis should be made for the milk data using a nonlinear least-squares fitting of Eq. (17a), neglecting μ_3 and all higher order terms. Report the $\overline{\Gamma}$, $\overline{D}$, and $\overline{R}$ values as well as the second cumulant μ_2 and the polydispersity index Q.

DISCUSSION

Compare your DLS value of R for PS with the value of the particle radius given by the supplier of the sample. Do these R values agree, given your estimated uncertainty value and the error limit quoted for R by the supplier? Report also any polydispersity information given by the PS supplier.

Predict the changes that would occur if the PS spheres were studied in a more viscous solvent like propanol. In particular, how would the time scale needed for data acquisition change? Estimate the diffusive relaxation time τ for your PS spheres in propanol at 25°C (for which $\eta = 1.945$ cP and $n = 1.385$).

For the milk sample, make a rough estimate of the width of the distribution in decay rates by *assuming* that $f(\Gamma)$ is a Gaussian distribution, that is, $f(\Gamma) = C \exp[-(\Gamma - \overline{\Gamma})^2/2\sigma^2]$ Normalize this, so that $\int_0^\infty f(\Gamma)\,d\Gamma = 1$ (i.e., find the value of C), and substitute this $f(\Gamma)$ into Eq. (16) to obtain an expression for μ_2. *Hint.* You will need to use error functions. Equate the resulting expression to your experimental value $\mu_2 \equiv \overline{\Gamma}^2 Q$, and calculate a value of the width σ. Then calculate radii corresponding to $\overline{\Gamma}(\text{eff})$ values of $\overline{\Gamma} + \sigma$ and $\overline{\Gamma} - \sigma$ to get an estimate of the size range in the milk sample.

In fact, in some systems, $f(\Gamma)$ is known to be a log-normal distribution,[15] for which $\overline{\Gamma}$ and μ_2 can be transformed into a mass-mean radius and a geometric standard deviation. As an optional exercise, carry out such an evaluation.

SAFETY ISSUES

The laser power delivered to the scattering cell is very low (few mW). At this power level, the incident beam is relatively harmless, but of course the beam should never be viewed directly. Further discussion of laser safety practices is given in Appendix C.

APPARATUS

Optical set-up similar to that shown in Fig. 3, assembled on a small optical table or on a 30″ × 30″ metal plate or breadboard set on a sturdy bench (available from Newport Corp.); 5- to 50-mW laser with a coherence length greater than a few mm—(suitable sources are HeNe or argon gas lasers, 630- to 650-nm red diode lasers, and 532-nm Nd:YAG diode-pumped lasers (available from Melles Griot, Coherent Laser, CrystaLaser, and other manufacturers). Although not ideal because of their shorter coherence length, multimode 532-nm green laser modules from a variety of suppliers are adequate for this experiment; a 5-mW laser pointer will also work but a higher power source is better. Optical components, including a sheet dichroic polarizer, two 75-mm focal length lenses, a narrow band-pass filter at the laser wavelength, an iris aperture closing to 0.5 mm, and a 0- to 3-mm adjustable slit, along with mounts to hold these items and two *XYZ* stages (available from Newport Corp., Melles Griot, and other suppliers); photomultiplier detector with amplifier/discriminator for photon counting such as the Hamamatsu HC125 or Correlator.com PMT120-OP, or others described in the Method section; personal computer with PC pulse-counter board such as National Instruments M Series 622x and LabVIEW software with autocorrelation capability.

Glassware for diluting the PS suspensions: 100-μL syringe for extracting known volume from the commercial PS sample bottle; 20-mL syringe with Luer-lock fittings for filtering of saline solution into storage bottle; disposable 0.1-micron filters (Pall or Gilman) with Luer-lock features; 100-mL of 10-mM NaCl solution; skim milk (nonfat, not 1% milk); clean 25-mL dilution vial with screw cap to avoid dust and allow mixing by inversion; scattering cell (1-cm square glass or plastic cuvette free from scratches); thermometer.

Monodisperse polystyrene (latex) spheres having a radius value in the range 30–100 nm with a standard deviation of 8% or less characterizing the size uniformity. Such samples can be obtained from Duke Scientific Corp. as 3000 Series Nanospheres (www.dukescientific.com), Polymer Laboratories as PL-Latex Plain White (www.polymerlabs.com), and Polysciences, Inc., as Nanobead Traceable Size Standards (www.polysciences.com).

REFERENCES

1. W. Brown (ed.), *Dynamic Light Scattering,* pp. 100–112, 660–661, Clarendon Press, Oxford (1993).

2. B. J. Berne and R. Pecora, *Dynamic Light Scattering with Applications to Chemistry, Biology and Physics,* pp. 1–28, 62–65, 83–86, Wiley, New York (1976).

3. B. B. Weiner, "Particle Sizing Using Photon Correlation Spectroscopy," in H. G. Barth (ed.), *Modern Methods of Particle Size Analysis,* pp. 93–116, Wiley, New York (1984).

4. W. Tscharnuter, "Photon Correlation Spectroscopy in Particle Sizing," in R. A. Meyer (ed.), *Encyclopedia of Analytical Chemistry,* vol. 6, pp. 5469–85, Wiley, Chichester, UK (2000).

5. http://en.wikipedia.org/wiki/Brownian_Motion. A simulation video of Brownian motion is available at http://galileo.phys.virginia.edu/classes/109N/more_stuff/Applets/brownian/brownian.html.

6. D. A. McQuarrie, *Statistical Mechanics,* sec. 22-3, University Science Books, Sausalito, CA (2000).

7. H. Z. Cummins et al., *Biophys. J.* **9,** 518 (1969); R. Pecora, *J. Chem. Phys.* **40,** 1604 (1964).

8. P. N. Pusey et al., *Physica A* **235,** 1 (1997).

9. A. P. Y. Wong and P. Wiltzius, *Rev. Sci. Instrum.* **64,** 2547 (1993); L. Cipelletti and D. A. Weitz, *Rev. Sci. Instrum.* **70,** 3214 (1999).

10. I. N. Levine, *Physical Chemistry,* 6th ed., sec. 15.4, McGraw-Hill, New York (2009).

11. R. Xu, *Particle Characterization: Light Scattering Methods,* chap. 5, Kluwer Academic, Dordrecht, The Netherlands (2000).

12. G. D. J. Phillies, *Anal. Chem.* **62,** 1049A (1990).

13. P. M. Kampmeyer, *J. Appl. Phys.* **23,** 99 (1952).

14. D. C Harris, *J. Chem. Ed.* **75,** 119 (1998); R. de Levie, *J. Chem. Ed.* **76,** 1594 (1999).

15. J. C. Thomas, *J. Coll. Interf. Sci.* **117,** 187 (1987).

GENERAL READING

B. J. Berne and R. Pecora, *Dynamic Light Scattering with Applications to Chemistry, Biology and Physics.* Wiley, New York (1976).

B. Chu, *Laser Light Scattering: Basic Principles and Practice,* 2d ed., chap. 4, Academic Press, New York (1991).

G. D. J. Phillies, *Anal. Chem.* **62,** 1049A (1990).

W. Tscharnuter, *op. cit.*

XIV

Spectroscopy

EXPERIMENTS

34. Absorption Spectrum of a Conjugated Dye

35. Raman Spectroscopy: Vibrational Spectrum of CCl_4

36. Stimulated Raman Spectra of Benzene

37. Vibrational–Rotational Spectra of HCl and DCl

38. Vibrational–Rotational Spectra of Acetylenes

39. Absorption and Emission Spectra of I_2

40. Fluorescence Lifetime and Quenching in I_2 Vapor

41. Electron Spin Resonance Spectroscopy

42. NMR Determination of Keto–Enol Equilibrium Constants

43. NMR Study of Gas-Phase DCl–HBr Isotopic Exchange Reaction

44. Solid-State Lasers: Radiative Properties of Ruby Crystals

45. Spectroscopic Properties of CdSe Nanocrystals

EXPERIMENT 34

Absorption Spectrum of a Conjugated Dye

Absorption bands in the visible region of the spectrum correspond to transitions from the ground state of a molecule to an excited electronic state that is 170 to 300 kJ mol^{-1} above the ground state. In many substances the lowest excited electronic state is more than 300 kJ mol^{-1} above the ground state and no visible spectrum is observed. Those compounds that are colored (i.e., absorb in the visible) generally have some weakly bound or delocalized electrons such as the odd electron in a free radical or the π electrons in a conjugated organic molecule. In this experiment we are concerned with the determination of the visible absorption spectrum of several symmetric polymethine dyes and with the interpretation of these spectra using the "free-electron" model.

THEORY

The visible bands for polymethine dyes arise from electronic transitions involving the π electrons along the polymethine chain. The wavelength of these bands depends on the spacing of the electronic energy levels. Bond-orbital and molecular-orbital calculations have been made for these dyes, but the predicted wavelengths are in poor agreement with those observed. We shall present here the simple free-electron model first proposed by Kuhn;[1,2] this model contains some drastic assumptions but has proven reasonably successful for molecules like conjugated dyes.

As an example, consider a dilute solution of 1,1'-diethyl-4,4'-carbocyanine iodide (cryptocyanine):

The cation can "resonate" between the two limiting structures above, which really means that the wavefunction for the ion has equal contributions from both states. Thus all the bonds along this chain can be considered equivalent, with bond order 1.5 (similar to the C—C bonds in benzene). Each carbon atom in the chain and each nitrogen at the end is involved in bonding with three atoms by three localized bonds (the so-called σ bonds). The extra valence electrons on the carbon atoms in the chain and the three remaining electrons on the two nitrogens form a mobile cloud of π electrons along the chain (above and below the plane of the chain). We shall assume that the potential energy is constant along the chain and that it rises sharply to infinity at the ends; i.e., the π electron system is replaced by free electrons moving in a one-dimensional box of length L. The quantum mechanical solution for the energy levels of this model[3] is

$$E_n = \frac{h^2 n^2}{8mL^2} \qquad n = 1, 2, 3, \ldots \tag{1}$$

where m is the mass of an electron and h is the Planck constant.

Since the Pauli exclusion principle limits the number of electrons in any given energy level to two (these two have opposite spins: $+\frac{1}{2}$, $-\frac{1}{2}$), the ground state of a molecule with N π electrons will have the $N/2$ lowest levels filled (if N is even) and all higher levels empty. When the molecule (or ion in this case) absorbs light, this is associated with a one-electron jump from the highest filled level ($n_1 = N/2$) to the lowest empty level ($n_2 = N/2 + 1$). The energy change for this transition is

$$\Delta E = \frac{h^2}{8mL^2}(n_2^2 - n_1^2) = \frac{h^2}{8mL^2}(N + 1) \tag{2}$$

Since $\Delta E = h\nu = hc/\lambda$, where c is the speed of light and λ is the wavelength,

$$\lambda = \frac{8mc}{h} \frac{L^2}{N+1} \tag{3}$$

Let us denote the number of carbon atoms in a polymethine chain by p; then $N = p + 3$. Kuhn assumed that L was the length of the chain between nitrogen atoms plus one bond distance on each side; thus $L = (p + 3)l$, where l is the bond length between atoms along the chain. Therefore

$$\lambda = \frac{8mcl^2}{h} \frac{(p+3)^2}{p+4} \tag{4}$$

Putting $l = 1.39$ Å $= 0.139$ nm (the bond length in benzene, a molecule with similar bonding) and expressing λ in nanometers, we find

$$\lambda \text{ (in nm)} = 63.7\frac{(p+3)^2}{p+4} \tag{5}$$

If there are easily polarizable groups at the ends of the chain (such as benzene rings), the potential energy of the π electrons in the chain does not rise so sharply at the ends. In effect this lengthens the path L, and we can write

$$\lambda \text{ (in nm)} = 63.7\frac{(p+3+\alpha)^2}{p+4} \tag{6}$$

where α should be a constant for a series of dyes of a given type. If such a series is studied experimentally, this empirical parameter α may be adjusted to achieve the best fit to the data; in any event, α should lie between 0 and 1.[1]

In order to compare the results of the free-electron model with calculations based on simple bond-orbital or molecular-orbital schemes, let us use Eq. (5), which assumes that $\alpha = 0$, to calculate the wavelength λ for cryptocyanine (in which $p = 9$) and compare that value with those obtained from orbital calculations:

Free electron	$\lambda = 707$ nm
Bond orbital (case 1)[4]	$\lambda = 3900$ nm
Bond orbital (case 2)[4]	$\lambda = 2900$ nm
Molecular orbital[4]	$\lambda = 2700$ nm
Hückel model[5]	$\lambda = 475$ nm

Note that only the free-electron and Hückel models predict an absorption band in the visible in agreement with observation. While the orbital calculations in Ref. 4 are poor for polymethine dyes, they are in principle a superior approach and have given excellent results for saturated hydrocarbons. Indeed unsaturated molecules like the polymethines can be very well modeled by the semiempirical configuration interaction (CIS) method.

METHOD

Many commercial visible–UV spectrophotometers are suitable for this experiment. These instruments range from simple single-beam devices such as the Spectronic model 20 to high-performance double-beam scanning spectrophotometers such as various Varian–Cary models. The components and operational principles of these instruments are discussed in Chapter XIX, and this material should be reviewed prior to undertaking the experiment.

It is necessary to define several terms commonly used in spectrophotometry. Absorption spectra are often characterized by the *transmittance T* at a given wavelength; this is defined by

$$T \equiv \frac{I}{I_0} \tag{7}$$

where I is the intensity of light transmitted by the sample and I_0 is the intensity of light incident on the sample. When the sample is in solution and a cell must be used, I is taken to be the intensity of light transmitted by the cell when it contains solution, while I_0 is taken to be the intensity of light transmitted by the cell filled with pure solvent. Another way of describing spectra is in terms of the *absorbance A*, where

$$A \equiv \log\frac{I_0}{I} \tag{8}$$

A completely transparent sample would have $T = 1$ or $A = 0$, while a completely opaque sample would have $T = 0$ or $A = \infty$.

The absorbance A is related to the path length d of the sample and the concentration c of absorbing molecules by the Beer–Lambert law,[6]

$$A = \varepsilon c d \tag{9}$$

where ε is called the *molar absorption coefficient* when the concentration is expressed in moles per unit volume.† The quantity ε is an intrinsic property of the absorbing material that varies with wavelength in a characteristic manner; its value depends only slightly on the solvent used or on the temperature. The SI unit for ε is $\text{mol}^{-1}\,\text{m}^2$, but a more practical and commonly used unit is $\text{mol}^{-1}\,\text{L}\,\text{cm}^{-1}$, which corresponds to using the concentration c in $\text{mol}\,\text{L}^{-1}$ and the path length d in cm.

For quantitative measurements it is important to calibrate the cells so that a correction can be made for any small difference in path length between the solution cell and the solvent cell. For analytical applications, one must check the validity of Beer's law, since slight deviations are often observed and a calibration curve of absorbance versus concentration is then required. Such corrections will not be necessary in the present work.

EXPERIMENTAL

For the present experiment, a manually scanned spectrophotometer with a resolution of 10 to 20 nm is adequate, although an autoscanning instrument is, of course, more convenient. Instructions for operating the spectrophotometer will be made available in the laboratory. Turn on the instrument as instructed, and allow it to warm up for a few minutes in order to achieve stable, drift-free performance.

Several polymethine dyes should be studied, preferably a series of dyes of a given type with varying chain length. In addition to the 1,1′-diethyl-4,4′-carbocyanine iodides mentioned previously, 1,1′-diethyl-2,2′-carbocyanine iodides and 3,3′-diethyl thiacarbocyanine iodides are suitable. Other possible compounds can be found in the literature.[7]

---

†The quantity ε was previously called the *extinction coefficient,* and this name is still frequently used in the scientific literature.

Choose any one of the available dyes, and prepare 10 mL of a solution using methyl alcohol as the solvent. The concentration should be approximately 10^{-3} M. (If the solutions are to be kept for a long time, they should be stored in dark-glass bottles to prevent slow decomposition by daylight.)

Follow the spectrophotometer operating instructions carefully, and obtain the spectrum of this initial solution. Take absorbance readings at widely spaced intervals throughout the spectrophotometer range until the absorption band is located. Then take readings at much closer intervals throughout the band. Dilute the initial solution and redetermine the spectrum. Repeat this procedure until a spectrum is obtained with an absorbance reading of about 1 at the peak (transmittance ~ 0.1). The band shape may change at high concentration, since many dyes dimerize. At the final concentration used in this experiment, no dimer contribution is expected.

Make up solutions (10 mL) of the other dyes at approximately the same molar concentration that gave the best results previously. Obtain their spectra in the same way.

CALCULATIONS

Present all the spectra obtained (plotting A versus λ if data were obtained manually); label each plot clearly with the name of the compound and the concentration used. Determine λ_{max}, the wavelength at the peak, for each dye studied.

Using Eq. (6), calculate λ from the free-electron model for each compound. If a series of dyes of a single type has been studied, choose α to give the best fit for this series. Alternatively, one could fix α by fitting any one of the series and use this value for all others in the series.

Report in a table your experimental and theoretical values of λ_{max}.

Theoretical Calculations.   As a project exercise, the measured energy separations of the upper and ground electronic states of cryptocyanine might be compared with values calculated using quantum theory. Semiempirical methods using the *configuration interaction singles* (CIS) method can be done reasonably rapidly with the program HyperChem, which also gives intensities for the transitions. Hehre et al. describe a similar application using the program Spartan.[8] Calculations of excited state energies using the ab initio program Gaussian at the CIS level are described by Foresman and Frisch[9] and are more time intensive. In general, theoretical values for excited-state properties of molecules are less accurate than ground-state values.

SAFETY ISSUES

None.

APPARATUS

Spectrophotometer, such as a Bausch & Lomb Spectronic or Beckman DU series; four Pyrex or polystyrene sample cells; lens tissue; several 10-mL volumetric flasks; wash bottle.

Reagent-grade methyl alcohol (150 mL); several polymethine dyes, preferably a series such as 1,1′-diethyl-4,4′-cyanine iodide, -carbocyanine iodide, and -dicarbocyanine iodide or 1,1'-diethyl-2,2'-cyanine iodide, -carbocyanine iodide, and -dicarbocyanine iodide (a few milligrams of each is sufficient). (Gallard-Schlesinger Chemical Mfg. Corp., 1001 Franklin St., Garden City, NY, and K & K Laboratories, Inc., 121 Express St., Plainview, NY 11803, are possible but expensive suppliers of an entire series of dyes. Aldrich Chemical Co. and Kodak Laboratory & Research Products supply some of these dyes at lower cost. Check the current *Chem Sources* for other suppliers.)

REFERENCES

1. H. Kuhn, *J. Chem. Phys.* **17,** 1198 (1949); *Fortsch. Chem. Organ. Naturstoffe* **17,** 404 (1959).

2. I. N. Levine, *Quantum Chemistry,* 5th ed., Prentice-Hall, Upper Saddle River, NJ (2000).

3. P. W. Atkins, and J. de Paula, *Physical Chemistry,* 8th ed., p. 279, Freeman, New York (2006).

4. K. F. Herzfeld and A. L. Sklar, *Rev. Mod. Phys.* **14,** 294 (1942).

5. J. J. Farrell, *J. Chem. Educ.* **62,** 351 (1985).

6. R. J. Silbey, R. A. Alberty, and M. G. Bawendi, *Physical Chemistry,* 4th ed., Wiley, New York (2005).

7. L. G. S. Brooker, *Rev. Mod. Phys.* **14,** 275 (1942); N. I. Fisher and F. M. Hamer, *Proc. Roy. Soc.,* Ser. A, **154,** 703 (1936); B. D. Anderson, *J. Chem. Educ.* **74,** 985 (1997).

8. W. J. Hehre, A. J. Shusterman, and J. E. Nelson, *The Molecular Modeling Workbook for Organic Chemistry,* Wavefunction, Irvine, CA (1998).

9. J. B. Foresman and A. Frisch, *Exploring Chemistry with Electronic Structure Methods,* 2d ed., chap. 9, Gaussian, Pittsburgh, PA (1996).

GENERAL READING

P. W. Atkins and J. de Paula, *Physical Chemistry,* 8th ed. Freeman, New York (2006).

H. H. Willard, L. L. Merritt, Jr., J. A. Dean, and F. A. Settle, *Instrumental Methods of Analysis,* 7th ed., Wadsworth, Belmont, CA (1988).

EXPERIMENT 35
Raman Spectroscopy: Vibrational Spectrum of CCl$_4$

In Exps. 37 and 38, the vibrational frequencies of a molecule are determined by measuring the direct absorption of infrared radiation due to transitions between vibrational energy levels. Such changes in molecular state can also be caused by inelastic scattering of higher-energy visible photons. The energy (or frequency) *shifts* in the scattered radiation then give a direct measure of the vibrational frequencies of the molecule. Although the intensity of this scattered light is quite low, Raman was able to observe such scattering in 1928, and the effect was later named after him. With the advent of extremely intense monochromatic laser sources, Raman spectroscopy has become much faster and more sensitive, and instrumentation for such measurements is now reasonably common. Some of the advantages and unique features of Raman spectroscopy will be demonstrated in this experiment on liquid CCl$_4$.

THEORY

In Exp. 29, the effect of a static or low-frequency electric field on a molecule was considered and it was seen that a dipole moment is induced because of movement of the charged electrons and nuclei. At high optical frequencies ($\sim 10^{15}$ Hz), the nuclei cannot respond rapidly enough to follow the field, but polarization of the electron distribution can occur. For an isolated molecule, an oscillating radiation field of intensity **E** will induce a dipole moment of magnitude

$$\boldsymbol{\mu}_{\text{ind}} = \alpha \mathbf{E} \tag{1}$$

where α is the molecular polarizability. The field **E** oscillates at the frequency ν_0 of the light,

$$\mathbf{E} = \mathbf{E}_0 \cos 2\pi\nu_0 t \tag{2}$$

and hence the induced dipole will also change at this frequency. According to classical electromagnetic theory, any oscillating dipole will radiate energy; hence light of frequency ν_0 is emitted in all directions (except that parallel to the dipole). Such elastically scattered† light is termed Rayleigh scattering and typically one out of a million incident photons will be so scattered.

Inelastic or Raman scattering of light can be understood classically as arising from modulation of the electron distribution, and hence the molecular polarizability, because of vibrations of the nuclei. For example, for a diatomic molecule, α can be represented adequately by the first two terms of a power series in the vibrational coordinate Q:

$$\alpha = \alpha_0 + \left(\frac{d\alpha}{dQ}\right)_0 Q \tag{3}$$

Here the subscript zero indicates that the parameters α_0 and $(d\alpha/dQ)_0$ are evaluated at the equilibrium position. In the harmonic-oscillator model, Q for mode i oscillates at the vibrational frequency ν_i according to the relation

$$Q = A \cos 2\pi\nu_i t \tag{4}$$

where A is the amplitude of vibration. Substitution of Eqs. (2) to (4) into Eq. (1) reveals the time dependence of $\boldsymbol{\mu}_{\text{ind}}$:

$$\boldsymbol{\mu}_{\text{ind}} = \alpha_0 \mathbf{E}_0 \cos 2\pi\nu_0 t + \left(\frac{d\alpha}{dQ}\right)_0 A\mathbf{E}_0 \cos 2\pi\nu_0 t \cos 2\pi\nu_i t$$

$$= \alpha_0 \mathbf{E}_0 \cos 2\pi\nu_0 t + \frac{1}{2}\left(\frac{d\alpha}{dQ}\right)_0 A\mathbf{E}_0[\cos 2\pi(\nu_0 + \nu_i)t + \cos 2\pi(\nu_0 - \nu_i)t] \tag{5}$$

The first term is responsible for Rayleigh scattering at ν_0, while the second and third terms produce inelastic Raman scattering at frequencies that are higher (anti-Stokes component) and lower (Stokes component), respectively, than the incident-light frequency. The frequency ν_0 usually corresponds to light in the visible region (typically the 514.5-nm line of the argon-ion laser), and the Raman-shifted light then occurs in the visible region but with an intensity that is 10^{-8} to 10^{-12} times that of the incident light.

The quantum mechanical treatment of light scattering[1] predicts the same general results but more explicitly involves the vibrational energy levels and wavefunctions of the molecule. Figure 1 shows the usual energy-level representation of the scattering process. The virtual states are not real energy levels but serve as convenient intermediate levels for picturing the overall scattering process, which occurs in a time less than the period of a molecular vibration ($\sim 10^{-13}$ s). Should the incident radiation have a frequency that

---

†Elastic scattering means no energy change in the scattering process, hence no frequency shift.

FIGURE 1

Energy levels and transition
frequencies involved in
Raman spectroscopy.

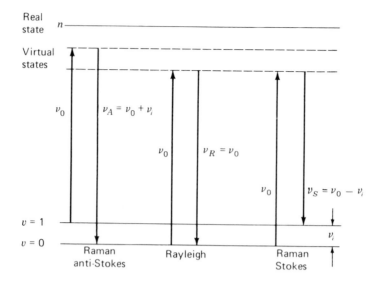

causes actual absorption to a real state n, the lifetime of the upper state is much longer
($\sim 10^{-8}$ s) and the resultant intense emission is termed *fluorescence*. In the absence of such
a resonance, the ratio of the Raman intensity of the Raman anti-Stokes and Stokes lines is
predicted[2] to be

$$\frac{I_A}{I_S} = \left(\frac{\nu_0 + \nu_i}{\nu_0 - \nu_i}\right)^4 \exp\left(\frac{-h\nu_i}{kT}\right) \tag{6}$$

Because of the ν^4 dependence, blue light is scattered more efficiently than red (a fact that
accounts for the blue color of the sky). However, the Boltzmann exponential factor is
dominant in Eq. (6), and the anti-Stokes features are always much weaker than the corresponding Stokes lines.

Raman Selection Rules. For polyatomic molecules a number of Stokes Raman
bands are observed, each corresponding to an allowed transition between two vibrational
energy levels of the molecule. (An allowed transition is one for which the intensity is not
uniquely zero owing to symmetry.) As in the case of infrared spectroscopy (see Exp. 38),
only the fundamental transitions (corresponding to frequencies $\nu_1, \nu_2, \nu_3, \ldots$) are usually
intense enough to be observed, although weak overtone and combination Raman bands are
sometimes detected. For molecules with appreciable symmetry, some fundamental transitions may be absent in the Raman and/or infrared spectra. The essential requirement is that
the transition moment P (whose square determines the intensity) be nonzero; i.e.,

$$P_{\text{ind}}(1 \leftarrow 0) = \int \psi_1 \mu_{\text{ind}} \psi_0 \, dQ = E \int \psi_1 \alpha \psi_0 \, dQ \neq 0 \tag{7}$$

for the Raman case, and

$$P(1 \leftarrow 0) = \int \psi_1 \mu \psi_0 \, dQ \neq 0 \tag{8}$$

for the infrared case. Here ψ_0 and ψ_1 are wavefunctions for states with vibrational quantum
numbers $\upsilon = 0$ and $\upsilon = 1$, and the integration extends over the full range of the coordinate
Q. Since these two constraints are not the same, infrared and Raman spectra generally differ in the number, frequencies, and relative intensities of observed bands. The prediction

of allowed transitions for both cases represents an elegant example of the role of symmetry and group theory in chemistry. Although a discussion of this topic is beyond the scope of this book, the application of group theory to vibrational spectroscopy is presented in several standard texts (e.g., Refs. 2 to 4). A summary of the conclusions reached for a few simple molecular structures is provided in Table 1. Here, the general form of the vibrational coordinates is shown with the mode numbering and the infrared and Raman activity for the fundamental vibrations. Each molecular type has a unique set of symmetry elements (rotation axes, reflection planes, etc.). Associated with these elements are operations (such as rotation about symmetry axes, reflections, etc.) that transform the molecule from one configuration in space to another that is indistinguishable. This set of operations defines the point group, which is indicated in the table. The point group is associated with a unique set of symmetry species (denoted Σ_g^+, A_1, B_1, E, F_2, etc.), each of which behaves differently under the symmetry operations of the molecule. Every vibrational coordinate (and wavefunction) belongs to one of these species, as indicated in Table 1; it is knowledge of this symmetry that permits the prediction of the infrared and Raman activity of a given fundamental transition.

Raman Depolarization Ratio.   A more complete discussion of vibrational selection rules must take into account the vector properties of the dipole moment and the tensor character of the polarizability α. Thus Eq. (1) should be written

$$\begin{pmatrix} \mu_x \\ \mu_y \\ \mu_z \end{pmatrix}_{\text{ind}} = \begin{pmatrix} \alpha_{xx} & \alpha_{xy} & \alpha_{xz} \\ \alpha_{yx} & \alpha_{yy} & \alpha_{yz} \\ \alpha_{zx} & \alpha_{zy} & \alpha_{zz} \end{pmatrix} \begin{pmatrix} E_x \\ E_y \\ E_z \end{pmatrix} \tag{9}$$

Such considerations are of considerable importance in the study of oriented molecules, as in a crystal, but are less important for liquids and gases, in which the molecular orientation is random. However, in the case of Raman spectroscopy, not all information about the components of α is lost by orientational averaging. In particular one finds that it is possible to distinguish totally symmetric from other molecular vibrations from a measurement of the depolarization ratio,

$$\rho_l \equiv \frac{I_\perp}{I_\parallel} \tag{10}$$

of the Raman bands. Here, $I_\perp$ and $I_\parallel$ are the intensities of scattered light with polarization perpendicular ($\perp$) and parallel ($\parallel$) to the polarization of the linearly polarized exciting light. Theory shows that ρ_l is related to combinations of the tensor components of α and that $\rho_l = \frac{3}{4}$ for any vibration that is *not* totally symmetric and $0 \leq \rho_l \leq \frac{3}{4}$ for those that are totally symmetric.[5] Raman bands for which $\rho_l = \frac{3}{4}$ are called depolarized, and those with $\rho_l < \frac{3}{4}$ are called polarized. Totally symmetric vibrations are those in which the symmetry of the molecule does not change during the vibrational motion (e.g., the Σ_g^+, A_1', and A_1 vibrations of Table 1). For such vibrations, ρ_l values near zero are observed for highly symmetric molecules, such as CO_2, SF_6, and CCl_4. Thus the measurement of ρ_l will be used in this experiment to aid in assigning the frequency of the totally symmetric ν_1 vibration of CCl_4.

Valence-Force Model.   The simplest harmonic oscillator model for the potential energy U of a tetrahedral molecule such as CCl_4 can be written as

$$U = \tfrac{1}{2}k(r_1^2 + r_2^2 + r_3^2 + r_4^2) + \tfrac{1}{2}k_\delta(\delta_{12}^2 + \delta_{13}^2 + \delta_{14}^2 + \delta_{23}^2 + \delta_{24}^2 + \delta_{34}^2) \tag{11}$$

where r_i is the change in the length of C—Cl bond i from its equilibrium value l and δ_{ij} is the change in the angle between bonds i and j. From classical mechanics one obtains[6] the following relations between the vibrational frequencies and the force constants k and k_δ/l^2;

$$4\pi^2\nu_1^2 = \frac{k}{m_{Cl}} \tag{12}$$

$$4\pi^2\nu_2^2 = \left(\frac{3}{m_{Cl}}\right)\frac{k_\delta}{l^2} \tag{13}$$

$$4\pi^2(\nu_3^2 + \nu_4^2) = \left(1 + \frac{4m_{Cl}}{3m_C}\right)\frac{k}{m_{Cl}} + \left(1 + \frac{8m_{Cl}}{3m_C}\right)\left(\frac{2}{m_{Cl}}\right)\frac{k_\delta}{l^2} \tag{14}$$

$$16\pi^4\nu_3^2\nu_4^2 = \left(1 + \frac{4m_{Cl}}{m_C}\right)\left(\frac{2}{m_{Cl}}\right)\frac{k_\delta}{l^2}\frac{k}{m_{Cl}} \tag{15}$$

TABLE 1 Normal modes of vibration for several molecular types

Type	Point group	Vibrational modes,[a] Symmetries, Infrared-Raman activities, and Raman polarizations[b]			
B—A—B	$D_{\infty h}$	ν_1 Σ_g^+ (Rp)	ν_{2x} ν_{2y} Π_u (IR)		ν_3 Σ_u^+ (IR)
B—A—B (C_{2v})	C_{2v}	ν_1 A_1 (IR, Rp)	ν_2 A_1(IR, Rp)		ν_3 B_1 (IR, Rdp)
B–A—B (tri)	D_{3h}	ν_1 A_1' (Rp)	ν_2 A_2'' (IR)	ν_3 E' (IR, Rdp)	ν_4 E' (IR, Rdp)
B–A(B)–B	C_{3v}	ν_1 A_1 (IR, Rp)	ν_2 A_1 (IR, Rp)	ν_3 E (IR, Rdp)	ν_4 E (IR, Rdp)
AB$_4$	T_d	ν_1 A_1 (Rp)	ν_2 E (Rdp)	ν_3 F_2 (IR, Rdp)	ν_4 F_2 (IR, Rdp)

[a] Only one of two E or three F modes in the respective degenerate sets is shown for the AB$_3$ and AB$_4$ cases. Each mode of a degenerate set oscillates at the same frequency, and the description of the nuclear motion associated with each component of a degenerate mode is somewhat arbitrary.

[b] Rp indicates an allowed Raman band that is polarized; Rdp indicates an allowed depolarized Raman band.

If one expresses the frequencies in wavenumbers $\tilde{\nu}$ (cm^{-1}) and uses appropriate *isotopic* masses (in amu) $m_C = 12.000$ and $m_{Cl} = 34.969$, $4\pi^2$ should be replaced by $4\pi^2 c^2/10^3 N_0 = 5.8918 \times 10^{-5}$ to obtain k and k_δ/l^2 in units of N m^{-1}.

Since there are two force constants and four frequencies, one can test the quality of this force field by determining k and k_δ/l^2 separately from Eqs. (12) and (13) and from Eqs. (14) and (15). However, for some tetrahedral molecules, it is found that the latter two equations yield imaginary values for k and k_δ, and this is an indication that stretch–stretch, stretch–bend, and bend–bend interaction terms may be important in U. Unfortunately addition of these terms leads to more complex expressions than Eqs. (12) to (15) and to more force constants than can be determined from the four observed frequencies. Use of frequencies of isotopic forms of the molecule is often helpful in such cases and can permit more accurate determination of potential functions.[5,6]

METHOD

There are many different Raman spectrometers in current use, almost all of which are suitable for this experiment. A general description of Raman instruments with grating monochromators and laser sources is given in Chapter XIX. Before beginning the experiment, you should review the pertinent material in Chapter XIX and read carefully any available information about the actual instrument to be used in the laboratory. **Use the spectrometer with care;** if in doubt, ask the instructor for advice about proper procedures.

In order to measure depolarization ratios, a polaroid analyzer and a scrambler are placed in front of the entrance slit of the spectrometer. Since the laser source is polarized, ρ_l is obtained by measuring $I_\parallel$ and $I_\perp$ when the polaroid is oriented to pass only light whose polarization is parallel or perpendicular, respectively, to the polarization of the exciting light (see Fig. 2). The scrambler serves to completely depolarize the light passed by the polaroid; this is necessary since the reflection efficiency of the gratings in the spectrometer is polarization dependent. It should be mentioned that the theoretical value of $\frac{3}{4}$ for depolarized Raman bands applies strictly only for 90° scattering. Since the exciting source is focused and the collection angle is usually large, measured values of ρ_l can exceed $\frac{3}{4}$. If very precise values are desired, a collimated excitation beam and a smaller collection angle near 90° are used in conjunction with calibration by bands of known ρ_l value.

EXPERIMENTAL

Detailed instructions for the use of the Raman spectrometer will be provided in the laboratory. *It is particularly important to follow all safety instructions regarding the use of the excitation laser, since serious eye damage can occur if the beam, or even a small reflection from it, should strike the eye.* **Safety goggles must be worn when inserting samples in the beam and when adjusting the optics for maximum signal.**

A convenient liquid cell for Raman spectroscopy is a 1-cm glass or quartz cuvette, with a clear bottom if the excitation beam enters from below. Alternatively melting-point capillaries can serve as inexpensive cells; these have the added virtue when studying a scarce or expensive sample in that very little sample volume is required. A 10-cm length of thin-walled capillary tubing is melted at the center and separated to form two 5-cm tubes. Any excess glass is removed and the end is sealed while twirling to form a hemispherical bottom. This tube can be used as is, or a flat bottom can be achieved by inserting a flat stainless steel or tungsten rod and pressing the heated bottom against a flat carbon plate.

FIGURE 2

Polarization arrangements for Raman depolarization measurements.

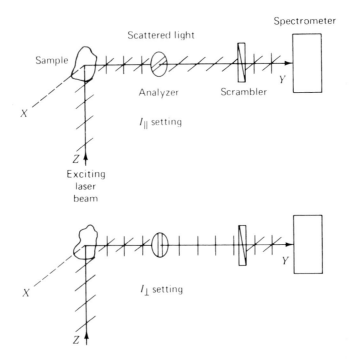

A hypodermic syringe can be used to add sample to the capillary up to a height of 1 to 2 cm. The top can then be sealed if the sample is sensitive to air or is volatile.

The scattering cylinder of the focused laser beam must be aligned vertically to be parallel to the entrance slit of the spectrometer. The collection lens is used to image this cylinder onto the slit. The sample is then inserted at the laser focal point with its long axis parallel to the beam and a quick survey scan is taken until a Raman band is found. The sample, the excitation beam, and the collection lens are then carefully adjusted for maximum signal at this spectrometer setting.

Scan the Stokes Raman spectrum of CCl_4 for shifts of 150 to 1000 cm^{-1} from the exciting line.[†] Note that the frequency in wavenumbers is given by $\tilde{\nu}$ (cm^{-1}) $\equiv 1/\lambda = \nu/c$, where c is the speed of light in cm s^{-1}. Record both parallel and perpendicular polarization scans so that you can determine the depolarization ratio of all bands. Indicate the spectrometer wavelength or wavenumber reading on the chart at several points in the scan to provide reference points for the determination of the Raman shifts for all bands.

Block the laser beam or spectrometer entrance slit and adjust the spectrometer to an anti-Stokes shift of 1000 cm^{-1}. **Caution:** *Exposure of the sensitive phototube to the intense Rayleigh scattering line can seriously damage the detector.* Scan the anti-Stokes spectrum from -1000 to -150 cm^{-1} in the parallel polarization configuration and, using appropriate sensitivity expansion,[‡] measure the ratio of anti-Stokes to Stokes peak heights for each band.

[†]For reference, suitable Stokes and anti-Stokes Raman spectra of CCl_4 can be found in the review of Raman spectroscopy in Ref. 7. A high-resolution scan of the 460-cm^{-1} band of liquid and solid CCl_4 is available in Ref. 8. Raman spectra of several molecules are described in Refs. 9 and 10.

[‡]Known scale expansions should be chosen that give peaks that are nearly full scale, so that the heights can be measured accurately. Some errors will arise in the measurement of Stokes and anti-Stokes intensities because of variation with frequency of spectrometer transmission and detector efficiency. These effects should be unimportant for bands involving small Raman shifts, and calibration corrections can be made if necessary for very accurate measurements.

You will note some structure in the Stokes band near 460 cm^{-1}, which is due to chlorine isotopic frequency shifts. (See Exp. 37 for a discussion of isotope effects in diatomic molecules.) Rescan this region at higher resolution (1 cm^{-1} or less) with an expansion of the chart display and measure the frequencies of each of the components. Record the ambient temperature near the Raman cell.

CALCULATIONS AND DISCUSSION

Assign the fundamentals of CCl_4 using the fact that the ν_1 and ν_3 CCl_4 stretching motions are expected to occur at higher frequencies than ν_2 and ν_4 CCl_4 bending motions. The depolarization ratios can be used to determine which of the two high-frequency bands should be assigned as ν_1. To decide on the ν_2 and ν_4 assignments, use the fact that the only infrared band observed below 600 cm^{-1} is a feature near 300 cm^{-1}.

The ν_3 vibration will be observed as a doublet due to Fermi resonance splitting, a process that can occur if an overtone or combination energy level happens to fall near the energy level of a fundamental vibration and if both levels have the same symmetry. The ν_3 frequency can be taken as the mean of the doublet peaks. From your values of ν_1, ν_2, and ν_4, deduce which combination band is involved in the Fermi resonance with ν_3.

Tabulate your Stokes and anti-Stokes frequencies and also the intensity ratios I_A/I_S wherever possible. According to Eq. (6), the intensity ratio for each mode can be used to determine the vibrational temperature characterizing the Boltzmann distribution. In the present experiment, the vibrational temperature should agree with the ambient temperature of the laboratory. Does it? This technique of using information about spectral intensities to determine T is a very convenient method for finding the temperature of flames, shock waves, and plasmas.

Calculate the valence-force constants k and k_δ/l^2 from Eqs. (12) and (13), and then compare these values with those obtained from Eqs. (14) and (15), if possible. How important are the neglected interaction terms in the potential function of CCl_4?

From the natural abundance of ^{35}Cl (75.5 percent) and ^{37}Cl (24.5 percent), calculate the relative intensities expected for the various components of the band near 460 cm^{-1} and compare these with your observations. Suggest a reasonable modification of one of Eqs. (12) to (15) that would permit you to predict the frequency spacing of these components.

Alternative or additional liquids suitable for Raman investigation include CS_2, PCl_3, $CHCl_3$, and benzene. Comparisons of similar molecules (e.g., CCl_4, $CHCl_3$, $CDCl_3$, or C_6H_6, C_6D_6, and substituted benzenes) can provide a basis for frequency assignments and can be used to test the transferability of force constants. Note that several of these samples are toxic and must be handled with care. Capillary samples can be prepared in a hood, frozen with liquid nitrogen, and then sealed with a flame (or simply sealed off with soft wax).

Aqueous solutions can also be studied, since the Raman spectrum of water is weak and quite broad. Possible samples would be 1 M to 3 M solutions of Na_2CO_3, $NaNO_3$, Na_2SO_3, or Na_2SO_4. Such samples are difficult to study using infrared spectroscopy because of solvent absorption and the solubility of most infrared cell windows.

Theoretical Calculations. The harmonic vibrational frequencies and infrared/Raman intensities for CCl_4 can be calculated using an ab initio program such as Gaussian, but this may be time consuming for large basis sets due to the number of electrons on the chlorine atoms. Calculations at the lower PM3 level are recommended instead; and if done with programs with visualization options, such as HyperChem, animated displays of the different normal modes can be seen.

SAFETY ISSUES

As described on p. 197, CCl_4 is a toxic substance and must be used with care. Clean up any spillage, and dispose of waste materials properly. Laser beams can cause serious eye damage, and safety goggles must be worn.

APPARATUS

Raman spectrometer with argon- or krypton-ion laser source; liquid-sample cell or melting-point capillaries; reagent-grade CCl_4; safety goggles (such as those available from Glendale Optical Co., Woodbury, NY 11797).

REFERENCES

1. D. A. Long, *Raman Spectroscopy,* McGraw-Hill, New York (1977).

2. E. B. Wilson, Jr., J. C. Decius, and P. C. Cross, *Molecular Vibrations,* p. 51, McGraw-Hill, New York (1955), reprinted in unabridged form by Dover, New York (1980).

3. D. C. Harris and M. D. Bertolucci, *Symmetry and Spectroscopy,* Dover, Mineola, NY (1989).

4. F. A. Cotton, *Chemical Applications of Group Theory,* 3d ed., Wiley-Interscience, New York (1990).

5. E. B. Wilson, Jr., J. C. Decius, and P. C. Cross, *op. cit.,* pp. 47, 131.

6. G. Herzberg, *Molecular Spectra and Molecular Structure II: Infrared and Raman Spectra of Polyatomic Molecules,* p. 182, reprint ed., Krieger, Melbourne, FL (1990).

7. R. S. Tobias, *J. Chem. Educ.* **44,** 2, 70 (1967).

8. H. J. Sloane, *Appl. Spectrosc.* **25,** 430 (1971).

9. F. P. DeHaan, J. C. Thibeault, and D. K. Ottesen, *J. Chem. Educ.* **51,** 263 (1974).

10. L. C. Hoskins, *J. Chem. Educ.* **52,** 568 (1975); **54,** 642 (1977).

GENERAL READING

P. F. Bernath, *Spectra of Atoms and Molecules,* chaps. 7–8, Oxford Univ. Press, New York (1995).

F. A. Cotton, *Chemical Applications of Group Theory,* 3d ed., chaps. 1–5, 9, Wiley-Interscience, New York (1990).

J. R. Ferraro and K. Nakamoto (eds.), *Introductory Raman Spectroscopy,* Academic Press, San Diego, CA (1994).

G. Herzberg, *Molecular Spectra and Molecular Structure II: Infrared and Raman Spectra of Polyatomic Molecules,* chap. II, reprint ed., Krieger, Melbourne, FL (1990).

D. A. Long, *Raman spectroscopy,* McGraw-Hill, New York (1977).

EXPERIMENT 36
Stimulated Raman Spectra of Benzene

In Exp. 35, spontaneous Raman scattering of laser radiation by a sample produced new, shifted frequencies of scattered light, with the energy shifts corresponding to changes between molecular energy levels. The intensity of such Raman scattered light is low and increases linearly as the intensity of the exciting laser increases. However, at much higher laser intensities, new interactions can occur between the light and the molecules that cause dramatic nonlinear increases in the intensity of scattered light of certain frequencies and in certain directions. This effect, first observed in 1962 by Woodbury and Ng[1] and termed *stimulated Raman scattering,* will be studied in this experiment on benzene and its deuterated form C_6D_6. The angular characteristics of the stimulated light will be examined, and from a measurement of the wavelengths emitted, the frequencies and force constants will be determined for the totally symmetric CC and CH (CD) stretching vibrations in benzene.

THEORY[2–5]

The interaction of light with matter is the perturbation of the charged electrons and nuclei by the electric field **E** associated with the radiation. This results in the production of a polarization **P** in the sample that can be expressed as

$$\mathbf{P}_{\text{ind}} = \mathbf{X}^{(1)}\mathbf{E} + \mathbf{X}^{(2)}\mathbf{EE} + \mathbf{X}^{(3)}\mathbf{EEE} \tag{1}$$

Here $\mathbf{P}_{\text{ind}}$ is the induced dipole moment per unit volume, and $\mathbf{X}^{(1)}$, $\mathbf{X}^{(2)}$, and $\mathbf{X}^{(3)}$ are the first-, second-, and third-order electric susceptibilities of the sample, where the order refers to the power of electric field (not of **X**). Equation (1) represents the macroscopic (bulk) form for the polarization; in terms of single molecule properties, the induced dipole moment per molecule is written as

$$\mu_{\text{ind}} = \alpha E + \beta EE + \gamma EEE \tag{2}$$

where α, β, and γ are the molecular polarizability, hyperpolarizability, and second hyperpolarizability constants, respectively. These and the **X**s are in fact tensors and μ_{ind}, P_{ind}, and E are vectors, see Eq. (35-9), but for simplicity they will be treated as scalar quantities in the following discussion.

The first-order or linear term in electric field is thus seen to involve the polarizability α, whose value fluctuates at all the characteristic molecular vibrational and rotational frequencies. The electric field E oscillates at the light wave frequency ν_0 and, as discussed in Exp. 35, the interaction with a molecular vibration ν_i via α produces fluctuating dipole moments at frequency ν_0 and at the beat frequencies $\nu_0 + \nu_i$ and $\nu_0 - \nu_i$. According to Maxwell's equations, these oscillating dipoles act as source terms to generate fields (light waves) that correspond to scattered light at the Raleigh (ν_0), first Stokes Raman $(\nu_{S1} = \nu_0 - \nu_i)$, and first anti-Stokes Raman $(\nu_{A1} = \nu_0 + \nu_i)$ frequencies. The label 1 is added here since multiple shifts can occur, as discussed later. An energy level diagram depicting these processes is shown in Fig. 35-1, and the Stokes case is illustrated in Fig. 1a.

The ratios $\beta E/\alpha$ and $\gamma E/\beta$ are about 10^{-10}. Thus the linear α term is sufficient to describe the scattering caused by ordinary light sources and by continuous wave (cw) lasers of low intensity. However, this is not the case for pulsed lasers where, when a beam is focused, the electric field values can be comparable to that produced on electrons in a molecule by the charged nuclei (about 100 V/nm) and ionization can occur. At somewhat lower intensities and fields, the higher-order terms manifest themselves through the production

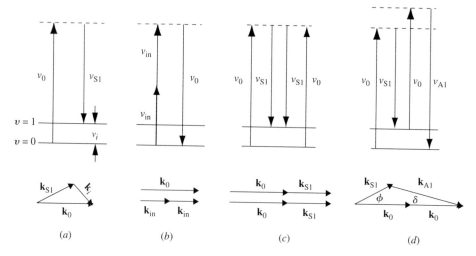

FIGURE 1

Energy level and momentum conditions for various mixing processes: (*a*) spontaneous Raman scattering; (*b*) second harmonic generation (which occurs in KDP to create frequency doubling of the 1064-nm Nd:YAG laser output); (*c*) Stimulated Raman Scattering (SRS); (*d*) Coherent Anti-Stokes Raman Scattering (CARS). Raman signals are generated in benzene via processes *a, c,* and *d* when it is exposed to an incident laser beam of frequency ν_0, but processes *c* and *d* require high intensity. Processes *a* and *c* involve energy transfer to the molecule and are termed nonparametric. The dashed levels do not correspond to real molecular states but are *virtual* levels that can be regarded as intermediate states during the scattering event.

of light scattered at various new beat frequencies. For example, the β hyperpolarizability term can give rise to second harmonic generation of light, a useful method of generating a higher-energy photon from two low-energy ones, as depicted in the energy level diagram of Fig. 1*b*. This method is commonly employed in crystal doublers such as KDP (KH_2PO_4) to efficiently (20–40%) convert 1064-nm radiation (ν_{in}) from a Nd:YAG laser to more useful coherent 532 nm radiation ($\nu_0 = 2\nu_{in}$) in the visible. In this case the crystal sample serves only to combine the light waves and no energy or momentum is transferred to it. This process is termed *parametric,* to distinguish it from a *nonparametric* process, in which the energy state and the momentum of the sample are changed as a result of the interaction between the sample and the light.

If not one but two light beams, at frequencies ν_0 and ν_1, impinge upon a crystal, a $\beta E(\nu_0)E(\nu_1)$ term can produce beats at $\nu_0 \pm \nu_1$, and at appropriate crystal angles this interaction can be used to generate intense coherent radiation at a sum frequency in the visible or ultraviolet region or at the difference frequency in the infrared.

Momentum Considerations. The energy and momentum (or phase) matching conditions for doubling are that

$$\nu_0 = \nu_{in} + \nu_{in} \quad \text{and} \quad \mathbf{k}_0 = \mathbf{k}_{in} + \mathbf{k}_{in} \tag{3}$$

where the momentum of a wave at frequency ν_j traveling through a material with index of refraction n_j is proportional to the wavevector magnitude $k_j = 2\pi n_j \nu_j/c$, c being the velocity of light. The collinear momentum condition, shown as vector addition in Fig 1*b*, can be written as

$$n_0\nu_0 = n_{in}\nu_{in} + n_{in}\nu_{in} \tag{4}$$

and this can be achieved only at a certain *index-matching* angle in KDP. The result of satisfying this momentum or phase matching condition is the *coherent* addition of the fields of the second harmonic photons to give a laserlike beam of intensity $I(\nu_0) = c\varepsilon_0[\Sigma_j E_j (\nu_0)]^2/2$ that propagates along the original laser direction. Typically the overlapping ν_{in} and ν_0 beams are separated by mirrors coated to transmit ν_{in} and reflect ν_0. In the current experiment, the latter isolated beam serves as the primary pump source in all the Raman processes studied.

Momentum also plays a role in ordinary spontaneous Raman spectroscopy. When the pump radiation at 532 nm is passed through a sample, the αE term of Eq. (2) produces scattering and, for the first Stokes case shown in Fig. 1a, the frequency is $\nu_{S1} = \nu_0 - \nu_i$, where ν_i is the Raman-active vibration excited in the sample. It should be noted that there is an exchange between the radiation field and molecule not only of energy but also of momentum, represented by the vector $\mathbf{k}_i$. The direction and magnitude of $\mathbf{k}_i$ are determined by the photon-scattering direction, which is random for this spontaneous event. The result is scattering in all directions so that there is no coherent addition of photon amplitudes, as expressed in the summation $I(\nu_{S1}) = c\varepsilon_0 \Sigma_j[E_j(\nu_{S1})]^2/2$. The net intensity from this *incoherent*, nonparametric process is low and hence recording of spontaneous Raman spectra requires efficient light collection and sensitive detectors.

Stimulated Raman Scattering (SRS). If three light frequencies are incident upon a sample with a molecular resonance ν_i, new components at all combinations $\nu_0 \pm \nu_1 \pm \nu_2 \pm \nu_i$ can occur via a third-order term in Eq. (1). This is termed four-wave mixing since three field waves serve, via P_{ind}, to generate a fourth. Of these various combinations, two are of particular interest in the current experiment, since they correspond to the generation of Stimulated Raman Scattering (SRS) and Coherent Anti-Stokes Raman Scattering (CARS), *coherent* processes that give information about the Raman-active transitions in molecules.

The energy level diagram representing the SRS process is shown in Fig. 1c. It is a third-order nonparametric process involving a mixing of fields $E(\nu_0)$ and $E(\nu_{S1})$ to induce a polarization at $\nu_{S1} = 2\nu_0 - \nu_{S1} - 2\nu_1 = \nu_0 - \nu_i$

$$P_{ind}(\nu_{S1}) = X^{(3)}E(\nu_0)E(\nu_0)E(\nu_{S1}) \tag{5}$$

As shown in texts on nonlinear spectroscopy[2-5], this polarization can cause the magnitude of the field $E(\nu_{S1})$, and hence the intensity $I(\nu_{S1})$, to grow exponentially with propagation distance z:

$$I(\nu_{S1}) = I_0(\nu_{S1}) \exp (Gz) \tag{6}$$

In this expression, the gain coefficient G is

$$G = K\bar{N}(d\sigma/d\nu_S)I(\nu_0)/\Delta\nu_S \tag{7}$$

where K is a proportionality constant and $\bar{N}$ is the sample number density. The quantities $d\sigma/d\nu_S$ (which is proportional to $X^{(3)}$) and $\Delta\nu_S$ are the scattering cross-section and frequency width, respectively, for the Raman transition. The requirement for SRS is that the pump laser intensity $I(\nu_0)$ be high enough to produce sufficient Raman gain to overcome other absorption or scattering losses of $I(\nu_{S1})$. The initial intensity $I_0(\nu_{S1})$ is that of a photon scattered by ordinary Raman scattering in the direction of propagation of the ν_0 laser beam. When the intensity of ν_0 is above some threshold value, amplification of the ν_{S1} field occurs and an intense, coherent beam at this frequency is generated along the propagation direction. For each photon generated at ν_{S1}, one photon of ν_0 is annihilated and one molecule in the sample is excited to the upper level in a Raman-active transition. Only the strongest Raman transitions in a molecule can satisfy the gain requirement, so typically only one or two vibrations in a molecule give rise to stimulated Raman scattering. For transitions with a high Raman cross-section and narrow linewidth, the process can be quite efficient, with greater than 10 percent conversion of the ν_0 beam into coherent radiation at the first Stokes frequency.

If the intensity of the Raman Stokes beam at ν_{S1} is high enough, this beam itself can generate a new beam at the *second* Stokes frequency $\nu_{S2} = \nu_{S1} - \nu_i = \nu_0 - 2\nu_i$, in a manner analogous to the first Stokes generation depicted in Fig 1c. Again, this SRS beam is coherent and travels along the original propagation direction. Note that the molecular transition is still from $v = 0$ to $v = 1$. At even higher intensities, this process repeats, with the generation of additional Stokes-shifted beams satisfying the wavelength condition

$$1/\lambda_m = \nu_m/c = \nu_0/c - m\nu_i/c \tag{8}$$

where m indexes the net number of Raman shifts from ν_0. Thus one can generate a rainbow of colors whose wavelengths can be measured and used to deduce the ν_i transition frequency.

Coherent Anti-Stokes Raman Scattering (CARS). In addition to the nonparametric SRS process to generate Stokes-shifted beams that propagate collinearly with the ν_0 pump radiation, a second *parametric* process can occur that generates noncollinear beams at frequencies $\nu_{A1} = 2\nu_0 - \nu_{S1} = \nu_0 + \nu_i$ and $\nu_{S1} = \nu_0 - \nu_i$. The energy level diagram for this case is depicted in Fig. 1d, along with the corresponding momentum-matching condition. This process results in the production of cones of anti-Stokes and first Stokes radiation, where the cone angles ϕ and δ are determined by the wavelength variation of the index of refraction of light in the sample according to the relations

$$2\mathbf{k}_0 - \mathbf{k}_{S1} - \mathbf{k}_{A1} = \Delta\mathbf{k} = 0 \tag{9}$$

$$2n_0\nu_0 - n_{S1}\nu_{S1}\sin\phi - n_{A1}\nu_{A1}\sin\delta = 0 \tag{10}$$

The basic requirement is that the momentum mismatch $\Delta\mathbf{k}$ must be zero for efficient generation of light. For the index dispersion in benzene, this results in blue anti-Stokes light with a cone angle of about 4 degrees.[3] A similar cone is produced at the Stokes frequency, although this may be difficult to distinguish from the intense collinear beam produced by SRS scattering.

In an analogous way, the mixing of these off-axis beams with each other and with the collinear beams can yield a complex pattern of Stokes and anti-Stokes beams at various frequencies and angles. These additional beams become more apparent at higher ν_0 pump laser powers but their directional characteristics will be examined only qualitatively in this experiment, where the emphasis is on wavelength measurement. A more complete discussion of all these wave-mixing processes, along with beautiful color prints of the resultant rings of light, can be found in Refs. 3 and 6.

Benzene Vibrations. With $N = 12$ atoms, benzene has $3N - 6 = 30$ vibrational modes, but 10 are degenerate (E type) due to symmetry so that only 20 unique vibrational frequencies are predicted. Using group theory[7,8], these are classed according to the symmetry species of the D_{6h} point group of benzene as

$$\underbrace{2A_{1g} + E_{1g} + 4E_{2g}}_{\text{Raman}} + \underbrace{A_{2u} + 3E_{1u}}_{\text{Infrared}} + \underbrace{A_{2g} + 2B_{2g} + 2B_{1u} + 2B_{2u} + 2E_{2u}}_{\text{Inactive}}$$

Since benzene has a center of symmetry, the "rule of mutual exclusion" applies; i.e., bands that appear in the IR are absent in the Raman and vice versa. Thus, just from symmetry considerations, one predicts four infrared-active and seven Raman-active fundamental transitions, with no required coincidences among these frequencies.

As for most molecules, Raman-active modes that retain the full symmetry of the molecule have the highest cross section for scattering since these modes produce the largest change in the polarizability during the course of the vibrational motion. Hence in benzene the two polarized A_{1g} vibrations (ν_1 and ν_2) are the most intense in the Raman spectrum and the in-plane ring-breathing mode (ν_2) is especially effective in producing stimulated Raman scattering. Appropriate vibrational coordinates of A_{1g} symmetry are

$$S_1 = (s_1 + s_2 + s_3 + s_4 + s_5 + s_6)/(6)^{1/2} \tag{11}$$

$$S_2 = (t_1 + t_2 + t_3 + t_4 + t_5 + t_6)/(6)^{1/2} \tag{12}$$

where s_i and t_i are bond extensions of one of the six CH and CC bonds, respectively. Keeping only a single interaction constant of the form k_{st}, the contribution of these two A_{1g} coordinates to the potential energy of the molecules is

$$U(A_{1g}) = \tfrac{1}{2}k_s S_1^2 + \tfrac{1}{2}k_t S_2^2 + k_{st} S_1 S_2 \tag{13}$$

and the force constants are related to the vibrational frequencies and atomic masses through the relations[7]

$$4\pi^2 \nu_1^2 + 4\pi^2 \nu_2^2 = k_s[(1/m_H) + (1/m_C)] - 2k_{st}/m_C + k_t/m_C \tag{14}$$

$$(4\pi^2 \nu_1^2)(4\pi^2 \nu_2^2) = (k_s k_t - k_{st}^2)/m_H m_C \tag{15}$$

See Exp. 35 for a discussion of the units to be used in these equations.

It is apparent from Eqs. (14) and (15) that the two measured A_{1g} frequencies for C_6H_6 are not sufficient to determine the three force constants k_s, k_t, and k_{st}. However, because the force constants are invariant to isotopic substitution, insertion of the C_6D_6 frequencies into Eq. (14) gives a third equation so that one can solve for the three force constants. Note that Eq. (15) for C_6D_6 *adds no new constraint,* since the force constants cancel out on taking a ratio of this equation for the two isotopic forms. This ratio is an example of the *Product Rule* discussed in Ref. 7.

EXPERIMENTAL

The pulsed laser beams in this experiment are of high power and can cause eye damage. Read the safety discussion about lasers in Appendix C and pay close attention to directions given by your instructor. Remove all potentially reflective watches, bracelets, and rings. Goggles should be worn except when it is necessary to remove them for making wavelength measurements, in which case the instructor should ensure that all reflections from optics are accounted for and blocked. Do not look directly along the axis of the laser or signal beams and avoid staring at any intense spots in making measurements.

Figure 2 shows a suitable arrangement for the experiment. This experiment requires only a small ν_0 pulse energy of about 1 mJ in a 532-nm beam of about 10 ns pulse duration, but this corresponds to a peak power of 100 kW. This energy is easily obtained from the doubled output of a Q-switched Nd:YAG laser and such a source is recommended. Alternatively a comparable output of a pulsed dye laser, pumped by a nitrogen, excimer, or Nd:YAG laser, can be employed. The instructor will describe the laser to be used and assist in setting up the experiment. If it is necessary to run the laser at higher energies for best pulse-to-pulse stability, the beam should be attenuated with suitable optical density

FIGURE 2

Experimental setup for the measurement of stimulated Raman spectra. The distance x is measured from the grating to the wall or viewing screen. The diffracted beams show the pattern of the beams involving the $\tilde{\nu}_2$ ring stretching vibration of benzene; not shown are additional weak features that may be seen involving the $\tilde{\nu}_1$ CH or CD stretching vibrations.

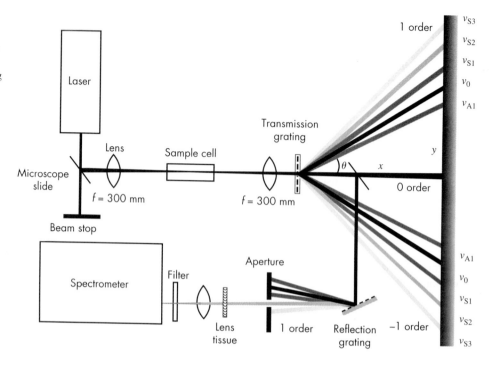

filters or a microscope slide can be used as a beam splitter to pick off 1–2 mJ for the experiment, as shown in Fig. 2.

Using a hood, fill a 100-mm optical cell with reagent or spectroscopic quality benzene and seal it with Teflon stoppers. The initial alignment of the laser beam and optical components should be done at very low laser power, in the non-Q-switched mode of operation of the laser, where the pulse is about 1 ms in duration and the peak power is only a few watts. Direct the faint green beam through the center of a 300-mm focal length lens with the benzene cell positioned so that the middle of the cell is at the beam focus. Do not place the windows of the cell near the focus since the beam intensity when in the short pulse Q-switched mode is high enough to damage the glass. Also tilt the cell slightly so that the beam reflection from the entrance window to the cell is blocked by the lens mount and does not come to a focus in the lens itself. Center and adjust the position of the second lens so that the green beam comes to a focus on a wall or suitable screen that is perpendicular to the beam.

To disperse the radiation and obtain an approximate measure of the wavelength of the generated beams, a transmission grating can be used. This should be attached to a mirror mount so that orthogonality to the beam can be achieved in the diffraction plane by adjusting the direct back-reflection from the grating just above (or below) the original beam. Under these conditions, a simple distance measurement as shown in Fig. 2 can be used to determine the diffraction angle and the wavelength from the Bragg equation; i.e.,

$$\tan \theta_m = y_m/x \tag{16}$$

$$n\lambda_m = d \sin \theta_m \tag{17}$$

Here the index m has values 0 for the 532-nm measurements, +1 for the first anti-Stokes (blue-shifted) line, and -1, -2 for the first, second Stokes (red-shifted) lines, etc.; n is the diffraction order (0, ± 1, ± 2); and d is the groove spacing of the grating, supplied by the manufacturer. A suitable grating with $d = 1333.3$ nm (750 lines per mm) is described in the Apparatus section. The wavelength of the doubled Nd:YAG laser is 532.260 nm, and this can be used to confirm the d spacing.

After the alignment is complete, turn the laser energy to the lowest setting and activate the Q-switch mode of operation. Increase the energy slowly until yellow Stokes radiation is seen in the diffracted light. Continue to increase the energy very slightly and observe the appearance of orange, red, and on the anti-Stokes side of the green source, blue light. Adjust the position of the second lens along the propagation axis and note the faint blue ring that corresponds to the cone of radiation produced by the parametric process of Fig. 1d. Similar rings should be discernible on the Stokes side. Increase the laser power only to the point that the new beam intensities are moderately stable. (At higher laser powers, breakdown of benzene occurs at the sample focus with the production of soot that can produce burns on the windows. If the amount of soot is appreciable, the sample should be filtered or replaced with new benzene.) If an energy meter is available, measure the threshold for generation of the light and the energy at which a stable pattern is seen. Typically this will be 0.5 to 2 mJ/pulse. At the higher energy a series of beams will be seen that are produced by the CC ring-stretching vibration of benzene, Stokes-shifted in each case by about 1000 cm^{-1}. Just to the red of the third Stokes beam, there will be a brighter spot that corresponds to a Raman shift by the symmetric CH vibration near 3050 cm^{-1}. Note that, if the intensity of the latter is high, it may produce a ν_{CC}-shifted spot about 1000 cm^{-1} to the red; this should not be assigned as the fourth ν_{CC}-shifted Stokes beam.

To measure wavelengths, adjust the second lens position to focus the beams on the wall (or screen). Tape sheets of paper to the wall at the 0 and ± 1 order diffraction positions. Attenuate (or block) the brighter beams, and taking care to avoid the diffracted beams, mark the location of all spots with a nonreflecting pencil. Block the pump laser beam before the cell and measure the x and y_m values of the marks with a meter stick or tape measure as accurately as possible. Other variants of the experiment are possible; for example, a reflection rather than transmission diffraction grating could be employed as the dispersing element for these simple wavelength measurements.

Repeat the operation with a second cell filled with deuterated benzene, C_6D_6. In this case the Stokes shift for the CD stretching mode is much smaller, and this beam is expected to occur between the second and third Stokes spots produced by the CC stretch. This feature is less intense than its CH counterpart in C_6H_6, but it should be observable, along with the fourth ν_{CC}-shifted Stokes feature.

If possible, make more accurate measurements of the wavelengths using a high-resolution grating spectrometer as indicated in Fig. 2. The instructor will advise on the most appropriate measurement procedure for the particular setup employed. The spectrometer should have 0.1-nm resolution, or better if possible, and should be equipped with a photomultiplier or charge-coupled detector (CCD). An instrument used for Raman or fluorescence measurements would be quite suitable.

Since the beam intensities are high, care should be taken not to damage the detector. Initially the slits should be set near the minimum, and the beam being measured should be attenuated by suitable colored or neutral density filters. To measure the wavelengths of the weaker beams, the attenuation can be reduced and/or the slits opened slightly. The beams should be imaged onto a piece of lens tissue or translucent paper/glass placed in front of the lens that focuses light onto the entrance slit of the spectrometer. The purpose of this is not just to further attenuate the beam but also to reduce the coherence of the light passing through the slits, since the narrow slits themselves can diffract a coherent beam of a single color. As a result, a coherent beam might produce sidelobes, which can appear as spurious lines in the spectrum.

With a photomultiplier detector, the measurement can be done simply by observing the signal maximum with an oscilloscope as the wavelength is scanned by hand or electronically. Alternatively a pulse integration system can be employed to record the spectra. The use of a CCD detector for such measurements is advantageous since the detector itself serves to integrate the pulsed signal. Such measurements are described in Ref. 9.

CALCULATIONS AND DISCUSSION

Assemble a table with your measured quantities for both C_6H_6 and C_6D_6. Compute the wavelengths and the corresponding wavenumber values, $\tilde{\nu}_m(\text{cm}^{-1}) = 1/\lambda_m(\text{cm})$. Estimate the uncertainty in these from your uncertainty estimate in the x and y_m measurements. Most of the lines are a progression in the symmetric ring-stretching mode $\tilde{\nu}_2$ at about 1000 cm^{-1}. Do a linear regression of your measured wavenumber $\tilde{\nu}_m$ values for the lines in this progression using Eq. (8) in the form

$$\tilde{\nu}_m = \tilde{\nu}_0 + m\tilde{\nu}_2 \tag{18}$$

where the index m is as described for Eq. (17). Compare your resulting $\tilde{\nu}_0$ value with the expected value of 18787.8 cm^{-1} and also your $\tilde{\nu}_2$ result with the literature value of 991.6 cm^{-1}.[8] Perform a separate linear regression of the data for C_6D_6. Do your $\tilde{\nu}_0$ values agree within the standard error given by the linear regressions? Are the $\tilde{\nu}_2$ values significantly different for the two isotopic forms? From the Stokes measurements of the CH and CD stretching features, calculate and report the $\tilde{\nu}_1$ vibrational frequencies for C_6H_6 and C_6D_6. Literature values can be found in Ref. 8.

If more accurate wavelength measurements have been made with a spectrometer, repeat the calculations and compare the uncertainties with those of the simpler diffraction experiment described above. Using your best frequency results, solve the three force constant equations for k_s, k_p, and k_{st}. You will find two possible sets of these constants since the solution of the equations yields a quadratic expression. Choose between these two by recognizing that the interaction constant k_{st} is usually much smaller than k_s or k_t. (*Note:* if the frequencies used are appreciably in error, the solution may yield imaginary roots.) Typically the force constants for single, double, and triple CC bonds are about 500, 1000, and 1500 N m^{-1}, respectively. What does your value of k_t imply about the CC bond type in benzene?

This experiment can also be done with other liquid samples that generate stimulated Raman beams; examples are acetonitrile[9] (CH_3CN and CD_3CN) and CCl_4. Solid calcite crystals $CaCO_3$ (feldspar) can be used to produce intense stimulated Raman emission via the symmetric CO_3 carbonate stretching mode.[3]

Theoretical Calculations. The vibrational frequencies and infrared and Raman intensities can be calculated for benzene using programs such as Gaussian/Gaussview and HyperChem. These programs also permit animated visualization of the normal modes such as the symmetric CC and CH stretches studied in this experiment.

SAFETY ISSUES

As noted in the text, special care should be taken to avoid stray reflections of the laser and signal beams generated in this experiment. The pulse energies are high enough to cause serious eye injury and safety goggles should be worn at all times except when the instructor has determined it is safe to remove them to make measurements. The filling of the benzene cells should be done in a hood, and the sealed cell should be handled carefully. Benzene is a carcinogenic substance and it should be disposed of properly.

APPARATUS

Nd:YAG laser with second harmonic generator (doubler) capable of generating at least 2 mJ 532-nm pulses of 5 to 10 ns duration (e.g., Polaris model from New Wave, Minilite model from Continuum); laser safety goggles (available from Lase-R Shield, Glendale Optical, Kentek, etc.); reagent or spectral grade benzene and deuterated benzene; 100-mm cell with optical glass windows and Teflon stoppers (available from McCarthy Scientific, Helma, Starna, etc.); mirrors and mounts; 300-mm convex lenses; lens tissue; colored filters; neutral density filters (available from Newport, Oriel, Edmunds Scientific, etc.); microscope slides; gratings; meter stick or tape measure with fine markings; spectrometer with photomultiplier or CCD detector and a resolution of about 0.1 nm; oscilloscope.

An inexpensive and very efficient holographic transmission grating with 750 lines/ mm is available from Learning Technologies, 40 Cameron Avenue, Somerville, MA 02144; inexpensive reflection gratings can be obtained from Edmunds Scientific.

REFERENCES

1. E. J. Woodbury and W. K. Ng, *Proc. IRE* **50,** 2367 (1962).

2. Y. R. Shen, *The Principles of Nonlinear Optics,* chap. 15, Wiley, New York (1984).

3. G. S. He and S. H. Liu, *Physics of Nonlinear Optics,* chap. 8, World Scientific, River Edge, NJ (1999).

4. R. W. Boyd, *Nonlinear Optics,* chap. 9, Academic Press, San Diego, CA (1992).

5. J. W. Nibler and G. V. Knighten, "Coherent Anti-Stokes Raman Spectroscopy," chap. 7 in A. Weber (ed.), *Topics in Current Physics,* Springer-Verlag, New York (1978).

6. D. A. Long, *Raman Spectroscopy,* chap. 8, McGraw-Hill, New York (1977).

7. E. B. Wilson, J. C. Decius, and P. C. Cross, *Molecular Vibrations,* chap. 10, McGraw-Hill, New York (1955), reprinted in unabridged form by Dover, New York (1980).

8. G. Herzberg, *Molecular Spectra and Molecular Structure II: Infrared and Raman Spectra,* pp. 362–369, reprint ed., Krieger, Melbourne, FL (1990).

9. C. A. Grant and J. L. Hardwick, *J. Chem. Educ.* **74,** 318 (1997).

GENERAL READING

J. L. McHale, *Molecular Spectroscopy,* Prentice-Hall, Upper Saddle River, NJ (1999).

S. Mukamel, *Principles of Nonlinear Laser Spectroscopy,* Oxford Univ. Press, Oxford (1995).

G. S. He and S. H. Liu, *Physics of Nonlinear Optics,* chap. 8, World Scientific, River Edge, NJ (1999).

R. W. Boyd, *Nonlinear Optics,* Academic Press, San Diego, CA (1992).

EXPERIMENT 37

Vibrational–Rotational Spectra of HCl and DCl

The infrared region of the spectrum extends from the long-wavelength end of the visible region at 1 μm out to the microwave region at about 1000 μm. It is common practice to specify infrared frequencies in wavenumber units: $\tilde{\nu}(\text{cm}^{-1}) = 1/\lambda = \nu/c$, where c is the speed of light in cm s^{-1} units. Thus this region extends from 10,000 cm^{-1} down to 10 cm^{-1}. Although considerable work is now being done in the far-infrared region below 400 cm^{-1}, the spectral range from 4000 to 400 cm^{-1} has received the greatest attention because the vibrational frequencies of most molecules lie in this region.

This experiment is concerned with the rotational fine structure of the infrared vibrational spectrum of a linear molecule such as HCl. From an interpretation of the details of this spectrum, it is possible to obtain the moment of inertia of the molecule and thus the internuclear separation. In addition the pure vibrational frequency determines a force constant that is a measure of the bond strength. By a study of DCl also, the isotope effect can be observed.

THEORY

Almost all infrared work makes use of absorption techniques in which radiation from a source emitting all infrared frequencies is passed through a sample of the material to be studied. When the frequency of this radiation is the same as a vibrational frequency of the molecule, the molecule may be vibrationally excited; this results in loss of energy from the radiation and gives rise to an absorption band. The spectrum of a polyatomic molecule generally consists of several such bands arising from different vibrational motions of the molecule. This experiment involves diatomic molecules, which have only one vibrational mode.

The simplest model of a vibrating diatomic molecule is a harmonic oscillator, for which the potential energy depends quadratically on the change in internuclear distance. The allowed energy levels of a harmonic oscillator, as calculated from quantum mechanics,[1] are

$$E(v) = h\nu(v + \tfrac{1}{2}) \tag{1}$$

where v is the vibrational quantum number having integral values 0, 1, 2, . . . ; ν is the vibrational frequency; and h is the Planck constant.

The simplest model of a rotating diatomic molecule is a rigid rotor or "dumbbell" model in which the two atoms of mass m_1 and m_2 are considered to be joined by a rigid, weightless rod. The allowed energy levels for a rigid rotor may be shown by quantum mechanics[1] to be

$$E(J) = \frac{h^2}{8\pi^2 I} J(J + 1) \tag{2}$$

where the rotational quantum number J may take integral values 0, 1, 2, ... The quantity I is the moment of inertia, which is related to the internuclear distance r and the reduced mass $\mu = m_1 m_2/(m_1 + m_2)$ by

$$I = \mu r^2 \tag{3}$$

Since a real molecule is undergoing both rotation and vibration simultaneously, a first approximation to its energy levels $E(v, J)$ would be the sum of expressions (1) and (2). A more complete expression for the energy levels of a diatomic molecule[2] is given below, with the levels expressed as *term values* T in cm^{-1} units rather than as energy values E in joules:

$$T(v, J) = \frac{E(v, J)}{hc} = \tilde{\nu}_e(v + \tfrac{1}{2}) - \tilde{\nu}_e x_e(v + \tfrac{1}{2})^2 + B_e J(J + 1)$$
$$-D_e J^2(J + 1)^2 - \alpha_e(v + \tfrac{1}{2})J(J + 1) \tag{4}$$

where c is the speed of light in cm s^{-1}, $\tilde{\nu}_e$ is the frequency in cm^{-1} for the molecule vibrating about its equilibrium internuclear separation r_e, and

$$B_e = \frac{h}{8\pi^2 I_e c} \tag{5}$$

The first and third terms on the right-hand side of Eq. (4) are the harmonic-oscillator and rigid-rotor terms with r equal r_e. The second term (involving the constant x_e) takes into account the effect of anharmonicity. Since the real potential $U(r)$ for a molecule differs from a harmonic potential U_{harm} (see Fig. 1), the real vibrational levels are not quite those given by Eq. (1) and a correction term is required. The fourth term (involving the constant D_e) takes into account the effect of centrifugal stretching. Since a chemical bond is not truly rigid but more like a stiff spring, it stretches somewhat when the molecule rotates. Such an effect is important only for high J values, since the constant D_e is usually very small. The last term in Eq. (4) accounts for interaction between vibration and rotation. During a vibration the internuclear distance r changes; this changes the moment of inertia and affects the rotation of the molecule. The constant α_e is also quite small, but this term should not be neglected.

Selection Rules.   The harmonic-oscillator, rigid-rotor selection rules[2] are $\Delta v = \pm 1$ and $\Delta J = \pm 1$; that is, infrared emission or absorption can occur only when these "allowed" transitions take place. For an anharmonic diatomic molecule, the $\Delta J = \pm 1$ selection rule is still valid, but weak transitions corresponding to $\Delta v = \pm 2, \pm 3$, etc. (overtones) can now be observed.[2] Since we are interested in the most intense absorption band (the "fundamental"), we are concerned with transitions from various J'' levels of the vibrational ground

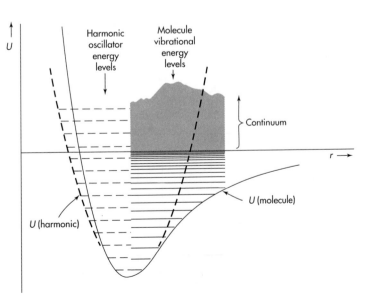

FIGURE 1

Schematic diagram showing potential energy U as a function of internuclear separation r for a diatomic molecule. The harmonic potential U_{harm} is indicated by the dashed curve. The vibrational levels are also shown

state ($v'' = 0$) to J' levels in the first excited vibrational state ($v' = 1$). From the selection rule we know that the transition must be from J'' to $J' = J'' \pm 1$. Since $\Delta E = h\nu = hc\tilde{\nu}$, the frequency $\tilde{\nu}$ (in wavenumbers) for this transition will be just $T(v', J') - T(v'', J'')$. When $\Delta J = +1$ ($J' = J'' + 1$) and $\Delta J = -1$ ($J' = J'' - 1$), we find, respectively, from Eq. (4) that

$$\tilde{\nu}_R = \tilde{\nu}_0 + (2B_e - 3\alpha_e) + (2B_e - 4\alpha_e)J'' - \alpha_e J''^2 \qquad J'' = 0, 1, 2, \ldots \qquad (6)$$

$$\tilde{\nu}_P = \tilde{\nu}_0 - (2B_e - 2\alpha_e)J'' - \alpha_e J''^2 \qquad\qquad\qquad J'' = 0, 1, 2, \ldots \qquad (7)$$

where the D_e term has been dropped and $\tilde{\nu}_0$, the frequency of the *forbidden* transition from $v'' = 0, J'' = 0$ to $v' = 1, J' = 0$, is

$$\tilde{\nu}_0 = \tilde{\nu}_e - 2\tilde{\nu}_e x_e \qquad (8)$$

The two series of lines given in Eqs. (6) and (7) are called R and P branches, respectively. These allowed transitions are indicated on the energy-level diagram given in Fig. 2. If α_e were negligible, Eqs. (6) and (7) would predict a series of equally spaced lines with separation $2B_e$ except for a missing line at $\tilde{\nu}_0$. The effect of interaction between rotation and vibration (nonzero α_e) is to draw the lines in the R branch closer together and spread the lines in the P branch farther apart as shown for a typical spectrum in Fig. 3. For convenience let us introduce a new quantity m, where $m = J'' + 1$ for the R branch and $m = -J''$ for the P branch as shown in Fig. 3. It is now possible to replace Eqs. (6) and (7) by a single equation:

$$\tilde{\nu}(m) = \tilde{\nu}_0 + (2B_e - 2\alpha_e)m - \alpha_e m^2 \qquad (9)$$

FIGURE 2

Rotational energy levels for the ground vibrational state ($v'' = 0$) and the first excited vibrational state ($v' = 1$) in a diatomic molecule. The vertical arrows indicate allowed transitions in the R and P branches; numbers in parentheses index the value J'' of the lower state. Transitions in the Q branch ($\Delta J = 0$) are not shown since they are not infrared active.

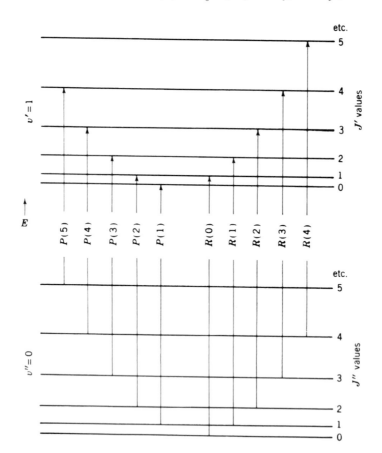

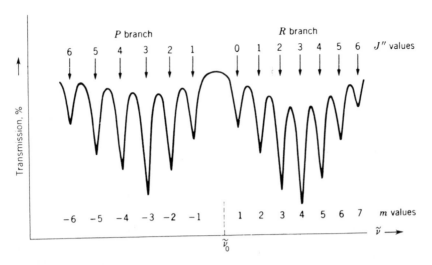

where m takes all integral values and $m = 0$ yields the frequency $\tilde{\nu}_0$ of the forbidden "purely vibrational" transition. If one retains the D_e term of Eq. (4) (which assumes $D'' = D' = D_e$), Eq. (9) takes the form

$$\tilde{\nu}(m) = \tilde{\nu}_0 + (2B_e - 2\alpha_e)m - \alpha_e m^2 - 4D_e m^3 \qquad (10)$$

Thus a multiple linear regression can be performed to determine $\tilde{\nu}_0$, B_e, α_e, and D_e.

Isotope Effect.   When an isotopic substitution is made in a diatomic molecule, the equilibrium bond length r_e and the force constant k are unchanged, since they depend only on the behavior of the bonding electrons. However, the reduced mass μ does change, and this will affect the rotation and vibration of the molecule. In the case of rotation, the isotope effect can be easily stated. From the definitions of B_e and I, we see that

$$\frac{B_e^*}{B_e} = \frac{\mu}{\mu^*} \qquad (11)$$

where an asterisk is used to distinguish one isotopic molecule from another.

For a harmonic oscillator model, the frequency $\tilde{\nu}_e$ in wavenumbers is given by

$$\tilde{\nu}_e = \frac{1}{2\pi c}\left(\frac{k}{\mu}\right)^{1/2} \qquad (12)$$

which leads to the relation

$$\frac{\tilde{\nu}_e^*}{\tilde{\nu}_e} = \left(\frac{\mu}{\mu^*}\right)^{1/2} \qquad (13)$$

The ratio $\tilde{\nu}_0^*/\tilde{\nu}_0$ differs slightly from this harmonic ratio due to deviation of the true potential function from a quadratic form, as depicted in Fig. 1. A closer approximation to the solid curve can be had by adding cubic and higher anharmonic terms to $U(r)$, viz.,

$$U(r) = \tfrac{1}{2}k(r - r_e)^2 + c(r - r_e)^3 + d(r - r_e)^4 + \cdots \qquad (14)$$

Although somewhat complicated, it can be shown[2,3] that the c and d terms yield, as the first correction to the energy levels, precisely the $-\tilde{\nu}_e x_e(v + \tfrac{1}{2})^2$ term given in Eq. (4).

A similar conclusion is reached if $U(r)$ is taken to have the Morse potential form given by Eq. (9) of Exp. 39. In both cases, the mass dependence of $\tilde{\nu}_e x_e$ is found to be greater than for $\tilde{\nu}_e$ and is

$$\frac{\tilde{\nu}_e^* x_e^*}{\tilde{\nu}_e x_e} = \frac{\mu}{\mu^*} \tag{15}$$

Equations (13) and (15) are useful in obtaining the $\tilde{\nu}_0^*$ counterpart of Eq. (8),

$$\tilde{\nu}_0^* = \tilde{\nu}_e^* - 2\tilde{\nu}_e^* x_e^* = \tilde{\nu}_e \left(\frac{\mu}{\mu^*}\right)^{1/2} - 2\tilde{\nu}_e x_e \frac{\mu}{\mu^*} \tag{16}$$

and it is seen that a measurement of $\tilde{\nu}_0$ for HCl and DCl suffices for a determination of $\tilde{\nu}_e$ and $\tilde{\nu}_e x_e$. Alternatively of course the latter constants can be determined from overtone vibrations ($\Delta v > 1$) of a single isotopic form (see Exp. 39). However, such overtones generally have low intensity and the transitions may fall outside the range of many infrared instruments, so the isotopic shift method is used in the present experiment.

Since HCl gas is a mixture of $H^{35}Cl$ and $H^{37}Cl$ molecules, a chlorine isotope effect will also be present. However, the ratio of the reduced masses is only 1.0015: therefore high resolution is required to detect this effect. HCl is predominantly $H^{35}Cl$, and for this experiment we shall assume that the HCl bands obtained are those of $H^{35}Cl$. If deuterium is substituted for hydrogen, the ratio of the reduced masses, $\mu(D^{35}Cl)/\mu(H^{35}Cl)$, is 1.946 and the isotope effect is quite large.

Vibrational Partition Function.[4,5]   The thermodynamic quantities for an ideal gas can usually be expressed as a sum of translational, rotational, and vibrational contributions (see Exp. 3). We shall consider here the heat capacity at constant volume. At room temperature and above, the translational and rotational contributions to C_v are constants that are independent of temperature. For HCl and DCl (diatomic and thus linear molecules), the molar quantities are

$$\tilde{C}_v(\text{trans}) = \tfrac{3}{2}R$$

$$\tilde{C}_v(\text{rot}) = R \tag{17}$$

The vibrational contribution to $\tilde{C}_v$ varies with temperature and can be calculated from the vibrational partition function q_{vib} using

$$\tilde{C}_v(\text{vib}) = R\frac{\partial}{\partial T}\left(T^2 \frac{\partial \ln q_{\text{vib}}}{\partial T}\right) \tag{18}$$

The partition function q_{vib} of HCl or DCl is well approximated by the harmonic-oscillator partition function q^{HO}. Since the energy levels of a harmonic oscillator are given by $(v + \tfrac{1}{2})h\nu$, one obtains[4]

$$q^{HO} = \sum_{v=0}^{\infty} \exp\left[\frac{-(v+\tfrac{1}{2})h\nu}{kT}\right] = \frac{e^{-h\nu/2kT}}{1 - e^{-h\nu/kT}} \tag{19}$$

Combining Eqs. (18) and (19), we find

$$\tilde{C}_v(\text{vib}) = R\frac{u^2 e^{-u}}{(1 - e^{-u})^2} \tag{20}$$

where $u = h\nu/kT = hc\tilde{\nu}/kT = 1.4388\tilde{\nu}/T$.

EXPERIMENTAL

A general description of infrared instruments is given in Chapter XIX. Medium to high resolution is required for this experiment; a grating or FTIR instrument with at least 2 cm^{-1} resolution is desirable. In addition grating spectrometers require careful calibration. CO and CH_4 are suitable gases for calibration purposes. Detailed instructions for operating the spectrometer will be given in the laboratory. Before beginning the experiment, the student should review the pertinent material. **Use the spectrometer carefully;** if in doubt ask the instructor.

The infrared gas cell is constructed from a short (usually 10-cm) length of large-diameter (4 to 5 cm) Pyrex tubing with a vacuum stopcock attached. Infrared-transparent windows are clamped against O-rings at the ends of the cell or are sealed on the ends with Glyptal resin. For studies concentrating on the region 4000 to 700 cm^{-1}, NaCl windows will suffice. When the spectrum of interest extends down to ~400 cm^{-1}, KBr windows are needed. Both types of salt windows become "foggy" on prolonged exposure to a moist atmosphere and should be protected (e.g., stored in a desiccator) when not in use. In the present experiment, inert sapphire windows with transmission down to 1600 cm^{-1} are particularly convenient and relatively inexpensive.

Filling the Cell and Recording the Spectra.   An arrangement for filling the cell is given in Fig. 4. Attach the cell at D. With stopcock C open and B closed, open stopcock A and pump out the system. Make sure that the needle valve of the HCl cylinder is closed. Then open B and continue pumping.

Close A and **slowly** open the valve on the HCl cylinder. Fill the system to a pressure recommended by the instructor (50 to 500 Torr, depending upon the resolution of the spectrometer). Close the valve on the HCl cylinder and then close B and C. Remove the cell and take a spectrum at the highest available resolution.

Preparation of DCl.   If DCl gas is not available in a commercial cylinder or bulb, it must be prepared in the laboratory. Deuterium chloride gas can be synthesized readily by the reaction between benzoyl chloride and heavy water:

$$C_6H_5COCl + D_2O \rightarrow C_6H_5COOD + DCl(g)$$
$$C_6H_5COOD + C_6H_5COCl \rightarrow (C_6H_5CO)_2O + DCl(g)$$

(21)

An arrangement for carrying out this reaction is shown in Fig 5; a somewhat simpler apparatus for a smaller-scale synthesis is described in Exp. 43. Approximately 2.5 mL (0.14 mol) of D_2O in the separating funnel S is added slowly to 70 g (0.5 mol) of benzoyl

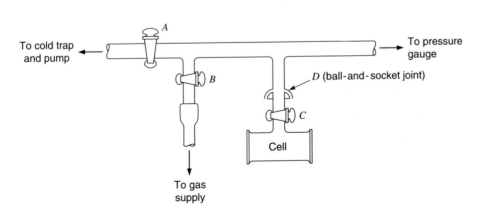

FIGURE 4

Gas-handling system for filling infrared cell.

FIGURE 5

Apparatus for preparing
deuterium chloride.

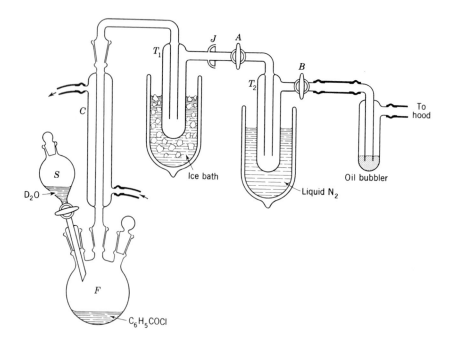

chloride in flask *F*. With a water-cooled reflux condenser *C* attached and either ice or Dry
Ice and trichloroethylene in trap T_1, the flask is gently heated. Stopcocks *A* and *B* are left
open, and the system is swept out by allowing some DCl gas to escape through the oil
bubbler into a hood. Stopcock *B* is then closed, and trap T_2 is cooled with liquid nitrogen.
After 20 min remove the heat from flask *F*. Wait a few minutes, then close stopcock *A* and
disconnect the ball-and-socket joint *J*. Keep trap T_2 in liquid nitrogen, and attach it to the
system used to fill the cell at stopcock *B* shown in Fig. 4. After the line is pumped out, the
cell can be filled by allowing the DCl in the trap to warm up slowly until the desired pressure is achieved.

A similar synthesis can be used to generate DBr from benzoyl bromide if it is desired
to study DBr and HBr in addition or as an alternative to DCl and HCl. In fact, if an
equimolar mixture of the benzoyl halides is used along with D_2O, which is about 95 percent
D, the product mix will contain all isotopic forms with reasonable intensity ratios for an
infrared scan of all four species in a single cell. Such a mixture can also be used for the
determination of equilibrium constant K_p for the H–D exchange reaction

$$DCl + HBr \rightleftarrows DBr + HCl \tag{22}$$

Unfortunately, the infrared intensities do not give accurate concentrations due to pressure-
broadening effects and the narrowness of the spectral lines compared to the spectral resolution. A preferable and more accurate measurement of K_p can be made with an NMR
spectrometer, as described in Exp. 43.

CALCULATIONS

Select your best HCl and DCl spectra, and index the lines with the appropriate *m* values as shown in Fig. 3. If $^{35}Cl/^{37}Cl$ splitting is seen, index the stronger ^{35}Cl lines. Make a
table of these *m* values and the corresponding frequencies $\tilde{\nu}(m)$. Express the frequencies in

cm^{-1} units to tenths of a cm^{-1} if possible. Then list the differences between adjacent lines $\Delta\tilde{\nu}(m)$, which will be roughly $2B_e$ but should vary with m. Plot $\Delta\tilde{\nu}(m)$ against m, draw a straight line through the points, and check any points that seem out of line. Then carry out a multiple linear least-squares fit to the data with Eq. (9) to determine $\Delta\tilde{\nu}_0$, B_e, and α_e and their standard errors. Repeat this fitting procedure using Eq. (10), noting that high m transitions will be the most important ones in determining D_e due to its m^3 dependence. Use an F test as described in Chapters III and XXI to determine if the value of D_e obtained in this second fit is significant at the 95 percent confidence level. An example of a spreadsheet calculation of this type is given in Fig. III-3. Use your values of $\tilde{\nu}_0$ for HCl and DCl to determine $\tilde{\nu}_e$ and $\tilde{\nu}_e x_e$ for HCl. From $\tilde{\nu}_e$, calculate k for HCl.

Calculate I_e, the moment of inertia, and r_e, the internuclear distance, for both HCl and DCl. The masses (in atomic mass units) are H = 1.007825, D = 2.014102, ^{35}Cl = 34.968853, and ^{37}Cl = 36.965903. If HBr and DBr are examined, it is unlikely that the two nearly equal ^{79}Br and ^{81}Br spectral lines will be resolved, and it is appropriate to use the average atomic mass (79.904) in similar calculations. Tabulate all of your results, along with your estimates of the experimental uncertainty. Compare your results with literature values, which can be found in Ref. 2.

Using your value of $\tilde{\nu}_0$ for HCl, calculate $\tilde{C}_v(\text{vib})$ at 298 K and at 1000 K from Eq. (20). Compare the spectroscopic value $\tilde{C}_v = 2.5R + \tilde{C}_v(\text{vib})$ with the experimental $\tilde{C}_v$ value obtained from directly measured values[5,6] of $\tilde{C}_p$ and the expression $\tilde{C}_v = \tilde{C}_p - R$: $\tilde{C}_v = 20.80$ J K^{-1} mol^{-1} at 298 K and 23.20 J K^{-1} mol^{-1} at 1000 K.

DISCUSSION

Compute the ratio B_e^*/B_e and compare with the rigid-rotor prediction of Eq. (11). How constant is r_e for HCl and DCl? Compute $B_v = B_e - \alpha_e(v + \frac{1}{2})$ for the $v = 0, 1$, and 2 levels of HCl and DCl and from these obtain average r_v values for these levels. Comment in your report on the changes in these distances. Compare your $\tilde{\nu}_0^*/\tilde{\nu}_0$ ratio with the ratio $(\mu/\mu^*)^{1/2}$ expected for a harmonic oscillator. How anharmonic is the HCl molecule: i.e., how large is x_e? Use your values of $\tilde{\nu}_e$ and $\tilde{\nu}_e x_e$ and Eq. (4) to predict the frequencies of the first overtone transitions of HCl and DCl (ignore the rotational terms). Did you see any indication of these overtones in your spectra? Do your spectra show any evidence of a ^{35}Cl–^{37}Cl isotope effect? Use Eq. (10) to calculate the splitting expected for this effect for several P and R branch transitions in HCl and in DCl.

Theoretical Calculations.   A class project to calculate the physical properties of HCl using an ab initio program such as Gaussian is described in Chapter III. Compare your experimental values of r_e and $\tilde{\nu}_e$ with values deduced from these theoretical calculations.

SAFETY ISSUES

Lecture bottles of HCl and CO/CH$_4$ gas must be strapped or chained securely to the laboratory bench. Work on a vacuum system requires preliminary review of procedures and careful execution in order to avoid damage to the apparatus and possible injury from broken glass; in addition the liquid nitrogen used for cold traps must be handled properly (see Appendix C). Safety glasses must be worn. Benzoyl chloride, which is a potent lachrymator, and other waste chemicals must be disposed of properly.

APPARATUS

Medium- or high-resolution infrared grating or FTIR spectrometer; gas cell with sapphire, NaCl, or KBr windows; vacuum line (with pressure gauge) for filling cell; cylinder of HCl gas with needle valve; three-neck round-bottom flask; reflux condenser; glass-stoppered dropping funnel; two traps, one with a stopcock on each arm; oil bubbler; exhaust hood; heating mantle; CO and CH_4 gas for calibration check (optional).

Heavy water (2.5 mL at least 95 percent D_2O); benzoyl chloride (70 g); ice or Dry Ice and isopropanol; liquid nitrogen; or a 5-L flask of DCl gas (available from Cambridge Isotope Laboratories, 20 Commerce Way, Woburn, MA 01801, and other suppliers of isotopically substituted compounds).

REFERENCES

1. P. W. Atkins and J. de Paula, *Physical Chemistry,* 8th ed., chap. 13, Freeman, New York (2006).

2. G. Herzberg, *Molecular Spectra and Molecular Structure I: Spectra of Diatomic Molecules,* 2d ed., chap. III, reprint ed., Krieger, Melbourne, FL (1989); K. P. Huber and G. Herzberg, *Molecular Spectra and Molecular Structure IV: Constants of Diatomic Molecules,* Van Nostrand Reinhold, New York (1979). Although the latter book is now out of print, its contents are available as part of the *NIST Chemistry WebBook* at http://webbook.nist.gov.

3. I. N. Levine, *Molecular Spectroscopy,* chap. 4, Wiley-Interscience, New York (1975).

4. I. N. Levine, *Physical Chemistry,* 6th ed., sec. 21.6, McGraw-Hill, New York (2009); D. A. McQuarrie, *Statistical Thermodynamics,* reprint ed., University Science Books, Sausalito, CA (1985).

5. G. N. Lewis and M. Randall (revised by K. S. Pitzer and L. Brewer), *Thermodynamics,* 2d ed., pp. 60 and 419*ff.,* McGraw-Hill, New York (1961).

6. H. M. Spencer, *Ind. Eng. Chem.* **40,** 2152 (1948).

GENERAL READING

P. F. Bernath, *Spectra of Atoms and Molecules,* chap. 7, Oxford Univ. Press, New York (1995).

G. Herzberg, *Molecular Spectra and Molecular Structure I: Spectra of Diatomic Molecules,* 2d ed., reprint ed., Krieger, Melbourne, FL (1989).

J. M. Hollas, *Modern Spectroscopy,* 3d ed., Wiley, New York (1996).

J. I. Steinfeld, *Molecules and Radiation,* 2d ed., MIT Press, Cambridge, MA (1985).

EXPERIMENT 38
Vibrational–Rotational Spectra of Acetylenes

In this experiment, several vibrational-rotational infrared bands of C_2H_2 and C_2D_2 will be recorded at medium to high resolution (~ 1 cm^{-1}). These spectra will be analyzed to extract rotational constants for use in the calculation of accurate values for the C—H and C—C bond lengths. The role of symmetry and nuclear spin in determining the activities and intensity patterns of the spectral transitions is also examined. From such consider-ations, the infrared bands can be assigned to specific modes of vibration and values can be deduced for the fundamental vibrational frequencies of C_2H_2 and C_2D_2.[1,2]

THEORY

Vibrational Levels and Wavefunctions.   Acetylene is known to be a symmet-ric linear molecule with $D_{\infty h}$ point group symmetry and $3N - 5 = 7$ vibrational *normal modes,* as depicted in Table 1. Symmetry is found to be an invaluable aid in understanding the motions in polyatomic molecules, as discussed in detail in Refs. 3 through 9. Group theory shows that each vibrational coordinate and each vibrational energy level, along with its associated wavefunction, must have a symmetry corresponding to one of the symmetry species of the molecular point group. The $D_{\infty h}$ symmetry species corresponding to the dif-ferent types of atomic motion in acetylene are indicated in the table. Motions that retain the center of inversion symmetry, such as the ν_1, ν_2, and ν_4 modes of Table 1, are labeled g (*gerade,* German for even), while those for which the displacement vectors are reversed on inversion are labeled u (*ungerade,* odd). Modes involving motion along the molecular axis (z) are called parallel vibrations and labeled Σ, while those involving perpendicular motion are labeled Π and are doubly degenerate since equivalent bending can occur in either x or y directions. From the appearance of the nuclear displacements, it can be seen that only the ν_3 and ν_5 modes produce an oscillating change in the zero dipole moment of the molecule and hence give rise to infrared absorption.

From the harmonic-oscillator model of quantum mechanics, the term value G for the vibrational energy levels for a linear polyatomic molecule can be written as[5]

$$G(v_1, v_2, \ldots) = \sum_{i=1}^{3N-5} \tilde{\nu}_i (v_i + \tfrac{1}{2}) \tag{1}$$

TABLE 1   Fundamental vibrational modes of acetylene

Normal mode		Symmetry species	Description	Activity, band type[a]	Frequency $(cm^{-1})^b$	
					C_2H_2	C_2D_2
H—C≡C—H	ν_1	Σ_g^+	Sym. CH stretch	Rp, ‖	3372.8	2705.2
H—C≡C—H	ν_2	Σ_g^+	CC stretch	Rp, ‖	1974.3	1764.8
H—C≡C—H	ν_3	Σ_u^+	Antisym. CH stretch	IR, ‖	3294.8	2439.2
H—C≡C—H / H—C≡C—H	ν_4	Π_g	Sym. bend (gerade)	Rdp, ⊥	612.9	511.5
H—C≡C—H / H—C≡C—H	ν_5	Π_u	Antisym. bend (ungerade)	IR, ⊥	730.3	538.6

[a] The designations IR or R indicate that the fundamental transition for the mode is infrared- or Raman-active, respectively, and the labels p and dp give the polarization of the Raman band (see Exp. 35 for a detailed discus-sion). Parallel bands (‖) have PR branches, while perpendicular bands (⊥) show PQR branches.

[b] The frequencies are from compilations in Refs. 1 and 2.

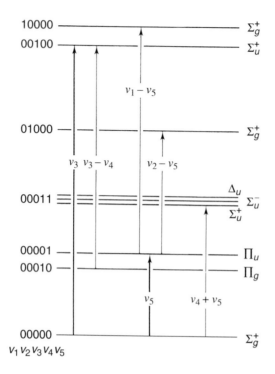

where $\tilde{\nu}_i = \nu_i/c$ is the vibrational frequency of mode i in cm^{-1} when c is the speed of light in cm s^{-1}. Additional anharmonicity corrections, analogous to $\tilde{\nu}_e x_e$ for diatomic molecules (see Exps. 37 and 39), can be added; but these are usually small (1 to 5 percent of $\tilde{\nu}_i$) and will be neglected in this discussion. The energy levels of some of the states of acetylene are shown in Fig. 1. Each level is characterized by a set of harmonic oscillator quantum numbers $v_1 v_2 v_3 v_4 v_5$, shown at the left of the figure. The *fundamental* transitions from the ground state are those in which only one of the five quantum numbers increases from 0 to 1; the two infrared-active fundamentals ν_3 and ν_5 are indicated with bold arrows in the figure.

The set of quantum numbers of a level also serves to define the corresponding wave-function, which in the usual approximation is written as a product of one-dimensional harmonic oscillator functions,[3,5,7]

$$\psi_{v_1 v_2 v_3 \cdots} = \phi_{v_1}(Q_1)\phi_{v_2}(Q_2)\phi_{v_3}(Q_3) \cdots \tag{2}$$

The latter have the form

$$\phi_{0_i}(Q_i) = \left(\frac{\gamma_i}{\pi}\right)^{1/4} \exp\left(\frac{-\gamma_i Q_i^2}{2}\right) \tag{3a}$$

$$\phi_{1_i}(Q_i) = \left(\frac{4\gamma_i}{\pi}\right)^{1/4} \exp\left(\frac{-\gamma_i Q_i^2}{2}\right) \gamma_i^{1/2} Q_i \tag{3b}$$

$$\phi_{2_i}(Q_i) = \left(\frac{\gamma_i}{4\pi}\right)^{1/4} \exp\left(\frac{-\gamma_i Q_i^2}{2}\right)(2\gamma_i Q_i^2 - 1) \quad \text{etc.} \tag{3c}$$

where $\gamma_i = k_i/hc\tilde{\nu}_i$ and k_i is the quadratic force constant for the mode of frequency $\tilde{\nu}_i$. The function ϕ is an even or odd polynomial in the *normal coordinate* Q_i, depending on the evenness or oddness of the quantum number v_i. In general Q_i is a combination of bond-stretching and angle-bending coordinates that all oscillate in phase and at the characteristic frequency ν_i. The precise combination is obtained by solution of Newton's equations as described in Refs. 4 through 9, but here we restrict our analysis to a consideration of the role of symmetry in limiting the possible mix of coordinates and the transitions between vibrational levels.

Symmetry Relations. Each normal coordinate Q_i, and every wavefunction involving products of the normal coordinates, must transform under the symmetry operations of the molecule as one of the symmetry species of the molecular point group. The ground-state function in Eq. (3a) is a Gaussian exponential function that is quadratic in Q, and examination shows that this is of Σ_g^+ symmetry for each normal coordinate, since it is unchanged by any of the $D_{\infty h}$ symmetry operations. From group theory the symmetry of a product of two functions is deduced from the symmetry species for each function by a systematic procedure discussed in detail in Refs. 4, 5, 7, and 9. The results for the $D_{\infty h}$ point group applicable to acetylene can be summarized as follows:

$$g \times g = u \times u = g \qquad g \times u = u \times g = u$$
$$\Sigma^+ \times \Sigma^+ = \Sigma^- \times \Sigma^- = \Sigma^+ \quad \Sigma^+ \times \Sigma^- = \Sigma^-$$
$$\Sigma^+ \times \Pi = \Sigma^- \times \Pi = \Pi \qquad \Sigma^+ \times \Delta = \Sigma^- \times \Delta = \Delta \quad \text{etc.} \qquad (4)$$
$$\Pi \times \Pi = \Sigma^+ + \Sigma^- + \Delta \qquad \Delta \times \Delta = \Sigma^+ + \Sigma^- + \Gamma$$
$$\Pi \times \Delta = \Pi + \Phi$$

Application of these rules shows that the product of two or more Σ_g^+ functions has symmetry Σ_g^+, hence the product function for the ground-state level (00000) is of Σ_g^+ symmetry.

From Eq. (3b) it is apparent that the symmetry species of a level with $v_i = 1$ is the same as that of the coordinate Q_i. In the case of a degenerate level such as (00001), there are two wavefunctions involving the degenerate Q_{5x}, Q_{5y} pair of symmetry Π_u. The symmetry of combination levels involving two *different* degenerate modes is obtained according to the above rules, and for example, for the (00011) level, one obtains $\Sigma_g^+ \times \Sigma_g^+ \times \Sigma_g^+ \times \Pi_g \times \Pi_u = \Sigma_u^+ + \Sigma_u^- + \Delta_u$. Thus one sees that the product of two degenerate functions gives rise to multiplets of different symmetries. For *overtone* levels of *degenerate* modes, a more detailed analysis[7,9] is necessary in which it is found that levels such as (00020), (00003), and (00004) consist of multiplets of symmetry $\Sigma^+ + \Delta$, $\Pi + \Phi$, and $\Sigma^+ + \Delta + \Gamma$, respectively.

From such considerations the symmetry species of each wavefunction associated with an energy level is determined, and these are indicated at the right in Fig. 1. It is important to realize that this symmetry label is the correct one for the true wavefunction, even though it is deduced from an approximate harmonic-oscillator model. This is significant because transition selection rules based on symmetry are *exact* whereas, for example, the usual harmonic-oscillator constraint that $\Delta v = \pm 1$ is only approximate for real molecules.

Selection Rules. The probability of a transition between two levels i and j in the presence of infrared radiation is given by the transition moment P_{ij} [see Eq. (35-8)],

$$P_{ij} = \int \psi_i \mu \psi_j \, d\tau \qquad (5)$$

For a given molecule, P_{ij} is a physical quantity with a unique numerical value that must remain unchanged by any molecular symmetry operation such as rotation or inversion. Hence, to have a nonzero value, P_{ij} must be totally symmetric: that is, $\Gamma(\psi_i) \times \Gamma(\mu) \times \Gamma(\psi_j) = \Sigma_g^+$, where $\Gamma(\psi_i)$ denotes the symmetry species of ψ_i, etc. The dipole moment component μ_z

and the equivalent pair μ_x, μ_y are of symmetries Σ_u^+ and Π_u, respectively, for the $D_{\infty h}$, point group and are usually indicated in point-group (or character) tables.[3-9] From this and the rules of Eq. (4), it follows that, for a transition between two levels to be infrared allowed, it is necessary that the symmetry species of the product of the two wavefunctions be the same as one of the dipole components. Thus, from the Σ_g^+ ground state of acetylene, transition to the Σ_u^- or Δ_u members of the (00011) multiplet is forbidden while that to the Σ_u^+ level is allowed by the μ_z dipole component. Transitions involving μ_z are termed *parallel* bands, while those involving μ_x, μ_y are called *perpendicular* bands.

In the case of a Raman transition, the same symmetry arguments apply, except that the dipole function μ must be replaced by the polarizability tensor elements α_{zz}, α_{xy}, etc. [see Eq. (35-7)]. For molecules of $D_{\infty h}$ symmetry, these elements belong to the symmetry species Σ_g^+, Π_g, and Δ_g so that the condition for a Raman-active transition is that the product $\Gamma(\psi_i) \times \Gamma(\psi_j)$ include one of these species. Thus, from the Σ_g^+ ground state of acetylene, Raman transitions to the (10000) Σ_g^+, (01000) Σ_g^+, and (00010) Π_g levels are allowed and can be used to determine the ν_1, ν_2, and ν_4 fundamental frequencies, respectively. As can be seen in Table 1, these three modes do not produce a dipole change as vibration occurs, and thus these transitions are absent from the infrared spectrum. This is an example of the "rule of mutual exclusion," which applies for IR/Raman transitions of molecules with a center of symmetry.[4-9]

Although direct access to the (10000), (01000), and (00010) levels from the (00000) ground-state level by infrared absorption is thus rigorously forbidden by symmetry, access from molecules in the (00010) or (00001) levels can be symmetry allowed. For example $\Gamma(00001) \times \Gamma(10000) = \Pi_u \times \Sigma_g^+ = \Pi_u = \Gamma(\mu_{x,y})$, and so the transition between these levels, termed a *difference band*, $\nu_1 - \nu_5$, is not formally forbidden. As can be seen in Fig. 1, the frequency $(\nu_1 - \nu_5)$ can be added to the fundamental frequency ν_5 to give the *exact* value of ν_1, the (10000)–(00000) spacing. Similarly, the $\nu_2 - \nu_5$ and $\nu_3 - \nu_4$ difference bands are infrared-active and can be combined with ν_5 and ν_3 to deduce ν_2 and ν_4, respectively. Such difference bands are detectable for acetylene but will of course have low intensity because they originate in excited levels that have a small Boltzmann population at room temperature. The intensity of such bands increases with temperature, hence they are also termed "hot-band" transitions.

Other nonfundamental bands often appear in infrared spectra and can be used to obtain an estimate of the fundamental frequencies. For example, from the ground state of acetylene, an infrared transition to the (00011) level is permitted and is termed the $\nu_4 + \nu_5$ *combination* band. The difference $(\nu_4 + \nu_5) - \nu_5$ can be used as an estimate of ν_4, but it should be noted that this is actually the separation between levels (00011) and (00001) and not the true ν_4 separation between (00010) and (00000). Because of anharmonicity effects, these two separations are *not identical* and hence the determination of fundamental frequencies from difference bands is preferred.

Force Constants of Acetylene. From the vibrational frequencies of the normal modes, one can calculate the force constants for the different bond stretches and angle bends in the C_2H_2 molecule. In the most complete valence-bond, harmonic-oscillator approximation, the potential energy for C_2H_2 can be written as[5,8]

$$U = \tfrac{1}{2}k_r(r_1^2 + r_2^2) + \tfrac{1}{2}k_R R^2 + \tfrac{1}{2}k_\delta(\delta_1^2 + \delta_2^2) + k_{rr}r_1 r_2 + k_{rR}R(r_1 + r_2) + k_{\delta\delta}\delta_1\delta_2 \quad (6)$$

where r and R refer, respectively, to the stretching of the CH and CC bonds and δ represents bending of the H—C—C angle from its equilibrium value. The interaction constants k_{rr}, k_{rR}, and $k_{\delta\delta}$ characterize the coupling between the different vibrational coordinates and are usually small compared to the principal force constants k_r, k_R, and k_δ.

The normal modes are combinations of r, R, and δ coordinates that provide an accurate description of the atomic motions as vibration takes place. These combinations must be chosen to have a symmetry corresponding to the symmetry species of the vibration. Consequently, for example, there is no mixing between the orthogonal axial stretches, and the perpendicular bending modes and U contains no cross terms such as $r\delta$ and $R\delta$. The process of finding the correct combination of coordinates, termed a normal coordinate analysis, involves the solution of Newton's equations of motion. This solution also gives the vibrational frequencies in terms of the force constants, atomic masses, and geometry of the molecule.[4–9]

Such an analysis yields the following relations for acetylene:[5,8]

$$4\pi^2 \nu_1^2 + 4\pi^2 \nu_2^2 = (k_r + k_{rr})\left(\frac{1}{m_H} + \frac{1}{m_C}\right) + \frac{2(k_R - 2k_{rR})}{m_C} \tag{7a}$$

$$4\pi^2 \nu_1^2 \times 4\pi^2 \nu_2^2 = \frac{2[(k_r + k_{rr})k_R - 2k_{rR}^2]}{m_H m_C} \tag{7b}$$

$$4\pi^2 \nu_3^2 = (k_r - k_{rr})\left(\frac{1}{m_H} + \frac{1}{m_C}\right) \tag{7c}$$

$$4\pi^2 \nu_4^2 = (k_\delta - k_{\delta\delta})\left[\frac{1}{R_{CH}^2 m_H} + \left(\frac{1}{R_{CH}} + \frac{2}{R_{CC}}\right)^2 \frac{1}{m_C}\right] \tag{7d}$$

$$4\pi^2 \nu_5^2 = (k_\delta + k_{\delta\delta})\frac{1}{R_{CH}^2}\left(\frac{1}{m_H} + \frac{1}{m_C}\right) \tag{7e}$$

When C_2D_2 frequencies are used, m_H should be replaced by m_D. The force constants for acetylene can be calculated from these relations using the measured vibrational frequencies, and the bond lengths can be determined from the rotational analysis described below. If one expresses the frequencies in cm^{-1} units and the masses in appropriate *isotopic* mass units, the factors $4\pi^2$ should be replaced by $4\pi^2 c^2 10^{-3} N_0^{-1} = 5.8918 \times 10^{-5}$ (this includes a factor of 10^{-3} kg/g mass conversion). This substitution gives the force constants k_r, k_R, k_{rr}, and k_{rR} in N m^{-1} units and the bending constants k_δ and $k_{\delta\delta}$ in units of N m.

Rotational Levels and Transitions. The vibrational–rotational energy levels for a linear molecule are similar to those for a diatomic molecule and to a good approximation are given in cm^{-1} units by the sum $G(\nu_1\nu_2 \ldots) + F_\nu(J)$, where[5]

$$F_\nu(J) = B_\nu[J(J + 1) - l^2] - D_\nu[J(J + 1) - l^2]^2 \tag{8}$$

The general label ν characterizes the set $\nu_1\nu_2\nu_3 \ldots$ and is added to F_ν to account for the fact that the rotational constant B and centrifugal distortion constant D change slightly with vibrational level. B_ν is related to the moment of inertia I_ν by the equation

$$B_\nu = \frac{h}{8\pi^2 c I_\nu} \tag{9}$$

where

$$I_\nu = \sum_{i=1}^{N} m_i r_i^2 \tag{10}$$

and the sum is over all atoms in the molecule, having mass m_i and located a distance r_i from the center of mass of the molecule. The quantum number l characterizes the vibrational angular momentum about the linear axis and is 0, 1, 2, . . . for levels of symmetries

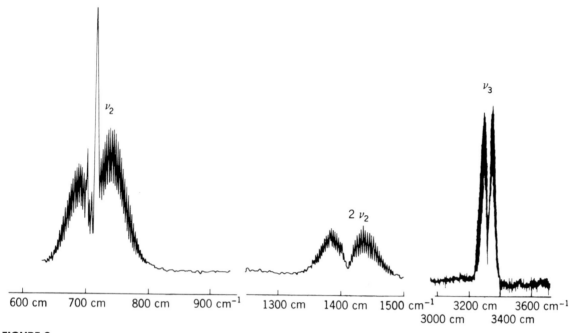

FIGURE 2
Portions of the infrared absorption spectrum of HCN. The ν_2 bending vibration is a perpendicular band and therefore has allowed P, Q, and R branches. The ν_3 [C–H stretching] vibration is a parallel band with P and R branches only.

$\Sigma, \Pi, \Delta, \ldots$, respectively. This angular momentum derives from a rotary motion produced about the linear axis by a combination of the degenerate x and y bending motions. For acetylene there are two bending modes, requiring l_4 and l_5 quantum numbers, which are sometimes shown as superscripts to the ν_4 and ν_5 labels.

The allowed changes in the rotational quantum number J are $\Delta J = \pm 1$ for parallel (Σ_u^+) transitions and $\Delta J = 0, \pm 1$ for perpendicular (Π_u) transitions.[3,5,7,8] Parallel transitions such as ν_3 for acetylene thus have $P(\Delta J = -1)$ and $R(\Delta J = +1)$ branches with a characteristic minimum between them, as shown for diatomic molecules such as HCl in Fig. 37-3 and for the HCN ν_3 mode in Fig. 2. However, perpendicular transitions such as ν_5 for acetylene and ν_2 for HCN (Fig. 2) have a strong central Q branch ($\Delta J = 0$) along with P and R branches. This characteristic PQR-versus-PR band shape is quite obvious in the spectrum and is a useful aid in assigning the symmetries of the vibrational levels involved in the infrared transitions of a linear molecule.

The individual lines in a Q branch are resolved only under very high resolution, but the lines in the P and R branches are easily discerned at a resolution of 1 cm^{-1} or better. As discussed in Exp. 37, it is possible to represent both P and R transition frequencies with a single relation:

$$\tilde{\nu}_m = \tilde{\nu}_0 + B''l''^2 - B'l'^2 + (B' + B'')m + (B' - B'')m^2 - 4D_e m^3 \qquad (11)$$

Here $\tilde{\nu}_0$ is the rotationless transition frequency corresponding to ΔG, the spacing between the two vibrational levels with $J = 0$. B' and B'' are the rotational constants of the upper and lower states, respectively, and the index $m = -J$ for P branch lines, $J + 1$ for R branch lines. The centrifugal distortion constants are extremely small (typically 10^{-6} cm^{-1}), and it is assumed that $D_e' = D_e'' = D_e$.

Intensities and Statistical Weights. The absolute absorption intensity of a vibrational–rotational transition is proportional to the square of the transition moment P_{ij} times the population in the lower state. P_{ij} varies only slightly for different rotational levels, so the principal factors determining the relative intensity are the degeneracy and the Boltzmann weight for the lower level,

$$I_J \propto g_l g_J \exp\left[\frac{-hcBJ(J+1)}{kT}\right] \qquad (12)$$

The rotational degeneracy g_J is $2J + 1$, and the nuclear-spin degeneracy g_l varies with rotational level only when the molecule contains symmetrically equivalent nuclei.

A complete discussion of the factors that determine g_l is beyond the scope of this book but can be found in Refs. 5 and 7. Briefly, the total wavefunction Ψ_{tot} for molecules with equivalent nuclei must obey certain symmetry requirements upon exchange, as determined by the Pauli principle. Exchange of nuclei with half-integral spin, such as protons ($I = \frac{1}{2}$), must produce a sign change in Ψ_{tot}. Such nuclei are termed *fermions* and are distributed among energy levels according to Fermi–Dirac statistics. Nuclei with integral nuclear spin, such as deuterium ($I = 1$), obey Bose–Einstein statistics and are called *bosons;* for these the sign of Ψ_{tot} is unchanged by interchange of the equivalent particles. The total wavefunction can be written, approximately, as a product function,

$$\Psi_{tot} = \psi_{elec}\psi_{vib}\psi_{rot}\psi_{ns} \qquad (13)$$

For the ground vibrational state of acetylene, $\psi_{elec}\psi_{vib}$ is symmetric with respect to nuclear exchange, so $\psi_{rot}\psi_{ns}$ must be antisymmetric for C_2H_2, symmetric for C_2D_2. For linear molecules the ψ_{rot} functions are spherical harmonics that are symmetric for even J, antisymmetric for odd J.[5,7] The ψ_{ns} spin-product functions for two protons consist of three that are symmetric ($\alpha\alpha, \alpha\beta + \beta\alpha, \beta\beta$) and one that is antisymmetric ($\alpha\beta - \beta\alpha$), where α and β are the functions corresponding to M_I values of $+\frac{1}{2}$ and $-\frac{1}{2}$ (see Exp. 32). Thus for C_2H_2, it follows that g_l is 1 for even J, 3 for odd J, and the P and R branch lines will alternate in intensity. For C_2D_2, with spin functions α, β, γ representing the M_I values of $+1, 0, -1$, there are six symmetric nuclear spin combinations ($\alpha\alpha, \beta\beta, \gamma\gamma, \alpha\beta + \beta\alpha, \alpha\gamma + \gamma\alpha, \beta\gamma + \gamma\beta$) and three that are antisymmetric to exchange ($\alpha\beta - \beta\alpha, \alpha\gamma - \gamma\alpha, \beta\gamma - \gamma\beta$). Consequently the *even J* rotational lines are stronger in this case. The experimental observation of such intensity alternations confirms the $D_{\infty h}$ symmetry of acetylene, and in the present experiment serves as a useful check on the assignment of the J values for the P and R branch transitions.

EXPERIMENTAL

An infrared grating or Fourier-transform (FTIR) spectrometer covering the spectral region from 600 to 4000 cm^{-1} is sufficient for this experiment, although extension to 400 cm^{-1} is desirable if the ν_5 band of C_2D_2 at about 540 cm^{-1} is to be studied. Table 2 indicates the spectral regions of interest and the approximate pressures that give satisfactory intensities. These pressures may require some adjustment depending on the resolution capabilities of the instrument, since the peak absorbance of a narrow line increases as the spectral resolution improves. For the survey scan, a resolution of 4 cm^{-1} is adequate to permit rapid data collection at reasonable signal-to-noise ratio. The regions to be studied in detail should be examined at an expanded scale to permit accurate frequency measurements. A resolution of at least 1.5 cm^{-1} is needed to resolve the rotational structure of the

TABLE 2   Infrared regions to be scanned for acetylenes

	C_2H_2	C_2D_2
	Scans at ~300 Torr	
Survey scan	400–4000 cm^{-1}	400–4000 cm^{-1}
$\nu_2 - \nu_5$	1235–1255 cm^{-1}	1220–1235 cm^{-1}
$\nu_1 - \nu_5$	2635–2650 cm^{-1}	2160–2175 cm^{-1}
$\nu_3 - \nu_4$	2675–2690 cm^{-1}	1920–1935 cm^{-1}
	Scans at ~25 Torr	
Survey scan	400–4000 cm^{-1}	400–4000 cm^{-1}
ν_5	720–740 cm^{-1}	530–550 cm^{-1}
$\nu_4 + \nu_5$	1230–1410 cm^{-1}	980–1120 cm^{-1}
ν_3	3275–3325 cm^{-1}	2425–2450 cm^{-1}

acetylene bands, and a value of 0.5 cm^{-1} or better is desirable. (The effect of resolution on the ν_5 mode of C_2H_2 can be seen in Fig. XIX-28.) Detailed instructions for operating the spectrometer will be given in the laboratory.

The C_2H_2 sample can be taken from a commercial gas cylinder† in the manner described in Exp. 37 or a sample can be synthesized as described below for C_2D_2. **Acetylene is flammable; there should be no flames in the filling or synthesis area,** which, if feasible, should be in a hood. A 10-cm cell fitted with KBr windows should be filled to a pressure of about 300 Torr for a survey scan and for expanded traces of the weak difference band regions as indicated in Table 2. The cell pressure should then be reduced to about 25 Torr so that the strong ν_3 and ν_5 fundamentals and the $\nu_4 + \nu_5$ combination band have a more reasonable intensity. Expanded scans of the latter bands are recorded according to Table 2, preceded by a second survey scan.

C_2D_2 can be synthesized by addition of D_2O to calcium carbide using the apparatus shown in Fig. 3. About 2.5 g of calcium carbide is placed in flask F, which is then evacuated to remove traces of H_2O; 0.5 mL of D_2O is added with a syringe to flask F through the rubber septum, and the entire system up to the vacuum stopcock V is allowed to "cure" at room temperature for about 5 to 10 min to allow deuterium exchange with H_2O adsorbed on the walls of the system. The pressure should be monitored and kept below 1 atm during this period, reducing it if necessary by opening and closing V.

The system is then evacuated, after which the cold traps are put in place and 0.5 mL of D_2O is added to flask F. The pressure will rise and then drop as the C_2D_2 is condensed in the storage vessel cooled by liquid nitrogen. When the pressure drops to a few Torr, another increment of D_2O is added and the procedure is repeated until a total of 4 mL has been added. When gas evolution has slowed or stopped, stopcock A is closed and the reaction flask and water trap are placed in the hood, open to allow any further reaction to occur harmlessly.

---

†Commercial acetylene is widely used for welding purposes and is shipped dissolved in acetone, in which it is extremely soluble. The acetone is retained by a porous filler material within the cylinder so that the discharged acetylene is typically >99 percent. If desired, residual traces of acetone can be eliminated by passage through a Dry Ice/isopropanol trap. In its free state, acetylene may decompose violently; the stability decreases at higher pressures. At pressures below 1 atm, the sampling conditions of this experiment, the gas can be handled safely but one should of course wear safety glasses and exercise reasonable judgment. Unalloyed copper, silver, and mercury should never be used in direct contact with acetylene, particularly when wet, owing to the possible formation of explosive acetylides.

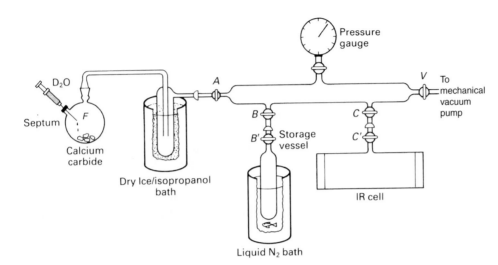

FIGURE 3

Apparatus for the synthesis of deuterated acetylene.

To fill the infrared cell, the system is evacuated with the liquid nitrogen trap still in place. Stopcock V is then closed and the nitrogen Dewar is lowered to allow the storage vessel to warm slowly until the pressure is about 300 Torr. The stopcocks C and C' leading to the infrared cell are then closed, and as warming continues, the system pressure is monitored and adjusted with stopcock V if necessary to keep it below 1 atm. When room temperature is reached, stopcock B' can be closed to save some residual C_2D_2 as "insurance" until all cell pressure adjustments and spectral measurements are completed. The spectral regions of C_2D_2 indicated in Table 2 should be scanned in the same manner as for C_2H_2.

Do not open stopcock B' to air with the storage vessel in liquid nitrogen, since liquid oxygen will condense on top of the acetylene, forming a potentially explosive mixture. At the end of the experiment, the acetylene in the infrared cell and in the storage vessel can be disposed of by simply exhausting it through the roughing pump of the vacuum system.

CALCULATIONS AND DISCUSSION

Vibrational Assignments and Fundamental Frequencies. Examine your survey spectra for C_2H_2 and C_2D_2 and note the striking difference between parallel and perpendicular bands. Determine the frequencies of as many of the transitions shown in Fig. 1 as your data allow. In doing this take the Q branch maximum of each expanded spectrum of the ν_5, $\nu_1 - \nu_5$, $\nu_2 - \nu_5$, and $\nu_3 - \nu_4$ perpendicular bands as a measure of these vibrational frequencies. The Q branch of the C_2D_2 $\nu_2 - \nu_5$ band may be difficult to detect, since it overlaps some of the rotational structure of the $\nu_4 + \nu_5$ band of C_2HD, an inevitable impurity in C_2D_2.†

The parallel bands are expected to show a gap between the P and R branches at the position of the missing Q branch as in Fig. 37-3, but in fact such a gap is not seen for the

---

†A higher D/H ratio can be achieved by adding the acetylene to a storage bulb containing 5 mL of 99+ percent D_2O and about 1 g of basic alumina. The latter serves to promote the exchange of acidic protons on C_2H_2 with the D_2O and thereby improves the D content of the acetylene. For best results the protons on the basic alumina should first be exchanged by adding a few milliliters of D_2O and evacuating the storage bulb prior to addition of more D_2O and acetylene. After exchange at room temperature for a few hours, the storage bulb can be cooled in a Dry Ice/isopropanol bath and the enriched C_2D_2 distilled into the infrared cell.

acetylenes owing to overlap with combination and difference bands. For example, in the expanded trace of the ν_3 region of C_2D_2, a weak feature seen at the Q branch position is *not* a consequence of a violation of selection rules but rather is due to overlapping $R(J'' = 2)$ branch lines of the difference bands $(\nu_3 + \nu_4) - (\nu_4)$ and $(\nu_3 + \nu_5) - (\nu_5)$.[10] This line happens to be the one of *minimum* intensity between the P and R branches and may be taken as a good approximate value of ν_3 for C_2D_2. The corresponding ν_3 region for C_2H_2 is more complicated[11] owing to additional overlapping absorption by a combination band $\nu_2 + \nu_4 + \nu_5$, and the ν_3 value given in Table 1 can be used for subsequent calculations.

Use your data to obtain as many of the fundamental transition frequencies of acetylene as possible, and compare the results with the literature values listed in Table 1. Use these values to assign other combination or difference bands that you observe in the C_2H_2 spectra, using band shapes and symmetry arguments as a guide. Draw a vibrational energy-level diagram from 0 to 4000 cm^{-1} and show all the vibrational transitions you observe for C_2H_2. For C_2D_2, such an assignment task is more difficult because of C_2HD impurities, for which all the fundamental transitions are allowed because of the lower symmetry. One clue serving to identify the transitions of the latter species is the absence of intensity alternation in the P and R branches, since there is no longer exchange symmetry for the protons.

Rotational Analysis.   The $\nu_4 + \nu_5$ parallel combination bands of C_2H_2 and C_2D_2 should be analyzed to obtain the ground-state B values for each species.† Note the alternation of line intensities and use the intensity predictions from the nuclear spin statistics as an aid in assigning an m value to each line in the P and R branches. The feature that appears at the Q branch position is due to overlapping R lines of the difference bands $(2\nu_4 + \nu_5) - (\nu_4)$ and $(2\nu_5 + \nu_4) - (\nu_5)$.[11,12] These overlapping branches also cause the alternating intensity ratios to differ somewhat from the values of 3:1 and 6:3 predicted for C_2H_2 and C_2D_2.

Tabulate the transition frequencies and fit them to Eq. (11) using a least-squares method. Since this is a parallel transition between two Σ states, l' and l'' are zero in this equation. (If a perpendicular fundamental such as ν_5 is to be analyzed, a value of $l' = l'_5 = 1$ should be substituted.) Compare your rotational constants for the ground state with the literature values $B''(C_2H_2) = 1.176608$ cm^{-1}, $B''(C_2D_2) = 0.847887$ cm^{-1} cited in Refs. 1 and 2. Assume that the structure is unchanged by deuteration and, using Eqs. (9) and (10), calculate the C—H and C—C bond lengths. Use the uncertainties from the least-squares analysis to calculate the uncertainty in these bond lengths and compare your results with values of $R_{CH} = 1.0625$ Å, $R_{CC} = 1.2024$ Å that correspond to the *equilibrium* positions of the atoms on the potential energy surface.[2]

Force-Constant Determination.   Calculate the force constants for acetylene using Eqs. (7) with the fundamental frequencies and the C—H and C—C bond lengths that you have determined. The bending force constants k_δ and $k_{\delta\delta}$ have units of energy, N m, when the angular displacements are in (dimensionless) radians. Compare values of k_δ and $k_{\delta\delta}$ obtained with C_2D_2 frequencies with those calculated for C_2H_2: how good is the assumption that the force constants are independent of isotopic substitution?

Compute the stretching force constants (k_r, k_R, k_{rr}, and k_{rR}) and discuss their magnitudes in terms of the strengths of the chemical bonds and the likely interactions among these. Two independent determinations of the quantity $k_r - k_{rr}$ are obtained using the isotopic data and Eq. (7c), but the calculation of $k_r + k_{rr}$, k_R, and k_{rR} requires the combined

---

†The ν_3 band of C_2D_2 and the ν_5 perpendicular band for both isotopic species are also suitable for analysis. In addition the ν_2, $2\nu_2$, and ν_3 bands of HCN seen in Fig. 2, along with the DCN counterparts, can serve as alternatives for a similar vibrational–rotational study.

solution of Eqs. (7*a*) and (7*b*) for both isotopic species. In fact, Eq. (7*b*) for C_2D_2 is redundant and places no new constraint on the force constants, since these factor out of the ratio of Eq. (7*b*) for the two isotopes:

$$\frac{\nu_1^2(D)\nu_2^2(D)}{\nu_1^2(H)\nu_2^2(H)} = \frac{1/m_D + 1/m_C}{1/m_H + 1/m_C} \tag{14}$$

Equation (14) is an example of a relation derived from a general *product rule*[5,8] that provides a useful method of checking frequency assignments without doing a detailed normal-coordinate analysis.

Theoretical Calculations.   Because of its simple structure and limited number of electrons, acetylene is a good candidate for an ab initio quantum mechanical computation using a program such as Gaussian. Examples of such calculations at various levels are described in Foresman and Frisch[13] and in Hehre et al.[14] Compare your values for the CH and CC bond lengths with those obtained from ab initio calculations. In making comparisons of vibrational frequencies, it should be noted that the quantum calculations yield *harmonic* values that the molecule would have for motion about the equilibrium position. For C_2H_2 the harmonic frequencies, derived by correcting the experimental values for anharmonicity, are 3497, 2011, 3415, 624, and 747 cm^{-1} for modes 1 to 5, respectively.[7] Theoretical values for these from ab initio calculations are generally high by about 10 percent, but agreement with experiment improves as the basis set and level of calculation increase. The vibrational frequencies can also be calculated using semiempirical methods such as PM3 in the program HyperChem, which also provides an animated display of the normal modes.

The spectroscopic value of the $C_2H_2(g)$ heat capacity $\widetilde{C}_v = 2.5R + \widetilde{C}_v(\text{vib})$ can be calculated if desired by using the vibrational partition function and the resulting $\widetilde{C}_v(\text{vib})$ harmonic-oscillator expression given in Eqs. (37-19) and (37-20). Since C_2H_2 has seven normal modes, $\widetilde{C}_v(\text{vib})$ is of course given by an appropriate sum of Eq. (37-20) over the seven values of $u = hc\widetilde{\nu}/kT$. The experimental value of $\widetilde{C}_p$ for $C_2H_2(g)$ at 298.15 K is 43.93 J K^{-1} mol^{-1}; thus $\widetilde{C}_v = \widetilde{C}_p - R = 35.62$ J K^{-1} mol^{-1}.

SAFETY ISSUES

The acetylene cylinder must be chained securely to the wall or laboratory bench. Work on a vacuum system requires a preliminary review of procedures and careful execution in order to avoid damage to the apparatus and possible injury from broken glass; in addition, the liquid nitrogen used for cold traps must be handled properly (see Appendix C). Safety glasses must be worn. Acetylene is flammable; no flames can be permitted in the synthesis area. Take great care not to allow oxygen condensation to occur in the presence of acetylene (potentially explosive). Carry out as many operations as possible in a fume hood. Dispose properly of excess CaC_2.

APPARATUS

Infrared-grating or FTIR instrument with a resolution of 1.5 cm^{-1} or better; 10-cm gas cell with KBr windows; vacuum line with pressure gauge for synthesis and for filling cell, located in a hood if feasible; cylinder of acetylene.

Round-bottom flask (250 mL) with septum port; syringe; 5 mL D_2O (99+ percent); calcium carbide (3 g); D_2O trap and 1-L storage flask with stopcocks; two Dewars; Dry Ice/isopropanol slurry; liquid nitrogen; basic alumina (optional).

REFERENCES

1. E. Kostyk and H. L. Welsh, *Can. J. Phys.* **58**, 534 (1980).

2. *Ibid.,* p. 912.

3. P. W. Atkins and J. de Paula, *Physical Chemistry,* 8th ed., chaps. 12 and 13, Freeman, New York (2006).

4. D. C. Harris and M. C. Bertolucci, *Symmetry and Spectroscopy,* chaps. 1–3, Dover, Mineola, NY (1989).

5. G. Herzberg, *Molecular Spectra and Molecular Structure II: Infrared and Raman Spectra of Polyatomic Molecules,* chaps. II–III, reprint ed., Krieger, Melbourne, FL (1990).

6. J. M. Hollas, *Modern Spectroscopy,* 3d ed., chaps. 6–7, Wiley, New York (1996).

7. I. N. Levine, *Molecular Spectroscopy,* chaps. 1, 4–6, 9, Wiley-Interscience, New York (1975).

8. J. J. Steinfeld, *Molecules and Radiation,* 2d ed., chaps. 6–8, MIT Press, Cambridge, MA (1985).

9. E. B. Wilson, J. C. Decius, and P. C. Cross, *Molecular Vibrations,* chaps. 5–7, McGraw-Hill, New York (1955), reprinted in unabridged form by Dover, New York (1980).

10. S. Ghersetti and K. N. Rao, *J. Mol. Spectrosc.* **28**, 27 (1968).

11. K. F. Palmer, M. E. Mickelson, and K. N. Rao, *J. Mol. Spectrosc.* **44**, 131 (1972).

12. S. Ghersetti, J. Pliva, and K. N. Rao, *J. Mol. Spectrosc.* **38**, 53 (1971).

13. J. B. Foresman and A. Frisch, *Exploring Chemistry with Electronic Structure Methods: A Guide to Using Gaussian,* chap. 4, Gaussian, Pittsburgh, PA (1993).

14. W. J. Hehre, L. Radom, P. v.R. Schleyer, and J. A. Pople, *Ab Initio Molecular Orbital Theory,* pp. 156, 238, Wiley, New York (1986); out of print but available from Gaussian, Inc.; see http://www.gaussian.com/allbooks.htm.

GENERAL READING

P. F. Bernath, *Spectra of Atoms and Molecules,* chaps. 7–8, Oxford Univ. Press, New York (1995).

EXPERIMENT 39
Absorption and Emission Spectra of I_2

Although the electronic spectra of condensed phases are typically quite broad and unstructured, the spectra of small molecules in the gas phase often reveal a wealth of resolved vibrational and rotational lines. Such spectra can be analyzed to give a great deal of information about the molecular structure and potential energy curves for ground and excited electronic states.[1,2] The visible absorption spectrum of molecular iodine vapor in the 490- to 650-nm region serves as an excellent example,[3–5] displaying discrete vibrational bands at moderate resolution and extensive rotational structure[6] at very high resolution. The latter structure is not seen at a resolution of $\sim$0.2 nm, a common limit for commercial ultraviolet–visible spectrophotometers, but the vibrational features can be easily discerned in both absorption and emission measurements. In this experiment the absorption spectrum of I_2 will be used to obtain vibrational frequencies, anharmonicities, bond energies, and other molecular parameters for the ground $X^1\Sigma_g^+$ and excited $B^3\Pi_{0u}^+$ states involved in this electronic transition. As an additional option, emission spectra[7,8] can be used to measure many more vibrational levels of the X state and hence to get improved values of the ground-state parameters.

THEORY

The relevant potential energy curves for I_2 are depicted in Fig. 1, which also shows some of the parameters to be determined from the spectra. The spacings between levels in the two electronic states can be measured by either absorption or emission spectroscopy. Emission occurs following an absorption event if the upper state is not relaxed by a nonradiative collisional process (called *quenching*). The emission is termed *fluorescence,* and the transition between two states is said to be spin allowed if the states have the same spin multiplicity (e.g., both are singlets or both are triplets). Fluorescence intensities are usually high, and the lifetime of the emitting state is short ($\sim$10^{-8} s). If the multiplicity changes in the transition, the emission is termed *phosphorescence*. In that case the intensity is lower and the lifetime is longer ($\sim$10^{-3} s), since the transition is "forbidden" by the spin-selection rules (which are only approximate owing to electron spin–orbit interactions). There is no strict selection rule for the change Δv in vibrational quantum number during an electronic transition; thus sequences of transitions are observed. Each band in the sequence contains rotational structure, which, for I_2, is subject to the selection-rule constraint that $\Delta J = \pm 1$.[9]

The total energy of a diatomic molecule may be separated into translational energy and internal energy. We are concerned here with the internal energy E_{int}, which can be expressed to a good approximation by $E_{int} = E_{el} + E_v + E_r$, where E_{el} is the electronic

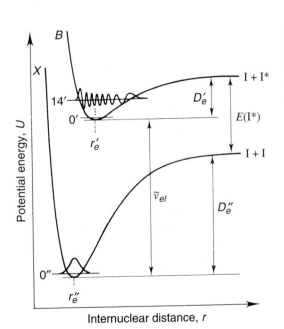

FIGURE 1
Potential-energy diagram for molecular iodine. The energy zero has been arbitrarily set at the minimum of the ground-state potential.

energy, E_v is the vibrational energy, and E_r is the rotational energy. This electronic energy E_{el} refers to the minimum value of the potential curve for a given electronic state. The zero of energy is arbitrarily taken as the minimum in the potential curve for the lowest electronic state (ground state). It is convenient to divide E_{int} by the quantity hc, where c is expressed in units of cm s^{-1}, to get the so-called *term value*, $T_{int}(cm^{-1}) = E_{int}/hc = T_{el} + G + F$, where the vibrational and rotational term values E_v/hc and E_r/hc are given their conventional symbols G and F, respectively. The advantage of this change is that the frequency $\tilde{\nu}$ (expressed in cm^{-1}) for a transition between two electronic states can be simply expressed by

$$\tilde{\nu} = T'_{el} - T''_{el} + G(v') - G(v'') + F(J') - F(J'') \tag{1a}$$

$$\simeq \tilde{\nu}_{el} + G(v') - G(v'') \tag{1b}$$

where $\tilde{\nu}_{el} = T'_{el} - T''_{el} = T'_{el}$ since $T''_{el} = 0$ for the ground electronic state. $G(v)$ is the vibrational term value, which for an anharmonic oscillator is

$$G(v) = \tilde{\nu}_e (v + \tfrac{1}{2}) - \tilde{\nu}_e x_e (v + \tfrac{1}{2})^2 + \tilde{\nu}_e y_e (v + \tfrac{1}{2})^3 + \ldots \tag{2}$$

The rotational-term difference $F(v', J') - F(v'', J'')$ will be ignored, since the rotational structure is not resolved in this experiment. The cubic term in $G(v)$ is also small and can be neglected in obtaining the transition frequency

$$\tilde{\nu}(v', v'') = \tilde{\nu}_{el} + \tilde{\nu}'_e (v' + \tfrac{1}{2}) - \tilde{\nu}'_e x'_e (v' + \tfrac{1}{2})^2 - \tilde{\nu}''_e (v'' + \tfrac{1}{2}) + \tilde{\nu}''_e x''_e (v'' + \tfrac{1}{2})^2 \tag{3}$$

If the quantum numbers v' and v'' are known, the measured frequencies in an absorption or emission spectrum can then be used with a multiple linear least-squares technique (see Chapter XXI) to determine the parameters $\tilde{\nu}_{el}, \tilde{\nu}'_e, \tilde{\nu}'_e x'_e, \tilde{\nu}''_e$, and $\tilde{\nu}''_e x''_e$.

An alternative analysis procedure that is often used concentrates on the determination of $\tilde{\nu}_e, \tilde{\nu}_e x_e$ parameters within each electronic state. Differences between levels in the upper state are obtained from

$$\Delta\tilde{\nu}(v') \equiv \tilde{\nu}(v' + 1, v'') - \tilde{\nu}(v', v'') \simeq \tilde{\nu}'_e - 2\tilde{\nu}'_e x'_e (v' + 1) \tag{4}$$

A plot of $\Delta\tilde{\nu}(v')$ versus v', termed a Birge–Sponer plot, will thus have a slope of $-2\tilde{\nu}'_e x'_e$ and an intercept of $\tilde{\nu}'_e - 2\tilde{\nu}'_e x'_e$. The values of $\Delta\tilde{\nu}(v')$ for all v'' values are combined in this plot, so the two methods should give the same $\tilde{\nu}'_e$ and $\tilde{\nu}'_e x'_e$ parameters. A similar treatment can be used for lower-state differences $\Delta\tilde{\nu}(v'')$ to yield $\tilde{\nu}''_e$ and $\tilde{\nu}''_e x''_e$. The electronic spacing $\tilde{\nu}_{el}$ is then determined using these parameters and the observed frequencies in Eq. (3). This alternative procedure has the virtue of providing a visual representation of the data so that discordant points can be examined and the data can be fitted with a single least-squares treatment that is easily done on a personal computer. The multiple linear regression technique is preferred however, since it uses all the data with equal weighting and has minimum opportunity for calculational error in forming differences. Such regressions are easily performed with spreadsheet programs, as discussed in Chapter III.

Dissociation Energies. Because of the anharmonicity term, the spacing between adjacent vibrational levels decreases at higher v values, going to zero at the point of dissociation of the molecule into atoms. From Eq. (4), the value of $v = v_{max}$ at which this occurs is $v_{max} = (1/2x_e) - 1$. Substitution of this into Eq. (2) gives an expression for the energy D_e required to dissociate the molecule into atoms:

$$D_e = G(v_{max}) = \frac{\tilde{\nu}_e (1/x_e - x_e)}{4} \tag{5}$$

The energy D_0 to dissociate from the $v = 0$ level is smaller than D_e by the zero-point energy $G(0) = \tilde{\nu}_e/2 - \tilde{\nu}_e x_e/4$, so

$$D_0 = \frac{\tilde{\nu}_e(1/x_e - 2)}{4} \tag{6}$$

The expressions used in Eqs. (3) to (6) assume that $\tilde{\nu}_e y_e$ and higher-order anharmonicity terms can be neglected, an approximation that is good for the B state of I_2 but more typically leads to D_e values that are high by 10 to 30 percent. The error for the X ground electronic state is particularly large if only the absorption data are used to deduce $\tilde{\nu}''_e$, $\tilde{\nu}''_e x''_e$, and D''_e, since only the $v'' = 0, 1, 2$ levels are appreciably populated at room temperature. Extension to higher levels, v'' up to $\sim$30, is possible using the emission spectrum, so that improved values of $\tilde{\nu}''_e$ and $\tilde{\nu}''_e x''_e$ are obtained. The value of D''_e remains poorly determined however, since even the $v'' = 30$ level is less than halfway to the dissociation limit.

A more accurate value of D''_e can be obtained by combining $\tilde{\nu}_{el}$ and D'_e values with $E(I^*)$, the difference in electronic energy of the iodine atoms produced by dissociation from the X and B states. The value of $E(I^*)$ is known to be 7603 cm^{-1} from atomic spectroscopy,[10] so that, as seen in Fig. 1,

$$D''_e = \tilde{\nu}_{el} + D'_e - E(I^*) \tag{7}$$

Potential Functions.   Near the minimum in the potential-energy curve of a diatomic molecule, the harmonic-oscillator model is usually quite good. Therefore the force constant k_e can be calculated from the relation

$$k_e = \left(\frac{\partial^2 U}{\partial r^2}\right)_{r_e} = \mu(2\pi c \tilde{\nu}_e)^2 \tag{8}$$

where μ is the reduced mass and c is the speed of light in cm s^{-1} units. The constant k_e is the curvature of the potential curve at the minimum distance r_e and, like the dissociation energy, serves as a measure of the bond strength.

At large displacements from the equilibrium position, the harmonic representation of the potential energy is invalid and a more realistic model is necessary. One simple function that is often employed is the Morse potential,

$$U(r - r_e) = D_e\{\exp[-\beta(r - r_e)] - 1\}^2 \tag{9}$$

which has the desired values of 0 at $r = r_e$ and D_e at $r = \infty$. The parameter β is determined by equating k_e to the curvature of the Morse potential at $r = r_e$, yielding

$$\beta = \left(\frac{k_e}{2hcD_e}\right)^{1/2} \tag{10}$$

This three-parameter function provides a very good approximation to the real potential energy curve at all distances except $r \ll r_e$, a region of no practical significance.

Rotational Structure.   Although rotational structure is not resolved in the present I_2 absorption experiment, each vibrational hand consists of P ($\Delta J = -1$) and R ($\Delta J = +1$) branches as discussed in Exp. 37. For vibrational changes *within* a given electronic state, such as those measured for HCl in Exp. 37, the P and R branches are distinct, with a pronounced dip between them that characterizes the missing Q branch frequency for the "pure" vibrational transition (see Fig. 37-3). The spacing between lines in each branch is not constant, a slight asymmetry arising from a quadratic term [see Eqs. (37-9, 37-10; 38-11)]:

$$\tilde{\nu} = \tilde{\nu}_0 + (B' + B'')m + (B' - B'')m^2 \tag{11}$$

This is a general equation for the transition frequencies in which $m = -J$ for the P lines and $m = J + 1$ for the R lines. For $B' < B''$, the m^2 term causes a decrease (increase) in line spacing in the $R(P)$ branch at high J values. The resultant asymmetry is small for HCl, since $B' - B''$ is small.

If the upper and lower levels of a transition correspond to *different* electronic states, $B' - B''$ is generally much larger and the corresponding quadratic term in Eq. (11) will often cause a frequency maximum ($B' < B''$) in the R branch or a frequency minimum ($B' > B''$) in the P branch. This reversal in the progression of lines at low values of J produces a sharp *band head*, which in the case of I_2 occurs on the R branch edge at a J value as low as $J = 2$. The R branch thus folds back and merges with the P branch so that only a single band is seen for each transition to a vibrational level. A transition frequency measured at the intensity maximum of this band will be *lower* than the "pure" vibrational transition frequency $\tilde{\nu}_0$ assumed in Eq. (3). This error is not constant, varying from 20 to 50 cm^{-1} for I_2 as ν' increases from 0 to the dissociation limit. In contrast the difference $\tilde{\nu}_{head} - \tilde{\nu}_0$ is quite small, varying from 0 to 0.13 cm^{-1}. For this reason, in the present experiment, band-head frequencies rather than band maxima will be measured to obtain the best values of the transition frequencies and the vibrational spacings.

The *emission* of bands of I_2 will also contain many rotational lines if the spectral width of the excitation source is broad enough to populate many upper-state levels. However, if the source is *monochromatic*, excitation to a single ν', J' level can occur and the resultant spectral emission is greatly simplified. Assuming that there is no change to another level in the upper state owing to collisions, the emission to a given lower ν'' level will consist of only the two transitions corresponding to $\Delta J = -1$ and $\Delta J = +1$. Since there is no restriction on $\Delta \nu$, one will observe sequences of doublets whose large spacings give the vibrational-level separations in the ground electronic state. The small spacing corresponds to $2B''(2J' + 1)$, the separation between the $J'' = J' + 1$ and $J'' = J' - 1$ levels in the lower ν'' state. If a doublet of known J' value can be resolved, the splitting can be used to determine the rotational constant $B''(\nu'')$.

EXPERIMENTAL

Absorption Spectrum.   The absorption spectrum of I_2 vapor is easily obtained with any commercial visible spectrometer having a resolution of about 0.2 nm or better; see Fig. 2. A general description of such spectrometers is given in Chapter XIX, and the instrument manual of the instrument to be used should be consulted for specific operational details. Follow the guidelines provided by the instructor in recording the spectra at the highest resolution possible with the instrument. Calibration corrections to the wavelength readout should be provided or made as described in Chapter XIX. Unless these are quite variable over the 450- to 650-nm range, a single correction value is sufficient.

Crystals of I_2 can be placed in a conventional glass cell of 100-mm length, which is then closed with a Teflon stopper. A usable spectrum can be obtained at room temperature (vapor pressure of $I_2 \sim 0.2$ Torr), although the absorption is much more intense if the cell is wrapped with heating tape to raise the temperature to $\sim 40°C$ (vapor pressure ~ 1 Torr). In this case, to avoid condensation of I_2, the windows should be heated to a higher temperature by wrapping the ends of the cell with extra coils of heating tape.

Emission Spectrum.   Several sources are suitable for exciting the emission spectrum of I_2. In previous editions of this text, the use of a low-pressure mercury discharge lamp was described, in which the green Hg line at 546.074 nm causes a transition from

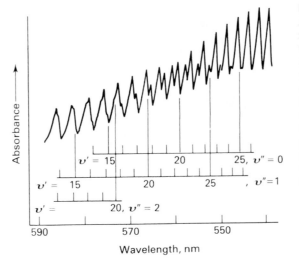

FIGURE 2

A portion of the medium-resolution spectrum of the visible $B \leftarrow X$ iodine absorption spectrum with assignments for the overlapping progressions for $v'' = 0, 1, 2$. The upper-state v' values are indicated at the estimated band-head positions on the short-wavelength side of each transition; the band maxima are at the *top* of the figure.

$v'' = 0$, $J'' = 33$ in the ground state to the $v' = 25$, $J' = 34$ level in the excited state. In conventional spectroscopic notation, this is designated as the 25-0 $R(33)$ transition, with the J'' value of the *lower state* indicated in parentheses and the letter R or P given to indicate the change in J in going to the upper state. As discussed earlier, emission from the upper state would yield doublet sequences in v'' that would be labeled 25-v'' $R(33)$, 25-v'' $P(35)$.

Alternatively laser sources in the green to red region can be used to produce more intense emission spectra with a simpler optical arrangement for excitation. Examples of suitable sources and the I₂ transitions caused by these are listed in Table 1.

The three green sources are especially effective, since in these cases excitation is from the ground $v'' = 0$ level. The two red sources are less efficient because the excitation occurs from high v'' levels, thus in order to obtain reasonable signal levels, heating of the sample is necessary to increase the vapor pressure and, of lesser importance, to improve the relative Boltzmann populations.

The use of a doubled Nd:YAG laser is particularly appealing, since this is becoming increasingly available as for example a relatively inexpensive green "laser pointer." Figure 3 shows the overlap of the I₂ absorption lines with the doubled output of a pulsed Nd:YAG laser as its frequency was varied by tuning the temperature of a single frequency

TABLE 1 Laser excitation wavelengths suitable for excitation of I₂[a]

Laser	λ (nm air)	ν (cm^{-1} vac.)	Assignment
Argon ion	514.5	19429.81	43-0 $P(13)$, 43-0 $R(15)$
Krypton ion	520.8	19194.61	40-0 $R(76)$
Nd:YAG	532.1	18788.45	32-0 $P(53)$, 34-0 $P(103)$
(doubled)		18788.34	32-0 $R(56)$
		18787.80	33-0 $P(83)$
Krypton ion	647.1	15449.50	11-7 $R(98)$, 12-7 $P(138)$
Helium-Neon	632.8	15797.99	6-3 $P(33)$, 11-5 $R(127)$

[a] In the table, the laser wavelengths are air values and the wavenumbers for the I₂ transitions are vacuum values from Ref. 11. The assignments are based on a calculation of the transition wavenumbers using the molecular parameters in Ref. 12.

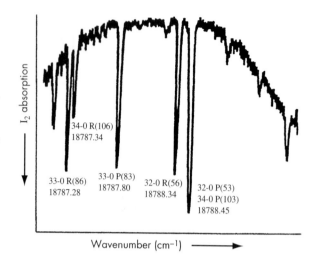

I_2 absorption

34-0 R(106)
18787.34

33-0 R(86)
18787.28

33-0 P(83)
18787.80

32-0 R(56)
18788.34

32-0 P(53)
34-0 P(103)
18788.45

Wavenumber (cm^{-1}) ⟶

seed laser.[13] An unseeded Nd:YAG multimode source will produce light with a width of 0.5 to 1 cm^{-1} so that several I_2 states will be excited. This will lead to a more complex mixture of emission doublets, but this multimode source is still suitable for this experiment.[14]

An argon-ion laser causes excitation to two upper state levels of the I_2 B state, from which an extended emission progression to many levels of the X state results. Under high resolution triplet structure is observed owing to overlap of the P, R doublets expected for each originating J' level. If it is available, the 520.832-nm green line of a krypton-ion laser is an especially good source, since it excites mainly one rotational level, the $J' = 77$ level for $v' = 40$. The emission thus consists of a sequence of resolved P, R doublets extending from $v'' = 0$ to 41. The emission also shows a pronounced alternation of intensities for transitions to even and odd v'' values, and this can be reproduced quite nicely by a Franck–Condon calculation for a Morse oscillator, as discussed in Chapter III. The combination of an argon-ion or krypton-ion laser with a Raman spectrometer is ideal for this experiment, since such instruments are designed to collect scattered light efficiently and to measure intensities that are much lower than the I_2 emission signals. A photomultiplier tube with "extended red" response is desirable for detection of the long-wavelength emission to high v'' levels.

For laser excitation, a 50-mm cylindrical glass cell with two flat end windows is used to contain the I_2; see Fig. 4. The focused laser beam enters and leaves through the windows, traversing the cell parallel with the entrance slit (to optimize the collection efficiency) and near the cell wall facing the spectrometer (to minimize reabsorption of the emitted light by I_2).

FIGURE 4

Fluorescence cell for use with laser excitation.

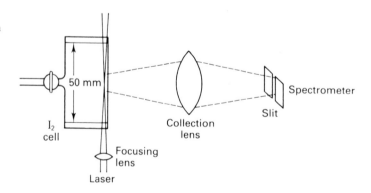

50 mm

Spectrometer

Slit

I_2 cell

Collection lens

Focusing lens

Laser

The cell is prepared by adding several crystals of I_2, after which it is pumped down to 10^{-3} Torr or less residual air pressure and sealed off. Heating of the cell is not necessary, but it is essential that the cell should not leak, since the addition of air serves to quench the emission intensity very efficiently. The track of the laser beam through the cell should be quite visible to the eye. (One can also use a cell with a greaseless stopcock, which permits reevacuation if necessary.)

CALCULATIONS AND DISCUSSION

Absorption Spectrum.   Assign the vibrational quantum numbers of the absorption bands using the numbering indicated in Fig. 2. This numbering choice is *not* obvious and was established, after some controversy, from considerations of intensity distributions[10] and isotopic frequency shifts.[15] Note that there are three overlapping progressions, since there is appreciable population in the $v'' = 0$, 1, and 2 levels at room temperature. As a measure of the band-head position, take the *minimum* on the *short* wavelength side of each peak, estimating this as best you can for overlapping peaks from different progressions. Correct for any spectrometer calibration error.

Once the vibrational transitions in the spectrum have been assigned and the frequencies measured, it is convenient to organize the values of $\tilde{\nu}(v', v'')$ into a *Deslandres table*. As an example, a portion of such a table for PN vapor[1] is presented in Table 2. Note that the frequencies along diagonals (which correspond to bands with the same Δv) lie close together. Now consider the first two rows in the table. Since the frequencies are for transitions from two adjacent upper vibrational states ($v' = 0$ and $v' = 1$) to common lower vibrational states ($v'' = 0$, 1, 2, or 3), there should be a constant separation between the rows—a separation corresponding to the upper-state vibrational separation $(G'_{v=1} - G'_{v=0})$, as seen from Eq. (1b). The separation between the rows $v' = 1$ and $v' = 2$ will give $(G'_{v=2} - G'_{v=1})$ and so forth. The separation between successive rows should decrease as v' increases owing to the effect of anharmonicity; see Eq. (2). In exactly the same way, the separation between adjacent columns will give information about the vibrational levels of the lower electronic state. Agreement among the separations between corresponding frequencies of two rows (or two columns) is a definite check on the correctness of the entries in the Deslandres table. Inspection of the differences given in Table 2 shows a slight variation, which is however greater than the experimental error. This is caused by the use of band-head frequencies $\tilde{\nu}_{head}$ rather than band origins $\tilde{\nu}_0$, as discussed above in the Rotational Structure section.

TABLE 2 Deslandres table of PN bands[1] (Band-head frequencies in cm⁻¹. Differences between the entries in rows 0 and 1, 1 and 2 are in parentheses.)

v' \ v''	0	1	2	3	Average differences
0	39 698.8	38 376.5	37 068.7	—	
	(1 087.4)	(1 090.7)	(1 086.8)		(1 088)
1	40 786.2	39 467.2	38 155.5	36 861.3	
	(1 072.9)	(1 069.0)		(1 071.6)	(1 071)
2	41 859.1	40 536.2	—	37 932.9	
3	—	41 597.4	40 288.3	—	

Emission Spectrum. The emission spectrum should form a regular progression to the red of the exciting line, with $v'' = 0$ corresponding to the exciting wavelength. Measure the wavelengths of as many emission bands as are observable, including any calibration correction if a scanning spectrometer is used. If multiplets are resolved, use the wavelength of the most intense member. For photographic recording, a dispersion curve for the spectral region between 500 and 700 nm is obtained by carefully measuring mercury and neon calibration lines with a comparator microscope. A compilation of permanent neon lines is given in Chapter XIX.

Interpretation and Discussion. Use a multiple linear least-squares technique and Eq. (3) to determine the parameters $\tilde{\nu}_{el}, \tilde{\nu}_e', \tilde{\nu}_e' x_e', \tilde{\nu}_e'', \tilde{\nu}_e'' x_e''$ from the absorption data. Calculate D_e and D_0 for both states using Eqs. (5) and (6) and compare the lower state D_e'' value with the more accurate value obtained from Eq. (7). If emission data have been recorded, analyze these to get improved values for the X-state parameters. If a Birge-Sponer plot of these data shows curvature, you might see whether inclusion of a $\tilde{\nu}_e'' y_e''$ term improves your fit.

Compare your constants with literature values.[5,12,16] Note that the $\tilde{\nu}_e$ and $\tilde{\nu}_e x_e$ values change if $\tilde{\nu}_e y_e$ and higher anharmonicity terms are included in the analysis of the same data set,[5] so close agreement may not be obtained with the literature results given in Refs. 12 and 16. Discuss other possible errors in the experiment or analysis that would add to the uncertainties obtained from the least-squares treatment.

Calculate the harmonic force constant k_e and the Morse parameter β for the two I_2 states. Using the known r_e'' value (0.2666 nm),[16] plot the Morse curve for the ground electronic state of I_2. Compare this with the harmonic potential calculated from k_e.

To include the upper-state potential curve on the same graph, it is necessary to know r_e', which can be estimated from the observed intensities of the absorption spectrum in the following way. According to the Franck–Condon principle,[9,17] the intensity of an electronic transition is related to the overlap of the vibrational wavefunctions of the two states by

$$I(v', v'') \propto \left| \int \psi_{v'}(r) \psi_{v''}(r)\, dr \right|^2 \tag{12}$$

This overlap will be the greatest when $\psi_{v'}(r)$ and $\psi_{v''}(r)$ have their maximum values at the same distance r. The maximum in $\psi_0''(r)$ occurs at $r_0'' \simeq r_e''$ for a harmonic-oscillator wavefunction, but for higher vibrational levels, this maximum approaches the classical turning-point limits of the potential. Since r does not change during the transition (the heavy nuclei take time to move, whereas the light electrons redistribute "instantly"), the transition is said to be *vertical*. From Fig. 1 it can be seen that the $v'' = 0$ transition of greatest intensity, $\tilde{\nu}(v^{*'})$, intersects the upper-state potential curve at $r' = r_e''$, so one can write

$$U'(r_e'' - r_e') = D_e'\{\exp[-\beta'(r_e'' - r_e')] - 1\}^2 + \tilde{\nu}_{el}$$

$$= \tilde{\nu}(v^{*'}) + \tfrac{1}{2}\tilde{\nu}_e'' - \tfrac{1}{4}\tilde{\nu}_e'' x_e'' \tag{13}$$

Determine $\tilde{\nu}(v^{*'})$ from your spectrum and then $r_e'' - r_e'$ from this expression. Compare the resultant value of r_e' with the literature value of 0.3025 nm obtained by analysis of the rotational structure of the electronic spectrum.[16] Include the Morse potential curve for the upper state on your plot for the X state and comment on the differences in the various parameters determined for the two states.

Theoretical Calculations. Because of the number of electrons and limited basis-set options, ab initio quantum mechanical calculations of I_2 properties are not as accurate as those for molecules such as HCl, studied in Exp. 37. It may be instructive however to calculate a ground-state potential-energy curve for comparison with the Morse form deduced in the experiment.

A Mathematica calculation of Franck–Condon factors that determine electronic transition intensities of I$_2$ is presented in Chapter III, and program statements for this are illustrated for I$_2$ in Fig. III-6. In this figure, note the dramatic differences between the intensity patterns predicted for the harmonic oscillator and Morse cases and compare these patterns with those seen in your absorption spectra. If you have access to this software, you might examine the changes in the harmonic-oscillator and Morse-oscillator wavefunctions for different v', v'' choices. A calculation of the relative emission intensities from the $v' = 25$, 40, or 43 level could also be done for comparison with emission spectra obtained with a mercury lamp or with a krypton- or argon-ion laser. In contrast to the smooth variation in the intensity factors seen in the absorption spectra, wide variations are observed in relative emission to v'' odd and even values, and this can be contrasted with the calculated intensities. Note that, if accurate relative comparisons are to be made with experimental intensities, the theoretical intensity factor from the Mathematica program for each transition of wavenumber value v should be multiplied by v for absorption and v^4 for emission.[1]

SAFETY ISSUES

Solid iodine is corrosive to the skin and also stains badly. Handle the I$_2$ crystals with a spatula or tweezers. If a laser is used as the excitation source, safety goggles must be worn, since accidental exposure to a laser beam can cause serious eye damage. Care should also be taken in the use of the vacuum system and liquid nitrogen cold trap while preparing the cell.

APPARATUS

Medium-resolution absorption spectrometer; emission spectrometer with red-sensitive photomultiplier or CCD detector; laser excitation source such as listed in Table 1 (or medium pressure mercury arc such as described in earlier editions of this text); neon calibration lamp and power supply (available from, e.g., Oriel Corp., Stratford, CT); reagent-grade iodine; 100-mm glass cell with Teflon stoppers for absorption studies; heating tape with controlling Variac; 50-mm cell for emission studies; vacuum system, preferably with a diffusion pump and cold trap, for pumping down emission cell.

REFERENCES

1. G. Herzberg, *Molecular Spectra and Molecular Structure I: Spectra of Diatomic Molecules*, reprint ed., Krieger, Melbourne, FL (1989).

2. G. Herzberg, *Molecular Spectra and Molecular Structure III: Electronic Spectra and Electronic Structure of Polyatomic Molecules*, reprint ed., Krieger, Melbourne, FL (1991).

3. F. E. Stafford, *J. Chem. Educ.* **39**, 626 (1962).

4. R. D'alterio, R. Mattson, and R. Harris, *J. Chem. Educ.* **51**, 283 (1974).

5. I. J. McNaught, *J. Chem. Educ.* **57**, 101 (1980); see also E. L. Lewis, C. W. P. Palmer, and J. L. Cruickshank, *Am. J. Phys.* **62**, 350 (1994).

6. J. D. Simmons and J. T. Hougen, *J. Res. Natl. Bur. Std.* **81A**, 25 (1977).

7. J. I. Steinfeld, *J. Chem. Educ.* **42**, 85 (1965).

8. J. Tellinghuisen, *J. Chem. Educ.* **58**, 438 (1981).

9. See G. Herzberg, Vol. I, *op. cit.,* chap. 4.

10. J. I. Steinfeld, R. N. Zare, J. M. Lesk, and W. Klemperer, *J. Chem. Phys.* **42**, 15 (1965).

11. S. Gerstenkern and P. Luc, *Atlas du Spectra d'Absorption de la Molecule d'Iode,* Centre National de la Recherche Scientifique, Paris (1978).

12. P. Luc, *J. Mol. Spectrosc.* **80**, 41 (1980).

13. M. Leuchs, M. Crew, J. Harrison, M. F. Hineman, and J. W. Nibler, *J. Chem. Phys.* **105**, 4885 (1996).

14. J. S. Muenter, *J. Chem. Educ.* **73**, 576 (1996).

15. R. I. Brown and T. C. James, *J. Chem. Phys.* **42**, 33 (1965).

16. K. P. Huber and G. Herzberg, *Molecular Spectra and Molecular Structure IV: Constants of Diatomic Molecules,* p. 332, Van Nostrand Reinhold, New York (1979). Although out of print, its contents are available as part of the *NIST Chemistry WebBook* at http://webbook.nist.gov.

17. E. U. Condon, *Amer. J. Phys.* **15**, 365 (1947).

GENERAL READING

P. F. Bernath, *Spectra of Atoms and Molecules,* chaps. 7 and 9, Oxford Univ. Press, New York (1995).

G. Herzberg, *Molecular Spectra and Molecular Structure I: Spectra of Diatomic Molecules,* 2d ed., chaps. 2–4, 8, reprint ed., Krieger, Melbourne, FL (1989).

J. M. Hollas, *Modern Spectroscopy,* 3d ed., chaps. 6–7, Wiley, New York (1996).

J. I. Steinfeld, *Molecules and Radiation,* 2d ed., chap. 5, MIT Press, Cambridge, MA (1985).

EXPERIMENT 40
Fluorescence Lifetime and Quenching in I_2 Vapor

The vibrational energy levels of the $B\,^3\Pi_{0u}^+$ electronic state of I_2 were studied by absorption spectroscopy in Exp. 39. In the present experiment, selected vibrational–rotational levels of this state will be populated using a pulsed laser. The fluorescence decay of these levels will be measured to determine the lifetime of excited iodine and to see the effect of fluorescence quenching caused by collisions with unexcited I_2 molecules and with other molecules. In addition to giving experience with fast lifetime measurements, the experiment will illustrate a Stern–Volmer plot and the determination of quenching cross-sections for iodine. Student results for different quenching molecules will be pooled and the dependence of the cross sections on the molecular properties of the collision partners will be compared with predictions of two simple models.

THEORY

Absorption Process. The 532-nm doubled output of a pulsed Nd:YAG laser is a convenient excitation source for this experiment. Alternatively, a pulsed dye laser can be used;[1] in this case the instructor should determine which I_2 levels are excited and modify

TABLE 1

I_2 atlas line no.[2]	Wavenumber	Assignment
1109	18787.80 cm^{-1}	$v'' = 0, J'' = 83 \rightarrow v' = 33, J' = 82$
1110	18788.34 cm^{-1}	$v'' = 0, J'' = 56 \rightarrow v' = 32, J' = 57$
1111	18788.45 cm^{-1}	$v'' = 0, J'' = 53 \rightarrow v' = 32, J' = 52$

the following discussion as appropriate. As shown in Fig. 39-3, the gain curve of a doubled Nd:YAG laser extends over about 2 cm^{-1}, but normal multimode operation yields laser output only over the central portion of the gain curve, yielding a 532-nm linewidth of about 1 cm^{-1}. This is sufficient to excite mainly the central three $X(v'', J'') \rightarrow B(v', J')$ I_2 transitions in Fig. 39-3, as shown in Table 1.

Of these absorptions the latter two produce most of the emission intensity so that we are concerned mainly with the $v' = 32$ excited vibrational level. Detailed studies[3-5] of single vibrational–rotational states show only slow variation of the relaxation constants with upper state J'. Thus in this experiment it will be assumed that a single decay constant is sufficient to describe the average relaxation. This assumption has been validated by using a Nd:YAG laser that had a single-frequency output, which was tunable to any of the three transitions; the decay times vary by less than 10 percent among these three upper states.[6]

Fluorescence Decay.   The fluorescence is mainly from the $v' = 32$ vibrational level of state B to various v'' levels of the ground state, each consisting of a rotational doublet as discussed in Exp. 39. This emission will be to the red (long wavelength or Stokes) side of the 532-nm excitation source; thus an orange or red filter is used to block green and pass red light. It is not necessary to resolve the emission into individual transitions since the decay rate of each is assumed to have the same dependence on the upper B state concentration, I_2^*.

Following excitation, an excited state can relax by radiative and/or nonradiative processes. The latter may or may not require a collision. We can distinguish four processes as follows:

a. $I_2^* \rightarrow I_2 + h\nu_f$ $dI_2^*/dt = -k_r(I_2^*)$ fluorescence

b. $I_2^* \rightarrow I + I$ $dI_2^*/dt = -k_{nr}(I_2^*)$ nonradiative decay by predissociation (unimolecular)

c. $I_2^* + I_2 \rightarrow I + I + I_2$ $dI_2^*/dt = -k_S(I_2)(I_2^*)$ collisional predissociation (bimolecular, self-quenching)

d. $I_2^* + Q \rightarrow I + I + Q$ $dI_2^*/dt = -k_Q(Q)(I_2^*)$ collisional predissociation (bimolecular, added quencher Q)

The unimolecular predissociation of process **b** is thought to occur by I_2^* crossover from the bound B state to an unbound (repulsive) $^1\Pi_{lu}$ state, which crosses the inner part of the B potential curve at low v levels.[7,8] The evidence for such predissociation is the spectroscopic observation of unexcited ground-state I atoms following excitation of I_2 molecular beams at energies below the dissociation limit of the B state.[9] Collisions such as indicated in processes **c** and **d** also serve to promote predissociation, but in these cases it is believed that the crossover is to a second repulsive curve corresponding to a state of symmetry $^3\Pi_{0g}^+$. This state mixes with the B state during the symmetry-destroying collision due to the van der Waals interaction of I_2^* and the collision partner. Steinfeld gives a plot of these potential curves and further discussion of the mechanism for relaxation.[7]

It is found that the lifetimes change as v' and J' vary in the upper state. For example, Capelle and Broida[3] found that the lifetime decreased from 1420 to 690 ns as v' decreased from 40 to 21, a drop that can be accounted for by more favorable overlap of ground and excited state wavefunctions (Franck–Condon factors) for lower v' levels.[7] The effect of a change in rotational state is less; Castaño, Martínez, and Martínez[5] found for the $v' = 25$ level that the lifetime decreased from 745 ns to 625 ns as J' increased from 0 to 106. Such a shortening of the lifetime is consistent with enhanced predissociation at higher rotational levels due to bond lengthening by the increased centrifugal force.

Collisions can also cause small changes in v' and J' levels within the B state, an effect that can lead to nonexponential decay curves since the emission rates vary somewhat with vibrational and rotational level. In the present experiment, these effects of vibrational and rotational relaxation should be minor since the total emission is measured and the pressure of collision partners is kept low. At higher collision pressures however, clear deviations from single exponential decay curves can be observed and the simplified analysis presented here is inadequate.

The emission intensity is proportional to I_2^* so that, from the integrated rate equation for these first-order decay processes, the fluorescence intensity will decrease according to the relation

$$I_f(t) = I_{f0} \exp(-kt) = I_{f0} \exp(-t/\tau) \tag{1}$$

where I_{f0} is the intensity at time $t = 0$. The experimental fluorescence decay rate k can thus be obtained from the slope of a plot of $\ln I_f(t)$ versus t. From rate processes **a** to **d**, it is seen that

$$k = 1/\tau = k_r + k_{nr} + k_S p(I_2)/k_B T + k_Q p(Q)/k_B T \tag{2}$$

where we have used the ideal gas law to convert from gas concentrations to pressures (k_B is the Boltzmann constant). Using hard-sphere gas kinetic theory,[10] the quenching constant k_S can be interpreted in terms of a cross-section ($\pi d_S^2 = \sigma$) for an I_2—I_2^* collision:

$$k_S = \pi d_S^2 c_{\rm rel} = \sigma c_{\rm rel} = \sigma \sqrt{\frac{8k_B T}{\pi \mu_S}} \tag{3}$$

Here $d_S = [d(I_2) + d(I_2^*)]/2$ is the effective mean diameter for quenching, $c_{\rm rel}$ is the relative collisional velocity, and $\mu_S = m(I_2)/2$ is the reduced mass of the two colliding molecules. It is important to note that self-quenching cross sections σ_S given in the literature[1,3,4,7] differ from the above conventional gas kinetic definition of σ by a factor of π, i.e., $\sigma = \pi \sigma_S = \pi d_S^2$. An expression identical to Eq. (3) results for k_Q but with an effective cross-section $\sigma_Q = d_Q^2$ serving as a measure of the efficiency of quenching of I_2^* by Q and with $\mu_Q = m(I_2)m(Q)/(m(I_2) + m(Q))$ as the reduced mass of the I_2, Q pair. Using these definitions of σ_S and σ_Q, it follows that

$$k = 1/\tau = k_0 + \sigma_S \sqrt{\frac{8\pi}{\mu_S k_B T}} p_{I_2} + \sigma_Q \sqrt{\frac{8\pi}{\mu_Q k_B T}} p_Q \tag{4}$$

where $k_0 = k_r + k_{nr} = 1/\tau_0$, with τ_0 the lifetime in the absence of collisions.

Assuming that only I_2 vapor is present (no Q), the last term in Eq. (4) can be dropped and τ_0 and σ_S can be deduced from the intercept and slope of a plot of k versus p_{I_2}. Such a display is termed a Stern–Volmer plot. Alternatively, if the pressure of I_2 is fixed and the pressure of an added quencher is varied, a Stern–Volmer plot of k versus p_Q gives the first two terms of Eq. (4) as the intercept and the effective quenching cross-section σ_Q can be calculated from the slope.

EXPERIMENTAL

The pulsed laser beam in this experiment is of high power and can cause eye damage. Read the safety discussion about lasers in Appendix C and pay close attention to directions given by your instructor. Remove all potentially reflective watches, bracelets, and rings. Goggles should be worn, and the instructor should ensure that all reflections from optics are accounted for and blocked. The energy of the 532-nm laser beam need only be about 1 mJ/pulse. If it is necessary to run the laser at higher energies for best pulse-to-pulse stability, the beam should be attenuated with suitable optical density filters or else one or more microscope slides can be used as beam splitters to pick off about 1 mJ for the experiment. Block all unused laser beams.

Students should work in groups of two or three. Figure 1 shows the apparatus for this experiment, with the sample cell consisting of a glass bulb containing a few I₂ crystals in equilibrium with I₂ vapor. The green 532-nm laser beam is passed through the cell and the red fluorescence is detected by a photomultiplier at 90 degrees using an orange or red filter (e.g., Kodak Wratten Red #25) to block any scattered green light. The main body of the bulb should be wrapped with black electrical tape, with a small opening for the laser beam and about a 3-cm opening for detection of the fluorescence. This path should be shielded from stray light from the laser and from room lights, which are best turned off when the photomultiplier voltage is on.

The energy of the laser beam should be adjusted to give an undistorted decay curve on the oscilloscope with a minimum of no more than -100 mV at a photomultiplier voltage in the range -500 to -800 V. (The photomultiplier voltage is negative, with the initial negative signal pulse decaying to zero at long times.) A digital oscilloscope with frequency response of 100 MHz or greater is needed and the signal input should be terminated by

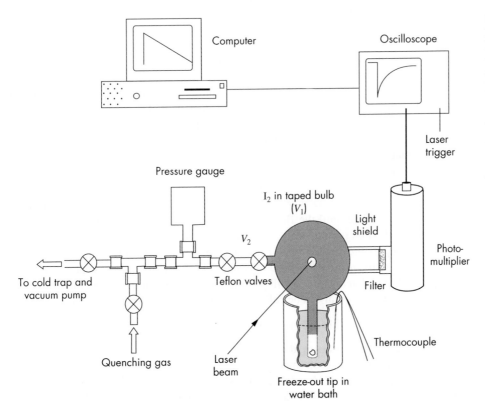

FIGURE 1

Experimental setup for I₂ fluorescence lifetime measurement. The computer display is generated using Eq. (5) after transfer of the averaged decay curve shown on the oscilloscope. The vacuum system can be constructed using Swagelok or Ultra-Torr fittings and quarter-inch glass or Teflon tubing. The pressure gauge can be a 0-to-1 bar capacitance or reluctance manometer or a mercury manometer. The volume ratio $r = V_2/(V_1 + V_2)$ should be about 0.0025 to 0.01 for accuracy in adding quenching gas.

50 Ω at the oscilloscope to ensure fast time response. This can be checked during setup of the experiment by placing a card at the cell position to scatter a small amount of green light toward the photomultiplier; the system time response can be judged from the shape of the 5 to 10 ns 532-nm laser pulse. The oscilloscope should be triggered by the Q-switch sync output provided by most lasers.

For each decay time measurement, 256 or more measurements should be averaged and the decay curve transferred to a computer. The means to accomplish this transfer can be provided by software from the oscilloscope manufacturer, LabVIEW, or a custom program developed for this purpose. Once the decay curve data are available, they may be imported into a spreadsheet for determination of k using

$$\ln[I_f/I_{max}] = \ln[(I_{obs} - I_\infty)/I_{max}] = \ln[I_{f0}/I_{max}] - kt = a - kt \qquad (5)$$

For convenience in making plot comparisons, the intensity in Eq. (5) is scaled to the maximum measured intensity I_{max}. The I_∞ value is to correct the measured I_{obs} intensity values (voltages recorded on the digital scope) for any nonzero value at long times. Note that, for a 20-Hz laser, the interval between laser shots is 50 ms, an "infinite" time compared to the decay times of about a microsecond. Thus I_∞ can be obtained by averaging the "pretrigger" part of the digitized signal that is automatically recorded by the digital oscilloscope just *before* each laser pulse. It is recommended that a spreadsheet template be prepared to make the plotting and least-squares determination of k more efficient.

The iodine pressure can be controlled by varying the temperature of solid I_2 in equilibrium with the vapor. It is important to minimize the presence of any other quenching gases Q, and these should be removed as much as possible at the beginning of the experiment. This can be done by pumping on the cell for several minutes with the sample at room temperature. As solid I_2 evaporates, it will "flush" any remaining air or adsorbed water vapor from the cell. A roughing pump with a nitrogen or Dry Ice cold trap is adequate for this purpose. Alternatively, several flushes of the cell with a few Torr of He, which has a low quenching cross-section, can expedite this operation. Repeat as necessary until the decay curve on the scope shows a maximum decay time. Close the greaseless valve to the cell and freeze the I_2 into the lower extension of the glass bulb ("freeze-out tip") using liquid nitrogen or Dry Ice. As the I_2 pressure decreases, observe with the oscilloscope the decrease in fluorescence signal and the concomitant increase in lifetime. When all the I_2 is condensed, confirm that any remaining signal due to scattered 532-nm light is very small.

Place the freeze-out tip of the sample cell in a Dewar of ice and water at 0.0°C. Use a thermocouple to measure the bath temperature here and subsequently. Watch the fluorescence signal to judge when the pressure has stabilized, and adjust the oscilloscope time base and amplitude to get a reasonable picture of the decay curve. Stir the ice–water mix and average 256 laser shots with the computer, noting the bath temperature at the time of data collection. Replace the Dewar with one containing water at about 5°C and allow the temperature to stabilize until the fluorescence intensity is again constant. (During this 5 to 10 min equilibration period, analyze the data just collected.) Make a series of measurements with temperatures within one or two degrees of the following values: 0, 5, 10, 15, 20°C, recorded to the nearest 0.1°C. Do not exceed room temperature, which should also be recorded. The bath temperature will govern the vapor pressure of I_2 within the cell, and the iodine pressure in pascal can be calculated using the Clausius–Clapeyron equation:

$$\ln p(\text{Pa}) = 28.89129 - 7506/T(\text{K}) \qquad (6)$$

which is based on vapor pressure data in Ref. 11.

Finally, each student group should investigate the relative quenching efficiency of one of the following gases: Q = He, Ne, Ar, H_2, N_2, O_2, or CO_2. This measurement should be done with the freeze-out tip of the sample cell held at 0.0°C in an ice bath. Thus the second term on the right-hand side of Eq. (4) will be constant and the last term can be examined by measuring the decay curves as a function of added gas pressures. For He and H$_2$, make five pressure additions of about 100 Pa to cover the range 0 to 500 Pa, so that the lifetime shortens by about a factor of 5. For the other gases, also make five additions, but the overall range should be about 0 to 150 Pa. After each addition wait 5 to 10 min for equilibration to occur. To give improved accuracy for the pressure measurement, the addition can be done by filling a small volume V_2 outside the cell to a relatively high pressure p_2, followed by expansion into the much larger cell volume V_1. From the volume ratio $r = V_2/(V_1 + V_2)$ to be provided by the instructor, each pressure increment added to the cell is given by $\delta p = r(p_2 - p_c)$, where p_c is the pressure in the cell before the given addition.

CALCULATIONS AND DISCUSSION

Examine your logarithmic decay plots for linearity and choose an appropriate range for least-squares determination of k based on Eq. (5). (Typically a range going from 10 to 90 percent of the decay curve is reasonable.) In your report provide a summary plot showing the logarithmic decay curves for different iodine pressures, along with a table of your k results. Use your decay data to make a Stern–Volmer plot of k versus $p(I_2)$, and from a least-squares determination of intercept and slope, calculate τ_0 and σ_S. Provide similar information for your assigned quencher, and from the slope of the Stern–Volmer plot of k versus $p(Q)$, determine the quenching cross-section σ_Q. Note that the temperature T to be used in Eq. (4) is room temperature since the collisions occur in the warm part of the glass cell.

In the literature (Refs. 3–5, 7), the τ_0 collision-free lifetimes are reported to range from about 600 ns for low B state vibrational levels up to 5000 to 9000 ns at v' levels greater than 60, near the dissociation limit. The more rapid decay at lower v' values is generally consistent with the belief that the potential energy curve crossing that leads to dissociation occurs at low quantum numbers near $v' = 3$. An intermediate τ_0 value can be expected for the present experiment, where excitation is mainly to the $v' = 32$, $J' = 52$, 57 levels. Paisner and Wallenstein[4] report an average lifetime of 1090(30) ns for excitation into $v' = 32$, $J' = 9$, 14 levels. A somewhat lower value would be expected for higher J' levels since high rotational excitation generally enhances the rate of predissociation of a given vibrational level. It should be noted that all of the measured lifetimes are less than the purely radiative lifetime, which has been found from absorption measurements to range from about 1000 to 10,000 ns as v' varies from 0 to 60.[3]

Class Project.   The quenching results can be shared and form the basis for a class project to examine which molecular properties of the quenching molecule are important in causing relaxation of I_2^*. According to one simple model proposed by Rössler[12] and discussed by Steinfeld[7], the quenching cross-section $\sigma_Q = d_Q^2$ should be proportional to the polarizability α_Q of the gas molecule and to the duration of the collision. Since the latter is inversely proportional to the relative collision velocity c_{rel}, one predicts

$$\sigma_Q \propto \alpha_Q \mu_Q^{1/2} \tag{7}$$

This is only an approximate relation correlating the cross-section to quencher properties since it ignores Franck–Condon and other effects. Nonetheless, a plot of $\ln(\sigma_Q)$ versus $\ln(\alpha_Q \mu_Q^{1/2})$ is found to be reasonably linear.[2,13]

Selwyn and Steinfeld[7,13] later derived an expression for σ_Q based on a van der Waals interaction between the excited molecule and the collision partner. This model gives a slightly more complicated prediction of a quenching correlation relation:

$$\sigma_Q \propto \alpha_Q \mu_Q^{1/2} I/R_c^3 \tag{8}$$

TABLE 2 Molecular properties of some collision partners

	m	d	α'	I
	amu	Å	Å^3	ev
He	4.00	2.58	0.21	24.6
Ne	20.18	2.79	0.40	21.6
Ar	39.95	3.42	1.67	15.8
Kr	83.80	3.61	2.54	14.0
Xe	131.30	4.06	4.18	12.1
H_2	2.02	2.92	0.83	15.4
N_2	28.02	3.68	1.78	15.5
O_2	32.00	3.43	1.61	12.2
CO_2	44.00	3.90	2.71	13.7
SF_6	146.07	5.51	4.57	19.3
I_2	253.80	4.98	13.03	9.0

d = Lennard–Jones collision diameters from Ref. 14, where the symbol σ is used for d. Using a gas viscosity value at 124°C in Ref. 15, 1st ed, Vol. V, p. 2, a somewhat larger value of 6.8 Å is obtained for I_2.

$\alpha' = \alpha/4\pi\varepsilon_0$ = volume polarizabilities deduced from index of refraction values in Ref. 15, 6th ed. Vol. II-8, pp. 871–74.

I = ionization potentials from Ref. 15, 6th ed. Vol. I-1, p. 211 and Vol. I-3, p. 359.

Here I is the ionization potential of the quenching molecule and $R_c = [d(Q) + d(I_2^*)]/2$ is the distance of closest approach of the collision pair, where the d values are taken as Lennard–Jones collision diameters deduced from viscosity measurements. Thus a plot of $\ln(\sigma_Q)$ versus $\ln(\alpha_Q \mu_Q^{1/2} I/R_c^3)$ would be predicted to be linear. This model also predicts some variability for different v' vibrational levels due to Franck–Condon effects, but this can be ignored in the present experiment where mainly the $v' = 32$ level is excited by the 532-nm source.

To explore the validity of these two predicted relations, the class results should be pooled and examined. Table 2 contains values of relevant molecular properties that can be used to make comparisons and to make appropriate logarithmic plots. In computing R_c values, $d(I_2^*)$ can be taken to be $d(I_2) + 0.70$ Å, where the difference is that for the bond lengths of $B(v' = 32)$ and $X(v'' = 0)$ states (calculated from the rotational constants given in Ref. 2). The logarithmic form of plotting is convenient because of the large range of σ values. Note in particular that the slope of such a plot is expected to be unity if the model gives the correct power dependence on the molecular parameters. (Proportionality constants determine only the intercept so that, for example, volume polarizabilities $\alpha' = \alpha/4\pi\varepsilon_0$ can be used in place of α in the plots.) Perform least-squares fits of the data for both models and compare the slopes and the R^2 correlation coefficients. What conclusions do you reach about the relative merits of these two models?

As noted earlier, the quenching cross-section is usually reported as $\sigma_Q = d_Q^2$, i.e., as a distance squared. Another measure of molecular diameter can be obtained from the volume polarizability α' using the relation $\alpha' = 4\pi(d_\alpha/2)^3/3$. In your report, compare the diameters so obtained with your d_Q values and with the Lennard–Jones collision diameters listed in Table 2.

Theoretical Calculations.   The program Gaussian can be used at the configuration interaction (CIS) level using the STO-3G basis set to calculate the energies of the ground and low-lying electronic states as the I_2 bond length is varied. Do any of the repulsive curves cross the bound *B* state curve and, if so, how far above the minimum of the latter? Note how the dissociation limits at large bond length value vary for the various states, depending upon whether the product I atoms are in their ground or first excited electronic states.

SAFETY ISSUES

As noted in the text, special care should be taken to avoid stray reflections of the laser beam used in this experiment (see Appendix C also). The pulse energies are high enough to cause serious eye injury and safety goggles should be worn at all times except when the instructor has determined that it is safe to remove them. Iodine is corrosive to the skin and should be handled with a spatula or tweezers in adding a few crystals to the fluorescence cell.

APPARATUS

Nd:YAG laser with second harmonic generator (doubler) capable of generating at least 1-mJ 532-nm pulses of 5 to 10 ns duration (e.g., Polaris model from New Wave, Minilite model from Continuum); laser safety goggles (available from Lase-R Shield, Glendale Optical, Kentek, etc.); reagent grade iodine; cell consisting of a 1-L round bottom flask to which has been added two Teflon stopcocks and a 10-cm freeze-out extension of about 12-mm diameter (the body of the cell should be wrapped with black electrical tape as a safety measure in case of implosion and also to minimize the amount of background light that reaches the detector; the cell can be made by the instructional staff or can be purchased as a special order from optical cell manufacturers such as Helma, Starna, etc.); vacuum system with liquid nitrogen or Dry Ice cold trap; pressure gauge; two 1-L Dewars; thermocouple with readout for temperature measurements; colored filters such as Kopp (formerly Corning) 2-61 or Kodak Wratten Red #25; mirrors, mounts, neutral density filters (available from Newport, Oriel, Edmunds Scientific, etc.); glass microscope slides; photomultiplier with nanosecond time response (e.g., Oriel model 77348); digital oscilloscope with frequency response of 100 MHz or greater.

REFERENCES

1. G. Henderson, R. Tennis, and T. Ramsey, *J. Chem. Educ.* **75,** 1139 (1998).

2. S. Gerstenkorn and P. Luc, *Atlas du Spectra d'Absorption de la Molecule d'Iode,* Centre National de la Recherche Scientifique, Paris (1978).

3. G. A. Capelle and H. P. Broida, *J. Chem. Phys.* **58,** 4212 (1973).

4. J. A. Paisner and R. Wallenstein, *J. Chem. Phys.* **61,** 4317 (1974).

5. F. Castaño, E. Martínez, and M. T. Martínez, *Chem. Phys. Lett.* **128,** 137 (1986).

6. T. Masiello and J. W. Nibler, unpublished results.

7. J. I. Steinfeld, *Accts. of Chem. Res.* **3,** 313 (1970).

8. K. L. Duchin, Y. S. Lee, and J. W. Mills, *J. Chem. Educ.* **50,** 858 (1973).

9. G. E. Busch, R. T. Mahoney, R. I. Morse, and K. R. Wilson, *J. Chem. Phys.* **51**, 837 (1969).

10. I. N. Levine, *Physical Chemistry,* 6th ed., sec. 14.7 and 22.1, McGraw-Hill, New York (2009).

11. L. J. Gillespie and L. H. D. Fraser, *J. Amer. Chem. Soc.* **58**, 2260 (1936).

12. F. Rössler, *Z. Phys.* **96**, 251 (1935).

13. J. E. Selwyn and J. I. Steinfeld, *Chem. Phys. Lett.* **4**, 217 (1969).

14. J. O. Hirschfelder, C. F. Curtiss, and R. B. Bird, *Molecular Theory of Gases and Liquids,* Table I-A, Wiley, New York (1964).

15. *Landolt-Börnstein Physikalisch-chimische Tabellen,* Springer, Berlin.

GENERAL READING

G. Henderson, R. Tennis, and T. Ramsey, *J. Chem, Educ.* **75**, 1139 (1998).

J. I. Steinfeld, *Accts. of Chem. Res.* **3**, 313 (1970).

J. E. Selwyn and J. I. Steinfeld, *Chem. Phys. Lett.* **4**, 217 (1969).

G. A. Capelle and H. P. Broida, *J. Chem. Phys.* **58**, 4212 (1973).

J. A. Paisner and R. Wallenstein, *J. Chem. Phys.* **61**, 4317 (1974).

EXPERIMENT 41
Electron Spin Resonance Spectroscopy

Electron spin resonance (ESR), also called electron paramagnetic resonance (EPR), is a form of magnetic resonance spectroscopy that is possible only for molecules with unpaired electrons. Despite this restriction this sensitive technique has proven useful in the study of the electronic structures of many species, including organic free radicals, biradicals, triplet excited states, and most transition-metal and rare-earth species.[1-3] Important biological applications include the use of "spin labels" as probes of molecular environment in enzyme active sites and membranes. ESR has also been used to examine interior defects in solid-state chemistry and to study reactive chemical species on catalytic surfaces. The present experiment, on several benzosemiquinone radical anions, provides an introduction to the theoretical principles and experimental techniques used in ESR investigations. The utility of simple Hückel molecular orbital theory in interpreting the experimental spectra is also demonstrated.[3-6]

THEORY

As indicated in Eqs. (31-4) and (31-5), a single unpaired electron has spin angular momentum characterized by the quantum number $S = \frac{1}{2}$ and a magnetic moment of magnitude

$$\mu(\text{electron}) = g_e\mu_B[S(S + 1)]^{1/2} \tag{1}$$

where g_e, the electron g factor, equals 2.0023 for a free electron and $\mu_B = eh/4\pi m_e$, the Bohr magneton, equals 9.274×10^{-24} J T^{-1}. For atoms and molecules with many electrons, the total electron spin angular momentum can be as high as the sum of the spin angular momenta for each electron. Since spin is a vector quantity, the total spin is usually less because some of the individual spins are oriented so that they cancel each other. In fact most of the electrons in molecular species are *paired,* with zero angular momentum per pair, because this arrangement gives a lower energy for the system. In practice the species of greatest interest are those in which the total spin is a small integral multiple of $\frac{1}{2}$: $S = \frac{1}{2}$ (radicals and some transition metal species), $S = 1$ (biradicals, triplets, and some transition metals), and $S > 1$ (some transition metals).

The energy of interaction of the electronic magnetic moment with a magnetic field (the Zeeman energy) is given by[1-3]

$$E = g_e \mu_B M_S B \tag{2}$$

where B is the magnetic induction and M_S is the quantum number that measures the component of the spin angular momentum along the field direction (z). The difference in sign for this expression and the comparable expression for the nuclear case, cf. Eqs. (32-2) and (42-1), is due to the opposite charges of electrons and nuclei. For a single electron, $M_S = -\frac{1}{2}$ or $+\frac{1}{2}$, corresponding to two degenerate levels that split in the presence of the field, as depicted by the dashed lines in Fig. 1.

As for NMR spectroscopy, the selection rule for magnetic transitions is $\Delta M_S = \pm 1$ so that transition can take place between the $M_S = -\frac{1}{2}$ and $M_S = +\frac{1}{2}$ spin levels upon excitation by radiation of frequency

$$\nu = \frac{\Delta E}{h} = \frac{g_e \mu_B B}{h} \tag{3}$$

This relation is identical to Eq. (32-8) for a proton NMR transition except that, at a given B value, the splitting between the ESR levels is larger by the ratio $g_e \mu_B / g_N \mu_N (1 - \sigma_i) \cong 658$, where σ_i is the shielding constant. As a consequence, microwave radiation at a frequency of about 10 GHz is commonly employed for ESR spectroscopy, in contrast to the lower-energy radio-frequency radiation (about 100 MHz) used in NMR.

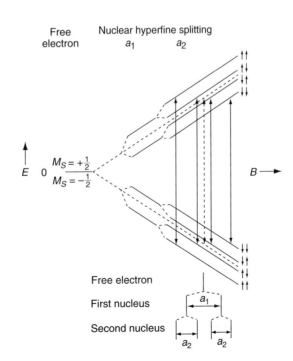

FIGURE 1

Energy levels and spectral line pattern (stick spectrum) for an unpaired electron interacting with two nonequivalent protons whose spin orientations are indicated at the right. The dashed levels and dashed transition arrow indicate the case for an uncoupled free electron.

To chemists, the principal aspect of interest in electron spin resonance is that the electron spin energies are sensitive to the molecular environment, as in the NMR case. The ESR counterpart to the NMR chemical shift is a variation in the g_e factor for an electron in a compound. For systems with orbital angular momentum, such as transition-metal atoms and ions, g_e can deviate substantially from 2.0023 if there is appreciable coupling between the electron spin and orbital motion. For most organic radicals however, g_e changes only very slightly and its variation does not give nearly as much chemical information as does the NMR chemical shift for a nucleus.† Instead, the primary value of the ESR measurement comes from the fact that the unpaired electron is generally not localized on a given atom but samples the magnetic environment over much if not all of the molecule. If the molecule contains nuclei with magnetic moments, especially protons, the electron–nuclear interaction produces characteristic splitting patterns in the ESR spectrum that can be used to deduce the number of different types of nuclei and their geometric symmetry.

More specifically, the total energy in a magnetic field can be written, to first order, as a sum of three contributions:

$$E = \begin{matrix} \text{electron} \\ \text{Zeeman} \\ \text{energy} \end{matrix} + \begin{matrix} \text{nuclear} \\ \text{Zeeman} \\ \text{energy} \end{matrix} + \begin{matrix} \text{electron-} \\ \text{nuclear} \\ \text{coupling} \end{matrix}$$

$$E = g_e \mu_B M_S B - \sum_i g_{Ni} \mu_{Ni} M_{Ii}(1 - \sigma_i)B + \sum_i a_i M_S M_{Ii} \tag{4}$$

The hyperfine splitting constant a_i has units of energy and is a characteristic parameter for the interaction of the unpaired electron with a nucleus of type i. It is the analog of the NMR J spin–spin constant describing the magnetic coupling between two nuclei.

The nuclear Zeeman term in Eq. (4) can be omitted in the following presentation, since it does not change for levels involved in an ESR transition ($\Delta M_I = 0$ if $\Delta M_S = \pm 1$). Thus, for example, the energy levels for an unpaired electron interacting with two different protons can be written as

$$E = M_S(g_e \mu_B B + a_1 M_{I1} + a_2 M_{I2}) \tag{5}$$

giving rise to the pattern and transitions depicted in Fig. 1. Here the usual experimental procedure is assumed, in which radiation of fixed frequency impinges on the system as the magnetic field is varied. As indicated in the figure, the consequence of the electron–nuclear coupling in Eq. (5) is to split the free-electron levels by an amount $\pm \frac{1}{2}a_1 \pm \frac{1}{2}a_2$ and to produce a quartet of lines whose spacings yield the hyperfine splitting parameters directly. The latter are usually reported in gauss units ($1 \text{ gauss} = 10^{-4} \text{ tesla}$),

$$a_i \text{ (gauss)} \equiv \frac{a_i \text{ (joule)}}{g_e \mu_B} \times 10^{-4} \tag{6}$$

so they can be extracted directly from the spectrum. This statement is not changed by a more exact, second-order treatment[1–3] of the energy levels, which produces only a slight common shift in all transitions such that the hyperfine splittings remain the same.

If the two protons are equivalent ($a_1 = a_2$), the two central transitions in Fig. 1 merge and a triplet hyperfine pattern is produced with intensity ratios 1:2:1. In general n equivalent protons give a spectrum of $n + 1$ hyperfine lines equally spaced by the hyperfine splitting constant a_H. The intensities are proportional to the degeneracy of the lower energy

---

†The g_e factor is actually a tensor quantity, whose average reduces to a single value for isotropic media such as gases, liquids, or solutions such as those studied in this experiment. For anisotropic samples such as crystals and impurities in solids, measurement of the components of g_e can give added information about the local molecular environment.

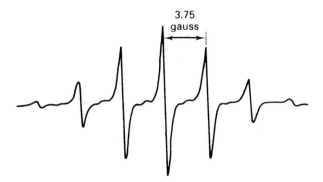

FIGURE 2
Proton hyperfine splitting
pattern in the ESR spectrum
of the benzene anion radical.

level involved in the transition; the relative values can be determined from the coefficients in a binomial expansion $(1 + x)^n$, which is equivalent to the relation

$$I_M = \frac{n!}{(\frac{1}{2}n + M)!(\frac{1}{2}n - M)!} \qquad M = -\tfrac{1}{2}n, -\tfrac{1}{2}n + 1, \ldots, \tfrac{1}{2}n - 1, \tfrac{1}{2}n \qquad (7)$$

For the benzene radical anion, one thus expects a seven-line pattern with intensity ratios of 1:6:15:20:15:6:1, in good agreement with the ESR spectrum shown in Fig. 2.

In the present experiment, we are concerned with the hyperfine structure of the benzosemiquinone radical anions. The delocalized unpaired π electron is of course distributed over the entire molecular frame of six C atoms and two O atoms. With R = H, by symmetry, it is clear that the four protons are all equivalent in the *para* species; hence five hyperfine lines with relative intensities 1:4:6:4:1 are expected in the ESR spectrum of this radical. By contrast, when R is not a proton, the three ring protons are not related by symmetry, and thus each may be expected to possess a different splitting constant. A hyperfine structure pattern of eight unequally spaced lines of equal intensity is expected. The line splittings and relative intensities in ESR spectra thus convey information about the geometric arrangement of the atoms.

The magnitudes of the hyperfine splitting parameters also yield information about the electron distribution in the molecule. The theory of the electron–nuclear coupling interaction was first worked out by Fermi, who showed that the constant a depended on the electron density at the nucleus. For a free hydrogen atom, a is given by the Fermi contact interaction in the form[1-3]

$$a = \left(\frac{8\pi}{3}\right) g_e \mu_B g_N \mu_N \rho(0) \qquad (8)$$

where $\rho(0) = |\psi(0)|^2$ is the unpaired electron density at the H nucleus. The wavefunction for the ground state of the H atom is well known, namely $\psi = (\pi a_0^3)^{-1/2} \exp(-r/a_0)$ with

$a_0 = 0.0529$ nm equal to the Bohr radius. The resultant value for a in gauss is 507 for this "pure" s orbital, while that for $p, d, f,$ and all other orbitals is zero since these have a node at the nucleus.

For the molecular case, the essential conclusion is that the orbital must have some s (or σ) character for the impaired electron to interact with a magnetic nucleus. Consider however the case of the benzene radical anion, in which the electron is usually described as being in a π orbital with a node in the molecular plane. As a consequence no coupling with the proton nuclei is expected, a prediction clearly in conflict with the hyperfine splitting of 3.75 gauss seen in the ESR spectrum of this species as shown in Fig. 2. How, then, does the unpaired π electron density appear at the H nucleus?

The answer is that the electrons cannot be so neatly labeled as σ or π type, and part of the unpaired π-electron density is transferred through the C—H sigma bonding electrons to the H nucleus through *exchange interactions*.[1-3] In the case of the π electron in the planar methyl radical, this process, termed spin polarization, results in a hyperfine constant of -23 gauss, about 5 percent of the limiting value of 507 gauss for an electron isolated completely on the H atom. The negative sign of a is not determined directly in the ESR experiment but is given by the theory. Thus one might say that the unpaired electron polarizes the CH bonding pair such that there is a net "negative spin excess of 5 percent" about the proton. For the benzene anion radical, the electron is equally distributed over six carbon atoms and one would expect a to be about $-23/6 = -3.83$ gauss, a value in good agreement with the experimental splitting of 3.75 gauss.

In the case of organic free radicals, McConnell[7] has shown that a simple empirical proportionality can be used to relate the observed hyperfine structure constant a_H and the unpaired electron spin density on the nearest carbon atom:

$$a_H = Q\rho_\pi \tag{9}$$

The constant Q is of the order of -20 to -30 gauss for aromatic hydrocarbons, and the benzene anion value of $Q = -6 \times 3.75 = -22.5$ gauss is commonly used. This relation may also be applied to give the hyperfine constant a_H for splitting arising from protons on the first carbon of a substituent attached to a carbon in an aromatic system, e.g., each of three methyl hydrogens in the toluene radical cation. Again Q is in the range of -20 to -30 gauss, and a value of -28 is usually assumed[8] when an independent experimental determination cannot be made.

The hyperfine structure constant thus allows us to probe the electron distribution in radicals. Theoretically calculated values of the spin densities can then be compared with the experimental values obtained from Eq. (9). One of the simplest methods for calculating electron density in an aromatic hydrocarbon is to use Hückel molecular orbital theory as discussed later.

The present experiment involves an investigation of the ESR spectra of the unsubstituted *ortho-* and *para-*benzosemiquinone radical anions, along with one or more of the methyl and *t*-butyl derivatives. Aspects of interest include the elucidation of splitting constants from complex spectra, the examination of substituent effects in ESR spectra, and the interpretation of spectra using molecular-orbital (MO) calculations.

EXPERIMENTAL

A schematic of an ESR spectrometer is shown in Fig. 3; more detailed discussion of construction and operation of the instrument can be found in Ref. 1 and in the citations therein, as well as in the manuals accompanying the spectrometer to be used. The microwave source is a vacuum-tube Klystron or a solid-state Gunn diode, which provides

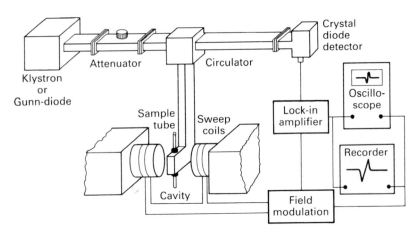

FIGURE 3
Schematic diagram of an ESR spectrometer.

tunable monochromatic radiation at about 10 GHz at a power level adjustable with an attenuator from zero to a few tenths of a watt. Radiation of this frequency is conveniently directed to the sample cavity by rectangular tubing, termed X-band waveguide. The sample is placed in a quartz tube and inserted into a rectangular cavity assembly whose dimensions are such as to produce at the microwave frequency a standing wave with maximum magnetic field and minimum electric field at the sample position. This arrangement provides the most efficient coupling of radiation into the sample for the magnetic dipole transitions of interest while at the same time minimizing energy absorption by nonresonant electric-dipole absorption (dielectric loss or "microwave cooking").

Typically the source is tuned with the sample in place and then locked to match the cavity resonance frequency so as to achieve maximum energy storage and minimum reflected power. This reflected power is directed through a "one-way" coupler called a circulator to a crystal diode detector to convey information about sample absorption in the cavity. An iris opening to the cavity is adjusted to match the impedance of the cavity to that of the source so as to produce minimum reflection of radiation from the cavity. This condition gives maximum sensitivity for the impedance mismatch produced when sample absorption occurs in the cavity.

Resonance is achieved by varying the current in the electromagnet to sweep the field a few gauss about a typical center value of 3500 gauss. On resonance, energy is absorbed in the cavity and the amount of radiation reflected back to the detector is altered. To achieve high sensitivity, the magnetic field is modulated (typically at 30 to 100 kHz) by an amount chosen to be small compared to the transition line width. The resultant ac component from the detector is sent to a phase-sensitive lock-in amplifier, producing a derivative spectrum, the common mode of display in ESR spectroscopy.

Details of the operation of the spectrometer will be provided by the instructor. Familiarity with the instrument can be gained by some preliminary experimentation with a stable radical sample such as solid α,α'-diphenyl-β-picrylhydrazyl (DPPH), which is often used as an ESR calibration standard.

The semiquinone radicals are produced by base-induced oxidation of 1,4-dihydroxybenzene (hydroquinone) or 1,2-dihydroxybenzene (catechol) by molecular oxygen, present in dissolved form. Radical concentration will increase over a period of time as the oxidation reaction proceeds and then decay as radical–radical reaction and other processes destroy the anions. Rates for these processes will depend on temperature, concentration of the dihydroxybenzene, and other parameters, so some experimentation may be necessary to obtain optimal spectra.

Prepare 5 to 10 mL of concentrated solution (1 *M* or greater) of the hydroquinones in methanol (ethanol or acetonitrile can also be used as solvent). A basic NaOH solution in methanol (1 *M* or greater) can be made by adding 1 g of NaOH to 25 mL of alcohol. With a syringe, place 1 to 2 mL of the hydroquinone solution into a small beaker and then add a drop of basic methanol. Stir until the solution turns yellow, then transfer to a quartz ESR tube and record the spectrum. A scan range of 10 to 20 gauss and a modulation amplitude of 0.01 gauss would be suitable starting parameters. The spectra can be recorded in conventional 5-mm-OD quartz ESR tubes, although the high dielectric constant of methanol makes the cavity more difficult to tune. This problem can be reduced by inserting the tube so that the solution extends only partially into the cavity region or, better, by using a 2-mm-OD ESR tube. Special ESR tubes for aqueous solutions are also available that have a flat rectangular sample section that can be oriented to maximize the sample volume at the central plane of the cavity where the electric field has a node, giving lower dielectric loss. **Special care should be taken to wipe all liquid off the outside of the tubes to avoid contamination of the cavity.** The cells should also be rinsed carefully with methanol or acetone prior to storage at the end of the experiment.

The parent *para*-benzosemiquinone radical anion may gradually increase in concentration and will last about 2 h; the methyl- or *t*-butyl-substituted radicals form more quickly and have lifetimes of about 15 min. The color serves as some guide to the optimal concentration; a reddish color suggests that too much base has been added, while a brown-black color indicates that radical–radical combination has occurred.

A satisfactory spectrum for the *ortho*-benzosemiquinone anion is more difficult to obtain by the above method because the radical–radical reaction is much faster. One simple way to promote the oxidation of the parent to provide a reasonable anion concentration is to increase the surface area of the solution to give greater access to oxygen in the air. This can be done by adding one drop each of catechol solution and base to a 5-mm-OD ESR tube. Turn the tube to form on the tube wall a *film* of solution, which should be green; a brown color is the result of radical–radical reaction. Shake any excess solution from the tube and record the spectrum immediately. It may be necessary to experiment a bit to obtain optimal spectra.

Spectra for substituted catechols can be obtained in the same way as for the hydroquinones, but these are much less stable and interference by other radical intermediates makes the interpretation of these spectra more difficult. Thus it is recommended that the student obtain spectra for the parent *ortho*- and *para*-benzosemiquinones and for one or more of the substituted methyl-, 2,3-dimethyl-, or *t*-butyl-*para*- forms. All the expected lines in the parent and dimethyl anion spectra should be clearly resolved. For the methyl compound, it may be difficult to obtain a recording showing all 32 expected lines clearly separated, but at least 20 separate lines should be readily distinguishable. Resolution of all the hyperfine structure of the *t*-butyl species will depend on the instrument and the sampling conditions. It may be possible to sharpen the lines somewhat by varying the concentration and by minimizing any excess dissolved oxygen gas, which can cause line broadening since it is paramagnetic.

CALCULATIONS AND DISCUSSION

Analyze the spectrum of each of the benzosemiquinone radicals to obtain the hyperfine splitting constants a_H. If you have at least 20 separate lines in the spectrum of the methyl derivatives, it should be possible to deduce all the splitting constants for this system. Note that the separation between the two outermost lines in the spectrum is given by a simple linear combination of the constants. After you have a set of constants for a spectrum, refine the values to give the best fit to all the lines in the spectrum. Construct a predicted stick spectrum derived from your constants for comparison with your actual ESR spectrum.[9]

Theoretical Calculations.   According to the McConnell relation, Eq. (9), the proton hyperfine splitting constants are proportional to ρ_π, the unpaired π electron density of the carbon atom bonded to the proton. Using quantum mechanics ρ_π can be calculated at different levels of approximation. The approach outlined here is one of the simplest, the Hückel molecular orbital (HMO) method.[2-6] This model assumes that a π molecular orbital, delocalized over n atoms, can be written as a linear combination of n atomic p_z orbitals (z is perpendicular to the molecular plane),

$$\psi = \sum_i c_i p_i \tag{10}$$

The corresponding one-electron π density at atom i is given by

$$\rho_{i\pi} = c_i^2 \tag{11}$$

The energy E and the coefficients c_i for the ground state are determined by use of the *variation principle*,[6] which says that one should choose the coefficients such that

$$E = \frac{\displaystyle\int \psi H \psi \, d\tau}{\displaystyle\int \psi^2 \, d\tau} = \text{minimum} \tag{12}$$

Here the Hamiltonian H is an effective one-electron energy operator whose explicit form need not be written in the HMO approximation. The variation principle ensures that the lowest-energy state is as close as possible to the true energy, and the coefficients are obtained from the minimization relations

$$\frac{\partial E}{\partial c_i} = 0 \quad i = 1, \dots, n \tag{13}$$

This results in n equations of the form

$$c_1(H_{11} - S_{11}E) + c_2(H_{12} - S_{12}E) + \cdots + c_n(H_{1n} - S_{1n}E) = 0$$
$$\vdots \qquad\qquad\qquad \vdots \tag{14}$$
$$c_1(H_{n1} - S_{n1}E) + c_2(H_{n2} - S_{n2}E) + \cdots + c_n(H_{nn} - S_{nn}E) = 0$$

In these equations $H_{ii} = \int p_i H p_i \, d\tau \equiv \alpha_i$, called the *Coulomb integral,* is the energy of an electron in a $2p$ orbital, while $H_{ij} = \int p_i H p_j \, d\tau \equiv \beta_{ij}$ the *resonance* or *bond integral,* represents the interaction energy of the two atomic orbitals. Both α_i and β_{ij} are negative energy quantities. $S_{ij} = \int p_i p_j \, d\tau$ is the overlap integral, which in the simplest approximation is taken to be 1 if $i = j$ and 0 otherwise. For a π system involving only carbon atoms, $\alpha_i \equiv \alpha$ and Eqs. (14) take the form

$$c_1(\alpha - E) + c_2\beta_{12} + \cdots + c_n\beta_{1n} = 0$$
$$\vdots \qquad\qquad \vdots \tag{15}$$
$$c_1\beta_{n1} + c_2\beta_{n2} + \cdots + c_n(\alpha - E) = 0$$

TABLE 2 Hückel molecular-orbital calculations for *ortho*-benzosemiquinone

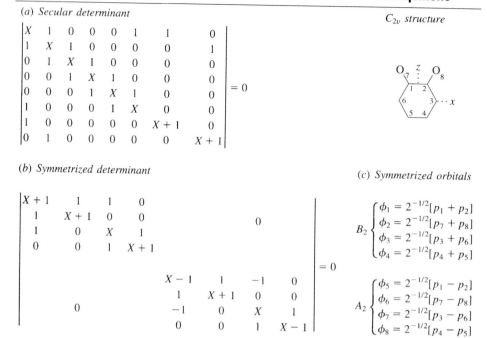

(a) Secular determinant

C_{2v} structure

$$
\begin{vmatrix}
X & 1 & 0 & 0 & 0 & 1 & 1 & 0 \\
1 & X & 1 & 0 & 0 & 0 & 0 & 1 \\
0 & 1 & X & 1 & 0 & 0 & 0 & 0 \\
0 & 0 & 1 & X & 1 & 0 & 0 & 0 \\
0 & 0 & 0 & 1 & X & 1 & 0 & 0 \\
1 & 0 & 0 & 0 & 1 & X & 0 & 0 \\
1 & 0 & 0 & 0 & 0 & 0 & X+1 & 0 \\
0 & 1 & 0 & 0 & 0 & 0 & 0 & X+1
\end{vmatrix} = 0
$$

(b) Symmetrized determinant

(c) Symmetrized orbitals

$$
\begin{vmatrix}
X+1 & 1 & 1 & 0 \\
1 & X+1 & 0 & 0 \\
1 & 0 & X & 1 \\
0 & 0 & 1 & X+1
\end{vmatrix}
\quad 0
$$

$$
0 \quad
\begin{vmatrix}
X-1 & 1 & -1 & 0 \\
1 & X+1 & 0 & 0 \\
-1 & 0 & X & 1 \\
0 & 0 & 1 & X-1
\end{vmatrix} = 0
$$

$$
B_2 \begin{cases}
\phi_1 = 2^{-1/2}[p_1 + p_2] \\
\phi_2 = 2^{-1/2}[p_7 + p_8] \\
\phi_3 = 2^{-1/2}[p_3 + p_6] \\
\phi_4 = 2^{-1/2}[p_4 + p_5]
\end{cases}
$$

$$
A_2 \begin{cases}
\phi_5 = 2^{-1/2}[p_1 - p_2] \\
\phi_6 = 2^{-1/2}[p_7 - p_8] \\
\phi_7 = 2^{-1/2}[p_3 - p_6] \\
\phi_8 = 2^{-1/2}[p_4 - p_5]
\end{cases}
$$

(d) Energies

Energies	Symmetry	Molecular orbitals
$X_8 = 2.127$	A_2	$\psi_8 = 0.664\phi_5 - 0.212\phi_6 + 0.536\phi_7 - 0.476\phi_8$
$X_7 = 1.197$	A_2	$\psi_7 = 0.579\phi_5 - 0.263\phi_6 - 0.149\phi_7 + 0.757\phi_8$
$X_6 = 1.095$	B_2	$\psi_6 = 0.475\phi_1 - 0.226\phi_2 - 0.768\phi_3 + 0.366\phi_4$
$X_5 = -0.262$	B_2	$\psi_5 = 0.505\phi_1 - 0.685\phi_2 + 0.312\phi_3 - 0.423\phi_4$
$X_4 = -0.748$	A_2	$\psi_4 = 0.129\phi_5 - 0.512\phi_6 - 0.737\phi_7 - 0.422\phi_8$
$X_3 = -1.477$	B_2	$\psi_3 = 0.226\phi_1 + 0.475\phi_2 - 0.366\phi_3 - 0.768\phi_4$
$X_2 = -1.576$	A_2	$\psi_2 = 0.455\phi_5 + 0.790\phi_6 - 0.383\phi_7 - 0.149\phi_8$
$X_1 = -2.356$	B_2	$\psi_1 = 0.685\phi_1 + 0.505\phi_2 + 0.423\phi_3 + 0.312\phi_4$

change) between each pair of adjacent atoms. Examine the nodal surfaces for the other wavefunctions and comment on the correlation between energy and number of nodes.

For the benzosemiquinone radical anion, one has nine π electrons, with two each in the four lowest levels and the unpaired electron residing in ψ_5. Using the HMO results in the tables, calculate ρ_π for the carbons attached to the protons and compare with the values obtained from your experimental results and the McConnell relation [Eq. (9)] with $Q = -22.5$ gauss. Note that the calculations provide a basis for assignment of the hyperfine splitting constants to specific ring protons in *ortho*-benzosemiquinone. Draw out the possible valence-bond resonance structures for both *ortho* and *para* compounds and discuss the relative importance of these.

Improvements in the simple HMO theoretical model are of course possible. For the *para*-benzosemiquinone anion, Vincow and Fraenkel[10] suggest that better parameters for oxygen are $\alpha_O = \alpha_C + h\beta$, $\beta_{CO} = k\beta_{CC} = k\beta$, with $h = 1.2$ and $k = 1.56$; values of $h = 2$ and $k = 2^{1/2}$ are also common choices for oxygen in C=O groups.[4,11] The student may find it interesting to repeat the calculations of Table 1 using these assumptions. For the *ortho*

TABLE 3   Calculated Hückel molecular-orbital spin densities for substituted benzosemiquinone anions

	ρ_2	ρ_3	ρ_4	ρ_5	ρ_6
2-Methyl-p-[a]	0.120	0.0786	. . .	0.111	0.112
2-t-Butyl-p-[a]	0.107	0.0704	. . .	0.132	0.091
2,3-Dimethyl-p-[b]	—	—	. . .	0.0884	0.0884
4-Methyl-o-[a]	. . .	~0	0.175	0.158	0.0388
4-t-Butyl-o-[a]	. . .	0.0110	0.171	0.163	0.045

[a] Ref. 12, [b] Ref. 4.

compound, higher-level self-consistent-field (SCF) calculations give very good agreement with experiment.[12]

Molecular-orbital calculations have also been made for the methyl- and t-butyl-substituted benzosemiquinones,[4,13] and the resultant spin densities are listed in Table 3. In these cases, the transfer of unpaired spin density from the π system to the proton is explained in terms of hyperconjugation.[2,13] Use these theoretical results as an aid in assigning your experimental hyperfine splitting constants to specific protons. Can you rationalize the charge distributions in these species and the changes from the distributions in the parent benzosemiquinones?

SAFETY ISSUES

Dispose of waste chemicals as instructed.

APPARATUS

ESR spectrometer (a relatively inexpensive teaching instrument is available from Micro-Now Instruments, Skokie, IL); 2.5-mm-OD ESR tube (or special ESR tube for aqueous samples) (Wilmad); 5-mL syringe; five 10-mL beakers.

1,4-Dihydroxybenzene (hydroquinone), 2-methyl-, 2,3-dimethyl-, and 2-t-butyl-hydroquinones, 1,2-dihydroxybenzene (catechol), (~1 g each); methanol; NaOH pellets.

REFERENCES

1. J. A. Weil, J. R. Bolton, and J. E. Wertz, *Electron Paramagnetic Resonance: Elementary Theory and Practical Applications,* Wiley-Interscience, New York (1994).

2. A. Carrington and A. D. McLachlan, *Introduction to Magnetic Resonance,* 2d ed., Chapman & Hall, New York (1989).

3. R. S. Drago, *Physical Methods in Chemistry,* 2d ed., chaps. 1–3, 9, 13, Saunders, New York (1992).

4. A. Streitwieser, Jr., *Molecular Orbital Theory for Organic Chemists,* Wiley, New York (1961).

5. K. Higasi, H. Baba, and A. Rembaum, *Quantum Organic Chemistry,* Interscience, New York (1965).

6. P. W. Atkins and J. de Paula, *Physical Chemistry,* 8th ed., pp. 386–392 and 549–552, Freeman, New York (2006).

7. H. M. McConnell, *J. Chem. Phys.* **24,** 764 (1956).

8. A. D. McLachlan, *Mol. Phys.* **1,** 233 (1958).

9. A PC program for simulating and plotting first-derivative ESR spectra for simple radicals is ESR, by Ronald D. McKelvey, *J. Chem. Educ.* **64,** 497 (1987). It can be obtained as program PC4102 from Project SERAPHIM, Department of Chemistry, University of Wisconsin–Madison. This program is especially useful in deducing coupling constants from experimental spectra by itera- tive comparison with spectra calculated for assumed hyperfine constants. See the website http:// ice.chem.wisc.edu/seraphim/.

10. G. Vincow and G. K. Fraenkel, *J. Chem. Phys.* **34,** 1333 (1961).

11. A useful program for Hückel MO calculations for up to 22 atom molecules, including N and O atoms, is Hückel, by Ronald D. McKelvey. This program runs on a PC and is available from Project SERAPHIM as program PC4001; see Ref. 9.

12. G. Vincow, *J. Chem. Phys.* **38,** 917 (1963).

13. C. Trapp, C. A. Tyson, and G. Giacometti, *J. Amer. Chem. Soc.* **90,** 1394 (1968).

GENERAL READING

N. J. Bunce, *J. Chem. Educ.* **64,** 907 (1987).

A. Carrington and A. D. McLachlan, *Introduction to Magnetic Resonance,* 2d ed., Chapman & Hall, New York (1989).

R. S. Drago, *Physical Methods in Chemistry,* 2nd ed., Saunders, New York (1992).

J. W. Nibler and R. Beck, *J. Chem. Educ.* **66,** 253 (1989).

A. Streitwieser, Jr., *Molecular Orbital Theory for Organic Chemists,* Wiley, New York (1961).

J. A. Weil, J. R. Bolton, and J. E. Wertz, *Electron Paramagnetic Resonance,* Wiley-Interscience, New York (1994).

A useful compilation of EPR data is J. A. Pedersen (ed.), *Handbook of EPR Spectra from Natural and Synthetic Quinones and Quinols,* CRC Press, Boca Raton, FL (1985).

EXPERIMENT 42
NMR Determination of Keto–Enol Equilibrium Constants

In this experiment proton NMR spectroscopy is used in evaluating the equilibrium compo- sition of various keto–enol mixtures. Chemical shifts and spin–spin splitting patterns are employed to assign the spectral features to specific protons, and the integrated intensities are used to yield a quantitative measure of the relative amounts of the keto and enol forms. Solvent effects on the chemical shifts and on the equilibrium constant are investigated for one or more β-diketones and β-ketoesters.

THEORY

Chemical Shifts. In Exp. 32, the Zeeman energy levels of a nucleus in an external applied field were given as

$$E_N = -g_N \mu_N M_I B_{\text{loc}} \tag{1}$$

where B_{loc} is the magnetic induction ("local field") at the nucleus. As a result of the $\Delta M_1 = \pm 1$ selection rule, a transition will occur at frequency

$$\nu_i = \left(\frac{g_N \mu_N}{h} \right) B_{i,loc} = \left(\frac{g_N \mu_N}{h} \right) B(1 - \sigma_i) \tag{2}$$

for a nucleus i; see Eq. (32-8). The chemical shift in parts per million (ppm) of this nucleus relative to a reference nucleus r is defined by

$$\delta_i \equiv \frac{\nu_i - \nu_r}{\nu_r} \times 10^6 = \frac{\sigma_r - \sigma_i}{1 - \sigma_r} \times 10^6 \approx (\sigma_r - \sigma_i) \times 10^6 \tag{3a}$$

Here the definition is based on the resonant frequencies for a fixed external induction (field) B. A second (nearly equivalent) relation is based on the alternative experimental case, where B is varied to achieve resonance at a fixed instrumental frequency ν. In this case $B_i(1 - \sigma_i) = B_r(1 - \sigma_r)$ and

$$\delta_i = \frac{B_r - B_i}{B_r} \times 10^6 = \frac{\sigma_r - \sigma_i}{1 - \sigma_i} \times 10^6 \approx (\sigma_r - \sigma_i) \times 10^6 \tag{3b}$$

Tetramethylsilane (TMS) is usually used as the proton reference, since it is chemically inert and its 12 equivalent protons give a single transition at a field B_r, higher than the field B_i found in most organic compounds. Thus δ is generally positive and increases when substituents are added that attract electrons and thereby reduce the shielding about the proton. This shielding arises because the electrons near the proton are induced to circulate by the applied field B (see Fig. 1a). This electron current produces a secondary field that *opposes* the external field and thus reduces the local field at the proton. As a result resonance at a fixed frequency such as 60 MHz requires a higher external field for protons with larger shielding. This shielding effect is generally restricted to electrons localized on the nucleus of interest, since random tumbling of the molecules causes the effect of secondary fields due to electrons associated with neighboring nuclei to average to zero. Nuclei such as ^{19}F, ^{13}C, and ^{11}B have more local electrons than hydrogen, hence their chemical shifts are much larger.

Long-range *deshielding* can occur in aromatic and other molecules with delocalized π electrons. For example, when the plane of the benzene molecule is oriented perpendicular to B, circulation of the π electrons produces a ring current (see Fig. 1b). This ring current induces a secondary field at the protons that is *aligned parallel* to B and thus increases

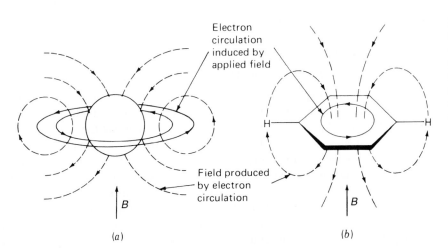

Electron circulation induced by applied field

Field produced by electron circulation

B

(a)

B

(b)

FIGURE 1

Shielding and deshielding of protons: (a) shielding of proton due to induced diamagnetic electron circulation; (b) deshielding of protons in benzene due to aromatic ring currents.

TABLE 1 Typical proton chemical shifts δ

CH₃ protons		Acetylenic protons	
$(CH_3)_4Si$	0.0	$HOCH_2C\equiv CH$	2.33
$(CH_3)_4C$	0.92	$ClCH_2C\equiv CH$	2.40
CH_3CH_2OH	1.17	$CH_3COC\equiv CH$	3.17
CH_3COCH_3	2.07	Olefinic protons	
CH_3OH	3.38	$(CH_3)_2C\!=\!CH_2$	4.6
CH_3F	4.30	Cyclohexene	5.57
CH₂ protons		$CH_3CH\!=\!CHCHO$	6.05
Cyclopropane	0.22	$Cl_2C\!=\!CHCl$	6.45
$CH_3(CH_2)_4CH_3$	1.25	Aromatic protons	
$(CH_3CH_2)_2CO$	2.39	Benzene	7.27
$CH_3COCH_2COOCH_3$	3.48	C_6H_5CN	7.54
CH_3CH_2OH	3.59	Naphthalene	7.73
CH protons		α-Pyridine	8.50
Bicyclo[2.2.1]heptane	2.19	Aldehydic protons	
Chlorocyclopropane	2.95	CH_3OCHO	8.03
$(CH_3)_2CHOH$	3.95	CH_3CHO	9.72
$(CH_3)_2CHBr$	4.17	C_6H_5CHO	9.96

the local field at the protons. This induced field changes with benzene orientation but does not average to zero, since it is not spherically symmetric. Because of this net deshielding effect, the resonance of the benzene protons occurs at a relatively low external field. The proton chemical shift δ for benzene is 7.27, greatly downfield from the value $\delta = 1.43$ that is observed for cyclohexane, in which ring currents do not occur. Similar deshielding occurs for olefinic and aldehydic protons because of the π electron movement. Typical values of δ for different functional groups are shown in Table 1, and additional values are available in Refs. 1 to 3. Although the resonances change somewhat for different compounds, the range for a given functional group is usually small and δ values are widely used for structural characterization in organic chemistry.

Spin-Spin Splitting. High-resolution NMR spectra of most organic compounds reveal more complicated spectra than those predicted by Eq. (2), with transitions often appearing as multiplets. Such *spin–spin splitting patterns* arise because the magnetic moment of one proton (A) can interact with that of a nearby nucleus (B), causing a small energy shift up or down depending on the relative orientations of the two moments. The energy levels of nucleus A then have the form

$$E_A = -g_{N_A}\mu_N M_{I_A}(1 - \sigma_A)B + hJ_{AB}M_{I_A}M_{I_B} \tag{4}$$

and there is a similar expression for E_B. The spin–spin interaction is characterized by the coupling constant J_{AB}, and the effect is to split the energy levels in the manner illustrated for acetaldehyde in Fig. 2. It is apparent from this diagram that the external field B does not effect the small spin–spin splitting that is characterized by the coupling constant J. The quantity J is a measure of the strength of the pairwise interaction of the proton spin with the spin of another nucleus. Since there are only proton–proton interactions in acetaldehyde, the same splitting occurs for both CH and CH₃ resonances.

The total integrated intensity of the CH and CH₃ multiplets follows the proton ratio of 1:3. However, the intensity distribution within each multiplet is determined by the relative population of the lower level in each transition. Since the level spacing is much less

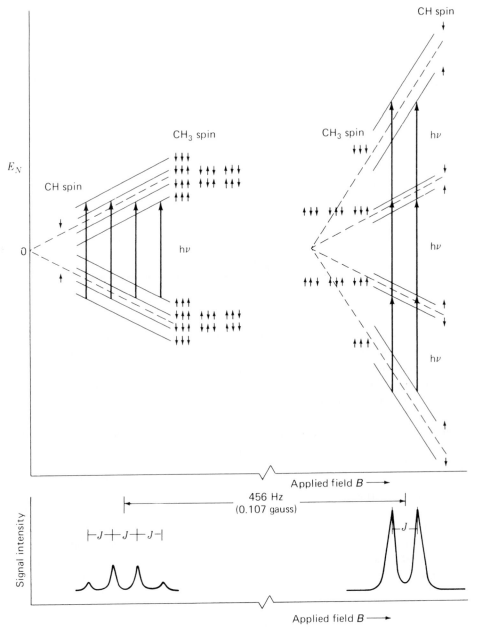

FIGURE 2
Energy levels, transitions, and 60-MHz NMR spectrum for acetaldehyde (CH_3CHO). The coupling $J = J_{CH_3} = J_{CH} = 2.2$ Hz ($= 5.2 \times 10^{-4}$ gauss $= 5.2 \times 10^{-8}$ T). constant For CH, the quantum number $M_I = -\frac{1}{2}$ or $\frac{1}{2}$. For the CH_3 group, $M_I = -\frac{3}{2}, -\frac{1}{2}, +\frac{1}{2}, +\frac{3}{2}$. The dashed lines represent the level spacing that would occur in the absence of the spin–spin interaction. The slopes of the energy levels are greatly exaggerated in the figure. Also, to be correct, all dashed lines should extrapolate to a common $E_N = 0$ at $B = 0$.

than kT, the Boltzmann population factors are essentially identical for these levels. However, there is some degeneracy because rapid rotation of the CH_3 group around the C—C bond makes the three protons magnetically equivalent. The number of spin orientations of the CH_3 protons that produce equivalent fields at the CH proton determine the degeneracy. The eight permutations of the CH_3 spins shown in Fig. 2 thus lead to a predicted intensity ratio of 1:3:3:1 for the CH multiplet. Similarly the CH_3 doublet peaks will be of equal intensity, with a total integrated intensity three times that of the CH peaks. In a more general sense, it can be seen that n equivalent protons interacting with a different proton

FIGURE 3
NMR spectrum of highly
purified ethanol obtained at
100 MHz.

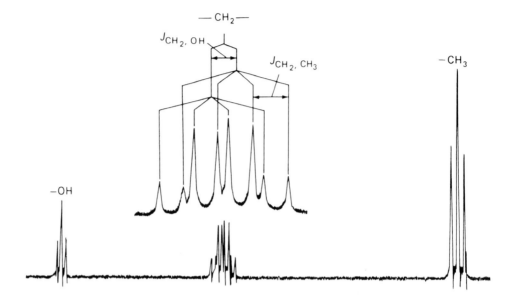

will split its resonance into $n + 1$ lines whose relative intensities are given by coefficients of the terms in the binomial expansion of the expression $(\alpha + \beta)^n$. Equivalent protons also interact and produce splitting in the energy levels. However, these splittings are symmetric for upper and lower energy states, so no new NMR resonances are produced.

If a proton is coupled to more than one type of neighboring nucleus, the resultant multiplet pattern can often be understood as a simple stepwise coupling involving different J values. For example, the CH_2 octet that occurs for pure CH_3CH_2OH (Fig. 3) arises from OH doublet splitting ($J = 4.80$ Hz) of the quartet of lines caused by coupling ($J = 7.15$ Hz) with CH_3. It should be mentioned that such regular splitting and intensity patterns are expected for two nuclei A and B only if $|\nu_A - \nu_B| \gtrsim 10 J_{AB}$. The spectra for this weakly coupled case are termed *first order*. Since the difference $\nu_A - \nu_B$ (in Hz) increases with the field while J_{AB} does not, NMR spectra obtained with a high-field instrument (400 MHz) are often easier to interpret than those from a low-field spectrometer (60 MHz). However, even if the multiplets are not well separated, it is still possible to deduce accurate chemical shifts and J values using slightly more involved procedures, which are outlined in most texts on NMR spectroscopy.[1–5] Such an exercise can be done as an optional part of this experiment, although it will not be necessary for the determination of equilibrium constants.

The mechanism of spin–spin coupling is known to be indirect and to involve the electrons in the bonds between interacting nuclei. The spin of the first nucleus A is preferentially coupled antiparallel to the nearest bonding electron via the so-called Fermi contact interaction, which is significant only when the electron density is nonzero at the first nucleus. (Such is the case only for electrons in *s* orbitals, since *p, d,* and *f* orbital wavefunctions have zero values at the nucleus.) This electron-spin alignment information is transmitted by electron–electron interactions to the second nucleus B to produce a field that thus depends on the spin orientation of the first nucleus (Fig. 4). Since the strength of this interaction falls rapidly with separation, only neighboring groups produce significant splitting. A few typical spin–spin coupling constants are given in Table 2 and these, along with the chemical shifts, serve to identify proton functional groups. As mentioned above, the multiplet intensities also give useful information about neighboring groups. Thus NMR spectra can provide detailed structural information about large and complex molecules.

(a) low-energy case   *(b)* high-energy case

FIGURE 4
Illustration of nuclear spin–
spin interaction transmitted
via polarization of bonding
electrons. The two electrons
about each carbon will tend
to be parallel, since this
arrangement minimizes the
electron–electron repulsion
(Hund's rule for electrons in
degenerate orbitals).

Keto–Enol Tautomerism. It is well known that ketones such as acetone have an isomeric structure, which results from proton movement, called the enol tautomer, an unsaturated alcohol:

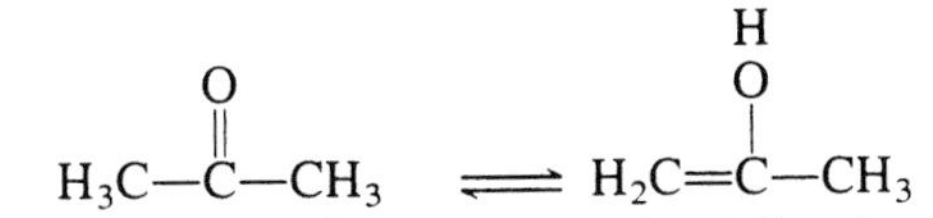

$$H_3C-\overset{\overset{\displaystyle O}{\|}}{C}-CH_3 \rightleftharpoons H_2C=\overset{\overset{\displaystyle O}{|}}{\underset{}{C}}-CH_3 \qquad (5)$$

acetone (keto form)   (enol form)

For acetone and the majority of cases in which this keto–enol tautomerism is possible, the keto form is far more stable and little if any enol can be detected. However, with β-diketones and β-ketoesters, such factors as intramolecular hydrogen bonding and conjugation increase the stability of the enol form and the equilibrium can be shifted significantly to the right.

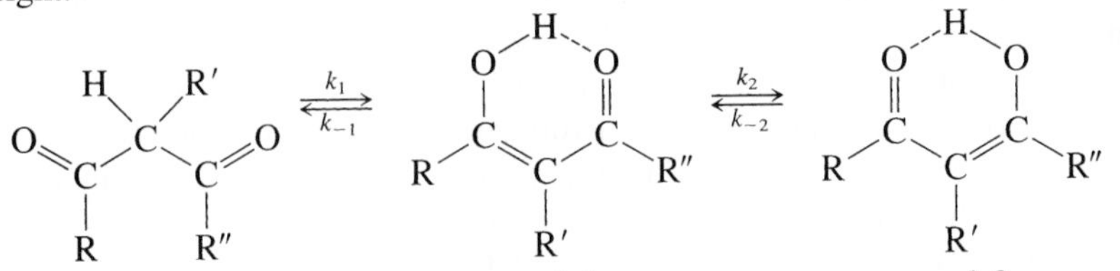

keto   enol 1   enol 2   (6)

The proton chemical environments are quite different for the keto and enol tautomers, and the interconversion rate constants k_1 and k_{-1} between these forms are small enough

TABLE 2 Typical proton spin–spin coupling constants

Coupling	J(Hz)	Coupling	J(Hz)
	−20 to +5		0 to 3.5
$\ce{CH-CH}$	2 to 9	$C=C$	6 to 14
$\ce{CH-(C)_n-CH}$	0		
ortho- meta- para-	6 to 9 1 to 3 1	$C=C$	11 to 19

that distinct NMR spectra are obtained for both forms. In principle, the two enols are also distinguishable when $R' \neq R''$. However, the intramolecular OH proton transfer is quite rapid at normal temperatures, so that a single (average) OH resonance is observed. In general, such averaging occurs when the conversion rates k_2 and k_{-2} (in Hz) exceed the frequency separation $\nu_1 - \nu_2$ (also in Hz) of the OH resonance for the two enol forms.[2] The magnetic field at the OH proton is thus averaged and resonance occurs at $(\nu_1 + \nu_2)/2$. Similarly, rapid rotation about the C—C bonds of the keto form explains why spectra due to different keto rotational conformers are not observed. Thus distinct spectra are expected only for the two tautomers, and these can be used to determine the equilibrium constant for keto-to-enol conversion:

$$K_c = \frac{(\text{enol})}{(\text{keto})} \tag{7}$$

where parentheses denote concentrations in any convenient units.

The keto arrangement shown in Eq. (6) is the configuration which is electrostatically most favorable, but the steric repulsions between R and R'' groups will be larger for this keto form than for the enol configuration. Indeed, experimental studies have confirmed that the enol concentration is larger when R and R'' are bulky.[4] This steric effect is less important in the β-ketoesters, in which the $R \cdots R''$ separation is greater. For both β-ketoesters and β-diketones, α substitution of large R' groups results in steric hindrance between R' and R (or R'') groups, particularly for the enol tautomer, whose concentration is thereby reduced. Inductive effects have also been explored; in general, α substitution of electron-withdrawing groups such as —Cl or —CF$_3$ favor the enol form.[4]

The solvent plays an important role in determining K_c. This can occur through specific solute–solvent interactions such as hydrogen bonding or charge transfer. In addition the solvent can reduce solute–solute interactions by dilution and thereby change the equilibrium if such interactions are different in enol–enol, enol–keto, or keto–keto dimers. Finally the dielectric constant of the solution will depend on the solvent and one can expect the more polar tautomeric form to be favored by polar solvents. Some of these aspects are explored in this experiment.

EXPERIMENTAL

The general features of a CW NMR spectrometer were described briefly in Exp. 32, and details about Fourier-transform NMR instruments are given in Exp. 43. For the spectrometer you are to use, more specific operating instructions will be provided by the instructor. Obtain several milliliters each of acetylacetone (CH$_3$COCH$_2$COCH$_3$, $M = 100.13$, density = 0.98 g cm^{-3}) and ethyl acetoacetate (CH$_3$CH$_2$OCOCH$_2$COCH$_3$, $M = 130.14$, density = 1.03 g cm^{-3}). Prepare small volumes of two solvents and three solutions.

Solvent A: Carbon tetrachloride, spectrochemical grade ($M = 153.83$, density = 1.58 g cm^{-3}) with 5 percent-by-volume tetramethylsilane (TMS) added. Prepare in a 10-mL volumetric flask.

Solvent B: Methanol, spectrochemical grade ($M = 32.04$, density = 0.791 g cm^{-3}) with 5 percent-by-volume TMS added. Prepare in a 5- or 10-mL volumetric flask.

Solution 1: 0.20 mole fraction of acetylacetone in solvent A.

Solution 2: 0.20 mole fraction of acetylacetone in solvent B.

Solution 3: 0.20 mole fraction of ethyl acetoacetate in solvent A.

For FT-NMR instruments that require a deuterated sample for frequency-locking purposes, $CDCl_3$ and CD_3OD can be used as solvents A and B.

Use a 1-mL pipette graduated in 0.01-mL increments to measure out 0.010 mol of solute, and use a 2-mL graduated pipette to then add the correct amount (0.040 mol) of solvent. You may neglect the presence of the 5 percent TMS when determining the necessary volumes of solvent. **Warning:** All work with TMS should be carried out in a hood. All containers or samples containing TMS should be tightly sealed and stored at low temperatures because of its volatility.

Prepare an NMR tube containing about 1 in. of solvent A and another containing solvent B, and record both NMR spectra, setting the TMS signal at the chart zero. Repeat for solutions 1 to 3, taking care to scan above $\delta = 10$ ppm since the enol OH peak is shifted substantially downfield. Determine which peaks are due to solute and measure chemical shifts for all solute features. Integrate the bands carefully at least three times, expanding the vertical scale by known factors as necessary in order to obtain accurate relative intensity measurements.

CALCULATIONS

Assign all spectral features using Table 1 and other NMR reference sources.[2,3,5] Tabulate your results and use your integrated intensities to calculate the percentage enol present in solutions 1 to 3. If possible, use the total integral corresponding to the sum of methyl (or ethyl), methylene, methyne, and enol protons. If this proves difficult because of overlap with solvent bands, indicate clearly how you used the intensities to calculate the percentage enol.

For both the enol and the keto forms, compare experimental and theoretical ratios of the integrated intensities for different types of protons (e.g., methyl to methylene protons in the keto form).

Using Eq. (7), calculate K_c and the corresponding standard free-energy difference ΔG^0 for the change in state keto → enol in each solution.

DISCUSSION

Discuss briefly your assignments of chemical shifts and spin–spin splitting patterns of acetylacetone and ethyl acetoacetate. Which compound has a higher concentration of enol form, and what reasons can you offer to explain this result? What changes would you expect in the NMR spectra of these two compounds if the interconversion rate between enol structures were much slower?

Compare the value of K_c for acetylacetone in CCl_4 with that in CH_3OH. What does your result suggest regarding the relative polarity of the enol and keto forms? Which form is favored by hydrogen bonding and why?

Compare your values of ΔG^0 with those for the gas phase ($\Delta G^0 = -9.2 \pm 2.1$ kJ mol^{-1} for acetylacetone, $\Delta G^0 = -0.4 \pm 2.5$ kJ mol^{-1} for ethyl acetoacetate).[6] What solvent properties might account for any differences you observe?

Additional compounds suitable for studies of steric effects on keto–enol equilibria include α-methylacetylacetone ($CH_3COCHCH_3COCH_3$), diethylmalonate (CH_3CH_2OCO-$CH_2COOCH_2CH_3$), ethyl benzoylacetate ($C_6H_5COCH_2COOCH_2CH_3$), and t-butyl acetoacetate (CH_3COCH_2COOt-Bu). Some other possible compounds are listed in Refs. 4 and 5.

Further aspects of this equilibrium that could be studied include the effects of concentration, temperature, and solvent dielectric constants on K_c.[5]

SAFETY ISSUES

Carbon tetrachloride and tetramethylsilane (TMS) are both toxic chemicals; see p. 197 for details about CCl_4. TMS is volatile and must be kept in a tightly sealed container. Carry out all solution preparations in a fume hood. Use a pipette bulb; do not pipette by mouth. Dispose of waste chemicals as instructed.

APPARATUS

NMR spectrometer with integrating capability; several 5- and 10-mL volumetric flasks; precision 1-mL and 2-mL graduated pipettes; pipetting bulb; NMR tubes; spectrochemical-grade CCl_4 and CH_3OH (or $CDCl_3$ and CD_3OD); tetramethylsilane, acetylacetone, and ethyl acetoacetate; fume hood.

REFERENCES

1. J. C. Davis, Jr., *Advanced Physical Chemistry: Molecules, Structure, and Spectra,* Wiley-Interscience, New York (1965).

2. J. A. Pople, W. G. Schneider, and H. J. Bernstein, *High Resolution Nuclear Magnetic Resonance,* McGraw-Hill, New York (1959); C. P. Slichter, *Principles of Magnetic Resonance,* 3d ed., Vol. I in P. Fulde (ed.), *Solid State Sciences,* Springer-Verlag, New York (1996).

3. R. M. Silverstein and F. X. Webster, *Spectrometric Identification of Organic Compounds,* 6th ed., Wiley, New York (1997); C. J. Pouchert and J. Behnke, *The Aldrich Library of ^{13}C and ^1H FT NMR Spectra,* Vols. 1–3: *300-MHz ^1H / 75-MHz ^{13}C,* Aldrich Chemical, Milwaukee, WI (1993).

4. J. L. Burdett and M. T. Rogers, *J. Amer. Chem. Soc.* **86,** 2105 (1964).

5. M. T. Rogers and J. L. Burdett, *Can. J. Chem.* **43,** 1516 (1965).

6. M. M. Folkendt, B. E. Weiss-Lopez, J. P. Chauvel, Jr., and N. S. True, *J. Phys. Chem.* **89,** 3347 (1985).

GENERAL READING

R. J. Abraham, J. Fisher, and P. Lofthus, *Introduction to NMR Spectroscopy,* Wiley, New York (1991).

R. Drago, *Physical Methods for Chemists,* 2d ed., chaps. 7, 8, Saunders, Philadelphia (1992).

R. K. Harris, *Nuclear Magnetic Resonance Spectroscopy,* Longman, London (1986).

J. K. M. Sanders and B. K. Hunter, *Modern NMR Spectroscopy: A Guide for Chemists,* 2d ed., Oxford Univ. Press, New York (1993).

EXPERIMENT 43

NMR Study of Gas-Phase DCl–HBr Isotopic Exchange Reaction

The development of pulsed Fourier–transform NMR spectrometers has greatly increased the sensitivity of NMR measurements, allowing spectra to be obtained for ^{13}C and other nuclei at natural abundances and low sample concentrations. In this experiment this enhanced capability is utilized in a low-density gas-phase measurement of the equilibrium constant K_p for H–D exchange in the reaction

$$DCl + HBr \rightleftharpoons DBr + HCl \tag{1}$$

The measured value of K_p is compared with the value calculated using statistical thermodynamics and vibrational–rotational spectroscopic information of the type obtained in Exp. 37.

THEORY

As shown in most physical chemistry texts (e.g., Refs. 1, 2; see also the discussion in Exp. 48), the molecular partition function q for the reactants and products determines the equilibrium constant K_p of a gas-phase reaction. For reaction (1) the relation is

$$K_p = \frac{p_{DBr}\, p_{HCl}}{p_{DCl}\, p_{HBr}} = \frac{q_{DBr}\, q_{HCl}}{q_{DCl}\, q_{HBr}} \exp\left(\frac{-\Delta E^0}{kT}\right) \tag{2}$$

Here each q consists of a product of terms

$$q = q_{trans} \cdot q_{rot} \cdot q_{vib} \cdot q_{elec} \tag{3}$$

where, in the rigid-rotor harmonic-oscillator approximation,

$$q_{trans} = V\left(\frac{2\pi MkT}{N_0 h^2}\right)^{3/2} \qquad \text{(Sackur–Tetrode equation)} \tag{4}$$

with V = volume, M = molar mass, and N_0 = Avogadro's number; and

$$q_{rot} = \sum_{J=0}^{\infty} (2J + 1) \exp\left\{-\left[\frac{hcBJ(J + 1)}{kT}\right]\right\} \tag{5}$$

$$\approx \int_0^{\infty} (2J + 1) \exp\left\{-\left[\frac{hcBJ(J + 1)}{kT}\right]\right\} dJ = \frac{kT}{hcB} \left(\text{if } \frac{hcB}{kT} \text{ is small}\right) \tag{6}$$

$$q_{vib} = \sum_{v=0}^{\infty} \exp\left(\frac{-hc\tilde{\nu}_0 v}{kT}\right) = \left[1 - \exp\left(\frac{-hc\tilde{\nu}_0}{kT}\right)\right]^{-1} \tag{7}$$

$$q_{elec} = \sum_{i=0}^{\infty} \exp\left[\frac{-(E_i - E_0)}{kT}\right] = 1 + (\approx 0 \text{ for higher electronic states}) \tag{8}$$

Equations (5) to (8) involve the partition function for the energy levels *internal* to a molecule, and the zero of energy for each molecule is the $v = 0$ level. The exponential term of Eq. (2) is then required to provide a common *external* energy reference point for all four molecules. A convenient reference state for the four-molecule system is that corresponding to complete dissociation to free atoms with no kinetic energy, as shown in

FIGURE 1

Potential-energy diagram relevant to the DCl-HBr exchange reaction. The $v = 0$ vibrational levels are shown for each molecule, and the zero-energy point of the system is taken to be the dissociated molecules limit. The bond dissociation D_0 and zero-point energy G_0 quantities are displayed for DCl.

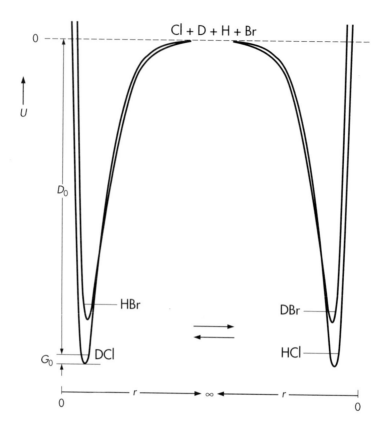

Fig. 1. From this figure, it is apparent that $\Delta E^0 \ (= \Delta U^0)$ for the reaction is determined by the bond-energy difference

$$\Delta E^0 = E^0_{DBr} + E^0_{HCl} - E^0_{DCl} - E^0_{HBr} = -D_{0DBr} - D_{0HCl} + D_{0DCl} + D_{0HBr} \quad (9)$$

This is also equal to the difference in the zero-point energies

$$\Delta E^0 = hc(G_{0DBr} + G_{0HCl} - G_{0DCl} - G_{0HBr}) \quad (10)$$

where

$$G_0 = \tfrac{1}{2}\tilde{\nu}_e - \tfrac{1}{4}\tilde{\nu}_e x_e \quad (11)$$

Equation (10) follows from the fact that, to a very good approximation, the sum $D_0 + G_0 = D_e$ is isotopically invariant. Equation (10) is useful because it allows one to use accurate G_0 values rather than the more poorly determined D_0 bond energies in calculating ΔE^0, and it is seen that the only input data needed to compute K_p are isotopic masses and vibrational–rotational parameters. Since the q's appear as ratios in Eq. (2), most of the constants in fact cancel and K_p consists of four factors (the electronic contribution is essentially unity):

$$K_p = \left(\frac{M_{DBr}M_{HCl}}{M_{DCl}M_{HBr}}\right)^{3/2} \cdot \left(\frac{B_{DCl}B_{HBr}}{B_{DBr}B_{HCl}}\right) \cdot \exp\left(\frac{-\Delta E^0}{kT}\right) \quad (12)$$

$$\cdot \left(\frac{[1-\exp\,(-hc\,\tilde{\nu}_{0DCl}/kT)]\,[1-\exp\,(-hc\,\tilde{\nu}_{0HBr}/kT)]}{[1-\exp\,(-hc\,\tilde{\nu}_{0DBr}/kT)]\,[1-\exp\,(-hc\,\tilde{\nu}_{0HCl}/kT)]}\right)$$

TABLE 1 Molecular parameters $(cm^{-1})^3$

Molecule	M (g/mol)	$B(v = 0)$	$\tilde{\nu}_e$	$\tilde{\nu}_e x_e$	G_0	$\tilde{\nu}_0$
$H^{35}Cl$	35.97668	10.43983	2990.95	52.82	1482.27	2885.31
$D^{35}Cl$	36.98296	5.39215	2145.16	27.18	1065.79	2090.80
$H^{81}Br$	81.92412	8.34824	2648.98	45.22	1313.18	2558.54
$D^{81}Br$	82.93039	4.24560	1884.75	22.72	936.70	1839.31

Literature values for some of these constants are given in Table 1, but the calculation could also be done using parameters determined previously from analysis of the infrared spectra of the four diatomic species as in Exp. 37. It should be noted that it does not matter much which halogen isotopes are used in the calculations, since the parameters differ only slightly for ^{35}Cl–^{37}Cl and ^{79}Br–^{81}Br pairs,[3] and changes in K_p are even smaller.

METHOD

The ratios p_{DBr}/p_{DCl} and p_{HCl}/p_{HBr}, and hence K_p, are to be determined from the relative integrated peak areas of the deuteron and proton NMR spectra. A sensitive Fourier-transform NMR instrument with multinuclei capability is thus required, ideally at 200-MHz proton frequency or higher. Here we outline the essentials of a pulsed NMR experiment; more detailed discussions can be found in Refs. 4 to 7.

The basic components of a Fourier-transform NMR (FT-NMR) instrument are the same as those in the continuous-wave (CW) spectrometer discussed in Exp. 32. In modern instruments of greater than 100 MHz frequency for protons, a high static magnetic field (B_0) is utilized, which is generated by superconducting magnets cooled to 4 K. As in the CW experiment, the sample tube is inserted into a probe consisting of a room-temperature transmitter and detector coils positioned between the magnet pole faces and perpendicular to the field direction z. In one form of the CW proton experiment, the sample is subjected to a radio wave whose frequency is slowly tuned across the proton resonances in the sample; the signal induced in the detector coil is measured to give the NMR spectrum directly. In a 200-MHz FT-NMR instrument, monochromatic radiation at 200 MHz is *gated* to produce a frequency-broadened pulse that serves to excite all proton resonances simultaneously. The frequency spread $\Delta\nu$ of the pulse is inversely proportional to the duration of the gate τ_p and is $\Delta\nu \approx (2\pi\tau_p)^{-1}$. A typical pulse of 10 μs thus produces a spread of about 60,000 Hz, corresponding to a chemical-shift range of 300 ppm. This width greatly exceeds the 0- to 10-ppm proton range of interest, so the intensity of the exciting radiation is nearly constant for each proton resonance—an important feature for quantitative measurements as in the present experiment.

The effect of the transmitter radiation pulse on the sample can be understood by picturing the changes that occur in the orientation of the net sample magnetization vector M for a sample containing a single proton type (Fig. 2). Initially, according to classical mechanics, each individual nuclear magnetic moment precesses on a cone about B_0 at the Larmor frequency $\nu_L = \gamma B_0/2\pi$ and only the z component of the magnetic moment is stationary in time. In a quantum mechanical description, this corresponds to the fact that only the M_z component of the magnetic moment is quantized, and the uncertainty about M_x and M_y implies that the moment is randomly distributed along a cone oriented *with* (α) or *against* (β) B_0. In both descriptions the net macroscopic sample magnetization (due to the Boltzmann excess of α spins) is accordingly aligned along z in a laboratory axis system

FIGURE 2

(*a*) Magnetic-moment distribution at the start of the nuclear induction experiment. The excess of $\alpha(M_I = \frac{1}{2})$ spin moments are randomly positioned on the precessional cone. The transmitter coil will generate a circularly polarized B_1 component, rotating with the precessing nuclei.
(*b*) Distribution at resonance $\nu = \nu_L$ after rotation of M through an angle θ. The α nuclei are "bunched" on the cone to produce a rotating $M_{y'}$ component that induces a signal in the detector coil.

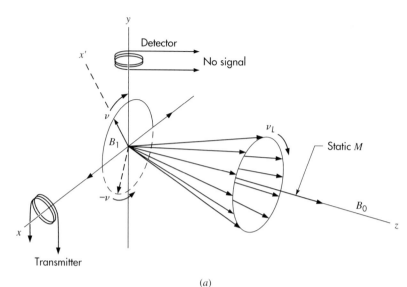

(*a*)

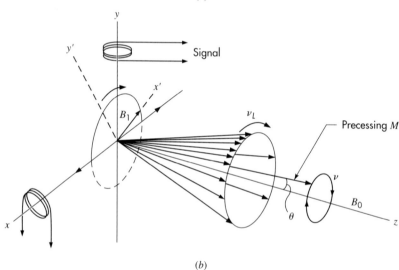

(*b*)

(x,y,z) and has magnitude $M = \gamma \hbar (N_\alpha - N_\beta)$ given by the vector sum of the microscopic z components. Thus, initially, $M_x = M_y = 0$, $M_z = M$, as shown in Fig. 2*a*.

The direction of M will change if a static field B_1 is applied perpendicular to B_0. That the nuclei experience an effective static field from the transmitter radiation when it oscillates at the resonant frequency ν_L can be seen by recognizing that the transmitter coil produces a fluctuating B field component along the coil axis (x). The resultant radiation emitted along z is thus x polarized and, like any plane-polarized wave, can be decomposed into two circularly polarized components rotating about z in opposite directions. If the radiation is at the Larmor frequency, one of these circular components rotates at the same frequency as the precessing α spins and, in this rotating (x', y', z) axis system, appears to be a static field component, B_1. (The other counterrotating component averages to zero for the α nuclei.) The effect of this static B_1 field is to exert a torque about the x' axis, resulting in a rotation of the M vector in the $y'z$ plane by an angle (in radians) of $\theta = \gamma B_1 \tau_p$; see Fig. 2*b*. For a given B_1 intensity, the τ_p time necessary for a 90° rotation is termed a

$\pi/2$ pulse, after which the M vector is aligned along the y' axis. Since y' (and hence M) is rotating at the Larmor frequency about the z axis, a large signal is induced in the detector coil which measures the M_y component in the laboratory framework. After a $\pi/2$ pulse, $M_{y'} = M$, $M_{x'} = M_z = 0$ ($N_\alpha = N_\beta$), and, in the laboratory frame, $M_y = M \sin(2\pi\nu_L t)$.

At the end of the excitation pulse (plus a delay to allow the electronics to quiet), the detector signal dies away exponentially as the individual nuclei reorient their moments back to the B_0 z direction. This decreasing signal is termed the free-induction decay (FID), and in the case of exact resonance ($\nu = \nu_L$), the FID curve corresponds to the dashed line in Fig. 3a. If the radiation frequency is changed slightly off resonance, the B_1 field appears to the precessing nuclei to be rotating at a frequency $\nu - \nu_L$ and the FID is thereby modulated to produce the solid line in Fig. 3a. Typically the FID data from many pulses are averaged and the Fourier transform is taken to obtain the frequency-domain spectrum, which in this example consists of a single Lorentzian line. In the nonresonant case, the effect of the modulation is to shift the Lorentzian peak an amount $\nu - \nu_L$, as seen in Fig. 3b. For a sample with two or more nuclei with different Larmor frequencies, the detector measures a sum of FID signals but the Fourier transform operation unravels this sum to yield the desired frequency spectrum.

As shown in Fig. 3, the decay time T_2^* of the FID is related to the half-width of the Lorentzian peak. The quantity $1/T_2^*$ gives a measure of the rate at which the M_y component decays due to randomization of the "bunched" individual magnetic moments on the precessional cone. This dephasing involves no energy change and is caused by spin–spin

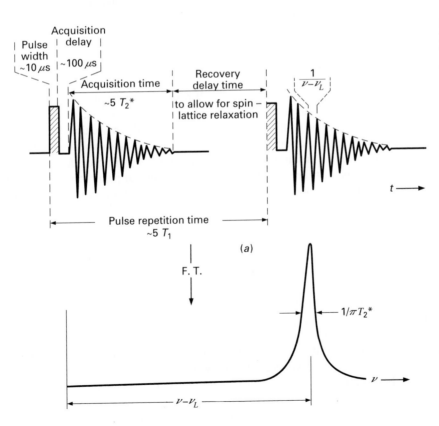

FIGURE 3

Pulse sequence in FT-NMR. The FID data in (*a*) are averaged and Fourier transformed to yield the frequency-domain spectrum shown in (*b*).

exchange processes (T_2) and by the residual magnetic field inhomogeneity ΔB. The relationship among these quantities is

$$\frac{1}{T_2^*} = \frac{1}{T_2} + \gamma \Delta B \qquad (13)$$

Although it is not sensed by the detector coil, the return of the M_z component to the equilibrium value is also an exponential process, characterized by a longitudinal relaxation time T_1. This time is also called the *spin-lattice relaxation time*, since it involves energy exchange with the surroundings (lattice) as $N_\alpha - N_\beta$ returns to the original value. Special multiple pulse sequences are usually used to determine T_1 and T_2.[4-7] Both relaxation processes result from fluctuating magnetic field components caused by collisions and rotational motion, and $T_1 \geq T_2$ since the spin-lattice energy exchange also contributes to the dephasing process. For liquid solutions $T_1 \sim T_2^*$ and is of the order of 1 s.

These decay times are important in determining both the spectral resolution and the rate at which data can be collected in the FT-NMR experiment. The resolution increases with the total time of FID data acquisition, but this is usually limited to about $5T_2^*$, at which point the signal magnitude is 0.7 percent of the original value and approaches the noise level. However, a further delay before the next excitation pulse is necessary to ensure that the spin-lattice relaxation is also 99.3 percent complete. This results in a repetition period of $5T_1$ and long measurement times if many pulses must be averaged for a good signal-to-noise (S/N) ratio. Such delays are required if accurate relative intensity measurements are to be made for two nulcei with significantly different T_1 times. However, for two protons (or deuterons) with comparable T_1 relaxation rates, as for HCl and HBr in the current experiment, adequate data can be obtained using a shorter pulse repetition period of about $2T_1$.

EXPERIMENTAL

For the equilibrium mixture of HCl, HBr, DCl, and DBr at a total pressure of about 1 bar, the T_1 times are about 30, 30, 250, and 250 ms, respectively, while the T_2^* times are typically an order of magnitude smaller. The latter includes relaxation due to the magnetic field inhomogeneity, and some "reshimming" of the NMR instrument with the gas sample may be necessary to ensure narrow lines. The instructor will provide guidance as to the proper use of the instrument and the choice of operational parameters such as number of pulses to average and optimal acquisition and delay times for accurate peak intensity integrations. It should be noted that the D/H magnetic moment ratio of 0.857/5.585 implies a lower sensitivity factor of 0.15 in measuring D signals. In fact the practical experimental D/H sensitivity factor cited by NMR manufacturers is significantly smaller (0.01) and much longer measurement times are needed for D than for H spectra. This problem is alleviated somewhat in the present experiment by using a sample with a D content of about 95 percent.

Since the data-acquisition time for a given S/N scales inversely with the square of the number of molecules sampled, high pressures and large tube volumes are desirable. NMR tubes are produced in a range of diameters, and the largest of these that is compatible with the instrument probe design should be chosen. Tubes of 5-, 10-, and 15-mm diameter with an in-line Teflon valve, which permits easy attachment to a vacuum line via $\frac{1}{4}$-in. or $\frac{3}{8}$-in. Ultra-Torr fittings and which can be spun in the normal way, are available from various suppliers.

Fill the NMR tube to 1 bar with a gas mixture prepared from the reaction of D_2O (95 percent D) with an equimolar mixture of benzoyl chloride and bromide. The reaction

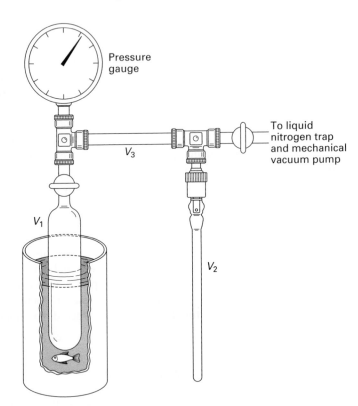

FIGURE 4
Apparatus for small-scale
synthesis of hydrogen halides
and for transfer to an NMR
tube fitted with a concentric
valve.

and one possible synthesis set up are shown in Exp. 37. Figure 4 shows an alternative simpler apparatus consisting of a reaction tube of volume V_1 connected by polyethylene tubing and $\frac{3}{8}$-in. Ultra-Torr fittings to a pressure gauge, a vacuum port, and the valved NMR tube of volume V_2. The halides are powerful lachromators, so care should be exercised in handling them. Work in a fume hood, and use microsyringes to add through the valve opening of the reaction tube the stoichiometric amounts of reactants necessary to produce 0.5 bar total gaseous products (the air will be pumped out as described below). For $V_1 = 1$ L, this is 0.21 mL of D_2O (95 percent D) and 1.2 mL each of benzoyl chloride ($M = 140.6$, $\rho = 1.21$) and bromide ($M = 185.0$, $\rho = 1.57$); for other V_1 sizes, scale these amounts accordingly. **Care should be taken to avoid excessive pressures in this synthesis—if in doubt about the reaction amounts or the procedures, consult the instructor.**

The reaction is slow at room temperature, so it is safe to close the valve on the reaction flask and then freeze the sample mixture in a Dewar filled with liquid nitrogen. With the reaction flask still in the Dewar, attach the flask to the vacuum line and pump out the air. Close the reaction valve, warm the flask to room temperature, and then repeat this freeze–pump cycle to eliminate any residual air trapped in the initially frozen solid. Now remove the sealed flask and place it in an oil bath on a hot plate in the hood. Gradually raise the bath temperature from room temperature to 145°C and hold this temperature until bubbling of the acid halides ceases. Then reattach the reaction flask to the vacuum line, cool it in liquid N_2, evacuate any residual air, close the valve, and replace the N_2 Dewar with an ice bath to allow the frozen gases to evaporate and mix. Keep the ice bath on the sample during the following filling operation to minimize the vapor pressure of the organic reaction products and of any residual D_2O.

To fill the valved NMR cell to 1 bar, it is necessary to know the volume ratio $r = V_2/(V_2 + V_3)$, where V_3 is the connection volume including the pressure gauge (but not V_1). An r ratio of about 0.1 to 0.2 is convenient, and it can be easily determined by measuring the pressure ratio when 1 bar of air in V_2 is expanded into the evacuated volume V_3 (plus V_2). Now evacuate the air, open the reaction flask valve carefully to fill volume $V_2 + V_3$ to a pressure of r bar, and then close the reaction flask valve. Condense the gas into the NMR tube by immersing it in liquid N_2, and close the tube valve. This filling procedure yields a 1-bar pressure of the sample mix in the NMR tube. Initial fillings of $V_2 + V_3$ to a pressure of 2 or 3 r can be used if higher pressures of 2 or 3 bar are desired in order to reduce the NMR data-collection time. *In this case, the tube should first be carefully tested by pressurizing to about 5 bar by this filling procedure (or by using a N_2 cylinder equipped with a pressure regulator).*

Record the proton and deuteron NMR spectra according to the directions of the instructor. The chemical shifts (relative to liquid tetramethylsilane) are -0.43 and -4.33 ppm for HCl and HBr, respectively, and a similar chemical-shift ordering is seen for DCl and DBr. Obtain expanded traces and integrate the peaks. Be consistent in choosing the integration limits; the same ranges (about five peak widths) should be used for all four peaks. Repeat the integration procedure several times to obtain an estimate of the uncertainty in the peak area ratios for H and D spectra. Record the sample probe temperature.

CALCULATIONS AND DISCUSSION

From the relative NMR areas obtained from the instrumental integration, calculate the value of K_p and its uncertainty. If possible obtain the spectrum as an ASCII intensity file in order to obtain improved integration values by doing a nonlinear least-squares fit of the data. A suitable function based on Lorentzian line shapes is

$$I(x) = \frac{a_1}{1 + 4W_1^{-2}(x - x_1)^2} + \frac{a_2}{1 + 4W_2^{-2}(x - x_2)^2} + b + cx \qquad (14)$$

Here a_i is the peak amplitude, W_i is the full width at half-maximum, and x_i is the center of peak i, while the terms b and cx provide a baseline correction. The variable x can be in ppm, frequency units, or arbitrary units, since these cancel in the area ratio $a_1 W_1/a_2 W_2$ for the two Lorentzian peaks (the integrated area of a Lorentzian peak is $\pi a_i W_i/2$). An example of the use of spreadsheets for nonlinear fitting applications is given in Chapter III. Alternatively other nonlinear least-squares programs or commercial line-fitting software, such as Peakfit from Jandel Scientific, can be used. Such fitting procedures generally give a more accurate measure of the peak areas than does the instrumental integration and are instructive in their own right.

Use of a spreadsheet is also helpful for the statistical calculation of K_p from Eqs. (2) to (12). Comment in your report on the relative importance of the various factors in Eq. (12) in determining the equilibrium constant. Compare the calculated value of K_p with your measured NMR value. Compute the free-energy change $\Delta G^0 = -RT \ln K_p$ for the isotopic exchange reaction.

By setting up the calculations in a spreadsheet mode, it is also easy to explore the effect of various changes in the parameters. For example the assumption that the choice of halogen isotope is not important can be tested by repeating the computations using spectroscopic data[3] for diatomics containing ^{37}Cl instead of ^{35}Cl. The effect of approximating the zero-point energy as $G_0 \approx \tilde{\nu}_0/2$ ($\tilde{\nu}_0$ = vibrational fundamental) rather than as the expression in Eq. (11) can be examined. Another calculation might explore the accuracy of the rotational partition function approximation of Eq. (6) by performing the actual sum over the rotational levels.[1,2] Finally the effect of temperature on K_p can be calculated to judge whether a temperature-dependence study to obtain ΔH^0 and ΔS^0 would be feasible.

SAFETY ISSUES

Avoid pressures in excess of 1 bar during the synthesis work and in excess of 5 bar for the NMR tube. Work on a vacuum system requires preliminary review of procedures and careful execution in order to avoid damage to the apparatus and possible injury; in addition the liquid nitrogen used for cold traps must be handled properly (see Appendix C). Benzoyl chloride and bromide are toxic chemicals and potent lachrymators; they must be disposed of properly.

APPARATUS

FT-NMR spectrometer, with 10- or 15-mm-diameter probe capability if possible; gas NMR tube with concentric Teflon valve (e.g., J. Young NMR-10); 1-L reaction flask with greaseless stopcock; 0- to 1-bar pressure gauge; vacuum line; Ultra-Torr fittings; microsyringes; hot plate and oil bath with 0 to 200°C thermometer; liquid N_2 and ice baths; benzoyl chloride and bromide, D_2O (95 percent); fume hood.

REFERENCES

1. R. J. Silbey, R. A. Alberty, and M. G. Bawendi, *Physical Chemistry,* 4th ed., secs. 16.2–16.8, Wiley, New York (2005).

2. P. W. Atkins and J. de Paula, *Physical Chemistry,* 8th ed., chaps. 16, 17 and pp. 611–618, Freeman, New York (2006).

3. K. P. Huber and G. Herzberg, *Molecular Spectra and Molecular Structure IV: Constants of Diatomic Molecules,* Van Nostrand Reinhold, New York (1979); although this book is now out of print, its contents are available as part of the *NIST Chemistry WebBook* (http://webbook.nist.gov).

4. R. J. Abraham, J. Fisher, and P. Lofthus, *Introduction to NMR Spectroscopy,* Wiley, New York (1991).

5. R. K. Harris, *Nuclear Magnetic Resonance Spectroscopy,* Longman, London (1986).

6. R. Drago, *Physical Methods for Chemists,* 2d ed., chaps. 7, 8, Saunders, Philadelphia (1992).

7. R. W. King and K. R. Williams, *J. Chem. Educ.* **66,** A213, A243 (1989); **67,** A93, A100, A125 (1990).

GENERAL READING

R. J. Abraham, J. Fisher, and P. Lofthus, *Introduction to NMR Spectroscopy,* Wiley, New York (1991).

P. W. Atkins and J. de Paula, *Physical Chemistry,* 8th ed., Freeman, New York (2006).

R. K. Harris, *Nuclear Magnetic Resonance Spectroscopy,* Longman, London (1986).

J. W. Nibler, P. Minarik, W. Fitts, and R. Kohnert, *J. Chem. Educ.* **73,** 99 (1996).

EXPERIMENT 44

Solid-State Lasers: Radiative Properties of Ruby Crystals

The first optical laser, the ruby laser, was built in 1960 by Theodore Maiman. Since that time lasers have had a profound impact on many areas of science and indeed on our every-day lives. The monochromaticity, coherence, high-intensity, and widely variable pulse-duration properties of lasers have led to dramatic improvements in optical measurements of all kinds and have proven especially valuable in spectroscopic studies in chemistry and physics.[1,2] Because of their robustness and high power outputs, solid-state lasers are the "workhorse" devices in most of these applications, either as primary sources or, via non-linear crystals or dye media, as frequency-shifted sources. In this experiment the 1064-nm near-infrared output from a solid-state Nd:YAG laser will be frequency doubled to 532 nm to serve as a fast optical pump of a ruby crystal. Ruby consists of a dilute solution of chromium 3^+ ions in a sapphire (Al_2O_3) lattice and is representative of many metal ion-doped solids that are useful as solid-state lasers, phosphors, and other luminescing materials. The radiative and nonradiative relaxation processes in such systems are important in determining their emission efficiencies, and these decay paths for the electronically excited Cr^{3+} ion will be examined in this experiment.

THEORY

The two general requirements for lasing action in a medium are (1) the production of a population inversion between two levels of an optical transition and (2) cavity feedback to stimulate emission of further photons in a coherent (in-phase) fashion. Figure 1 shows the excitation cycle for a general four-level laser. For the cases of interest here, optical pumping into a high electronic state of a metal ion is followed by rapid intersystem crossing (ISC) in a few picoseconds to nanoseconds to a metastable state from which radiative return to the ground state is slow due to spin or symmetry selection rules. The concentration of excited ions in the metastable state thus builds up so that the required population inversion can be produced between the two levels involved. The rates of the excitation and relaxation processes are critical in determining the energy output and efficiency of a laser.

FIGURE 1
Pumping–emission cycle for a typical four-level laser.

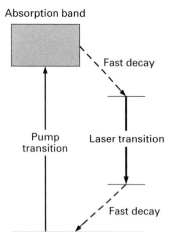

Absorption band

Fast decay

Pump transition

Laser transition

Fast decay

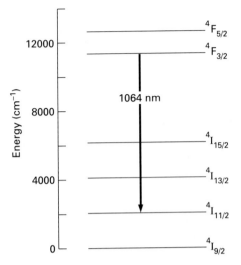

FIGURE 2
Energy-level diagram for the
Nd:YAG system.

In the subsequent discussion, the wavelengths of transitions are cited, as is conventional in describing laser lines. The frequency $\tilde{\nu}$ in cm^{-1} units is given by $\tilde{\nu} = 10^7/\lambda$, where λ is the wavelength in nm.

Nd:YAG System.   Figure 2 shows a somewhat simplified energy-level diagram for Nd^{3+} doped in an yttrium-aluminum garnet (YAG = $Y_3Al_5O_{12}$) crystal. Inversion is relatively easy to achieve in this four-level system because the ${}^4I_{11/2}$ lower state has little initial thermal population (the Boltzmann factor is $\exp(-E/kT) \simeq 10^{-6}$) and relaxation to the ${}^4I_{9/2}$ ground state is very fast. Typically the YAG rod is 50 to 100 mm in length by 3 to 9 mm OD and is pumped by a flashlamp whose visible output excites a number of higher electronic states of the Nd^{3+} ion. These then relax rapidly to the upper ${}^4F_{3/2}$ level of the lasing transition, leading to overall energy conversion values of 1 to 2 percent. Some improvement in this efficiency is possible by pumping with semiconductor diode lasers designed to emit at a strong Nd^{3+} absorption line near 808 nm.

The excited Nd^{3+} ions in the crystal can either emit spontaneously ($\tau_{spon} = 0.55$ ms) or they can be stimulated to emit faster in the electric field produced by other photons propagating back and forth in on optical cavity. The wavelength reflectivity of the cavity mirrors and their spacing L determine which wavelengths satisfy the requirement that the waves be in phase after a cavity round-trip distance of $2L$. In a free-running laser, one gets repeated spikes of emission during the flashlamp pumping duration of 0.2 to 0.3 ms as population builds up and then the system is induced to emit as a lasing threshold is crossed. To obtain a *single* pulse of higher energy, a so-called Q switch is added to the cavity to prevent lasing and to increase the energy stored in the rod. Typically this is a crystal of potassium dihydrogen phosphate (KDP), which has the property of rotating the polarization of the light when a voltage of up to a few kilovolts is applied to it. A polarizer plate is added to the cavity so that only light of a particular polarization angle can propagate in the cavity. By changing the voltage within a few nanoseconds, the polarization rotates rapidly until the switch opens, resulting in all the energy stored in the rod being dumped in a few nanoseconds. The Q-switched pulse has energies of millijoules to joules and peak powers (energy/time) of megawatts or higher.

The output from a Nd:YAG laser has a wavelength of 1064 nm, which corresponds to a photon energy that is too small for electronic excitation of most molecules. However, the high peak power makes it possible to achieve efficient conversion to the green second-harmonic light at 532 nm. This "frequency-doubling" process occurs because the high electric field of the 1064-nm light wave can induce an appreciable dipole μ in a medium due to nonlinear terms in the relation

$$\mu = \alpha E + \beta E^2 + \chi E^3 + \cdots \tag{1}$$

Periodic oscillations in this dipole can act as a source term in the generation of new optical frequencies. Here α is the linear polarizability discussed in Exps. 29 and 35 on dipole moments and Raman spectra, while β and χ are the second- and third-order dielectric susceptibilities, respectively. The quantity β is also called the hyperpolarizability and is the material property responsible for second-harmonic generation. Note that, since $E \sim \cos \omega t$, the E^2 term can be expressed as $\frac{1}{2}(1 + \cos 2 \omega t)$. The next higher nonlinear term χ is especially important in generating sum and difference frequencies when more than one laser frequency is incident on the sample. In the case of coherent anti-Stokes Raman scattering (CARS), χ gives useful information about vibrational and rotational transitions in molecules.

For symmetry reasons, β is zero for isotropic materials such as gases and liquids and for high-symmetry cubic crystals, but it is not zero for anisotropic crystals such as tetragonal potassium dihydrogen phosphate (KDP). By orienting the crystal at the proper angle, the index of refraction for the fundamental and second-harmonic waves can be chosen so that the waves remain in phase over distances of a few centimeters. Under these "phase-matched" conditions, 25 to 50 percent of the fundamental 1064-nm energy can be converted to 532-nm light. This mixing process can be repeated successively in other KDP crystals oriented at other angles to generate the 1064-, 532-nm sum frequency at 355 nm and the 532 second harmonic at 266 nm. A single-pulsed Nd:YAG laser can thus be used to generate four different laser outputs, which span the range from the near-infrared to the ultraviolet region of the spectrum. The 532- and 355-nm outputs are often used to pump dye lasers to provide tunable lasers, which find widespread application in physical chemistry.

Cr^{3+}:Al_2O_3 Ruby System. The transition-metal Cr^{3+} ion has a d^3 electron configuration, and when substituted for an Al^{3+} ion in the Al_2O_3 lattice, the oxygens produce a near-octahedral crystal field that splits the five degenerate d levels into e_g and t_{2g} levels. The lowest energy states derive from the $(t_{2g})^3$ and $(t_{2g})^2 e_g$ configurations of the three d electrons of Cr^{3+}, yielding the level pattern shown in Fig. 3. In the ground $^4A_2$ state, the three electron

FIGURE 3

Energy-level diagram for the Cr^{3+}:Al_2O_3 ruby system.

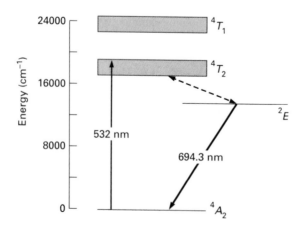

spins are unpaired, but pairing of two of the electrons yields low-lying doublet states in which the multiplicity ($2S + 1$) is 2. Since the site symmetry of Cr^{3+} is actually lower than octahedral due to a slight distortion along one of the cube diagonals in the Al_2O_3 lattice, optical selection rules allow excitation into both the $^4T_1$ and $^4T_2$ levels, corresponding to broad absorption bands at 400 and 550 nm, respectively. This is followed by prompt nonradiative relaxation to the metastable ^2E level within a few nanoseconds. Due to the distortion of the lattice symmetry, this level actually consists of two sublevels split by 29 cm^{-1}, and two lines, termed R_1 and R_2, are seen in the emission (phosphorescence) as the ions return to the ground state. The stronger of these, R_1, corresponds to the famous ruby-laser emission at 693.4 nm.

Since the lower state of the transition is the ground state, ruby is an example of a three-level laser. Achieving a population inversion is more difficult than for a four-level system, and strong pumping of the ruby crystal is necessary. Inversion and lasing is possible for ruby because the radiative lifetime of the ^2E level is quite long, about 10 times that of the metastable state of Nd^{3+} in YAG. This lifetime is also quite temperature dependent and becomes shorter at higher temperatures due to thermal population of the shorter-lived $^4T_2$ electronic state and also due to nonradiative relaxation involving host vibrational modes (phonons). Both aspects will be studied in this experiment.

In the simplest model of the temperature dependence of the lifetime, only the three low-lying $^4A_2$, ^2E, and $^4T_2$ electronic states of Cr^{3+} need be considered. Potential-energy curves of these states in terms of a single "configurational" coordinate are shown schematically in Fig. 4. In this figure, the horizontal coordinate has no precise meaning but roughly represents the displacement of the Cr^{3+} ion from its equilibrium position in the lattice. The relative positions and shapes of the curves have been chosen to match the observed transition frequencies and intensities using Franck–Condon concepts employed for diatomic molecular spectra. Details are given by Fonger and Struck.[3] In comparison to the ground state, the minimum in the $^4T_2$ curve lies to the right, since this state derives from the $(t_{2g})^2 e_g$ configuration in which the Cr–O ligand interaction is presumably weaker. The curvature or shape of the potentials can be related to the effective "vibrations" of the chromium ion, which are of the order of 500 cm^{-1}, i.e., are similar to lattice vibrations (phonons) of the Al_2O_3 host, which range from 200 to 700 cm^{-1}.

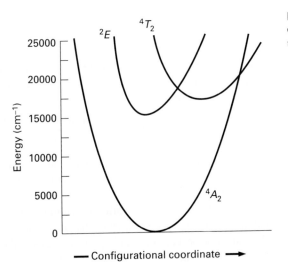

FIGURE 4

Configurational curves for the Cr^{3+} states in ruby.

Following excitation at 532 nm into the $^4T_2$ state at room temperature, most of the excited Cr^{3+} ions fall to the ^2E level, thereafter emitting at 694.3 nm. The resultant phosphorescence intensity decays as

$$\frac{I}{I_0} = \exp\left(\frac{-t}{\tau_E}\right) \tag{2}$$

and the lifetime τ_E gives a direct measure of the radiative decay rate constant A_E, since $A_E = (\tau_E)^{-1}$. At higher temperatures, rapid thermal equilibrium between the nearby ^2E and $^4T_2$ states results in an appreciable fraction of the emission coming from the latter state. Since the approximate $\Delta S = 0$ selection rule is not violated in this case, the radiative rate constant A_T is larger than A_E. Broad fluorescence from the $^4T_2$ level to higher phonon levels of the ground state occurs from 600 to 800 nm and can be distinguished from the sharper 694.3-nm phosphorescence line from the ^2E state. It is significant, however, that the same lifetime is observed for both emission bands,[3] as would be expected if the two upper states are in thermodynamic equilibrium. In this event the relative populations are given by the state degeneracies and a Boltzmann factor:

$$\frac{n_T}{n_E} = \frac{g_T}{g_E} \exp\left(-\frac{E_T - E_E}{kT}\right) \tag{3}$$

In Eq. (3) the number densities are designated n_T and n_E for Cr^{3+} in $^4T_2$ and ^2E states. The degeneracy of a state is the product of the $2S + 1$ spin multiplicity times the electronic degeneracy (1, 2, or 3 for A, E, or T state designations, respectively). Thus the degeneracy is $2 \times 2 = 4$ for the $^2E_2$ state. The degeneracy would be $4 \times 3 = 12$ for the $^4T_2$ state in an octahedral field; however, as mentioned earlier, the actual symmetry is lower due to lattice distortion. The consequence is that the 12 T states divide into two levels of degeneracy 8 and 4, with the latter so-called π states lying 500 cm^{-1} higher than the 8 states denoted as σ. Since only the σ states are important in the relaxation processes,[3] the effective degeneracy ratio g_T/g_E is 2.

From measurements of the optical absorption and emission spectra as a function of temperature, the energy separation $E_T - E_E$ (in cm^{-1}) is $2350 - 0.97T$, with the latter factor accounting for the relative shifts of the E and T states as the lattice expands with increasing temperature.[3] The relative populations are thus given by

$$\frac{n_T}{n_E} = \frac{8}{4} \exp\left[-\frac{hc}{kT}(2350 - 0.97T)\right] = 8.11 \exp\left(-\frac{3380}{T}\right) \tag{4}$$

Both the E and T states can radiate to the ground $^4A_2$ level, so the rate of radiative emission per unit volume will be $A_E n_E + A_T n_T$. Since the total number density of excited atoms is $n_T + n_E$, the radiative decay time is

$$\tau = \frac{n_E + n_T}{A_E n_E + A_T n_T} = \frac{1 + n_T/n_E}{A_E + A_T(n_T/n_E)} \tag{5}$$

At room temperature, n_T/n_E is negligibly small and A_E can be taken as τ^{-1}. This value of A_E can then be used with Eqs. (4) and (5) to calculate A_T from lifetimes measured at higher temperatures.

The extent to which A_T varies with temperature at higher temperatures is an indication that nonradiative processes may contribute to the relaxation of the higher electronic states. This can arise because of coupling of the states to the lattice phonons so that the excitation energy is dissipated in the form of heat. Evidence exists that such a nonradiative process occurs at temperatures above about 200°C, where a crossing from the $^4T_2$ potential curve to that of the ground state is believed to become important.[3] The latter process results in nonradiative cascading down the phonon levels of the ground state and causes a shortening of the overall radiative lifetime. The rate $N(T)$ at which this nonradiative decay process occurs is temperature dependent and can be included in the model by adding it to the A_T term in Eq. (5), yielding

$$\tau = \frac{1 + n_T/n_E}{A_E + [A_T + N(T)](n_T/n_E)} \tag{6}$$

Fonger and Struck[3] have developed a rather complicated model to describe the temperature dependence of $N(T)$, but we will examine the suitability of a much simpler Arrhenius-type expression:

$$N(T) = \mathscr{A} \exp \left[\frac{-hcE_a(\text{cm}^{-1})}{kT} \right] \tag{7}$$

As a physical interpretation of this expression, we can associate the parameter $\mathscr{A}$ with a frequency factor that might approach the frequency of the lattice vibrations, while the E_a term measures an activation energy necessary for ions in the $^4T_2$ state to make the crossing to the $^4A_2$ potential-energy curve.

EXPERIMENTAL

A 10- to 20-Hz Nd:YAG laser is very convenient as an excitation source for this experiment, since the doubled output at 532 nm is near the broad 550-nm ruby absorption and the laser pulse is short (5 to 10 ns) compared to the excited Cr^{3+} radiative lifetime. The student should note the cavity construction of such a laser and, following the directions of the laboratory instructor, adjust the doubling crystal for optimum green output. A 532-nm output of 0.1 to 1 mJ is adequate for the experiment, so a small and relatively inexpensive flashlamp- or diode-pumped pulsed laser is sufficient. Also suitable as an excitation source would be a dye laser operated near 550 nm and pumped by a nitrogen or excimer laser. (An incoherent pulsed source such as a strobe light can also be used if the pulse is about 10 μs or less and if appropriate band-pass filters are used.) **For all laser experiments, safety goggles must be worn to minimize hazard due to the high intensity of these sources.** The instructor will provide instructions about any special features of the lasers and their safe operation.

Rectangular ruby pieces of a few millimeters dimension and with polished sides can be obtained from a variety of laser-rod manufacturers. Also suitable at lower cost are 3- to 5-mm polished ruby ball lenses and small inexpensive ruby-bearing endstones, which are readily available. The Cr^{3+} concentration should be relatively low (0.05 weight percent or less), since energy transfer between nearby chromium ions can affect the radiative decay rate at high concentrations.[4] The color should be pink to pale red and, if desired, the Cr^{3+} concentration can be determined from an absorption measurement using the extinction coefficients in Ref. 5.

Figure 5 shows an aluminum (or copper) sample holder that can serve to hold a 5-mm ruby ball for measurements at temperatures up to 450°C. A snug hole contains a 110-V, 150-W cartridge heater (Omega) that is controlled by a Variac. A type-K thermocouple can be stuffed with glass wool in the hole above the ruby ball to monitor the temperature. A 1-in. steel set screw serves to thermally isolate the block from a support post. A 3-mm hole is drilled through for access of the excitation laser beam, and a second 3-mm hole on

FIGURE 5

Ruby sample holder and experimental arrangement for radiative-lifetime measurements.

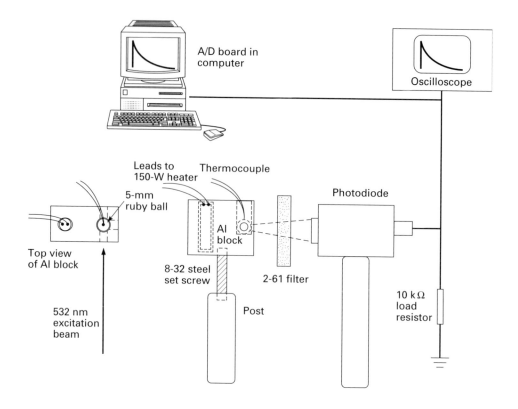

the side face allows the ruby emission to be detected. A red Kopp (formerly Corning) 2-61 optical filter serves to block scattered green light from the detector.

For excitation energies of 0.1 to 1 mJ, the red ruby emission is easily visible to the eye and can be measured by any photomultiplier or by photovoltaic or photoconductive detectors. Care should be taken to operate in a linear range of the detector and to ensure that the detection electronics are fast enough to give an accurate measurement of the decay curves. A time response of a few microseconds is suitable, and this can be obtained by using a 5- to 10-kΩ load resistor in parallel with the detector signal. The response time of the detector and all the electronics can be checked by removing the 2-61 filter and observing the intense pulse of scattered 532-nm light.

If a digital oscilloscope is available, it can be used to average 100 to 1000 decay curves, which can then be transferred to a computer for analysis. However, since the lifetimes are fairly long, in the range from 1 to 10 ms, the ruby decay curves are ideal for measurement with a simple analog-to-digital (A/D) conversion board with a data-acquisition rate of 40 kHz or better. An example is the 100-kHz DAS-1600 board from Keithley, which is easily programmed in Visual BASIC on a personal computer to collect data sequentially at 10-μs intervals. A variety of other A/D boards and software packages is also available. Useful summaries of current hardware and compatible software are also provided in manufacturer catalogs.

After checking the response time of the system, adjust the $t = 0$ signal level to be at about 75 percent of the A/D range and ensure that the baseline is 0 when the green laser is blocked. The decay data should be collected about every 30°C from room temperature to 450°C. Since the decay time shortens as the temperature increases, the data time interval and collection duration should be varied as needed to obtain a good representation of the decay curve. After the series of runs at a given temperature is averaged, adjust the Variac slightly and transfer the data to a spreadsheet for analysis while waiting for the temperature of the block to increase. It is not necessary to wait for complete temperature equilibration if the rate of change is slow compared to the data-collection time and if the temperature is recorded at about the midpoint of the data-averaging period.

CALCULATIONS

For each set of decay data, make a plot of $\ln I$ versus t and examine it for linearity. Typically the decay at early times may show deviations due to scattered green light, and the data at long times will be noisy due to low signal levels. Choose the central linear portion of the curve and do a linear regression to obtain the decay time τ at each temperature. Make a plot of τ versus T.

Use the τ values from room temperature to 100°C to calculate an average value of the radiative rate A_E on the assumption that all emission comes from the ^2E state. Then use this average with Eqs. (4) and (5) to calculate A_T at each temperature from 100 to 200°C. Values of both A_E and A_T will show some variation due to inadequacies in the relaxation model. An alternative, and preferable, option is to fit these low-temperature data to Eq. (5) using the "Optimizer" feature of spreadsheets such as Quattro Pro to obtain the "best-fit" values of A_E and A_T, which give the lowest sum for the residuals $(\tau_{exp} - \tau_{calc})^2$ for this data range. In either case use your average or best-fit values of A_E and A_T in Eq. (5) and plot the predicted τ-versus-T line on your graph of τ_{exp} data. A marked deviation should be apparent for the data above 200°C. Using Eq. (6) and your average or best-fit values for A_E and A_T, calculate $N(T)$ for all temperatures higher than 200°C. Carry out a linear regression of $\ln N(T)$ versus $1/T$ and determine the parameters in the Arrhenius expression. Use these parameters to generate $N(T)$ values in Eq. (6) and add the curve predicted by Eq. (6) to your data plot.

DISCUSSION

Comment in your report on the extent to which the simple model presented above is or is not consistent with your data. How constant are A_E and A_T, and what effect would a nonradiative decay from the ^2E level directly to the ground state have on the lifetimes? Do your high-temperature data support the picture of an Arrhenius temperature dependence for the nonradiative decay from the $^4T_2$ state? Compare your value of $\mathcal{A}$ in this expression with a phonon vibrational frequency corresponding to 500 cm^{-1}. Is your value of E_a of reasonable magnitude?

Extension of the lifetime measurements to 77 K can be done easily by immersing the ruby sample in liquid nitrogen in a clear glass Dewar. In addition, spectral studies of the absorption and emission of ruby can be used to determine the Cr^{3+} dopant levels[3,5] and, at higher concentrations, the spectra of chromium-chromium pairs.[4]

SAFETY ISSUES

All laser sources should be treated with caution, since a stray reflection into an eye can result in serious damage. This is especially true with pulsed lasers, where the natural tendency to close the eyelid to a bright source is much too slow to provide protection. Laser goggles must be worn during this experiment, and care should be taken to eliminate or block all reflected beams.

APPARATUS

Nd:YAG laser with second-harmonic output of 1 mJ and a pulse duration of 5 to 10 ns (e.g., MiniLite model from Continuum, Inc.) or dye laser system with similar output near 550 nm; laser safety goggles; ruby samples such as 5-mm ball lenses from Edmonds Scientific or ruby endstones from the Swiss Jewel Company; aluminum or copper sample holder as in Fig. 5; 150-W cartridge heater (Omega) with Variac; type-K thermocouple with digital readout (Omega); photomultiplier (e,g., RCA 1P28 or equivalent) or photodiode detector (e.g., model PDA50 with built-in preamplifier and power supply, available from Thor Labs, Newton, NJ 07860); digitizing oscilloscope (e.g., Tektronix DPS 310 or equivalent) or analog oscilloscope plus analog-to-digital conversion board (e.g., Keithley MetraByte DAS series) in a microcomputer; data-collection and spreadsheet software.

REFERENCES

1. G. R. Van Hecke and K. K. Karukstis, *A Guide to Lasers in Chemistry,* Jones & Bartlett, Sudbury, MA (1998).

2. D. K. Evans, *Laser Applications in Physical Chemistry,* Marcel Dekker, New York (1989).

3. W. H. Fonger and C. W. Struck, *Phys. Rev.* **B11,** 3251 (1975).

4. G. F. Imbusch, *J. Lumin.* **53,** 465 (1992).

5. D. C. Cronmeyer, *J. Amer. Opt. Soc.* **56,** 1703 (1966).

GENERAL READING

D. L. Andrews, *Lasers in Chemistry,* 2d ed., Springer-Verlag, Berlin (1990).

J. K. Steehler, *J. Chem. Educ.* **67,** A37, A65 (1990).

G. R. Van Hecke and K. K. Karukstis, *A Guide to Lasers in Chemistry,* Jones & Bartlett, Sudbury, MA (1998).

R. N. Zare, B. H. Spencer, D. S. Springer, and M. P. Jacobson, *Laser Experiments for Beginners,* University Science Books, Sausalito, CA (1995).

EXPERIMENT 45
Spectroscopic Properties of CdSe Nanocrystals

The extremely small crystallites known as nanocrystals (NC), and sometimes called quantum dots, present a wide range of interesting scientific issues and have a considerable variety of practical applications, including biological labeling.[1,2] Although there are some metallic NCs of interest, the overwhelming majority is semiconductors. Upon excitation with radiation, semiconductor NCs produce electron-hole pairs, which, when they recombine, emit light. Thus, one can observe interesting absorption and fluorescence emission peaks in the spectra. The wavelengths of these peaks lie in the visible, and the color (wavelength) of the light will vary for particles of 1- to 4-nm radius, a size regime where quantum effects become apparent.

One of the most widely studied types of semiconductor NCs is CdSe. The hexagonal wurtzite structure that pertains for large CdSe crystals is shown in Fig. 1*a,* and a model of a nanocrystal is pictured in Fig. 1*b.* The size of such NCs can be controlled by reaction

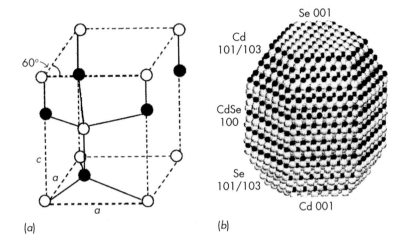

(a)

(b)

FIGURE 1
(*a*) The wurtzite crystal
structure of bulk CdSe, which
exhibits tetrahedral fourfold
coordination for both the
Cd and Se atoms, has the
hexagonal unit cell shown
here with lattice parameters
$a = 0.43$ nm and $c = 0.70$ nm
and containing two Cd and
two Se atoms. (*b*) Model of
a CdSe nanocrystal based
on transmission electron
micrographs, with Miller
indices shown for the low-
energy faces (from Ref. 1).
For core-shell NCs, this CdSe
core is coated by a shell of
one or more epitaxial layers
of ZnS.

kinetics and, to prevent agglomeration once formed, the surfaces are typically passivated by attaching organic ligands or by coating the particle with an epitaxial layer of ZnS or CdS.[3,4] The latter *core-shell* NCs exhibit better quantum yields (brighter luminescence) than ligand-capped *core* NCs and have improved photostability. However, in the present experiment, simple ligand-capped CdSe nanocrystals will be prepared with radii of 1 to 2 nm. These NCs contain 75–600 CdSe ion pairs and produce well-defined spectral peaks with wavelengths λ of 450–585 nm.

THEORY

In Exp. 34 on the absorption spectra of conjugated dyes, the model of a free electron in a one-dimensional box of length L is used to obtain the energy level expression

$$E_n = \frac{h^2 n^2}{8 m_e L^2} \tag{1}$$

Extension of this model to an electron confined to a three-dimensional rectangular box gives a similar expression for each of the x, y, and z coordinates, with the total energy given by the sum. For a spherical box, the angular part of the solution is identical to that of the hydrogen atom, but, since the potential energy is zero inside the sphere and infinite outside, the radial part of the wavefunction must go to zero at the surface of the sphere. The solution for an electron in a spherical box involves spherical Bessel functions and Kauzmann[5] gives the energies of the two lowest levels as

$$E_1 = \frac{h^2}{8 m_e a^2} \quad \text{and} \quad E_2 = 2.04 \frac{h^2}{8 m_e a^2} \tag{2}$$

where a is the radius of the sphere. The wavefunction for the lowest state is like that of an *s* orbital but with a radial part that goes smoothly to zero at the surface. That for the second level is a *p*-type orbital, which has a nodal plane and also goes to zero at the surface. It is triply degenerate, as is the case for the *p* orbitals in the hydrogen atom. Some higher levels have *d*, *f*, . . . angular shapes and some also have radial nodes as in the 2*s*, 3*s* hydrogen cases. Note that the above models predict an increase in the level spacing proportional to $1/a^2$ as the size a decreases.

For a more accurate description of the energy levels of CdSe NCs, a very brief review of the band theory for bulk semiconductors[6] is useful. All semiconductors have a band gap

FIGURE 2

Schematic parabolic band structure for CdSe, which has a band gap E_g of 1.75 eV. The conduction band is labeled C, and several valence bands (V_i) are shown. The filled and open circle symbols indicate the position of quantized k values $n\pi/a_1$ allowed for the $n = 1$ and $n = 2$ states of an NC with radius a_1. The solid arrow shows the $n = 1$ transition in which an electron is excited and a hole is created (open circle). The dashed arrow shows how the position of this $n = 1$ transition would change for a nanocrystal of smaller radius a_2. (Adapted from Ref. 7.) This simple diagram is for the cubic zinc blend structure; the hexagonal wurtzite structure has a small gap at $k = 0$ between the V_1 and V_2 bands.

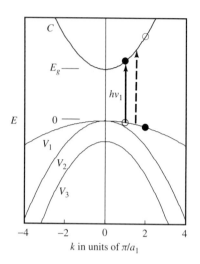

E_g between the valence band and the conduction band. The band structure of a direct II–VI intrinsic semiconductor like CdSe can be represented reasonably well by a parabolic band model like that shown schematically in Fig. 2. Here, $k = \pi/r$ is the wave vector and r is the radial distance from an arbitrary origin in the center of the crystal. The kinetic energy of the electron is proportional to k^2 and the energy minimum of the conduction band and the maxima of the valence bands occur at $k = 0$ (corresponding to $r = \infty$ in a bulk sample).

The valence bands, which arise from p atomic orbitals, have sixfold degeneracy and contain at 0 K the six p-orbital Se valence electrons. Due to spin-orbit coupling, this degeneracy is split at $k = 0$ into a fourfold degenerate $J = 3/2$ band and a twofold degenerate $J = 1/2$ "split-off" valence band V_3, where J is the total unit cell angular momentum. For $k \neq 0$, the $J = 3/2$ band splits into two doubly degenerate components: the heavy hole valence band V_1 and the light hole valence band V_2.

If one sets the zero of energy at the top of the V_1 valence band, then, according to band theory, the dispersion relations (i.e., the $E - k$ relations) for the conduction band $E_C(k)$ and the valence band $E_V(k)$ are

$$E_C(k) = E_g + \frac{h^2}{8\pi^2 m_e^*} k^2 \tag{3}$$

$$E_V(k) = -\frac{h^2}{8\pi^2 m_h^*} k^2 \tag{4}$$

Here m_e^* and m_h^* are the *effective masses* of the electron and the positive hole created when an electron is excited from the valence band to the conduction band. In bulk CdSe, m_e^*/m_e and m_h^*/m_e have been determined to be about 0.12 and 0.5, respectively.[8] The reduced mass $\mu = m_e^* m_h^* /(m_e^* + m_h^*) = 0.097 m_e$ is thus much smaller than the mass of an electron. It is seen that the form of equations (3) and (4) is similar to that for a free particle in a sphere but with a negative sign for the energies of the hole.

What will be the effect of the finite NC size on the electronic states of CdSe? The electronic structure of a CdSe NC can be semiquantitatively represented by a simple effective mass parabolic-band model, in which the finite NC size quantizes the allowed k values to $k_{NC} = n \pi/a$ with $n = 1, 2, \ldots$. A smaller NC radius shifts the first allowed state to a larger k value and increases the separation between conduction and valence states, as indicated in Fig. 2. Thus, the fundamental ($n = 1$) frequency of a photon that excites an electron into the conduction band is given by the parabolic-band model as

$$hv_1 = \Delta E = E_e - E_h = E_g + \frac{h^2}{8m_e^*a^2} + \frac{h^2}{8m_h^*a^2} = E_g + \frac{h^2}{8\mu a^2} \qquad (5)$$

Ignored in the above quantum confinement model is the fact that when an electron is excited into the conduction band and a positive hole is left behind in the valence band, the hole and electron are coulombically attracted. The pair can be treated as a well-defined quasiparticle called an *exciton*, a hydrogen-like system for which a bulk exciton Bohr radius $a_B = h^2\kappa\varepsilon_0/\pi\mu e^2$ can be defined. Here κ is the dielectric constant (10.2 for bulk CdSe[8]) and ε_0 is the permittivity of space (8.854×10^{-12} $C^2m^{-1}J^{-1}$). The quantity a_B is a measure of the size of the electron-hole pair orbital (just like the $n = 1$ orbital size of the electron-proton pair in a H atom). The value of a_B is 5.6 nm for a bulk CdSe exciton, versus 0.053 nm for the orbital size of a hydrogen atom.

For a nanocrystal, the basic model for an electron-hole exciton was first proposed by Brus[1,9] and independently by Efros.[10] A number of simplifying assumptions are made in this model: (1) the NC is taken to be a sphere of radius a, that is, the actual shape shown in Fig. 1b is simplified; (2) the interior of the NC is assumed to be a uniform dielectric medium with dielectric constant κ, that is, the excited electron and the hole are taken to be the only charged particles in this sphere; and (3) the potential energy is assumed to jump from zero inside the NC to infinity outside. The Schrödinger equation for the Brus strong confinement model can be solved analytically to yield hydrogen-like wavefunctions and the energy needed to create the exciton in its ground ($n = 1$) state.[1,9] Expressing all terms in wavenumber units, the resulting transition energy is

$$\tilde{v}_1 = \Delta E/hc = \Delta\tilde{E} = \tilde{E}_g + \tilde{E}_{pol} - \frac{1.8e^2}{4\pi hc\varepsilon_0\kappa a} + \frac{h}{8c\mu a^2} \qquad (6)$$

where $\tilde{v}_1(cm^{-1}) \equiv 1/\lambda(cm) = 10^7/\lambda(nm)$ is the wavenumber of the transition frequency for the absorption or luminescence fundamental (i.e., $n = 1$) band of an NC. The e^2/a term is an approximate coulombic interaction term, as calculated in first-order perturbation theory, and the term $\tilde{E}_{pol}$ is a small polarization energy whose size dependence can be ignored.[1] For NCs of radius a much smaller than $a_B = 5.6$ nm, one could also ignore the size variation of the coulombic contribution so that Eq. (6) would then reduce to the approximate form

$$\tilde{v}_1 \approx \tilde{E}_0 + h/8c\mu a^2 \qquad (7)$$

with $\tilde{E}_0 = \tilde{E}_g + \tilde{E}_{pol} - (1.8e^2/4\pi hc\varepsilon_0\kappa)<1/a>$, where $<1/a>$ is an average over the range of NC radii studied. The term $\tilde{E}_0$ can be treated as a constant independent of the NC radius a since its value is mostly due to $\tilde{E}_g$ and the weak $1/a$ variation of the coulombic term has been suppressed over a short range of a values. Note that Eq. (7) has the same $1/a^2$ dependence as the parabolic-band model, Eq. (5), but the constant terms differ.

A more exact treatment of the electronic states of NCs requires nonparabolic bands that can be coupled to each other, a treatment of the energy dependence of the effective masses, consideration of both the nonsphericity of the NCs and the leakage of the wavefunction out of the confines of the NC, and inclusion of electron-hole exchange.[4,11] Although these are important refinements, the remarkable thing is the fact that many essential features of NC spectroscopy can be captured by models as simple as the effective-mass parabolic band and Brus strong confinement descriptions.

It will be noted experimentally that the peak wavelength in the luminescence emission spectrum is typically about 4–5 nm larger than that for the $n = 1$ exciton absorption peak. This Stokes shift is generally attributed to the spin fine structure of exciton states;[12] similar but generally smaller Stokes shifts are seen in molecular electronic spectra.

Equations (6) and (7) predict a transition whose width will be small if the NC sample is monodisperse (i.e., all particles have the same a value). However, even samples with a narrow size distribution show transition line widths of $\sim$30 nm FWHM (full width half maximum), which is significantly broader than expected. This line broadening is due to both homogeneous broadening related to phonon excitations in the NC and inhomogeneous broadening due to interactions with the surrounding environment.[13]

EXPERIMENTAL

Synthesis. CdSe nanocrystals of various sizes are available commercially but can also be synthesized easily from CdO and elemental Se using a kinetic growth method, where the size is controlled by time of reaction.[14,15] A stock solution of Se can be prepared in advance by adding 60 mg of Se and 10 mL octadecene to a 50-mL round-bottom flask over a stirrer hot plate. **Warning: all operations with Se and CdO should be done in a fume hood using gloves. Cd compounds are classified as potential carcinogens and should not be ingested or inhaled.** Measure by syringe 0.8 mL of trioctylphosphine (a coordinating ligand to enable dissolution of the Se and to coat the surface of the final NCs) and add this to the flask, along with a magnetic stir bar. Warm and stir the solution as necessary to dissolve the selenium completely, then cool to room temperature. This stock solution is sufficient for five preparations and can be stored in a sealed container at room temperature for at least a week.

Each student group should prepare a fresh sample of Cd precursor by placing 26 mg of CdO into a 50-mL round-bottom flask clamped in a heating mantle in a fume hood. To the same flask, add 20 mL of octadecene and add by pipette or syringe 1.2 mL of oleic acid (a coordinating ligand). Insert a thermometer capable of measuring up to 225°C, and heat the cadmium solution. As soon as the temperature reaches 225°C, use a clean and dry pipette to quickly transfer 2 mL of the room-temperature selenium solution into the hot cadmium solution, stir the mixture quickly with a glass rod, and start timing. As the CdSe particles grow in size, remove approximately 2-mL samples quickly (at about 10-second intervals) using a 9-inch glass Pasteur pipette. Quench each sample by placing it into a 2-mL vial at room temperature. After the first five samples have been removed, take five more at longer times such as 65, 85, 110, 140, and 180 seconds. (There should be a noticeable color change between samples.) It is useful to have a team of two persons, one to record time and the other to quench the samples. The time of removal should be recorded for each of the ten samples removed within 3 minutes of the initial injection. The sizes of CdSe nanocrystals prepared in this manner will range from 1 to 2 nm in radius.

As an alternative to synthesizing the NCs as described above or as a way to augment the size range obtained from the synthesized particles, commercially available CdSe nanocrystals[2] can be used. Such samples, which can be either core type (ligand capped) or core-shell type (with a ZnS epitaxial layer coating the CdSe core), are obtained as dilute solutions, or actually stable dispersed suspensions, in an organic solvent, which is usually toluene. If commercial samples are used to augment the size range of home-grown particles, core NCs should be used. Two or three such samples with radii a in the range 1.6–3.0 nm are recommended. If only commercial samples are to be studied, five or six samples with radii spanning the range from $\sim$1 to $\sim$3 nm should be studied and either core or core-shell NCs can be used. The procedure for handling commercial samples is the same as that described below except that the solvent is changed from octadecene to toluene.

Spectroscopy. For absorption spectra, the concentration of the NC solutions should be such that a 1-cm path length glass cuvette has sufficient absorbance to show a well-defined maximum at the $n = 1$ exciton peak. Thus, the peak absorbance $A = \log(I_0/I)$ should be in the range 0.3 to 0.5; if necessary, dilute with octadecene solvent as needed, making use of the Beer-Lambert law, which shows that A is proportional to the concentration c. For the fluorescence spectra, dilute each solution so that A at the peak in the absorption spectrum is close to 0.1. This dilution is necessary so that photons emitted by a CdSe nanocrystal will not be reabsorbed by other nanocrystals nearby. If commercial samples are to be studied, dilute with the appropriate solvent (typically toluene).

Any suitable absorption and emission spectrometers can be used for these measurements. An Agilent 8453 diode array instrument is convenient for absorption spectra, and a Perkin Elmer LS50B spectrometer is useful for the emission spectra. Ocean Optics instruments such as the USB 2000 can also be used. Detailed operating procedures will be provided by the instructor, but a few general points will be discussed here. The spectrometer source lamp should be turned on at least 10 minutes prior to use so that it will achieve intensity stability. Absorption spectra should be obtained over the 350–700 nm wavelength range. The first step is to obtain a reference or background absorption spectrum of the solvent using a glass cuvette; do not use a plastic cuvette since it will be attacked by the solvent. This background will be subtracted from the absorption spectrum obtained for each of the NC samples. The fluorescence spectra should be measured over the 400–700 nm range using suitable bandwidths and integration or scan times to achieve a decent signal-to-noise ratio. The excitation radiation wavelength should be less than 400 nm; the exact wavelength is not critical but, if convenient, some experimentation to produce maximum emission can be made.

Size determination.   If commercial samples are used, the mean size of the nanocrystals will typically be provided by the supplier. The size of nanocrystal particles can be determined by scanning electron microscopy, transmission electron microscopy, dynamic light scattering studies, small-angle x-ray scattering, or scanning tunneling microscopy. From such measurements, Yu et al.[14] have determined the following empirical relation between the radius a (in nm) of CdSe nanocrystals and the wavelength λ (in nm) of the first (reddest) absorption maximum:

$$a = 0.8061 \times 10^{-9}\lambda^4 - 1.3288 \times 10^{-6}\lambda^3 + 0.8121 \times 10^{-3}\lambda^2 - 0.2139\lambda + 20.79 \quad (8)$$

You may use this empirical relation to determine the NC size for each of your synthesized samples. Note that such a values are *not* based on the quantum models we wish to test.

CALCULATIONS

From the absorption spectra of the various NC samples studied, determine the wavelengths λ of the $n = 1$ exciton peaks. It will be observed that the exciton peak lies on a sloping baseline due to the presence of other absorption processes that are dominant at shorter wavelengths. In order to obtain the best wavelength value for the maximum of the exciton peak, this sloping baseline should be subtracted from the spectrum. Then use Eq. (8) to obtain values for the radii a. Plot the transition energy $\tilde{\nu}_1$ $(\text{cm}^{-1}) \equiv 1/\lambda(\text{cm}) = 10^7/\lambda(\text{nm})$ for the $n = 1$ fundamental band versus $1/a^2$ and carry out a least-squares fit to these data with Eq. (7). To test the importance of retaining the explicit $1/a$ dependence of the coulombic term shown in Eq. (6), calculate this term for each a value and subtract it from $\tilde{\nu}_1$, yielding

$$\tilde{\nu}_i + \frac{1.8e^2}{4\pi hc\varepsilon_0 \kappa a} = \tilde{E}_g + \tilde{E}_{pol} + \frac{h}{8c\mu a^2} \quad (9)$$

and then do a least-squares fit of the left-hand side of Eq. (9) versus $1/a^2$. Report the values of the fit parameters with their standard deviations for both Eqs. (7) and (9). Do the fits clearly favor the use of one equation over the other?

From the emission spectra, determine the wavelength of the emission peak and the full width at half maximum (FWHM) for each sample. List these λ_{em} values in a table together with λ_{abs} values obtained from the $n = 1$ absorption peak, and list also the differences $\lambda_{em} - \lambda_{abs}$. Plot $\tilde{\nu}_{em}$ versus $1/a^2$, using the same a values as determined from the absorption maxima, and fit these emission data with Eqs. (7) and (9).

DISCUSSION

Compare the emission spectrum fit parameters with the absorption spectrum fit parameters. Do they agree within the 95 percent confidence limits obtained?

For the range of sizes studied, compare the contribution of the coulombic term with that from the $1/a^2$ quantum confinement term. From the $h/8c\mu$ values obtained by fitting with Eqs. (7) and (9), calculate μ/m_e values and compare these values with the value 0.097 for an exciton in bulk CdSe.

According to Eq. (9), the intercept value is $\tilde{E}_g + \tilde{E}_{pol}$. Estimate $\tilde{E}_{pol}$ using the bulk CdSe band gap value of 1.75 eV (14,110 cm^{-1}). From the two-parameter fit with Eq. (7), the intercept can be taken to be $\tilde{E}_0 = \tilde{E}_g + \tilde{E}_{pol} - (1.8e^2/4\pi hc\varepsilon_0\kappa)<1/a>$, where $<1/a>$ is an average over the range of NC radii studied. Using a suitable $<1/a>$ average value, calculate $\tilde{E}_{pol}$ and compare it with the value obtained from Eq. (9). Is the assumption that the polarization energy is small compared to $\tilde{E}_g$ justified?

It should be noted that spectroscopic data obtained over a wide range of a values show that $\tilde{\nu}_1$ is not quite linear when plotted against $1/a^2$, and the observed curvature is greater than that due to the coulombic term shown explicitly in Eq. (9).[16] Such curvature is due to nonparabolic band contours and to a breakdown of the effective mass approximation. As a result, the μ/m_e values obtained with Eqs. (7) and (9), and to a lesser extent the $\tilde{E}_{pol}$ value, will depend on the range of a values used in the fits.

As clusters become smaller, one would expect surface effects to become more important. Explore this by calculating the number of CdSe pairs contained in the smallest NC studied. What fraction of the CdSe pairs are on the surface of that particle?

SAFETY ISSUES

Cadmium compounds are classified as potential carcinogens and should not be ingested or inhaled. All operations with Se and CdO must be done in a fume hood using gloves.

APPARATUS

Suitable absorption/emission spectrometers such as Ocean Optics UV-Vis USB2000, Agilent 8453 diode array, Perkin Elmer LS50B; several glass cuvettes. Selenium; cadmium oxide; trioctylphosphine; oleic acid; 1-octadecene.

Various pipettes or syringes from 1 mL to 20 mL; two 50-mL round-bottom flasks; stir bar; two heating mantles; 9-inch glass Pasteur pipette with bulb; thermometer or thermocouple for measurements up to 225°C; 2-mL vials. Commercial suspensions of core and core-shell CdSe nanocrystals of different sizes are available from Evident Technologies (www.evidenttech.com), Nanomaterials and Nanofabrication Laboratories (www.nn-labs.com), or Invitrogen (http://probes.invitrogen.com).

REFERENCES

1. T. Kippeny, L. A. Swafford, and S. J. Rosenthal, *J. Chem. Educ.* **79,** 1094 (2002).

2. E. M. Boatman, G. C. Lisensky, and K. J. Nordell, *J. Chem. Educ.* **82,** 1697 (2005); L. D. Winkler, J. F. Arceo, W. C. Hughes, B. A. DeGraff, and B. H. Augustine, *J. Chem. Educ.* **82,** 1700 (2005). See also information on websites of commercial sources, e.g., www.evidenttech.com; www.nn-labs.com; http://probes.invitrogen.com.

3. X. Peng et al., *J. Am. Chem. Soc.* **119,** 7019 (1997); B. O. Dabbousi et al., *J. Phys. Chem. B* **101,** 9463 (1997).

4. M. Nirmal and L. Brus, *Acc. Chem. Res.* **32,** 407 (1999).

5. W. Kauzmann, *Quantum Chemistry,* p. 188, Academic Press, New York (1957); see also R. S. Berry, S. A. Rice, and J. Ross, *Physical Chemistry,* 2d ed., p. 84, Oxford University Press, New York (2000).

6. R. Dalven, *Introduction to Applied Solid State Physics,* 2d ed., pp. 1–24, Plenum, New York (1990).

7. C. B. Murray, C. R. Kagan, and M. G. Bawendi, *Annu. Rev. Mater. Sci.* **30,** 545 (2000).

8. *Landolt-Bornstein Numerical Data and Functional Relationships in Science and Technology. New Series,* vol. III-17b, pp. 202, 219, Springer-Verlag, New York (1982).

9. L. Brus, *J. Chem. Phys.* **79,** 5566 (1983); *loc. cit.* **80,** 4403 (1984).

10. Al. L. Efros and A. L. Efros, *Sov. Phys. Semicond.* **16,** 772 (1982).

11. Al. L. Efros and M. Rosen, *Annu. Rev. Mater. Sci.* **30,** 475 (2000).

12. D. J. Norris, Al. L. Efros, M. Rosen, and M. G. Bawendi, *Phys. Rev. B* **53,** 16347 (1996).

13. S. Empedocles and M. G. Bawendi, *Acc. Chem. Res.* **32,** 389 (1999).

14. W. Yu, L. Qu, W. Guo, X. Peng, *Chem. Mater.* **15,** 2845 (2003). See also www.mrsec.wisc.edu/Edetc/nanolab/CdSe/index.html.

15. E. M. Boatman, G. C. Lisensky, and K. J. Nordell, *J. Chem. Educ.* **82,** 1697 (2005).

16. D. J. Norris and M. G. Bawendi, *Phys. Rev. B* **53,** 16338 (1996).

GENERAL READING

T. Kippeny, L. A. Swafford, and S. J. Rosenthal, *J. Chem. Educ.* **79,** 1094 (2002).

XV

Solids

EXPERIMENTS

46. Determination of Crystal Structure by X-Ray Diffraction
47. Lattice Energy of Solid Argon
48. Statistical Thermodynamics of Iodine Sublimation

EXPERIMENT 46
Determination of Crystal Structure by X-Ray Diffraction

The object of this experiment is to determine the crystal structure of a solid substance from x-ray powder diffraction patterns. This involves determination of the symmetry classification (cubic, hexagonal, etc.), the type of crystal lattice (simple, body-centered, or face-centered), the dimensions of the unit cell, the number of atoms or ions of each kind in the unit cell, and the position of every atom or ion in the unit cell. Owing to inherent limitations of the powder method, only substances in the cubic system can be easily characterized in this way, and a cubic material will be studied in the present experiment. However, the recent introduction of more accurate experimental techniques and sophisticated computer programs make it possible to refine and determine the structures of crystals of low symmetry from powder diffraction data alone.

Knowledge of the crystal structure permits determination of the coordination number (the number of nearest neighbors) for each kind of atom or ion, calculation of interatomic distances, and elucidation of other structural features related to the nature of chemical bonding and the understanding of physical properties in the solid state.

THEORY

A perfect crystal constitutes the repetition of a single very small unit of structure, called the *unit cell,* in a regular way so as to fill the volume occupied by the crystal (see Fig. 1).

A crystal may for some purposes be described in terms of a set of three *crystal axes* **a, b,** and **c,** which may or may not be of equal length and/or at right angles, depending on the symmetry of the crystal. These axes form the basis for a coordinate system with which the crystal may be described. An important property of crystals, known at least a century before the discovery of X rays, is that the crystal axes for any crystal can be so chosen that all crystal faces can be described by equations of the form

$$hx + ky + lz = \text{positive constant} \tag{1}$$

where x, y, and z are the coordinates of any point on a given crystal face, in a coordinate system with axes parallel to the assigned crystal axes and with units equal to the assigned axial lengths a, b, c; and where h, k, and l are *small integers,* positive, negative, or zero. This is known as the law of rational indices. The integers h, k, and l are known as the *Miller indices* and, when used to designate a crystal face, are ordinarily taken relatively prime (i.e., with no common integral factor). Each crystal face may then be designated by three Miller indices hkl, as shown in Fig. 1. The law of rational indices historically formed the strongest part of the evidence supporting the conjecture that crystals are built up by repetition of a single unit of structure, as shown in Fig. 1.

The crystal axes for a given crystal may be chosen in many different ways; however, they are conventionally chosen to yield a coordinate system of the highest possible symmetry. It has been found that crystals can be divided into six possible systems on the basis of the highest possible symmetry that the coordinate system may possess as a result of the symmetry of the crystal. This symmetry is best described in terms of symmetry restrictions governing the values of the axial lengths a, b, and c and the interaxial angles α, β, and γ.

The crystal systems are as follows:[1]

Triclinic system: No restrictions.

Monoclinic system: No restriction on lengths a, b, c; however, $\alpha = \gamma = 90°; \beta \neq 90°$.

Orthorhombic system: No restrictions on a, b, c; however, $\alpha = \beta = \gamma = 90°$.

Tetragonal system: $a = b \neq c; \alpha = \beta = \gamma = 90°$.

Hexagonal system: *Hexagonal division:* $a = b = b' \neq c; \alpha = \beta = 90°, \gamma = \gamma' = 120°$ (there being three axes **a, b,** and **b'** in the basal plane, at $120°$ angular spacing; the **b'** axis is redundant). *Rhombohedral division:* $a = b = c; \alpha = \beta = \gamma \neq 90°$.

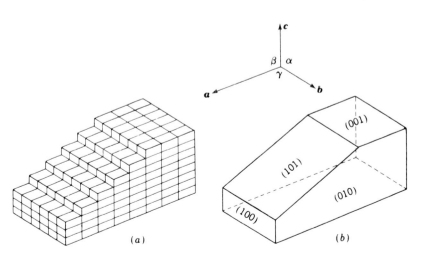

FIGURE 1
(*a*) Schematic diagram of a crystal, showing unit cells; (*b*) same crystal, showing axes and Miller indices.

planes. If, however, there is a very large (or "infinite") number of reflecting planes, the reflection from any given plane is canceled by that from another plane a considerable distance away. Reflection should be observed therefore only for values of θ extremely close to those that satisfy the Bragg equation, and this is experimentally found to be the case.

We shall now modify the treatment slightly by taking note of the fact that a second-order reflection ($n = 2$) from the planes $Q, R, S, T, \ldots$ corresponds to a hypothetical first-order reflection from the planes $Q, U, R, V, S, W, T, X, \ldots$, only half of which contain lattice points. By inserting the required number of additional equidistant parallel planes containing no lattice points, we can dispense with the order n and write the Bragg equation in the following form, which is the form that will be used henceforth in this discussion,

$$\lambda = 2d \sin \theta \tag{3}$$

where d is now the interplanar spacing of the new set of planes. We shall now say that Eq. (3) is the necessary condition for reflection of X rays of wavelength λ from a set of crystallographic planes with interplanar spacing d. A *set of crystallographic planes* is defined as an infinite array of equidistant and finitely spaced parallel planes so constructed that every point of the given crystal lattice lies on some plane of the set (though it is not necessarily true that all planes in the set contain lattice points). It is evident that sets of crystallographic planes may be constructed for a given crystal lattice in many ways (indeed, an infinite number of ways). The equations for such a set of planes are

$$
\begin{aligned}
&\vdots \\
hx + ky + lz &= -2 \\
hx + ky + lz &= -1 \\
hx + ky + lz &= 0 \\
hx + ky + lz &= 1 \\
hx + ky + lz &= 2 \\
&\vdots \\
hx + ky + lz &= N \\
&\vdots
\end{aligned}
\tag{4}
$$

where x, y, and z are coordinates as previously defined and where, when the coordinate system corresponds to a primitive unit cell, the coefficients h, k, and l may take on any integral values: positive, negative, or zero. They may be called the *Miller indices* of the set of crystallographic planes and are related to Miller indices as applied to crystal faces. (However, they need not be taken relatively prime.) The different integral values of N, ranging from $-\infty$ to ∞, define different planes in the set. That any given lattice point must lie on one of the planes, in the case of a coordinate system corresponding to a primitive unit cell, is easily seen from the fact that its coordinates xyz must be integers mnp; since the Miller indices are integers, the quantity $hx + ky + lz$ must be an integer, and one of the equations in the set of Eqs. (4) is satisfied.

If the coordinate system is not chosen in correspondence to a primitive unit cell, not all the lattice points have coordinates that are integers and all lattice points will lie on planes in the set only if certain restrictions are placed on the combinations of values that the Miller indices hkl may assume.

For a body-centered lattice I, some of the lattice points have coordinates that are expressible as integers mnp and some have coordinates that must be expressed as half integers $m + \frac{1}{2}, n + \frac{1}{2}, p + \frac{1}{2}$. For the latter,

$$hx + ky + lz = hm + kn + lp + \tfrac{1}{2}(h + k + l)$$

and for this quantity to be an integer, it is necessary that $h + k + l$ be even.

For a face-centered lattice F, some of the lattice points are at $xyz = mnp$; some are at $m, n + \frac{1}{2}, p + \frac{1}{2}$; some are at $m + \frac{1}{2}, n, p + \frac{1}{2}$; and some are at $m + \frac{1}{2}, n + \frac{1}{2}, p$; where $m, n,$ and p are in each case any three integers. In order that $hx + ky + lz$ be an integer, it is evidently necessary that $k + l, h + l,$ and $h + k$ simultaneously be even. An equivalent restriction is easily seen to be that $h, k,$ and l must be either all even or all odd.

When these restrictions are not obeyed, no reflections can be obtained from the set of crystallographic planes under consideration, for there will be lattice points lying between the planes and scattering out of phase with those in the planes, resulting in complete cancellation due to destructive interference. By observing experimentally what sets of planes reflect X rays, one can deduce what the restrictions are and thereby deduce the lattice type.

The interplanar distance d is determined by the Miller indices hkl. For the cubic system it is easy to show by analytical geometry that

$$d = \frac{a_0}{\sqrt{h^2 + k^2 + l^2}} = \frac{a_0}{M} \tag{5}$$

where a_0 is the length of the edge of the unit cube and

$$M^2 \equiv h^2 + k^2 + l^2 \tag{6}$$

Lattice Type.   From the angles at which X rays are diffracted by a crystal, it is possible to deduce the interplanar distances d using Eq. (3). To determine the lattice type and compute the unit-cell dimensions, it is necessary to deduce the Miller indices of the planes that show these distances. In the case of a powder specimen (where all information concerning orientations of crystal axes has been lost), the only available information regarding Miller indices is that obtainable by application of Eqs. (5) and (6).

To find which of the three types of cubic lattice is the correct one, we make use of some interesting properties of integers. From Eq. (5) we see that

$$\left(\frac{1}{d}\right)^2 = \left(\frac{1}{a_0}\right)^2 M^2 = \left(\frac{1}{a_0}\right)^2 (h^2 + k^2 + l^2) \tag{7}$$

so that, if we square our reciprocal spacings, it should be possible to find a numerical factor that will convert them into a sequence of integers, which we shall find convenient to make *relatively prime*. We shall see that the type of lattice is determined by the character of the integer sequence obtained.

For a simple cubic (primitive) lattice, all integral values are independently possible for the Miller indices $h, k,$ and l. Note that it is possible to express most, but not all, integers as the sum of the squares of three integers. In Table 1 the various possible values of M^2 are listed in the column under P, and it is seen that there are gaps where the integers 7, 15, 23, 28, 31, 39, 47, and 55 are absent. (Other gaps occur at higher values of M^2.) For those values of M^2 that are possible, the first column of the table gives the Miller indices the sum of whose squares yield the M^2 values. In some cases it is seen that there is more than one possible choice.

For a simple (primitive) cubic lattice P, there are no restrictions on the Miller indices and therefore none on M^2 except as noted above. In the case of a body-centered cubic lattice, only those values of M^2 can be allowed that arise from Miller indices whose sum is even. This has the effect of requiring M^2 to be even, as seen in the first of the two columns under I. We can then divide them by 2 and thereby reduce them to a *relatively prime* sequence, shown in the second column under I, for comparison with the sequence obtained from the $(1/d)^2$ values. We note immediately that the relatively prime sequence obtained differs from that for a primitive cubic lattice in having gaps at different places. By use of this fact, it is almost always possible to distinguish between a primitive cubic lattice and a

TABLE 1 Possible values of M^2 for cubic lattices

hkl	M	P, M^2	I M^2	I $M^2/2$	F, M^2
100	1.0000	1			
110	1.4142	2	2	1	
111	1.7321	3			3
200	2.0000	4	4	2	4
210	2.2361	5			
211	2.4495	6	6	3	
		—			
220	2.8284	8	8	4	8
300, 221	3.0000	9			
310	3.1623	10	10	5	
311	3.3166	11			11
222	3.4641	12	12	6	12
320	3.6056	13			
321	3.7417	14	14	7	
		—			
400	4.0000	16	16	8	16
410, 322	4.1231	17			
411, 330	4.2426	18	18	9	
331	4.3589	19			19
420	4.4721	20	20	10	20
421	4.5826	21			
332	4.6904	22	22	11	
		—			
422	4.8990	24	24	12	24
500, 430	5.0000	25			
510, 431	5.0990	26	26	13	
511, 333	5.1962	27			27
		—	—	—	
520, 432	5.3852	29			
521	5.4772	30	30	15	
		—			
440	5.6569	32	32	16	32
522, 441	5.7446	33			
530, 433	5.8310	34	34	17	
531	5.9161	35			35
600, 442	6.0000	36	36	18	36
610	6.0828	37			
611, 532	6.1644	38	38	19	
		—			
620	6.3246	40	40	20	40
621, 540, 443	6.4031	41			
541	6.4807	42	42	21	
533	6.5574	43			43
622	6.6332	44	44	22	44
630, 542	6.7082	45			
631	6.7823	46	46	23	
		—			

hkl	M	P, M^2	I			F, M^2
			M^2	$M^2/2$		
444	6.9282	48	48	24		48
700, 632	7.0000	49				
710, 550, 543	7.0711	50	50	25		
711, 551	7.1414	51				51
640	7.2111	52	52	26		52
720, 641	7.2801	53				
721, 633, 552	7.3485	54	54	27		
642	7.4833	56	56	28		56

body-centered cubic lattice on the basis of a powder photograph. For the face-centered cubic lattice, application of the restriction that the indices must be all even or all odd produces the characteristic sequence 3, 4, 8, 11, 12, 16, . . . given in the column under F. If most or all of the numbers in this sequence are present, and if none of the excluded numbers is present, the lattice is evidently face-centered cubic.

If no relatively prime sequence of integers can be found to within the experimental uncertainty of the measurements, the crystalline substance presumably does not belong to the cubic system.

When the cubic lattice type has been deduced, the unit-cell dimension a_0 can be calculated. From the unit-cell volume, the measured crystal density, and the formula weight, the number of formulas in one unit cell can be calculated.

Deduction of the Structure. The arrangement of the atoms or ions in the unit cell is at least partly determined by symmetry considerations, but in most cases it is necessary to take account of the *intensities* of the Bragg reflections. The way this is done in present crystallographic practice is far too complicated to describe here.[2] We shall here illustrate only by a simple example how it is possible to use qualitative arguments based on intensity.

If the substance is a binary compound AB, and if its unit cell is simple cubic P with one formula (one atom of A and one of B) per cubic cell, the relative positions of the two atoms are fixed by symmetry. This is true of the salt cesium chloride, CsCl, the structure of which is shown in Fig. 4. One of the ions, Cs^+ say, may without loss of generality be placed at the origin. The other ion, Cl^-, must be at the center of the unit cell; if it is in any other position, the structure will lack the threefold rotational axes of symmetry that are always present along all four body diagonals of the unit cell in the cubic system.

In many cases, including even some with one formula of a binary compound for each lattice point, the positions of the atoms or ions are not necessarily given uniquely by

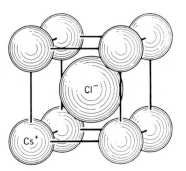

FIGURE 4
Unit cell of cesium chloride. Note that this structure is not body-centered cubic but would be if the two ions were identical.

symmetry, though the number of possible choices may not be large. The solution of the structure in some cases may lie in a simple clue from the intensities. Let it be supposed, for example, that a certain class of powder lines are relatively weak and that omitting these lines from the calculation leads to a "pseudo-lattice" with a smaller number of atoms per lattice point than in the case of the true lattice. A possible hypothesis might then be that the weak lines owe their weakness to destructive interference between two kinds of atoms or ions and that, if the chemical difference between these two kinds of atoms or ions could somehow be removed, the interference would become complete and the pseudo-lattice would become a true lattice. Knowledge of the pseudo-lattice (and pseudo-cell) may then show where the atoms or ions must be placed (irrespective of kind), and knowledge of the true lattice will then show which atoms are of each kind.

As an example of this kind of clue, let us deduce the relative positions of the ions in cesium chloride, bromide, and iodide (which all have the same structure) from intensity considerations, without appealing to any arguments based on symmetry. Putting the $(1/d)^2$ on a relatively prime basis, we obtain the integers 1, 2, 3, 4, 5, 6, —, 8, 9, 10, 11, 12, 13, 14, —, 16, . . . , which clearly indicates a primitive lattice. From the density we find that there is one formula per lattice point. However, we see that all the odd-numbered lines are somewhat weak in CsCl, much weaker in CsBr, and very faint or absent in CsI. If we neglect them, the sequence of integers obtained (on dividing by 2 to obtain relative primes) indicates a body-centered cubic "pseudo-lattice" with one-half formula, or one ion (irrespective of kind), at each pseudo-lattice point. In other words, if we were unable to distinguish between the two kinds of ions, the structure would look like one with a single atom of a single kind at each point of a body-centered cubic lattice. This shows clearly that, in the real structure, an ion of one kind is located at (000) and an ion of the other kind is located at $(\frac{1}{2}\frac{1}{2}\frac{1}{2})$. It may be mentioned that the virtually complete obliteration of the odd lines in CsI is due to the fact that the Cs^+ ion and the I^- ion are *isoelectronic* (that is, have the same number of electrons, namely 54, which is the number in xenon); if any odd lines are observed at all, it is owing to the fact that, because of the different nuclear charges of the two ions, the sizes of the electron clouds of the two ions are slightly different.†

When there are more than two atoms per lattice point, the structure determination will be more complicated. One frequently used procedure is trial and error, in which a number of "model" structures are successively proposed and tested by calculation of intensities and comparison with experiment, until a structure is found that yields satisfactory agreement. When the number of possible structures is too large for the practical application of trial-and-error methods, methods must be used that are too advanced for description here. With their use, crystal structures have been found in which there are thousands of atoms in a unit cell. In such cases and in most ordinary work, however, powder data would be inadequate for determining the structure, and diffraction patterns must be obtained from single crystals.

METHOD

There are several experimental techniques for realizing the diffraction conditions, the most powerful of which depends on having a single crystal of the substance to be studied;[2] see Exp. 46. In the present experiment we are concerned only with the Debye–Scherrer method (often called the powder method), which does not make use of a single crystal but rather of a powder obtained by grinding up crystalline or microcrystalline material.[3] This powder contains crystal particles of a few micrometers in size.

---

†The argument here depends upon the fact that it is the electrons, rather than the nuclei, that scatter X rays.

The diffraction pattern of a powdered material can be obtained by several different experimental methods. The one that we will illustrate in the present experiment is the *powder-photograph* method using the Straumanis technique.[4] Other specialized powder photograph techniques exist[3] for extremely precise work. It is also possible to record powder diffraction directly in digital form or on a strip-chart recorder with a *powder diffractometer.* In a powder diffraction, a radiation detector (a proportional counter or a crystal scintillation counter) with narrow entrance slits moves at a constant angular speed over the desired range of Bragg angles, picking up x-ray photons diffracted by the powder specimen. Each x-ray photon is converted by the detector into an electrical pulse. The stream of pulses is processed by an electronic circuit and is usually converted to a pulse rate (number of photons per unit time), which is stored in digital form on a computer and can be displayed on the monitor as a continuous curve having peaks that correspond to lines on a powder photograph. The present experiment can be done with such an instrument if it is available. Close supervision and instruction by qualified personnel is essential.

In the present method, the specimen for diffraction is obtained by sticking some of the powdered material onto a fine (0.1-mm) glass fiber with a trace of Vaseline or filling a very thin-walled glass capillary tube (0.2-mm diameter) with the powder. A narrow beam of parallel monochromatic X rays, about 0.5 mm in diameter, impinges on this specimen at right angles to its axis. The source of X rays is usually a Coolidge-type x-ray tube with a copper (or molybdenum) target, equipped with a filter (of nickel foil, in the case of a copper target) to remove all spectral components except the desired $K\alpha$ line. The narrow beam is formed by a *collimator,* which consists basically of a conical tube 5 or 6 cm long with a pinhole or slit at each end. On the opposite side of the specimen is a conical receptacle similar to the collimator but with only an entrance pinhole or slit and no exit; this is the *beam stop,* in which the undiffracted beam is trapped. The diffracted radiation is detected by a strip of photographic film bent into a cylinder and held firmly against the inside wall of a cylindrical camera, coaxial with the specimen. The arrangement of the collimator, specimen, and photographic film is shown schematically in Fig. 5. When the beam impinges on the randomly oriented particles in a stationary powder specimen, most of the particles will not diffract the X rays at all. Only those particles that happen by chance to be so oriented that Eq. (3) holds for Bragg reflection will diffract X rays from some set of

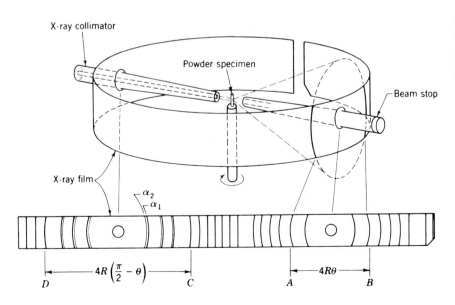

FIGURE 5
Schematic diagram illustrating x-ray powder method (Straumanis arrangement[4]). Note splitting of back-reflection lines.

crystallographic planes. The direction of the diffracted rays will then deviate from the direction of the incident beam by twice the Bragg angle. Since orientation is completely random with respect to the beam axis, the rays diffracted at a given scattering angle may lie with equal probability anywhere on a right circular cone with apex angle 4θ. In practice the specimen is usually rotated about its own axis, so that during a single rotation all or nearly all the particles present will have an opportunity to reflect X rays from any given set of crystallographic planes, the resulting diffracted radiation being distributed rather evenly over the cone. Where the cone intersects the photographic film, the latent image of a *powder line* is formed. When the film is removed from the camera and developed, fixed, washed, and dried, it is found to have on it a number of lines, each of which is due to reflection from one or more sets of crystallographic planes.

The positions of the lines on the film can be measured with a comparator or microphotometer, but adequate measurements can be made with a good millimeter scale. From measurements of the positions of the lines or distances between them, the Bragg angles can be calculated. The determination of these angles is a purely geometric problem. The "effective radius" R of the film in the camera can be obtained by picking two sets of lines, A, B and C, D as in Fig. 5, and measuring the four distances, $\overline{AC}$, $\overline{AD}$, $\overline{BC}$, and $\overline{BD}$:

$$4R = \frac{1}{\pi}(\overline{AC} + \overline{AD} + \overline{BC} + \overline{BD}) \tag{8a}$$

If it is desired to calculate angles in degrees rather than in radians, it is convenient to calculate

$$4R° = \frac{\pi}{180}(4R) = \frac{1}{180}(\overline{AC} + \overline{AD} + \overline{BC} + \overline{BD}) \tag{8b}$$

Then

$$\theta(\text{rad}) = \frac{\overline{AB}}{4R} \qquad \theta' = \frac{\pi}{2} - \theta(\text{rad}) = \frac{\overline{CD}}{4R} \tag{9a}$$

$$\theta(\text{deg}) = \frac{\overline{AB}}{4R°} \qquad \theta' = 90 - \theta(\text{deg}) = \frac{\overline{CD}}{4R°} \tag{9b}$$

One can then calculate $(1/d)$ from Eq. (3). In processing back-reflection data, it is unnecessary to convert from $(\pi/2 - \theta)$ to θ, since $\sin \theta \equiv \cos(\pi/2 - \theta) \equiv \cos \theta'$. Thus

$$\frac{1}{d} = \frac{2}{\lambda} \sin \theta = \frac{2}{\lambda} \cos \theta' \tag{10}$$

When the measurements are made and $(1/d)$ is computed, account should be taken of the fact that the $K\alpha$ spectral line generally used as a monochromatic source of X rays is not truly a single line but is actually a closely spaced doublet. The doublet is ordinarily not resolved in the forward-reflection part of the film, but in the back-reflection part of the film the lines are usually observably split into two components, known as $K\alpha_1$ and $K\alpha_2$ in order of increasing wavelength (and therefore in order of increasing θ), in the intensity ratio 2:1. When the two components are not resolved, the position taken for the line should be an estimate of the position of the "center of gravity," and the value of the wavelength λ used in the calculation should be the weighted mean wavelength:

$$\lambda_{\text{mean}} = \tfrac{1}{3}(2\lambda_{\alpha_1} + \lambda_{\alpha_2}) \tag{11}$$

When the components are sufficiently well resolved to make possible the measurement of the positions of the two components separately, such measurements should be made and the actual values of the wavelengths should then be used in the calculations. Each component will then yield a separate value of $1/d$; the two values obtained should be in good agreement and may be averaged for the ensuing calculations.

For copper ($K\alpha$) radiation,

$$\lambda_{\alpha_1} = 1.54050 \text{ Å}$$
$$\lambda_{\alpha_2} = 1.54434 \text{ Å} \tag{12}$$
$$\lambda_{\text{mean}} = 1.5418 \text{ Å}$$

In present-day x-ray crystallography, powder photography is only rarely used for complete structure determinations. It is, however, frequently used in the precise determination of unit-cell dimensions. Its most common technical use is as an analytical tool; powder patterns of thousands of crystalline substances are known.

EXPERIMENTAL

Preparation of Powder Specimen.   A specimen of the material to be studied is finely pulverized in an agate mortar with an agate pestle. The powder is then loaded into a Pyrex or Lindemann glass capillary tube, 0.2 to 0.3 mm OD, having a very thin wall (0.01 mm or less), so as to obtain a densely filled specimen about 1 cm in length. If the capillary tube is flared out at one end, it is easier to load; even without a flared-out end, however, the powder can be picked up by scraping one end of the capillary against the surface of the mortar and, with this end uppermost, agitating the powder by very gentle rasping with a file. One end of the capillary tube should be sealed off with a flame or with wax before filling, and the other end with wax after filling. The capillary tube is then affixed with wax to the end of a short ($\frac{3}{8}$-in. or less) length of $\frac{1}{8}$-in. brass rod; see Fig. 6.

Installation of the Specimen.   A powder camera of the Straumanis design is shown in Fig. 7. It is a valuable instrument and must be handled carefully. The camera should not be used until the procedures for using it have been demonstrated to the student by an experienced user. To install and line up the specimen, proceed as follows.†

Remove the circular cover, and lay it down, inside surface upward, in a *safe place*. Remove the two slits (collimator and beam stop), and lay them down *inside the cover*.

By means of tweezers or long-nose pliers, insert the brass pin supporting the specimen into its receptacle, where it is held by friction. Carefully replace the collimator, being sure not to force it or to drop it.

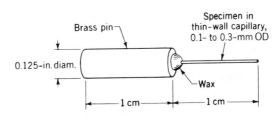

FIGURE 6
Mounted powder specimen.

†The procedure given is that applicable to the 114.6-mm camera manufactured by the North American Philips Co.

FIGURE 7

Philips powder camera
(shown with cover not in
place).

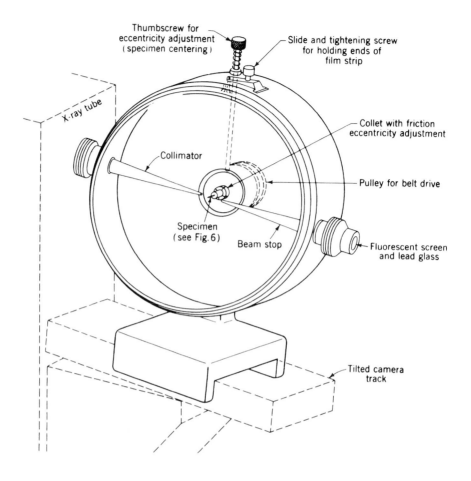

Place the lens cap over the outside collimator opening, and set the camera down with the beam-stop hole facing a well-lighted surface (beam stop *not* in place). By hand, turn the pulley wheel on the outside of the camera. As it goes around, the silhouette of the specimen should appear to go up and down. Stop it in an "up" position, and by means of the screw on top of the camera push it down to the center of the visual field. Turn the screw back up, and rotate again to see if further adjustment is needed; if so use the screw as before. Continue until the specimen appears to undergo no motion as the spindle is rotated.

Remove the lens cap, and replace the beam stop. As in the case of the collimator, do not force it or drop it. Replace the cover.

Loading the Camera with Film.   Take the camera into the darkroom. Remove the cover and the collimator and beam stop. Turn off the room light, and under a safelight cut, punch, and insert the film in accordance with specific instructions given. (It is advisable to rehearse the procedure ahead of time with the light on, using a piece of exposed film.) Insert the collimator and beam stop and replace the cover.

Making the X-ray Exposure.   This should be done under the constant supervision of a qualified person; detailed instructions cannot be given here. **Be very careful to avoid any x-ray exposure to any part of the body;** make sure that all x-ray ports are covered when the unit is in operation. The optimum exposure time depends on the composition and size of the specimen, the tube current and operating potential, and the dimensions of the collimating system. Typical exposures are 4 to 20 h. In the absence of a recommended exposure time, make a short trial exposure and a second longer exposure if necessary.

Developing the Film.   Take the camera into the darkroom and remove the film. Attach it to a clip or other support and develop for 3 to 5 min, depending on the temperature of the developer. Agitate the film, but do not allow its surface to rub against the walls or bottom of the tank or tray. Wash the film for 30 s and place it in the fixing bath. After the film has cleared, the room light may be turned on. The film should be in the fixing bath for at least twice as long as it takes to clear and should then be washed in running water for at least 30 min. It may then be carefully wiped with clean fingers, rinsed again with distilled water, and hung up to dry.

Measurement of Crystal Density.   The density of the substance can be determined with a pycnometer or with a 5-mL (or smaller) volumetric flask used as a pycnometer. The liquid used should be one in which the substance is insoluble; for a water-soluble inorganic salt, medium-boiling petroleum ether is convenient. The procedure is described in Chapter I (Sample Report). Make two determinations.

CALCULATIONS

If it is not possible to take the x-ray powder pattern in the laboratory, a contact print of a powder photograph will be furnished and the x-ray equipment will be discussed and shown. The instructions below should be followed with either the negative taken or the print provided.

Measure $\overline{AB}$ or $\overline{CD}$ (see Fig. 5) for as many lines as possible; estimate each distance to ± 0.1 mm. Enter the measurements, together with estimated intensity, into a table. Make a separate measurement for each component of a resolved doublet.

Pick a pair of sharp lines A, B in the forward-reflection part of the film and another pair C, D in the back-reflection region. Measure the four distances $\overline{AC}$, $\overline{AD}$, $\overline{BC}$, $\overline{BD}$ as accurately as possible. Calculate $4R$ or $4R°$ from Eq. (8).

Calculate θ or θ' for each line or resolved component of a line [Eq. (9)] and then calculate $1/d$ from Eq. (10), averaging the values obtained for the two components of each resolved doublet. Calculate $(1/d)^2$.

Find a factor that will reduce the $(1/d)^2$ values to relatively prime integers within experimental error. Refer to Table 1 and identify the lattice type. Also from Table 1, obtain the Miller indices hkl for each line.

The most precise values of a_0 are obtained from the lines in the back-reflection part of the film, where θ is close to 90°. This can be seen by combining Eqs. (3) and (5), solving for a_0, and differentiating:

$$da_0 = -\frac{M\lambda}{2 \sin^2 \theta} \cos \theta \, d\theta$$

$$= -a_0 \cot \theta \, d\theta$$

The experimental uncertainty in measuring θ is proportional to that of measuring the distances between pairs of lines and is approximately constant over the film unless lines in the back-reflection region are unduly faint or broad. However, even if the uncertainty of measurement is a little larger in the back-reflection region, the effect is ordinarily far outweighed by the cot θ factor. In fact lines very close to the collimator hole should always be used if at all possible, even if they are broad and diffuse. Another good reason for using only lines in the back-reflection region is that they are much more free of shifts due to absorption of X rays by the specimen. From a few well-chosen lines in the back-reflection region, preferably resolved doublets, calculate a_0 from Eq. (5) and average the values obtained.

From the measured density and the a_0 value determined as above, calculate the number of atoms per unit cell and per lattice point. Report the lattice type, the value of a_0, and the number of atoms per unit cell.

DISCUSSION

Determine the crystal structure, if possible, by methods similar to those described for CsCl (see Theory). Draw a diagram of the cubic unit cell showing the positions of all atoms or ions.

SAFETY ISSUES

The individuals who are to perform this experiment must be thoroughly informed by the staff member in charge of the facility, especially in matters of radiation safety. The apparatus must be checked with a radiation monitor to ensure that there is no significant stray radiation in the surrounding space. Any adjustment of the camera or the x-ray ports must be done with the x-ray power off or set to the minimum x-ray intensity required for the adjustment. Avoid loitering in the x-ray room during operation of the apparatus.

APPARATUS

Access to an x-ray diffraction facility must be under the active supervision and control of an individual trained and qualified in its use. The facility should be in a room by itself. It should contain an x-ray generator, typically consisting of an x-ray tube mounted in a radiation-shielding tube stand over a high-voltage power supply. A portable radiation monitor is needed. The x-ray tube is typically one with a solid copper anode that, with a nickel foil filter, yields Cu–Kα radiation; other commonly used anode materials are molybdenum and tungsten. The tube stand is configured to fit the entrance pinhole of the camera or other diffraction apparatus so as to prevent radiation from leaking into the surrounding space, and it should have some provision for accommodating the metal foil used for filtering, as well as a remotely operated shutter for turning the x-ray beam to the camera on and off. The diffraction camera needed in this experiment is a Debye–Scherrer powder camera (usually a 114.6-mm powder camera), with a film cutter and punch to match. Also needed are thin-walled Lindemann glass capillary tubes (available from Caine Scientific Sales Co., Chicago), agate mortar and pestle, small file, x-ray film (Eastman no-screen 35-mm double-coated continuous strip), photographic darkroom equipped with water-cooled developer and fixer tanks or adequate trays, developer and fixer solutions, timer, thermometer, good millimeter scale, pycnometer or 5-mL volumetric flask, small pipette or eyedropper. Alternatively, the entire experiment can be done with an x-ray powder diffractometer, used under the supervision and instruction of qualified personnel.

Small quantity of crystalline material of cubic structure for study [e.g., alkali halides, alkaline earth oxides, cuprous or silver halides, simple metals such as aluminum or copper (finely powdered with a file)]; liquid (medium boiling petroleum ether) for density work.

REFERENCES

1. C. Giacovazzo (ed.), *Fundamentals of Crystallography,* chap. 1, Oxford Univ. Press, Oxford (1992).

2. G. H. Stout and L. H. Jensen, *X-Ray Structure Determination,* 2d ed., Wiley, New York (1989).

3. B. D. Cullity and S. R. Stock, *Elements of X-Ray Diffraction,* 3d ed., chaps. 6 and 7, Prentice-Hall, Upper Saddle River, NJ (2001).

4. M. J. Buerger, *Am. Miner.* **21,** 11 (1936); *J. Appl. Phys.* **16,** 501 (1945); M. Straumanis and A. Ievins. *Z. Phys.* **98,** 461 (1936).

GENERAL READING

D. L. Bish and J. E. Post (eds.), *Modern Powder Diffraction,* Mineralogical Society of America, Washington, DC (1989).

C. Giacovazzo, (ed.), *Fundamentals of Crystallography,* Oxford Univ. Press, Oxford (1992).

J. P. Glusker and K. N. Trueblood, *Crystal Structure Analysis: A Primer,* 2d ed., Oxford Univ. Press, New York (1985).

R. Jenkins, *J. Chem. Educ.* **78,** 601 (2001).

R. A. Young (ed.), *The Rietveld Method,* Oxford Univ. Press, Oxford (1993).

EXPERIMENT 47
Lattice Energy of Solid Argon

Accurate measurements of the sublimation pressures of solid argon are to be made as a function of temperature in the range 65 to 78 K. Applying both a second- and third-law thermodynamic treatment to these equilibrium measurements, one obtains the heat of sublimation of argon. From this heat of sublimation and the Debye theory of lattice vibrations, the lattice energy of solid argon can be determined. Representing the Ar–Ar interaction by a Lennard–Jones 6,12 potential, the lattice energy will be calculated using potential parameters obtained from the room-temperature transport properties of argon gas. This "theoretical" value is then compared with the "experimental" thermodynamic value.

THEORY

The change in state involved here is simply

$$\text{Ar}(s) = \text{Ar}(g) \qquad (\text{const } p, T) \tag{1}$$

The molar enthalpy of sublimation $\Delta \widetilde{H}$ is rigorously related to the slope dp/dT of the sublimation pressure curve by the Clapeyron equation, which has been discussed in Exp. 13. The exact expression is

$$\Delta \widetilde{H} = T \frac{dp}{dT} (\widetilde{V}_g - \widetilde{V}_s) \tag{2}$$

If $\widetilde{V}_s$ is assumed to be negligible in comparison to $\widetilde{V}_g$ and if the vapor is assumed to be an ideal gas, an approximate form of the Clapeyron equation is obtained,

$$\Delta \widetilde{H} \simeq -R \frac{d \ln p}{d(1/T)} \tag{3}$$

TABLE 1   Second virial coefficient of argon[a] (B in units cm^3 mol^{-1})

T(K)	B	dB/dT	T(K)	B	dB/dT
66	-497	23.9	76	-335	11.2
68	-454	19.8	78	-314	9.9
70	-418	17.0	80	-295	8.8
72	-386	14.7	83.81	-265	7.1
74	-359	12.8	87.30	-242	5.9

[a] Extrapolated from measurements above 84 K; see M. A. Byrne, M. R. Jones, and L. A. K. Staveley, *Trans. Faraday Soc.* **64**, 1747 (1968).

which will yield an approximate value of $\Delta\tilde{H}$ directly from the slope of a plot of ln p versus $1/T$. This approximate Clapeyron equation is very nice for examination questions or the quick analysis of crude data, but it should never be used to represent accurate experimental measurements. Fortunately the equation of state of Ar(g) has been determined experimentally down to the triple point of argon. Table 1 gives values of B, the second virial coefficient, and dB/dT as a function of temperature. Also, the molar volume of the solid can be taken to have the constant value 25 cm^3 mol^{-1} over the investigated temperature range. Thus one can use the exact Clapeyron equation to determine second-law values of $\Delta\tilde{H}$.

On the basis of the third law, the thermodynamic properties of a pure substance [such as $S^0, (H^0 - H_0^0)/T$ and $(G^0 - H_0^0)/T$, where $S^0 = \int_0^T C_p^0 \, d \ln T$, $H^0 - H_0^0 = \int_0^T C_p^0 \, dT$ and $G^0 = H^0 - TS^0$] can be tabulated as functions of temperature. Table 2 gives these functions for Ar(s, cubic), Ar(l), and Ar(g). In the preparation of this table, experimental values

TABLE 2   Thermodynamic properties of argon[a] (All quantities are in units of J K^{-1} mol^{-1})

T, K	$\tilde{S}^0$	$\dfrac{\tilde{H}^0 - \tilde{H}_0^0}{T}$	$\dfrac{\tilde{G}^0 - \tilde{H}_0^0}{T}$	$\tilde{S}^0$	$\dfrac{\tilde{H}^0 - \tilde{H}_0^0}{T}$	$\dfrac{\tilde{G}^0 - \tilde{H}_0^0}{T}$
		Ar(s, l)			Ar(g)	
0	0	0	0	0	0	0
10	1.092	0.828	-0.264	84.275	20.786	-63.489
20	6.259	4.418	-1.841	98.684	20.786	-77.898
30	12.611	8.234	-4.377	107.111	20.786	-86.325
40	18.556	11.339	-7.217	113.094	20.786	-92.308
50	23.870	13.845	-10.025	117.730	20.786	-96.944
60	28.652	15.912	-12.740	121.521	20.786	-100.735
70	33.024	17.698	-15.326	124.726	20.786	-103.940
72	33.870	18.037	-15.833	125.312	20.786	-104.526
74	34.710	18.384	-16.326	125.880	20.786	-105.094
76	35.547	18.723	-16.824	126.433	20.786	-105.647
78	36.376	19.066	-17.310	126.972	20.786	-106.186
80	37.208	19.405	-17.803	127.500	20.786	-106.714
83.81(s)	38.798	20.075	-18.723	128.466	20.786	-107.680
83.81(l)	53.003	34.280	-18.723			
87.30	54.815	34.681	-20.134	129.316	20.786	-108.530

[a] The entries for Ar(s) and Ar(l) were calculated from data in Refs. 1 and 2; the entries for Ar(g) are statistical thermodynamic values for a monatomic ideal gas.

of $\widetilde{C}_p^0$ for the solid and liquid phases and the experimental enthalpy of fusion were used. For the gas phase, $\widetilde{C}_p^0 = 5R/2$ (except at extremely low T) and $\widetilde{S}^0$ were calculated from the Sackur-Tetrode equation. We can now proceed to evaluate third-law values of the enthalpy of sublimation. For the change in state (1), $\Delta \widetilde{G}^0 = -RT \ln K$, where the equilibrium constant $K = f_g / a_s$. At low pressures, only the second virial coefficient need be considered, and a satisfactory approximation for $\Delta \widetilde{G}^0$ is

$$\Delta \widetilde{G}^0 = -RT \ln p - Bp - \widetilde{V}_s^0 (1 - p) \tag{4}$$

Indeed, the last term in the above expression is required only for data of the highest accuracy. From experimental $\Delta \widetilde{G}^0$ values given by Eq. (4) and interpolated values of $(\widetilde{G}^0 - \widetilde{H}_0^0)/T$ from Table 2, we may now determine $\Delta \widetilde{H}_0^0$, the enthalpy of sublimation at absolute zero:

$$\frac{\Delta \widetilde{G}^0}{T} = \left(\frac{\widetilde{G}^0 - \widetilde{H}_0^0}{T} \right)_g - \left(\frac{\widetilde{G}^0 - \widetilde{H}_0^0}{T} \right)_s + \frac{\Delta \widetilde{H}_0^0}{T} \tag{5}$$

A value of $\Delta \widetilde{H}_0^0$ may be calculated from each experimental value of the sublimation pressure, and these $\Delta \widetilde{H}_0^0$ values must be constant within the limits of experimental error. The value of $\Delta \widetilde{H}^0$ at any temperature can be calculated from $\Delta \widetilde{H}_0^0$ and the entries in Table 2. It is important to recognize that $\Delta \widetilde{H}^0$ refers to the change in state

$$\text{Ar}(s) = \text{Ar(perfect gas)} \qquad (1 \text{ bar, const } T) \tag{6}$$

whereas $\Delta \widetilde{H}$ refers to change in state (1) involving the real gas at pressure p. If the enthalpy of the solid phase is assumed to be independent of pressure for low pressures, then third-law values of $\Delta \widetilde{H}$ can be calculated from

$$\Delta \widetilde{H} = \Delta \widetilde{H}^0 + \left(B - T \frac{dB}{dT} \right) p \tag{7}$$

If all the experimental data were perfect, the second-law value of $\Delta \widetilde{H}$ calculated from Eq. (2) and the third-law value calculated from Eqs. (4), (5), and (7) would be identical. However, in this case, the third-law result should be the more reliable, since it is difficult to determine dp/dT with very high accuracy.

An independent value for $\Delta \widetilde{H}$ has been reported by two different investigators from direct calorimetric measurements of the enthalpy of vaporization.[1-3] The calorimetric measurements, which are in good agreement with each other, give a $\Delta \widetilde{H}_0^0$ value that is about 63 J higher than the third-law value. This discrepancy although small is disturbing because it is well outside the limits of error claimed for the various experimental results. These studies of the thermodynamic properties of argon were made prior to the discovery in 1964 of a metastable hexagonal phase of argon, and the earlier investigators assumed without proof that their solid phase was the face-centered cubic phase. Until more is known concerning the hexagonal phase, it is impossible to say how the presence of this phase might have affected the earlier measurements.†

---

†Face-centered cubic argon is the stable solid phase of pure argon, but upon the addition of as little as 1 percent of N_2 to Ar, the solid phase crystallizes with the close-packed hexagonal structure. A study of the sublimation pressures of solid solutions of Ar with small amounts of N_2 would give valuable information concerning the hexagonal phase, and such a study could be made with the apparatus used in this experiment.

Lattice Energy. If we measure the energy content of argon on an arbitrary scale which defines $\widetilde{E}^0(g)$ to be zero at 0 K, then the total energy of the crystal lattice at 0 K is $-\Delta\widetilde{E}_0^0$ which for all practical purposes is the same as $-\Delta\widetilde{H}_0^0$. There are two additive contributions to this absolute-zero energy content of solid argon: the *lattice energy* Φ_0, which is the potential energy of the argon atoms *at rest* in the lattice relative to a zero of energy for these atoms in the gas at infinite separation, and the zero-point vibrational energy, which arises from the quantum mechanical vibrational motion of the atoms about their equilibrium positions.[4] In terms of the Debye characteristic temperature Θ_D, the zero-point vibrational energy equals $\frac{9}{8}R\Theta_D$. Thus the lattice energy is given by

$$\Phi_0 = -\Delta\widetilde{H}_0^0 - \tfrac{9}{8}R\Theta_D \tag{8}$$

where $\Theta_D = 93$ K for argon.[2]

It is possible to calculate a theoretical value of the lattice energy for a molecular crystal if data are available on the potential energy between atoms as a function of their separation. A commonly used form for the interatomic potential (see Fig. 1) is due to Lennard–Jones:[4,5]

$$U(r) = 4\varepsilon\left[\left(\frac{\sigma}{r}\right)^{12} - \left(\frac{\sigma}{r}\right)^6\right] \tag{9}$$

where $U(r)$ is the potential energy for *two* atoms at a distance r. The r^{-12} term is an empirical function to describe the repulsion at short distances, and the r^{-6} term represents the r dependence of the potential energy found by London to describe the attraction at large distances due to induced dipole–dipole interaction. The Lennard–Jones potential for a given atom is characterized by the two constants ε and σ (which are shown in Fig. 1). These parameters can be evaluated from an analysis of gas data (second virial coefficient, Joule–Thomson effect, or gas viscosity); the best values for argon[6] are

$$\frac{\varepsilon}{k} = 119.5\,\mathrm{K} \qquad \sigma = 3.405\ \text{\AA} \tag{10}$$

For the solid it is assumed that the total potential energy (i.e., lattice energy) is the sum of all pair potentials $U_{ij}(r_{ij})$. The result of this summation for a face-centered cubic lattice (such as argon) is[7]

$$\Phi_0 = 2N_0\varepsilon\left[12.132\left(\frac{\sigma}{d}\right)^{12} - 14.454\left(\frac{\sigma}{d}\right)^6\right] \tag{11}$$

FIGURE 1

Lennard–Jones potential $U(r)$ as a function of interatomic distance r. The characteristic parameters ε and σ determine this potential curve; see Eq. (9).

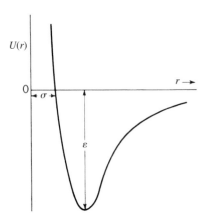

where d is the distance between nearest neighbors. Note that almost all the repulsion part of the potential comes from nearest-neighbor interactions (nearest neighbors alone would give 12 for both coefficients). Since r^{-6} falls off much more slowly than r^{-12}, the coefficient of the second term in Eq. (11) is considerably greater than 12. A knowledge of the lattice spacing for solid argon and the parameters of Eq. (10) will permit calculation of the lattice energy for comparison with the experimental value obtained from Eq. (8).

EXPERIMENTAL

Temperatures below 77 K can be achieved with a liquid-nitrogen bath by reducing the pressure over the nitrogen by pumping, thus lowering its boiling point. It is possible to lower the bath temperature to 63 K before liquid nitrogen begins to solidify. Measurement of the temperature can be made by using a resistance thermometer or a thermocouple (see Chapter XVII) or by measuring the vapor pressure of nitrogen.

The low-temperature cell, shown in Fig. 2, consists of a copper block containing a small chamber A for solid argon and a large chamber B that can be used as an N_2 vapor-pressure bulb. If a thermocouple or a platinum resistance thermometer (PRT) is used, it should be mounted in the copper cell near the argon chamber. The cupronickel (or stainless-steel) connecting tubes are soldered through a cap with a side-arm, This cap fits over the top of a tall glass Dewar flask and is attached to the Dewar with a rubber sleeve.† The side-arm is connected by heavy-wall vacuum hose to a high-capacity mechanical vacuum pump.

If the vapor pressure of liquid nitrogen is used for the temperature measurement, one allows N_2 gas to condense in chamber B and the N_2 pressures can be read directly on a pressure gauge. If a copper–Constantan thermocouple or a platinum resistance thermometer is used, it must be well calibrated, since accurate absolute temperatures are needed. If chamber B is not used, all further instructions concerning it may be disregarded.

Procedure. The apparatus should be assembled by connecting the necessary pressure gauges, filling lines, and pumping lines; see Fig. 3. Evacuate each chamber by closing clamp or stopcock C and stopcocks $A2$ and $B2$ and opening stopcocks $A1$ and $B1$. Then close stopcocks $A1$ and $B1$ and test for leaks in either chamber by waiting approximately 15 min to see if the manometers indicate any increase in pressure. The rate of increase in pressure should not exceed 5 Torr/h.

Filling chambers A and B with gas should be done with care. If a simple reducing valve is used, the high-pressure valve on the storage cylinder should never be opened when gas is being admitted to the system. Allow the gas contained in the section between the two valves to expand into the chamber; then close the needle valve and refill this section. Repeat this procedure several times (if necessary) until the chamber is full. Flushing the chamber once or twice with the appropriate gas will reduce the possibility of contamination.

Fill chamber A with argon at 760 Torr and close stopcock $A2$. **Slowly** raise the Dewar filled with liquid nitrogen until the copper cell is almost completely immersed (do *not* immerse the connecting tubes above the cell). The argon pressure should drop to about 200 Torr. **Slowly** allow N_2 gas to enter chamber B until the pressure is constant (about 760 Torr), and then close stopcock $B2$. Now raise the Dewar all the way up, and attach the cap to the top of the Dewar. The copper cell should be sufficiently far below the liquid-nitrogen level to ensure that the cell will remain submerged throughout the run. (If necessary, additional liquid nitrogen may be added through the filling port in the cap at this time.)

More argon may now be added to chamber A, although this is not necessary if the measurements are to be made on warming. Experience has shown that the argon pressures

†The top of a rubber surgical glove makes an excellent sleeve.

FIGURE 2
Low-temperature vapor-pressure apparatus. (Metal parts to be well tinned and soft soldered, except where otherwise indicated.)

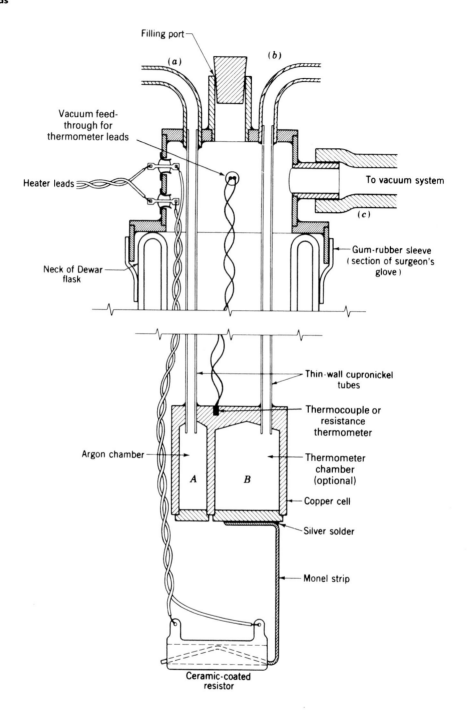

will not be reproducible unless the argon is pumped very briefly prior to any measurement made after the cell has been cooled. To avoid the possibility of pumping off too much argon, it is convenient to make the measurements when the cell is warming rather than cooling. The careful student will want to confirm the fact that reproducible results can be obtained on both warming and cooling.

Close the liquid-nitrogen filling port with a tightly fitting rubber stopper and begin pumping on the refrigerant by opening clamp or stopcock *C*. After the cell has cooled to approximately 65 K, the Ar pressure should be measured at some arbitrary constant

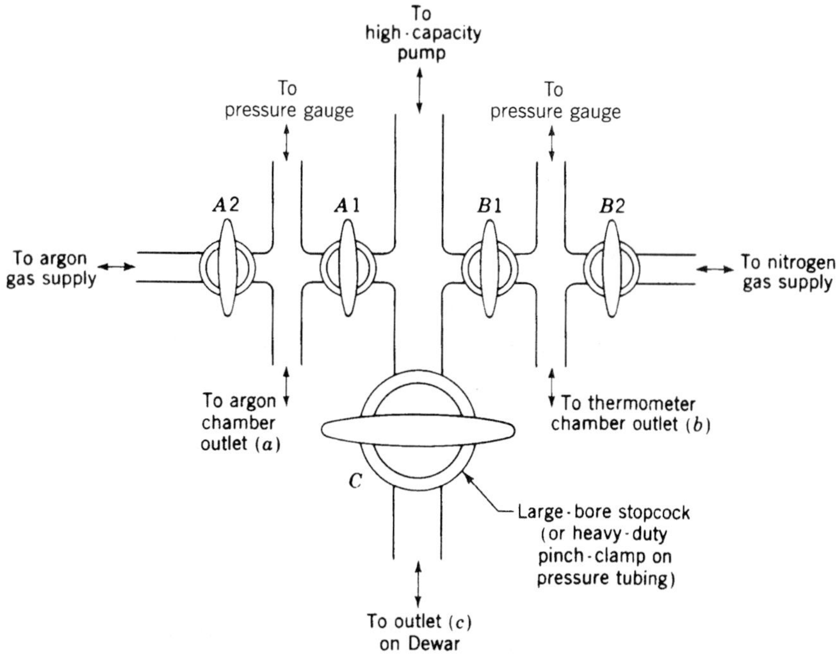

FIGURE 3
Vacuum and gas connections for the apparatus shown in Fig. 2.

temperature. Then pump chamber A briefly and measure the Ar pressure again at the same constant temperature. If the two pressures differ, repeat the procedure. Usually the pressure will be reproducible after the second pumping. No further pumping should be necessary at higher temperatures, although it is advisable to check for reproducibility. At each of a series of constant temperatures, record the vapor pressure of solid argon and measure the temperature. If the vapor pressure of liquid nitrogen in chamber B is being used as the thermometer, the constancy of this pressure is a good indication that the temperature is stable. As an aid to rapidly converting pressure readings into *approximate* bath temperatures, some nitrogen vapor pressures are listed in Table 3. The Ar and N_2 pressure readings should be estimated to within ± 0.2 Torr.

If only external heat leaks served to warm up the system, the rate of warmup would be inconveniently slow. Intermittent heating with an electric heater element immersed directly in the liquid nitrogen will provide a suitable rate of heating between readings. Emphasis should be placed on obtaining good equilibrium data rather than a large number of data points. Four or five good experimental points are sufficient to establish the sublimation curve.

TABLE 3 Vapor pressure of liquid nitrogen[a] (Equilibrium temperatures are given corresponding to the specified vapor pressure)

p, Torr	T, K	p, Torr	T, K	p, Torr	T, K	T, K
100	63.500	240	68.757	500	73.975	
120	64.521	280	69.785	560	74.860	
140	65.412	320	70.703	620	75.675	
160	66.207	360	71.535	680	76.431	
180	66.926	400	72.298	740	77.138	
200	67.584	440	73.004	800	77.803	

[a] Based on data in Ref. 8.

Warning: A closed system containing a liquid with a normal boiling point well below room temperature can cause the destruction of a vacuum system and injuries to personnel. *Make certain that the apparatus is properly pumped out or vented before it is allowed to warm up.* As you should at all times in the laboratory, be sure to wear safety glasses.

CALCULATIONS

Convert all pressure gauge readings to pressures in Torr. Compute the temperatures and tabulate them along with the corresponding argon sublimation pressures. If N_2 vapor pressures were used for thermometry, one can obtain T from the Antoine-type empirical equation given by Armstrong:[8]

$$\log p_{N_2} (\text{Torr}) = 6.49594 - \frac{255.821}{(T - 6.600)} \tag{12a}$$

or

$$T(K) = 6.600 + \frac{255.821}{6.49594 - \log p \, (\text{Torr})} \tag{12b}$$

If a thermocouple or PRT was used, convert its readings to temperatures with an appropriate calibration equation. It may be wise to check the calibration by measuring the bath temperature at atmospheric pressure. Calculate this bath temperature from Eq. (12) using the pressure determined with a barometer. If necessary, one can then apply a constant additive correction to all PRT resistance or thermocouple emf readings.

Represent the argon sublimation pressures by an equation of the form

$$\ln p \, (\text{Torr}) = -\frac{A}{T} + b \tag{13}$$

Only two constants are used because the temperature range of the measurements is quite limited. After the constants A and b are determined, calculate the deviation $p_{obs} - p_{calc}$ for each of the experimental points and add this to your table of p and T values. These deviations will serve as an estimate of the precision of the measurements.

Employing the second-law approach, calculate $\Delta \widetilde{H}$ at 66 K, 70 K, and 76 K from your experimental data, using both the exact and approximate Clapeyron equations. Estimate the experimental error in $\Delta \widetilde{H}$ and comment on any differences between the values obtained with Eqs. (2) and (3).

Utilizing the third-law approach, calculate $\Delta \widetilde{H}_0^0$ from each of your experimental sublimation pressures. Then calculate $\Delta \widetilde{H}^0$ at 66 K, 70 K, and 76 K, using what you consider to be the best value of $\Delta \widetilde{H}_0^0$ (justify this choice). From Eq. (7) obtain $\Delta \widetilde{H}$ values at the corresponding temperatures and compare these values with those obtained through the second-law treatment.

The report should show all numerical values of the various terms involved in the thermodynamic equations used and all appropriate graphs. Also compare your $\Delta \widetilde{H}$ values with those reported in the literature.[2,9]

Finally, use Eq. (8) to determine the "experimental" value of the lattice energy of argon at 0 K. X-ray diffraction data[10] give 5.30 Å for the cubic unit-cell parameter of solid argon at 4 K. Find the nearest-neighbor distance d, and use Eqs. (10) and (11) to calculate a "theoretical" value of Φ_0. Compare the theoretical and experimental values.

SAFETY ISSUES

Gas cylinders must be chained securely to the wall or laboratory bench (see pp. 644–646 and Appendix C). The handling of liquid nitrogen must be carried out properly. Do not allow any low-temperature liquid (A_r or N_2) to heat up in a confined chamber, since large vapor pressures can develop, leading to an explosive rupture of the system (see Appendix C).

APPARATUS

Tall Dewar flask (preferably with vertical unsilvered strip to permit viewing the cell and the liquid level); support for Dewar; copper cell and cap assembly as shown in figure; rubber stopper to fit filling port; rubber sleeve; eight lengths of heavy-wall rubber tubing; high-capacity mechanical vacuum pump; gas-handling assembly with four stopcocks (see figure); heavy-duty hose clamp; two pressure gauges; Variac for heater supply. If a copper–Constantan thermocouple is to be used, a Dewar flask for ice-water reference junction and either a potentiometer circuit or a digital voltmeter are needed.

Supplies of argon and dry nitrogen gas at 1 atm; liquid nitrogen.

REFERENCES

1. P. Flubacher, J. Leadbetter, and J. A. Morrison, *Proc. Phys. Soc.* (*London*) **78,** 1449 (1961).

2. R. H. Beaumont, H. Chihara, and J. A. Morrison, *Proc. Phys. Soc.* (*London*) **78,** 1462 (1961).

3. K. Clusius, *Z. Phys. Chem.* **31B,** 459 (1936); A. Frank and K. Clusius, *Z. Phys. Chem.* **42B,** 395 (1939).

4. J. O. Hirschfielder, C. F. Curtiss, and R. B. Bird, *Molecular Theory of Gases and Liquids,* pp. 1035–1044, Wiley, New York (1964[1954]).

5. J. A. Beattie and W. H. Stockmayer, "The Thermodynamics and Statistical Mechanics of Real Gases," in H. S. Taylor and S. Glasstone (eds.), *A Treatise on Physical Chemistry.* Vol. 2, *States of Matter,* 3d ed., chap. 2, pp. 305–306, Van Nostrand, Princeton, NJ (1951).

6. E. Whalley and W. G. Schneider, *J. Chem. Phys.* **23,** 1644 (1955).

7. J. A. Beattie and W. H. Stockmayer, *op, cit.,* p. 309.

8. G. T. Armstrong, *J. Res. Natl. Bur. Std.* **53,** 263 (1954).

9. A. M. Clark, F. Dim, J. Robb, A. Michels, T. Wassenaar, and Th. Zwietering, *Physica* **17,** 876 (1951).

10. O. G. Peterson, D. N. Batchelder, and R. O. Simmons, *Phys. Rev.* **150,** 703 (1966).

EXPERIMENT 48
Statistical Thermodynamics of Iodine Sublimation

This experiment is in some respects similar to two other experiments concerning enthalpy changes attending phase transformations, namely, Exps. 13 and 47. However, it differs from them in that the experimental data, which are vapor pressures of solid iodine at several temperatures, are obtained from optical absorption measurements. As in the other experiments mentioned, the enthalpy change (here the heat of sublimation of solid iodine) can be calculated with the Clausius–Clapeyron equation, which requires the values of vapor pressures at two or more temperatures.

The system $I_2(s)$–$I_2(g)$ also provides an opportunity for the application of statistical mechanics to derive thermodynamic information from spectroscopic data. For the gas phase, the vibrational frequency of the I_2 molecule, needed in formulating the vibrational partition function, can be obtained from the absorption spectrum in the visible region (see Exp. 39); the rotational partition function in the gas phase will be calculated from the known internuclear distance in the iodine molecule. For the crystalline phase, published phonon dispersion curves, obtained by inelastic neutron-scattering spectroscopy, will be used to determine the vibrational frequencies. With the above information and statistical mechanical theory, the molar energy difference $\Delta \widetilde{E}_0^0$ between the vibrational ground states of crystalline and gaseous iodine can be determined from a measurement of vapor pressure at *one* temperature. From the fully defined partition functions for both crystalline and gaseous iodine, the entropy and enthalpy changes attending sublimation of iodine can be calculated at any temperature T. The value of $\Delta \widetilde{H}_{\text{sub}}$ obtained in this way will then be compared with the value obtained with the Clausius–Clapeyron equation.

THEORY

This section will not be concerned with the Clausius–Clapeyron equation, which is discussed adequately in Exps. 13 and 47. The discussion here will focus on the application of statistical mechanics to the phase equilibrium

$$I_2(s) = I_2(g) \tag{1}$$

It is required for equilibrium that the chemical potential of I_2 be the same in the two phases:

$$\mu_s = \mu_g \tag{2}$$

The basic question is: *How can these chemical potentials be determined?*

Statistical Mechanical Background. We will present a brief review of the basic statistical mechanical concepts needed in this experiment, because standard textbooks in physical chemistry vary widely in their approach.

Given the canonical partition function Q for a system,

$$Q = \sum_i e^{-E_i/kT} \tag{3}$$

where E_i is the energy of the ith quantum state of the entire system, the Helmholtz free energy is given by the equation

$$A = -kT \ln Q \tag{4}$$

For a one-component system, the chemical potential per mole is given by[1]

$$\mu = \left(\frac{\partial A}{\partial n} \right)_{T,V} = -kT \left(\frac{\partial \ln Q}{\partial n} \right)_{T,V} = -RT \left(\frac{\partial \ln Q}{\partial N} \right)_{T,V} \tag{5}$$

where n is the number of moles in the system and N is the number of molecules in the system ($= N_0 n$, where N_0 is Avogadro's number).

In order to evaluate the canonical partition function Q for a gas, we shall consider the system to be composed of an aggregate of essentially independent particles (molecules). As we shall see later, a crystal may be considered to a good approximation as an aggregate of independent harmonic oscillators. Each of these has its own microcanonical partition function:

$$q_i = \sum_j e^{-\varepsilon_j^{(i)}/kT} \tag{6}$$

where $\varepsilon_j^{(i)}$ is the energy of the ith oscillator in the jth quantum state of that oscillator. Since the oscillators are only very weakly interacting, the canonical partition function of the solid is a simple product of the microcanonical partition functions of the individual oscillators:

$$Q_s = \prod_i q_i \qquad \ln Q_s = \sum_i \ln q_i \tag{7}$$

Since many of these oscillators differ from each other in the values of their frequencies, energy levels, and partition functions, it is convenient to define a new quantity q_s which is the geometric mean of all of the q_i for the crystal:

$$q_s \equiv \left[\prod_{i=1}^{M} q_i \right]^{1/M} \qquad \ln q_s \equiv \frac{1}{M} \sum_{i=1}^{M} \ln q_i \tag{8}$$

where M is the number of oscillators. Then, for the crystal,

$$\ln Q_s = M \ln q_s = 3tN \ln q_s \tag{9}$$

where t is the number of atoms in a molecule. Since $\ln q_s$ can be shown to be independent of N, we find from Eq. (5) that

$$\mu_s = -3tRT \ln q_s \tag{10}$$

For a one-component ideal gas, the microcanonical partition function for an individual molecule is q_g. Therefore, under all ordinary conditions, we may write for a gas

$$Q_g = \frac{q_g^N}{N!} \tag{11}$$

where the division by $N!$ takes into account the fact that the individual molecules are indistinguishable. With the aid of the Sterling approximation for $\ln(N!)$, we obtain

$$\ln Q_g = N \ln q_g - N \ln N + N \tag{12}$$

Using Eq. (5) again, we obtain

$$\mu_g = -RT \ln \frac{q_g}{N} \tag{13}$$

We will now develop expressions for the microcanonical partition functions q_s and q_g to substitute into Eqs. (10) and (13).

Gaseous I$_2$. The partition function q_g is very well approximated as a product of terms arising from translational, rotational, vibrational, and electronic degrees of freedom:

$$q_g = q_{\text{trans}} q_{\text{rot}} q_{\text{vib}} q_{\text{el}} \tag{14}$$

The translational partition function is given by[2]

$$q_{\text{trans}} = \left(\frac{2\pi mkT}{h^2} \right)^{3/2} V \tag{15}$$

where m is the molecular mass, k is the Boltzmann constant, T is the absolute temperature, h is the Planck constant, and V is the volume within which the molecule is constrained to move.

For a molecule as massive as I_2, the rotational energy levels are very closely spaced and the partition function has the simple form[3]

$$q_{\text{rot}} = \frac{kT}{\sigma hcB_0} = \frac{T}{\sigma \Theta_{\text{rot}}} \tag{16}$$

Here σ is the symmetry number of the molecule, c is the velocity of light, and B_0 is the rotational constant (conventionally expressed in units of cm^{-1} with c expressed in cm s^{-1} units) defined by

$$B_0 \equiv \frac{h}{8\pi^2 Ic} \tag{17}$$

where I is the moment of inertia of the molecule,

$$I = \mu r_0^2 \tag{18}$$

The reduced mass μ (not to be confused with chemical potential) is defined by

$$\mu = \frac{m_1 m_2}{m_1 + m_2} \tag{19}$$

where m_1 and m_2 are the respective atomic masses. In I_2 the interatomic distance r_0 is 0.2667 nm, and the rotational constant B_0 is 0.037315 cm^{-1}.[4] The quantity Θ_{rot} is the *rotational characteristic temperature,* given by

$$\Theta_{\text{rot}} = \frac{hcB_0}{k} \tag{20}$$

The factor hc/k has the value 1.43877 cm K. Since the I_2 molecule is end-for-end symmetric, $\sigma = 2$.

For the vibrational partition function the molecule is regarded as a quantum mechanical harmonic oscillator, for which[5]

$$q = (1 - e^{-h\nu_0/kT})^{-1} = (1 - e^{-\Theta_{\text{vib}}/T})^{-1} \tag{21}$$

where ν_0 is the molecular vibration frequency and Θ_{vib} is the *vibrational characteristic temperature,*

$$\Theta_{\text{vib}} = \frac{h\nu_0}{k} = \frac{hc\tilde{\nu}_0}{k} \tag{22}$$

For the I_2 molecule, $\tilde{\nu}_0$ has the value 213.3 cm^{-1}.[4]

Equation (21) as written applies to an oscillator for which the reference energy is the energy of the vibrational ground state ($v = 0$); i.e., the $v = 0$ state in a gas molecule has been assigned zero energy. For the present situation, in which I_2 molecules in the vapor phase are in equilibrium with crystalline iodine, it is more convenient to take the reference energy to be that of an I_2 molecule in the crystal when the *crystal* is in its ground vibrational state.† Accordingly the energy of the vibrational ground state of an I_2 molecule in the ideal gas phase is taken to be $\Delta\varepsilon_0$, which is the energy required to remove a molecule from the crystal at absolute zero temperature. Thus we should write for the I_2 molecule in the gas phase

$$q_{vib} = (1 - e^{-\Theta_{vib}/T})^{-1} e^{-\Delta\varepsilon_0/kT} \tag{23}$$

It remains to deal with q_{el}. The excited electronic states of I_2 are separated from the ground electronic state by an energy difference that is very large compared to kT. Therefore

$$q_{el} = 1 \tag{24}$$

Let us now introduce $\Delta\widetilde{E}_0^0 = N_0\Delta\varepsilon_0$, the energy needed to sublime 1 mol of crystalline I_2 into the ideal gas phase at absolute zero, and replace V by its ideal-gas equivalent NkT/p. We can then combine Eqs. (13) to (16), (23), and (24) to obtain

$$\mu_g = \Delta\widetilde{E}_0^0 - RT \ln\left[\left(\frac{2\pi mkT}{h^2}\right)^{3/2} \frac{kT}{p} \frac{T}{\sigma\Theta_{rot}} (1 - e^{-\Theta_{vib}/T})^{-1}\right] \tag{25}$$

Crystalline I_2.   The partition function for the crystalline state of I_2 consists solely of a vibrational part; the crystal does not undergo any significant translation or rotation, and the electronic partition function is unity for the crystal as it is for the gas.

The geometric mean partition function for the crystal can be expressed as

$$q_s = \left[\prod_{i=1}^{M} (1 - e^{-\Theta_i/T})^{-1}\right]^{1/M} \tag{26}$$

where Θ_i is defined in terms of $\widetilde{v}_i$ in the same way as Θ_{vib} is defined in terms of $\widetilde{v}_0$ in Eq. (22). Since the number of iodine atoms is $2N$ for a crystal containing N molecules of I_2 and since each atom contributes three degrees of freedom, the number of modes of vibration for the crystal is

$$M = 3 \times t \times N - 6 = 6N - 6 \cong 6N \tag{27}$$

The subtracted number 6 represents the 3 translational and 3 rotational degrees of freedom of the crystal as a whole and will henceforth be ignored.

We now present a brief discussion of the vibrations occurring in a crystal.[6,7] The crystal can be thought of as a gigantic molecule with a huge number of normal modes, and the student may find it useful to review the discussion of normal modes for small molecules given in Exps. 35 and 38. In the case of the I_2 crystal, each primitive (smallest) unit cell contains two molecules.[8] Figure 1 shows that these two molecules are distinguished easily because their spatial orientations are different. As a consequence of this crystal structure, there are 3×4 atoms = 12 mechanical degrees of freedom associated with each unit cell. In the gas phase there would be three translations, two rotations, and one vibration for each of the two I_2 molecules. In the crystal, however, only vibrations occur: six lattice modes, four librational modes, and two internal vibration (bond-stretching) modes.

Let us consider first the center-of-mass motions for each of the two I_2 molecules in a unit cell. These types of motion account for six degrees of freedom and give rise to two kinds of lattice vibration. When both I_2 molecules *in a given cell* move in phase with each other (say for example both are displaced in the $+x$ direction at the same time), there are

---

†It should be noted that the location of the energy zero is arbitrary; a different, but equally reasonable, choice is made in Exp. 47. All that matters is a consistent choice for the two phases in equilibrium—here gaseous and crystalline I_2.

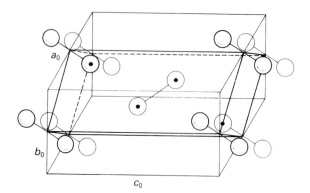

three so-called *acoustic* vibrations. When the two I_2 molecules *in a given cell* move out of phase (say one is displaced in the $+x$ direction while the other is displaced in the $-x$ direction), there are three *optic* vibrations.†

The four librations (torsional oscillations or rocking motions) arise because the crystal-field potential prevents the I_2 molecule from rotating as it would in the gas phase. There are some special crystals, called plastic crystals, in which symmetric molecules that interact weakly can still undergo hindered rotation in the solid phase, but $I_2(s)$ is not one of these. The librational motions for each I_2 occur about two axes (α, β) perpendicular to the I—I bond direction. The librations of the two I_2 molecules in the same unit cell are coupled—giving rise to SL_α, AL_α and SL_β, AL_β vibrations, where SL denotes symmetric libration (angle displacements in phase) and AL denotes antisymmetric libration (angle displacements out of phase).

Finally, there are two I—I bond-stretching vibrations that are essentially the same as the gas-phase stretching mode. As expected, these vibrations are coupled to produce an SS (symmetric stretch) in-phase vibration and an AS (antisymmetric stretch) out-of-phase vibration. In the latter case, one I_2 bond is stretching while the other is being compressed. As a result of interactions in the crystalline phase,[8,9] these SS and AS vibrations have lower frequencies than the gas-phase vibration at 213.3 cm^{-1}.

Now we must consider the fact that the motions of the I_2 molecules in any given unit cell are coupled to those of the molecules in other unit cells. An entire crystal of $N/2$ unit cells has $12 \times (N/2) = 6N$ degrees of freedom. Thus it would seem necessary to solve a $6N \times 6N$ secular determinant to obtain the normal-mode frequencies. However, symmetry and the periodicity of the lattice can be used to greatly simplify the problem,[6,7] and we can talk about 12 vibrational modes associated with each of $N/2$ discrete values of a *wave vector* **k.** This wave vector has a magnitude

$$k = \frac{2\pi}{\lambda} \tag{28}$$

and a direction that specifies the propagation direction of a *traveling wave* (i.e., of the "crests and troughs" of the periodic displacements). The vibrational wave motion in the crystal can be represented by traveling-wave equations of the general form

$$A_j(\mathbf{r}, t) = A_{j0}\cos(2\pi\nu_j t - \mathbf{k} \cdot \mathbf{r}) \tag{29}$$

---

†The name *optic mode* comes from the behavior of ionic crystals such as Na$^+$Cl$^-$. When Na$^+$ and Cl$^-$ in a given cell move out of phase with each other, there is an oscillating electric dipole. Optical absorption will occur for light having frequency equal to that of the optic lattice mode.

where A_j is the instantaneous amplitude of a displacement of type j ($j = 1$ to 12) in the cell at point $\mathbf{r}$. Equation (29) describes the 12 normal modes associated with a given $\mathbf{k}$, i.e., with a given wavelength and direction for the periodic displacements of molecules in *different cells*. All allowed $\mathbf{k}$ values lie inside a *Brillouin zone* (*BZ*), a region in $\mathbf{k}$ space ("reciprocal space") bounded by a polyhedron that is centered around k_x, k_y, $k_z = 0, 0, 0$.[6,7] As $k \rightarrow 0$, adjacent cell displacements approach being in phase and $\lambda \rightarrow \infty$; when $k \rightarrow k_{max}$ at the Brillouin-zone boundary, $\lambda \rightarrow \lambda_{min}$, which is the minimum wavelength for the given $\mathbf{k}$ direction.

The ν-versus-k curves, called phonon dispersion curves,[6,7] are shown in Fig. 2 for the **a**-axis direction in an I_2 crystal. These and similar dispersion curves in other directions were obtained by Smith et al.[9] using the technique of inelastic neutron scattering.[10,11] The frequencies of internal stretching and libration are not affected greatly by the coupling between unit cells; i.e., each ν_j is roughly constant for all $\mathbf{k}$ values for these modes. In contrast the center-of-mass motion is strongly affected, especially for the acoustic branches TA_1, TA_2, and LA and moderately for the optic branches TO_1, TO_2, and LO. The acoustic lattice vibrations are three-dimensional analogs of the one-dimensional vibrations of a violin string or the air in an organ pipe and the two-dimensional vibrations of a drum head. In the continuum (long-wave) limit, they represent three-dimensional vibrations in a bowl of Jello. Such acoustic frequencies range from 0 at the *BZ* center (point Γ) to $\sim 1-2$ THz at the *BZ* edge [1 terahertz (1 THz) $= 10^{12}$ Hz $= 33.3$ cm^{-1}]. The notation *TA* means transverse (shear) acoustic, and *LA* means longitudinal (compression–rarefaction) acoustic.

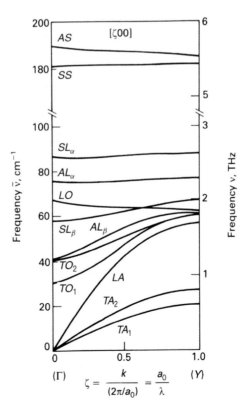

FIGURE 2

Phonon dispersion curves for $I_2(s)$ in the **a**-axis direction from the center of the *BZ* (Γ) to the boundary (Y) at 77 K. Adapted by permission from Ref. 9b.

TABLE 1   Discrete (representative) phonon frequencies in I$_2$ crystals[a]

Mode no. j	Type of mode	Representative frequency $\tilde{\nu}_i$, cm^{-1}	Frequency range	
			cm^{-1}	Point(s) in BZ[b]
1	TA_1	21.0	0–40.5	Γ, T
2	TA_2	26.5	0–53.1	Γ, T
3	LA	33.0	0–56.2	Γ, Y
4	TO_1	41.0	30.7–60.0	Γ, Y
5	TO_2	49.0	41.0–60.0	Γ, Y
6	AL_β	51.5	41.7–61.0	Γ, Y
7	SL_β	58.0	57.7–66.5	Γ, Y
8	LO	59.0	65.4–46.5	Γ, Z
9	AL_α	75.4	flat	Γ
10	SL_α	87.4	flat	Γ
11	SS	180.7	flat	Γ
12	AS	189.5	flat	Γ

[a] Estimated from Ref. 9.
[b] Points T and Z are not shown in Fig. 2; they are elsewhere on the surface of the BZ. See Ref. 9.

In order to assign frequency values $\tilde{\nu}_j$ to each of the 12 branches, we average the available values[9] over the Brillouin zone. The resulting values are given in Table 1, where limiting values at the zone center (point Γ) and zone edge (points Y, T, or Z) are also given. The choice of a single frequency for each mode corresponds to a version of the Einstein model for a solid.[1,6] This is quite reasonable for all branches except the three acoustic branches. For those three modes, the Debye model[1,6] would provide a better approximation. However, the simpler Einstein approximation for TA_1, TA_2, and LA is adequate for the present purposes.

We can now formulate the desired expressions for q_s and μ_s. Using Eqs. (8) to (10), together with the fact that $t = 2$ for I$_2$ and each unit cell contains two I$_2$ molecules, we find

$$\ln q_s = \frac{1}{6N} \sum_{i=1}^{6N} \ln q_i = \frac{1}{6N} \sum_{i=1}^{12(N/2)} \ln q_i \tag{30}$$

The number of discrete **k** values is $N/2$, the number of primitive unit cells in the crystal. Each of these is assumed to yield the *same* set of 12 branch frequencies ν_j. Thus we can simplify Eq. (30) to

$$\ln q_s = \frac{1}{6N} \frac{N}{2} \sum_{j=1}^{12} \ln q_j = \tfrac{1}{12} \sum_{j=1}^{12} \ln q_j \tag{31}$$

where 12 is the number of degrees of freedom per unit cell. Finally, for I$_2(s)$, we obtain

$$\ln q_s = -\tfrac{1}{12} \sum_{j=1}^{12} \ln(1 - e^{-\Theta_j/T}) \tag{32}$$

and

$$\mu_s = -6RT \ln q_s = \frac{RT}{2} \sum_{j=1}^{12} \ln(1 - e^{-\Theta_j/T})$$

$$= \frac{RT}{2} \ln\left[\prod_{j=1}^{12} (1 - e^{-\Theta_j/T}) \right] \tag{33}$$

Equilibrium Between Crystal and Gas. On substituting the expressions of Eqs. (25) and (33) into Eq. (2) and doing some rearranging and simplifying, we obtain

$$\ln p - \ln\left[\frac{T^{7/2}\Pi_{j=1}^{12}(1 - e^{-\Theta_j/T})^{1/2}}{(1 - e^{-\Theta_{\text{vib}}/T})}\right] = \ln\left[\left(\frac{2\pi mk}{h^2}\right)^{3/2}\frac{k}{\sigma\,\Theta_{\text{rot}}}\right] - \frac{\Delta\widetilde{E}_0^0}{\text{R}T} \qquad (34)$$

If the value of p is determined at *one* temperature, this equation can be solved for $\Delta\widetilde{E}_0^0$, the value of which is needed (along with Θ_{rot} and Θ_{vib}) to determine the chemical potential of gaseous I_2. Once $\mu_s(T)$ and $\mu_g(T)$ are both known, one can calculate $\Delta\widetilde{S}_{\text{sub}}$ and $\Delta\widetilde{H}_{\text{sub}}$. By contrast, the Clausius–Clapeyron equation, given by

$$\ln p = \text{constant} - \frac{\Delta\widetilde{H}_{\text{sub}}}{RT} \qquad (35)$$

in its approximate integrated form, requires at least two values of p at different temperatures in order to obtain a value of $\Delta\widetilde{H}_{\text{sub}}$.

Equation (35) has obvious similarities to Eq. (34). This correspondence can be enhanced by replacing $\Delta\widetilde{H}_{\text{sub}}/RT$ with $\Delta\widetilde{E}_{\text{sub}}/RT + 1$, which is equivalent since $\Delta(p\widetilde{V}) \cong RT$ is an excellent approximation under the conditions of the present experiment. However, $\Delta\widetilde{E}_{\text{sub}}$ is temperature dependent and refers to the energy of sublimation at the temperature of the experiment rather than at absolute zero. This temperature dependence is reflected in the statistical treatment by the variation with T of the second term on the left-hand side (LHS) of Eq. (34).

If p values have been measured at several temperatures, the LHS of Eq. (34) can be plotted against $1/T$, and the value for $\Delta\widetilde{E}_0^0$ can be determined from the slope of a straight line fitted graphically or by least squares. In addition the intercept can be compared with the predicted value of the constant term on the right-hand side (RHS) of Eq. (34). Alternatively, it is possible to calculate a $\Delta\widetilde{E}_0^0$ value from each p, T data point and see how well these values agree.

Entropy and Enthalpy of Sublimation. Since we have a system of only one component, the chemical potentials for I_2 in crystalline and gaseous forms, given in Eqs. (33) and (25), respectively, are equivalent to the molar Gibbs free energies $\widetilde{G}_s$ and $\widetilde{G}_g$, aside from an additive constant. The entropies of the two phases can be obtained by differentiating with respect to temperature. The expressions obtained are

$$\widetilde{S}_s = -\left(\frac{\partial\widetilde{G}_s}{\partial T}\right)_p = -\left(\frac{\partial\mu_s}{\partial T}\right)_p$$

$$= \frac{R}{2}\sum_{j=1}^{12}\left[\frac{\Theta_j/T}{e^{\Theta_j/T} - 1} - \ln(1 - e^{-\Theta_j/T})\right] \qquad (36)$$

$$\widetilde{S}_g = -\left(\frac{\partial\widetilde{G}_g}{\partial T}\right)_p = -\left(\frac{\partial\mu_g}{\partial T}\right)_p$$

$$= \frac{\Delta\widetilde{E}_0^0 - \mu_g}{T} + \frac{7}{2}R + R\frac{\Theta_{\text{vib}}/T}{e^{\Theta_{\text{vib}}/T} - 1} \qquad (37)$$

The heat of sublimation at temperature T is

$$\Delta\widetilde{H}_{\text{sub}} = T\Delta\widetilde{S}_{\text{sub}} = T(\widetilde{S}_g - \widetilde{S}_s) \qquad (38)$$

EXPERIMENTAL

The determination of the vapor pressure of solid iodine at temperatures from 20 to 70°C in steps of about 10°C is accomplished through spectrophotometric measurements of the absorbance A of the iodine vapor in equilibrium with the solid at the absorption maximum (520 nm), and also at 700 nm, where the molar absorption coefficient of iodine vapor is so small as to be negligible. On the short-wavelength side of the maximum, there is no accessible wavelength at which the absorption is negligible; hence the baseline can be drawn only from the long-wavelength side. In most instruments the absorbance is indicated directly; at every wavelength it is related to the incident beam intensity I_0 and the transmitted beam intensity I by the equation

$$A_\lambda = \log\left(\frac{I_0}{I}\right)_\lambda \tag{39}$$

The absorption peaking at 520 nm is for the transition

$$X^1\Sigma_g^+ \rightarrow B^3\Pi_{0^+u} \tag{40}$$

(see Exp. 39). There is also a transition

$$X^1\Sigma_g^+ \rightarrow A^3\Pi_{1u} \tag{41}$$

with a broad absorption peaking at about 700 nm, but its extinction is quite small in comparison to the other band and can be neglected.[12]

The absorption cell containing iodine in the solid and vapor states must also contain air or nitrogen at about 1 atm to provide pressure broadening of the extremely sharp and intense absorption lines of the rotational fine structure (which can be individually resolved only by special techniques of laser spectroscopy). The reason lies in the logarithmic form of Eq. (39). Within the slit width or resolution width of the kinds of spectrophotometers that may be used in this experiment, low-pressure $I_2(g)$ exhibits many very sharp lines separated by very low background absorption. The instrument effectively averages transmitted intensity I, not absorbance A, over the sharp peaks and background within the resolution width, but the logarithm of an average is not the average of the logarithm. If the extremely sharp lines are so optically "black" that varying the concentration has little effect on the amount of light transmitted in them, the absorbance is controlled mainly by the background between the lines, and the contribution of the lines to the absorbance is largely lost. Increasing the concentration of gas molecules increases the number of molecular collisions and thus decreases the time between them. This can greatly broaden the lines and lower their peak absorbances, causing them to overlap and smooth out the spectrum over the resolution width so that the absorbance readings are meaningful averages over that range. This effect is readily demonstrated experimentally by comparison of spectra taken of I_2 vapor with and without air present.[13]

Figure 3 shows the absorption spectrum of I_2 vapor over the range of interest at the vapor pressure of iodine at 27°C, at moderate resolution and at low resolution. A low-resolution spectrum, obtained with wide slits, is preferable for this experiment, since the vibrational structure is averaged out, facilitating the determination of the absorbance at 520 nm. An even lower resolution may be entirely satisfactory for this experiment.

The spectrometer to be used need cover only the visible portion of the electromagnetic spectrum. Preferably it should be a double-beam instrument, to allow for compensation for absorption by the cell windows. However, a single-beam instrument may be used if the absorption spectrum of an equivalent set of cell windows is obtained separately, so that the absorbances of the windows can be subtracted from those of the iodine-containing cell. The spectrophotometer must have a cell compartment large enough to contain the

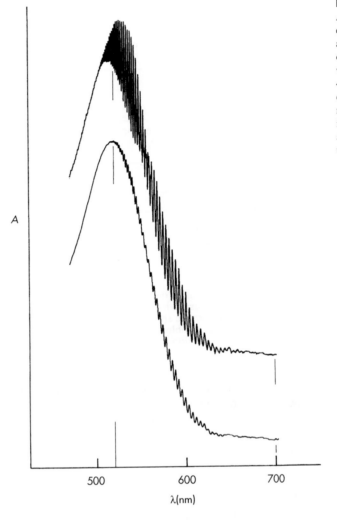

A

500   600   700

λ(nm)

FIGURE 3
Absorbance of $I_2(g)$, in equilibrium with the solid at 26°C and in the presence of air at about 1 atm, for wavelengths ranging from 470 to 700 nm. The top curve is at moderately high resolution. The bottom curve is at relatively low resolution, about the highest suitable for this experiment.

water-jacketed absorption cell that must be used for measurements at several temperatures. Further details concerning spectrophotometers are given in Chapter XIX and in the operating manuals of the specific instrument to be used.

A suitable water-jacketed absorption cell is shown in Fig. 4. The central iodine-containing cell should be constructed of Pyrex or optical glass. The end windows, which must be planar and polished, must be sealed (glass-blown) to the body of the cell with a torch, not cemented; cements are unsatisfactory because of the corrosive action of iodine vapor. The cell can be fabricated with commercially available windows by an experienced glassblower, or made by modifying a commercially available spectrophotometer cell, or fabricated on special order by a commercial vendor (e.g., NSG Precision Cells, Inc., Farmingdale, NY, or the Harrick Scientific Corporation, Ossining, NY). The optical path between the end windows should be about 4 or 5 cm and should be accurately known. At least 10 mg of solid iodine crystals (for a cell of about 15 cm³ volume) must be introduced into the cell through a side tube before the side tube is sealed off close to the body of the cell with a torch, with about 1 atm of air inside. The details of the water jacket design are very flexible but must ensure that the end windows of the inner cell are bathed in the circulating water to prevent condensation of iodine crystals on the windows at the higher

FIGURE 4

Suggested design for water-jacketed Pyrex spectrophotometer cell containing iodine. The recommended material for the main part of the jacket is Delrin, which can be easily machined. The outer windows may be either glass or clear plastic.

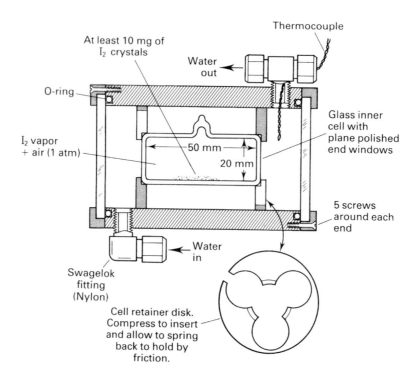

temperatures of the experiment. A calibrated thermocouple should be provided for measuring the temperature inside the water jacket. Plastic tubing is used to connect the water jacket to a water thermostat bath equipped with a variable temperature control and a circulating pump. The cell, with its water jacket constructed as shown, can be used year after year with little or no maintenance.

Note: A separate set of windows, both for the inner cell and for the water jacket, should be available for optical compensation as indicated above.

For each temperature, sufficient time must be allowed for the contents of the inner cell to come to equilibrium with the circulating water. The spectrophotometer can be set to 520 nm and used as a monitor to follow the approach to equilibrium. When the absorbance and the thermocouple reading are no longer changing, the temperature should be noted and the absorption spectrum should be recorded from 700 to 450 nm. It may be necessary to experiment with different settings of the absorbance scale, the slit and/or gain controls, the scanning speed, and the damping time to obtain optimum performance.

CALCULATIONS

For each temperature, determine the net absorbance as the difference between the absorbances at 520 and 700 nm:

$$A = A_{520} - A_{700} \tag{42}$$

The net absorbance is related to the concentration c and the pressure p of iodine vapor in the cell as follows (assuming the perfect-gas law):

$$A = \varepsilon dc = \frac{\varepsilon d}{RT} p \tag{43}$$

TABLE 2 Molar absorption coefficients of iodine vapor at $\lambda = 520$ nm^a

t, °C	ε, L mol^{-1} cm^{-1}
20.0	691
30.0	682
40.0	672
50.0	663
60.0	654
70.0	646

^{a} Based on an equation derived by Sulzer and Wieland (Ref. 12). The function "Tg" in their equation is the hyperbolic tangent. They tested this equation in the range 423 to 1323 K (150 to 1050°C). Their cells apparently contained no gas other than I_2 vapor, but at those temperatures the vapor pressures were apparently high enough to obtain the needed pressure broadening without the addition of another gas. Their equation satisfactorily fits lower-temperature data obtained with a cell containing about 1 atm of air.

where c is the concentration of I_2 vapor, p is the partial pressure of I_2 vapor, d is the optical path length inside the inner absorption cell, and ε is the molar absorption coefficient for I_2 vapor at 520 nm and temperature T.

The value of the molar absorption coefficient ε at each temperature can be found from Table 2 by interpolation. Then p may be calculated with Eq. (43). In the calculations of this experiment, considerable care must be taken with units. It is desirable to obtain p in pascals for the statistical mechanical calculations; accordingly ε should be converted into units of m^2 mol^{-1}, d should be in m, and R should be in units of J K^{-1} mol^{-1}.

To determine $\Delta \widetilde{E}_{\text{sub}}$ from the approximate Clausius–Clapeyron equation [Eq. (35)], plot ln p against $1/T$ and determine the slope of the best straight-line fit to the data by graphical or least-squares methods.

The values of p and T can now be used for the statistical mechanical calculations. In order to calculate the rotational characteristic temperature Θ_{rot} with Eq. (20), use the literature value[4] for the rotational constant $\widetilde{B}_0 = 0.037315$ cm^{-1} [or calculate $\widetilde{B}_0$ from the internuclear distance in the molecule, $r_0 = 0.2667$ nm, with Eqs. (17) to (19)]. From the literature value of the molecular vibrational frequency in the gas phase,[4] $\widetilde{\nu}_0 = 213.3$ cm^{-1}, calculate the vibrational characteristic temperature Θ_{vib} with Eq. (22). From the phonon dispersion data in Table 1, calculate the 12 vibrational characteristic temperatures Θ_j.

Calculate $\Delta \widetilde{E}_0^0$ with Eq. (34) at each temperature. A particularly easy and rapid way to do these calculations on a computer is with a spreadsheet program.

Do the values obtained for $\Delta \widetilde{E}_0^0$ agree satisfactorily? If not, check the calculations and/or consider possible systematic errors. Plot the LHS of Eq. (34) against $1/T$, and determine both $\Delta \widetilde{E}_0^0$ and the constant term graphically or by least squares. Does this value of $\Delta \widetilde{E}_0^0$ agree with the average of the values obtained by direct application of Eq. (34)? Does the constant term agree with the theoretical value?

Calculate the molar entropies $\widetilde{S}_s$ and $\widetilde{S}_g$ of the crystalline and vapor forms of I_2 at 320 K with Eqs. (36) and (37), and obtain the molar heat of sublimation $\Delta \widetilde{H}_{\text{sub}}$ with Eq. (38). Compare it with the value obtained by the Clausius–Clapeyron method and with any literature values that you can find.[14]

DISCUSSION

Of the two methods of determining $\Delta \widetilde{E}_0^0$ with Eq. (34), which do you judge gives the more *precise* value? Which gives the more *accurate* value? Which provides the better test of the overall statistical mechanical approach? Compare this approach with the purely thermodynamic method using the integrated Clausius–Clapeyron equation, taking into account the approximations involved in the latter (see Exp. 13). State the average temperature corresponding to your Clapeyron value of $\Delta \widetilde{H}_{\text{sub}}$.

Comment on the choice of representative values of $\widetilde{v}_j$ for the 12 vibrational modes of the crystal. How much would reasonable changes (say, 10 to 20 percent) in these values affect the results of the calculations? If possible, comment on the effect of using the Debye approximation for the acoustic lattice modes instead of the Einstein approximation.

Other experimental and theoretical methods have been developed for the determination of the heat of sublimation of solid iodine; these too are suitable for undergraduate laboratory experiments or variations on this experiment. Henderson and Robarts[15] have employed a photometer incorporating a He–Ne gas laser, the beam from which (attenuated by a $CuSO_4$ solution) has a wavelength of 632.8 nm, in a "hot band" near the long-wavelength toe of the absorption band shown in Fig. 3. Stafford[16] has proposed a thermodynamic treatment in which a free-energy function (*fef*), related to entropy, is used in calculations based on the third law of thermodynamics. In this method either heat capacity data or spectroscopic data are used, and as in the present statistical mechanical treatment, the heat of sublimation can be obtained from a measurement of the vapor pressure at only one temperature.

SAFETY ISSUES

If a filled and sealed cell is made available, there are no safety hazards. If the cell is to be filled and sealed by the student, the iodine must be handled carefully with a spatula or tweezers, since it is corrosive to the skin and also stains badly.

APPARATUS

Visible spectrophotometer (preferably double beam); water-jacketed absorption cell containing solid I_2 and air or N_2 (see text and Fig. 4); water thermostat with adjustable temperature regulator and circulating pump; plastic tubing to connect circulating pump with waterjacket of absorption cell; calibrated thermocouple and associated instrumentation (see Chapter XVII).

REFERENCES

1. I. N. Levine, *Physical Chemistry,* 6th ed., sec. 21.2, McGraw-Hill, New York (2009).

2. P. W. Atkins and J. de Paula, *Physical Chemistry,* 8th ed., p. 592, Freeman, New York (2006).

3. *Ibid.,* pp. 592–596.

4. Adapted from R. P. Huber and G. Herzberg, *Molecular Spectra and Molecular Structure IV: Constants of Diatomic Molecules,* p. 332, Van Nostrand Reinhold, New York (1979). Although this book is now out of print, its contents are available as part of the *NIST Chemistry WebBook* at http://webbook.nist.gov.

5. P. W. Atkins and J. de Paula, *op. cit.,* pp. 596–597.

6. C. Kittel, *Introduction to Solid State Physics,* 6th. ed., Wiley, New York (1995).

7. N. W. Ashcroft and N. D. Mermin, *Solid State Physics,* Harcourt, Orlando, FL (1976).

8. F. van Bolhuis, P. B. Koster, and T. Migghelsen, *Acta Crystallogr.* **23,** 90 (1967).

9. (a) H. G. Smith, M. Nielsen, and C. B. Clark, *Chem Phys. Lett.* **33,** 75–78 (1975); (b) H. G. Smith, C. B. Clark, and M. Nielsen, in J. Lascombe (ed.), *Dynamics of Molecular Crystals,* pp. 4411–4446, esp. fig. 2, Elsevier, Amsterdam (1987).

10. C. Kittel, *op. cit.,* pp. 120–121.

11. N. W. Ashcroft and N. D. Mermin, *op. cit.,* pp. 470–474.

12. P. Sulzer and H. Wieland, *Helv. Phys. Acta* **25,** 653 (1952).

13. J. G. Calvert and J. N. Pitts, Jr., *Photochemistry,* p. 184, ref. 424 in chap. 5, Wiley, New York (1966).

14. D. A. Shirley and W. F. Giauque, *J. Am. Chem. Soc.* **31,** 4778 (1959).

15. G. Henderson and R. A. Robarts, Jr., *Am. J. Phys.* **46,** 1139 (1978).

16. F. Stafford, *J. Chem. Educ.* **40,** 249 (1963).

GENERAL READING

N. W. Ashcroft and N. D. Mermin, *Solid State Physics,* chaps. 4, 5, 7, 22–24, Harcourt, Orlando, FL (1976).

D. A. McQuarrie, *Statistical Mechanics,* reprint ed., University Science Books, Sausalito, CA (2000).

C. Kittel, *Introduction to Solid State Physics,* 6th ed., chap. 4, Wiley, New York (1995).

XVI
Electronic Devices and Measurements

The rapid development of solid-state electronic devices in the last two decades has had a profound effect on measurement capabilities in chemistry and other scientific fields. In this chapter we consider some of the physical aspects of the construction and function of electronic components such as resistors, capacitors, inductors, diodes, and transistors. The integration of these into small operational amplifier circuits is discussed, and various measurement applications are described. The use of these circuit elements in analog-to-digital converters and digital multimeters is emphasized in this chapter, but modern integrated circuits (ICs) have also greatly improved the capabilities of oscilloscopes, frequency counters, and other electronic instruments discussed in Chapter XIX. Finally, the use of potentiometers and bridge circuits, employed in a number of experiments in this text, is covered in the present chapter.

CIRCUIT ELEMENTS

Figure 1 shows the symbols of common elements used in electronic circuits. These can be classed as either *passive* components, such as resistors, capacitors, inductors, and diodes, or *active* components, such as bipolar and field-effect transistors, and silicon-controlled rectifiers (SCRs). Some of the key features and physical characteristics of these devices are summarized in the first two sections of this chapter.

Resistors. Resistors are available in three commercial forms. The most common type, used for noncritical applications, is the graphite *composition resistor*, a mixture of graphite and silica in which their ratio determines the resistance. These resistors are generally molded with a binder in the shape of a cylinder, with two electrodes bonded at the ends. *Film resistors* are ceramic insulating rods coated with a thin conducting film of carbon, metal, metal oxide, or cermet, a mixture of glass and metal alloys. The film is then cut with a spiral groove to increase the effective length so as to produce the desired resistance. Electrodes are attached at each end and the unit is encased in a glass or ceramic material. *Wire-wound resistors* are made similarly but thin wire of an appropriate length is wrapped around a ceramic insulator, which is then coated. The wire is formed from alloys such as manganin or constantan whose resistance changes only slightly with temperature,

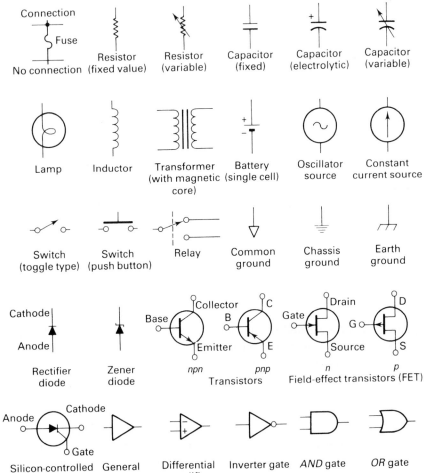

FIGURE 1
Common circuit component symbols.

about 10 ppm/°C versus 100 to 1000 ppm/°C for film or composition resistors. The color code for the labeling of fixed resistors is shown in Fig. 2.

Precision wire-wound resistors are used for critical applications in which low noise and resistance stability are important, but since they are coils, they have an inductive reactance that must be taken into account for ac circuits. By folding the insulated wire in the middle and carefully forming parallel windings, the inductive component can be reduced by about a factor of 100. Precision standard resistors with an accuracy of about 50 ppm in the range of 1 Ω to 10 MΩ are available from Electro Science Industries and other companies. These are often combined in the form of decade resistance boxes to provide a wide range of fixed values.

Resistors whose values can be varied are termed *potentiometers*. They are made of the same materials as fixed resistors but have a movable wiper to contact a coil of resistance wire or a strip of resistive film at any point along the resistor. Potentiometers used only occasionally to adjust a circuit are called *trimmers,* while those employed for high-wattage applications such as control of heating mantles and ovens are called *rheostats.* Precision potentiometers have played an important role in the measurement of electrical quantities and are considered in detail in a later section.

Capacitors.   A capacitor consists of two conductors separated by an insulator. The capacitance C gives a measure of the charge separation Q produced when a potential V is

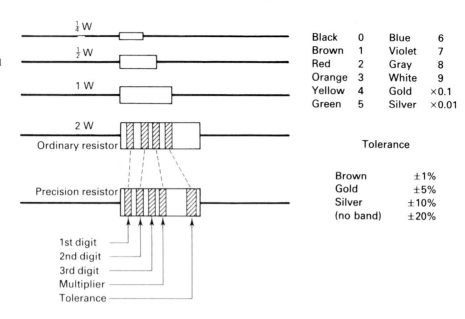

applied across the electrodes and is given by $C = Q/V$. C is determined by the geometry and the dielectric component of a capacitor. For the simplest case of two plates of area a separated by a distance d, the capacitance is

$$C = \varepsilon_0 \kappa_d \frac{a}{d} \tag{1}$$

where $\varepsilon_0 = 8.854 \times 10^{-12}$ F m^{-1} is the permittivity of free space and κ_d is the dielectric constant of the insulating material between the plates. For a cylindrical capacitor, such as is sometimes used for dipole-moment determinations, the cell capacitance is

$$C = \frac{2\pi\varepsilon_0 \kappa_d l}{\ln(r_2/r_1)} \tag{2}$$

where l is the length of concentric cylinders of radii r_1 and $r_2 = r_1 + d$ with the dielectric material in between. This reduces to Eq. (1) when $d/r_1 \ll 1$, where d is the gap between the cylinders.

Capacitors are produced in several forms. *Film* capacitors are commonly made of two thin metal foils, separated by a dielectric material such as paper or plastic. The foils are rolled into a tubular form with conducting leads added at each end, yielding capacitance values in the range of 0.01 to 50 μF. Often the conducting film is coated on one side of the dielectric film to give a very compact structure. Plastic dielectrics include Mylar, Teflon, polystyrene, polycarbonate, and other materials. The best film capacitors are Teflon capacitors, which have a very low current leakage and temperature dependence and are not greatly affected by humidity. Polycarbonate capacitors also have low leakage but are subject to high-frequency signal losses.

Mica and *ceramic* disk capacitors consist of stacks of the rigid dielectric material, interleaved with metal foils, deposited metallic films, or conducting silver paste. These are encapsulated in epoxy or a fired glass to produce a very durable device that can withstand high temperatures and mechanical shock. The ceramic material is usually titanium dioxide (TiO_2) or barium titanate ($BaTiO_3$). Ceramic capacitors are also produced in tube form with plated electrodes on the inside and outside of the tube. Ceramic capacitors have a range of 10 pF to 1 μF, while mica capacitors cover the range of 5 pF to 10 μF.

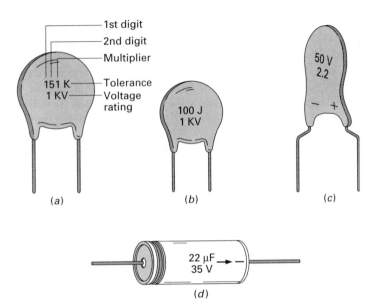

FIGURE 3
Typical labeling schemes for common capacitors. (*a*) and (*b*) are ceramic capacitors of 150 and 10 pF values, respectively. Tolerances are often indicated with letters, with lower values meaning less uncertainty, e.g., J = ±5%, K = ±10%. (*c*) and (*d*) are tantalum and aluminum electrolytic capacitors of values 2.2 and 22 μF, respectively. Polarity is irrelevant for ceramic capacitors but is indicated and must be maintained for electrolytic capacitors.

For applications requiring high capacitance and where the polarity is fixed, as in dc power supplies, inexpensive *electrolytic* capacitors are commonly used. These employ an aluminum or tantalum film that is electrolyzed to form a thin oxidized layer that serves as the dielectric material. A low-conductance paste or solution is added between layers, and because the films are extremely thin, large capacitance values are possible. These range from 0.01 μF up to 100,000 μF, with the largest values applicable only in cases where the voltage does not exceed a few volts. Electrolytic capacitors are usually manufactured in tubular form with the metal electrode indicated as positive. This polarity must be maintained in use, since a negative potential causes electrolytic destruction of the oxide film, leading to shorting and sometimes explosive rupture of the sealed capacitor.

Capacitor values are generally indicated on the body of the capacitor. If no units are shown and the value is less than 1, the unit is μF; if the value is greater than 1, the units are in pF. Some common ceramic and electrolytic capacitors are shown in Fig. 3.

Variable capacitors are available with values up to a few hundred picofarads. These are commonly formed from sets of interleaved plates, one fixed and the other attached to a shaft. Rotation changes the effective area and thereby the capacitance. Arrangements with sliding cylinders are also used and dielectrics include air, mica, and ceramic. *Varacter diodes, p–n* junction diodes in which the capacitance is determined by the reverse bias voltage, are now finding increasing use in circuits, because their capacitance can be actively controlled by other electronic circuit elements.

Inductors and Transformers. Inductance is related to the fact that a moving charge has associated with it a magnetic field. Changes in this field arising from variations in the charge current result in the generation of a voltage in the circuit that is given by

$$V = L\frac{dI}{dt} \tag{3}$$

where the proportionality constant L is called the inductance. The induced voltage acts to oppose the current change and its magnitude increases with the frequency of oscillation of the current. The unit of inductance is the *henry* (H), after Joseph Henry, an early American investigator of inductive effects.

Inductors, sometimes termed *chokes,* are usually made of coils of wire wound adjacent to one another to concentrate the magnetic flux produced by each coil. Ten to 100 loops about a hollow support serve to produce inductances in the μH to mH range, which is suitable for most high-frequency circuits. Large inductance values require hundreds of turns, usually about a core of ferromagnetic material such as iron. This core serves to greatly concentrate the magnetic flux in the region of the coil so as to yield inductance values up to several hundred henrys. Large inductors are useful principally in power applications and as filter elements.

Inductors are never pure, since there is always resistance in and capacitance between the coil windings. Because of this and because of the relatively large size and cost of inductors, inductive reactance is not used as much as capacitive reactance for ac control in electronic circuits.

The combination of two or more inductors arranged so that the coils are inductively coupled to each other is a *transformer.* The coupling can be through air or a metallic core. Power transformers are widely used to provide various voltages from the standard 60-Hz, 115-V power lines, the scaling being given simply by the turns ratio of the output (secondary) to input (primary) inductors. Smaller pulse transformers also find use in coupling high-frequency ac signals while isolating dc levels of primary and secondary circuits.

OPERATIONAL AMPLIFIERS

The use of semiconductor devices has been greatly facilitated by the incorporation of many discrete elements on a single microchip of specific purpose, such as amplification, memory storage, switching, time delay, and so on. An example is the operational amplifier (op amp), a device whose function is easy to understand and whose internal construction has been carefully designed to provide nearly ideal characteristics of high input resistance, high gain over a wide frequency range, and low output resistance. While it is not necessary to know the full details of the inner operation of the simple triangular symbol used to represent the operational amplifier in order to make effective use of it, some knowledge of its basic circuit elements is helpful in appreciating its specifications and limitations. A more extensive discussion of these aspects can be found in Refs. 1 to 6.

Figure 4 shows a simplified view of the three basic stages of an LF351 operational amplifier. The input stage consists of two matched *p*-channel JFETs used in a differential mode. This arrangement provides a very high input resistance and efficient cancellation of any noise voltages common to both inputs. The second stage consists of several BJT amplifiers to provide an overall current gain of about 10^6. The final output stage serves to provide current with an output resistance of a few ohms or less. The device actually has many more transistors, not indicated in the figure, which serve to provide appropriate biasing levels and current sources.

Typical specifications of several common operational amplifiers are provided in Table 1. The LM741 BJT amplifier consists of 20 transistors incorporated in a single chip, which is produced in high volume at a cost of less than $1.00. The 741 was the first general-purpose op amp and is still used since it is inexpensive, robust, and adequate for many routine applications at frequencies below about 10 kHz. However, more recent bipolar versions offer improved specifications; e.g., the LM11 op amp has a much lower input bias current, which is important in minimizing errors in many op amp applications.

Operational amplifiers that contain both bipolar and MOS or FET transistors on the same chip are termed BiMOS or BiFET op amps. These families have much higher input resistances ($\sim10^{12}$ versus $\sim10^6$ Ω for BJT op amps) and a frequency response generally

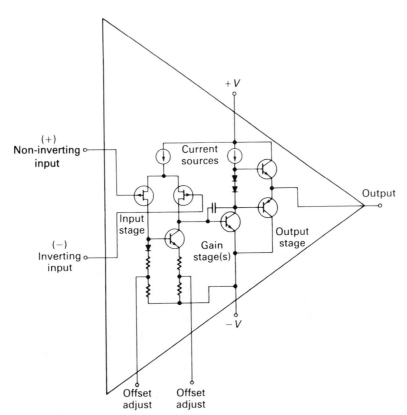

FIGURE 4
Simplified schematic of an
LF351 BiFET integrated
circuit operational amplifier
with FET input stage and BJT
amplification stages. (From
Ref. 4, with permission.)

TABLE 1 Some comparative specifications of common operational amplifiers[a]

Device	Bipolar		BiMOS		BiFET	
	LM741	LM11	CA3140A	TLC27M2A	LF351	LFT356
Input offset voltage V_{OS}, mV (max)	6	0.3	5	5	10	0.5
Input voltage drift $(dV_{OS}/dT)_{max}$, μV °C^{-1}	15	3	6	2	10	3
Input bias current I_b, nA (max)	200	0.05	0.04	0.001	0.2	0.05
Input offset current I_{OS}, nA (max)	200	0.01	0.02	0.001	0.1	0.01
Input noise voltage V_n, nV (Hz)$^{-1/2}$	20–30	150	40	38	16	12
Common-mode rejection ratio CMRR, dB (typ)	90	110	70	70	100	100
Unity-gain bandwidth f_u, MHz	1	0.5	3.7	0.7	4	4
Slew rate $(dV_O/dt)_{max}$, V μs^{-1}	0.5	0.3	7	0.6	13	12
Output current $(I_O)_{max}$, mA	25	2	10	10	20	25

[a] LM, LF: National Semiconductor; CA: Harris; TLC: Texas Instruments.

FIGURE 5

Representation of an operational amplifier.

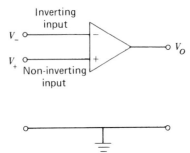

extending above 1 MHz when used in a circuit with a voltage gain of 1 (*unity-gain* condition; for a gain-of-10 circuit the frequency response would be reduced by a factor of 10). The 3140 op amp is pin compatible with the 741. In addition to these standard units, a number of manufacturers produce premium op amps with significantly improved performance at costs ranging from $5 to $200. For laboratory and research applications, where the component cost is usually small compared to design and construction costs, it is almost always best to choose these premium devices. Full specifications and helpful applications literature can be obtained from Analog Devices, Burr-Brown, Texas Instruments, National Semiconductor, Motorola, and other companies. A useful summary of many more op amps, with recommendations for different applications, can be found in Ref. 1.

In its overall function, an op amp is basically a high-gain amplifier of voltage differences, and it is usually represented by a triangular symbol as shown in Fig. 5. The plus and minus signs on the input terminals refer to the polarity of the output voltage produced by a positive voltage at that input with respect to the other. All voltages are referenced to the ground line, and this is often omitted from the symbol. The output voltage V_0 is related to the input difference

$$V_o = A(V_+ - V_- + \Delta V) \approx A(V_+ - V_-) \tag{4}$$

where A is the amplifier open-loop gain (10^6 to 10^8), which is ideally the same for both inputs. The quantity ΔV is a small voltage offset (~1 mV or less) that is characteristic of the amplifier and can be canceled for many op amps by adjustment of an external nulling potentiometer.

For stable operation, the operational amplifier is generally used with a negative-feedback loop that can involve a resistor, a capacitor, or a diode. The essential function of the amplifier is to produce whatever voltage is required at the output to hold, by means of the feedback loop, the inverting input at a voltage close (within 1 μV or so) to that of the noninverting input. When the feedback is achieved with a simple connector as shown in Fig. 6*a*, the output acts as a *voltage follower* with $V_o = V_i$. Such a circuit is commonly used when a voltage source cannot supply sufficient current to drive a voltage-measuring device or other load. The input resistance is that of the op amp (ranging from 10^5 to 10^{14} Ω) and the output resistance is small (10^{-1} to 10^3 Ω).

If amplification of a small input voltage is desired, either of the circuits shown in Fig. 6*b* and 6*c* can be used. In the latter case, the noninverting input is grounded and the amplifier acts to hold the inverting ($-$) terminal at a *virtual ground*, i.e., $V_- \cong 0$. (Actually, V_- approaches zero as the gain A approaches infinity.) The input resistance relative to ground potential is determined by the input resistor R_i, and a current of

$$I = \frac{V_i - V_-}{R_i} \cong \frac{V_i}{R_i} \tag{5}$$

FIGURE 6
Some circuit configurations
with operational amplifiers.

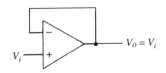

(a) Voltage follower

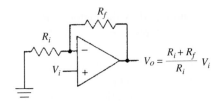

(b) Non-inverting voltage
amplifier

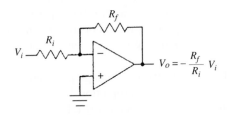

(c) Inverting voltage
amplifier

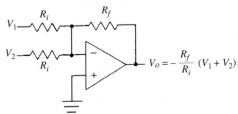

(d) Voltage summing
amplifier

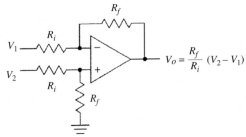

(e) Voltage difference
amplifier

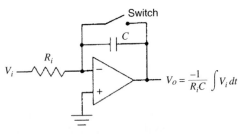

(f) Voltage integrator

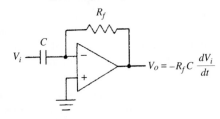

(g) Voltage differentiator

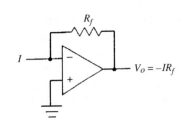

(h) Current-to-voltage
amplifier

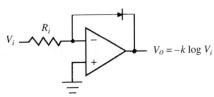

(i) Log amplifier

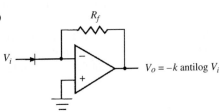

(j) Antilog amplifier

flows into the inverting input terminal. Since the input resistance to the amplifier is very high, essentially all of this input current must flow out through the feedback resistor R_f, corresponding to a potential at the output of

$$V_0 = V_- - IR_f \cong -\frac{R_f V_i}{R_i} \qquad (6)$$

The polarity is thus reversed and the output voltage is amplified by the factor R_f/R_i. This gain must be less than the open-loop gain A, and the feedback resistance R_f must be small compared to the op amp input resistance. If several voltage inputs are applied to the inverting input, as in Fig. 6d, the amplifier can be used to perform a summing operation with simultaneous amplification. Feedback arrangements to perform voltage subtraction, integration, and differentiation are shown in Fig. 6e to 6g. Other operations such as multiplication, division, and logarithmic manipulations are also possible by using diodes or transistors in the input or feedback circuit, as shown in Fig. 6i and 6j. Analog computation integrated circuits containing several precision circuits, such as those in Fig. 6b, 6d, 6e, 6i, 6j, are available from many manufacturers. Such chips make it easy for the experimental scientist to use operational amplifiers to perform analog mathematical operations such as multiplying or dividing voltages, squaring voltages, taking the square roots of voltages, and taking the logarithm of voltages.

For many applications in spectroscopy, electrochemistry, chromatography, and other areas, a current instead of a voltage is generated that is directly proportional to a property of interest. For example, electron emission from a photocathode is linearly related to variations in the incident-light intensity. In a photomultiplier, this cathode current is amplified and the anode current I (typically 10^{-9} to 10^{-3} A) can be converted to a voltage $V_0 = IR_L$ by passage through a load resistor R_L. However, if R_L is large, the potential at the anode varies significantly with anode current. As a result, the gain of the tube changes with current so that the anode current (and the voltage across R_L) is no longer linearly related to light intensity. A current-to-voltage amplifier, such as that depicted in Fig. 6h, can be used to solve this problem. This arrangement offers no resistance to current flow into the virtual ground at the inverting input and the anode potential is constant. Of course no current actually flows into the amplifier; instead the output terminal changes potential to ensure that all of the photomultiplier current flows through the feedback resistor R_f. The same magnitude of voltage signal is produced as with a simple load resistor but with opposite polarity. The amplifier circuit in effect acts as a zero-resistance device while generating a large voltage across R_f, which can then be read with a meter or recorder of relatively low input resistance connected between the op amp output and ground. Summation, integration, and other operations involving current sources can be done with the circuits shown in Fig. 6a to 6f if the input resistance R_i is simply eliminated.

In principle any voltage difference amplifier, including an ideal operational amplifier, produces an output that is proportional only to the differential voltage $V_+ - V_-$ and is independent of the *common-mode voltage* [CMV $= \frac{1}{2}(V_+ + V_-)$]. The extent to which this is true of a real amplifier can be judged by the *common-mode rejection ratio*:

$$\text{CMRR} = \frac{\text{gain for }(V_+ - V_-)}{\text{gain for CMV}} \qquad (7)$$

Typical values of CMRR for ordinary difference amplifiers are 10^3 to 10^5. Frequently, gain and CMRR factors are given in decibels, which is 20 times the base-10 log of the numerical factor. A CMRR factor of 10^5 corresponds to 100 dB. High values of CMRR are particularly important for applications where common noise signals (such as 60-Hz pickup from power lines) occur on both inputs to the amplifier. Examples include signal amplification of

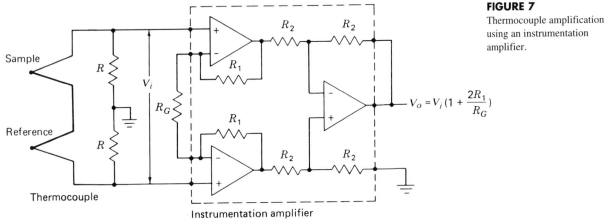

outputs from transducers such as thermocouples, thermistors, strain-gauge bridges, and other devices. For such applications instrumentation or isolation amplifiers are generally used in place of the simple amplifier circuits shown in Fig. 6.

An instrumentation amplifier consists of three or more operational amplifiers and is carefully designed to achieve gains in the range 1 to 10^4 with very high values of input impedance (10^8 to 10^{10} Ω) and CMRR (10^4 to 10^6). Examples in common use are AD524 and AD610 amplifiers from Analog Devices. Figure 7 shows a simplified circuit diagram for an instrumentation amplifier used to amplify a thermocouple output. With such an arrangement, the two thermocouple leads can have common voltages as large as ± 10 V. For common-mode voltages in excess of 10 V, an isolation amplifier is used in which the input is electrically isolated from the output and from the power supply. With such devices, CMRR values of 10^8 or more are possible for common-mode voltages that exceed 1000 V.

ANALOG-TO-DIGITAL CONVERSION

Most voltage or current measurements in physical chemistry involve conversion from an analog form to a digital number at some recording stage, so that subsequent numerical analysis can be performed. For example, the student acts as an analog-to-digital (A/D) converter in measuring and then recording a Chromel–Alumel thermocouple output as say 10.24 mV. This digital number then might be used with a reference table to deduce that the temperature at the sensing point is 252°C. This same analog-to-digital voltage conversion (and even the conversion to temperature) can also be done with integrated-circuit (IC) chips consisting of combinations of operational amplifiers acting in a linear mode for amplification and in a nonlinear fashion for switching purposes and interfacing to a display or a computer. These elements form the heart of digital voltmeters (DVMs), which are readily available with three to six display digits at accuracies of 0.1 to 0.003 percent of full scale. With the development of large-scale ICs, the cost of such meters dropped dramatically, and they are widely used in instrumentation and instructional laboratories.

A variety of methods are used for the digitizing operation, but only the two most common techniques will be discussed here. The dual-slope integration technique (Fig. 8) is used with most display meters when millisecond conversion times are adequate. The conversion involves two steps. In the first step the signal is applied to an operational amplifier, which provides high input resistance plus gain for low-level inputs. Often an initial voltage follower stage, as in Fig. 6a, is used to achieve even higher input resistance. At the start of a measurement cycle, control circuits simultaneously switch the amplifier output

FIGURE 8

Dual-slope type A/D converter. For calibration, either V_{ref} or the buffer gain G can be adjusted to give the correct display for a known input voltage.

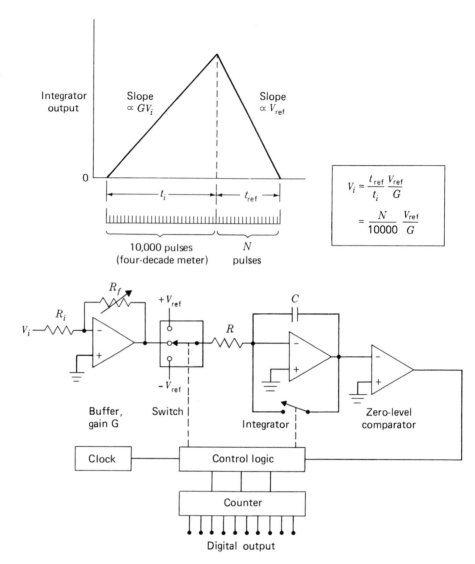

$$V_i = \frac{t_{ref}}{t_i} \frac{V_{ref}}{G}$$

$$= \frac{N}{10000} \frac{V_{ref}}{G}$$

to the integrator input, remove the short from the integrating capacitor, and open a gate to pass clock pulses to a counter. The integrator output, which begins from an initial value of zero, produces a linear ramp with slope and direction proportional to the instantaneous amplitude and polarity of the input voltage (cf. Fig. 6f). The integration continues until a fixed number of counts have accumulated (10^4 for a four-digit meter). At this time, in the second step of the cycle, the integrator input is switched to a reference voltage whose polarity will cause the capacitor C to discharge toward zero. The counter is reset to zero and again begins to count while the reference voltage drives the integrator back to zero. The slope of this second ramp is proportional to the reference voltage, and the time to reach zero is exactly proportional to the input voltage. When the integrator reaches zero, the counter gate is closed and, with calibration, the count displayed is numerically equal to the unknown input voltage.

The dual-slope method has a number of important features. Conversion accuracy is independent of the effect of temperature or aging on the clock frequency and integrating capacitor value, since they need to remain stable only during a conversion cycle. Accuracy is thus determined by the accuracy and stability of the reference source. Resolution is limited primarily by the analog resolution of the converter. Because of the integration operation, the converter gives excellent high-frequency noise rejection and acts to average the signal over the integration period. Moreover, by choosing the clock frequency to be a multiple of 60 Hz, such as 600 kHz, it is possible to achieve nearly complete rejection of 60-Hz ac fluctuations on the input signal, since this ac noise averages to zero when one (or more) full 60-Hz cycle equals the 10^4 count period. The minimum period for this process is 16.67 ms, so this advantage of the dual-slope method of integration is not obtained at high digitizing rates.

For applications requiring high resolution and high speed, the successive-approximation A/D conversion method is used. This method is illustrated in Fig. 9 and consists of comparing the unknown input against a precisely generated internal voltage

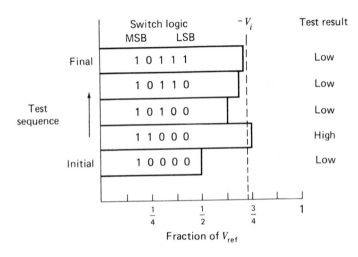

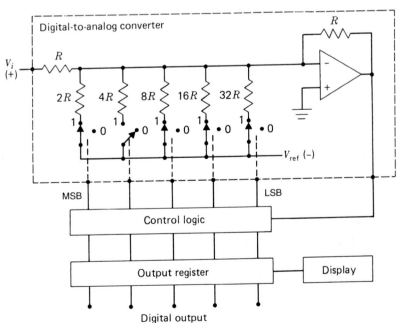

FIGURE 9

Successive-approximation-type A/D converter. A 5-bit converter is shown for simplicity of illustration, but commercial instruments typically use 10 to 16 bits.

from a digital-to-analog(D/A) converter. The D/A converter operates by switching in various binary fractions of a reference voltage (V_{ref} = full scale, typically 1 to 10 V) to the summing point (−) of an operational amplifier. The polarities of V_{ref} and V_i are opposite, so that subtraction occurs when V_i is also introduced into the summing point. The output polarity of the operational amplifier thus indicates the relative magnitudes of V_i and the total fraction of the reference voltage passed by the control logic. At the start of a conversion cycle, the D/A converter's most significant binary bit (MSB), which is one-half full scale, is compared with the input. If it is smaller than the input, the MSB is left on and the next bit (which is one-fourth full scale) is added. If the MSB is larger than the input, it is turned off when the next bit is turned on. The procedure is analogous to that used in placing weights on one side of a two-pan analytical balance. The process of comparison is continued down to the least significant bit (LSB), which is 2^{-n} of full scale for a converter of n bits. At this point the output register contains the complete output digital number, which can be sent to a computer or other processing device. The bit-comparison operation can be done in less than 10 ns, so a 10-bit conversion (with an accuracy of 1 part in $2^{10} = 1024$) can be done in 0.1 μs. In practice, conversion times of 0.1 to 100 μs are typical for high-resolution converters, which generally have a 10-bit to 16-bit digitizing capability.

DIGITAL MULTIMETERS

An A/D converter is basically a dc voltage-measuring device, but with the appropriate signal preprocessing, it can also serve to measure resistance and dc current as well as ac voltage and current. A block diagram of a digital multimeter (DMM) is shown in Fig. 10. The A/D unit, commonly based on the dual-slope A/D conversion method, will have a fixed input range corresponding to the most sensitive scale. The voltage attenuator is just a resistance divider network (Fig. 11a) that allows one to measure larger voltages. On some DMMs with autoranging capability, the full-range setting is changed automatically to ensure the highest resolution without overranging. The circuitry also senses the polarity of the input, which is displayed along with the voltage, the correct decimal position, and the appropriate voltage unit.

For current measurements, the input current can be routed through precision resistors to produce a proportionate dc voltage (Fig. 11b). A separate input is sometimes provided for high currents to avoid damage to sensitive elements. For ac measurement of voltage and current, the attenuated signal is usually rectified and filtered to present a dc voltage to the A/D converter. The ac converter may be either a true rms (root-mean-square) or an averaging type. Both types of meters display the same rms result for sine-wave inputs but

FIGURE 10

Block diagram of a digital multimeter (DMM).

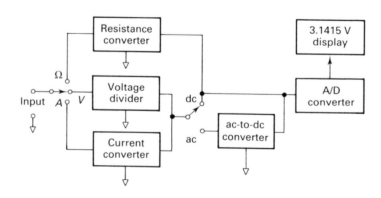

for example the averaging meter gives a value that is high by 11.1 percent for a square wave and low by 3.8 percent for a triangular wave.

To determine resistance, the DMM is equipped with an operational amplifier constant-current source that provides a known current I in the unknown resistor R_u (Fig. 11c). The resultant voltage drop is then measured by the A/D circuit. Some meters provide more accurate resistance measurement by providing leads for the current source that are separate from those used for measurement of the IR drop across the unknown resistor. Since very little current passes through the latter leads, this eliminates any error due to the voltage drop across the measuring leads and connections, which can be significant in the simpler two-lead meter.

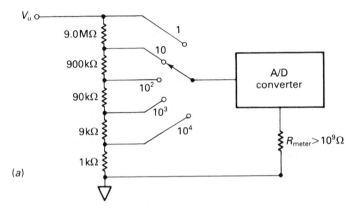

(a)

FIGURE 11
(a) Voltage divider network for DMM. For accurate measurement of an unknown voltage V_u, the input resistance of the meter should be much larger than the resistor network sum. (b) Shunt network to convert current to voltage for a DMM measurement of an unknown current I_u. (c) Circuit for measurement of an unknown resistance R_u with a DMM.

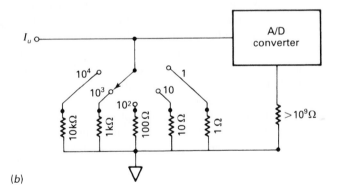

(b)

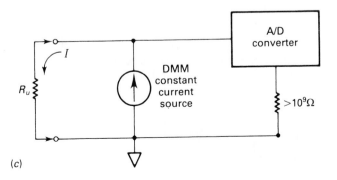

(c)

POTENTIOMETER CIRCUITS

For most measurements of electrical quantities, a precision DMM provides fast and accurate results, and the historical use of potentiometers for this purpose is no longer common in physical chemistry laboratories. If, due to "economic inertia" (potentiometers do not wear out), potentiometers are to be used, specific instructions on them can be found in earlier editions of this book. The following discussion will focus on the nulling principle used with potentiometer circuits, since nulling is a common method for improving the accuracy of physical measurements. For example, a five-decade DVM with a most sensitive scale of 20 mV provides a resolution of 1 μV, but on the 2-V scale the voltage can be read to only 100 μV. Using a potentiometer to accurately cancel out most of the voltage permits the use of the more sensitive scale, enabling one to obtain very accurate measurements over a wide voltage range. This nulling procedure also has the advantage that, at balance, no (or very little) current is drawn from the voltage source being measured. Thus, for cell potential (emf) measurements, as in Exp. 19, one gets the reversible emf value. The nulling principle is also used in dc and ac bridge circuits, discussed later and employed for several of the electrochemistry experiments in this text.

Basic Circuit.   As the name implies, the potentiometer is a device for measuring electrical potential (voltage) difference. The principle of the potentiometer can be described in terms of the simple "slide-wire" potentiometer shown schematically in Fig. 12. With a constant direct current flowing through the uniform-resistance slide-wire AB, the potential difference between A and C should be accurately proportional to the length AC. Thus the potentiometer is a device for selecting with high precision any desired voltage that is less than a certain maximum value. This maximum value depends on the construction of the slide-wire and the voltage of the battery. The voltage across part of the slide-wire (AC) may be compared with another voltage by connecting the negative pole of the external voltage V with the negative end of the slide-wire (at A) and connecting its positive pole with the positive slide-wire contact C through a null detector and a tapping key. When this key is depressed, a nonzero indication on the null detector will be observed if the voltage V and that between A and C are not equal. The sign of the imbalance will obviously depend on whether the potential at C is higher or lower than that at the positive pole of V. If the position of the sliding contact is adjusted until a zero indication is observed upon depressing the tapping key, the voltage V must be the same as that between A and C. Indeed, a linear scale marked off directly in volts can be placed alongside the slide-wire, and the voltage between A and C can be read directly from this scale, provided that the current through the slide-wire has been properly adjusted by manipulation of the rheostat R.

FIGURE 12

Schematic diagram of a simple slide-wire potentiometer.

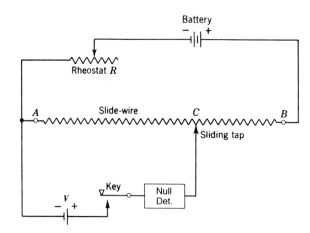

The adjustment of the current through the slide-wire is accomplished by connecting a source of known voltage, such as a *standard cell,* in place of the potential V. The slider C is then placed at the setting on the slide-wire scale corresponding to the known potential, and the rheostat R is adjusted until a null indication is observed on depressing the tapping key. This standardization of the potentiometer against a standard cell should be repeated frequently during the experimental work, because the battery voltage usually shows a tendency to drift slightly.

The accuracy of a voltage measurement depends most of all on the design and the quality of the potentiometer unit used. For a precision potentiometer, the relative accuracy of measurement (ability to measure very small changes in voltage) will depend a great deal on the null detector sensitivity, while the absolute accuracy will depend more on the accuracy of the standard-cell voltage value and the linearity of the slide-wire. The most accurate potentiometers replace the slide-wire by discrete resistors and switches.

For many years, the standard null detector for voltage comparisons was the D'Arsonval or moving-coil galvanometer. More recently, galvanometers have been replaced by stable, high-gain operational amplifiers of high sensitivity and very high input resistance. Two commercial nanovoltmeters that can be used for this purpose are the Keithley model 2182A (which provides 1-nV resolution and 40-nV accuracy on its most sensitive 10-mV scale) and at somewhat higher cost the Agilent model 34420A (which provides 0.1-nV resolution and 20-nV accuracy on its most sensitive 1-mV scale).

For most applications involved in the experiments in this book, a precision digital multimeter, operated on its lowest voltage range, serves as a suitable null detector. For example, the Keithley model 2000 DMM and the Agilent model 34401A DMM provide 0.1-μV resolution and 3-μV accuracy on a 100-mV scale, and the Agilent model 34405A DMM provides 1-μV resolution and 25-μV accuracy on its 100-mV scale. Such multimeters also serve as very flexible instruments for the measurement of resistance as well as ac–dc voltage and current.

Standard Cell. The electrochemical cell used as a voltage standard is the Weston cell, shown in Fig. 13. The voltage of this cell changes only slightly with temperature and is given in absolute volts by

$$V_W = 1.01865 - 4.1 \times 10^{-5}(t - 20) - 9.5 \times 10^{-7}(t - 20)^2 \text{ volts} \qquad (8)$$

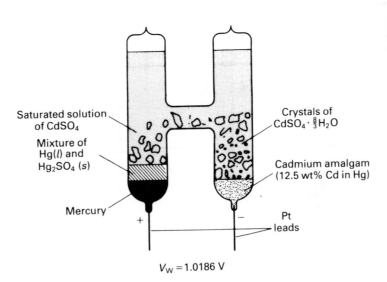

FIGURE 13
The saturated standard Weston cell.

Saturated solution of $CdSO_4$

Mixture of Hg(l) and Hg_2SO_4 (s)

Mercury

Crystals of $CdSO_4 \cdot \frac{8}{3}H_2O$

Cadmium amalgam (12.5 wt% Cd in Hg)

Pt leads

$V_W = 1.0186$ V

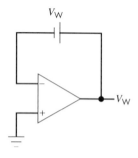

FIGURE 14

Operational amplifier circuit to deliver current at a standard Weston cell voltage.

where the temperature t is in degrees Celsius.[7] In actual practice an unsaturated cell is often used in which the $CdSO_4$ solution is made saturated at 4°C and the solution is slightly unsaturated at the operating temperature. This cell has the advantage of a lower temperature coefficient (1×10^{-5} V °C^{-1}) but the voltage at 20°C can vary from 1.0185 to 1.0195 for different cells. However, the calibrated value for a given cell is reproducible to about 10 μV. For highest accuracy, standard cells should be calibrated against a reliable reference every few years and, in use, one should *never* pass more than 10^{-4} A through the cell. A simple op amp circuit (Fig. 14), with the Weston cell in the feedback loop, can be used to deliver current at the standard voltage without significant loading of the standard cell.

WHEATSTONE BRIDGE CIRCUITS

Direct-Current Wheatstone Bridge. The dc Wheatstone bridge circuit provides a simple means of accurately determining an unknown resistance by comparing it to a known resistance. As shown in Fig. 15, an arbitrary dc potential drop is established across the bridge from A to C and a null detector with tapping key serves as a detector of current flow from B to D. Since direct current is involved, all arms of the bridge are treated as purely resistive elements. When the bridge is balanced (i.e., zero current through the detector when the tapping key is closed), the potential at B must be the same as that at D and it follows that

$$\frac{R_1}{R_3} = \frac{R_2}{R_4} \tag{9}$$

Commercial Wheatstone bridges are available in a form in which R_2/R_4 can be set by a step switch to accurate decimal ratios from 10^{-3} to 10^3, and R_3 is a precision four- or five-decade resistance box. The use of the Wheatstone bridge with platinum resistance thermometers is discussed in Chapter XVII.

Alternating-Current Wheatstone Bridge.[8-10] A Wheatstone bridge can be operated with an alternating- as well as a direct-current source. This is advantageous even when measuring an unknown element that is a pure resistance, since high-sensitivity lock-in-detection techniques can be used for the null detection (see Chapter XIX). For measurements such as those in Exp. 17 on the conductance of electrolyte solutions, the use of alternating current is necessary to prevent polarization of the electrodes in the conductance cell. The basic circuit for an ac bridge is the same as that shown in Fig. 15 except that the

FIGURE 15

General schematic diagram of the dc Wheatstone bridge.

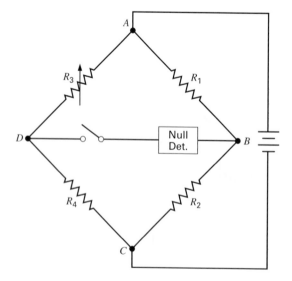

voltage source is an oscillator, operating usually at 1 kHz (or any convenient frequency in the range of 0.5 to 10 kHz) and the detector is an oscilloscope or lock-in detector.

In general one must treat each arm of the bridge as a complex impedance Z:

$$Z = |Z| \exp(i\phi) = |Z| \ (\cos \phi + i \sin \phi) \qquad (10)$$

where ϕ is the phase angle by which the voltage vector is advanced with respect to the current vector. One can also express Z in the form

$$Z = R + iX \qquad (11)$$

where R is the resistance and X is the reactance caused by the capacitance and inductance. It then follows that

$$|Z| = (R^2 + X^2)^{1/2} \qquad \text{and} \qquad \tan \phi = \frac{X}{R} \qquad (12)$$

The general bridge balance condition, where no ac voltage is observed by the null detector, is

$$Z_1 Z_4 = Z_2 Z_3 \qquad \text{or} \qquad |Z_1||Z_4| \exp[i(\phi_1 + \phi_4)] = |Z_2||Z_3| \exp[i(\phi_2 + \phi_3) \qquad (13)$$

which requires that

$$|Z_1||Z_4| = |Z_2||Z_3| \qquad (14)$$

$$\phi_1 + \phi_4 = \phi_2 + \phi_3 \qquad (15)$$

Equations (14) and (15) represent the necessity to balance *both* the real (resistive) and the imaginary (reactive) elements. If one starts from the Z expression given in Eq. (11) instead of Eq. (10), the resulting balance conditions can be written in the form

$$(R_1 R_4 - R_2 R_3) = (X_1 X_4 - X_2 X_3) \qquad (16)$$

$$(X_1 R_4 - X_2 R_3) = (X_3 R_2 - X_4 R_1) \qquad (17)$$

In general, two adjustable elements are sufficient to satisfy relations (16) and (17). In practice, it is desirable to have the adjustments in the resistances independent of those in the reactances. This can be achieved in a *ratio bridge*.[8] Let us put the unknown element in arm 1 of the bridge, allow arm 3 to have a variable impedance Z_3, and choose the ratio arms 2 and 4 so that $Z_2/Z_4 = K$, where K is some real positive number (i.e., $R_2/R_4 = X_2/X_4 = K$). This is equivalent to having the same phase shift in legs 2 and 4, i.e., $\phi_2 = \phi_4$. It follows from Eq. (13) that the balance condition for this ratio bridge is

$$(R_1 + iX_1) = (R_3 + iX_3)K \qquad (18)$$

Equating real and imaginary parts yields

$$R_1 = R_3 K = \frac{R_3 R_2}{R_4} \qquad (19)$$

$$X_1 = X_3 K = \frac{X_3 X_2}{X_4} \qquad (20)$$

and, from Eq. (17),

$$\frac{X_1}{R_1} = \frac{X_3}{R_3} \qquad \frac{X_2}{R_2} = \frac{X_4}{R_4}$$

$$\phi_1 = \phi_3 \qquad \phi_2 = \phi_4 \qquad (21)$$

Equation (19) is the familiar Wheatstone resistance balance condition, and Eq. (20) is a new condition that is needed when phase shifts can occur. Together, Eqs. (19) and (20) ensure that the alternating potential at points B and D in Fig. 15 will be equal in amplitude and exactly in phase at all times.

Since the reactance can be inductive or capacitive and can be parallel or in series with the resistance, there are many variations of the ac bridge. An arrangement suitable for conductivity measurements is discussed in Exp. 17, and an ac bridge used for capacitance determinations is shown in the following section. A more extensive treatment of ac bridges can be found in Refs. 8–10.

REFERENCES

1. P. Horowitz and W. Hill, *The Art of Electronics,* 2d ed., Cambridge Univ. Press, Cambridge (1989).

2. A. J. Diefenderfer and B. E. Holton, *Principles of Electronic Instrumentation,* 3d ed., Saunders, Philadelphia (1994).

3. H. V. Malmstadt, C. E. Enke, and S. R. Crouch, *Microcomputers and Electronic Instrumentation: Making the Right Connections,* American Chemical Society, Washington, DC (1994).

4. H. V. Malmstadt, C. E. Enke, and S. R. Crouch, *Electronics and Instrumentation for Scientists,* W. A. Benjamin, Menlo Park, CA (1981).

5. J. H. Moore, C. C. Davis, and M. A. Coplan, *Building Scientific Apparatus,* 3d ed., chap. 4, Perseus, Cambridge, MA (2002).

6. J. J. Brophy, *Basic Electronics for Scientists,* 5th ed., McGraw-Hill, New York (1989).

7. J. J. Lingane, *Electroanalytical Chemistry,* 2d ed., Interscience, New York (1958).

8. J. R. Duff and S. L. Herman, *Alternating Current Fundamentals,* 6th ed., Delmar, Albany, NY (1999).

9. L. Hartshorn, *Radio-Frequency Measurements by Bridge and Resonance Methods,* Wiley, New York (1941).

10. K. B. Klaasen, *Electronic Measurement and Instrumentation,* Cambridge Univ. Press, New York (1996).

GENERAL READING

J. J. Brophy, *Basic Electronics for Scientists,* 5th ed., McGraw-Hill, New York (1989).

A. J. Diefenderfer and B. E. Holton, *Principles of Electronic Instrumentation,* 3d ed., Saunders, Philadelphia (1994).

H. V. Malmstadt, C. E. Enke, and S. R. Crouch, *Microcomputers and Electronic Instrumentation,* American Chemical Society, Washington, DC (1994).

J. H. Moore, C. C. Davis, and M.A. Coplan, *Building Scientific Apparatus,* 3d ed., Perseus, Cambridge, MA (2002).

E. H. Piepmeier, "Analog Electronics," in P. J. Elving (ed.), *Treatise on Analytical Chemistry,* 2d ed., part 1, vol. 4, chap. 2, Wiley, New York (1984).

XVII

Temperature

Temperature is one of the most important variables in physical chemistry. In this chapter we are concerned with the methods and instruments that are used in measuring and controlling temperature.

The first mercury-in-glass thermometer was invented in 1714 by an instrument maker named Gabriel Fahrenheit, who followed Newton's suggestion of defining the temperature scale by fixing two points. Fahrenheit chose 0° for the coldest temperature that could be produced with an ice–salt bath and 96° for the temperature of human blood. Later it was discovered that the freezing and boiling points of water were attractively reproducible values. Since these were near 32° and 212° on the original scale, the Fahrenheit scale was redefined to make these "fixed points" exactly 32°F and 212°F. The Swedish scientist Celsius proposed in 1742 a scale based on 0°C and 100°C for these fixed-point temperatures, and the same proposal was made independently in 1743 by the Frenchman Christin, who called it the centigrade scale. This scale, now defined in a different way as described below, is very widely used and is officially known as the Celsius temperature scale although it is often still called the centigrade scale.

TEMPERATURE SCALES

The general concept of temperature scales is discussed briefly in Exp. 1. The *thermodynamic temperature scale,* based on the second law of thermodynamics, embraces the Kelvin (absolute) scale and the Celsius scale, the latter being defined by the equation

$$t(\text{Celsius}) \equiv T(\text{Kelvin}) - 273.15 \tag{1}$$

The size of the kelvin, the SI temperature unit with symbol K, is defined by the statement that the triple point of pure water is exactly 273.16 K. The practical usefulness of the thermodynamic scale suffers from the lack of convenient instruments with which to measure absolute temperatures routinely to high precision. Absolute temperatures can be measured over a wide range with the helium-gas thermometer (appropriate corrections being made for gas imperfections), but the apparatus is much too complex and the procedure much too cumbersome to be practical for routine use.

TABLE 1   Fixed points of the International Temperature Scale of 1990 (ITS-90)

Number	T_{90} (K)	Substance	Kind[a]	$W_r(T_{90})$[b]
1	3 to 5	He	V	
2	13.8033	$e\text{-}H_2$[c]	T	0.00119007
3	17	$e\text{-}H_2$ (or He)	$V(G)$	
4	20.3	$e\text{-}H_2$ (or He)	$V(G)$	
5	24.5561	Ne	T	0.00844974
6	54.3584	O_2	T	0.09171804
7	83.8058	Ar	T	0.21585975
8	234.3156	Hg	T	0.84414211
9	273.16 exact	H_2O	T	1.00000000
10	302.9146	Ga	M	1.11813889
11	429.7485	In	F	1.60980185
12	505.078	Sn	F	1.89279768
13	692.677	Zn	F	2.56891730
14	933.473	Al	F	3.37600860
15	1234.93	Ag	F	4.28642053
16	1337.33	Au	F	
17	1357.77	Cu	F	

[a] V = vapor pressure, G = gas thermometry, T = triple point (gas, liquid, and solid in equilibrium), M/F = melting/freezing at 1 atm.

[b] $W_r(T_{90})$ is a reference function for platinum thermometers.

[c] $e\text{-}H_2$ is an equilibrium mixture of *ortho-* and *para*-hydrogen.

The *International Practical Temperature Scale* of 1968 (IPTS-68) has been replaced by the *International Temperature Scale* of 1990 (ITS-90).[1] The ITS-90 scale is basically arbitrary in its definition but is intended to approximate closely the thermodynamic temperature scale. It is based on assigned values of the temperatures of a number of defining fixed points and on interpolation formulas for standard instruments (practical thermometers) that have been calibrated at those fixed points. The fixed points of ITS-90 are given in Table 1.

The major change from IPTS-68 to ITS-90 has been the elimination of the normal (i.e., 1-atm) boiling points that were previously used as fixed points. This change was made because the temperature of a boiling point is much more sensitive to the ambient pressure than that of a freezing point. The latter is defined as the equilibrium temperature of coexisting pure solid and liquid at one standard atmosphere (101 325 Pa), and corrections can be made for small pressure deviations (the effect is only about 5 mK per atm).[1]

An unofficial list of secondary reference points is given in Table 2. These represent previously specified points that have been abandoned but are still useful for practical work of lower accuracy. Their IPTS-68 values[2] have been corrected to the appropriate new values on the ITS-90 scale.

The ITS-90 scale extends from 0.65 K to the highest temperature measurable with the Planck radiation law (~6000 K). Several defining ranges and subranges are used, and some of these overlap. Below ~25 K, the measurements are based on vapor pressure or gas thermometry. Between 13.8 K and 1235 K, T_{90} is determined with a platinum resistance thermometer, and this is by far the most important standard thermometer used in physical chemistry. Above 1235 K, an optical pyrometer is the standard measurement instrument. The procedures used for different ranges are summarized below.

TABLE 2 Unofficial list of secondary reference points

T_{90} (K)	Substance	Kind[a]
63.152	N_2	Triple point
77.356	N_2	B
194.686	CO_2	S
273.150	H_2O	F
373.124	H_2O	B
600.612	Pb	F
2042	Pt	F
3695	W	F

[a] B = boiling, S = sublimation, F = freezing point (all at 1 atm).

1. Range 0.65 to 5.0 K. The value of T_{90} is defined in terms of the vapor pressures of liquid ^{3}He and ^{4}He. The empirical vapor pressure equations and necessary parameter values are given in Ref. 1.

2. Range 3.0 to 24.5561 K. Over this range, T_{90} is defined in terms of a constant-volume helium-gas thermometer. If ^{4}He is used above 4.2 K, the basic equation is

$$T_{90} = a + bp + cp^2 \tag{2}$$

where p is the pressure in the gas thermometer and the coefficients a, b, and c are determined by calibration at three defining fixed points. For a ^{3}He-gas thermometer or a ^{4}He thermometer used below 4.2 K, nonideality must be included explicitly via the appropriate second virial coefficient; see Ref. 1 for details.

3. Range 13.8033 to 1234.93 K. The standard instrument for this range is the platinum resistance thermometer, although in practice no single thermometer is likely to be usable over the entire range. Details about the types of Pt thermometers available and their optimal ranges of operation are given in Ref. 3. Unfortunately the ITS-90 interpolation formulas are quite complicated. Temperatures are determined from the measured ratio $W(T)$ of the resistance R at temperature T to that at the triple point of water:

$$W(T) = \frac{R(T)}{R(273.16 \text{ K})} \tag{3}$$

Note that IPTS-68 and many simpler interpolation schemes still in use, such as the Callendar–van Dusen equation described below, use the ratio $R(T)/R(273.15 \text{ K})$ based on the ice point as the reference temperature.

For T above (below) 273.16 K, the value of T on the ITS-90 scale is given as a 10-term (16-term) power series involving the reference function $W_r(T_{90})$. These power series and tabulated values of the necessary coefficients are given in Ref. 1. The quantity $W_r(T_{90})$ is defined by

$$W_r(T_{90}) = W(T) - \Delta W(T) \tag{4}$$

where the deviation function $\Delta W(T)$ is given by a series expansion in $W(T)$ with unknown coefficients that are determined by the calibration procedure. See Ref. 1 for numerous details.

4. Range Above 1234.93 K. Above the freezing point of silver, an optical pyrometer is used to measure the emitted radiant flux (radiant excitance per unit wavelength interval) M_λ of a blackbody at wavelength λ. The defining equation is the Planck radiation law in the form

$$\frac{M_\lambda(T)}{M_\lambda(T_X)} = \frac{\exp(c_2/\lambda T_X) - 1}{\exp(c_2/\lambda T) - 1} \tag{5}$$

where T_X refers to the silver ($T_{Ag} = 1234.93$ K), gold ($T_{Au} = 1337.33$ K), or copper ($T_{Cu} = 1357.77$ K) freezing points. If λ is expressed in nanometer units, one should use $c_2 = 1.43877 \times 10^7$ nm K. Practical details about the use of pyrometers are given in Ref. 3.

Effect of T_{90} Scale. Although ITS-90 conforms to the thermodynamic temperature scale more closely than its predecessors (IPTS-68 and IPTS-48), the distinction between these scales is not of great importance except for very precise work or work above 1500 K. The differences ($T_{90} - T_{68}$) lie within ± 0.015 K over the range 14 to 330 K, vary from -0.015 K at 330 K to -0.089 K at 800 K, and lie within ± 0.35 K over the range 800 to 1500 K. Above 1500 K this difference varies from -0.31 K to -2.58 K at 4300 K.[1,4] The changes in calorimetric values of heat capacity, enthalpy, and entropy caused by conversion from IPTS-68 to ITS-90 have been evaluated for several solids, and the changes all lie within the experimental uncertainty in the measured values for the range 16 to 2800 K.[4] Indeed even the differences ($T_{90} - T_{48}$) are quite small over the range 90 to 880 K, less than ± 0.045 K,[4] and this includes the effect of changes in the thermodynamic temperature values of the fixed points and the inadequacies in the older interpolation formulas. Thus, for most purposes, the introduction of ITS-90 has a negligible effect on property values.

Callendar–van Dusen Equation. The complexity of the ITS-90 equations, especially for temperatures below 273.16 K, creates an awkward situation. For practical use, the Callendar–van Dusen equation, which was the basis for IPTS-48, is still a very convenient form. This is especially true if one wishes to determine the resistance R of a Pt thermometer digitally under computer control and convert it into a temperature with a reasonably simple algorithm. The general form of the Callendar–van Dusen equation is

$$R_t = R_0\left\{1 + \alpha\left[t - \delta\left(\frac{t}{100}\right)\left(\frac{t}{100} - 1\right) - \beta\left(\frac{t}{100} - 1\right)\left(\frac{t}{100}\right)^3\right]\right\} \tag{6}$$

where R_0 is the resistance at the ice point, t is in degrees Celsius, and α, β, δ are calibration constants. The constant β is zero by definition above 0°C. It is obvious from Eq. (6) that

$$\alpha = \frac{R_{100} - R_0}{100R_0} \tag{7}$$

Thus α equals the average value over the range 0 to 100°C of the resistivity coefficient α_r, a quantity defined later in Eq. (11). A more convenient form of Eq. (6) is

$$\frac{R_t}{R_0} = 1 + At + Bt^2 + Ct^3(t - 100) \tag{8}$$

where $A = \alpha(1 + \delta/100)$, $B = -\alpha\delta/10^4$, $C = -\alpha\beta/10^8$ below 0°C, and $C = 0$ above 0°C. Note that the constant A equals the value of α_t at $t = 0$°C. A calibration at 0°C and at two points above 0°C will yield R_0, A, and B (or α and δ); one more calibration point below 0°C will then determine C (or β).

The typical uncertainties in temperature values based on a four-point Callendar–van Dusen calibration over the range -180 to $+260$°C vary from $\pm(10 - 20)$ mK over this range but rise to ±80 mK at 350°C because an extrapolation beyond the fitted calibration range is involved. A crude calibration can be achieved on the basis of only two points at 0°C and 100°C. These two points will determine R_0 and α; see Eq. (7). One can then *assume* that $\alpha\delta = 0.0058755$ and $\beta = 0.11$ for $t < 0°C$.[5] The resulting calibration formula can be used to interpolate reasonably well between 0°C and 100°C (errors of about ±50 mK at 50°C) and to extrapolate roughly down to −50°C and up to 150°C (errors grow to about ±200 mK).

Another approach is to use a carefully selected standard platinum alloy for which the constants in Eqs. (6) and (8) are well known. In the mid-1980s, the International Electrotechnical Commission (IEC) recommended that ITS-90 be based on the Callendar–van Dusen interpolation formula and proposed constants for a Pt resistor with $R_0 = 100\ \Omega$: $A = 3.90802 \times 10^{-3}$, $B = -5.802 \times 10^{-7}$, $C = -4.27350 \times 10^{-12}$ (or $\alpha = 0.00385$, $\delta = 1.50701$, $\beta = 0.111$). The platinum wire used in platinum resistance thermometers that conform to this proposed standard is a platinum alloy containing small amounts of several different elements (mostly noble metals) adjusted so as to achieve the required $\alpha = 0.00385\ K^{-1}$. This alloy is now widely used in Europe and by some American manufacturers of resistance thermometers; note, however, that other American firms use a wire for which $\alpha = 0.00392\ K^{-1}$. In spite of the fact that this IEC proposal was not adopted, the Callendar–van Dusen constants given above are a guide to appropriate values, which can always be checked by calibration.

TRIPLE-POINT AND ICE-POINT CELL

Since the triple point of water is defined to be exactly 273.16 K on the thermodynamic temperature scale, this is an especially important fixed point. It is also a point that can be reproduced with exceptionally high accuracy. If the procedure of inner melting (described below) is used, the temperature of the triple point is reproducible within the accuracy of current techniques (about ±0.00008 K). This precision is achieved by using the *triple-point cell* shown in Fig. 1. This cell, which is about 7.5 cm in outer diameter and 40 cm in overall length, has a well of sufficient size to hold all thermometers that are likely to be calibrated.†

The procedure for using a triple-point cell is as follows. Cool the cell by placing it in a bath of crushed ice and distilled water. Fill the thermometer well with powdered Dry Ice in order to form a clear mantle of ice around the well. When this mantle is about 0.5 cm thick, remove the Dry Ice and place a warm tube in the well just long enough to free the mantle and provide a thin layer of liquid next to the wall. If the mantle is free, it will rotate about the well when the cell is given a quick twist about the vertical axis. This "inner melting" provides an additional step of purification. As the water freezes to form the mantle, impurities are left behind in the liquid near the outer wall. When the mantle is partly melted, the water near the inner well is exceptionally pure and this water does not mix rapidly with the water outside the mantle. The equilibrium between the inner water layer, the ice mantle, and the water vapor in the top of the cell occurs at the defined triple-point temperature. In order to ensure good thermal contact between the wall of the well and the thermometer being calibrated, the well is filled with ice water. It is also wise to precool the thermometer in an ice bath before it is placed in the well.

Although the triple point of water is the most fundamental and stable of all fixed points, the *ice point* also serves as a convenient fixed point for less demanding calibrations

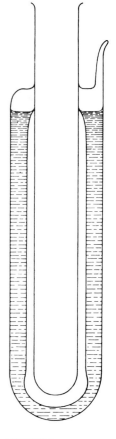

FIGURE 1

Triple-point cell for water (NBS/NIST design). Note the clear mantle of ice formed around the inner well.

†The design shown in Fig. 1 was developed at the National Bureau of Standards (NBS), which has now been renamed the National Institute of Standards and Technology (NIST).

of practical thermometers. The ice point refers to an ice and pure water mixture in equilibrium with air saturated with water vapor at a total pressure of 1 atm. Procedures for preparing an ice bath are given in the section on temperature control. Commercial ice-point devices that eliminate the need for frequent attention are also available.[6] Such devices consist of a sealed chamber containing distilled water; the outer walls are cooled with a thermoelectric element to create a shell of ice. The increase in volume that accompanies the conversion of liquid into solid is detected by the expansion of a bellows, which operates a microswitch controlling the cooling element. The alternating cycle of freezing and thawing maintains a constant temperature and eliminates the slow drift in temperature that will occur in an unattended ice bath as melting proceeds. An ice-point device will maintain a reference temperature that is constant to within ± 0.04 K, but the absolute value may differ from 273.15 K $\equiv$ 0°C by a somewhat larger amount.

THERMOMETERS

In this section, the design and operation of familiar liquid thermometers, thermocouples, platinum resistance thermometers, thermistors, and optical pyrometers are discussed in detail. Briefer descriptions are also given of a variety of special thermometric devices such as quartz thermometers, germanium resistance thermometers, and silicon-diode thermometers.

Liquid-in-Glass Thermometers. Major emphasis is given below to the mercury thermometer, although any liquid with a thermal expansion coefficient significantly greater than that of glass can be used. The familiar household thermometer uses an organic liquid such as toluene or alcohol containing a little red dye to enhance visual contrast. Although liquid-in-glass thermometers are inexpensive, compact, and reliable over long periods of use, there are problems discussed below in obtaining both precision and absolute accuracy. Furthermore, such thermometers do not lend themselves to computer automation of the measurement process, which is important for frequently repeated procedures.

1. Mercury Thermometers.[7,8] In the past, mercury thermometers were by far the most common type of laboratory thermometer. Concerns about health hazards associated with mercury have now reduced their role. However, when high precision is required (calorimetry and freezing-point depressions, for example), oil-in-glass thermometers are not suitable since the use of fine capillaries is not feasible with oil. Furthermore, the safety hazards involved in the laboratory use of mercury thermometers are quite low, as discussed at the end of this section. For these reasons, a description of the use of mercury thermometers is still pertinent.

A mercury thermometer is based on the differential volume thermal expansion of liquid mercury (about 1.8×10^{-4} K^{-1}) and glass (about 0.2×10^{-4} K^{-1}). In the manufacture of such thermometers, the stem, a capillary of uniform bore, is usually marked at two points (say, 0°C and 100°C) and then graduated uniformly in between, on the tacit assumption that the volume of a fixed mass of mercury in glass is a linear function of the temperature. The error resulting from this assumption (about +0.12°C at 50°C for a 0°C to 100°C mercury thermometer made with Corning normal thermometer glass) is ordinarily smaller than that due to variations in the bore of the capillary.

Laboratory thermometers are commonly available in two types: solid-stem and enclosed-scale. The former has a stem of solid glass, with a scale engraved on the outside surface; the latter has a slender capillary and a separate engraved scale, both enclosed in an outer glass shell. The enclosed-scale type is preferable for thermometers with extremely fine threads, largely because it suffers less from parallax in reading.

Since the mercury in the thread, as well as that in the bulb, is susceptible to thermal expansion, it is important in precise work to take account of the temperature of the thermometer stem. Most thermometer calibrations, especially those for enclosed-stem types, are for total immersion—it is assumed that the thread is at the same temperature as the bulb. Other thermometers are meant to be used with partial immersion, often to a ring engraved on the stem, and the remainder of the stem is assumed to be at room temperature (say, 25°C). For precise work, stem corrections should be made if the stem temperatures differ significantly from those assumed in the calibration. The correction that should be *added* to the thermometer reading is given by the equation

$$\Delta t_{corr} = -0.00016(\Delta t_{stem})(\Delta L_{stem}) \tag{9}$$

where Δt_{stem} is the amount by which the temperature of the stem *exceeds* that for which the calibration applies (or the amount by which it exceeds the thermometer reading itself, if the calibration is for total immersion) and ΔL_{stem} is the length of mercury thread, expressed in degrees Celsius, for which the temperature is different from that assumed in the calibration. To obtain Δt_{stem} a second thermometer may be positioned near the first, with its bulb near the midpoint of ΔL_{stem}.

For most purposes a partial-immersion thermometer need not be stem corrected because of a few degrees variation in room temperature or a few degrees error in the immersion level. On the other hand, it is usually worthwhile to apply stem corrections to readings of a total-immersion thermometer when it is used in partial immersion, particularly when reading temperatures well removed from room temperatures.

Other important sources of error in mercury thermometers are parallax and sticking of the mercury meniscus. The former can largely be avoided by careful positioning of the eye when reading or by use of a properly designed magnifier. It can be eliminated entirely by use of a cathetometer (see p. 230). Sticking of the mercury meniscus is due to the fact that the contact angle of mercury to glass (see Fig. 25-5) varies depending on whether the mercury surface is advancing or receding, and thus the capillary pressure due to surface tension is variable. This combines with the small but finite compressibility of the mercury and the elasticity of the glass to yield a small variability in meniscus position, especially for very sensitive thermometers. *Gentle tapping of the stem before taking readings usually leads to reproducible results.*

When precision of better than about 1 percent of full scale is required, it is necessary to use a calibration chart that gives corrections to be added to or subtracted from the readings. Detailed procedures for calibrating liquid-in-glass thermometers are given in Ref. 8. The best calibration method is to calibrate the test thermometer against a platinum resistance thermometer, since such thermometers have exceptional reproducibility and long-term stability. The calibrations of mercury thermometers should be checked at a single point from time to time, as significant irreversible changes may take place in the glass, particularly if the thermometer is used above 150°C.

Special thermometers are made for calorimetric and cryoscopic work, where it is desired to measure very accurately (to 0.01 or even 0.001 K) a temperature *difference* of the order of a few kelvin. For these thermometers the fineness of scale graduation has little to do with the accuracy with which the thermometer measures a single temperature. The scale may be in error by several tenths of a kelvin, but this error cancels out in taking differences. A typical thermometer for bomb calorimetry has a range of 19 to 35°C, with graduations of 0.02°C. For measuring freezing-point depressions with water or cyclohexane as solvent, a range of -2 to $+7°C$ with graduations of $0.01°$ is convenient. Such thermometers require careful handling. Not only are they relatively fragile, but they are susceptible to certain malfunctions (separation of the mercury column, bubbles in the bulb) arising principally from the extreme fineness of the thread. Whenever possible, keep these thermometers upright; *avoid overly rapid heating or cooling.* If the mercury thread separates (which often happens when thermometers are shipped), cool the bulb in an ice–salt mixture to bring the mercury entirely into the bulb and tap if necessary to bring any bubbles to the top of the bulb. Then allow the thermometer to warm to room temperature in an upright position.

In recent years, there has been considerable stress on the potential health hazards of oral mercury thermometers used for medical purposes. In fact, the risk of mercury poisoning from a broken mercury thermometer in a physical chemistry experiment is extremely low. The key safety rule if such a thermometer were to break is the prompt and efficient cleanup of any spilled mercury; see Appendix C for further details. Note that essentially no mercury will be spilled if the stem rather than the bulb is broken. In conclusion, mercury thermometers should be handled with reasonable care, but they are useful and precise scientific instruments of considerable value for research applications.

2. Special Liquid Thermometers. For low temperatures, one can use several kinds of liquid-in-glass thermometers. Toluene thermometers may be used down to $-95°C$, and pentane thermometers will operate as low as $-130°C$. However, it is usually more convenient, as well as more accurate, to use thermometric devices of other types, especially thermocouples or resistance thermometers.

Thermocouples.[2,6,9] Thermocouples provide one of the most convenient and versatile means of measuring temperatures over a wide range from very low ($-250°C$) to very high ($2300°C$) values. All that is required is two kinds of fine wire made from appropriate metals or alloys, a suitable precision voltage-measuring instrument (a good potentiometer or digital voltmeter), and a constant-temperature bath (almost always an ice–water bath) for the reference junction. Thermocouples are simple to use and, being small, have a rapid response to temperature changes. Although their sensitivity is limited, temperature resolution of 0.001 K can be achieved with very good voltmeters, and much better resolution is possible with multijunction thermocouples called thermopiles (see p. 629). The principal disadvantage of thermocouples is the need for very careful calibration to achieve good absolute accuracy and some tendency for the calibration to change with time, especially at high temperatures.

When two dissimilar metals are placed in contact, a transfer of electrons from one to the other takes place and a double charge layer forms at the junction surface. As in the case of a junction between a metal electrode and an electrolyte solution, the resulting electrical potential must be measured in the presence of a reference junction, which in the present instance is a junction of the same two metals at a known temperature. As shown in Fig. 2,

FIGURE 2

Schematic diagram of a two-junction thermocouple setup.

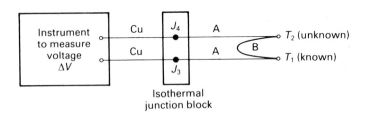

the thermoelectric potential difference ΔV is measured between the ends of two wires of the same kind. One wire of metal A leads to the junction at the reference temperature T_1, and the other wire of metal A leads to the junction at the unknown temperature T_2; these two junctions are connected directly by a wire of the second metal B, which completes the circuit. The potential drop ΔV is a measure of the temperature difference $\Delta T = T_2 - T_1$. In cases where ΔT is not too large, ΔV is roughly proportional to the difference between the unknown and the reference temperature:

$$\Delta V = \alpha \Delta T \qquad (10)$$

where α is called the Seebeck coefficient in honor of Thomas Seebeck, who discovered the thermoelectric effect in 1821. In fact ΔV is not exactly a linear function of ΔT, since the sensitivity α varies slowly with temperature. Thus a well-known reference temperature T_1 and calibration tables are needed to obtain values of T_2 from measured ΔV values. It is common practice to use the ice point (0°C) as the reference temperature; in this case ΔT equals the value of T_2 in degrees Celsius.

The presence of an isothermal junction block in Fig. 2 should also be noted. Almost all high-quality voltage-measuring instruments have input terminals made of copper. Unless metal A of the thermocouple pair is also copper, connection of the thermocouple to the voltage instrument will create two new bimetallic junctions J_3 and J_4. The presence of any temperature difference between the input terminals would then generate an unwanted extra thermoelectric potential drop and cause an error in the determination of T_2. In practice, this error voltage is usually very small and can often be neglected. However, for high-precision measurements this source of error can be eliminated by using an isothermal junction block. This block is made from an electrical insulator with high thermal conductivity, which ensures that J_3 and J_4 are at the same temperature, T_{block}. The value of T_{block} has no effect on the measured ΔV values.

The isothermal junction block is also used in commercial thermocouple devices that utilize only a single junction. Since T_{block} can have any value, it is made equal to the reference temperature T_1, as shown in Fig. 3a. Another simplification is the elimination of wire A in the junction block, so that two junctions (Cu–A and A–B) at the same temperature T_1 are replaced by one (a Cu–B junction) as shown in Fig. 3b. It can be shown empirically that

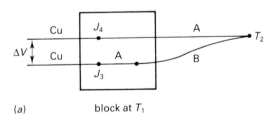

(a)   block at T_1

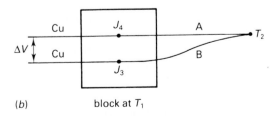

(b)   block at T_1

FIGURE 3

Configurations for a "one-junction" thermocouple arrangement utilizing an isothermal junction block held at a reference temperature T_1; see text for details.

TABLE 3 Thermocouple characteristics

ANSI type	Metals	Useful temperature range (°C)	Sensitivity (μV K^{-1})			Notes
			−200°C	25°C	300°C	
T	Copper Constantan	−200 to 370	15.8	40.7	58.1	R, I, V, O (mild)
J	Iron Constantan	0 to 750	21.8^a	51.7	55.4	R, I, V, O (moderate T)
E	Chromel Constantan	−200 to 900	25.1	60.9	77.9	I, O
K	Chromel Alumel	−200 to 1250	15.2	40.5	41.5	I, O, L
N	Nicrosil Nisil	−200 to 1250	9.9	26.8	35.4	I, O, L
S	Platinum Pt + 10%Rh	0 to 1450	—	6.0	9.1	I, O

Notes: Atmospheres in which bare-wire thermocouples can be used: R = reducing, O = oxidizing, I = inert, V = vacuum, L = limited use in vacuum or reducing atmospheres.

^{a} Not recommended for low temperatures.

this modification has very little effect on the measured value of ΔV. The isothermal junction block can be mounted in an ice–water bath so that $T_1 = 0°C$. Many commercial thermo-couple systems (voltage-measuring instruments plus thermocouple assembly) do not utilize an ice bath but employ software or hardware compensation schemes so that the reference temperature T_1 can have an unregulated value equal to the ambient room temperature.[6] For high-precision work, a well-regulated ice bath should always be used.

The most useful thermocouples for general work are listed in Table 3.[10] Both pure metals and various alloys are used. Many of these alloys are nickel alloys. *Constantan* (sold under the trade names Advance and Cupron) is available in two different compositions: 60%Cu + 40%Ni is designed for use with type J thermocouples, and 55%Cu + 45%Ni is used for both type T and type E thermocouples.† *Chromel* is the trade name of a 90%Ni + 10%Cr alloy, and *Alumel* is the trade name of a 95%Ni + 2%Mn + 2%Al + 1%Co alloy. Two other nickel alloys with trade names are *Nicrosil* (84.4%Ni + 14.2%Cr + 1.4%Si) and *Nisil* (95.5%Ni + 4.4%Si + 0.1%Mg).

The *copper–Constantan* thermocouple, called type T, is widely used. This has the important advantage that one of the metals (metal A in Fig. 3) is copper. Thus an iso-thermal junction block is not needed, and no stray thermoelectric potentials can occur at the input terminals of the voltage instrument even if a temperature difference does occur there. Furthermore, if a thermocouple is to be used without calibration, the possible sys-tematic errors are smaller for type T than for other types of thermocouples. However, Cu has a very high thermal conductivity and oxidizes above 350°C. *Iron–Constantan* (type J) thermocouples are popular because of their high sensitivity and the wide range of atmo-spheres in which they can be used (although Fe oxidizes at high temperatures). Individual thermocouples may show variations from the standard voltage–temperature conversion table values owing to trace impurities in the iron. Type J couples should never be used above 760°C, since a magnetic phase transition at that temperature can shift the calibration

†The letter codes used for thermocouples have been established by the American National Standards Institute (ANSI).

at lower temperatures after cycling above 760°C. A *Chromel–Constantan* (type E) thermo-couple is attractive for low-temperature use owing to its high sensitivity and low thermal conductivity. The *Chromel–Alumel* (type K) couple is convenient because it can be used over a very wide temperature range and the sensitivity is almost constant above 25°C (e.g., $\alpha = 36 \ \mu V \ K^{-1}$ at 1250°C). There is some calibration instability due to an order–disorder phase transition at ~500°C. The *Nicrosil–Nisil* (type N) thermocouple is a more recently developed couple for use at high temperatures. It is similar to type K with changes in the alloy composition to reduce problems associated with the order–disorder transition and to improve the resistance to oxidation at high temperatures. As a result, this couple has excel-lent long-term stability. *Platinum–platinum + 10% rhodium* (type S) thermocouples have low sensitivity but can be used at temperatures as high as 1700°C. The calibration stability is excellent, and the type S thermocouple was the standard device specified by IPTS-68 to measure temperatures in the range 630.74 to 1064.43°C. However, it was replaced in this role for ITS-90 by the platinum resistance thermometer.

Other thermocouples not listed in Table 3 include the Chromel–gold couple for cryogenic use below −200°C (useful range 4 K to above 100 K) and several tungsten–tungsten + rhenium couples that can be used up to 2300°C.

Although extensive tables are given in various handbooks for converting measured thermoelectric voltages into temperatures, the best tables are those published by the National Institute of Standards and Technology (NIST).[6,11] No matter what tables are used, thermocouple wires should be selected carefully and one or more specimens of each pro-duction lot (the spools are marked with this number) should be calibrated at a number of widely spaced temperatures. Such a calibration can be made using fixed points or by com-parison with a "standard" thermometer; see p. 573. For very precise work, each individual thermocouple should be calibrated. Deviations between the reading of a given thermo-couple and the entries in standard conversion tables are due to strains and small composi-tional variations that can occur during fabrication of the wires. Over the temperature range from −200 to +300°C, these systematic errors can vary from ±1°C to ±4°C for different thermocouples. As an example the uncertainty in the absolute accuracy of an uncalibrated type T thermocouple is the larger of ±0.8°C *or* ±0.75 percent of the Celsius temperature above 0°C (±1.5 percent of the Celsius temperature below 0°C). For type K, the analogous values are ±2.2°C or ±0.75 percent above 0°C (±2.0 percent below 0°C).

Junctions to be used only at low or moderate temperatures (below 170°C) may be joined with soft solder (be sure to use a noncorrosive flux such as rosin) or silver solder, but it is better to weld the two metals together with an electric arc or an oxygen–gas flame. The two wires are stripped of any insulating material and then twisted together tightly over a distance of ~0.5 cm. If a glass-blowing torch is used, the end of the junction should be dipped in paste flux and placed in the reducing region at the tip of the blue part of the flame. Remove the wires as soon as the metals melt and form a small bead at the end. This bead should be pinched with a pair of long-nosed pliers and examined carefully to make sure it is hard and metallic rather than a bead of oxide (which crumbles easily). The two wires must be electrically insulated from each other, and some consideration must be given to the temperature characteristics of the insulating materials. In particu-lar, at high temperatures nothing should come into contact with the wires, particularly Chromel-P and Alumel, that will form a liquid flux (low-melting eutectic) with the pro-tective oxide film.

Thermocouple wire is available commercially either as bare wire or as wire protected with a variety of insulators. The best choice of insulation for use up to 280°C is Teflon. Its resistance to abrasion, chemical reaction, solvents, and humidity is excellent, and its flex-ibility is good. Braided fiberglass is very good for use up to 480°C, but it has poor abra-sion resistance and quite high porosity. Special braided ceramic fibers are available for use as high as 1425°C. Bare wires and Alundum tubes can also be used above 450°C. The junctions of any thermocouple, and the wire itself when bare wire is used, must always be protected carefully from corrosion or mechanical damage. For use in liquids and solutions at moderate temperatures, a thermocouple is usually placed in a closed glass tube (~6 mm in diameter) with a small amount of nonvolatile oil or silicone grease at the bottom to improve the thermal contact. It is also possible to bond thermocouple junctions to a wide range of solid surfaces (metals, ceramics, glass, plastics) with high-thermal conductivity epoxy adhesives. Fully assembled thermocouple probes are also available commercially with both wires mounted in a stainless steel or Inconel protective sheath and insulated from each other with a compacted ceramic such as MgO, which is good up to 1650°C. The junctions can be obtained exposed as a bare butt weld or bare bead or enclosed in the sheath metal (grounded and ungrounded options are available).

A satisfactory environment for the 0°C reference junction is provided by a slushy mixture of ice and distilled water in a Dewar flask, with a ring stirrer and a monitoring mercury thermometer. Elaborate thermoelectric ice–water chambers are also available; these are convenient for prolonged periods of use but rather expensive. As mentioned previously many commercial thermocouple systems eliminate the ice bath by placing the cold junction on an isothermal block that is at room temperature and compensating for the resulting error. This is a convenient but less accurate procedure.

For ordinary work with thermocouple wires of large or medium diameter, the thermoelectric potential can be measured with a good digital voltmeter (DVM) or a dc potentiometer. If a continuous record of the temperature is required, the thermocouple can be connected to a millivolt strip-chart recorder; or a DVM with an interface to a computer can be used to store frequent periodic readings. For precision work a high-resolution ($6\frac{1}{2}$-digit) high-impedance DVM is required. Such equipment used with a calibrated copper–Constantan thermocouple having one measuring junction and one reference junction maintained carefully at the ice point permits one to measure the absolute temperature around room temperature with an uncertainty of about ±0.01 K. However, the precision is much better than this, and temperature differences can be measured to within ±0.002 K or better.

Several precautions should be observed in order to enhance the long-term stability of a thermocouple.

1. Use the largest diameter wire that is feasible without conducting too much heat away from the area of the measurement junction.
2. If very thin wires are required, use larger-diameter extension wires for regions where there is almost no temperature gradient (typically the room temperature leads that go to the voltage-measuring device).
3. Avoid mechanical stresses and vibrations that might cause strain distortions in the wires.
4. Avoid very large temperature gradients along the wire.
5. Operate the thermocouple only at temperatures below its design limit.
6. If operating in a hostile atmosphere, use an adequate external sheath to protect the wires.

Platinum Resistance Thermometers.[2,5,6,12] The platinum resistance thermometer (PRT) is capable of extremely high accuracy, owing to the high purity attainable for platinum and the high reproducibility of its temperature coefficient of resistivity. For a standard PRT used to determine absolute temperature values with the highest possible accuracy, the resistance element is a coil of pure platinum wire, carefully annealed both before and after winding and enclosed in a sealed sheath (glass or thin-walled metal tube) containing dry air or helium as a heat-transfer gas. The design of standard PRTs is aimed at minimizing errors due to heat conduction and radiation and maximizing stability and calibration reproducibility. Such PRTs are rather fragile and sensitive to mechanical vibration. They also have low-resistance elements, typically an ice-point resistance R_0 of 25 Ω, which means a sensitivity of only 0.1 Ω K^{-1}.

In most experimental applications, one uses "practical" PRTs that are less expensive, more compact, more rugged, and more sensitive (owing to the choice of R_0 values of 100 or 200 Ω) than standard PRTs.† There are of course some trade-offs since practical PRTs are slightly less stable and have poorer interpolation accuracy than standard PRTs. In practical PRTs the platinum wire is wound on a glass or ceramic bobbin, and the assembly is hermetically sealed in a glass or ceramic capsule. Considerable care is taken to match the thermal expansion coefficients of platinum and the bobbin in order to minimize strain-induced resistance changes. Since practical PRTs are quite small (typically 25 mm long and 1.5 to 4.5 mm in diameter), they have moderately fast responses to temperature changes. The 90 percent response times in stirred or flowing liquids are typically 1 to 3 s. A more recent development is the thin-film PRT, in which a metal slurry is deposited on a ceramic substrate. These devices are very small and have 90 percent response times of 0.3 s or less, but their stability is significantly less than that of wire-wound PRTs.

The basic principle underlying the operation of any resistance thermometer[2] is a finite temperature coefficient of resistivity:

$$\alpha_t = \frac{1}{R}\left(\frac{\partial R}{\partial T}\right)_P \cong \frac{1}{R_0}\frac{dR}{dt} \tag{11}$$

where t is the Celsius temperature and R_0 is the resistance at 0°C. For platinum, α_t is roughly constant over a wide temperature range and is very reproducible for different wire samples. Since $\alpha_t \cong 0.0039$ K^{-1} and R_0 is typically 100 Ω in a practical PRT, a change of 1 K will cause a resistance change of only 0.39 Ω. In many applications, especially when measuring low or high temperatures, rather long electrical leads are needed. Such leads are usually made of fine wire to reduce heat conduction and thus have appreciable resistance (say 5 to 10 Ω). Clearly any accurate determination of temperature requires a careful measurement of R_t for the PRT at temperature t free from any errors due to lead resistance. This can be accomplished by using a four-lead PRT, which has a pair of identical leads connected to each end of the resistance element (i.e., separate current leads and voltage leads).

Measurement of R_t can be made with a Wheatstone bridge or with a potentiometric method. A three-terminal bridge configuration is shown in Fig. 4, where leads a, b, A, B have resistance R_a, R_b, R_A, R_B. Since the two fixed resistance arms have been chosen to have the same values R_1, the balance condition is simply

$$R_t + R_B = R_2 + R_A \tag{12}$$

for the configuration in Fig. 4a, and

$$R_t + R_A = R_3 + R_B \tag{13}$$

for the configuration in Fig. 4b. There are special three-pole double-throw switches to allow one to switch easily from configuration a to b. Differences in lead resistances and also any thermal voltages (thermocouple effects at junctions between platinum and other metals such as copper) are canceled out by balancing the bridge in both configurations and calculating R_t from

$$R_t = \tfrac{1}{2}(R_2 + R_3) \tag{14}$$

---

†In the past, standard PRTs could be obtained from Leeds & Northrup, and many of these are still in use. Practical PRTs are currently available in a wide variety of formats from several companies, such as Rosemount Inc., Minneapolis, MN, and Omega Engineering Inc., Stamford, CT; and Omega will calibrate PRTs for a modest fee.

FIGURE 4

Measurement of a four-lead PRT with a three-terminal Wheatstone bridge. The resistance $R_t = (R_2 + R_3)/2$; see text for details. If the lead resistances R_A and R_B are equal and there are no disturbing thermal emfs at the lead junctions, R_t can be determined from either of the single bridge balance conditions since R_2 and R_3 are then the same.

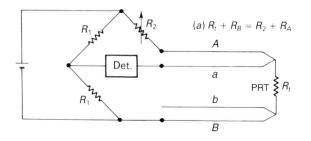

(a) $R_t + R_B = R_2 + R_A$

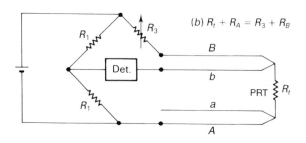

(b) $R_t + R_A = R_3 + R_B$

FIGURE 5

Potentiometric method of determining the resistance R_t of a four-lead PRT. The measured potential drops across R_t and a series standard resistance R_S are V_t and V_S, respectively. If thermal emfs at lead junctions are a problem, the PRT leads can be reversed as in Fig. 4b; in this case $R_t = R_S(V_t + V_{t,r})/(V_S + V_{S,r})$ is used to get R_t.

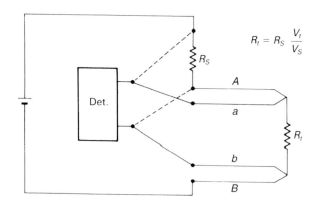

$$R_t = R_S \frac{V_t}{V_S}$$

Note that use of an ac bridge will also eliminate any errors due to thermal voltages, which are dc effects.

A potentiometric setup for measuring R_t is shown in Fig. 5. The detector is usually a high-resolution, high-impedance digital voltmeter. Voltage drops are measured across R_t and across a series standard resistance R_S that has a stable constant value, usually 100 Ω. The value of R_t is then given by

$$R_t = R_S(V_t/V_S) \tag{15}$$

In order to avoid self-heating (electrical $I^2 R_t$ heating caused by the measuring current I), one must use a DVM that operates at 1 mA or less on the appropriate resistance scale. The same limitation applies to the current flowing through R_t in the Wheatstone bridge. A current of 1 mA through a PRT of 100 Ω corresponds to a power dissipation of 0.1 mW, which will heat a typical PRT by ~50 mK in still air, by ~5 mK in flowing air, and by less than 1 mK in a stirred liquid.

For a PRT with $R_0 = 100 \, \Omega$, a temperature resolution of ± 1 mK requires that R_t be measured with a *precision* of $\pm 4 \times 10^{-4} \, \Omega$; i.e., a random error $\leq 4 \times 10^{-4}$ percent. Thus the measurement bridge and digital voltmeter must be high-precision instruments. In the case of the DVM, one needs at least six-digit resolution. It should be noted that, owing to decreasing sensitivity as T approaches 0 K, PRTs do not make attractive practical thermometers at very low temperatures.

The absolute accuracy of a PRT temperature value depends not only on the *accuracy* of the resistance ratio R_t/R_0 but also on the quality of the calibration. A calibration can be carried out by using fixed points[13] or by making a comparison with a standard PRT. The fixed-point method is more accurate if done carefully but is much more time-consuming. Over the range 0 to 232°C, a two(three)-point calibration will yield an accuracy of ± 1 mK (± 0.5 mK).[13] In the comparison method, both thermometers are placed in good thermal contact with an "isothermal" metal block, which can then be placed in an ice bath and in convenient stirred liquid baths over the range -150 to $+400$°C. For a calibration point below -150°C, liquid nitrogen at its normal boiling point is the best choice. The freezing points of several metals can be used for calibration above 400°C.

For PRT measurements of normal precision, the Callendar–van Dusen interpolation formula, described by Eqs. (6) to (8), is the most practical choice. Once the constants that appear in Eq. (6) or (8) have been determined, there are two ways to proceed in converting R_t readings into temperatures. One can prepare a conversion table of closely spaced R_t–T entries at even intervals of R_t and linearly interpolate between entries. Alternatively one can calculate t directly from Eq. (8). An analytic formula is used above 0°C since R_t/R_0 is quadratic in t; an iterative numerical method must be used below 0°C.

Thermistors.[2,6,14] Thermistors (a portmanteau contraction of "thermally sensitive resistors") are semiconductor devices consisting of sintered mixtures of metallic oxides such as NiO, Mn_2O_3, and Co_2O_3. They are usually extrinsic p-type materials containing excess oxygen above the stoichiometric amount. These ceramiclike resistors have large *negative* temperature coefficients of resistivity. The variation of the resistance R can be roughly approximated by

$$R = R_\infty \exp\left(\frac{\Delta E}{2kT}\right) \tag{16}$$

where ΔE is the electron energy gap. For typical thermistors, $\Delta E/2k \approx 3500$ K, which means a 4 percent change in resistance per kelvin at room temperature. Thus the magnitude of the temperature coefficient of resistivity $\alpha = R^{-1}(dR/dT)$ is about 10 times that of metallic resistors, and high resolution in a resistance thermometer can be attained with a less sensitive bridge or DVM than is required for a platinum thermometer. In addition the resistivity is so high that very small thermistor elements can still have high resistances. By using a thermistor with a high resistance in the temperature range of interest (say 10 to 100 kΩ), complications due to lead and contact resistances are eliminated and two-wire measurements are adequately precise.

Figure 6 shows a comparison of the temperature dependence of R_t/R_{25} for a platinum resistance thermometer and three typical thermistors. The greater thermistor sensitivity is obvious, but one should also note that the practical operating range for a given thermistor is relatively narrow. Let us assume that a resistance range from 2 to 20 kΩ is ideal for the measuring device to be used. Then a thermistor with $R_{25} = 20$ kΩ will operate over a range from 25°C to about 100°C, while one with $R_{25} = 200$ kΩ is useful from 100°C to about 200°C. Figure 6 also shows clearly that practical thermistors deviate from Eq. (16), the simple exponential expression valid for many intrinsic semiconductors. The standard

FIGURE 6
Resistance versus
temperature characteristics
for several typical
thermistors. The behavior
of a platinum resistance
thermometer (dashed line) is
shown for comparison.

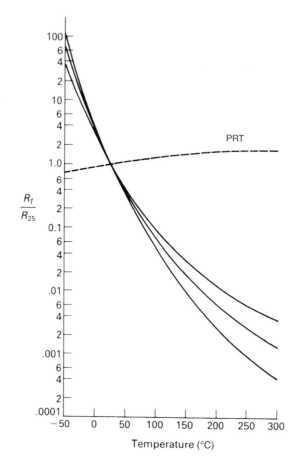

empirical curve-fitting technique used to represent the thermistor $R(T)$ variation is to consider $\ln R$ to be a polynomial in $1/T$ or vice versa. These equations are usually truncated at the cubic term, yielding

$$\ln R = A_0 + \frac{A_1}{T} + \frac{A_2}{T^2} + \frac{A_3}{T^3} \tag{17}$$

$$\frac{1}{T} = a_0 + a_1(\ln R) + a_2(\ln R)^2 + a_3(\ln R)^3 \tag{18}$$

It is common practice to drop the quadratic terms for the temperature range above 0°C; however, these quadratic terms must be retained in calibrating thermistors for use at low temperatures. In any event, neither Eq. (17) nor (18) provides a good representation of the $R(T)$ variation over a wide temperature range. Any given calibration is accurate to within ±0.1 K over a range of about 80 K. Two other disadvantages of thermistors in comparison with platinum resistance thermometers are their lower reproducibility and poorer long-term stability. A comparison of many performance characteristics of thermocouples, PRTs, and thermistors is given in Table 4.

The most stable and most useful type of thermistor is the glass-coated bead. In this design, a sintered oxide bead (0.075 to 1 mm in diameter) is sealed in glass, resulting in a probe bead with a diameter ranging from 0.125 to 1.5 mm. Important advantages of this small thermistor size are low probe heat capacity and rapid temperature response (0.1 to

TABLE 4 Comparison of performance characteristics for thermocouples, platinum resistance thermometers (PRTs), and thermistors

Characteristic	Thermocouple	PRT	Thermistor
Sensitivity	Moderate	Moderate	High (+)
Temperature range[a]	Very wide (+)	Wide (+)	Narrow (−)
Lowest temperature	4 K	15 K	77 K
Highest temperature[b]	2600 K	1000 K	575 K
Linearity	Fair	Good	Poor
Calibration stability	Very good (+)	Excellent (+)	Fair
Ease of measurement	Fair	Fair	Good
Measurement problems	Cold junction, Thermal emfs	Lead and contact resistance	Self-heating
Mechanical stability	Excellent (+)	Very good	Good
Size	Tiny (+)	Moderate	Very small (+)
Time response	Very fast (+)	Moderate	Fast (+)
Heat capacity of probe	Very low (+)	Moderate	Low (+)
Interchangeability	Good	Very good	Poor
Cost	Very low (+)	Moderate	Low

Note: A distinct advantage is indicated by (+) and a disadvantage by (−).

[a] For a given unit.

[b] For continuous operation.

1 s in still air and 5 to 15 ms in water). Such bead or microbead thermistors can be obtained commercially† with room-temperature resistances ranging from $10^3\ \Omega$ to $5 \times 10^6\ \Omega$.

The measurement of thermistor resistance can be carried out with two-lead bridge circuits or high-quality digital multimeters, as described in the preceding section on platinum resistance thermometers. The major concern is to limit the electrical power dissipation in the thermistor during the measurement. Small bead thermistors are sensitive to self-heating; i.e., the current flowing through a high-resistance thermistor can heat it above the temperature of the surroundings. Typical values range from 0.5 to 4 K per milliwatt of power dissipated. To avoid serious errors, this heating effect must be negligible compared with the desired accuracy of the temperature measurement. For a self-heating temperature rise of less than 1 mK, the power dissipation must be less than $\sim 1\ \mu$W, which implies, for example, a current of less than 7 μA through a 20-kΩ thermistor.

The calibration of a thermistor with either Eq. (17) or (18) requires at least three calibration points, even if a_2 and A_2 are set equal to zero. Preferably one should calibrate at five or six points distributed uniformly over the temperature range of interest. These equations should be considered as interpolation formulas: they do not provide a trustworthy extrapolation for temperatures outside the calibration range. One of the problems associated with calibrating thermistors is their fast response and the possibility of a thermal lag between the thermistor and the standard thermometer. If a thermistor and a mercury thermometer or a large PRT are positioned independently in a regulated constant-temperature bath, the thermistor will show larger and more rapid temperature fluctuations than the standard thermometer. The proper procedure is to mount both thermistor and standard thermometer in good thermal contact with a large metal block, which will damp out short-term fluctuations in the bath temperature. Calibration can also be carried out using the temperature fixed

†Microbead thermistors are available from Thermometrics, Edison, NJ; Fenwal Electronics, Inc., Framingham, MA; Victory Engineering Corp., Springfield, MA; YSI, Inc., Yellow Springs, OH; and many other firms.

points given in Tables 1 and 2. Since thermistors are often used between 0°C and 100°C, the obvious choices would be the water freezing point (or triple point), the water boiling point, and the gallium melting point.

Unfortunately, the stability of a thermistor resistance is not as good as that achieved with PRTs. The principal causes of instability in hermetically sealed bead thermistors are changes in lattice defect concentrations and changes in the electrical contact with the leads. Both of these changes can be caused by operating at high temperatures or by a transient electrical power surge. Contact resistance can also change during rapid cooling. As a result, thermistors "age" (i.e., exhibit a drift in resistance at constant temperature over a long time). This aging occurs even at room temperature and more rapidly at elevated temperatures. Typical drift rates for bead thermistors with $R_{25} = 10$ kΩ are only ~0.03 mK per day at 30 to 60°C but can reach 0.6 mK per day at 100°C and 2 to 5 mK per day at 200°C.[14,15]

Optical Pyrometers.[16,17]   The optical pyrometer can be used for the determination of temperatures above ~900 K, where blackbody radiation in the visible part of the spectrum is of sufficient intensity to be measured accurately. The blackbody emitted radiation intensity at a given wavelength λ in equilibrium with matter at temperature T is given by the Planck radiation law,

$$M_\lambda = (c/4)\rho_\lambda = \frac{c_1}{\lambda^5} \frac{1}{\exp(c_2/\lambda T) - 1} \tag{19}$$

where ρ_λ is the spectral radiant energy density, M_λ is the blackbody emitted radiant flux per unit wavelength interval, and c_1 and c_2 are universal constants ($c_1 = 2\pi hc^2 = 3.74177 \times 10^{-16}$ W m^2; $c_2 = hc/k = 0.0143877$ m K). Figure 7 shows the blackbody radiant flux per unit wavelength interval in units of W m^{-2} per micrometer.

FIGURE 7

Blackbody radiant excitance (emitted radiant flux) per unit wavelength as a function of wavelength and temperature. The effective wavelength of 655 nm for optical pyrometers is indicated by the vertical broken line. The position of maximum intensity λ_{max} at temperature T is given by the simple expression $\lambda_{max}T = 2898$ μm K, and this is shown by the dashed line.

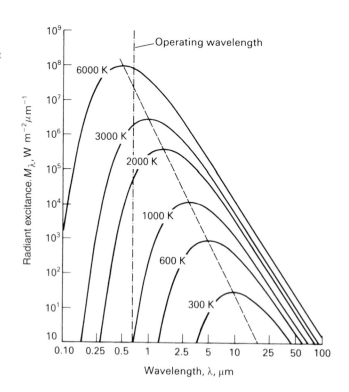

The optical pyrometer is essentially a photometer that measures the radiation intensity in a given wavelength interval. We shall here consider only the most common type of optical pyrometer, the disappearing-filament type. The object whose temperature is to be measured is viewed through a telescope containing, at an image plane, a lamp with a pure tungsten filament enclosed in an evacuated glass tube. The temperature of this filament can be varied by adjustment of the current passing through it. A red filter in the eyepiece and the insensitivity of the human eye to long-wavelength radiation create a narrow wavelength range (centered at 655 nm) for visual observation. When the current has been adjusted so that the filament becomes invisible against the blackbody radiation, the temperature is obtained from the measured current with a calibration curve or chart.

Fixed points for the calibration of the optical pyrometers are the silver, gold, and copper points and higher temperature secondary fixed points such as those given in Table 2. Calibration at other temperatures can be accomplished by use of a rotating sector or a filter of accurately known transmission factor between the fixed-point source and the pyrometer, in order to simulate a source of lower temperature in accordance with the Planck equation. Such sectors or filters are used also to permit the optical pyrometer to be used for the measurement of temperatures much above 2000 K.

Optical-pyrometer measurements are most reliable when the object being examined is the interior of a furnace or cavity of uniform temperature viewed through a small opening. Readings for an exposed surface are dependent upon the emittance ε of the substance concerned, which for an ideal "blackbody" is unity and for actual materials is less than unity. The emittance in the visible range is near unity for carbon ($\varepsilon \simeq 0.85$) and oxidized metals, but it is considerably less for platinum ($\varepsilon = 0.3$) and other unoxidized metals, especially when they are polished. The difference between the brightness temperature T_B obtained from the optical pyrometer and the actual temperature can be approximated by

$$\frac{1}{T} - \frac{1}{T_B} \cong \frac{\lambda}{c_2} \ln \varepsilon \tag{20}$$

and correction tables are available that take the emittance and any nonideal transmission of the viewing device into account. Under the best conditions, the optical pyrometer is accurate to about 0.2 percent of the absolute temperature (e.g., ± 4 K at 2000 K). It is also a convenient instrument for less precise measurements at high temperatures, such as routine measurement of furnace temperatures, etc.

Cryogenic Thermometers.[18]   The measurement of temperatures in the low-temperature region, below 77 K (liquid-nitrogen boiling point) and especially below ~30 K, requires special devices. Although the platinum resistance thermometer is the standard ITS-90 device down to 13.80 K, its sensitivity decreases rapidly at low temperatures and approaches zero as T approaches 0 K. As practical thermometers PRTs are limited to use above about 35 K. Special gold–Chromel thermocouples are available for low-temperature use. The alloy Au + 0.07%Fe yields the best sensitivity (≥ 15 μV K^{-1} above 10 K) and quite good stability, but the reproducibility of gold–Chromel couples is poor compared with that of the thermocouples listed in Table 3 for use at 77 K and above.

There are several special types of resistance thermometers for use in the range 1 K to 30 K: germanium resistors, carbon resistors, and carbon–glass resistors. Bulk germanium thermometers, which are widely used, have good stability (± 0.5 mK long-term) and excellent sensitivity at low T values ($d \ln R/dT \cong -2/T$). More recently germanium wafers have been developed with smaller heat capacities and better "high-temperature" sensitivities that allow them to be used up to 100 K. Problems with bulk and wafer germanium thermometers include high cost and difficulties with contact resistance. Carbon resistors are inexpensive and sensitive, but they are subject to poor long-term stability and require frequent recalibration. Carbon–glass resistance thermometers are significantly more reliable and easier to use than carbon and have characteristics comparable to those of germanium thermometers.

Several semiconductor devices (*p*-type GaAs and Si) are used at low temperatures; these diode thermometers provide an output voltage that increases as T is decreased. The Si-diode thermometer is more sensitive (typically -50 mV K^{-1} in the 2–30 K range and -3 mV K^{-1} from 30 to 80 K for a diode giving 1 V at 30 K) and more linear than germanium or carbon–glass resistance thermometers. It is probably the practical thermometer of greatest utility in the range 4.2 to 80 K.

Another attractive and widely used special thermometer is the quartz frequency thermometer, which is described in the next subsection, since it can be used over the very wide range from 4 K to 500 K.

Other Thermometric Devices. The vapor pressure of a pure liquid or solid is a physical property sensitive to temperature and thus suitable for use as a thermometer. The use of a liquid-nitrogen vapor-pressure thermometer is suggested for the range 64 to 78 K in Exp. 47. At very low temperatures (1 to 4.2 K), the vapor pressure of liquid helium can be used.

A recent innovation is the integrated-circuit temperature transducer, which is available in both current- and voltage-output versions.[6] This device provides an output linearly proportional to the absolute temperature (typically 1 μA K^{-1} or 10 mV K^{-1}). As semiconductor devices, they resemble thermistors in having limited range, self-heating problems, and mechanical fragility. However, they do provide a convenient way of obtaining an analog voltage proportional to T for various automated control devices.

Finally, the quartz thermometer should be emphasized as a versatile and high-precision device.[19] This thermometer makes use of the temperature dependence of the resonance frequency f_r of a quartz single crystal. This frequency variation is small (35 ppm K^{-1}) but very reproducible, and it can be measured with excellent precision. The temperature dependence of f_r depends strongly on the orientation of the quartz crystal slab. The frequency of a temperature-sensitive crystal beats against that of another crystal whose frequency is essentially independent of temperature. By counting the beat frequency, a resolution of $\pm 10^{-4}$ K can be achieved over the range 200 to 500 K and better than ± 0.02 K over the range 4 to 200 K. However, the sampling time is 10 s for $\pm 10^{-4}$ K resolution, which is inconveniently long; happily, it is only 0.1 s for ± 0.01 K resolution. The output is digital, and there are no lead-wire problems. Hysteresis and aging do occur, so calibrations are not better than ± 0.1 mK and a calibration drift of ~ 10 mK per month occurs. However, the quartz thermometer is generally a very convenient device for high precision as opposed to high absolute accuracy.

TEMPERATURE CONTROL

In addition to the measurement of temperature, it is often necessary to maintain a constant temperature. The importance of this type of control in experimental physical chemistry is illustrated by the fact that 30 of the 48 experiments described in this book require temperature control of some kind. Many physical quantities such as rate constants, equilibrium constants, and vapor pressures are sensitive functions of temperature and must be measured at a known temperature that is held constant to within ± 0.1 K or better. Certain physical techniques are even more demanding; for example, the measurement of the coexistence curve in Exp. 16 requires that the temperature be controlled to within ± 0.02 K.

There are two basic methods of achieving a constant temperature:[20]

1. Phase equilibrium is maintained at constant pressure between two phases of a pure substance or three phases of a two-component system (eutectic).
2. A temperature sensor provides a feedback signal to control the input of heat (or refrigeration cooling in some cases) in order to maintain the temperature close to any arbitrary desired value.

Method 1 is the simplest approach; method 2 is more flexible and generally more useful.

1. PHASE-EQUILIBRIUM BATHS

The greatest disadvantage of the phase-equilibrium method is the impossibility of attaining a desired *arbitrary* temperature unless a liquid–vapor system is used with a complicated manostat to maintain an arbitrary boiling pressure. In addition it is often difficult to maintain temperature control for very long periods of time with this method. There are however several advantages: economy and simplicity in operation, excellent temperature stability, and potentially high accuracy in the absolute temperature of the bath. Several commonly used systems of this type are listed below, together with brief comments on their use.

Liquid-Nitrogen Bath.   Liquid nitrogen at its normal boiling point (77.36 K = $-195.79°C$) provides a very convenient low-temperature bath. Since O_2 dissolved in liquid N_2 will raise the temperature of the bath, the mouth of the Dewar flask should be plugged loosely with glass wool or cotton to retard the slow condensation of atmospheric oxygen. The temperature of a liquid-nitrogen bath can be calculated on the assumption that the nitrogen vapor pressure is equal to the atmospheric pressure. Clearly temperature stability will depend on the absence of large changes in pressure; fortunately the boiling point of nitrogen changes by only 0.011 K Torr^{-1} near 77 K. Be careful in handling liquid nitrogen to avoid "burns" due to prolonged direct contact with bare hands or other exposed skin; see Appendix C.

Dry Ice Bath.   Solid carbon dioxide in equilibrium with CO_2 vapor at 1 atm will provide a temperature of 194.69 K = $-78.46°C$. Thus, Dry Ice, which is inexpensive and readily available, would seem to be very suitable for a constant-temperature bath at moderately low temperatures. Unfortunately, the use of Dry Ice alone is complicated by two difficulties: the problem of obtaining and maintaining the proper pressure of CO_2 gas; and the problem of achieving good thermal contact between the Dry Ice and the object to be cooled. Although these difficulties can be overcome by careful bath design and the use of a heater to cause a constant evolution of CO_2 gas, it is much easier to use a bath consisting of Dry Ice and a liquid that does not freeze at $-78.5°C$. This liquid provides good thermal contact throughout the bath and prevents air from diluting the CO_2 gas at the surface of the Dry Ice as it would at an exposed Dry Ice surface. The traditional choice of liquid is acetone, but acetone is flammable and should be handled with care. An alternative nonflammable liquid is isopropanol. In making up a Dry Ice bath, it is necessary to minimize the foaming that occurs owing to rapid evolution of gas when Dry Ice is placed in contact with liquid initially at room temperature. The Dry Ice should be pulverized. If a special grinder is not available, one can wrap chunks of Dry Ice in a towel and pound them with a mallet. **Be careful in handling pieces of Dry Ice; it can cause painful "burns" if held in the bare hand for more than a few seconds. The use of tongs or insulated gloves is strongly recommended.** A Dewar flask is first filled about two-thirds full with isopropanol or acetone, and then small quantities of very finely powdered Dry Ice are added

TABLE 5 Materials suitable for low-temperature slush baths

Compound	M.P. (°C)	Compound	M.P. (°C)
Isopentane (2-methyl butane)	−159.9	Ethyl acetate	−84
Methyl cyclopentane	−142.4	p-Cymene	−67.9
Allyl chloride	−134.5	Chloroform	−63.5
n-Pentane	−129.7	N-methyl aniline	−57
Allyl alcohol	−129	Chlorobenzene	−45.6
Ethyl alcohol	−117.3	Anisole	−37.5
Carbon disulfide	−110.8	Bromobenzene	−30.8
Isobutyl alcohol	−108	Carbon tetrachloride	−23
Acetone	−95.4	Benzonitrile	−13
Methyl alcohol	−93.9	p-Xylene	+13.3

slowly with a spatula. This Dry Ice will evaporate almost immediately, and there will be considerable foaming at the surface. Add more Dry Ice only after the foaming has subsided. After a while the liquid will have cooled to the point where the evaporation of Dry Ice is much slower, and some Dry Ice will begin to accumulate on the bottom of the Dewar. At this point, small lumps of Dry Ice can be added without causing serious foaming. Good temperature control with this bath is ensured only if it is well stirred and there is a slow but steady steam of CO_2 bubbles rising from the bottom.

Ice Bath. The ice–water equilibrium at 0°C provides an excellent constant-temperature bath. The ice should be washed, and distilled or deionized water must be used. To avoid thermal gradients between water at 4°C (maximum density) at the bottom of the Dewar and a 0°C liquid surface in which ice is floating, it is usually necessary to stir this bath. As a bath for the reference junction of a thermocouple, gradients can be eliminated by completely filling the Dewar with ice and adding only a small amount of cold distilled water; the weight of ice above the liquid will force ice down to the very bottom of the Dewar.

Organic Melting-Point Baths. These low-temperature baths can be made easily by cooling an organic liquid with either Dry Ice or liquid nitrogen until the bath liquid begins to freeze and a slush of solid and liquid is formed in a Dewar flask. **If you are using liquid nitrogen, be careful not to condense air, which could create an explosive mixture.** With frequent stirring of the slush but no special attention to purity or pressure control, one can achieve bath temperatures within ±1 K of the values given for each bath material in Table 5.

Vapor Baths. Boiling water (100.0°C) and naphthalene (218.0°C) can be used as vapor baths. In each case, the object whose temperature is to be controlled is immersed in the refluxing or condensing vapor. See Exp. 1 for details of a steam bath.

2. SYSTEMS UTILIZING TEMPERATURE CONTROLLERS

The most important method of achieving temperature control is to use a sensitive thermometer that generates an electrical signal, such as a resistance thermometer, thermistor, or thermocouple. Comparison of this signal with a reference signal that establishes the *set point* provides an error signal to a feedback circuit that controls the power to the heater. Thus any temperature deviation from the desired value detected by the sensor will be corrected automatically.

As in any system employing feedback to maintain a steady-state condition, certain design criteria must be met in order to obtain reasonably rapid response to environmental changes while avoiding excessive "hunting" or even uncontrolled oscillations. The performance of the system will depend upon such factors as the sensitivity and speed of response of the thermosensing element, the conduction or circulation of heat in the system (stirring, convection), and the fraction of the total energy input (heat input plus stirring work, etc.) that is being controlled by the regulator.

To provide damping of oscillations, a proportionating circuit is usually employed: in the vicinity of the desired temperature, the current or power to the heater is made roughly proportional to the difference between the actual temperature and a temperature setting that is slightly above the desired value. Such a circuit usually has an adjustment for providing an optimum range of proportionation, since a proportionating system provides stability at the expense of some precision of temperature control.

Design of Controllers. A brief review of the operating principles for temperature-control systems will be given below; more details are available in the literature.[21,22] The simplest type of controller is an on–off device, in which the heater is either fully on or fully off. This inexpensive and easily implemented control is suitable for moderate-quality control of large, sluggish baths and furnaces. A serious disadvantage of such a controller is its inherent cycling about the set point (i.e., the desired constant temperature value T_s). Figure 8 shows the time response after a step increase in the set point of an on–off controlled system. Not only does the temperature overshoot the desired value T_s when the set point is changed from a previous value T_s', but there is a persistent oscillation about T_s that never disappears. This long-term cycling can be eliminated by the use of proportional control.

With a proportional controller, the heater power supplied to the system is changed by an amount proportional to an error signal corresponding to the temperature difference $(T_s - T)$; see Fig. 9. The controller output can be either time proportionating or current proportionating;[22] only the latter will be considered here, as it is the more commonly used mode. Proportional controllers provide a fast response and show very little cycling about the set point, which means better temperature control. Two disadvantages are a steady-state *offset* from the set-point temperature and the presence of damped oscillations associated with a change in the set point, both of which are illustrated in Fig. 10. The offset can be reduced by decreasing the proportional band PB (i.e., increasing the controller gain), but there is a limit to this approach since instability in the form of undamped oscillations will occur at too high a gain.

The overshoot oscillations occurring after a change in set point can be reduced greatly by adding to the simple proportional control a second control signal proportional to the time derivative of the temperature. This feature "anticipates" the magnitude of future error

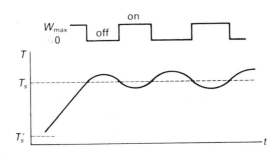

FIGURE 8

On–off controller behavior; the temperature variation observed after a step increase in the set-point temperature from T_s' to T_s. Note the long-term oscillation that persists about T_s. The heater power W can have only two values: 0 or W_{max}.

FIGURE 9
Heater power output W from
a proportional controller
as a function of sensor
temperature. The steeply
sloping linear ramp extends
over a temperature range
called the proportional band
(PB). The controller gain
is inversely proportional
to the width of this band:
$(1/W_{max})dW/dT = \mathrm{PB}^{-1}$. Note
that an on–off controller is a
limiting case of proportional
control with a proportional
band of zero width.

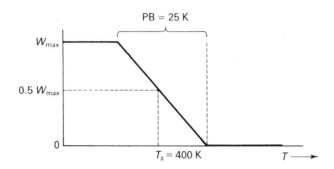

FIGURE 10
Response of a proportionally
controlled system to a step
increase in the set point from
T_s' to T_s. In this example the
gain is roughly one-half the
critical gain and the transient
oscillations are well damped.
Note the offset from the set-
point temperature.

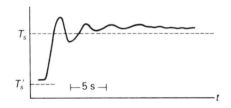

FIGURE 11
Response of a system with
(a) proportional + derivative
(PD) control and (b) proportional + integral + derivative
(PID) control when the gain is
15 percent of the critical gain.
Part (c) shows the response
with PID control when the
gain is 50 percent of the
critical gain, a practical limit
for good regulation. Compare
these responses to that
shown in Fig. 10 for simple
proportional (P) control.

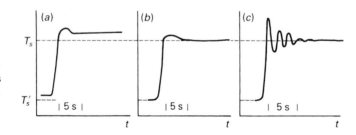

signals and smooths out overshoot oscillations. The offset problem could of course be
eliminated by a manual reset of the set-point value. However, this can be done automatically by adding an integrating control feature, for which the control signal is proportional to the integral of the error signal. As long as there is any difference between
the control temperature and the set-point temperature, a small corrective change in
the heater power will occur to slowly reduce this difference to zero. Figure 11 illustrates the behavior of systems with proportional + derivative (PD) or proportional +
integral + derivative (PID) controllers. Note that there are also proportional + integral
(PI) controllers.

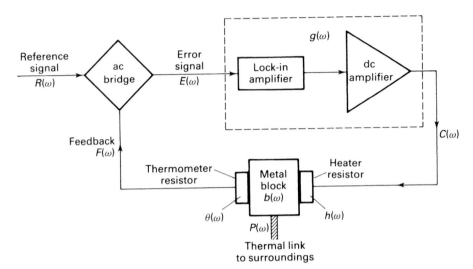

FIGURE 12
Schematic diagram of a
temperature-controlled
block at a set-point
temperature above the
ambient temperature of the
surroundings. The notation
is that of Forgan.[23] The dc
voltage V applied to the
heater is modified by the
small correction voltage
$C(\omega)$; the resulting heater
power output is $W(\omega) = V^2/R_h + 2VC(\omega)/R_h$, where R_h is
the electrical resistance of the
heater. See the text for other
details.

The proper design of constant-temperature systems requires judicious choice of sensor and controller and also careful attention to mechanical/thermal layout of the bath or block that serves as the heat bath. Since control problems are more acute in maintaining constant temperature at low temperatures, much of the pertinent literature is in cryogenic journals. However, the same general principles are equally valid for control at low, moderate, or high temperatures. A very clear mathematical analysis of the coupling between electrical and thermal responses has been given by Forgan,[23] and the conclusions of his analysis will be summarized here. A typical temperature-control system is shown schematically in Fig. 12. Frequency-domain responses are indicated; e.g., the Fourier transform of the temperature T as a function of time t gives an amplitude $T(\omega)$. The symbols $h(\omega)$, $b(\omega)$, and $\theta(\omega)$ denote the thermal responses of the heater to electrical power input, of the block to energy input from the heater, and of the thermometer to temperature variations of the block. The electrical gain $g(\omega)$ represents the response of the lock-in amplifier and the dc operational amplifier. The temperature $T(\omega)$ of the block is related to the control signal $C(\omega)$ and the perturbation $P(\omega)$, which is mostly heat leakage from the block to colder surroundings.

There are four time constants τ associated with the three thermal responses $h(\omega)$, $b(\omega)$, $\theta(\omega)$ and the electrical gain $g(\omega)$. The effect of a coupled set of four time constants on the behavior and stability of the system is a complex problem with rather messy solutions. However, a reasonably simple result is obtained for the limiting case in which

$$\tau_{max} = \tau_a > \tau_b > \tau_c > \tau_d = \tau_{min} \tag{21}$$

and all four τ values are well separated. In this case the best arrangement is to have the time constants of the block and of the lock-in amplifier (τ_a and τ_b) be long and the time constants of the heater and of the thermometer (τ_c and τ_d) be very short. This increases the critical open-loop dc gain and allows operation at high amplifier gain, which leads to good regulation control. Indeed one wants the time constants of the block, which characterize how rapidly the block cools when the heater is off, to be as long as possible, say 10 s or more. The time constant of the thermometer must be very short (1 to 10 ms is desirable), since the controller is actually controlling the temperature of the thermometer, which must therefore be closely coupled to the block. The heater should have a small thermal mass and be in very good contact with the block so that τ_c will be short. The absence of temperature gradients within the block (or bath) is also important; the heater should be distributed as uniformly as possible over the block (or the bath should be well stirred).

Commercial Controllers. There is a large variety of commercial temperature control units capable of low (± 0.2 to ± 1 K) or even moderate (± 0.05 to ± 0.1 K) resolution available from companies such as Minco, Omega, and Cole-Parmer. Many of these are simple on–off or proportional controllers, and most are designed to operate from the signal from an analog thermometer. Controllers capable of ± 5 to ± 10 mK regulation are available from Grant (Science/Electronics) and Lauda (Brinkmann Instruments). Even better constancy (± 1 mK or less) can be achieved with temperature controllers from Tronac. These instruments permit high-resolution selection of the sample temperature, adjustment of the gain to match the rate of heat loss, and adjustment of the time constant for the integrating circuit. Although the units described above are all analog controllers, some of them can be computer controlled and some even have LabView (National Instruments) software drivers. It is further possible to make modifications in order to convert an analog thermometer signal into digital form, operate on that signal with a computer, and then convert back to power an analog heater.

Digital control systems are intrinsically more attractive, since they are especially convenient for a computerized data acquisition system in which the temperature control is reset at programmed times or at preset intervals. Programmable digital controllers are available from Neslab (± 10 mK) and from Scientific Instruments and Lakeshore (± 1 mK). The latter two units are designed for cryogenic temperature control, but they can be used successfully at higher temperatures (up to ~ 450 K).

Stirred Water Baths. The water bath, equipped with stirrer, temperature sensor, control circuit, and heater, provides the most commonly used means of temperature control in the range 15 to 80°C. There are several commercial sources of regulated water baths with fluid volumes of 6 to 50 L.† A good design for a water bath for general student laboratory use consists of a large rectangular tank, perhaps 18 in. $\times$ 36 in. in horizontal area and 18 in. deep, with a water capacity of about 170 L. The tank should be constructed of welded stainless steel; the inclusion of glass windows in two or more sides is a convenience but complicates the design. The thermometer, temperature controller, and stirrer are best mounted in the middle; this arrangement leaves the ends free for experimental work. There should be adequate provision for mounting rods and clamps to support flasks, dielectric cells, etc. Depending on the experiment being done, two or four sets of apparatus can be operated in a single bath of this size. The bath should be provided with the following items.

Stirrer. The bath fluid must be well stirred in order to avoid temperature gradients within the bath. A large centrifugal water circulator, driven by a $\frac{1}{20}$-hp motor, provides adequate stirring. It should be positioned carefully so that the effluent stream will cause efficient circulation throughout the entire tank.

Heater.[24] A single copper-sheathed immersion heater ($\frac{1}{4}$ to $\frac{3}{8}$ in. in diameter) of 250-W capacity is adequate for temperature control up to about 30°C. It is recommended that the heater be positioned low in the bath and in the effluent stream from the stirrer. At higher temperatures a second heater may be needed. In this case one can be maintained at a constant power and the other can be regulated by the temperature controller, or both can be controlled, depending on their power ratings, the temperature desired, and other design factors. Blade heaters sheathed in stainless steel have longer time constants than copper-sheathed tube types and will not give as good temperature control, but they can be used as unregulated constant-power heaters.

---

†Grant constant-temperature baths are available from Science/Electronics, Inc., P.O. Box 986, Dayton, OH 45401; Lauda baths are available from Brinkmann Instruments Co., Cantiague Rd., Westbury, NY 11590.

Thermometers. The sensor probe should be placed in the bath in a central location removed from the heater and stirrer units. If a single cell or sample holder immersed in the bath is the only object of interest, the probe should be placed near this cell but not touching it. The measuring thermometer should be placed in good thermal contact with the cell or sample holder if there is only one or positioned close to the location of multiple cells.

Temperature controller. Large water baths are typically provided with controllers of moderate resolution (say ±0.1 to ±0.5 K), but much better constancy (±0.1 K or less) can be achieved. See further details given in the preceding section on commercial controllers.

Water-level and cooling control. When a water thermostat bath is operated for long periods, it is subject to loss of water by evaporation to a point that will interfere with its normal operation unless the water is replaced. Therefore the tank should have an overflow drain and should be provided with a continuous supply of cold water that can be controlled from a few drops per minute to a small, steady stream. This supply of water also performs an important control function. A $\frac{1}{20}$-hp motor delivers constantly about 37 W of mechanical energy to the bath, and on days when the room temperature is not far below the desired bath temperature, this amount of energy is itself sufficient to maintain the bath temperature above the desired level. A stream of cold water may compensate largely or entirely for the stirring heat. The adjustment of the water flow is often critical and must be checked frequently.

Miscellaneous. To reduce rusting of iron hardware in the bath, a zinc or aluminum electrode with an applied anodic potential of a few volts with respect to the tank as ground may be provided. This should be protected from accidental contact by a perforated plastic tube.

Low-temperature modifications.[25] For operation at temperatures in the range 5 to ~15°C (~20°C in the summer), one needs a supply of chilled cooling water. For work down to about -15°C, one needs a sealed refrigeration unit that can cool the bath sufficiently. The water in the bath must be replaced by water + ethylene glycol (antifreeze), and better insulation may be required.

Stirred Oil Baths. Open water baths are subject to substantial evaporation loss above 50°C, and even closed water baths present problems above ~80°C. In the past, "cylinder oil" was frequently used in the range 70 to 250°C, but there are safety concerns connected with fire hazards and toxicity. Commercial peanut oil is an inexpensive bath fluid that can be used up to about 225°C. It is nontoxic and even smells good, but fire hazards still remain. The best modern bath fluids for high-temperature use are high-molar-mass polyethylene glycol or polypropylene glycol (both of which are essentially nonflammable with "flash points" of 260°C) and Dow-Corning 510, 550, or 710 silicone fluids (nonflammable polydimethylsiloxanes with flash points of 310°C). Both types of materials are nontoxic and have low vapor pressures even at 200°C; the glycols can be used in closed baths up to 230°C, and the silicones can be used up to 300°C. Unfortunately, none of these materials is inexpensive, but they have long service lifetimes at high temperatures, especially the silicone fluids. For even higher temperatures, up to about 425°C, a eutectic mixture of sodium, potassium, and lithium nitrates (14, 56, and 30 wt%, respectively) can be used as the bath fluid, but furnaces are preferable to baths for temperatures above 300°C.

Stirred fluid baths designed to operate below -15°C require low-viscosity, low-freezing-point fluids such as Univis (a nonflammable oil) or Dow-Corning XLT silicone fluid. Large water-cooled compression refrigeration units and heavy-duty fluid stirrers are also needed. Many of the low-temperature units available commercially use circulator pumps instead of stirrers, as described below.

Circulating Liquid Baths.   These devices consist of a fluid reservoir, a heater, a temperature sensor, a controller, and a circulating pump. Commonly used general-purpose baths are made by Haake, Lauda, and Omega Engineering. Some of these have refrigeration units and are designed for operation at low temperatures (typically down to −20°C but sometimes to −35°C or even −70°C), as well as above room temperature.

Modern circulating baths are digital, and even programmable in some cases, with PRT sensors and PID control. They can achieve control stability of ±0.01°C (even ±0.005°C for the best ones) over the range 20 to 200°C without refrigeration units. These baths are easy to use, and no detailed information on their operation is given here. However, there are many simpler baths with manually set thermoregulators and proportional controllers.

Constant-Temperature Sample Blocks.   The design principles discussed at the beginning of this section apply just as well to the control of a metal block containing the sample of interest as to a fluid bath. This "dry design" usually involves a three-stage vacuum thermostat: an outer vacuum jacket that is immersed in a bath at roughly constant temperature, an inner vacuum can that serves as an adiabatic shield and radiation shield, and the sample block (usually copper). For low-temperature operation (4 to 300 K), the outer bath is a cryogenic liquid such as nitrogen or helium, and excellent commercial controllers are available from Lake Shore Cryotronics.[26] For operation in the moderate temperature range, from −40 to +70°C, the outer jacket is maintained at a temperature below the desired control temperature with a commercial circulating device, and microcomputer digital control is capable of maintaining a long-term temperature constancy of better than ±1 mK.[27] For operation at high temperatures (300 to 1300 K), a water jacket can serve as the outer bath, and digital control yields temperature stability of ±0.1 K at 1000 K.[28] Less elegant high-temperature control can be achieved with electric furnaces; see the next subsection.

3. SPECIAL SYSTEMS

A few constant-temperature systems of lower precision involving "air baths" are described below. In addition to ovens and furnaces, there are air thermostats, which are useful for complex gas-handling systems, but these will not be described.

Ovens.   Commercially available laboratory ovens can be used up to temperatures as high as 200 or 300°C (∼575 K). They are usually regulated to within ±(1–2) K with a bimetallic thermoswitch. A bimetallic strip, which consists of two dissimilar metal strips bonded together, undergoes differential thermal expansion. At some arbitrary temperature the two metals are of equal length and the strip is flat. As the temperature changes, one metal will expand or contract more than the other, thus causing the strip to bend or curl up. Although such thermoswitches are cheap, mechanically rugged, and capable of operating over a wide range, their accuracy is only about 1 percent and the contacts sometimes stick.

Air circulation in typical ovens is by convection, and the temperature inside the oven is usually nonuniform. To obtain better temperature uniformity and improved control, a lining made of thick sheet aluminum or copper can be installed.

Furnaces.   Electric furnaces are used for higher temperatures. These are obtainable commercially, but satisfactory ones can be made in the laboratory (see Fig. 24-1). A pregrooved Alundum core is wound with suitable heating wire (nichrome, Chromel A, Kanthal) of 12 to 20 gauge, and covered with a thick coating of Alundum cement. This is surrounded by several inches of powdered magnesia, for thermal insulation, in a large metal container. After assembly, the furnace is slowly heated to slightly above 1000°C to set the cement. A cover plug may be cut from firebrick. Internal metal radiation shields are needed if a high degree of temperature uniformity is required. Temperature is controlled with a Variac or similar transformer operated manually or by a proportionating control circuit with a thermocouple potentiometer or resistance thermometer bridge.

For temperatures above 1000°C, special high-temperature furnaces, including induction furnaces, vacuum furnaces, and arc furnaces, are used, but these are beyond the scope of this book.

REFERENCES

1. M. L. McGlashan, *J. Chem. Thermodynamics* **22,** 653 (1990); H. Preston-Thomas, *Metrologia* **27,** 3 and 107 (1990).

2. J. F. Schooley, *Thermometry,* CRC Reprint, Franklin, Elkins Park, PA (1986).

3. *Supplementary Information for the ITS-90,* International Bureau of Weights and Measures, Pavilion de Breteuil, Sevres, France (1990).

4. R. N. Goldberg and R. D. Weir, *Pure & Appl. Chem.* **64,** 1545 (1992).

5. *Temperature Calibration and Interpolation Methods for Platinum Resistance Thermometers.* Report 68023F, Rosemount, Inc., Minneapolis, MN (1980).

6. *The Temperature Handbook,* 5th ed., Omega Engineering, Stamford, CT (2005). Also available as a CD-ROM; see the omega.com website.

7. R. E. Bentley (ed.), *Resistance and Liquid-in-Glass Thermometry,* Springer, New York (1998).

8. J. A. Wise and J. F. Swindell, *Liquid-in-Glass Thermometry,* Natl. Bur. Stand. Monogr. 150, U.S. Government Printing Office, Washington, DC (1975).

9. *Manual on the Use of Thermocouples in Temperature Measurement,* ASTM Special Publ., ASTM, Philadelphia (1993).

10. D. D. Pollock, *Thermocouples: Theory and Practice,* CRC Press, Boca Raton, FL (1991).

11. *NIST ITS-90 Thermocouple Database* (NIST SRD60); see http://srdata.nist.gov/its90/main.

12. L. Michlaski, K. Eckersdorf, and J. Mcghee, *Temperature Measurement,* 2d ed., Wiley, New York (2000).

13. J. Ancsin and E. G. Murdock, *Metrologia* **27,** 201 (1990).

14. E. D. Macklen, *Thermistors,* Electrochemical Society/State Mutual Book and Periodical Service, Bridgehampton, NY (1980).

15. S. D. Wood, B. W. Mangum, J. J. Filliben, and S. B. Tillett, *J. Res. Natl. Bur. Stand.* **83,** 247 (1978).

16. D. P. DeWitt and G. D. Nutters (eds.), *Theory and Practice of Radiation Thermometry,* Wiley, New York (1988).

17. P. J. Dickerman (ed.), *Optical Spectroscopic Measurements of High Temperatures,* reprint ed., Books on Demand, Ann Arbor, MI (1961); R. E. Bentley (ed.), *Temperature and Humidity Measurement,* Springer, New York (1998).

18. L. G. Rubin, B. L. Brandt, and H. H. Sample, "Cryogenic Thermometry: A Review of Recent Progress," in J. F. Schooley (ed.), *Temperature: Its Measurement and Control in Science and Industry,* vol. 5, part II, p. 1333*ff,* American Institute of Physics, New York (1982).

19. A. Benjaminson and F. Rowland, "The Development of the Quartz Resonator as a Digital Temperature Sensor with a Precision of 10^{-4} K," in H. H. Plumb (ed.), *Temperature: Its Measurement and Control in Science and Industry,* vol. 4, part IV, p. 701*ff,* Instrument Society of America, Pittsburgh, PA (1972).

20. S. C. Greer, "Measurement and Control of Temperature," in J. H. Moore, C. C. Davis, and M. A. Coplan, *Building Scientific Apparatus,* 3d ed., Perseus, Cambridge, MA (2002).

21. J. Sandford, "Understanding Temperature Control," in *Instruments and Control Systems,* June 1977, pp. 37–42.

22. C. de Silva and M. A. Aronson, "Process Control," a series appearing in *Measurements and Control,* esp. part 3 (issue 106, Sept. 1984, pp. 165–170) and part 4 (issue 107, Oct. 1984, pp. 133–145). Much of this material has also appeared in *Principles of Temperature Control,* available from Gulton Industries, Inc., Service Division, Costa Mesa, CA.

23. E. M. Forgan, *Cryogenics* **14,** 207 (1974).

24. *The Electric Heaters Handbook,* "21st century" ed., Omega Engineering, Stamford, CT (2000).

25. G. I. Williams and W. A. House, *J. Phys. E: Sci. Instrum.* **14,** 755 (1981).

26. Lake Shore Cryotronics, Westerville, OH, has an extensive website www.lakeshore.com, which includes a reference manual on "Thermal Resistances of Cryogenic Temperature Sensors from 1–300 K."

27. R. B. Strem, B. K. Das, and S. C. Greer, *Rev. Sci. Instrum.* **52,** 1705 (1981).

28. W. M. Cash, E. E. Stansbury, C. F. Moore, and C. R. Brooks, *Rev. Sci. Instrum.* **52,** 895 (1981).

GENERAL READING

R. P. Benedict, *Fundamentals of Temperature, Pressure, and Flow Measurements,* 3d ed., Wiley-Interscience, New York (1984).

S. C. Greer, "Measurement and Control of Temperature," in J. H. Moore et al., *Building Scientific Apparatus,* 3d ed., Perseus, Cambridge, MA (2002).

G. K. McMillan, *Advanced Temperature Measurement and Control,* ISA, Research Triangle Park, NC (1995).

G. C. M. Meijer and A. W. Herwaarden (eds.), *Thermal Sensors,* Inst. of Physics Pub., Bristol, UK/ Philadelphia, PA (1994).

J. B. Ott and J. R. Goates, "Temperature Measurement with Application to Phase Equilibrium Studies," in B. W. Rossiter and R. C. Baetzold (eds.), *Physical Methods of Chemistry. Vol. VI: Determination of Thermodynamic Properties,* 2d ed., chap. 6, Wiley-Interscience, New York (1992).

T. J. Quinn, *Temperature,* 2d ed., Academic Press, New York (1991).

J. F. Schooley, *Thermometry,* CRC Reprint, Franklin, Elkins Park, PA (1986).

Temperature, Its Measurement and Control in Science and Industry, vol. 3 (C. M. Herzfeld, ed.), reprint ed., Krieger, Melbourne, FL (1972); vol. 4 (H. H. Plumb, ed.), Instrument Society of America, Pittsburgh, PA (1972) (Books on Demand reprint), parts 1 and 2 are especially valuable; vols. 5 and 6 (J. F. Schooley, ed.), American Institute of Physics, New York (1982 and 1993).

Vacuum Techniques

This chapter includes discussions of various topics of importance in the practice of handling gases at low pressures. Major emphasis is given to high-vacuum techniques, since they play an important role in a variety of physical chemistry research problems. For vacuum applications in this book, see Exps. 5, 24, 26, 47, and to a lesser extent 30, 37, 38, 40, and 43. Some of the material presented is also pertinent to the problem of pumping on refrigerant baths; for an application of this technique, see Exp. 47. This chapter is not intended as a comprehensive treatment. The appropriate theoretical background is given in Ref. 1, and more detailed technical information is available in the references cited in the General Reading list.

INTRODUCTION

The term *vacuum* refers to the condition of an enclosed space that is devoid of all gases or other material content. It is not experimentally feasible to achieve a "perfect" vacuum, although one can approach this condition quite closely. It is possible routinely to obtain a vacuum of 10^{-6} Torr and with more sophisticated techniques 10^{-10} Torr (1.3×10^{-13} bar); it is even possible by special techniques to obtain a vacuum of 10^{-15} Torr, or about 30 molecules per cubic centimeter. One *Torr,* the conventional unit of pressure in vacuum work, is the pressure equivalent of a manometer reading of 1 mm of liquid mercury; 1 Torr $\equiv \frac{1}{760}$ atm $= 1.333 \times 10^{-3}$ bar.

For Dewar flasks, metal evaporation apparatus, and most research apparatus, a vacuum of 10^{-5} to 10^{-6} Torr is sufficient; this is in the "high vacuum" range, while 10^{-10} Torr would be termed "ultrahigh vacuum." However, for many routine purposes a "utility vacuum" or "forepump vacuum" of about 10^{-3} Torr will suffice, and for vacuum distillations only a "partial vacuum" of the order of 1 to 50 Torr is needed.

PUMPS

The principal types of vacuum pumps will be discussed in this section. The water aspirator is a crude but useful pump for many routine operations. Rotary oil pumps are used for pumping on refrigerant baths and as the forepump for "backing" low-pressure pumps. For most high-vacuum work, diffusion pumps are utilized to achieve pressures of about 10^{-6} Torr; however, there are special types of pumps that can be used to achieve ultrahigh vacuums of 10^{-9} Torr or less.

Water Aspirator.   The water aspirator produces only a partial vacuum but one that is entirely adequate for a large number of simple chemical operations, such as filtration by suction, distillation under reduced pressure, and evacuation of desiccators.

This device operates through Bernoulli's principle. Water enters the aspirator from a pipe at mains pressure (about 4 atm) and leaves against 1 atm external pressure from a nozzle of lesser diameter. In an internal constriction of still smaller diameter, the pressure drops nearly to zero. The air from the system being evacuated enters at this constriction and is carried along with the water.

The ultimate vacuum achievable with a water aspirator is limited by the vapor pressure of the water itself; any significant drop in the pressure much below this will be reversed by the vaporization of water. The vapor pressure of water is 12.8 Torr at 15°C, 17.5 Torr at 20°C, and 23.8 Torr at 25°C. A lesser vacuum (higher pressure) can be achieved by reducing the flow of water into the aspirator and/or bleeding air into the system through a controlled leak.

When the water supply to an aspirator is shut off or greatly reduced while the system is under vacuum, the water in the aspirator can be sucked into the system. Modern aspirators often contain a check valve to prevent this, but this should not be relied upon; a trap bottle should always be provided between the aspirator and the system.

A word of **warning:** the flow rate of water through an aspirator is subject to change, often sudden change, as the water mains pressure fluctuates. This may produce changes in the pressure in the system being evacuated, or even cause "suckback" of water into the trap bottle. If a constant pressure (especially for reduced-pressure distillation) is required, one must be alert for such changes and be ready to make required adjustments.

Mechanical Pumps.   Perhaps the most common form of vacuum pump is a mechanical pump that operates with some sort of rotary action, with moving parts immersed in oil to seal them against back-streaming of exhaust as well as to provide lubrication. These pumps are used as forepumps for diffusion pumps. Other common laboratory applications are the evacuation of desiccators and transfer lines and distillation under reduced pressure. These pumps have ultimate pressures ranging from 10^{-4} to 0.05 Torr, and pumping speeds from 0.16 to 150 L s^{-1} or more, depending on type and intended application.

Two commonly used belt-driven pumps are the Hyvac 2 and the Welch Duo-Seal 1400. These are moderately priced laboratory pumps capable of good ultimate vacuum (10^{-4} Torr) but with relatively low pumping speed (0.4 L s^{-1} at 1 atm; about half that at 1 mTorr). At somewhat higher cost, the Hyvac 7 and the Duo-Seal 1402 provide both good ultimate vacuum and higher pumping speed (1.2 L s^{-1}). There are also less expensive all-purpose pumps that give reasonable pumping speed ($\sim$0.6 L s^{-1} at 1 atm) but a less good ultimate vacuum ($\sim$15 mTorr). Some models of Pressovac and pumps from Leybold can also be used as a source of positive pressures above 1 atm.

Direct-drive pumps, with an ultimate vacuum that is typically 5 mTorr, are increasingly becoming the dominant design since they are more compact and quieter. However, they are somewhat less durable, since the rotor operates at significantly higher speed than that of a belt-driven pump. A wide variety of pumps with different capabilities are available from Hyvac, Welch, Balzers, Kurt Lesker, Edwards, and Kinney.

One common design of the rotary pump is that of the Welch Duo-Seal pump, as shown in Fig. 1. In this type of pump, a rotating drum, tangent to the inside of the cylinder (with a clearance of only 0.003 mm), has deep slots that hold two (or more) vanes that slide along the cylinder wall, compressing the gas and carrying it to the exhaust port. Some belt-driven pumps tend to be noisy (hammering sound) when pumping a good vacuum;

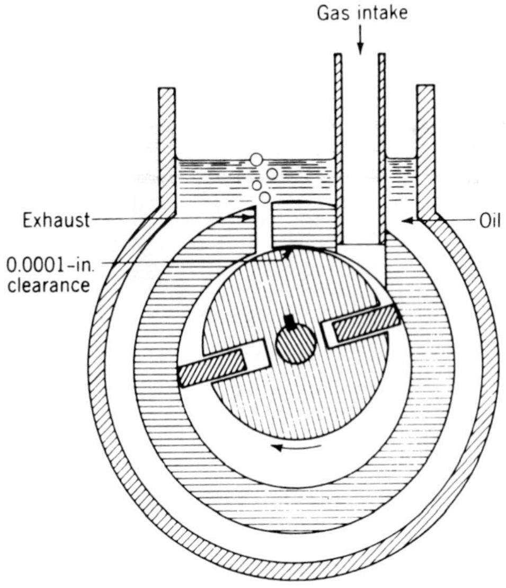

FIGURE 1
Schematic diagram of a
rotary oil pump similar to the
Welch Duo-Seal; based on
the Gaede oil pump design.

the hammer noises tend to be intermittent or muffled when the vacuum is less good. By contrast, the Duo-Seal and Edwards belt-driven pumps are most quiet when the vacuum is best; this is a definite advantage for continuous operation in a room where people are working. Direct-drive pumps are in general quite quiet.

There are considerable variations on this basic design. Many commercial types are compound pumps in which two single stages are mounted on the same shaft and are connected in series as a means of increasing the pumping speed and improving the ultimate vacuum. The pumping speed at high pressures depends on both the size and design of the pump.

An important limitation on the ultimate vacuum achievable with a rotary oil pump is the characteristics of the oil used. It must of course have a negligible vapor pressure and contain no significant quantity of dissolved volatile components. Accordingly, *care should be taken to prevent vapors, especially those of organic solvents, from entering the pump.* A Dry Ice–isopropanol trap is adequate to prevent this. The pump oil should be viscous enough to serve as a lubricant but fluid enough to permit air bubbles from the exhaust to rise rapidly to the oil surface, so as not to reenter the pump with the oil. *The oil must be clean;* dirt or metal particles suspended in the oil will cause abrasion and wear at points where tolerances are critical, and chemical fumes (e.g., hydrogen halides, sulfur dioxide, oxides of nitrogen) will cause corrosion. The oil level and cleanliness should be inspected at regular intervals, and the oil should be replaced whenever it is suspect. In the event of any serious contamination (e.g., corrosive fumes, water, or mercury), the pump must be disassembled by a competent mechanic, cleaned with solvents, dried, carefully inspected, reassembled (with due care for the extremely critical mechanical tolerances), and filled with fresh oil.

Although direct-drive pumps are manufactured, many rotary oil pumps are belt driven by an electric motor. The belt should be tight enough to prevent slippage but not so tight that it will produce wear on the motor and pump bearings and on the belt itself. Since the belt drive of a pump presents a mechanical hazard for fingers, loose clothing, and long hair, it should always be covered by a safety guard. With care and periodic oil changes, rotary pumps can be left on continuously and will operate well for years. When a rotary pump must be used under conditions in which significant amounts of condensable vapors—generally water vapor—pass into it, a pump with a *vented exhaust* should be used. In this pump a quantity of room air is introduced into the exhaust as a diluent to prevent condensation and

FIGURE 2
Roots blower.

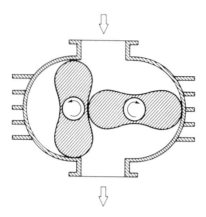

resultant contamination of the oil. It is inadvisable to permit organic vapors to enter even a vented exhaust pump, because of their high solubility in the oil; it is much better to freeze them out with a cold trap.

A mechanical pump giving very high pumping speeds, a few hundred to a few thousand liters per second in the 10 to 10^{-2} Torr range, is the *Roots pump*. As illustrated in Fig. 2, this pump, also known as a Roots blower, consists of a pair of two-lobed rotors that rotate in opposite directions. The rotors have a clearance of a few tenths of a millimeter and are synchronized by a gear drive so that they "push" the gas on the inlet side through to the exhaust outlet. There is no oil seal between the rotors, so compression ratios of only about 10:1 to 100:1 are achieved at typical rotational speeds of 3000 rpm. Because of this compression however, a rotary pump of lower pumping speed in series with the blower is adequate to prevent back-streaming at high throughput levels. Two-stage Roots pumps are also produced that can reduce the pressure to the 10^{-3} to 10^{-5} Torr range. Roots pumps are widely used for pumping large chambers.

A mechanical pump providing even lower vacuum levels is the *turbomolecular pump,* in which one or more balanced rotors (turbine blades) spin at 20,000 to 50,000 rpm. At these rotation rates, the periphery moves at a speed that exceeds the mean molecular speeds of most molecules, and gas–rotor collisions impart a momentum component to the gas in the direction of the exhaust. Compression ratios up to 10^6 can be achieved as long as the outlet pressure is kept below about 0.1 Torr by a forepump.

Balzers TPU 170 and the Welch 3134 are good-quality turbomolecular pumps. They have an ultimate vacuum limit of 10^{-9} Torr and pumping speeds of 300 L s^{-1} from 10^{-2} to 10^{-8} Torr. Other pumps with speeds up to 10,000 L s^{-1} are available, although these rates are reduced somewhat for light gases such as H_2 and He, which have high molecular velocities. The turbomolecular pump is a very clean pump, requires no trap between itself and the system, and is not sensitive to contamination (although in some cases the forepump should be protected by a cold trap). It is not widely used in routine laboratory applications however, as it is quite expensive and demands careful maintenance.

Diffusion Pumps.   The diffusion pump is the standard means of achieving high vacuum (about 10^{-6} Torr); with special care, ultrahigh vacuum (10^{-10} Torr) can be achieved. The operation of the diffusion pump is superficially similar to that of an aspirator, although the principles are quite different. The molecules of the gas being pumped diffuse into a jet of hot vapor, which carries them along by essentially viscous flow inside a condenser in which the vapor is gradually condensed to a liquid of low vapor pressure. The liquid falls back into a boiler where it is again vaporized and brought to the jet nozzle. The exhaust is essentially isolated from the intake by a sort of "plug" of viscous-flowing vapor, which contains the gas being pumped as a minority component; the very slight pressure difference against which the pump must operate (normally less than 10^{-1} Torr) is overcome by the momentum possessed by the vapor as it emerges from the jet nozzle. The gas being pumped is carried away by a forepump, which is usually a rotary oil pump.

Two very different kinds of pump fluids have been employed in diffusion pumps. For many years, mercury diffusion pumps, were used in small laboratory-bench glass vacuum systems. Mercury pumps are now seldom used owing to the health hazards associated with mercury and the high probability of contamination of the vacuum system with mercury unless a cold trap is used (the vapor pressure of mercury at room temperature is ~1.5 mTorr). The *oil diffusion pump* eliminates the safety hazard and can serve for both small glass and larger metal vacuum systems.

The essential features of an all-glass single-stage oil diffusion pump are shown in Fig. 3. Glass two- or three-stage oil pumps of slightly more complex construction are available from Ace Glass, Kontes, and other manufacturers. These are either air cooled or water cooled and have pumping speeds ranging from 8 to 30 L s^{-1} at 10^{-4} Torr. Before the heater of a water-cooled diffusion pump is turned on, it is important to make sure that the forepump has reduced the pressure below the limiting head pressure (about 10^{-1} Torr) and that a reliable flow of cooling water is flowing through the condenser. If water is turned on after the condenser has become hot, the glass will very likely crack, even if it is Pyrex. A reliable precaution against such damage is to power the heater through the contacts of a holding relay, with the current through the hold coil passing through a flow-operated switch in

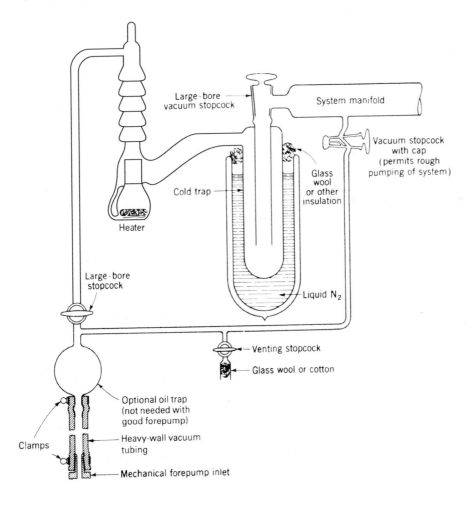

FIGURE 3

Typical all-glass air-cooled oil diffusion pump, shown with appropriate connections to the vacuum manifold and to the mechanical forepump. Cooling of the vapor condensation area can be achieved with a small electric fan or a compressed air line (not shown). The use of a liquid nitrogen trap is optional.

TABLE 1 Diffusion pump oils

Name	Composition	Vapor pressure at 25°C (Torr)
Apiezon A	Hydrocarbon	5×10^{-7}
Apiezon B	Hydrocarbon	1×10^{-7}
Neovac-SY	Phenyl ether	1×10^{-7}
Octoil-S	Hydrocarbon	5×10^{-8}
Convoil-20	Hydrocarbon	5×10^{-8}
DC-704	Silicone	2×10^{-8}
DC-705	Silicone	3×10^{-10}
Santovac-5	Polyphenyl ether	2×10^{-10}
Conalex-10	Polyphenyl ether	1×10^{-10}

the outlet side of the condenser, so that cessation of flow through the condenser—whether by water-main failure, blockage of the line, or a detached hose—will cut off the current to the heater. To turn the heater on again requires the positive act of resetting the relay. For protection against floods from a detached hose, a solenoid valve can be placed in the water inlet line.

The fluids used in oil diffusion pumps are usually hydrocarbon esters (e.g., di-*n*-octylphthalate) that are heated to 130 to 160°C in operation or silicone or polyphenyl ether fluids that operate at higher temperatures (180 to 280°C). The latter fluids are quite expensive but give lower ultimate vacuum levels (10^{-7} to 10^{-10} Torr) and greater resistance to oxidation by excessive amounts of air. Further information is given in Table 1. **Warning:** When hot, a diffusion pump should never be exposed to pressures above about 0.1 Torr; *never vent an oil diffusion pump to air when it is hot.*

Oil diffusion pumps of metal construction have the general characteristics shown in Fig. 4. The pump consists of a chamber with oil at the bottom that is heated to boiling at about 0.5 Torr. The streaming vapor is redirected to provide the pumping action by the control jet assembly. The limiting forepressure at which a single-jet diffusion pump will operate increases with decreasing annular spacing between the exit rim of the nozzles and the pump wall. Therefore high-speed diffusion pumps usually have two or more stages of jet nozzles and are referred to as multistage pumps. By the proper design of the size and spacing of these jets, the pump acts like several separate diffusion pumps connected in series. The smallest, fastest stage is positioned nearest to the system, and the largest, slowest stage (which operates at the highest forepressure) is nearest to the forepump, as shown in Fig. 4. Oil diffusion pumps have inlet diameters that range from 5 cm, with a pumping speed of 100 to 300 L s^{-1}, to an extreme of 1.2 m, with a speed of about 10^5 L s^{-1}. The latter enormous pump finds use in aerospace test chambers and other specialized cases; 5- to 15-cm pumps are more typical for laboratory applications.

An oil diffusion pump can be used without a trap if it is sufficiently well provided with baffles to stop back-streaming molecules, for the oils used usually have room-temperature vapor pressures less than 10^{-7} Torr when pure (see Table 1). However, pump oils generally undergo some decomposition in service, and some decomposition products may find their way into the system. Liquid-nitrogen cold traps are usually added when back-streaming and oil creep along the walls of the system must be prevented.

Ion Pump. A pump now much used in ultrahigh vacuum systems is the ion pump, of which the Varian Vac-Ion pump is an example. In this pump a "gas discharge" such as is commonly observed in air and other gases at about 10^{-2} Torr (e.g., neon signs, mercury and sodium lamps) is generated by a strong electric field and maintained all the way down

To vacuum
manifold

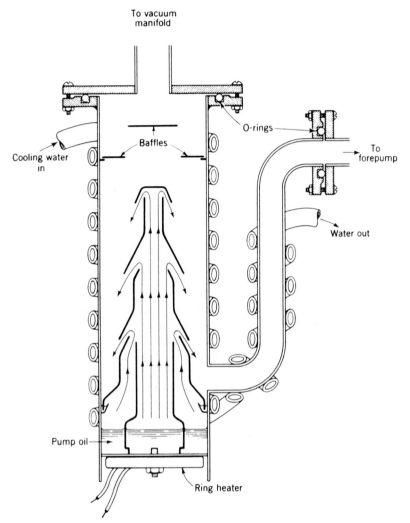

FIGURE 4
Schematic drawing of a
multistage, water-cooled, all-
metal oil diffusion pump. The
arrows indicate the flow of
oil vapor.

to 10^{-10} Torr by application of a strong magnetic field from an Alnico magnet. This field constrains the ions to move in circular or helical paths, thereby greatly increasing their path lengths between wall collisions and giving them the opportunity of colliding with neutral gas molecules and ionizing them so as to continue the discharge. The ions eventually strike certain parts of the specially designed titanium electrodes, eroding them by a process known as sputtering. The titanium atoms come to rest on other titanium pump surfaces, to maintain a fresh and very active titanium surface where the real pumping action takes place by surface adsorption. As the pumped molecules are adsorbed they are continually buried by additional titanium atoms, which refresh the surface.

The ion pump needs no forepump, but it is necessary to reduce the pressure initially to 10^{-2} or 10^{-3} Torr before turning on the pump. With care, this may be accomplished with a rotary oil pump, but it is often done instead with a "sorption pump" such as the Varian Vac-Sorb pump, in which the air in the system is adsorbed on a molecular sieve (synthetic zeolite, Linde type 5A) chilled with liquid nitrogen. This type of pump has the advantage of presenting no danger of contaminating the system with oil.

The ion pump has an ultimate vacuum capability of 10^{-10} Torr or better and may have a pumping speed for air of several hundred liters per second depending on its size.

The components of air are chemisorbed on the fresh titanium and are thus permanently fixed by chemical bonds. Noble gases (helium, argon, etc.) are pumped at a much slower rate, as they are only physically adsorbed on the surface and are held principally by being "plastered over" by titanium atoms. An ion pump that has pumped a considerable quantity of argon is subject to a condition called argon instability, in which bursts of argon are released at intervals into the system; other noble gases show a similar effect.

The ion pump is widely used in all-metal systems, which are usually fabricated of stainless steel with copper gaskets between machined stainless-steel flanges. With the ion pump in operation, these systems can be baked out for several hours at 250°C (or even as high as 400°C) to desorb surface gases (mainly water). The ion pump can be baked out at the same time, but its temperature should not exceed 250°C (to protect the magnet).

An important application of systems of this kind is to fundamental studies of ultra-clean surfaces and of chemisorbed molecular monolayers on such surfaces. Ultrahigh vacuum is necessary to provide ample time for the preparation and study of such surfaces. Assuming a molecular cross section for N_2 or O_2 of 10 Å^2 ($= 0.1$ nm^2) and a sticking probability of unity, it would take only about 5 s for the surface to be covered with a monolayer at 10^{-6} Torr; this time is extended to many hours at 10^{-10} Torr.

Cryopump. As the name cryopump or cryogenic pump implies, this pump operates at very low temperatures; in its most effective form it is a liquid-helium-chilled "cold trap." The pumping surface is at 4.2 K if the liquid helium is at 1 atm, or lower if the pressure over the coolant is reduced by pumping. Overly rapid evaporation of the liquid helium must be prevented by cooling the system walls and by the use of baffles chilled with liquid nitrogen (77 K), which will greatly reduce the input of energy in the form of blackbody radiation. Commercial cryopumps are available that utilize a closed-cycle helium refrigerator to cool condensing surfaces in a turnkey fashion. These pumps, which cool to 10 to 20 K, often include a cold cryosorbent material such as charcoal that absorbs noncondensable gases such as H_2, He, and Ne. The cryopump would normally be used in a system already pumped to a high or ultrahigh vacuum by some other pump. It would be useful for example for rapidly pumping down argon following argon-ion bombardment of a surface in a LEED apparatus having an ion pump. The liquid helium cryopump is effective under the best conditions down to 10^{-13} to 10^{-15} Torr for all gases except helium, hydrogen, and possibly neon.

PRESSURE GAUGES

A large variety of gauges is available for the measurement of low pressures, and the range of useful operation depends a great deal on the type used (see Table 2). This section contains a discussion of the principles of operation for the most commonly used vacuum gauges. More specific details of construction and operation are given in Refs. 1–4 and in the appropriate manufacturers' catalogs. Bourdon gauges and pressure transducers based on diaphragms are discussed in the section on gas-handling procedures in Chapter XX.

Mercury Manometers. A U-tube manometer filled with mercury is simple to construct, requires no calibration, and operates over a wide pressure range. It is now less frequently used due to concerns about safety hazards associated with mercury (see Appendix C) and its slow visual readout. However, it is a historically important device and it is directly related to barometers, which are discussed in Chapter XIX.

Most commonly, one arm is evacuated and the manometer indicates the total pressure directly. A temperature correction is necessary to obtain the absolute pressure. This correction allows one to convert the observed reading p(mm) in millimeters of mercury to Torr (1 Torr $\equiv$ 1/760 atm = 1 mm Hg at 0°C and standard gravity):

TABLE 2 Operational range of various vacuum gauges (in Torr)

Bourdon gauge	1–1000
Mercury manometer	1–1000
Oil manometer	0.03–10
Diaphragm gauges	
Capsule	1–1000
Piezoresistive	1–1000
Magnetic reluctance	0.1–1000
Capacitance	10^{-4}–1000
Thermocouple gauge	10^{-3}–3
Pirani gauge	10^{-4}–0.3
McLeod gauge	10^{-5}–1
Ionization gauges	
Alphatron	10^{-4}–100
Penning (cold cathode)	10^{-7}–10^{-2}
Thermionic	10^{-8}–10^{-3}
Bayard-Alpert	10^{-11}–10^{-3}
Redhead (magnetron)	10^{-12}–10^{-3}
Residual gas analyzer	10^{-14}–10^{-4}

$$p(\text{Torr}) = p(\text{mm})\left[1 - \frac{3\alpha t - \beta(t - t_s)}{1 + 3\alpha t}\right]$$

$$\approx p(\text{mm})(1 - 3\alpha t) = p(\text{mm})(1 - 1.8 \times 10^{-4}t) \qquad (1)$$

where 3α is the volume coefficient of expansion for mercury (average value of 3α between 0 and 40°C is $18.2 \times 10^{-5}\,\text{K}^{-1}$), β is the linear expansion coefficient of the scale, t is the Celsius temperature of the manometer, and t_s is the temperature at which the scale was graduated. Equation (1) does not contain any correction for the small effect due to the difference between local and standard values of gravity. A suitable manometer for medium-precision (± 0.5-mm) work is shown in Fig. 5. In this design the scale is constructed by carefully covering a smooth hardwood board with good-quality millimeter paper and the U-tube is rigidly attached to this board. Before being mounted and filled with vacuum-distilled mercury, the manometer should be thoroughly cleaned inside (with nitric acid), rinsed and dried, evacuated with a good forepump, and preferably flamed with a torch to release adsorbed water.

For pressures between 0.03 Torr and about 10 Torr, oil manometers are more accurate than mercury manometers, since oil has a much lower density. For example, dibutyl phthalate, which is often used, has a density of 1.046 g cm^{-3} at 20°C. The absolute pressure is given by

$$p(\text{Torr}) = \frac{\rho_{\text{oil}}}{13.59}\, h \qquad (2)$$

where h is the reading of the oil manometer in millimeters. However, oil-filled manometers have two drawbacks. Most gases are quite soluble in oil, and this may cause frothing or erratic behavior when the pressure changes suddenly. Also the wetting of the walls of the manometer by a viscous oil causes sluggish response.

Alternative Manometer Gauges. Alternatives to a mercury manometer are capacitance and magnetic reluctance gauges and also solid-state strain gauges. In a capacitance manometer, the pressure is sensed by the deflection of a membrane that is one

FIGURE 5
Design for a closed-tube laboratory mercury manometer. The capillary constriction acts to retard the mercury flow and thus prevents damage in case of sudden changes in pressure.

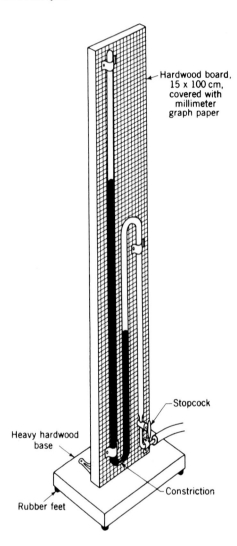

Hardwood board, 15 x 100 cm, covered with millimeter graph paper

Stopcock

Heavy hardwood base

Constriction

Rubber feet

element of a capacitor forming part of a capacitance bridge circuit. The gauge can be an absolute instrument, is independent of the kind of gas present, and is available in many ranges and sensitivities from manufacturers such as MKS Instruments and Setra Systems. For example the MKS model 622A gauge covers the 0 to 1000 Torr range with an accuracy of ±0.25 percent of full scale. At about three times the cost, the MKS model 690A has a full scale accuracy of ±0.05 percent, which is comparable to that of a mercury manometer. In its most sensitive form, a capacitance gauge can measure pressures in the neighborhood of 10^{-4} Torr to within a few percent.

A reluctance manometer consists of a thin metal diaphragm sandwiched between two coils. The diaphragm material is magnetically permeable, and the introduction of gas on one side causes a deflection that produces an imbalance in an ac inductive bridge circuit that contains the coils. The result is a small signal that is linear in the pressure of the gas. A model DP-15 gauge from Validyne Engineering covers the pressure range from 0 to 1000 Torr with an accuracy of ±0.25 percent. Reluctance gauges are somewhat less expensive than capacitance gauges and the diaphragms can be changed to allow coverage of different pressure ranges.

A third class of pressure-measuring devices is based on the change in resistance of thick-film strain gauges deposited on a flexible membrane. Silicon films are often used in a Wheatstone bridge arrangement that becomes unbalanced when the membrane moves as gas is introduced on one side. With calibration, this sensor yields a voltage output that is quite linear (0.05%) with pressure and that can give accuracies of ± 0.05 percent (e.g., Omega Engineering model PX01-5V). These are generally the least expensive of the alternatives to the mercury manometer.

McLeod Gauge.   The McLeod gauge has the advantages of simplicity of operation and a wide pressure range ($\sim 5 \times 10^{-6}$ to $\sim 10^{-1}$ Torr). In addition the readings depend only on the geometry of the gauge and not on the properties of the gas whose pressure is being measured. Its operation involves the compression by mercury of a known large volume V of gas at an unknown pressure p into a known small capillary volume v, where the final pressure p' can be measured. Both p' and p are quite low, and the perfect-gas law can be used in the form $pV = p'v$ since the compression is carried out at constant temperature. Detailed descriptions for the construction and operation of a McLeod gauge are given in earlier editions of this book and in standard reference texts.[1,2]

There are several significant disadvantages to the McLeod gauge that offset its attractive features. Since it contains mercury, there is a safety hazard in case of breakage and a cold trap (containing Dry Ice or liquid nitrogen) may be necessary between the gauge and the system to avoid unwanted mercury vapor inside the vacuum system. The McLeod gauge is also slow in operation and it cannot measure the pressure of a vapor that will condense when compressed into the capillary. Thus, this gauge is now used mostly to calibrate other direct-reading gauges.

Thermocouple and Pirani Gauges.   The thermocouple and Pirani gauges depend on the pressure variation of the thermal conductivity of a gas at low pressures. A fine wire or ribbon filament, often in the form of a coil for greater sensitivity, is sealed into a glass envelope that is connected to the vacuum system. The filament is made of a metal with a high temperature coefficient of resistance (such as platinum, nickel, or tungsten) and is heated electrically. For a fixed energy input, the temperature of the filament, and therefore its resistance, will depend on the gas pressure. In a Pirani gauge, the filament resistance is measured, while in a thermocouple gauge, the temperature of the filament is measured by a very fine wire thermocouple that is welded to the midpoint of the filament. The thermocouple output is usually measured with a low-resistance microammeter ($\sim 70\ \Omega$) connected in series with the thermocouple. Since the performance of such gauges depends on the geometry and design and on the nature of the gas, it is necessary to calibrate them against an absolute gauge. The standard method of calibration and operation of a thermocouple gauge is to maintain constant filament current and obtain the thermocouple emf as a function of pressure. The thermocouple gauge is a rugged, inexpensive instrument that is well suited for leak detection, since it is direct-reading and has a rapid response to pressure changes. However, ambient-temperature variations can cause erratic readings since thermal conduction depends on the temperature of the walls of the envelope as well as that of the filament.

Ionization Gauges.   Ionization gauges are of several types, having in common the production of positive ions from the molecules present and the measurement of the ion current to a cathode. In the *Alphatron gauge,* the ions are produced by the action of alpha rays emitted by a small amount of radioactive material. In the *Penning gauge,* a cold-cathode discharge produces the ions; the *Redhead* or *magnetron gauge* uses a magnetic field to constrain the ions to long spiral paths so that the cold-cathode "discharge" can be maintained as low as 10^{-13} Torr.

However, most ionization gauges are *thermionic gauges* that obtain positive ions through bombardment of the gas molecules with electrons produced thermionically from a heated filament and accelerated by a grid biased 100 to 200 V positive with respect to the filament. The ions are collected by a cathode that is typically about 20 to 50 V negative with respect to the filament, and the ion current in the cathode circuit is measured with a sensitive electrometer.

Thermionic ionization gauges are usually constructed like old-fashioned electron tubes (with a central filament, outside cathode plate, and anode grid in between) but with their glass envelopes provided with a stem of large-diameter tubing for connection to the system. In large metal systems, however, it is common to use the gauge in the form of a so-called *nude gauge;* that is, the gauge structure is supported from electrical feedthroughs in its mounting flange so that it protrudes, without any envelope, into the vacuum chamber itself. This substantially eliminates errors due to the limited conductance of the connecting tube, which is advantageous in view of the fact that the gauge itself often acts either as an ion pump or as a virtual leak (owing to outgassing of the filament). The filament of the gauge is usually tungsten, either bare or coated with a thoria activator that permits it to be operated at a lower temperature. Tungsten is notorious for emitting large amounts of CO until it has been thoroughly outgassed by many hours of incandescent heating under high vacuum. Iridium wire is sometimes used because it gives less gas and does not oxidize easily. Another satisfactory filament is rhenium wire activated with a coating of lanthanum boride; this operates at a much lower temperature than a tungsten filament. In any case, when a thermionic ionization gauge is employed, attention must be given to thoroughly outgassing the filament under vacuum before pressure readings are taken. The pumping or "cleanup" characteristics of the ionization gauge, which give rise to some uncertainties in measuring the system pressure, can be reduced to tolerable proportions by restricting application of high voltages to grid and collector to the very short interval needed to measure the electron and ion currents.

Thermionic ionization gauges cannot measure pressures lower than $\sim 10^{-8}$ Torr since the minimum cathode current is produced by a phenomenon altogether independent of the presence of gas molecules, namely, the photoelectric ejection of electrons from the cathode plate by soft X rays produced by the impact of the electrons on the anode grid. This led Bayard and Alpert to invent an "inverted" ionization gauge in which the cathode attracting the positive ions is a slender wire *inside* the anode grid and the thermionic filament is outside. The *Bayard–Alpert gauge* has an "x-ray limit" which is of the order of 10^{-10} or 10^{-11} Torr and is widely used in ultrahigh-vacuum work.

The ionization gauge has to be calibrated, and the calibration is in general different for every gas. The ion current for a given gas is however linear with the gas pressure over a very wide range, extending almost down to the x-ray limit.

The ionization gauge has a number of disadvantages in addition to those described in the preceding paragraphs. The hot filament may decompose organic vapors that may be present; it may be "poisoned" by some gases and thereby lose its emissivity; it can easily burn out if exposed to air while hot; and it gradually erodes in use through evaporation, slow oxidation, and ion bombardment. Many gauges (especially nude gauges) are so constructed that a burned-out or poisoned filament can easily be replaced in the laboratory. The cold-cathode gauge is preferred in cases where the pressure is generally greater than 10^{-5} Torr, such as in vacuum lines devoted primarily to synthesis or the handling of gases.

Residual Gas Analyzer. The residual gas or partial-pressure analyzer is a form of ionization gauge in which the positive ions are mass analyzed either by a conventional sector-mass-spectrometer arrangement (employing a permanent magnet) or by a quadrupole-mass-spectrometer arrangement (which requires no magnet). The ion current is amplified either inside by an electron multiplier arrangement or outside by a vacuum-tube electrometer or by both. Devices of this kind are sold by Balzers, Varian, and other companies and typically are capable of unit mass resolution for singly ionized molecules up to masses of a few hundred atomic units. The sensitivity is of the order of 10^{-10} Torr over most of the mass range. The partial-pressure analyzer is very useful in determining the actual composition of the residual gases inside a vacuum system and in hunting for leaks (with a helium "torch" outside the vacuum); it is virtually indispensable for surface adsorption and desorption studies under ultrahigh vacuum.

SAFETY CONSIDERATIONS

Vacuum apparatus probably presents fewer accident hazards than almost any other kind of laboratory apparatus of comparable complexity. However, these hazards are by no means entirely negligible, and we summarize the principal ones.

Implosion. This hazard is most important with glass apparatus and is ever-present when large glass bulbs (over 1 L in size) or flat-bottomed vessels (of any size) are evacuated. The force of atmospheric pressure makes dangerous missiles of glass fragments from imploding vessels. If at all possible, avoid evacuating Erlenmeyer flasks or other flat-bottomed vessels. Reduce the hazard of flying glass by placing strips of plastic electrician's tape on all large glass vessels that are to be evacuated. Wear safety glasses.

Explosion. If significant quantities of a gas have been liquefied or taken up by an adsorbent at low temperature, an explosion can result when the system warms up if adequate vents or safety valves have not been provided. An explosion of a different kind can take place if an oil diffusion pump (particularly a glass one) is vented to air while hot.

Liquid air. Do not refrigerate a trap with liquid nitrogen until the system pressure is reduced below a few Torr. (Liquid air boils at higher temperature than liquid nitrogen and can easily be condensed at 77 K.) Liquid air present in a trap may react explosively with any organic substances that condense there. The liquid air can also produce excess pressure and likely damage to the vacuum system should the trap warm without venting: this is a common mishap with vacuum systems.

Mercury poisoning. Mercury manometers pose hazards of mercury contamination in the event of fracture of a glass apparatus. See Appendix C.

REFERENCES

1. S. Dushman, *Scientific Foundations of Vacuum Technique,* 2d ed. (J. M. Lafferty, ed.), Wiley-Interscience, New York (1962); A. Roth, *Vacuum Technology,* 3d ed., North-Holland, Amsterdam/New York (1990).

2. H. Adam, "Foundations of Vacuum Technology," in B. W. Rossiter and J. F. Hamilton (eds.), *Physical Methods of Chemistry,* 2d ed., vol. I, chap. 3, Wiley-Interscience, New York (1986).

3. R. P. Benedict, *Fundamentals of Temperature, Pressure, and Flow Measurements,* 3d ed., Wiley, New York (1984).

4. M. H. Hablanian, *High-Vacuum Technology: A Practical Guide,* 2d ed., Dekker, New York (1997).

GENERAL READING

R. P. Benedict, *Fundamentals of Temperature, Pressure, and Flow Measurements,* 3d ed., Wiley, New York (1984).

M. H. Hablanian, *High-Vacuum Technology,* 2d ed., Dekker, New York (1997).

J. M. Lafferty (ed.), *Foundations of Vacuum Science and Technology,* Wiley, New York (1998).

A. Roth, *Vacuum Technology,* 3d ed., North Holland, Amsterdam/New York (1990).

XIX

Instruments

This chapter consists of brief descriptions and discussions of certain devices and instruments that are commonly used in experimental physical chemistry. Considerably greater detail can be found in the references cited. It should also be noted that most commercial scientific instruments are furnished with a detailed instruction manual. This should be reviewed and thoroughly understood *before* the instrument is used.

BALANCES

The determination of mass is one of the most common and important measurements in experimental chemistry. It is common practice to use the terms *mass* and *weight* interchangeably, but of course they have quite different meanings. Whereas the mass m of an object in kilograms is a measure of the amount of matter in that object, the weight W represents the gravitational force exerted on the object by the earth and should properly be expressed in newtons. Since $W = mg$ and g varies with geographical location, the weight of an object of a given mass will depend on where it is measured. The usage of such expressions as "a 10-g weight" to mean a mass whose weight equals that of a 10-g mass arises naturally from the common method of comparison weighing. Unless otherwise specified (as in Exp. 31), the term *weight* as used in this book actually means the mass in grams; this should be clear from the context and from dimensional analysis.

We shall assume that the reader has some prior experience with the use of an equal-arm analytical balance, with which an object is weighed by determining the "weights" which must be added to the right-hand side of the beam in order to make the rest point of the loaded balance the same as the zero point (rest point of unloaded balance). In Exp. 31, a Gouy balance for magnetic susceptibility measurements is described that incorporates a modified version of a two-pan (equal-arm) analytical balance (although the Gouy balance can be redesigned to utilize a modified single-pan balance instead). The design, construction, and operation of two-pan analytical balances are described in older textbooks on quantitative chemical analysis[1,2] and will not be discussed here.

In recent years, the use of single-pan "automatic" balances has become common because of their convenience and speed of operation. Below are brief descriptions of the operation of two balances of this type.

FIGURE 1

A single-pan mechanical
analytical balance.

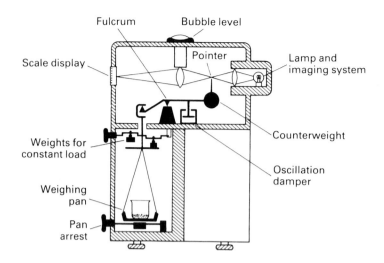

Single-Pan Mechanical Analytical Balances.   Figure 1 is a schematic drawing showing the key features of mechanical balances still in use in many university laboratories.[1-3] The weighing pan hangs on a hard knife edge of agate or sapphire on one end of the beam. The beam is supported at the center of gravity on another knife edge, and a counterweight is permanently fixed at the other end of the beam. The fixed weights are hung above the weighing pan and may be removed or replaced by operation of the weight-control knobs. The balance is damped by the motion of a piston in a cylinder (air damping) to prevent the beam from acting as a pendulum and oscillating for an inconvenient length of time. A well-adjusted balance usually swings just to equilibrium and stops, or it reaches equilibrium after one or two brief excursions. The rest position of the beam is displaced by putting an object on the pan. Sufficient weights are then removed with the weight-control knobs to bring the beam approximately back to its original position, and the residual displacement from the normal rest point is read on an optical scale that is calibrated directly in milligrams.

This "constant-load" condition gives optimal reproducibility over the entire mass range of the balance. Because of the delicacy of the knife edges and the sensitivity of the balance, the beam is always arrested and the knife edges lifted out of contact with the bearing surfaces whenever anything is being put on or removed from the pan or, for that matter, whenever a weighing is not actually in progress. Failure to observe this precaution results in rapid wear of the knife edges with resulting decrease in sensitivity of the balance and erratic behavior.

To carry out a weighing, one first checks that the balance is level and steady and that the pan is empty and clean. *Chemicals must never be weighed directly on the balance pan*—corrosion or contamination of the balance pan can contribute serious errors! With the balance case closed and the pan empty, the beam arrest is *gently* released and the rest point of the balance observed. If the empty pan does not give a reading of precisely zero on the optical scale, the optical scale should be adjusted to read zero. If a zero reading cannot be achieved with the adjustment knob, a more serious adjustment is required and the instructor should be notified. When the zero has been adjusted, the object to be weighed is picked up with tongs or a paper loop and placed carefully and gently in the center of the pan. The balance case is closed and the *partial beam release* is operated. The weight knobs are then manipulated until the approximate weight of the object is determined as shown by the optical scale indicating insufficient weight with one setting and overweight with the next individual weight added. With this last weight removed, the beam is then completely

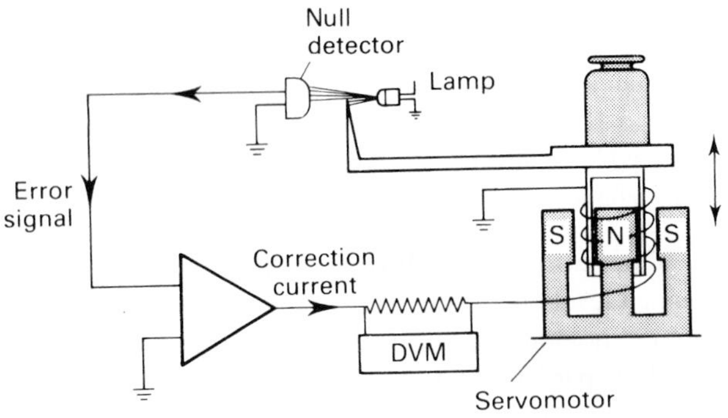

FIGURE 2

Simplified electromagnetic servo system for an electronic balance.

released, and after equilibrium has been established, the fractional weight is read on the optical scale. The beam is then arrested, the sample is removed, the weights are replaced, and the zero on the balance is checked.

Single-Pan Electronic Analytical Balances. Modern analytical balances commonly employ an electromagnetic force principle to achieve a reproducible balance point.[1,3,4] Figure 2 shows the basic electromagnetic servo system in which the current necessary to hold a pointer at the null position is proportional to the mass placed on the coil platform. Such a servo system is incorporated into balance enclosures as depicted in Fig. 3. Knife edges are replaced by a parallelogram arrangement with flexure points to provide a limited movement of the weighing pan. The null point is chosen to be the relaxed position of the flexure pivots, and the current necessary to maintain this position on addition of a mass object is measured and converted to a digital readout. Calibration is achieved by use of a standard weight and adjustment of the electronic circuitry to give the proper readout. Typical capacities are 30 to 200 g with a precision of 0.01 to 0.1 mg.

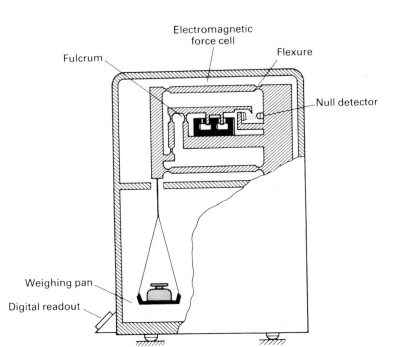

FIGURE 3

An electronic balance based on a parallelogram flexure support and an electromagnetic force cell. The flexure points allow limited movement about the null position and provide resistance to torsional distortion by off-center loading of the weighing pan. Drawing taken from Ref. 3 with permission.

In top-loading versions of the electronic balance, the weighing pan is supported on an upward extension of the moving bar of the parallelogram. This type of balance is very convenient for moderate precision weighing operations, and such balances have capacities of 0.1 to 10 kg with a precision of 1 mg to 1 g. Hybrid versions (electronic–mechanical balances) have also been produced, particularly for low-capacity (<10-g) micro and ultra-micro balances; these have a precision of 0.1 to 1 μg.

Operation of Balances. Any type of precision analytical balance should be mounted level on a sturdy bench, free from vibrations, in a room with a fairly stable temperature (no drafts or direct sunlight near the balance). Detailed instructions for the operation of the particular type to be used should be made available by the instructor. Indeed, if the student is not already familiar with the use of that particular design, demonstration by an experienced user and a practice weighing are strongly recommended. The most important general principle is the need for a sense of personal responsibility. Analytical balances are delicate instruments that are capable of excellent precision if they are used with respect and care. It is especially important to use beam and pan arrest controls properly, since they protect the components of the balance (especially the knife edges) from damage. If any difficulties arise, consult an instructor; **do not attempt repairs or adjustments.** After use, leave the balance clean and restored to the zero settings as a courtesy to other users.

Corrections and Errors. Buoyancy will affect the results of a weighing, since air will exert a buoyant effect both on the object and on the weights; in general these two effects will not cancel. The weight in vacuo W_v of an object can be obtained from the weight in air W_a by adding the weight of air displaced by the object and subtracting the weight of air displaced by the weights. Thus

$$W_v = W_a + V\rho - v\rho \tag{1}$$

where V is the volume of the object, v is the volume of the weights, and ρ is the density of air. This relation applies to electronic balances, since these are calibrated with standard metal weights. Analytical weights are usually made of stainless steel (density 7.8 to 8.0 g cm^{-3}) or brass that is lacquered or plated with gold, nickel, or chromium (density 8.4 g cm^{-3}). Taking 8.0 as a reasonable average density, we can replace v in Eq. (1) by $W_a/8.0$ to obtain

$$W_v = W_a\left(1 - \frac{\rho}{8.0}\right) + V\rho \tag{2}$$

Equation (2) is useful if V is known, but often it is not. However, if d_0 (the density of the object) is known, V can be replaced by W_v/d_0 to give

$$W_v = W_a\frac{1 - (\rho/8.0)}{1 - (\rho/d_0)} \cong W_a\left[1 + \left(\frac{1}{d_0} - \frac{1}{8.0}\right)\rho\right] \tag{3}$$

where the final approximation is excellent as long as $\rho/d_0 \ll 1$. Although the density of air varies with temperature, pressure, and moisture content, ρ can be taken to be about 0.0012 g cm^{-3} at any relative humidity over the range 15 to 30°C and 730 to 780 Torr. Buoyancy corrections are often neglected in the weighing of solids, but they are essential in weighing gases and are quite important in weighing large volumes of liquids (as when calibrating volumetric apparatus). For example W_v for water is 0.1 percent higher than W_a, and the buoyancy correction for 100 g of water would be about 100 mg (much greater than any other source of weighing error). Further discussion of buoyancy corrections can be found in Refs. 5 and 6.

The most common source of error in weighings, especially with two-pan balances, is due to inaccurate weights. Even in a good set, the weights are often in error by as much as 1 mg. It is recommended that a given set of weights be used only with a single balance and that these weights be calibrated on that balance against a good secondary standard set meeting the tolerance limits set by the National Institute of Standards and Technology (formerly the National Bureau of Standards, NBS). The calibration procedure is given in detail elsewhere.[2,7] With single-pan mechanical balances, the weights are inside the case, effectively protected from handling and dust. When delivered new or reconditioned, or after periodic servicing (which should include a check of the weights), the weights should be within the manufacturer's specifications. For the most precise work, it may be advisable for the user to make a prior check of the balance with a good secondary standard set of weights and to look carefully for any evidence of dulled knife edges (hysteresis, poor reproducibility).

Finally, there are several other weighing errors that may occur as a result of poor technique but that can usually be avoided. Volatile, hygroscopic, or efflorescent samples and samples that adsorb gases (e.g., CO_2 or O_2) should be kept in closed weighing bottles. **An object should never be weighed while warm,** since convective air currents will occur, causing the weighing to be in error. Weighings may also be in error because of the condensation of moisture on dry glass walls or the force produced by static charge caused by vigorous wiping of a glass surface.

BAROMETERS

The Fortin barometer is simply a single-arm, closed-tube mercury manometer equipped with a precise metal scale (usually brass). The bottom of the measuring arm of the barometer dips into a mercury reservoir that is in contact with the atmosphere. The mercury level in this reservoir can be adjusted by means of a knurled screw that presses against a movable plate (see Fig. 4). When the meniscus in the reservoir just touches the tip of a pointed indicator, the zero level is properly established and the pressure can be determined from the position of the meniscus in the measuring arm. Both the front and back reference levels on a sliding vernier are simultaneously lined up with the top of this meniscus in order to eliminate parallax error, and the height of the arm can be read to the nearest tenth of a millimeter using the vernier scale. A thermometer should be mounted on or near the barometer, since the temperature must be known in order to make a correction for thermal expansion. This correction is discussed in Chapter XVIII and the appropriate formula is given by Eq. (XVIII-1). The metal scale on most barometers is made of brass (linear coefficient of thermal expansion $1.84 \times 10^{-5} \text{ K}^{-1}$) and is usually graduated so as to read correctly at 0°C. A table of barometer corrections over the range 16 to 30°C and 720 to 800 mm is given in Appendix G. High-precision work also requires corrections for the effect of gravity, residual gas pressure in the closed arm, and errors in the zero position of the scale.[8,9] Since these usually amount to only a few tenths of a millimeter, they will not be discussed here.

OSCILLOSCOPES

The basic principles of dc and ac voltage measurements are discussed in Chapter XVI and in standard textbooks on electronics.[10,11] In many applications these measurements are carried out by complex electronic instruments designed to produce a visual record of the detected signal, either as a trace on a luminescent screen, a plot on paper, or a numerical

FIGURE 4
Detailed sketch of a Fortin
barometer. Uncorrected
reading shown is 79.02 cm.

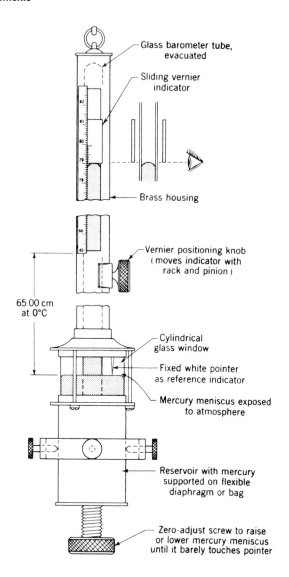

Glass barometer tube,
evacuated

Sliding vernier
indicator

Brass housing

Vernier positioning knob
(moves indicator with
rack and pinion)

65.00 cm
at 0°C

Cylindrical
glass window

Fixed white pointer
as reference indicator

Mercury meniscus exposed
to atmosphere

Reservoir with mercury
supported on flexible
diaphragm or bag

Zero-adjust screw to raise
or lower mercury meniscus
until it barely touches pointer

(digital) readout of the voltage measured. Analog and digital storage oscilloscopes are described briefly in this section, and other instruments for voltage measurements are discussed in Chapter XVI.

In an analog oscilloscope using a cathode-ray tube, electrons are emitted from a cathode and are then accelerated and focused by a series of special anodes to form a beam that impinges on the face of the tube. This face is coated with fluorescent materials such as ZnS, and the beam produces a sharp visible spot. Horizontal or vertical displacement of this spot from the center of the screen can be achieved by passing the electron beam through the electrostatic field between pairs of charged plates. Since high voltages are required across these plates to produce a rapid displacement of the beam, oscilloscopes have wide-band amplifiers to provide voltage amplification for the input signals. Amplifier bandwidths typically range from 10 to 20 MHz in inexpensive oscilloscopes to several GHz in costly research instruments.

The primary use of an oscilloscope is to display the shape of a voltage waveform (i.e., to plot out voltage vertically against a horizontal time scale). To accomplish this, a sweep voltage must be applied to the horizontal deflection plates that will cause the beam to move from left to right at a uniform rate and then return very rapidly to the starting point.

All oscilloscopes have an internal sawtooth waveform generator to produce this linear time-base sweep.

In recent years, analog oscilloscopes have been supplanted by digital storage oscilloscopes, in which a fast analog-to-digital conversion of the vertical input signal is done at regular timed intervals, as determined by an accurate clock and by the time scale chosen for the horizontal display. Typically, the vertical resolution is one part in 256 or 1024 (8- or 10-bit A/D) and the x-axis display consists of 1024 to 4096 points (2^{10} to 2^{12} time intervals). For some oscilloscopes, the x array stored in memory can be much larger—10^6 time intervals or more. The resultant x-y array of data is stored in memory internal to the oscilloscope and, once recorded, can be displayed at slower sweep speeds and can be plotted or transferred to a computer via RS-232, IEEE-488, or USB ports. Displays typically show a grid of ten x and y divisions, with a readout of the scales in volts/division and time/division. For repetitive signals, most digital storage oscilloscopes allow averaging for signal-to-noise improvement.

The operation of an oscilloscope can best be described by reference to Fig. 5, which shows a simplified layout of the controls of a commercial (Tektronix) digital instrument. The signal to be measured is applied to the input connector (BNC) of one of the vertical amplifier channels and must not exceed an upper limit of, typically, ± 400 volts if the scope input impedance is one megaohm and ± 5 volts for 50-ohm input impedance. The latter impedance is necessary for signal changes that occur rapidly, such as in the fluorescence decay measurements of Exps. 40 and 44. The lower limit of sensitivity is about 1 mV/division, so preamplification is sometimes needed if very low signal levels are to be measured.

The horizontal resolution is governed by the horizontal control knob and by the number of channels the oscilloscope can save. The maximum resolution is determined by the frequency response of the instrument and is usually about 5 ns/division or better. Measurements are typically made of 20 such divisions; any 10 of these can be scrolled across

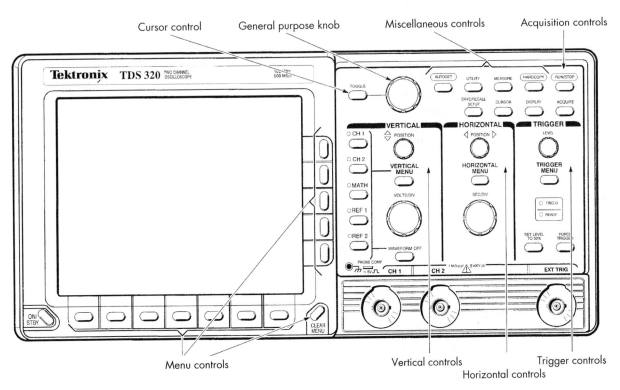

FIGURE 5

A typical digital oscilloscope and its controls.

the screen display. The start of the display is determined by the trigger controls, which can be adjusted so that the display begins when a sample voltage equals some preset level. This triggering can occur at positive or negative voltages and can also be set on rising or falling edges of a signal. Sometimes the trigger voltage is a separate source sent to the external input or is the 60-Hz line voltage signal.

The various menu controls provide for delayed triggering, expanded displays in time and voltage, storage of traces for comparisons, and many other useful functions such as measuring peak maxima, average voltages, and periods between peaks. One of the most useful features of a digital scope is its ability to average a repetitive signal, as is done for radiative lifetime measurements in Exps. 40 and 44. This information is usually transferred to a computer for processing; an example illustrating this application is discussed in the interfacing section of Chapter III. With many modern oscilloscopes, extensive calculations such as subtractions, multiplications, and fast Fourier transforms (FFTs) can be performed internally.

One application of an analog oscilloscope is in examining the relative phase relationship of two sinusoidal wave forms, one of which (of frequency f_V) is supplied to the vertical input (y) of the oscilloscope, the other (of frequency f_H) to the horizontal input (x). When the display mode is chosen to be x vs. y, the x voltage replaces the internal sawtooth sweep voltage and the oscilloscope becomes a sensitive device for comparing the frequencies of the sine waves. When the ratio of the frequencies is a rational fraction, a symmetrical closed pattern called a Lissajous figure will appear on the screen. The frequency ratio can be obtained from the form of the pattern by using the formula

$$\frac{f_H}{f_V} = \frac{n_V}{n_H} \tag{4}$$

where f_H and f_V are the frequencies applied at the horizontal and vertical inputs, n_H is the number of points at which the figure is tangent to a horizontal line, and n_V is the number of points of tangency between the figure and a vertical line. Several simple types of Lissajous figures are shown in Fig. 6a.

Shown in Fig. 6b is the Lissajous figure for several different values of the phase angle between two signals of the same frequency and amplitude; note that the pattern is a circle when the two sine waves are 90° out of phase and is a straight line tilted at 45° when the phase angle is 0 or 180°. This fact provides a convenient means of determining the phase shift in a circuit. For balancing an ac Wheatstone bridge, a signal from the ac power source is applied to the horizontal input, and the unbalance signal across the bridge is applied to the vertical input. The bridge is first balanced capacitively by obtaining a straight-line pattern. After this is done, the resistive balance is indicated by obtaining a horizontal line (zero vertical amplitude).

FIGURE 6

Lissajous figures: (*a*) several simple figures, with the ratios $f_H{:}f_V$ indicated; (*b*) the 1:1 figure for various values of the phase angle between signals of equal amplitude.

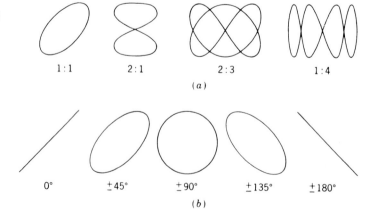

TABLE 1   Half-cell potentials of calomel reference electrodes

KCl Conc	Potential at 25°C, V
0.1 N	0.3338
1.0 N	0.2800
Saturated	0.2415

pH METERS

A pH meter is a special type of millivolt potentiometer designed to measure the emf of a cell in which the electrolyte contains hydrogen ions.[12] A "glass electrode" is employed as the measuring electrode, and a calomel electrode is used as the reference electrode.

Calomel Electrode.[13]   In addition to pH measurements, there are many other emf cell measurements for which it is convenient to use the calomel electrode as a reference electrode against which a measuring electrode is compared. Actually this "electrode" is really a half-cell that is connected via a KCl salt bridge to another half-cell containing the solution of interest. The *saturated* calomel electrode can be written as

$$Hg(l) + Hg_2Cl_2(s), K^+Cl^-(aq, \text{ sat.}), \text{aq. electrolyte} \qquad (5)$$

There are two other common versions of this half-cell: the *normal* and *tenth-normal* calomel electrodes, in which the KCl concentration is either 1.0 or 0.1 N. The saturated electrode is the easiest to prepare and the most convenient to use but has the largest temperature coefficient. The half-cell potential for each of the calomel electrodes has a different value relative to the standard hydrogen electrode; these emf values are given in Table 1. Calomel electrodes can be easily prepared in the laboratory and are also available commercially. Two typical calomel electrode designs are shown in Fig. 7.

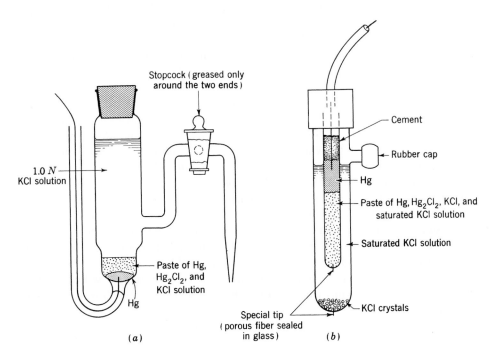

FIGURE 7
Two typical calomel electrode designs: (*a*) laboratory type, shown unsaturated; (*b*) commercial type, shown saturated.

Glass Electrode.[12] The glass electrode is usually a silver–silver chloride electrode, surrounded by a thin membrane of a special glass that is permeable to hydrogen ions. The glass membrane is essentially a special type of salt bridge—one in which the anions are immobile (have zero transference number), since they are part of the porous glass framework through which the H^+ cations can move. The glass electrode may be formulated as

aq. soln. containing H^+ at activity a_{H^+} |H^+ (glass)$^-$|H^+, Cl^- (aq, a_0)|AgCl(s)|Ag(s) (6)

The change in state per faraday for this half-cell is thus

$$AgCl(s) + H^+(a_{H^+}) + e^- = Ag(s) + Cl^-(a_0) + H^+(a_0) \qquad (7)$$

The activity a_0 has a definite constant value, usually obtained by using a phosphate buffer containing Cl^- ions as the solution inside the membrane. An important advantage of the glass electrode is that it can be used under many conditions for which the hydrogen electrode is subject to serious error.[14]

pH Measurement.[12,14] The overall emf for a cell with a calomel and a glass electrode dipping into an aqueous electrolyte solution is

$$\mathscr{E} = \mathscr{E}' - \frac{RT}{\mathscr{F}} \ln a_{H^+} = \mathscr{E}' + \frac{2.303\,RT}{\mathscr{F}}(\text{pH}) \qquad (8)$$

where pH $\equiv -\log a_{H^+}$ and $\mathscr{E}'$ is the difference between the half-cell potential of the calomel reference electrode and the "standard potential" of the glass electrode. Obviously $\mathscr{E}'$ will depend on the type of calomel electrode used and on the activity a_0 of hydrogen chloride in the inner solution of the glass electrode. Since these factors are kept constant, changes in emf are a direct indication of variations in pH.

Because of the high resistance of the glass membrane (10 to 100 MΩ), it is not practical to measure the emf directly. Instead, pH meters either use a direct-reading electronic voltmeter or electronically amplify the small current that flows through the cell and detect the voltage drop across a standard resistor potentiometrically. Both battery-operated and ac line-operated pH meters are available commercially from such firms as Beckman Coulter, Thermo Orion, and Corning. Such pH meters are calibrated to read directly in pH units, have internal compensation for the temperature coefficient of emf, and have provision for scale adjustments.

Since the operation of a pH meter is very simple but slightly different for each model, no detailed operational procedure will be given here. However, a few general remarks are necessary. If a glass electrode and a silver–silver chloride electrode were placed in an HCl solution for which a_{H^+} equals a_0, the emf of this cell would ideally be zero (i.e., there should be no potential difference across the glass membrane). However, there is always some small emf (1 or 2 mV) across the membrane under these conditions. This so-called *asymmetry potential* is presumably due to strains in the membrane and may change slowly with time or be temporarily changed by exposure of the electrode to very strong acid or base. Therefore it is necessary to compensate for this asymmetry potential by calibrating the pH meter frequently against a buffer solution of known pH. Also pH readings on solutions of pH greater than 12 may be in error owing to a significant contribution from sodium-ion transference in the glass at these low hydrogen-ion concentrations. This difficulty can be avoided by the use of special high-resistance glass membranes.

POLARIMETERS

The polarimeter (Fig. 8) is an instrument for measuring the optical rotation produced by a liquid or solution.[15] The *specific rotation* $[\alpha]_{\lambda}^{t}$ of a solute in solution at a given wavelength λ and Celsius temperature t is given by

$$[\alpha]_{\lambda}^{t} = \frac{100\alpha}{Lc} = \frac{100\alpha}{Lp\rho} \qquad (9)$$

where α is the angle in degrees through which the electric vector is rotated, L is the path length in *decimeters,* c is the concentration of solute in grams per 100 cm^3 of solution, p is the weight percent of solute in the solution, and ρ is the density of the solution in g cm^{-3}. The angle α is considered positive if the rotation of the electric vector as the light proceeds through the solution is in the sense of a left-hand screw or negative if it is in the sense of a right-hand screw.

Usually the optical rotation is measured with the sodium D yellow line (a doublet, 589.0 and 589.6 nm). For more precise work the 546.1-nm green mercury line may be used.

Light from the source (a sodium-vapor arc lamp or a mercury-vapor lamp with appropriate filter) is polarized by a Nicol prism, termed the polarizer, which consists of two prisms of calcite cemented together with canada balsam so that one of the two rays produced in double refraction (the "ordinary ray") is totally reflected at the interface and lost, while the other (the "extraordinary ray") is transmitted. The polarized light passes through the solution and then through a second Nicol prism, termed the analyzer, which can be rotated around the instrument axis. The normal position (zero rotation) is one in which the two Nicol prisms are at 90° to each other and no light passes through. When an optically rotating medium is introduced between the two Nicol prisms, light is transmitted; the observed rotation α is the angle in degrees through which the dial must be turned (clockwise with respect to the observer if α is positive, counterclockwise if negative) in order to restore the field to complete darkness.

As the dial is turned, the intensity of emergent light is proportional to $\sin^2(\alpha - \alpha_d)$, where α_d is the setting for complete darkness. Since this quantity behaves approximately quadratically near the dark position, the setting is of limited sensitivity if only a single polarizer and a single analyzer are used. In many instruments the field of view is divided into two equal parts with polarization angles differing by a few degrees. This is done either by use of a composite polarizer (two Nicol prisms cemented together side by side with their planes of polarization at a small angle) or by use of an added Nicol prism covering half of the field of the polarizer. In use the analyzer is adjusted so that the two fields appear equally bright.

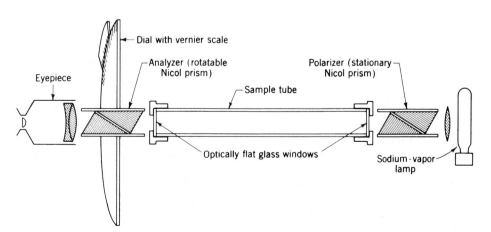

FIGURE 8

Schematic drawing of a polarimeter.

Although laboratory polarimeters generally use Nicol prisms as polarizers and analyzers, dichroic crystals (such as tourmaline) or dichroic sheet polarizers (such as Polaroid) may be used in the construction of special apparatus.†

The polarimeter is commonly used in organic and analytical chemistry as an aid in identification of optically active compounds (especially natural products) and in estimation of their purity and freedom from contamination by their optical enantiomers. The polarimeter has occasional application to chemical kinetics as a means of following the course of a chemical reaction in which optically active species are involved. Since the rotation α is a linear function of concentration, the polarimeter can be used in studying the acid-catalyzed hydrolysis of an optically active ester, acetal, glycocide, etc.

In modern organic chemistry optical-rotatory dispersion,[16] the variation with wavelength of optical-rotatory power (and certain related properties), is used in molecular-structure investigation as a means of identifying and characterizing chromophore groups. Automatic polarimetric spectrophotometers of high complexity have been developed for this purpose.

RADIATION COUNTERS

Electromagnetic radiations with wavelengths below about 1 nm are called X rays or gamma rays. These have a penetrating power that increases strongly as the wavelength decreases and is dependent on the atomic numbers of the atoms present rather than their state of chemical combination.

A 0.1-nm photon has about 8000 eV of energy, while ultraviolet radiation has only 3 or 4 eV. This much energy, absorbed in matter, may result in the virtually simultaneous production of many ions—enough that under optimum conditions the absorption of a single x-ray or gamma-ray photon can be detected with an efficiency approaching 100 percent. The measurement of x-ray and gamma-ray intensities thus amounts to a *counting* of discrete events wherein individual photons are detected.[17]

The best-known detector is the *Geiger–Müller counter*. This usually consists of a cylindrical container (the cathode) filled with an absorbing gas such as argon or krypton, with an insulated central wire (the anode) to serve as a collector for electrons. When an x-ray photon is absorbed, it produces ions and electrons, and the acceleration of the ions and electrons to the electrodes results in collisions resulting in more ionization; thus a cascade process develops that results in a general gas discharge. This continues until the fall of potential between the electrodes is sufficient to quench the discharge and allow the potential to be restored. The Geiger–Müller counter with its associated circuitry is relatively simple, but for many radiation-counting purposes suffers from a long "dead time" (0.1 to 1 ms) resulting from the complete discharge and required recharge. This dead time results in coincident counts and nonlinear response at counting rates higher than about 100 counts per second. Another limitation of the Geiger–Müller counter is that the size of the pulse generated is independent of the energy of the incoming particle.

To overcome these limitations, the self-quenching *proportional counter* has been developed. This counter is very much like the Geiger–Müller counter in construction. The electrons and positive ions from the primary ionizing event go to their respective electrodes, but the production of additional electrons through positive-ion bombardment of the

---

†In the phenomenon known as dichroism, the optical absorption depends strongly on the orientation of the plane of polarization with respect to the crystallographic axes (or axis of preferred orientation). Commercial sheet polarizers are made from acicular dichroic crystals, herapathite (iodoquinine) in the case of Polaroid, suspended in a viscous or plastic medium and aligned by extrusion or stretching. Dichroic polarizers and analyzers are inferior to Nicol prisms for use in polarimeters, because the transmission is considerably less than unity when the planes of polarization are parallel and the absorption is not quite complete when they are perpendicular.

wall is prevented by molecules of some organic compound (i.e., ethanol) that is present as a "quench gas." Thus the tube does not discharge completely. The pulse is of very short duration, of the order of 1 μs. Therefore counting rates of up to 10,000 counts per second are essentially linear with intensity. Since the pulse is very small, the proportional counter requires an exceedingly sensitive (high-gain) preamplifier, well shielded from electrical disturbances. Most important for many purposes is the fact that the pulse height depends upon the energy of the incident photon or other particle. The output of the preamplifier may be fed to an electronic pulse-height discriminator circuit, connected to two or more scaling and counting circuits, among which the pulses are distributed according to the height ranges in which they fall. The self-quenching proportional counter does not last indefinitely; after about 10^{11} counts the quench gas is entirely consumed, and the tube thereafter behaves like a Geiger–Müller counter.

Another commonly used detector is the *scintillation detector*. This makes use of a crystal that produces a scintillation (pulse of visible light) upon absorption of an x-ray photon. The visible light is detected by a photomultiplier tube and associated amplifier circuit, which is sensitive enough to detect nearly every scintillation. The scintillating crystal is usually sodium iodide doped with an activator such as thallous iodide.

RECORDING DEVICES AND PRINTERS

It is no longer common to use mechanical instruments such as strip-chart recorders or plotters for displaying on paper experimental data such as absorption intensity versus wavelength or temperature versus time. Rather, such data are usually displayed on a computer screen and stored in memory for later recall. For remote applications, small data loggers with internal memory and magnetic or optical storage capability can be used and this information is then transferred to a computer. Long-term storage of huge amounts of information is possible by using compact disks (CDs) or digital video display (DVD) disks, which have expected lifetimes of about 20 years. CDs are made from polycarbonate plastic coated with an aluminum film in which pits have been formed by using a focused diode laser beam, yielding a writing density of about 1 gigabyte for a disk of 120-mm diameter. DVDs have smaller pits and can have multiple layers on each side of the disk so the information density is about ten times larger.

When a paper display is needed or when textual information is to be printed, as in a report, it is common to use a digital printer such as a laser jet or inkjet printer. These devices are inexpensive and provide 600 to 2400 dots per inch, more than adequate for most applications. The laser jet devices use an electrostatic transfer of black toner material to paper to form the desired image, which is then fused to the paper by passage over a heater. Color versions are available but are more expensive because they require three toner cartridges, one for each of the primary colors. The inkjet printers use movable solid-state heads with many fine holes through which small droplets of ink are ejected by selective localized heating behind the orifices. The ink is drawn by capillary action from a reservoir chamber and, by use of three primary colors, full-color images can be printed at up to 2400 dots per inch.

REFRACTOMETERS

The term *refractometer* is applied principally to instruments for determining the index of refraction of a liquid, although instruments also exist for determining the index of refraction of a solid.[18] The index of refraction n for a liquid or an isotropic solid is the ratio of the phase velocity of light in a vacuum to that in the medium. It can be defined relative to a plane surface of the medium exposed to vacuum as shown in Fig. 9*a;* it is the ratio of

FIGURE 9
Reflection and refraction at
an interface: (a) $\phi_m < \phi_{crit}$;
(b) $\phi_m = \phi_{crit}$;
(c) $\phi_{m_1} = \phi_{m_2} > \phi_{crit}$.

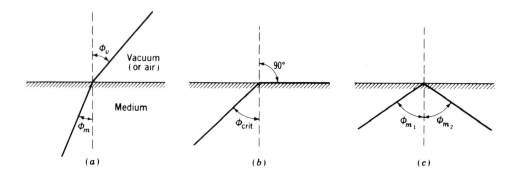

the sine of the angle ϕ_v a ray of light makes with a normal to the surface in vacuum to the sine of the corresponding ϕ_m in the medium:

$$n = \frac{c_v}{c_m} = \frac{\sin \phi_v}{\sin \phi_m} \tag{10}$$

It is common practice to refer the index of refraction to air (at 1 atm) rather than to vacuum for reasons of convenience; the index referred to vacuum can be obtained from that referred to air by multiplying the latter by the index of refraction of air referred to vacuum, which is 1.00027.

The index of refraction is a function of both wavelength and temperature. Usually the temperature is specified to be 20 or 25°C. The former is more in accord with past practice, but the latter is somewhat easier to maintain with a constant-temperature bath under ordinary laboratory conditions. The wavelength is usually specified to be that of the yellow sodium D line (a doublet, 589.0 and 589.6 nm), and the index is given the symbol n_D.

Most refractometers operate on the concept of the *critical angle* ϕ_{crit}; this is the angle ϕ_m for which ϕ_v (or ϕ_{air}) is exactly 90° (see Fig. 9b). A ray in the medium with any greater angle ϕ_{m_1} will be totally reflected at an equal angle ϕ_{m_2} as shown in Fig. 9c. The index of refraction is given in terms of the critical angle by

$$n = \frac{c_v}{c_l} = \frac{1}{\sin \phi_{crit}} \tag{11}$$

where c_l is the phase velocity of light in the liquid. In a refractometer the critical angle to be measured is that inside a glass prism in contact with the liquid, since the index of refraction of the glass is higher than that of the liquid. Therefore

$$\frac{c_l}{c_g} = \frac{c_l}{c_v}\frac{c_v}{c_g} = \frac{n_g}{n} = \frac{1}{\sin \phi_g}$$

where n_g is the index of refraction of the prism glass and ϕ_g is the critical angle in the glass. By trigonometry it can be shown that the index of refraction of the liquid is given by

$$n = \sin \delta \cos \gamma + \sin \gamma \sqrt{n_g^2 - \sin^2 \delta} \tag{12}$$

where γ is the prism angle (angle between the two transmitting faces) and δ is the angle of the critical ray in air with respect to the normal to the glass–air prism face (see Fig. 10). The most precise type of refractometer is the *immersion refractometer*. It contains a prism fixed at the end of an optical tube containing an objective lens, an engraved scale reticle, and an eyepiece. It also contains an Amici compensating prism (see below). In use, the instrument is dipped into a beaker of the liquid clamped in a water bath for temperature control. A mirror in the bath or below it reflects light into the bottom of

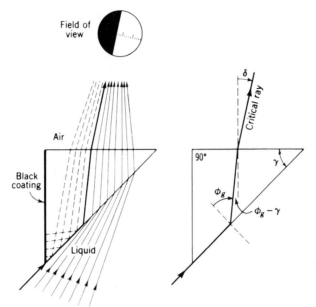

FIGURE 10
Essential features of an
immersion refractometer. The
behavior of the critical ray
is shown in detail, since this
represents the basic principle
of almost all refractometers.

the beaker at the requisite angle and with some angular divergence. The field of view
is divided into an illuminated area and a dark area, as shown in Fig. 10; the scale reading that corresponds to the boundary-line (critical-ray) position is read and referred to a
table to obtain the refractive index. This instrument is capable of measuring the refractive index to ± 0.00003. Its scale normally covers only a small range; a set containing
several refractometers or detachable prisms is required to cover the ordinary range of
refractive indexes for liquids (1.3 to 1.8).

The most commonly used form of refractometer is the *Abbe refractometer,* shown
schematically in Fig. 11. This differs from the immersion refractometer in two important
respects. First, instead of dipping into the liquid, the refractometer contains only a few
drops of the liquid held by capillary action in a thin space between the refracting prism and
an illuminating prism. Second, instead of reading the position of the critical-ray boundary
on a scale, one adjusts this boundary so that it is at the intersection of a pair of cross-hairs
by rotating the refracting prism until the telescope axis makes the required angle δ with the
normal to the air interface of the prism. The index of refraction is then read directly from a
scale associated with the prism rotation.

The Abbe refractometer commonly contains two Amici compensating prisms, geared
so as to rotate in opposite directions. An Amici prism is a composite prism of two different kinds of glass, designed to produce a considerable amount of dispersion but to produce
no angular deviation of light corresponding to the sodium D line. By use of two counterrotating Amici prisms, the net dispersion can be varied from zero to some maximum value
in either direction. The purpose of incorporating the Amici prisms is to compensate for
the dispersion of the sample so as to produce the same result with white light that would
be obtained if a sodium arc were used for illumination. This is achieved by rotating the
prisms until the colored fringe disappears from the field of view and the boundary between
light and dark fields becomes sharp. It should be borne in mind that the dispersion of a
sample is not always exactly compensated, for dispersion is not exactly defined by a single
parameter for all substances. The most precise results are obtained with illumination from
a sodium arc, the Amici prisms being set at zero dispersion.

The Abbe refractometer is less precise (± 0.0001) than the immersion refractometer
and requires somewhat less exact temperature control ($\pm 0.2°C$). For this purpose water

FIGURE 11
Schematic diagram of an
Abbe refractometer.

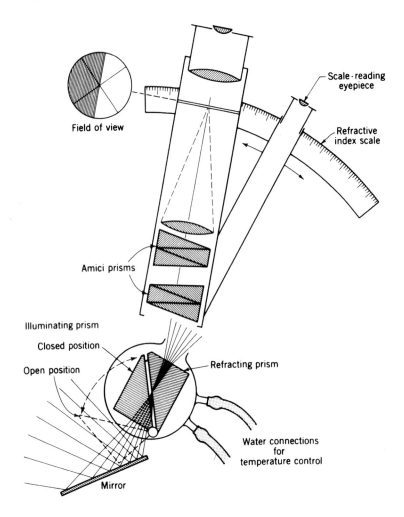

FIGURE 11
Schematic diagram of an
Abbe refractometer.

from a thermostat bath is circulated through the prism housings by means of a circulating pump. Alternatively, tap water is brought to the temperature of a thermostat bath by flow through a long coil of copper tubing immersed in the bath and is then passed once through the refractometer and down the drain.

The procedure for the use of an Abbe refractometer is as follows:

1. If a sodium arc is being used, check to see that it is operating properly. The sodium arc should be treated carefully and should be turned on and off as infrequently as possible. It should be turned on at least 30 min before use.
2. Check to see that the temperature is at the required value by reading the thermometer attached to the prism housing.
3. Open the prism (rotate the illuminating prism with respect to the stationary refracting prism). Wipe both prism surfaces *gently* with a *fresh* swab of cotton wool dampened with acetone. **Caution:** The prisms must be treated with great care, since scratches will decrease the sharpness of the boundary and permanently reduce the accuracy of the instrument.
4. When the prism surfaces are clean and dry, introduce the sample. In some models the refracting prism surface is horizontal and faces upward. In this case place a few drops of the liquid sample on the refracting prism and move the illuminating prism to the closed position. Other models have a prism configuration like that shown in Fig. 11. In that case bring the prisms into the closed position first and then squirt a small amount of sample into the filling hole.

5. Rotate the prism until the boundary between the light and dark fields appears in the field of view. If necessary, adjust the light source or the mirror to obtain the best illumination.

6. If necessary, rotate the Amici prisms to eliminate the color fringe and sharpen the boundary.

7. Make any necessary fine adjustment to bring the boundary between the light and dark fields into coincidence with the intersection of the cross-hairs.

8. Turn on the lamp (if any) that illuminates the scale, and read off the value of the refractive index.

9. Open the prism and wipe it *gently* with a clean swab of cotton wool, dampened with acetone. When it is dry, close the prism.

If the sample is very volatile, it may evaporate before the procedure is completed. In this case or in the event of drift, add more sample.

One of the worst enemies of the refractometer is *dust*. A gritty particle may scratch the prisms badly enough to require their replacement. The cotton wool used for wiping the prisms should be kept in a covered jar. Each swab of cotton wool should be used only once and then discarded. *Do not rub* the prisms with cotton wool, and do not attempt to wipe them dry; if streaks are left when the acetone evaporates, wipe again with a fresh swab dampened with fresh solvent. Do not use lens tissue on the prism surfaces. Finally, the instrument should be protected with its dust cover when not in use, and the table on which the instrument is used should be kept scrupulously clean.

For adjustment of the scale, a small "test piece" (a rectangular block of glass of accurately known index of refraction) is usually provided with the refractometer. The illuminating prism is swung out and the surface of both the refracting prism and the test piece are carefully cleaned. They are then carefully brushed with a clean camel's-hair brush (which is normally kept in a stoppered container) and inspected at grazing incidence to detect particles of dust or grit. A very small drop (ca. 1 mm^3) of a liquid (such as 1-bromonaphthalene or methylene iodide) that has a higher refractive index than the refracting prism is placed on the test piece, and the latter is then carefully pressed against the refracting prism and carefully moved around to spread the liquid. The reading of refractive index is made in the usual way. If it is not in agreement with the true value of the test piece, an adjustment of the instrument scale is made or a correction is calculated.

The procedure for determining the index of refraction of an isotropic solid sample is similar; like the test piece, it must have at least one highly polished plane face.

The refractometer is essentially an analytical instrument, used to determine the composition of binary mixtures (as in Exp. 14) or to check the purity of compounds. Its most common industrial application is in the food and confectionery industries, where it is used in "saccharimetry"—the determination of the concentration of sugar in syrup. Many commercially available refractometers have two scales: one calibrated directly in refractive index, the other in percent sucrose at 20°C.

The refractive index of a compound is a property of some significance in regard to molecular constitution. The *molar refraction,* defined by Eq. (29-14), is a constitutive and additive property; for a given compound it may be approximated by the sum of contributions of individual atoms, double bonds, aromatic rings, and other structural features.

SIGNAL-AVERAGING DEVICES

An important class of electronic measuring instruments is designed to retrieve weak voltage signals from accompanying noise.[10,11] Since the frequency spectrum of "white" noise is very wide, typically from 0.1 Hz to several MHz with a $1/f$ intensity distribution, much of it can be eliminated with the use of a *frequency-selective amplifier* that passes only a narrow bandwidth of the input at a specified frequency f_0. By modulating the dc signal to be measured at this same frequency, the signal-to-noise ratio is greatly improved. This modulation can be achieved in a number of ways: mechanically chopping a light beam, applying a modulated electric field to the sample in Stark-modulated microwave spectroscopy, or applying a modulated magnetic field in nuclear magnetic or electron paramagnetic resonance spectroscopy are a few examples. The combination of a frequency-selective amplifier with a *phase-sensitive detector,* which locks the amplifier input to a reference voltage, is known as a *lock-in detector.*

If the signal event is of short duration compared to the repetition time, a *gated integrator* provides better signal averaging than a lock-in amplifier. In this device, current from the signal source is allowed to pass through a fast, gated transistor switch to charge an integrating capacitor on a low-noise op amp (see Fig. XVI-6*f*). By making the gate overlap only the signal, noise outside this time period is excluded and the voltage across the capacitor reaches an average value representative of the input signal. Alternatively the voltage can be measured on each cycle, passed to a computer, and the capacitor discharged by a shorting transistor prior to the next input pulse; in this case signal averaging can be done by direct addition with the computer. A typical application of a gated integrator is in the study of transient signals produced by light sources or lasers with pulse durations of nanoseconds to microseconds but at a repetition rate of only 1 to 100 Hz.

When time resolution of a signal is required, the sampling period of a gated integrator can be made small and stepped or swept slowly across a repetitive waveform. A device with this capability, termed a *boxcar integrator,* is available from several manufacturers with time resolutions down to the nanosecond and even picosecond range. Alternatively one can use a *transient digitizer* or *digital oscilloscope* to capture a single-shot event in time by essentially doing many analog-to-digital (A/D) conversions in rapid succession. Such measurements can be done at rates up to about 100 kHz with relatively inexpensive A/D boards that plug into microcomputers, and at MHz and even GHz rates with more expensive commercial instruments. To handle the high data flow of such devices, signal averaging of repetitive sources is usually done by direct accumulation of the digitized signal in computer memory addresses assigned to each sampling interval.

SPECTROSCOPIC COMPONENTS

The interaction of electromagnetic radiation with matter serves as one of the most useful methods for studying the structures, energy levels, and dynamics of chemical systems. Figure 12 shows the energy transitions associated with different portions of the electromagnetic spectrum, and various experiments in this book utilize most of the types of spectroscopy associated with these changes. Some experimental details on magnetic resonance techniques are given in Exps. 21, 32, and 41–43, and the discussion here will center on the physical elements (sources, dispersion devices, and detectors) used for spectroscopy in the visible–ultraviolet (VIS–UV) and infrared (IR) regions.

Sources. The ultimate source for spectroscopic studies is one that is intense and monochromatic but tunable, so that no dispersion device is needed. Microwave sources such as klystrons and Gunn diodes meet these requirements for rotational spectroscopy, and lasers can be similarly used for selected regions in the infrared and for much of the visible–ultraviolet regions. In the 500 to 4000 cm^{-1} infrared region, solid-state diode and F-center lasers allow scans over 50 to 300 cm^{-1} regions at very high resolution (<0.001 cm^{-1}), but these sources are still quite expensive and nontrivial to operate. This is less true

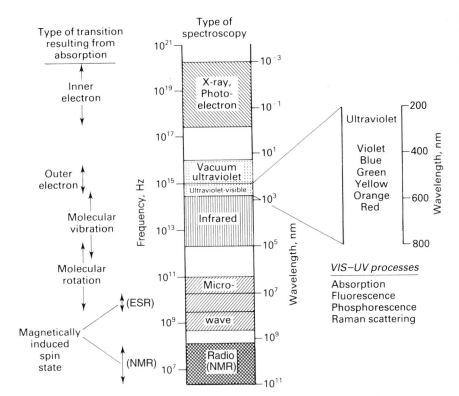

FIGURE 12

The electromagnetic spectrum and its applications in spectroscopy.

in the 5,000 to 10,000 cm^{-1} near-infrared region, where tunable diode lasers have been developed extensively because of applications to optical communication. Similarly red and blue diode lasers have become available, since these are widely used in CD devices and in other consumer products. Also for the visible region, dye lasers and solid-state lasers such as titanium-doped sapphire provide broad tunability for electronic spectroscopy, and the frequency of these can be doubled into the ultraviolet region by nonlinear interactions in crystals such as potassium dihydrogen phosphate (KDP). These sources can be pumped by either continuous argon/krypton-ion lasers or pulsed lasers, such as nitrogen or XeF excimer discharge lasers or by flashlamp-pumped (or diode-pumped) Nd:YAG solid-state lasers whose 1060-nm output is readily doubled to 532 nm by KDP. Again, these lasers are costly and, although widely used for research applications, are only slowly becoming available to instructional laboratories. This trend will surely increase in the future, and several of the experiments in this book make use of lasers (e.g., Exps. 15, 30, 33, 35, 36, 39, 40, 44, and 45). However, the most common sources still used in spectroscopy are incoherent lamps and discharges, and the remainder of the discussion in this section will focus on these devices.

In emission spectroscopy the molecule or atom itself serves as the source of light with discrete frequencies to be analyzed. In some cases, such as Exp. 39, which deals with the emission spectrum of molecular iodine vapor, excitation by a monochromatic or nearly monochromatic laser or mercury lamp is utilized. For other cases, such as the emission from N_2 molecules, electron excitation of nitrogen in a discharge tube provides an intense source whose spectrum is analyzed to extract information about the electronic and vibrational levels. Such low-pressure ($p < 10$ Torr) line sources are available with many elements, and lamps containing Hg, Ne, Ar, Kr, and Xe are often used for calibration purposes. The Pen-Ray pencil-type lamp is especially convenient for the visible and

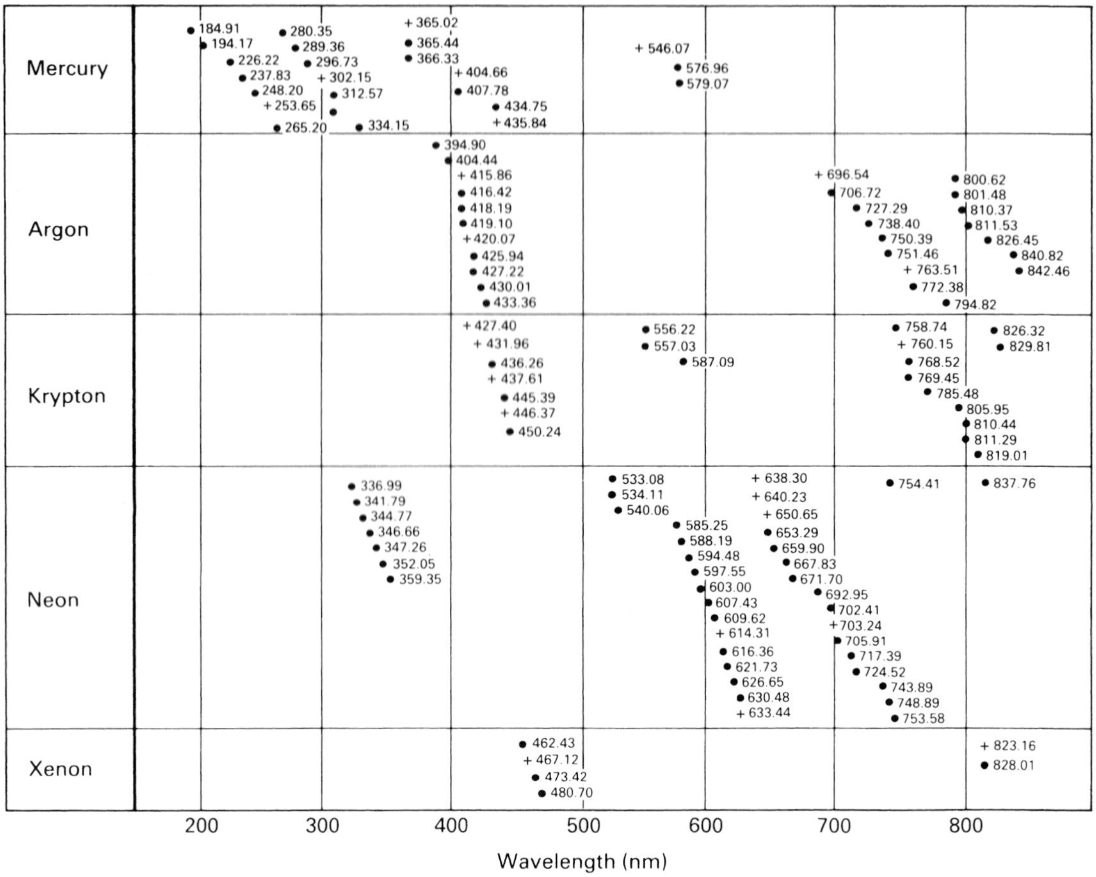

Element	Wavelengths (nm)
Mercury	• 184.91, • 194.17, • 226.22, • 237.83, • 248.20, + 253.65, • 265.20, • 280.35, • 289.36, • 296.73, + 302.15, • 312.57, • 334.15, + 365.02, • 365.44, • 366.33, + 404.66, • 407.78, • 434.75, + 435.84, + 546.07, • 576.96, • 579.07
Argon	• 394.90, • 404.44, + 415.86, • 416.42, • 418.19, • 419.10, + 420.07, • 425.94, • 427.22, • 430.01, • 433.36, + 696.54, • 706.72, • 727.29, • 738.40, • 750.39, • 751.46, + 763.51, • 772.38, • 794.82, • 800.62, • 801.48, • 810.37, • 811.53, • 826.45, • 840.82, • 842.46
Krypton	+ 427.40, + 431.96, • 436.26, + 437.61, • 445.39, + 446.37, • 450.24, • 556.22, • 557.03, • 587.09, • 758.74, + 760.15, • 768.52, • 769.45, • 785.48, • 805.95, • 810.44, • 811.29, • 819.01, • 826.32, • 829.81
Neon	• 336.99, • 341.79, • 344.77, • 346.66, • 347.26, • 352.05, • 359.35, • 533.08, • 534.11, • 540.06, • 585.25, • 588.19, • 594.48, • 597.55, • 603.00, • 607.43, • 609.62, + 614.31, • 616.36, • 621.73, • 626.65, • 630.48, + 633.44, + 638.30, + 640.23, + 650.65, • 653.29, • 659.90, • 667.83, • 671.70, • 692.95, • 702.41, + 703.24, • 705.91, • 717.39, • 724.52, • 743.89, • 748.89, • 753.58, • 754.41, • 837.76
Xenon	• 462.43, + 467.12, • 473.42, • 480.70, + 823.16, • 828.01

Wavelength (nm): 200   300   400   500   600   700   800

FIGURE 13

Calibration lines from low-pressure discharge lamps. The wavelengths in nm are given in air and the stronger lines are indicated with a plus sign (+).

ultraviolet regions, and some calibration wavelengths are given in Fig. 13. More complete compilations are to be found in Refs. 19 to 21.†

If a continuum source is needed for absorption spectroscopy, this can be provided by discharge lamps filled to higher densities, such that pressures can exceed 100 bar at operational temperatures. The result is a broad continuum emission with superimposed line spectra, as shown for several lamps in Fig. 14. In commercial spectrometers the deuterium lamp is commonly used for the UV region below 350 nm while the tungsten–halogen lamp is convenient for the 350 to 900 nm range. The latter is an example of a thermal source whose radiant excitance per unit wavelength M_λ closely approximates that predicted by the Planck formula for a blackbody radiator:[24,25]

†The wavelengths shown in Fig. 13 are those measured in air. Conversion to a vacuum wavelength or to a vacuum wavenumber value, $\tilde{\nu}_{vac} = 1/\lambda_{vac} = 1/n_{air}\lambda_{air}$, can be done using the index of refraction for dry air, CO_2-free, at 1 bar, 15°C:

$$(n_{air} - 1) \times 10^8 = 6431.8 + 2{,}949{,}330(146 - \tilde{\nu}^2)^{-1} + 25{,}536(41 - \tilde{\nu}^2)^{-1}$$

where $\tilde{\nu}$ must be inserted in units of μm^{-1}.[22,23] The correction $-\Delta\lambda/\lambda = \Delta\tilde{\nu}/\tilde{\nu} = \Delta n/n$ is about 0.028 percent over the visible–UV region; for the mercury green line, for example, $\lambda_{air} = 546.075$ nm corresponds to $\lambda_{vac} = 546.227$ nm. For more precise values, one must include a correction to the index of refraction n_{air} for the temperature, humidity, and CO_2 content of the air.[22,23] (The temperature correction can be obtained to adequate accuracy with the ideal-gas law.)

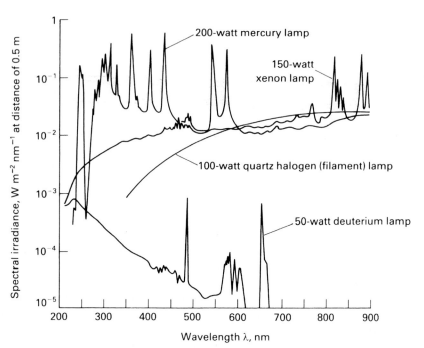

FIGURE 14

Typical output spectra of some discharge and filament lamps. Figure courtesy of Oriel Corp.

$$M_\lambda = (c/4)\rho_\lambda = \frac{2\pi hc^2}{\lambda^5[\exp(hc/\lambda kT) - 1]} \tag{13}$$

where ρ_λ is the spectral radiant energy density and M_λ has units of W m^{-3}, and its variation with temperature is shown in Fig. XVII-7. The wavelength of maximum excitance λ_m at temperature T obeys Wien's displacement law,

$$\lambda_m T = 2.8978 \times 10^6 \text{ nm K} \tag{14}$$

while the total energy emitted per unit time per unit area is given by the Stefan–Boltzmann equation,

$$M = \int_0^\infty M_\lambda d\lambda = \frac{2\pi^5 k^4}{15c^2 h^3} T^4 = \sigma T^4 \tag{15}$$

where $\sigma = 5.6705 \times 10^{-8}$ W m^{-2} K^{-4}.

A source whose spectral excitance differs from that of a blackbody only by a multiplicative constant is termed a *graybody*. The proportionality constant is called the emittance and denoted as ε. A tungsten source at 3000 K has an emittance of 0.43 ± 0.03 in the 350 to 800 nm spectral range of typical use. The lamp emission can also be characterized by specification of a *color temperature,* which is the temperature at which a blackbody would have a spectral emittance closest in shape to that of the lamp. Tungsten filaments in ordinary lightbulbs are operated at color temperatures of 2500 to 2600 K, while a xenon-arc lamp has plasma emission that corresponds to a color temperature of 6000 K. To obtain higher intensities, a tungsten source can be operated at higher temperatures, albeit at a cost of shorter filament life. In practice, iodine is often added in quartz–halogen–tungsten lamps to extend the filament life through the formation of volatile WI$_2$ near the quartz wall of the lamp. This compound then diffuses to the hot filament, where it decomposes and redeposits the tungsten. A coiled-filament lamp of this type serves as a NIST standard of spectral irradiance and is available from commercial sources such as Gamma Scientific and Optronics Laboratories.

FIGURE 15
Schematic diagram of a
simple prism spectrograph.
Short wavelengths are
refracted more than long
wavelengths. The collimating
lenses are multielement
achromats to provide a sharp
image at the exit slit for a
broad wavelength range.

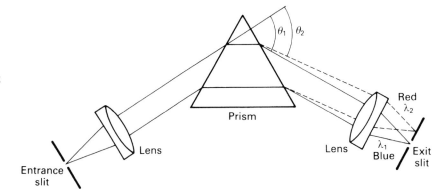

A thermal source that is commonly used for infrared spectroscopy is the *globar,* a 6-mm-diameter rod of silicon carbide, typically operated at 1000 to 1500 K with an emittance of 0.86 ± 0.04 in the 4000 to 600 cm^{-1} spectral region. Another resistively heated source is the *Nernst glower,* a smaller-diameter (1 to 3 mm) rod of semiconducting rare-earth materials such as ZrO_2, Y_2O_3, ThO_2, or CeO_2, which is operated at 1200 to 2000 K. Because of a negative temperature coefficient of resistance, it must be preheated to achieve conduction and must be used in series with a ballast resistor to prevent excessive currents and burnout. For the far-infrared region below 200 cm^{-1}, a mercury lamp with a fused-quartz envelope gives higher intensities. The hot envelope serves as the source down to about 35 cm^{-1}; below this the mercury plasma contributes most of the emission.

Wavelength Selection.   Although most modern spectrophotometers use a grating as a dispersion device, older visible–ultraviolet or infrared spectrophotometers employed prisms of quartz or flint glass or, for the infrared, salts such as LiF, NaCl, KBr, or CsI. A typical prism arrangement is shown in Fig. 15. The angular dispersion D_a of a polychromatic beam of radiation as it travels through a prism is due to the wavelength dispersion $dn/d\lambda$ of the index of refraction,

$$D_a = \frac{d\theta}{d\lambda} = \frac{d\theta}{dn}\frac{dn}{d\lambda} \tag{16}$$

The geometric factor $d\theta/dn$ is a function of the prism apex angle and, at a constant angle of incidence, changes only slightly with wavelength. Thus the resolving power of a prism is determined primarily by $dn/d\lambda$ as well as by the size (base length) of the prism. The refractive index variation for various materials is shown in Fig. 16.[26] It should be noted that the dispersion (slope) changes significantly with wavelength, with the greatest value occurring near the high- and low-wavelength limits of transmission determined by lattice absorption or by electronic absorption, respectively.

The dispersion of a grating is generally much higher than that of a prism and is more nearly constant with wavelength. The diffraction from a reflection grating is governed by the grating law

$$m\lambda = d(\sin\alpha - \sin\beta) = 2d\sin\theta\cos\phi \tag{17}$$

where d is the spacing between grating grooves and the angles are as defined in Fig. 17. The order of diffraction, m, is an integer equal to $\pm 1, \pm 2, \ldots$ for the dispersed orders of light; a value of $m = 0$ corresponds to undispersed specular reflection as for a plane

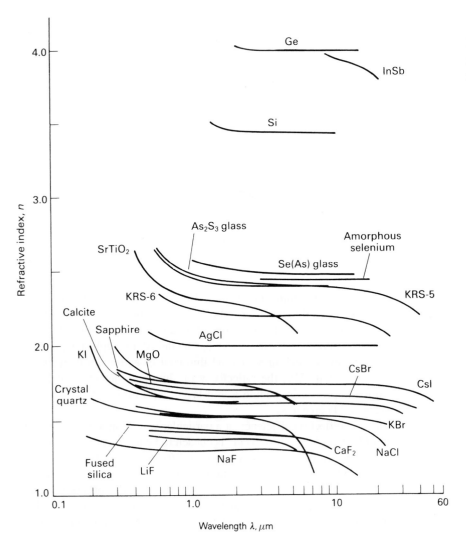

FIGURE 16
Refractive indexes for
various optical materials.
Figure from Ref. 26 with
permission.

mirror. For a fixed angle of incidence, the angular dispersion of the diffracted beam is
obtained by differentiation:

$$\frac{d\beta}{d\lambda} = \frac{-m}{(d \cos \beta)} \qquad (18)$$

The resolving power of a grating thus is better in higher orders and for small groove spacings. It is also linear in the total number of grooves illuminated; i.e., it is proportional to the width of the grating.

At a particular angle of diffraction β, wavelengths for several orders will satisfy Eq. (17), but since these are widely separated in magnitude, a simple bandpass filter of colored glass or other materials can be used to block undesired orders. Normally a grating is used in first order, since the reflection efficiency is highest there.

Ruled plane gratings are made by cutting regular grooves into an aluminum film on glass using a diamond tool. Line densities range from 20 grooves per millimeter for the far IR to 3600 or more grooves per millimeter for the VIS and UV ranges. Special interferometrically controlled ruling machines are required to produce master gratings over distances as great as 10 to 25 cm. Master gratings are expensive, and most instruments use

FIGURE 17

Cross section of a reflection grating with groove spacing d and blaze angle θ_0. Here α is the angle of incidence, β that of diffraction, and ϕ that of reflection. The blaze arrow, usually scribed on the back or top of a grating, indicates the direction (left or right) of enhanced diffraction of a beam incident along the grating normal.

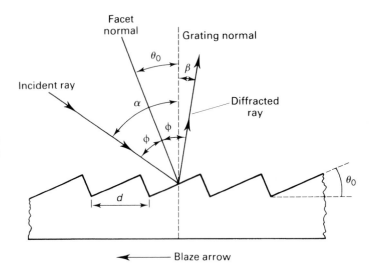

replica gratings that are of comparable quality. These are made by application of a parting agent such as silicone oil to a master, followed by evaporation of a film of aluminum. Epoxy resin is then poured onto the film, topped by a glass plate that bonds to the epoxy and provides a rigid support. After the epoxy has hardened, the replica is parted from the master, the oil is removed, and an additional thin coat of aluminum is evaporated onto the grating. In the vacuum UV, the reflectivity of aluminum decreases and the grating efficiency is often improved by adding a thin layer ($\sim$30 nm) of MgF_2 to increase the reflectivity to about 70 percent.

To increase the reflection efficiency at a particular wavelength, the grating groove is usually cut at a *blaze angle* θ_0 whose value is chosen to be $\theta_0 = (\alpha - \beta)/2$, so that specular reflection from the groove facet and first-order diffraction coincide. This concentrates most of the diffracted light at the corresponding blaze wavelength λ_B and provides high efficiency over a range from about $\frac{2}{3}\lambda_B$ to $\frac{3}{2}\lambda_B$. The reflectivity of a typical grating for the visible region is shown in Fig. 18, and it is seen that there is substantial variation depending upon the electric polarization of the incident light. To eliminate this dependence, the incident polarized light, such as occurs in Raman scattering, can be depolarized ("scrambled") by passing the radiation through a wedged quartz plate.

Because of the difficulty of controlling the groove spacing during ruling, extraneous lines (grating "ghosts") and other defects sometimes occur in mechanically ruled gratings. Since the late 1960s, an alternative method of production has involved the use of two collimated laser beams arranged to cause a pattern of interference lines on a substrate coated with a positive photoresist emulsion. After a few minutes exposure, the emulsion is developed to produce a periodic groove pattern that is then overcoated with aluminum. Groove densities up to 6000 lines mm^{-1} are commercially available. The production of a well-defined blaze angle for the groove is more difficult with these *holographic gratings,* so the overall reflection efficiency tends to be somewhat lower than that of a ruled grating (see Fig. 19). However, since all grooves are formed at the same time, ruling errors are nonexistent and these gratings exhibit very low scattered light levels as well. Another special advantage of holographic gratings is that the grooves do not have to be straight lines. By interfering spherical wavefronts, groove patterns can be placed on concave substrates so as to eliminate the need for collimating and focusing optics in a spectrometer. In addition the groove pattern can be designed to correct for spectrometer aberrations so that improved resolution with fewer optical elements becomes possible. Finally, we note that volume–phase gratings have recently been produced that diffract light by refractive

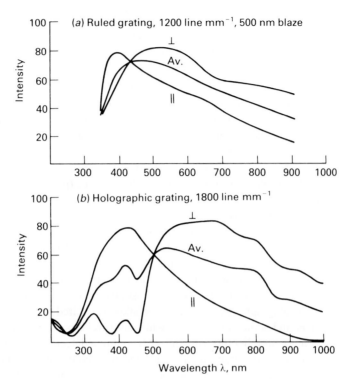

(a) Ruled grating, 1200 line mm^{-1}, 500 nm blaze

$\perp$

Av.

$\parallel$

Intensity

(b) Holographic grating, 1800 line mm^{-1}

$\perp$

Av.

$\parallel$

Intensity

Wavelength λ, nm

FIGURE 18
Diffraction efficiencies of ruled (a) and holographic (b) gratings for use in the visible region. The symbols $\parallel$ and $\perp$ correspond to light whose polarization is parallel (P type) or perpendicular (S type, S for senkricht) to the grating grooves. Also shown is the average, which is characteristic of unpolarized light. Figure courtesy of Spex Industries, Inc.

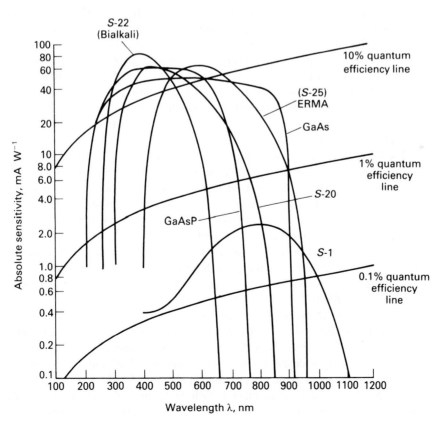

S-22 (Bialkali)

10% quantum efficiency line

(S-25) ERMA

GaAs

1% quantum efficiency line

S-20

GaAsP

S-1

0.1% quantum efficiency line

Absolute sensitivity, mA W^{-1}

Wavelength λ, nm

FIGURE 19
Wavelength dependence of the radiant sensitivity of various photocathode materials. S-1 = AgOCs, S-2 = Na$_2$KCsSb, S-22 = KCsSb (bialkali), S-25 = NaKCsSb, also denoted as ERMA (extended red multialkali).

index modulations within a thin layer of material sandwiched between two glass substrates. These gratings have high diffraction efficiencies, and a 750 lines/mm version for the visible region around 550 nm is available at modest cost from Learning Technologies, Somerville, MA.

Radiation Detectors.[26] Devices used for the detection of UV–VIS–IR electromagnetic radiation can be divided into two groups that are sensitive, respectively, to the *number* of incident photons (quantum detectors) or to the net *energy* of the beam (thermal detectors). In quantum detectors, photons interact directly with the electrons in the detector material and, if the energy of the photon is sufficient, cause a chemical reaction (photographic emulsion), a voltage or current change in the material (photovoltaic or photoconductive detectors), or actual ejection of an electron from the surface of the material (vacuum photocell or photomultiplier). Thermal detectors reveal the absorption of radiant energy through a change in a temperature-dependent property such as voltage, resistance, capacitance, or pressure and are generally slower and less sensitive than quantum detectors.

The use of photography in the 100 to 1000 nm spectral region has a long tradition in spectroscopy but is rarely used in modern instruments because of its low sensitivity and slow "readout" time with respect to most other photon detectors. We shall give no discussion of photography, as most of the techniques required for routine work are well known or are given in instructions supplied with commercial photographic materials.[27,28]

Photomultiplier tubes are the most common detectors used in VIS–UV spectrometers. These consist of a transparent evacuated tube containing a photoemissive material, the *photocathode,* plus a series of secondary electron emitters called *dynodes* that provide current gain. A simple tube with no dynodes is called a *phototube* or *photocell;* often gas is added to the tube to provide secondary electrons from collisions with the photoelectron as it is accelerated toward the anode by the tube potential. The photocathode is usually made of one or more alkali metals plus a group V element such as P, As, Sb, or Bi, and sometimes silver and/or oxygen. The formulation of the cathode is something of an art, the objective being to produce a material that absorbs all incident photons and that has the least resistance to photoemission. A very sensitive photocathode material of recent development employs GaAs and InP semiconductor substrates, coated with a thin film of Cs and CsO to lower the electron affinity below that of the substrate and hence to allow the electron to escape more easily. Figure 19 shows the spectral response curves of some typical photocathode materials.[29]

The first dynode in a photomultiplier is usually biased at a positive potential of about 75 to 150 V with respect to the photocathode, which has a potential of -1000 to -2000 V. An electron emitted from the photocathode is thus accelerated and, on impact with the dynode, generates additional electrons that are then accelerated toward the second dynode for further current amplification, as depicted in Fig. 20. The actual dynode shape and positioning are designed to give the most efficient collection and redirection of electrons along the chain to the anode that delivers the current pulse to the measuring electronics. Common dynode materials are Be–Cu and Cs–Sb alloys, which generate two to six electrons per incoming electron. Dynode materials such as GaP can give gains of up to 40 at higher biasing voltages of 800 V.

Typically, 5 to 15 dynode stages are used in photomultipliers to obtain gains that can be as high as 10^8. Each photon produces a current pulse of 5 to 20 ns duration because of the path differences traveled by the secondary dynode electrons. At pulse rates higher than 10^6 Hz, pulse coincidences become more frequent and this introduces counting errors. Thus direct-current measurement is preferred at high light levels. A photomultiplier should never be exposed to ambient or other high light levels with voltage on the tube, since destruction of the final dynode stages and even the photocathode can occur.

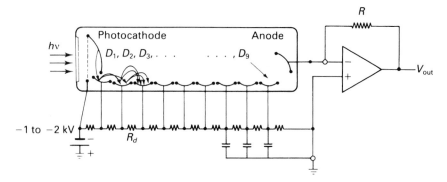

FIGURE 20

Typical connections for a photomultiplier tube. Secondary emission of electrons from 5 to 15 dynodes provides gains of up to 10^8. The dynode resistors are usually all the same and are about 50 kΩ. The capacitors on the last few dynodes store charge for improved operation with high current pulses. The output of the operational amplifier V_{out} is equal to the anode current i_a times the feedback resistance R.

Because of their extreme sensitivity, photomultipliers are critical elements in applications involving low light levels, such as astronomy, emission spectroscopy (Exps. 39, 40, 44), and Raman spectroscopy (Exp. 35). Single-photon detection is possible, and photon counting is the usual mode of operation under these signal-limited conditions. Even in complete darkness, some background thermal emission of electrons from the photocathode and the dynodes is always present, which can produce a dark count of 50 to 10^4 pulses per second. This emission decreases exponentially as the temperature is lowered and dark counts of 1 to 10 counts per second can be obtained for some tubes by cooling the photomultiplier to about $-40°C$ thermoelectrically or to lower temperatures with solid CO_2 or liquid nitrogen. Some residual noise comes from cosmic rays and from nearby radioactive materials such as ^{40}K in the glass housing.

The short-wavelength limit for a photomultiplier is determined by the transmission of the window material covering the photocathode. Quartz can be used for tubes to be operated down to about 170 nm, and tubes with LiF windows are available to permit operation down to 105 nm. In the 30 to 300 nm region, an ordinary photomultiplier can be used by coating the quartz or pyrex entrance window with a film of sodium salicylate that, upon absorption of UV light, fluoresces with nearly unit quantum efficiency at about 400 nm.

Although significantly less sensitive than photomultipliers, photodiodes also serve as useful detectors throughout the UV–VIS–near-IR regions. The most common photodiode consists of a $p–n$ junction formed on a silicon chip, as depicted in Fig. 21. The absorption of light with an energy in excess of the band gap creates electron–hole pairs, which owing to the internal energy barrier of the junction, separate to produce a potential difference across the junction. This produces a voltage proportional to the incident light intensity that can be measured directly, without the need for a power supply. Typical response curves for some common photovoltaic detectors are shown in Fig. 22. The quantum efficiency of silicon can be as high as 90 percent, but the dark noise and noise from any subsequent amplification generally exceeds that of a photomultiplier. Silicon photovoltaic cells can deliver relatively large currents and are also used as panels for solar energy conversion. Photovoltaic cells made of selenium have found use in camera exposure meters.

Photovoltaic detectors are employed for low-noise, low-frequency applications. For faster and more accurate intensity measurements, it is usually best to operate the photodiode in a reverse-biased photoconductive mode. Here a positive voltage is applied to the n junction and the p junction is negative, creating a charge-depletion layer that reduces the conductance of the junction to a very low value. The creation of electron–hole pairs by light absorption then causes a current flow that is very linear in light intensity and that gives the device improved detectivity. The time response of a photodiode is very fast, less than 1 ns for a *pin*-type diode, which has a very thin insulating layer i between the p- and n-type semiconductor materials to permit operation at higher bias voltages.

FIGURE 21

Cross section of a
p–n photodiode (*a*) and
potential energy diagram
or the junction region (*b*).
Absorption of a photon
with energy in excess of the
band gap produces charge
separation of the resultant
electron–hole pair to produce
a voltage across the depletion
layer.

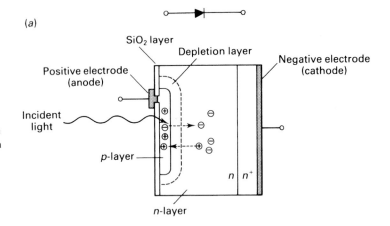

(a)

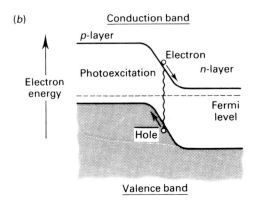

(b)

FIGURE 22

Typical detectivity D^* values
as a function of wavelength
for PbS photoconductive
and various photovoltaic
detectors. D^* is a figure of
merit defined as $A^{1/2}/NEP$,
where A is the detector
area and NEP is the *noise-
equivalent power*, the rms
radiant power in watts of a
sinusoidally modulated input
incident on the detector that
gives rise to an rms signal
equal to the rms dark noise in
a 1-Hz bandwidth. Data from
Hughes Aircraft Company.

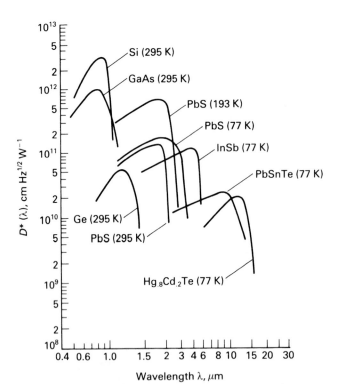

Detector arrays of up to 4000 silicon diodes are commercially available and are used in some modern UV–VIS spectrometers to give an entire spectrum in a few milliseconds. Often these include an intensifier stage that consists of a sensitive photocathode whose surface is imaged to the array through a *microchannel plate,* a bundle of hollow glass fibers subject to a large voltage drop from one end to the other. Each fiber has an interior resistive coating for secondary emission and produces a gain of 10^3 to 10^4 in the electron pulse, which strikes a phosphor to produce a strong visible pulse of light detected by one of the diode elements. This signal is then much larger than the inherently high dark noise found in the silicon detector.

Another very sensitive detector is the charge-coupled device (CCD), which is widely used in video cameras and in low-light-level applications such as occur in astronomy and in Raman and fluorescence spectroscopy. A CCD array consists of a two-dimensional grid of up to 2000×3000 anodes deposited on p-doped Si on an n-doped substrate. The basic detection element is similar to the single diode in Fig. 21a but differs in that the anode is deposited on top of the SiO_2 layer, thus forming an effective capacitor. Absorption of a photon in the p-doped region of one of the elements introduces an electron into the conduction band and a positive hole into the valence band. About 10^5 electrons can be stored beneath each anode before "spillover" into adjacent elements occurs, so the detector can be used to integrate a signal for improved sensitivity. By changing the potential on adjacent capacitors, the stored charge can be successively shifted from element to element until it is read out of an exit corner into a charge-measuring op amp. Both the photogeneration and the charge-transfer steps are very efficient, and when cooled to -50 to $-100°C$ to reduce background noise due to thermally generated electrons, CCD detectors can be up to 10 times more sensitive than photomultipliers. Further discussion of silicon and other multichannel detectors can be found in Ref. 30 and in manufacturers' current literature.

A number of other solid-state materials, both pure (intrinsic) and doped with impurities to reduce the effective band gap (extrinsic), have been developed for use in the near- to far-infrared regions. These generally require cooling to minimize noise from thermally generated charge carriers, especially for use in the far IR. Lead sulfide is the most sensitive photoconductive detector for the near-IR region from 1 to 3 μm and can be operated at room temperature. InSb, operated in a photovoltaic mode at 77 K, is useful for the 3 to 5.6 μm range, while mercury–cadmium–telluride (MCT) detectors at 77 K are often used in commercial infrared instruments covering the 2.5 to 25 μm range. Figure 22 shows the spectral response curves for these and a few other detector materials. Most of these detectors are used for specialized applications, where speed and sensitivity are critical. Examples include pulsed laser measurements and remote sensing of surface temperatures by orbital satellites.

The most common sensing devices found in commercial infrared spectrometers are thermal detectors. Although they are generally less sensitive than quantum detectors, these devices have broad, flat spectral response curves (see Fig. 23), and most can be operated at room temperature. One of the earliest types of IR detectors, the thermopile, is still in use. This consists of a multijunction thermocouple, each junction being composed of semiconductor elements or of metal pairs such as copper–Constantan, bismuth–silver, or antimony–bismuth. An equal number of junctions is kept in the dark in the detector housing to provide a reference to compensate for changes in the ambient temperature. The irradiated junctions are blackened to increase the absorption and have small leads in an evacuated housing to minimize cooling by heat conduction through the leads and residual gas. The time response is consequently slow, in the milliseconds range; but a temperature difference of 10^{-6} K can be detected. The far-IR limit of the detector is determined by the window material and is about 400 cm^{-1} for KBr and 200 cm^{-1} for CsI. Polyethylene is often used as a window for the 10 to 600 cm^{-1} far-IR region.

FIGURE 23

Typical detectivity $D*$ values for various thermal detectors assuming total absorption of incident radiation. See the legend of Fig. 22 for a definition of $D*$. The operating temperature, modulation frequency, and typical useful wavelength range are shown for each detector. Figure from Ref. 26 with permission.

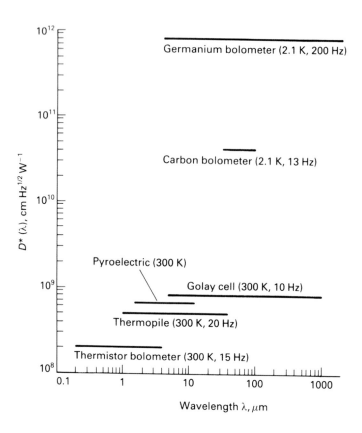

A more recent type of detector, offering greater sensitivity and much higher frequency response, is the pyroelectric detector. This is essentially a small capacitor made of a ferroelectric material such as triglycine sulfate (TGS), its deuterated form (DTGS), or lithium tantalate. When placed in an electric field, a surface charge results from alignment of the internal electric dipoles of these materials. Heating caused by absorption of radiant energy causes reorientation of the dipoles and a consequent change in surface charge that produces a capacitance change, sensed as a current pulse. The detector is inherently an ac device and can be used only with chopped or modulated sources. The output is a linear function of the input light intensity, and the time response can be as fast as a few picoseconds, since only charge reorientation is involved. In the infrared the detector is blackened so the time constant is longer due to the time required for heat transfer into the bulk material. The DTGS pyroelectric detector is the most common room-temperature detector used in modern Fourier transform infrared instruments.

The bolometer is another type of thermal detector that can offer extreme sensitivity for specialized applications.† This is essentially a resistance thermometer, usually with a platinum, nickel, carbon, or germanium element, although a semiconductor thermistor can also be used. Typically, two elements are used in a bridge circuit with one exposed to radiation and the other kept dark as a reference. The germanium bolometer provides exceptional

†The reputed sensitivity of the bolometer has inspired the following anonymous limerick:

Simon Langley invented the bolometer,

Which is really a kind of thermometer

That can measure the heat

    of a polar bear's seat

At a distance of half a kilometer.

sensitivity and is especially useful in the far-IR (see Fig. 23). Bolometer detectors have also found recent use in low-density molecular-beam experiments in which vibrational energy deposited in molecules by infrared laser absorption is detected as a temperature rise when the molecules strike the detector.

SPECTROSCOPIC INSTRUMENTS

Most spectroscopic measurements involve the use of an appropriate combination of source, dispersive device, and detector to analyze the absorption or emission spectrum of a sample. If only the wavelength or frequency of the radiation is measured, the resultant instrument is called a *spectrometer*. If the instrument provides a measure of the relative intensity associated with each wavelength, it is called a *spectrophotometer,* but this fine distinction is often ignored. Absorption spectra are often characterized by the *transmittance T* at a given wavelength; this is defined by

$$T \equiv \frac{I}{I_0} \tag{19}$$

where I is the intensity of light transmitted by the sample and I_0 is the intensity of light incident on the sample. When the sample is in solution and a cell must be used, I_0 is taken to be the intensity of light transmitted by the cell filled with pure solvent.

Another way of describing spectra is in terms of the *absorbance A,* where

$$A \equiv \log\left(\frac{I_0}{I}\right) = -\log T \tag{20}$$

The absorbance is related to the path length d of the sample and the concentration c of absorbing molecules by the Beer–Lambert law,

$$A = \varepsilon c d \tag{21}$$

where the proportionality constant ε is called the *absorption coefficient* (or, in older literature, the *extinction coefficient*). When the concentration is expressed in moles per liter, ε is called the molar absorption coefficient. The quantity ε is a property of the absorbing material that varies with wavelength in a characteristic manner; its value depends only slightly on the solvent used or on the temperature. Quantitative absorption measurements are widely used in chemistry, and accurate determinations require careful calibration of instruments and cells to confirm the validity of the Beer–Lambert law over the concentration range of interest.[4,30]

Visible–Ultraviolet Spectrophotometers. Many commercial spectrophotometers are suitable for the experiments in this text. Often, for instructional purposes, greater insight and more flexibility can be obtained by assembly of an instrument from modular components, available from such sources as Thermo Oriel, Optometrics USA, Ocean Optics, or other companies. More commonly an integrated instrument is used to provide greater reliability under use by relatively large numbers of students of varying backgrounds.

The simplest VIS–UV absorption instruments are relatively inexpensive single-beam devices such as the Thermo Spectronic 20 and Turner 350 models shown in Fig. 24. In a single-beam instrument, a reference detector is often used to provide feedback to hold the source intensity constant. The light from a tungsten source is focused through a slit and is dispersed by a grating (or by a prism in older instruments) whose rotation by an external manual knob governs the spectral wavelengths that are reimaged through an exit slit. The Spectronic 20 (Fig. 24*a*) has a particularly simple arrangement in which the beam is

FIGURE 24

Examples of single-beam spectrophotometers. (*a*) Spectronic 20 (available as Spectronics 20+ with photodiode detector from Thermo Spectronic); (*b*) HP 8452A multichannel diode-array spectrometer (available as model 8453 from Agilent Technologies).

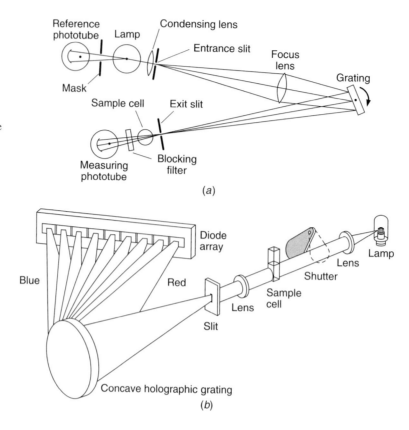

(*a*)

(*b*)

slightly convergent on a 600 line mm^{-1} grating, an economical compromise that provides a spectral bandwidth of 20 nm at the exit slit.

An alternative form of single-beam instrument is the multichannel diode-array spectrometer, such as the Hewlett-Packard 8452A shown in Fig. 24*b*. Here a concave holographic grating serves to both disperse and focus the light onto a photodiode array of 316 elements on a photosensitive area of about 18 mm × 0.5 mm, yielding a resolution of about 2 nm over the 190 to 820 nm range of the instrument. The grating is fixed and the instrument has no moving parts. The entire spectrum can be recorded in 0.1 s, making such instruments especially useful in kinetics studies. The cost of diode-array spectrometers varies considerably, depending upon the resolution and computer features desired.

The Varian-Cary double-beam instruments, whose optical schematic is shown in Fig. 25, are examples of a much more sophisticated double-pass absorption spectrophotometer with multiple sources and detectors to cover a 185 to 3150 nm scan range. A swing mirror is used to select between tungsten and deuterium source lamps. A filter wheel serves to eliminate overlapping spectral orders from the diffraction grating. The latter is two-sided and can be rotated 180° so as to provide better diffraction efficiency over the broad spectral range. All slits are curved for highest resolution (0.7 nm) in this *Czerny–Turner* monochromator configuration, and special optics match the source image to this shape. The exiting beam is sent alternately through sample and reference cells by a rotating half-sector mirror and is then detected by a phototube or, for the 900 to 3150 nm region, by a lead-sulfide detector. Simpler versions of such a double-beam instrument employ a single-pass monochromator with 1 to 2 nm resolution. A spectrum is acquired by recording the detector signal as the grating is rotated or, in some recent instruments, by holding the grating fixed and positioning a multielement detector array in the focal plane containing the dispersed radiation.

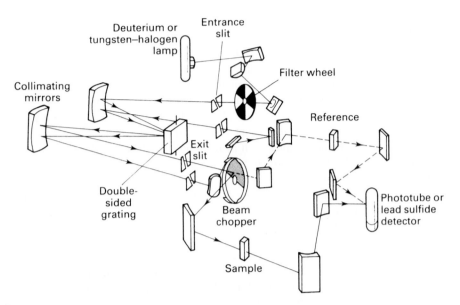

FIGURE 25
Optical diagram of Varian-Cary models 219 and 2000 series double-beam spectrophotometers (available as models 400 and 500 from Varian).

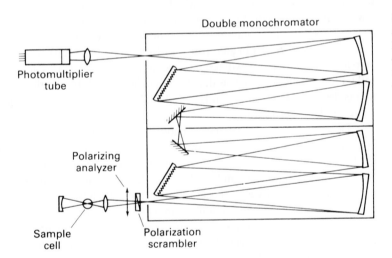

FIGURE 26
Schematic diagram of a Raman spectrometer. The exciting laser beam enters from below ($\perp$ to the plane of the drawing).

The Czerny–Turner configuration is also used in many emission or Raman instruments. A typical double monochromator arrangement for Raman spectroscopy is shown in Fig. 26. The excitation source is usually the 514.5-nm or the 488.0-nm line of an argon-ion laser at a typical power of 0.1 to 2 W in a spectral width of 0.25 cm^{-1}. The laser beam is focused into the sample by a lens (not shown in Fig. 26) to form a scattering cylinder approximately 0.1 mm in diameter and 10 mm in length. The scattered radiation is usually collected at an angle of 90° to the direction of the source beam by a large lens and is dispersed by the spectrometer. The gratings have 1800 to 2400 lines mm^{-1} for high dispersion and provide resolution down to 0.2 cm^{-1}. Since the Raman light intensity is quite low, stray Rayleigh light scattered off the slits, gratings, and mirrors of the spectrometer is minimized by the use of holographic gratings and a double (or triple) monochromator. Alternatively a single monochromator can be used in conjunction with a narrow "notch" or edge interference filter, which reflects the Rayleigh-scattered light at the exciting wavelength while passing Raman photons shifted by more than about 100 cm^{-1}. Special holographically produced filters that reduce the Rayleigh intensity by about 10^5 are available from Kaiser Optical Systems and Physical Optics.

Cooled CCD detectors or red-sensitive photomultiplier tubes are usually used as detection devices. CCD arrays permit signal integration and offer sensitivity comparable to or better than the photomultiplier operated in a photon-counting mode. In the latter case, a discriminator circuit can be used to reject noise pulses whose amplitude differs significantly from that produced by an electron ejected from the photocathode by a Raman-scattered photon. At 1 to 2 cm^{-1} resolution, count rates of 10^4 to 10^6 pulses s^{-1} are easily obtained with solids or with liquid samples such as benzene or carbon tetrachloride; for gases at 1 bar, count rates of 10^2 to 10^4 are more typical. Vibrational and/or rotational frequencies are deduced from the Raman spectrum by measuring the wavenumber values of each peak and finding the shift from the known exciting frequency. The spectrometer is easily calibrated[31] by using some of the emission lines shown in Fig. 13. Additional details of Raman sampling techniques and depolarization measurements can be found in Exp. 35 and in Ref. 31.

Infrared Spectrometers.   Infrared spectroscopy is one of the most powerful tools for quantitative and qualitative identification of molecules, and this led to the early development of prism and grating spectrophotometers.[32] Typically, these instruments cover the region from 400 to 4000 cm^{-1}, give a resolution of 1 to 4 cm^{-1}, and require calibration with polystyrene films or with standard gases such as H_2O, CO_2, CH_4, or NH_3.[33] This allows qualitative identification of materials and the measurement of vibrational frequencies of liquids, solids, and solutions of these in various nonaqueous solvents. Solids are often sampled as a dispersion in an oil such as Nujol (i.e., in a mull) or as a suspension in KBr, prepared by pressing a powdered mixture to form a pellet. Gases are examined in a simple cell with salt windows, such as that described in Exp. 37. If vibrational–rotational information is desired, as in Exp. 37 on HCl and DCl and in Exp. 38 on acetylene, a resolution of 1 cm^{-1} or better is desirable.

Descriptions of grating spectrophotometers can be found in earlier editions of this book, but such instruments have now been largely displaced by Fourier-transform infrared (FTIR) spectrometers.[34,35] Most FTIR instruments are based on the Michelson interferometer configuration depicted in Fig. 27. A collimated beam of light from an infrared source is divided into two halves by a beam splitter, typically KBr coated with germanium to give 50 percent reflectance. Reflections from a fixed mirror M_1 and a moving mirror M_2 are recombined and imaged onto a detector element that gives the net intensity. As one of the mirrors is moved, interference between the two beams occurs, which for a monochromatic source of wavelength λ, produces a periodic signal as depicted in the figure. The detector goes through one cycle for a mirror movement of $\lambda/2$, and hence the frequency f of this oscillation is given by $f = 2v/\lambda = 2v\tilde{\nu}$, where v is the mirror velocity. For infrared light in the 400 to 4000 cm^{-1} range (1.2 to 12 $\times$ 10^{13} Hz), a typical mirror velocity of 0.05 $cm\ s^{-1}$ gives $f = 40$ to 400 Hz, well within the response time of the pyroelectric and other infrared detectors commonly used in FTIR instruments.

For a polychromatic source, the signal is a sum of such cosine waves, with all adding constructively at the zero-path-difference point where the two mirrors are equidistant from the beam splitter. At other distances, the waves interfere and the detector signal, the *interferogram,* drops rapidly as shown in the figure. The interferogram is thus a sum of cosine waves, each of which has an amplitude and frequency proportional to the source intensity at a particular infrared frequency. Recovery of this desired information is achieved by performing a Fourier transform, a process greatly aided by the use of a computer and a fast-Fourier-transform algorithm developed in recent years (Cooley–Tukey procedure).[34] The resolution is determined by the total mirror travel L as illustrated in Fig. 28 and is essentially $\Delta\tilde{\nu} = 1/2L$; commercial research instruments (Bomem and Bruker) can provide a resolution of about 0.002 cm^{-1}.

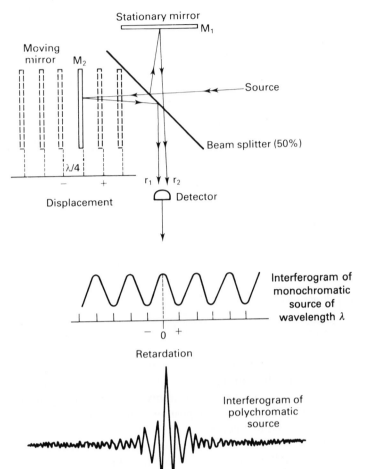

FIGURE 27
Schematic diagram of a
Michelson interferometer.
The detector signal variation
as a result of mirror motion
is displayed for the cases
of monochromatic and
polychromatic sources.

Stationary mirror

M_1

Moving
mirror

M_2

Source

$\lambda/4$

Beam splitter (50%)

$-$ $+$

r_1 r_2

Displacement

Detector

Interferogram of
monochromatic
source of
wavelength λ

$-$ 0 $+$

Retardation

Interferogram of
polychromatic
source

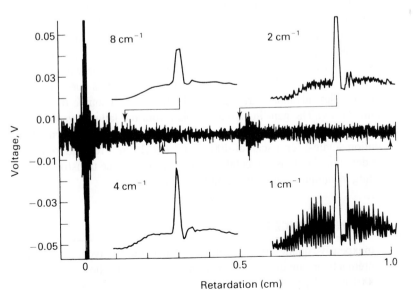

FIGURE 28
Interferogram and
transformed spectra for
acetylene in the 670 to 800
cm^{-1} ν_5 bending region. The
transformed spectra show the
effect of retardation distance
on the spectral resolution;
the 1-cm^{-1} case corresponds
to a total mirror travel of
0.5 cm. The vertical scale of
the interferogram is greatly
expanded; the peak-to-peak
voltage at zero retardation is
~4 V.

0.05

0.03

0.01

−0.01

−0.03

−0.05

Voltage, V

8 cm^{-1}

2 cm^{-1}

4 cm^{-1}

1 cm^{-1}

0

0.5

1.0

Retardation (cm)

FIGURE 29

Typical optical diagram of
the Mattson series of FTIR
instruments. Courtesy of
Mattson Instruments, Inc.

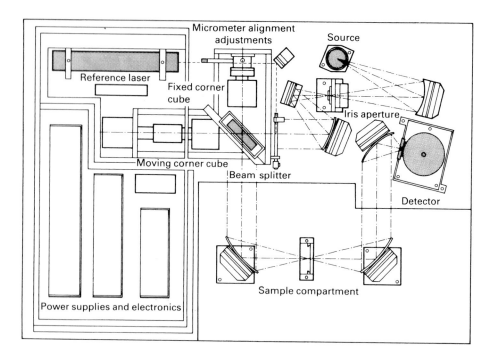

Figure 29 shows the optical configuration of a commercial FTIR instrument in which corner-cube mirror reflectors are used in place of plane mirrors. These "cat's-eye" reflectors serve to reduce the sensitivity of the mirror alignment and make it possible to use a simpler mirror drive. The position of the mirror is determined precisely by using the fringes produced by interference of a helium–neon laser reference beam. Since the frequency of the laser is accurately known, the infrared frequencies are determined to a few hundredths of a wavenumber and routine calibration with gas standards is unnecessary. This high precision also facilitates signal averaging over many scans and comparisons between spectra. Most FTIR instruments are single-beam devices, requiring separate measurements of the source background intensity and the signal when the sample is inserted. In some commercial instruments, a double-beam capability is provided by mirrors that oscillate to redirect the beams between the sample and reference cells.

Many FTIR spectrometers offer options to extend the coverage into the far-IR and, more recently, into the near-IR and even VIS–UV regions. In general the signal-to-noise ratio is much higher for FTIR instruments than for grating spectrometers, and this allows the use of shorter scan times. This improvement is due largely to two major advantages of an FTIR over a grating instrument: higher beam intensities due to the elimination of slits, and a multiplex advantage arising from the fact that all source frequencies are monitored simultaneously. The gain in signal to noise from the latter advantage is $N^{1/2}$, where N is the number of resolution elements to be examined. This is particularly helpful for high-resolution studies; for a scan of 4000 cm^{-1} at 0.1 cm^{-1} resolution, the advantage is 200. More detail on these instrumental aspects of FTIR, on the transform process itself, and on sampling techniques can be found in Refs. 34 and 35.

TIMING DEVICES

Inexpensive quartz-crystal digital stopwatches are convenient devices for time-interval measurements of durations ranging from 10 s to several hours, and they offer accuracies of 0.0001 to 0.01 percent. The principal error in the use of these is the reaction time of the experimenter, who is operating the start and stop buttons manually. Reaction times vary greatly from one individual to another, but a reasonable estimate of the error in a time interval for a (sober) operator would be ± 0.2 s.

More accurate time-interval measurements require electronic or mechanical triggering of the start–stop points. For example, the heating period in a calorimetric experiment can be timed automatically by using a fast double-pole switch to control both the timer and heater circuit simultaneously. Most electronic counters have a time-interval mode of operation in which successive electronic trigger pulses can turn a high-frequency (10 to 100 MHz) counter on and off with an accuracy of ± 1 count. For the most precise timing, a high-speed electronic frequency counter can be used to count the oscillations of an ultra-stable crystal-controlled oscillator. Agilent Technologies offers several instruments capable of making time measurements accurate to a few parts in 10^9. The international standard of time is Coordinated Universal Time (UTC), which is based on the NIST F1 "fountain" of cesium atoms, cooled to near 0 K by lasers. By stopping a vertical beam of atoms and causing it to fall slowly, observation times of a second or more are possible, resulting in a clock uncertainty of less than 2×10^{-15} (1 s in 20 million years!).

REFERENCES

1. D. A. Skoog and D. M. West, *Fundamentals of Analytical Chemistry,* 4th ed., Saunders, New York (1982).

2. I. M. Kolthoff, E. B. Sandell, E. J. Meehan, and S. Bruckenstein, *Quantitative Chemical Analysis,* 4th ed., chap. 19, Macmillan, New York (1969).

3. R. M. Schoonover, *Anal. Chem.* **54,** 976A (1982).

4. D. C. Harris, *Quantitative Chemical Analysis,* 6th ed., Freeman, New York (2002).

5. R. M. Schoonover and F. E. Jones, *Anal. Chem.* **53,** 900 (1981).

6. R. Batlino and A. G. Williamson, *J. Chem. Educ.* **61,** 51 (1984).

7. T. W. Lashof and L. B. Macurdy, "Precision Laboratory Standard of Mass and Laboratory Weights," *Natl. Bur. Stand. Cir.* **547,** U.S. Government Printing Office, Washington, DC (1954).

8. G. W. Thomson and D. R. Douslin, "Determination of Pressure and Volume," in A. Weissberger and B. W. Rossiter (eds.), *Techniques of Chemistry. Vol. 1: Physical Methods of Chemistry,* part V, chap. 2, Wiley-Interscience, New York (1971).

9. W. G. Brombacher, D. P. Johnson, and J. L. Cross, "Mercury Barometers and Manometers," *Natl. Bur. Stand. Monogr.* **8,** U.S. Government Printing Office, Washington, DC (1960).

10. P. Horowitz and W. Hill, *The Art of Electronics,* 2d ed., Cambridge Univ. Press, New York (1989).

11. A. J. Diefenderfer and B. E. Holton, *Principles of Electronic Instrumentation,* 3d ed., Harcourt, Fort Worth, TX (1994).

12. G. K. McMillan, *pH Measurement and Control,* 2d ed., ISA, Research Triangle Park, NC (1994).

13. D. J. G. Ives and G. J. Janz (eds.), *Reference Electrodes,* Academic Press, New York (1969); D. T. Sawyer et al., *Electrochemistry for Chemists,* 2d ed., Wiley, New York (1995).

14. H. H. Willard, L. L. Merritt, Jr., J. A. Dean, and F. A. Settle, Jr., *Instrumental Methods of Analysis,* 7th ed., chaps. 21, 22, Wadsworth, Belmont, CA (1988).

15. W. Heller and H. G. Curme, "Optical Rotation—Experimental Techniques and Physical Optics," in A. Weissberger and B. W. Rossiter (eds.), *Techniques of Chemistry. Vol. 1: Physical Methods of Chemistry,* part IIIC, chap. 2, Wiley-Interscience, New York (1972).

16. C. Djerassi, *Optical Rotary Dispersion,* McGraw-Hill, New York (1960); K. P. Wong, *J. Chem. Educ.* **51,** A573 (1974); **52,** A9, A89 (1975).

17. G. Friedlander, J. W. Kennedy, W. S. Macias, and J. M. Miller, *Nuclear and Radiochemistry,* 3d ed., Wiley, New York (1981).

18. S. Z. Lewin and N. Bauer, in I. M. Kolthoff and P. J. Elving (eds.), *Treatise on Analytical Chemistry,* vol. 6, part 1, chap. 70, Wiley-Interscience, New York (1965); A. Townshend (ed.), *Encyclopedia of Analytical Science,* vol. 7, pp. 4436–4443, Academic Press, San Diego, CA (1995).

19. A. G. Maki and J. S. Wells, *Wavenumber Calibration Tables from Heterodyne Frequency Measurements,* NIST Special Pub., 821, U.S. Government Printing Office, Washington, DC (1991); G. Guelachvila and K. N. Rao, *Handbook of Infrared Standards,* vol. I (1986), vol. II (1993), Academic Press, San Diego, CA.

20. G. R. Harrison, *M.I.T. Wavelength Tables,* M.I.T. Press, Cambridge, MA (1982 [1939]).

21. "NIST Atomic Spectra Database," NIST Standard Reference Database #78. See http://physics.nist.gov/cgi-bin/AtData/main_asd.

22. B. Edlin, *J. Opt. Soc. Am.* **43,** 339 (1953); H. Barrell and J. E. Sears, *Trans. Roy. Soc.* **A238,** 1 (1939).

23. G. Strey, *Spectrochim. Acta* **25A,** 163 (1969); K. Burns, K. B. Adams, and J. Longwell, *J. Opt. Soc. Am.* **40,** 339 (1950).

24. R. J. Silbey, R. A. Alberty, and M. G. Bawendi, *Physical Chemistry,* 4th ed., p. 340, Wiley, New York (2005).

25. P. W. Atkins and J. de Paula, *Physical Chemistry,* 8th ed., pp. 244–247, Freeman, New York (2006).

26. J. H. Moore, C. C. Davis, and M. A. Coplan, *Building Scientific Apparatus,* 3d ed., chap. 4, Perseus, Cambridge, MA (2002).

27. See, for example, Kodak publications CIS 51, *Kodak Materials for Emission Spectrography* (1982), and P-140, *Characteristics of Kodak Plates for Scientific and Technical Applications* (1984).

28. T. H. James (ed.), *The Theory of the Photographic Process,* 4th ed., Macmillan, New York (1977); B. H. Carroll, *Introduction to Photographic Theory,* Wiley, New York (1980).

29. M. J. Weber (ed.), *Handbook of Lasers,* CRC Press, Boca Raton, FL (1995).

30. J. D. Ingle and S. R. Crouch, *Spectrochemical Analysis,* Prentice-Hall, New York (1988).

31. J. R. Ferraro, K. Nakamoto, and C. W. Brown, *Introductory Raman Spectroscopy,* 2d ed., Academic Press, Boston, (2003).

32. J. E. Stewart, *Infrared Spectroscopy: Experimental Methods and Techniques,* reprint ed., Books on Demand, Ann Arbor, MI (1970); B. Stuart, *Infrared Spectroscopy,* 2d ed., Wiley, New York (1996).

33. A. R. H. Cole, *Tables of Wavenumbers for the Calibration of Infrared Spectrometers,* 2d ed., Pergamon, Oxford (1977). See also http://www.hitran.com/ (HITRAN is a compilation of spectroscopic parameters that a variety of computer codes use to predict and simulate the transmission and emission of infrared light in the atmosphere.)

34. P. R. Griffiths and J. A. de Haseth, *Fourier Transform Infrared Spectrometry,* Wiley-Interscience, New York (1986).

35. S. P. Davis, M. C. Abrams, and J. W. Brault, *Fourier Transform Spectrometry,* Academic Press, San Diego, CA (2001).

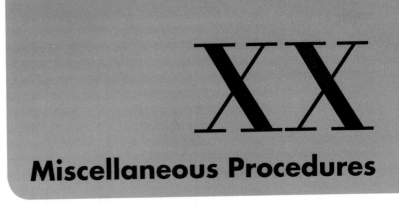

Miscellaneous Procedures

Physical chemistry laboratory work involves many manual arts, techniques, and procedures in addition to those described in earlier chapters. In this chapter we shall deal with some of the more important of these.

VOLUMETRIC PROCEDURES

A variety of physical chemistry experiments require either the preparation of solutions of precisely known concentrations or successive dilutions of a solution to obtain a series of solutions with known concentration ratios. Presented below is a summary of the more important aspects of volumetric methods. Considerably greater detail is available in many standard textbooks on quantitative chemical analysis.[1,2]

Volumetric Apparatus. Volumetric glassware is of three principal kinds: volumetric flasks, pipettes, and burettes; see Fig. 1.

The *volumetric flask* has a ring engraved around its neck; when the bottom of the liquid meniscus is level with this ring, at the designated temperature, the volume of liquid contained is that indicated by the engraved label. Frequently the letters TC, meaning "to contain," are present on the label. When such a flask is used "to deliver," it should be allowed to drain for at least 1 min; even so, its accuracy is not as high when used to deliver as when used to contain.

The technique for using a volumetric flask will be illustrated by a description of the proper way to make up a standard solution from a weighed amount of a solid. Once the solid has been introduced into the clean flask with the aid of a funnel, solvent is added to rinse out the funnel and rinse down the neck of the flask. A sufficient amount of the solvent is then added to dissolve the solid; shaking and swirling is more efficient when the flask is only about half full. When the solid has dissolved, additional solvent is added until the liquid level is about 2 cm below the calibration mark. The flask is stoppered, and the solution is mixed well by repeated inversion and shaking. The flask is then placed upright and the stopper is removed and rinsed down with a small amount of solvent. The flask is allowed to drain for a minute or two, and the neck is inspected to make sure that there are no clinging

FIGURE 1

Typical volumetric glassware: (*a*) volumetric flask, (*b*) transfer pipette, (*c*) measuring pipette, (*d*) burette.

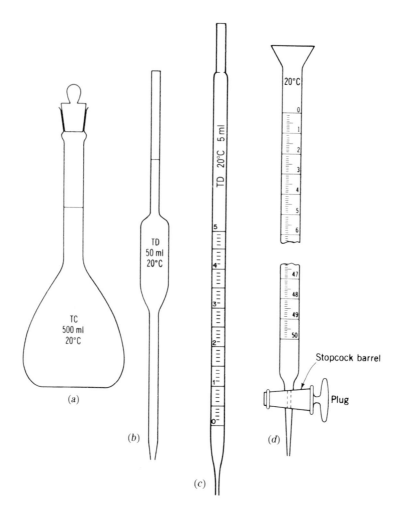

drops above the calibration mark. Finally the liquid level is raised until the bottom of the meniscus is even with the calibration mark by addition of solvent dropwise from a pipette or a dropper. The flask is then stoppered, and the solution is mixed thoroughly.

It should also be noted that *graduated cylinders* are often used for rough measurements of liquid volume. When they are used to deliver a given volume, the smaller sizes are not at all accurate. When they are used to contain, the larger graduated cylinders may be accurate to within ~3 percent. Graduated cylinders are much more convenient than volumetric flasks when solutions of approximate concentration are made up by dilution of concentrated reagents.

The *transfer pipette* has a single ring engraved on its upper stem, and its label frequently contains the letters TD, meaning "to deliver." The pipette is filled by drawing liquid well above the ring using a pipetting bulb. **As a safety precaution, never draw liquid into a pipette by mouth.** Slip off the pipetting bulb and close the top of the upper stem with a finger tip. By careful manipulation of the finger tip the meniscus is allowed to fall to the ring. On delivering the contents into a vessel, the pipette is held at an angle of about 35° from the vertical and the tip is held against the wall of the vessel to prevent spattering. When the meniscus stops falling, the pipette should be held for about 10 s to allow drainage and then withdrawn after again touching the tip to the wall of the receiving vessel but *without* blowing out the liquid held in the tip by capillary action.

Measuring pipettes are essentially small burettes without stopcocks. As with burettes, the need for adequate drainage time cannot be ignored. Because of the double reading involved in the use of measuring pipettes and the additional manipulation required, they are neither as precise nor as accurate as transfer pipettes, but they are quite satisfactory for semiquantitative operations.

Micropipettes are made in sizes from 0.1 μL to 500 μL (0.5 mL) in several designs. Some are simply capillaries that fill completely by capillary action and are blown or rinsed out. Others are capillaries with a single calibration mark that are filled beyond the mark by capillary action and then drawn down to the mark by touching the tip carefully to absorbent paper. Very often micropipettes are essentially miniature measuring pipettes fitted with a "pipette control," usually an uncalibrated hypodermic syringe or a threaded piston in a cylinder, which permits suction or pressure to be applied to adjust the level of the liquid with high precision. Micropipettes can be used with considerable precision if sufficient care is exercised, but the importance of droplets left on the outside of the tip, internal cleanliness, and uniformity of technique is even greater than with pipettes of ordinary size.

The *burette* is a cylindrical tube of uniform cross section, graduated in volume units along its length. A stopcock and drawn-out tip are attached at the bottom. Burette stopcocks are made in a variety of styles; those having a smooth glass barrel and a fitted Teflon plug are the most desirable. The set nut that holds the Teflon stopcock in position must be tightened sufficiently to avoid leakage but not overtightened lest the plug be deformed or the threads damaged. The traditional ground-glass stopcock requires careful lubrication with an appropriate stopcock grease. The use of a burette is described below in connection with titration.

For the delivery of small volumes, particularly of highly volatile materials or materials susceptible to the atmosphere, the use of a *hypodermic syringe* is recommended. Good-quality syringes may be read with reasonable accuracy, or the amount of liquid discharged may be found by weighing. Pyrex syringes are reasonably resistant to most reagents, but some care should be exercised with the choice of needles. The use of ordinary stainless-steel hypodermic needles as delivery tips with syringe burettes has a number of hazards. Although stainless steel tends to be quite resistant to many reagents, the needles are brazed into the shanks with an acid-susceptible alloy. Teflon needles are available commercially and are to be preferred.

Accuracy and Calibration. Specifications concerning the shape, dimensions, graduations, and capacity tolerances of precision glassware were established some time ago by the National Bureau of Standards (NBS), which has since been reorganized and renamed the National Institute of Standards and Technology (NIST). Various capacity tolerances are given in Table 1. Volumetric flasks that meet these specifications are designated "Class A" and are usually so marked. Precision pipettes, with a serial number and certificate, can also be obtained. The tolerances of most commercially available glassware are greater than those listed in Table 1 by approximately a factor of 2 or 3.

Whenever very reliable volumetric measurements are required, the glassware should be calibrated, preferably at 20°C. Volumetric flasks (TC) are best calibrated by weighing them empty and then filled with water. Pipettes are calibrated by filling with water, delivering their contents into a weighed glass-stoppered weighing bottle, and then reweighing. A burette is filled with water, and 10 to 20 percent at a time is run out into a weighed glass-stoppered weighing bottle or flask, which is stoppered and weighed after each addition

TABLE 1 Capacity tolerances of precision (class A) volumetric glassware

Capacity in mL less than and including	Volumetric Flasks		Pipettes	
	Calibrated to contain (TC) ± mL	Calibrated to deliver (TD) ± mL	Transfer (TD) ± mL	Measuring (TD) ± mL
1	0.01	—	0.006	0.01
2	0.015	—	0.006	0.01
5	0.02	—	0.01	0.02
10	0.02	0.04	0.02	0.03
25	0.03	0.05	0.03	0.05
50	0.05	0.10	0.05	
100	0.08	0.15	0.08	
200	0.10	0.20	0.10	
300	0.12	0.25		
500	0.20	0.40		
1000	0.30	0.60		
2000	0.50	1.00		
Above 2000	1 part in 4000	1 part in 2000		

(without being emptied in between); a calibration curve (required correction plotted against burette reading) is then prepared, analogous to a thermometer calibration curve. In each case the correct density of water at that temperature (0.99823 g cm^{-3} at 20°C) must be used in the calculations, and air-buoyancy corrections (see Chapter XIX) are applied. Further details on calibration procedures and calculations are given elsewhere.[1-3]

NIST has specified 20°C as the *normal temperature* for volumetric work. The cubical coefficient of expansion of Pyrex is about 0.9×10^{-5} K^{-1}; that of water at 20°C is about 21×10^{-5} K^{-1}. The expansion of glass will be of importance only in very precise work; that of water however will affect molar concentrations and will be of significance if the actual temperature is more than about 5°C removed from 20°C.

Cleaning of Glassware. Volumetric work of any quality depends upon glassware with a surface clean enough to be wetted uniformly by water and aqueous solutions; if the meniscus pulls away from areas of the glass leaving dry spots, the glass requires cleaning. If cleaning with an ordinary laboratory detergent is not sufficient, a commercial cleaning solution such as MICRO can be used. Chromic acid is probably the most effective cleaning solution for glass, but it should be handled with caution, using gloves and safety goggles. About 10 g of $Na_2Cr_2O_7$ is dissolved in the minimum quantity of hot water, and after cooling, about 200 mL of concentrated sulfuric acid is slowly added with stirring. This cleaning solution must be kept in a glass-stoppered bottle. If after much use it appears greenish, it should be replaced. **Contact of chromic acid cleaning solution with any organic material (e.g., wood, cloth) or with the skin must be carefully avoided.** Glassware to be cleaned is ordinarily filled with this solution, which may be moderately warm but should not be hot. The glassware should be reasonably dry to avoid dilution of the solution with water. The solution will attack most fillers and pigments used to fill graduations of burettes and other volumetric glassware; confine the solution to the inside surfaces as much as possible. After 15 to 30 min of contact, this solution should be poured back into the storage bottle and the glassware should be thoroughly rinsed with distilled water. Alternatives to chromic acid are hot 50–50 nitric + sulfuric acid (caution in handling also required) and alcoholic KOH. Clean glassware is usually allowed to drain if it is not to be used immediately; burettes however are often filled with distilled water and covered with an inverted beaker pending use.

Rinsing.   Volumetric glassware need not be dry before filling with the appropriate aqueous solution if the vessel is first rinsed two or three times with the solution. Several *small* portions are more effective than the same total volume of solution in a single portion.

Titration.   In titration with a standard solution, the amount of solution required to react quantitatively with a given sample is measured with a burette. The burette should be mounted vertically, with the graduations facing the operator and the stopcock handle on the right. A right-handed individual should learn to operate the stopcock with the left hand in order to leave the right hand free for swirling, manipulating a wash bottle, etc. The crook between the thumb and forefinger fits around the burette at the top of the barrel, and the handle is grasped with the thumb and first two fingers. With practice, very fine control can be obtained this way, and a fraction of a drop can be delivered if necessary.

The burette is filled with the required solution (after two or three rinsings with small portions of the same solution) to a point above the zero graduation. Solution is run out into a beaker until the meniscus has dropped to zero or some position on the scale. Be sure that no air bubbles are present, especially in the tip, when the initial reading is made. The burette is now ready to deliver solution. After running out any significant volume, allow at least 30 s for drainage from the burette walls before reading. Also remove any hanging drop by touching the tip against the wall of the receiving vessel. The volume delivered is the difference between the initial and final readings, subject to calibration corrections and temperature corrections if needed.

In reading the burette, avoid parallax error; the rings engraved all the way around the burette will help. Read the *bottom* of the meniscus to about a tenth of the smallest division. Some help in seeing the meniscus can be obtained from a white card containing a very broad horizontal black line, held behind the burette so that the line is reflected in the meniscus. Move the card upward until the top edge of the line and the bottom edge of its reflection become tangent in your field of view; read the position of the point of tangency.

In order to determine the equivalence point (the point at which exactly stoichiometric quantities of sample and titrant have been brought together), it is necessary to find a chemical or physical property that changes very rapidly at this point. Many properties have been used successfully, but the most common method is the visual observation of a color change in a chemical indicator present in very small concentration. This observable change takes place at the "end point," which must lie very close to the equivalence point. The technique of titration is concerned principally with approaching the end point with reasonable speed without "running over"; it is best learned by practice, but there are descriptions in the literature[1] that may be helpful.

Mixing of Solutions.   A special word should be said about *mixing,* particularly in the case of large volumes such as carboys of stock solutions for an entire class. It will not suffice merely to swirl the carboy for a few minutes. When the carboy is filled, enough air space (1 or 2 L) should be left to permit effective mixing. Most convenient for thorough mixing is an electric stirrer, comprising a motor with a long shaft and a swivel-mounted propeller that will go through the mouth of the carboy. This will produce adequate mixing in 15 to 20 min. If such a device is not available, the carboy should be stoppered tightly and turned onto its side on a table covered with a towel. It should then be *shaken vigorously* by a rapid back-and-forth rolling motion sufficient to distribute air bubbles throughout the volume. This should be continued for at least 1 or 2 min. After the carboy has been allowed to stand upright for a few minutes, samples should be withdrawn both from the bottom and from the top for titration.

PURIFICATION METHODS

Water.   Ordinary distilled water is pure enough for most purposes. In many situations, elimination of ionic impurities is the important goal, and the presence of neutral organic substances in trace amounts is not objectionable. Therefore, deionized water obtained from columns containing ion-exchange resins will serve well for almost all purposes. Dissolved air is objectionable for some applications and can be removed by boiling for a short period.

For conductance work, ions other than those resulting from the ionization of water itself must be reduced to the lowest possible concentrations. If distilled water is used, it should be triply distilled (ordinary distilled water distilled a second time from dilute acidified permanganate to oxidize organic impurities and a third time with a block-tin condenser from dilute barium hydroxide to remove volatile acids and CO_2). Conductivity water should be stored in polyethylene bottles in order to prevent leaching of soluble constituents from the surface of a glass bottle. Exposure to air should be minimized to prevent contamination with carbon dioxide.

Sodium Hydroxide.   For use as a reagent in acidimetric titration, sodium hydroxide must be freed from contamination by sodium carbonate, which forms rapidly on exposure of NaOH to air. Commercially obtainable reagent-grade sodium hydroxide needs no further purification if it is available in stick form, since any carbonate that forms on exposure to air can be quickly washed off with distilled water before dissolving the sticks in CO_2-free water (distilled water, reboiled if necessary) to make up the desired solution. This probably cannot be done effectively with the usual pellets. However, sodium carbonate is virtually insoluble in a saturated solution (15 to 18 M) of sodium hydroxide. Such a solution can be made up by stirring the solid hydroxide with cracked ice and allowing the resulting hot solution to cool. This syrupy liquid can be stored in a polyethylene bottle. When the insoluble carbonate has settled, the clear hydroxide solution can be drawn off as needed by pipette or siphon and diluted with CO_2-free water to the desired concentration. Sodium hydroxide solutions should not be stored for long in untreated glass flasks or bottles. Polyethylene bottles are satisfactory.

Handling of sodium hydroxide solutions at the high concentrations involved here entails extreme hazard of the destruction of the cornea if any solution gets into the eye. **Wear a face shield. Carefully avoid skin contact.**

GAS-HANDLING PROCEDURES

Many physical chemistry experiments involve the use of one or more gases such as oxygen, nitrogen, hydrogen, helium, argon, and carbon dioxide. We shall be concerned here with procedures for handling these gases.

Cylinders, Gas Regulators, Reducing Valves.   Although some gases can be prepared in chemical generators, see for example Exps. 37 and 43, it is far more convenient and desirable to obtain gases commercially in steel cylinders. These cylinders are available in several types and sizes. The large size is 51 in. in height and 9 in. in diameter; at 2200 psi and 70°F it contains approximately 280 moles of gas (~7000 L at STP). In addition, smaller cylinders and very small lecture bottles are useful for supplying special gases such as hydrogen chloride and acetylene. Note that many gas pressure gauges read in psi (pounds/square inch), where 14.7 psi equals 1 atm (and 14.5 psi equals 1 bar).

Each cylinder is delivered with a protective cap, which should be removed only when the cylinder has been chained against a laboratory table or a wall. **Cylinders should always be chained** to prevent upset, which has been known to cause violent release of the gas or even bursting of the cylinder, with serious consequences.

At the top of the cylinder are a needle valve and a threaded outlet. To clear the outlet of dust, the needle valve should be barely opened for an instant and reclosed. In most laboratory work a *regulator* is attached to the cylinder. This usually comprises a Bourdon gauge to indicate the cylinder pressure (up to 4000 psi), another gauge to indicate outlet pressure (ordinarily up to 60 psi), an adjustable diaphragm valve to regulate the outlet pressure, and an outlet needle valve to control the flow rate. In the diaphragm valve, the needle of a needle valve is attached to a flexible diaphragm. When the outlet pressure exceeds the desired value (controlled by a spring between the diaphragm and an adjusting screw), the motion of the diaphragm is such as to close this needle valve, shutting off the flow of gas from the high-pressure side; when the outlet pressure is too low, the diaphragm opens this needle valve.

The fittings and threads on the cylinder outlet are of several types, depending on the kind of gas, in order to prevent the wrong regulator or other fitting from being used with a given cylinder—e.g., an H_2 regulator on an O_2 cylinder.

The regulator fastens to the cylinder with a metal-to-metal contact, no gasket ordinarily being required except in the case of CO_2 cylinders. The connection is made tight with a large wrench. To verify that the seal is gas-tight, close the diaphragm valve, then open and close the cylinder valve and watch the cylinder pressure gauge for a few minutes.

Regulator Operation. When it is desired to use gas from a cylinder, follow the procedure given below:

1. Close the outlet needle valve and the diaphragm valve (turn the screw handle counterclockwise until it rotates freely).
2. Open the cylinder valve fully and check the cylinder pressure gauge to make sure that the cylinder is not empty.
3. *Make sure that the outlet needle value is closed,* and then slowly turn the handle on the diaphragm valve clockwise until the outlet gauge reads about 2 psi. Keep in mind that the outlet pressure gauge reads overpressure, i.e., psi *above* atmospheric pressure. 1 psi = 0.068 atm = 51.7 Torr.
4. In order to purge the system and the regulator chamber of air, open a vent valve and then *very slowly* open the outlet needle valve enough to achieve a reasonable flow rate. (If the gas is toxic or expensive, one can connect a vacuum pump to the vent valve and pump out the system and regulator.) After a brief purge period, close the vent valve, allow the system to fill, and then close the outlet needle valve. When one wishes to fill the system to an overpressure greater than 2 psi, *slowly* turn the diaphragm valve handle until the outlet pressure rises to the desired value.
5. When gas is no longer needed, close the outlet needle valve and the cylinder valve.

Do not allow a cylinder to be emptied down to 1 atm; there should be some over-pressure left when the cylinder is returned. Unneeded cylinders should be returned to the supplier to avoid needless demurrage (cylinder rental) charges.

A simple *reducing valve,* consisting of a needle valve with fittings to match the cylinder and a hose connector for the low-pressure side, can be used in place of a regulator for some purposes when gas flow through an open system is desired. This device is less expensive than a regulator, but its use requires more care because, if the system should accidentally be closed, the pressure will build up to whatever is required to burst it at its weakest point if this is less than the full cylinder pressure. It is advisable to connect the system to the reducing valve with a lightweight rubber tube that will easily blow off the hose connector if the pressure becomes too high, or to provide some other safety valve near the inlet end of the system.

Warning: Never connect a gas cylinder directly to a closed system that is not spe-cially designed to withstand high pressures (such as a "combustion bomb" for heats of combustion). Even when a regulator is used, remember that the diaphragm valve may slowly leak; arrange for some kind of safety pressure release, if only a rubber tubing con-nection that can be easily blown off.

Most gases undergo a Joule–Thomson cooling when they expand in a regulator or reducing valve. In the case of CO_2, the cooling at high flow rates is often large enough to be troublesome, causing frosting of apparatus or even clogging of the regulator or the reducing valve with solid. In addition to this effect, there may be a compressional heating of the gas if a significant pressure is built up quickly in some part of the system. If a gas must be maintained at a constant temperature, it should be passed through a long coil of copper tubing immersed in a constant-temperature bath. One hundred feet of $\frac{1}{4}$-in. copper tubing is adequate for flow rates up to about 5 L min^{-1}.

Needle Valves.   For control of gas flow at ordinary pressures, needle valves give much better control than stopcocks. The hard-steel tapered needle, at the end of a screw-threaded shaft, seats in a cylindrical hole so that the area of open space for gas flow is gradually increased or decreased on rotating the shaft. These usually do *not* provide a reli-able shut-off and are used for flow control only. Persons whose acquaintance with valves is limited to water faucets often damage needle valves by needlessly overtightening them when shutting off the flow. This results in a decrease in the sensitivity of control of the gas flow. *It is important not to exert any more force than necessary.*

Hoses.   Gases for open systems may be carried by ordinary $\frac{1}{4}$- or $\frac{5}{16}$-in. gum-rubber tubing. Closed systems may require heavy-wall rubber or plastic pressure tubing, which can safely be used with pressures up to several atmospheres. **Warning:** Do not subject glass apparatus containing bulbs more than 2 or 3 inches in diameter to internal pressures of more than 1 atm above the outside pressure.

Gas Purification.   Although in many cases the gas from the cylinder is sufficiently pure for direct use, for certain purposes it should be subjected to one or more purification procedures. Hydrogen is frequently contaminated with small amounts of oxygen, which should be removed if the gas is to be used in a hydrogen electrode. The most convenient procedure is to use a catalytic purifier such as Deoxo, which contains palladium; the oxy-gen combines with hydrogen to form water, which is subsequently removed with a drying tube if it is objectionable.

Another frequent contaminant is water vapor. This can be removed by passing the gas through a tube filled with a suitable drying agent. Gas flow should not be too fast, or the drying will be incomplete. Use of two or more drying tubes in series may provide better

TABLE 2 Common drying agents

Agent (Trade name)	Hydrated form	Comments
$Mg(ClO_4)_2$ (Anhydrone)	$Mg(ClO_4)_2 \cdot 6H_2O$	**Caution:** This is an oxidizing agent. Cloth and other organic materials impregnated with perchlorate can be flammable.
$MgSO_4$	$MgSO_4 \cdot 7H_2O$	Chemically inert. Commonly used for drying organic solvents; rapid and high capacity.
$CaSO_4$ (Drierite)	$CaSO_4 \cdot H_2O$	Inert, very rapid and efficient (H_2O vapor pressure 0.004 Torr above hydrate at 25°C); low capacity. Available as "indicating Drierite," containing some $CoCl_2$, which changes from blue to pink when hydrated. Relatively expensive but regenerated by heating in air to 250°C.
$CaCl_2$	$CaCl_2 \cdot 6H_2O$	Inexpensive drying agent with large capacity. Will liquefy on absorbing sufficient water.
P_2O_5	$(H_3PO_4)_n$	Rapid and efficient agent as long as P_2O_5 surface is exposed (not coated with H_3PO_4). **Caution:** P_2O_5 is toxic.
Molecular sieves (Linde type 5A)	Pores filled	Useful for drying gases. Can be reactivated by heating.

drying. A list of commonly used adsorbents for filling desiccators or drying tubes is given in Table 2.

Water Saturation. Gases to be used in systems containing water or aqueous solutions should be saturated with water before they are admitted to the sample cell. For this purpose a bubbler containing a fritted disk that disperses the gas in the form of very small bubbles is far superior to the ordinary laboratory bubbler. The temperature of the bubbler should be the same as that of the system, and the connection to the system should be as short as possible to avoid condensation of water from the gas before it enters the system.

Flow Meters. A wide variety of instruments are available for measuring the flow rates of liquids and gases in closed-tube systems.[4,5] Four general types are (1) differential-pressure devices, (2) variable-area devices, (3) velocity meters, and (4) mass meters. Important examples of each type are listed in Table 3, which contains information on the range of volume flow rates Q covered by a given style of instrument as well as the accuracy and range for any individual meter.

Differential pressure devices consist of two elements—one causes a *change* in the flow rate of a flowing fluid, which creates a pressure difference (Bernoulli effect) between two sections of the tube or pipe, and the second element measures the resultant Δp. Such a device is nonlinear, since Q is proportional to $(\Delta p)^{1/2}$. As a result, the range of Q values that can be covered by a given instrument is limited. A familiar example of a differential pressure device is the Venturi tube.

Variable area devices, such as rotameters, also involve a differential pressure; but they represent a distinct type of device, since the differential pressure is constant during operation. In the rotameter, a small "float" is contained in a vertical glass tube having a

TABLE 3   Flow meters for liquids and gases

Type	Range of Q (in L min^{-1})	$\dfrac{\text{Max}}{\text{Min}}$	Accuracy (reproducibility)	Viscosity effects	Gas use
1. Differential-pressure meter					
Venturi tube	100–10,000	4	1% FS	High	Y
Pitot tube	400–20,000	5	1–2% R (0.1% R)	Low	Y
2. Variable area meter					
Rotameter	0–600	10 (high Q) 100 (low Q)	2% R (0.5% R)	Medium	Y
3. Velocity meter					
Turbine/vane	0.25–2,500	15	1% FS (0.25% FS)	High	Y
Vortex	25–20,000	10	1% R (0.2% R)	Medium	Y
Electromagnetic	0–5,000	40	0.5–1% R	None	N
Ultrasonic	0–6,000	20	2–3% FS (0.2% FS)	None	N
4. Mass meter					
Thermal	—	10	1% FS (0.2% FS)	None	Y
Coriolis	—	10	0.4% R	None	N

Notes. These devices usually measure the volume flow rate Q. The mass flow rate $\dot{m}$ is given by ρQ and is measured directly by mass meters. The ratio max/min indicates the range of Q measurable with a given individual instrument. FS = full scale, R = reading, Y = yes, N = no.

tapered inside diameter (the cross-sectional area increasing with distance from the bottom of the tube).† When there is no flow, the float rests at the bottom of the tube. When gas or liquid is flowing upward in the tube, the float rises until the net force due to the fluid flow just balances the force of gravity. Thus the steady-state height h of the float depends on the rate of flow. This device can be designed to yield an almost linear response (i.e., $h \propto Q$), and rotameters are available in a wide variety of flow-rate ranges below 20 L min^{-1} for liquids or 600 L min^{-1} for gases.

Velocity meters measure the velocity v of fluid flow in a pipe of known cross section, thus yielding a signal linearly proportional to the volume flow rate Q. Mass meters provide signals directly proportional to the mass flow rate $\dot{m} = \rho Q$, where ρ is the mass density. Coriolis meters, which are true mass meters, can be used only for liquids. Thermal-type flow meters use a heating element and determine the rate of heat transfer, which is proportional to the mass flow rate. This type of device is used mostly for gas measurements, but liquid flow designs are also available.

Figure 2 shows a simple gas flow manometer that can be easily constructed in the laboratory. In this device a pressure difference, caused by viscous flow in a capillary tube (see Exp. 4), is measured by a simple manometer containing a colored organic liquid (such as dibutyl phthalate with added eosin). The response of this flow meter is very nearly linear over the range in which the gas flow in the capillary is laminar and may be roughly linear over a useful range even when the flow is turbulent. By variation of the capillary length and diameter and of the liquid density, a wide range of flow rates can be measured.

†A related simple device in which a plastic ball moves around a circular course is often used as a semiquantitative water flow indicator to monitor the flow of cooling water through a diffusion pump or bath circulating coil.

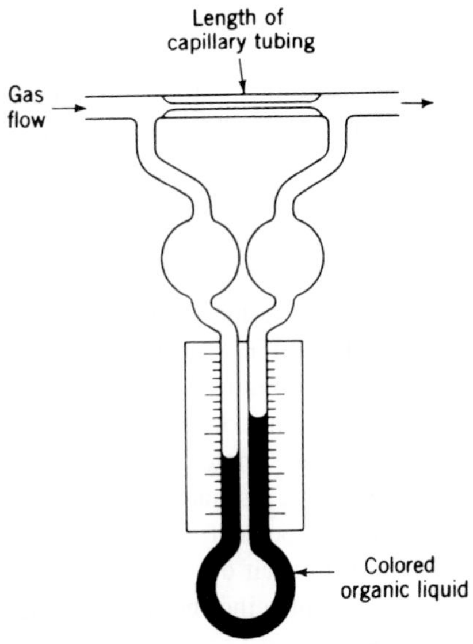

Length of
capillary tubing

Gas
flow

FIGURE 2
A typical flow manometer.

Colored
organic liquid

A steady gas flow calibrating a flow meter of this kind can be obtained from a needle valve attached to a regulator set to 5–10 psi gauge pressure or more (to avoid perturbations due to small variations in outlet pressure). At small flow rates the volume of gas flow over an interval of time measured with a timer can be determined with a gas burette; a three-way stopcock can be used to switch the gas burette in and out of the system. For larger flow rates a water-filled inverted graduated cylinder or volumetric flask, its mouth held under the surface of a water bath, can be used to collect gas from a rubber tube held underneath it for a time interval measured with a timer. For precise work, a correction should be made for the partial pressure of water vapor in the gas collected.

Pressure Gauges.[5,6] The most commonly used gauge for moderately precise measurement of pressures up to ~ 1000 bar is the *Bourdon gauge*. This gauge makes use of a curved metal tube of oval or flattened cross section; a higher pressure inside the tube than outside tends to straighten out the tube by a slight amount, producing a motion that is converted to a rotation of the indicating needle by a rack and pinion. In the best designs, the Bourdon tube is made of stainless steel and there is a mirror ring around the outer perimeter of the gauge dial so that the pointer can be read without parallax errors. Such gauges have an accuracy of ± 0.25 percent of full scale. The typical Bourdon gauge measures pressure relative to the ambient air pressure and has a scale in PSIG (pounds per square inch gauge). Thus a reading of 14.7 psi means an overpressure of 1 atm or an absolute pressure of 2 atm. Some gauges have scales indicating absolute pressure in psi (PSIA) or kilopascal (100 kPa = 1 bar = 0.987 atm). For low-pressure measurements (0–600 Torr) on dry gases, there are direct-reading gauges using flexible bronze diaphragms, usually corrugated, with appropriate mechanical linkages.

For more precise work, other devices may be used. The most familiar is the closed-tube mercury manometer (see Chapter XVIII). Clearly, the mercury manometer is convenient only for pressures that do not exceed ~ 1000 Torr (1.33 bar). Furthermore, manometer readings are time consuming and do not provide a convenient analog or digital output signal. Finally, concerns about health hazards associated with mercury (see Appendix C) have led to a reduction in the use of mercury manometers.

In recent years, electrical pressure transducers based on the use of strain gauges have been developed for use over ranges varying from 0–1 PSIG to 0–10,000 PSIG. The sensor is usually a thin diaphragm with a strain gauge mounted on it to detect deformations caused by a pressure differential across the diaphragm. The best stability is achieved using a stainless-steel diaphragm with a foil-type strain gauge bonded to it with epoxy cement (accuracy ±0.25 percent in the electrical output). This sensor is part of a complex compensation network needed to set the null (zero balance), correct for nonlinear characteristics of the strain gauge, and compensate for ambient temperature changes. The output can be displayed directly on an analog recorder or sent via an interface to a microcomputer. With microprocessor instruments, one can achieve conversion accuracy of ±0.05 percent full scale (including nonlinearity, hysteresis, and reproducibility errors) over ranges as small as 0–10 PSIG or as large as 0–10,000 PSIG. Moderate-cost instruments will perform almost as well, although with less flexibility in range selection and calibration checking. Piezoresistive silicon diaphragms have also been introduced. In these integrated-circuit devices, piezoresistors are diffused into a homogeneous single-crystal silicon diaphragm. A bend in this elastic diaphragm causes a change in the resistance of the embedded resistor, and the resulting accuracy can be as good as ±0.1 percent. Optical pressure transducers seem likely to become widely used in the future. In such devices the elastic deformation due to an applied pressure is detected optically, eliminating the friction-induced errors associated with conventional strain-gauge diaphragms.

Another type of electrical pressure transducer is a variable-reluctance device. The deflections of a diaphragm of magnetic stainless steel are sensed by two inductance pickup coils. With ac-bridge circuitry, this transducer delivers a full-scale output of 40 mV per volt for a variety of ranges from 0–4 Torr (0–0.08 PSIA) to 0–3200 PSIA.

For pressures in the range 500–10,000 bar, wire-wound *manganin resistance gauges* are widely used. The resistance change with pressure is rather low ($d \ln R/dp \cong 2.4 \times 10^{-6}$ bar^{-1}) but quite linear and stable. Absolute gas pressure measurements over a very wide range can be made with a *deadweight gauge*. This gauge measures the force exerted by a gas on a highly polished piston with a close sliding fit in a highly polished vertical cylinder, the mean diameter of which is known accurately. The effect of friction is largely eliminated by the presence of a lubricant and by a reciprocating mechanism that rotates the piston or the cylinder back and forth a few degrees around the vertical axis. The force due to gas pressure is balanced by weights placed on a pan supported by the floating piston. Pressures up to several thousand bar can be measured with high absolute accuracy (± 0.01 percent). The deadweight gauge is used mostly for the calibration of more convenient secondary pressure gauges. Other pressure devices, such as capacitance manometers (which are treated in Chapter XVIII), will not be discussed here.[5]

Manostats. A manostat is a device for maintaining the pressure in a system at a constant value. A gas regulator valve like that described on p. 645, operating in conjunction with an outlet leak, constitutes a simple manostat that prevents the system pressure exceeding a preset value. With the outlet open to the atmosphere, control pressures lie in the range 1–5 atm. More precise manostats, operating on the feedback principle, can be based on bellows (0–300 bar), diaphragms (0.3–100 bar), or pistons (30–200 bar) as the sensing element. The motion of the sensor is then coupled to an electrical control device. In the simplest designs this is a single-pole double-throw (on–off) switch, but proportionating controls can also be used.

Precautions. In addition to several general precautions given above for gas handling, certain particular precautions should be followed in handling hydrogen and oxygen.

Hydrogen is very flammable and also forms explosive mixtures with air over a wide range of compositions. If it is used in any quantity, the effluent hydrogen should be vented out of doors by a tube through a window. This is not essential for the slow rates of flow necessary with a hydrogen electrode, but good ventilation is important; *there should be no open flames.*

Oxygen is hazardous when in contact with flammable substances, particularly oil. *No oil or other organic substance should be allowed to come into contact with oxygen under pressure.* Oxygen lines, valves, and regulators must be kept scrupulously free of oil.

ELECTRODES FOR ELECTROCHEMICAL CELLS

We describe below the electrodeposition procedures required in the preparation of the platinum and silver–silver chloride electrodes used in electrochemical experiments. We shall not attempt to give a general treatment of electroplating or of electrode preparation. Further details can be found in various monographs.[7]

Platinum Electrodes. Platinum electrodes are used in conductivity cells and in hydrogen electrodes. For these purposes they are usually covered with a deposit of platinum black to increase the surface area. For certain other purposes, such as use in redox electrodes, this is usually not necessary.

Platinum can be used in the form of thin sheet or screen. Platinum wire can be welded to a small square of sheet or screen by placing it in position on a metal surface, heating to red heat with a torch, and striking lightly with a small ball peen hammer. The platinum wire can then be sealed into the end of a piece of soft-glass tubing. Electrical contact can be made by copper wire spot-welded to the platinum lead wire before the glass-to-metal seal is made.

Platinum can be plated onto other metals by use of a solution containing 1 g platinum diammino nitrate, 1 g sodium nitrate, 10 g ammonium nitrate, and 5 mL concentrated ammonium hydroxide in enough water to make 100 mL of solution. The cleaned metal is made the cathode, and a strip of platinum metal is the anode; a 4.5-V battery, a rheostat, and a milliameter are connected between the electrodes, and the plating is carried out at a current density of 50 to 100 mA per square centimeter of cathode surface.

For depositing platinum black, a solution of 3 g platinic chloride and 0.2 g lead acetate in 100 mL of distilled water is prepared. This can be kept in a glass-stoppered bottle and used repeatedly. The platinum electrode should be treated with warm aqua regia (one part concentrated HNO_3 to three parts concentrated HCl) to clean the surface and if necessary to remove old platinum black. It is then rinsed thoroughly with distilled water. While it is still wet, the electrode is immersed in (or the cell is filled with) the platinizing solution. If there is only one electrode to be treated, platinum wire will serve as the anode. A 3-V battery and a rheostat in series are connected between the electrodes, the electrode to be platinized being the cathode (negative). The rheostat is adjusted so that gas is produced only slowly. If both electrodes are to be platinized, the polarity is reversed every 30 s. The electrolysis should be stopped as soon as the electrodes are sooty black; an excessive deposit should be avoided. The platinic chloride solution is then returned to its stock bottle; the electrodes are rinsed thoroughly in distilled water, and electrolysis is continued with a very dilute solution of sulfuric acid in order to remove traces of chlorine. After a final washing with distilled water, the electrodes are ready for use. Pending use, they should be stored in contact with water; *they should never be allowed to dry out.*

Silver–Silver Chloride Electrodes. Silver–silver chloride electrodes may be made from thin sheet silver of high purity, but it is perhaps better to plate silver onto a clean square of platinum sheet or screen. This is made the cathode, and the anode is a strip of very pure sheet silver (at least 99.95 percent pure) in a plating bath containing 41 g silver cyanide, 40 g potassium cyanide, 11 g potassium hydroxide, and 62 g potassium carbonate per liter. Use a 4.5-V battery connected in series with a rheostat and milliammeter. Set the rheostat to about $1000\ \Omega$ and make electrical connections to the electrodes *before* immersion; then adjust the rheostat so as to obtain a current density of about 5 mA per square centimeter of electrode area. Plate for a few hours. Remove the electrodes, and wash very thoroughly with distilled water to remove all traces of cyanides.

The silver-plated electrode is "aged" in an acidified solution of silver nitrate and is then made the anode (and a platinum wire the cathode) in a 1 *M* HCl solution at a current density of 5 to 10 mA per square centimeter of electrode area. In a few minutes a brownish coat of silver chloride will appear and the electrolysis can be stopped. The electrode should be aged for a few days in distilled water before use, and *it should never be allowed to dry out*. After long periods the potential of the electrode changes, probably owing to crystal growth. The old silver chloride coating can be removed with ammonia or cyanide; after washing, a fresh coating can be prepared.

MATERIALS FOR CONSTRUCTION

"They should know enough not to build a bridge out of sodium."

Anonymous MIT Civil Engineering professor

This section describes a wide variety of materials that can be used successfully in the construction and repair of laboratory apparatus. Considerable details about the construction of scientific apparatus are given in monographs[8] and specialized journals.[9]

Glass.[10] The transparency, low thermal conductivity, and chemical inertness (except to HF and strongly basic solutions) of glass make it a valuable material for constructing many pieces of apparatus. Because of its low thermal expansion coefficient, Pyrex glass can be blown into many complex shapes without too great a danger of cracking due to internal strains. A major disadvantage of glass is its breakability, but this problem may be reduced by the application of Lexan polycarbonate resin to glass objects. Glasses do not have a sharp melting point but soften over a broad temperature range. The softening ranges for typical glassy materials are given in Table 4, together with other characteristics.

A glass object can be used for indefinite periods below its annealing temperature but only for very short intermittent periods at the high end of its softening range. Fused silica is obviously very useful for high-temperature apparatus; and Vycor, which contains a few percent of other oxides, is almost as good. At elevated temperatures fused quartz will devitrify, i.e., crystallize and become opaque.

Soft glass has a higher expansion coefficient than Pyrex and is much more subject to thermal stress. It is useful mainly as a glass that will produce a vacuum-tight seal with platinum. Pyrex does not make satisfactory glass-to-metal seals with most metals (tungsten is an exception if it is cleaned with sodium nitrite in a flame); however, vacuum-tight Pyrex–Kovar seals are readily available. Various thermosetting cements, such as quartz

TABLE 4   Characteristics of various glasses

Type	Composition	Softening range (°C)	Annealing temp. (°C)	Linear thermal expansion coef. (K^{-1})	Refractive index at 589 nm
Soft	Soda lime	400–700	510	9×10^{-6}	1.51
Pyrex	12% borate	500–800	560	3×10^{-6}	1.47
Vycor	3% borate	800–1500	890	8×10^{-7}	1.46
Fused quartz	Silica	1400–1700	1200	6×10^{-7}	1.46

TABLE 5   Mechanical properties of metals

Metal	Tensile strength (units of 10^3 lb/in.^{2})	Young's modulus of elasticity (units of 10^6 lb/in.^{2})	Melting point (°C)	Density (g/cm^3)
Aluminum (pure)	8.5	8–11	660	2.7
Duralumin	88.2	10	~640	3.0
Copper	33	18	1082	8.9
Brass (67 Cu–33 Zn)	66.8	13	940	8.4
Cast iron	16	15–20	1200	7.4
Mild steel	50–60	~20	1500	to
Carbon steel	200	28	1430	7.9
Stainless steel	250–300	28	1450	
Silver	40	11.2	961	10.5
Gold	36	11.4	1063	19.3
Platinum	48	24	1773	21.5
Monel	75–170	24–26	1350	8.9
Tantalum	50	27	2996	16.6
Tungsten	175	60	3410	19.3

cement and "solder glass" for soft glass, are also available commercially. Fused silica has high tensile strength and very low mechanical hysteresis; these properties make quartz fibers useful for certain applications. Glass transmits light from about 320 nm to 2 μm, thus covering the entire visible range. If ultraviolet transmittance is required, fused silica can be used for wavelengths down to 200 nm.

Metals.[11]   Hot-rolled *steel* is the least expensive metal for construction purposes and is available in sheet, strip, rod, girder, and also in angle form, all of which are useful in the laboratory for constructing stands and frameworks. To prevent rusting, it should be galvanized (zinc-dipped), cadmium-plated, or at least painted. Several commercial primers are effective for the retardation of rusting; they can be covered with additional coats of ordinary paint if desired. Cold-rolled (low-carbon) steel is also used for heavy-duty structural purposes, and owing to its excellent machining properties it has many varied uses in apparatus construction. It can be galvanized or plated with cadmium, copper, nickel, chromium, or any of several other metals. Copper should be plated on steel before nickel, and both before chromium. Steel parts can be joined by welding, silver soldering, soft soldering, or copper brazing (if previously copper- or nickel-plated). High-carbon steels, such as tool steels (e.g., drill rod), are less easily machinable but have the advantage that they can be hardened by heat-treatment. The mechanical properties of a number of commonly used metals are summarized in Table 5.

Stainless steels are in a special class, differing from other steels in being nonmagnetic and essentially free from rusting and corrosion. A typical stainless steel (type 316) contains 17 percent chromium, 12 percent nickel, and 3 percent molybdenum. The chemical resistance of stainless steel makes it very attractive for many purposes, but its cost and the difficulty of machining it limit its use somewhat. It can be silver-soldered or welded; soft soldering is difficult.

Aluminum and its alloys are excellent structural metals, with good machinability, fair corrosion resistance, good electrical and thermal conductivity. It is readily sand cast or die cast. Aluminum is seldom plated; it can be anodized (to produce a thick oxide layer) and then dyed or painted. Aluminum is difficult to weld owing to its flammability. Soldering is also difficult but can be aided by tinning with indium metal. The low melting point of aluminum (660°C) somewhat restricts its application.

Aluminum and its alloys are often used in the laboratory in sheet form for making electrical chassis and panels and in rod form for constructing frames for apparatus support. Aluminum foil is useful for heat-reflecting shields in low-temperature work. A common structural alloy is Duralumin (Dural), which contains about 4 percent copper and traces of manganese and magnesium. Duralumin that has been heated to 530°C and quenched in water is ductile for $\frac{1}{2}$ h or more and can be readily cold-worked; thereafter the alloy hardens and attains considerable strength. The hardening can be delayed for long periods (for rivets, etc.) by storage at Dry Ice temperatures.

Copper is used where its high electrical and thermal conductivity, its malleability and ductility, and its ease of soft soldering confer advantages. OFHC (oxygen-free, high-conductivity) copper should be used where highest conductivity is required or when employed in a vacuum system, particularly when soldering or welding is to be done in a hydrogen atmosphere. Soft copper is exceedingly difficult to machine. It is subject to oxidation and should not be heated above 100 to 200°C for any length of time except in a reducing atmosphere or unless adequately plated. Copper can be joined by brazing, silver soldering, or soft soldering. Copper amalgamates readily with mercury.

Brass is basically a copper–zinc alloy; *bronze* is a copper–tin alloy. In practice both often contain many other metals. Their high machinability, resistance to corrosion, and ease of soft soldering make them very useful in apparatus construction. Owing to the volatility of zinc, brass should not be used in vacuum components that must be baked out or operated hot. Certain bronzes such as phosphor bronze are useful for springs and diaphragms; beryllium copper is also useful in these applications.

Monel and *Inconel* are basically Ni−Cu−Co alloys containing small amounts of iron and manganese. These alloys have fair machinability and excellent corrosion resistance even at elevated temperatures; some are magnetic, others are not. They can be brazed, silver soldered, and soft soldered. Monel and cupronickel are very useful in the construction of apparatus for low-temperature work owing to their low thermal conductivities. Inconel can be used for heating elements; Nichrome (Ni−Cr or Ni−Cr−Fe−Mn) and Chromel (Ni−Cr−Fe) are also useful for this purpose.

Invar (64 percent iron, 36 percent nickel) has a very low coefficient of thermal expansion ($1 \times 10^{-6}\,K^{-1}$), which makes it a useful material for laser resonators and other devices in which dimensional stability is important. It is magnetic and only moderately corrosion resistant. *Kovar* (53.7 percent iron, 29 percent nickel, 17 percent cobalt, 0.3 percent manganese) is useful for vacuum-tight glass-to-metal seals, and such seals are available commercially in a variety of sizes.

Silver is an excellent conductor of heat and electricity and a good reflector of light. It is relatively immune to oxidation but becomes tarnished by exposure to sulfur compounds in exceedingly small concentrations. It is an excellent electroplating metal and can also be deposited in thin films by evaporation. In Dewar flasks and other vacuum glassware, it is deposited from an aqueous medium by the Brashear process. Silver is an excellent brazing material and an important constituent of "silver solder." The term silver is often applied to alloys of silver with copper; for example "Sterling" silver contains 7.5 percent copper. "Fine" silver is 99.9+ percent silver.

Platinum and *palladium* are useful because of their chemical inertness, electrical conductivity, high reflectivity, and high melting point; their very high cost restricts their use to applications in which only small amounts are employed: electrical contacts, suspension wires, heating elements, radiation shields, etc. They absorb hydrogen; palladium is very permeable to it and may be mechanically damaged by exposure to it. These metals easily spot-weld to themselves and to each other. *Gold* is also very useful because of its inertness. It is an excellent plating material and can be deposited by vacuum evaporation or chemical deposition.

Tungsten has the highest melting point of any known metal (3380°C) and is useful for heating elements and various vacuum-cell components. It is difficult to work and is usually handled in the form of wire or ribbon. Tungsten wires tend to have a fibrous structure; lead-through tungsten–Pyrex seals may not be vacuum-tight unless one end of the wire is coated with nickel. Tungsten spot-welds to itself and to nickel and tantalum. Tungsten wires will oxidize if heated in air. Molybdenum and tantalum are much more easily machinable and workable, are chemically resistant, and also have high melting points.

Polymeric Materials.[12]   There now exists a vast array of polymeric materials, both natural and synthetic. We shall concern ourselves here only with a few that are of particular usefulness in laboratory apparatus construction. Mechanical and chemical properties of a number of commonly used plastics are summarized in Table 6, and brief comments are given below.

Rubber is one of the most commonly encountered polymeric materials. In its vulcanized form it is used in rubber tubing and rubber stoppers. A sulfur coating may appear on the surface of such stoppers in the course of time. For tubing (other than pressure tubing), pure gum rubber is preferable, although it must be replaced more frequently. Gum rubber can be used up to ~100°C; vulcanized rubber is useful up to ~150°C.

Neoprene (du Pont) is a rubberlike material that is a polymer of 2-chloro-1,3-butadiene. Somewhat less flexible than natural rubber, it has greater resistance to oils, greases, hydrocarbon solvents, and other chemicals. Neoprene is useful for gaskets, O-rings, and tubing.

Tygon is a convenient substitute for rubber tubing in the laboratory. This transparent vinyl-plastic tubing is a compounding of polyvinylidene chloride with certain liquid plasticizers. This tubing is tough and flexible and makes a very good seal with glass fittings. It tends to become yellow with age, and after long exposure to water it may become somewhat milky. It is attacked by many organic solvents but has fairly good resistance to most other ordinary chemicals.

Lucite and *Plexiglas* (polymethyl methacrylate as marketed by du Pont and by Rohm and Haas, respectively) and *polystyrene* are transparent thermoplastic materials. Their machinability is fairly good but somewhat limited by their thermoplasticity. They are strongly attacked by solvents such as acetone. They can be cemented with solvents alone (trichloroethylene) or with such cements as Duco. Over long periods cracks may develop at points of strain, and discoloration may result from prolonged exposure to strong light.

Nylon (du Pont polyhexamethylenediamine adipic polyamide) is available in solid form as well as fiber and sheet. It has high strength and mechanical stability, excellent machinability, and low surface friction: it is excellent for bearings, small gears and cams, etc., in which it can be used with minimal lubrication or none at all. Nylon fibers and threads are useful in apparatus construction.

Teflon (du Pont polytetrafluoroethylene) is a somewhat more flexible solid that is virtually unsurpassed in chemical inertness, electrical insulating properties, and self-lubricating qualities. It is available in the form of rod, tube, tape, and sheet and is readily machinable. It is useful for gaskets and bushings, unlubricated vacuum seals for rotating shafts, etc. It can be used in dynamic high-vacuum systems if these are not baked out much above 130°C. **Caution:** When Teflon is heated to decomposition, it reportedly gives off fumes that are extremely toxic. After machining it clean up all chips and scraps at once. *Viton* (du Pont polydifluorodichloroethylene) has properties very similar to those of Teflon and is used for O-rings and greaseless stopcock bores.

TABLE 6 Properties of common plastics

	Plexiglas or Lucite	Saran	Polystyrene	Polyethylene	Teflon	Silicone rubber
Composition	Polymethyl-methacrylate	Polyvinylidene chloride	Polystyrene	Polyethylene	Polytetra fluoroethylene	Organo polysiloxane
High-temperature limit, °C	60	66	72	Conventional, 60 Linear, 120	260	320
Chemically resistant to:	Aqueous acids and alkalis, most oils	Aqueous acids and alkalis, most oils, some organic solvents	Aqueous acids and alkalis, alcohols	Aqueous acids and alkalis, most oils, hydroxylic solvents	Almost everything	Aqueous solutions, alcohols, ketones, some oils
Chemically sensitive to:	Alcohols, ketones, esters, aromatic hydrocarbons, halogenated hydrocarbons	Ketones, aromatic hydrocarbons, oxidizing acids	Most organic solvents	Hydrocarbon solvents, some ketones and esters	Almost nothing	Strong acids and alkalis, hydrocarbon solvents
Machinability	Good	Poor	Fair	Poor	Good	—
Tensile strength (units of 10^3 lb/in.^{2})	4–10	0.85–9	3–10	1.3	1.8	0.3–0.6
Young's modulus (units of 10^6 lb/in.^{2})	0.5	0.2–2.0	0.2–0.6	0.02	0.06	—
Other properties	High optical quality, high electrical resistivity, good mechanical strength	Fireproof (but gives off toxic HCl fumes), poor mechanical strength		Poor mechanical strength, flexible	Fireproof (but gives off toxic fumes), low friction coefficient	Poor mechanical strength (rubbery)

Saran (Dow polyvinylidene dichloride) is a tough, chemically resistant plastic available in a variety of forms that are useful in the laboratory. Saran pipe or tubing can easily be welded to itself or sealed to glass and is useful for handling corrosive solutions. Thin Saran film, available commercially as a packaging material, is useful for windows, support films, etc. *Mylar* (du Pont polyethylene terephthalate) film and other polyester films are also useful for these purposes. Mylar is chemically inert and has excellent electrical properties for electrical insulation and for use as a dielectric medium in capacitors. Much thinner than these are films that can be made in the laboratory by allowing a dilute ethylene dichloride solution of Formvar (polyvinyl acetal) to spread on a water surface and dry.

Delrin (du Pont polyformaldehyde) has excellent chemical, mechanical, and electrical properties. It is easily machineable, tough, and resilient even at low temperatures. Although there are no good adhesives for joining Delrin pieces, it is weldable and easily molded. Chemically stable enough for continuous use at temperatures up to 85°C, it is also inert to most organic solvents.

Polyethylene and *polypropylene* are somewhat similar to Teflon but are inferior in chemical resistance and many other respects. They are useful in the form of bottles, flasks, and beakers for containing such reagents as hydrofluoric acid, strong bases, etc., that attack glass. Polyethylene tubing is much less flexible than rubber or Tygon but more flexible than Saran; it can be used for handling caustics, corrosive gases, etc. Polyethylene film has better chemical resistance than Saran and Mylar but lower strength and poorer optical properties.

Thermal Insulating Materials. Vermiculite (expanded mica) is useful for insulating ovens and furnaces when the maximum temperatures are not too high. For high-temperature applications, alumina (Al_2O_3) or magnesia (MgO) in the form of compacted powders can be used for insulation up to 1550°C and 1675°C, respectively. For uncompacted powders, the upper limits are approximately 100°C higher. Braided fiberglass is suitable for use up to 480°C, and special high-temperature glass braids can be used up to 720°C. Teflon sleeves or tapes will provide insulation up to 260°C. Fabricated high-temperature parts can be made from boron nitride and other machinable ceramics.

At room temperature and below, foamed plastics are very effective insulators. Styrofoam, a rigid prefoamed material, is useful for making Dry-Ice cooling chambers and is very easily cut into any desired shape. Silicone foam is generally available in a soft and flexible form. Foam-in-place insulation is also handy for awkwardly shaped applications. Typically this is based on a polyurethane resin containing a "bubbling agent," i.e., a pair of reactants that, when mixed, release N_2 or CO_2 gas into the polymerizing resin and thus produce a foam.

At very low cryogenic temperatures, the best insulation is a vacuum jacket that is silvered to eliminate radiative heat leaks. Various kinds of "superinsulation" exist and are used to store and transport volatile cyrogens like liquid helium. In one type a fine insulating powder is placed in the vacuum jacket; in another type a metallized variety of Kapton film is used.

Lubricants and Coatings. Most of the lubrication problems encountered in the laboratory can be solved with SAE 10 or SAE 20 machine oil, except for ball bearings subject to heavy loads, for which cup grease should be used.

Teflon self-lubricating plastic bearings can be used in situations where a lubricant might contaminate the system. There are also a variety of nonoily lubricants such as silicone lubricant (usually dispensed as a spray) and molybdenum disulfide (suspended in petroleum distillates that evaporate to leave a dry coating). Molybdenum disulfide is an excellent alternative to graphite for lubrication of moving parts in a high-vacuum system.

Most of the materials that we have discussed—glass, stainless steel, aluminum, brass, plastics—require no protective coating. Parts made of mild steel, welded angle iron, or plywood should be given a coat of paint, which is most conveniently done using du Pont Krylon acrylic aerosol sprays. Clear polymer varnishes and Teflon coatings can also be applied in spray or liquid form.

SOLDERS AND ADHESIVES

The joining together of various parts of an experimental apparatus is a frequently encountered task. This section describes two major approaches to this task: the soldering together of metal parts and the cementing together with adhesives of a wide variety of materials.

Solders. Soldering is the technique of joining two metals with a fused metal of lower melting point that wets both surfaces.[11,13] *Soft solder* is a low-melting alloy of lead, tin, and often other metals (e.g., bismuth or antimony). It is commonly used in making electrical connections, making connections between copper or brass tubes or fixtures by means of "sweat fittings" (such as sleeves, elbows, or tees), assembling small parts of an apparatus, and occasionally assembling large parts (when reliance is not placed on it for much mechanical strength). It can be used for ordinary vacuum work but is not satisfactory for high-vacuum work, since its low melting point ($\sim 200°C$) is lower than most bakeout temperatures. Indium solders are especially handy for low-temperature use (below 150°C).

Silver solder is a common form of *hard solder;* it is an alloy of silver and copper, usually with zinc and sometimes tin or cadmium added. Silver solders melt in the range 600 to 700°C and have high mechanical strength.

The three requirements for successful soldering are clean surfaces, the correct flux, and sufficient heat. The first and crucial step is to clean thoroughly the parts that are to be soldered. A solder joint is easier to make and more reliable if the surfaces are initially clean and bright (free from dirt, oil, and oxide coatings). It is also important to use plenty of soldering flux. The function of this flux is to free the surface of any residual contamination and to protect both the surface and the solder from oxidation.

In electrical wiring soft solder is commonly used with noncorrosive rosin fluxes and soldering pastes. Soldering of electrical connections with rosin-core solder is easiest when both parts to be joined have been "tinned," i.e., wetted with a coating of solder. Hookup wire is often pretinned, as are the leads of resistors and capacitors, the lugs on terminal strips, etc. Copper wire usually requires no tinning but may require scraping to remove Formvar or other lacquer coatings (which are often invisible) or oxide and dirt. A tinned soldering iron or soldering gun carrying a drop of solder is applied to the connection and held there until the solder spreads over the metal by capillary action; more solder can be applied if necessary, but the minimum required amount should be used.

Soft-soldering of copper, iron, steel, and brass objects of large size is usually accomplished by the use of a burner or hand torch, with an acid flux—frequently a concentrated aqueous solution of zinc chloride and ammonium chloride (2:1 ratio)—brushed onto the hot metal concurrently with the addition of solder. Initially, heat should be applied around the area to be soldered but not directly on the region to be soldered. If the region to be soldered is heated too much without the application of flux, an oxide coating will form that can make proper surface adhesion impossible. The surfaces to be joined should be pretinned if possible, and excess solder is shaken off or wiped off with a cloth. The two surfaces to be joined are then placed in contact and heated with the torch until the solder begins to flow; more solder is then added as required, and the pieces are allowed to cool undisturbed. The finished work should be washed thoroughly with water to remove the flux.

Hard soldering, or silver soldering, requires the use of a hand torch and a soldering paste of borax and boracic acid. A fluoride flux, Aircosil, can also be used. The pieces to be joined are placed in contact, and the surfaces are brushed with the flux during heating. Solder is applied as soon as the work is hot enough to melt it and encouraged to spread if necessary by further application of flux. Silver solders containing zinc or cadmium should not be used in vacuum systems that must be baked out or operated hot. **Caution: Zinc and cadmium fumes are toxic; make sure there is adequate ventilation.**

The cleanest and most reliable joints in vacuum systems are made by *hydrogen furnace brazing.* The parts to be assembled are clamped together with a thin gasket or sheet of the brazing alloy (silver or gold solder) between them. They are heated in a hydrogen atmosphere until the brazing alloy melts, runs, and wets both metal surfaces. No flux is required.

Welding techniques are many and varied[13] and will not be discussed in detail. The ideal construction material and joining technique for vacuum apparatus and many other systems involves stainless-steel parts connected by heliarc welding (on the *inside* of the joint). However, stainless steel is difficult to machine, and the welding must be done by an experienced welder. A completely different welding technique, called cold welding, involves the use of pressure rather than heat. This requires a soft welding material that flows easily and "wets" the metal surface well. Indium, a soft metal that melts at 155°C, can be cold-welded by the application of modest loads with a handpress and makes a convenient vacuum-tight seal without heating the apparatus above room temperature. Tin is somewhat more difficult to cold-weld but can be used up to 230°C.

Adhesives. The use of adhesives to cement together pieces of apparatus has become a common technique for a wide variety of situations. A general class of adhesive material is *epoxy cement,* consisting of a reactive epoxide resin liquid that must be mixed with a "curing agent" activator (generally an alkylamine) before being applied. These cements set irreversibly to form tough, adherent solids that bond well to metals (including aluminum, which is difficult to solder), glass, wood, and some plastics. Epoxy seals are relatively inert to chemicals and solvents and are usually poor conductors of heat and electricity. However, many special varieties of epoxy are available: thermally conducting epoxies, electrically conducting epoxies, optical epoxies that are transparent, vacuum epoxies with very low residual vapor pressures, metal-filled epoxies that can be machined. A wide range of strength and hardness can be achieved by varying the resin/activator ratio and the curing time. It is advisable to roughen the surfaces to be joined slightly with emery paper prior to applying the adhesive. Two epoxy resin cements suitable for laboratory use are Araldite (Ciba Limited) and Omegabond 100 (Omega Engineering). **Caution:** Epoxy resins should be considered to be toxic until thoroughly set; skin contact should be avoided.

A special-purpose adhesive that forms an exceedingly strong bond to metals and many other materials, including careless experimenters, is cyanoacrylate *contact cement,* available as Eastman 910 Adhesive and Super Glue (Woodhill Chemical). It is expensive, but only small amounts are needed. In this cement, the material polymerizes rapidly without an added activator. It is not void filling, and a 0.001 in.-thick film yields the best bond. Do not buy more than you will need in a few months since it slowly deteriorates: store in the refrigerator.

For affixing devices such as heaters, strain gauges, and thermometers to experimental cells, one wants a thin adhesive layer that is stable over the temperature range of projected use but need not be very strong mechanically. *GE varnish* (General Electric product no. 7031) is an excellent adhesive that is widely used over the range from helium temperatures (4 K = −269°C) to above room temperature (∼150°C). A hot-curing epoxy adhesive Omegabond 200, available from Omega Engineering,[6] is useful over the range −270°C to

+310°C. Although it is expensive, it gives a strong and rigid bond that can meet the special needs of strains gauges.

Silicone rubbers, such as General Electric RTV, represent a useful type of flexible adhesive. They can be clear or opaque, are fairly inert chemically, and some are usable up to 300°C or above. These silicones are just poured or extruded from a tube, and no heat cure is required since they cure by air drying. The resulting seals are not mechanically strong but will provide protection against water and oil.

In cases where vacuum leaks cannot be properly repaired, sealants such as *Apiezon W* and *Glyptal* can be useful. Apiezon W (black waxes of petroleum origin) softens or "melts" on warming to 50 to 150°C, flows readily on warm surfaces, and sticks well to clean glass or metal surfaces. It has a fairly low vapor pressure, but its exposure on the surfaces of vacuum systems should be kept to a minimum. Glyptal (General Electric glycol phthalate), a lacquer with or without added pigment and with a solvent such as xylene, is useful for a wide variety of vacuum-sealing applications. It adheres well to clean glass or metal and has a relatively low vapor pressure after baking. It should not be heated above 150°C. Glyptal hardens very slowly because it forms a surface film that retards evaporation of the solvent from beneath the surface. It should be allowed to dry for several days at room temperature or for 12 h under an infrared lamp. For short-term emergency repair of a gross leak, the plumbing product Duct-Seal can save an experiment that might otherwise have to be terminated.

For high-temperature applications, *sauereisen cement* (Omega CC cement) and *zinc oxychloride* (dental cement) are useful irreversible cements. Sauereisen cement is made by suspending ceramic powders in sodium silicate solution ("water glass"). This cement sets very hard and withstands temperatures up to 1000°C. Zinc oxychloride is made by mixing calcined zinc oxide powder with concentrated zinc chloride solution. One can also use a ceramic putty (Omega CC high-temperature cement), which must be cured at 180°C and is then serviceable up to 850°C.

TUBING CONNECTIONS

It is often necessary to seal joints between dissimilar materials against leakage. A number of materials are available for doing this.

Metal tubing is often terminated in a pipe thread. This is an accurately tapered and threaded unit, and an assembly must be well tightened with a pair of wrenches in order to seat the joint. Small leaks can be sealed by wrapping the inner member with a thin film of Teflon tape, available as Teflon Tape (du Pont) or Tape Dope, before assembling the joint.

Various types of compression fittings are used as an alternative to pipe threads. The Swagelok fitting (Fig. 3) and Gyrolok fittings of similar design are more expensive than the standard types but can be disassembled more easily. They can also be used to couple wide varieties of tubing by using the combinations of materials shown in Table 7. It

FIGURE 3
Typical Swagelok
compression fitting.

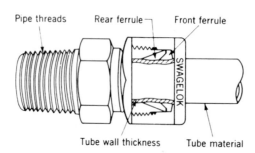

Pipe threads   Rear ferrule   Front ferrule

SWAGELOK

Tube wall thickness   Tube material

TABLE 7   Recommended materials for Swagelok fittings

Tubing	Threaded body	Front ferrule	Rear ferrule	Nut
Glass	Brass or stainless	Teflon	Nylon	Brass or stainless
Teflon	Teflon	Teflon	Stainless	Brass
Polyethylene	Brass	Nylon	Nylon	Brass
Tygon	(Either metal, Nylon, or polyethylene fittings can be used, but a special tube insert must be included.)			

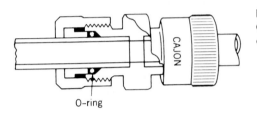

FIGURE 4
Cajon Ultra-Torr
compression fitting.

should be emphasized that inner and outer elements of the different types of fittings are *not* interchangeable.

Ultra-Torr fittings (Fig. 4) seal by means of an O-ring, usually made of Viton, and are especially useful in making glass–metal–plastic seals in simple vacuum systems. Viton is a rubberlike material with a very low vapor pressure, and Viton O-rings can be baked to ~200°C. Teflon is also occasionally used as a material for O-rings.

Glass pipe and fittings are available for applications where clean inert surfaces are required or where one wants to see what is going on inside the system. Glass tees, crosses, and unions with O-ring connectors (Ace Glass, Kontes) can be joined together or to metal apparatus to form vacuum systems and other apparatus. Viton and silicone O-rings are sufficiently inert for most applications, but Teflon O-rings can be employed when necessary.

SHOPWORK

Experimental work in physical chemistry is not limited to the assembly of standard pieces of commercial apparatus and the making of measurements with them. New or modified apparatus is frequently needed. This must be designed and must often be constructed by the experimenter. One must therefore be acquainted not only with the appropriate materials for constructing apparatus but also with many of the techniques for operating machine tools.

The most important machine tools are the drill press, the band saw, the grinding wheel, the lathe, and the milling machine. The first three of these are very simple basic tools that every experimenter should learn how to use. The lathe and the milling machine are more complex and more expensive tools of great value in constructing special apparatus; if possible, anyone interested in physical chemistry research work should learn how to operate these tools also. At the very least, it is necessary to understand the principles of their operation and to appreciate what they can and cannot do. Indeed, a familiarity with machine tools is a vital part of the ability to design complex apparatus properly. No attempt will be made here to describe the operation of any machine tools. Written descriptions are available elsewhere,[8] but it is of great importance to obtain basic instruction in the use of these tools from a qualified machinist in order to avoid risk of personal injury to the operator and damage to the machines.

REFERENCES

1. D. A. Skoog, D. M. West, and F. J. Holler, *Fundamentals of Analytical Chemistry,* 7th ed., Harcourt College, Fort Worth, TX (1996).

2. I. M. Kolthoff and P. J. Elving (eds.), *Treatise on Analytical Chemistry,* Wiley-Interscience, New York (1959 *et seq.,* a continuing series).

3. E. L. Peffer and G. C. Mulligan, *Testing of Glass Volumetric Apparatus,* U.S. Natl. Bur. Stand. Circ. 602, U.S. Government Printing Office, Washington, DC (1959).

4. *The Flow and Level Handbook,* "21st century" ed., Omega Engineering, Stamford, CT (2000); see the www.omega.com website.

5. R. P. Benedict, *Fundamentals of Temperature, Pressure, and Flow Measurements,* 3d ed., Wiley, New York (1984).

6. *The Pressure, Strain, and Force Handbook,* 8th ed., Omega Engineering, Inc., Stamford, CT (2008); see the www.omega.com website.

7. D. J. G. Ives and G. J. Janz (eds.), *Reference Electrodes,* Academic Press, New York (1961); D. T. Sawyer et al., *Electrochemistry for Chemists,* 2d ed., Wiley, New York (1995).

8. J. H. Moore, C. C. Davis, and M. A. Coplan, *Building Scientific Apparatus,* 3d ed., Perseus, Cambridge, MA (2002); R. F. Eisenberg, R. R. Kraybill, and R. B. Seymour, "Selection of Materials for Construction of Equipment," in E. S. Perry and A. Weissberger (eds.), *Techniques of Chemistry,* vol. XIII, chap. 1, Wiley-Interscience, New York (1979).

9. See for example the journals *Review of Scientific Instruments* and *Measurement Science and Technology.*

10. G. S. Brady and H. R. Clauser, *Materials Handbook,* 14th rev. ed., McGraw-Hill, New York (1996); N. P. Bansal and R. H. Doremus, *Handbook of Glass Properties,* Academic Press, Orlando, FL (1986).

11. J. R. Davis (ed.), *Metals Handbook Desk Edition,* 2d ed., American Society for Metals International, Materials Park, OH (1998).

12. C. A. Harper, *Handbook of Plastics, Elastomers and Composites,* 3d ed., McGraw-Hill, New York (1996); M. B. Ash and I. A. Ash (eds.), *Handbook of Plastic Compounds, Elastomers, and Resins,* Wiley, New York (1991).

13. *Welding,* 7th rev. ed., Deere & Co., East Moline, IL (1991); J. R. Davis, K. Ferjutz, and N. D. Wheaton, *ASM Handbook: Vol. 6, Welding, Brazing, and Soldering,* 10th ed., American Society for Metals International, Materials Park, OH (1983).

GENERAL READING

R. P. Benedict, *Fundamentals of Temperature, Pressure, and Flow Measurements,* 3d ed., Wiley, New York (1984).

J. H. Moore, C. C. Davis, and M. A. Coplan, *Building Scientific Apparatus,* 3d ed., Perseus, Cambridge, MA (2002).

J. Shackelford and W. Alexander (eds.), *CRC Materials Science and Engineering Handbook,* CRC Press, Boca Raton, FL (1991).

Least-Squares Fitting Procedures

Among the most common computational operations in experimental science is the fitting of experimental data to a theoretical function or model. When there is only one independent variable, as in fitting a straight line to a number of experimental points (x, y), the fitting can often be done graphically, with reliance on visual judgment for maximizing the quality of the fit. This is frequently sufficient, particularly when the experimental results are of only modest quality. However, when the number of variables or the number of fitting parameters is large or when it is important to make the best possible use of the data, the fitting must be done numerically. The computational method most commonly used is the method of least squares.

INTRODUCTION

The method of least squares has been used for many years, but the computations were once slow and onerous. Today, easy access to computers has made least-squares fitting a routine operation, especially using spreadsheets as discussed in Chapter III. Indeed, *least squares has become so easy to do that it is possible to be seduced into a false sense of security concerning the validity of the results obtained.* Least-squares fitting is not magic and is subject to the maxim of computer science, "garbage in, garbage out." It is important to realize that a set of experimental points can often be fitted quite well with an analytical expression that bears no relation whatsoever to any valid physical theory. There is nothing particularly wrong with fitting a data set with an artificial "mimic" function for purely empirical purposes, such as to obtain an analytical representation of the data that will be convenient for computational purposes (interpolation, differentiation, integration). For example, one might use a least-squares fit to the Gaussian error function, Eq. (II-11), in various contexts having little or nothing to do with the Gaussian error function in principle—atomic electron densities for instance. However, it is important to keep in mind which features of the model being tested are physically meaningful and which are not. One should resist cluttering up a theoretical model with features of doubtful physical significance merely to improve the fit. Furthermore, in trying to decide between two theoretical models on the basis of the "best" least-squares fit, one must be very careful in assessing the magnitude and the character of the experimental errors.

We will not undertake to present here a comprehensive and rigorous treatment of least squares. Our aim is to give a succinct and practical description of least squares as used in physical chemistry. More detail can be found in Ref. 1 and in the references listed under General Reading.

It must be kept in mind that the treatment of uncertainties given in this chapter is concerned with uncertainty in the data caused by random errors and not with that due to possible systematic errors. For a discussion of the latter, see Chapter II.

FOUNDATIONS OF LEAST SQUARES

The Maximum Likelihood Criterion. Let it be assumed that for m different values of an independent variable x (x_i; $i = 1, \ldots, m$) there are corresponding measured values of a dependent variable y (y_i; $i = 1, \ldots, m$). Let us further assume that there is a theoretical model predicting the way in which y is expected to depend on x and that this model may be represented by an analytical function f:

$$y = f(\alpha_1, \ldots, \alpha_n; x) \tag{1}$$

The quantities α_j ($j = 1, \ldots, n$)† are independently adjustable parameters that are initially unknown or known only approximately. The initial trial values of these parameters will be denoted as $\alpha_1^0, \ldots, \alpha_n^0$. It should be pointed out that we might just as easily have more than one independent variable; let there be p of them (x^k; $k = 1, \ldots, p$). The model is now represented by

$$y = f(\alpha_1, \ldots, \alpha_n; x^1, \ldots, x^p) \tag{2}$$

There are virtually no additional complexities or difficulties of any consequence resulting from having more than one independent variable, but for the sake of simplicity of expression we will assume only one in the treatment presented below.

Least-squares fitting deals with the problem of determining the "best" values of the n adjustable parameters $\alpha_1, \ldots, \alpha_n$ so as to maximize the agreement between the m observed y values and the values of y calculated with Eq. (1). In effect we are seeking the "best" (though generally not exact) solutions to a set of m simultaneous equations (called *observational equations* or *equations of condition*) in n unknowns:

$$y_i = f(\alpha_1, \ldots, \alpha_n; x_i) \tag{3}$$

The number m of observations exceeds (usually by an order of magnitude or more) the number n of adjustable parameters. Thus the mathematical problem is overdetermined. (For a situation in which $m = n$, the corresponding set of simultaneous equations could in principle be solved exactly for the parameters α_j. But this "exact" fit to the data would provide no test at all of the validity of the model.) In least squares as properly applied, the number of observations is made large compared to the number of parameters in order (1) to sample adequately a domain of respectable size for testing the validity of the model, (2) to increase the accuracy and precision of the parameter determinations, and (3) to obtain statistical information as to the quality of the parameter determination and the applicability of the model.

---

†In appropriate circumstances the index j can be chosen to run instead from 0 through $n - 1$, for example when the model function is a polynomial:

$$y = \alpha_0 + \alpha_1 x + \alpha_2 x^2 + \cdots + \alpha_{n-1} x^{n-1}$$

The least-squares criterion of best fit depends upon the concept known as *maximum likelihood:* the best set of parameters α_j is one that maximizes the *probability* function for the full set of measurements y_i. The probability P_i for making at $x = x_i$ a single measurement of y equal to y_i is [see Eq. (II-11)]

$$P_i = \frac{1}{\sqrt{2\pi}\sigma_i} \exp\left\{-\frac{[y_i - f(\alpha_1, \ldots, \alpha_n; x_i)]^2}{2\sigma_i^2}\right\}$$

assuming that the measurements y_i are normally distributed (i.e., are samples of a Gaussian distribution) with standard deviations σ_i. For making the entire set of measurements y_i, the probability is

$$P = \prod_i P_i = \left(\prod_i \frac{1}{\sqrt{2\pi}\sigma_i}\right) \exp\left\{-\sum_i \frac{[y_i - f(\alpha_1, \ldots, \alpha_n; x_i)]^2}{2\sigma_i^2}\right\}$$

Consider now the variation of P with respect to the α_j values. This affects only the sum in the exponent; maximizing P is the same thing as minimizing this sum. Thus maximum likelihood for P becomes the *least-squares principle:*

$$\delta_{\alpha 1}X^2 = \delta_{\alpha 2}X^2 = \cdots = \delta_{\alpha n}X^2 = 0 \tag{4}$$

where δ signifies variation with respect to infinitesimal and independent variations of the α_j,† and X^2 is defined by

$$X^2 \equiv \sum_i \left[\frac{y_i - f(\alpha_1, \ldots, \alpha_n; x_i)}{\sigma_i}\right]^2 \tag{5}$$

Actually, least squares is often applied in cases where it is not known with any certainty that measurements of y_i conform to a normal distribution or even when it is in fact known that they do *not* conform to a normal distribution. Does this destroy the applicability of the maximum-likelihood criterion? The answer is, not necessarily. The central-limit theorem is discussed briefly in Chapter II.[2] Simply stated, it says that the sum (or average) of a large number of measurements conforms very nearly to a normal distribution, regardless of the distributions of the individual measurements, *provided* that no one measurement contributes more than a small fraction to the sum (or average) and that the variations in the widths of the individual distributions are within reasonable bounds. (As we shall see, the average of a group of numbers is a special case of a least-squares determination.)

The factor $1/\sigma_i^2$ in Eq. (5) has the significance of the weight of the ith observation; i.e., it determines how heavily that observation contributes to the sum. However, it is not always possible to know even approximately the values of σ_i at the time the measurements are made. In this case, it is important and usually possible to assign *relative* weights w_i^R, which are estimates related to the *true* weights $w_i^T \equiv \sigma_i^{-2}$ by

$$w_i^R \approx kw_i^T = k\sigma_i^{-2} \tag{6}$$

where k is an unknown scale factor. Thus we will rewrite Eq. (5) as

$$X^2 = \sum_i w_i[y_i - f(\alpha_1, \ldots, \alpha_n; x_i)]^2 \tag{7}$$

---

†The validity of Eq. (4) depends on the function f being well behaved, i.e., possessing no discontinuities in the function itself or in the first derivative with respect to any α.

and use this form in what follows on the assumption that true weights w_i^T are available. If relative weights w_i^R are used, this will only introduce an unknown constant multiplicative factor into X^2 and will have no effect on minimizing this quantity.

If, by reason of difficulty or lack of time, problems of weighting are to be ignored, all factors w_i can be set equal to 1 in the equations that follow. In that case however, subsequent sections dealing with evaluation of uncertainties and goodness of fit lose much of their validity unless applied to a case in which all the observations just happen to be of equal weight. The assumption $w_i = 1$ is implicit in standard linear regression operations of spreadsheet programs.

Normal Equations. For X^2 to be a minimum, we require that Eqs. (4) hold:

$$\frac{\partial X^2}{\partial \alpha_1} = \frac{\partial X^2}{\partial \alpha_2} = \cdots = \frac{\partial X^2}{\partial \alpha_n} = 0$$

For the derivative with respect to α_1 we have

$$-2 \sum_i w_i [y_i - f(\alpha_1, \ldots, \alpha_n; x_i)] \frac{\partial f}{\partial \alpha_1} = 0 \tag{8}$$

For the general case that f is not a linear function of the α_j, let us expand it in a Taylor series around the values

$$y_i^0 \equiv f(\alpha_1^0, \ldots, \alpha_n^0; x_i)$$

calculated with the trial values α_i^0 of the parameters. *Keeping only terms to first order,*

$$f(\alpha_1, \ldots, \alpha_n; x_i) = y_i^0 + \frac{\partial f_i}{\partial \alpha_1} \Delta\alpha_1 + \frac{\partial f_i}{\partial \alpha_2} \Delta\alpha_2 + \cdots + \frac{\partial f_i}{\partial \alpha_n} \Delta\alpha_n \tag{9}$$

we obtain, after rearranging Eq. (8),

$$\sum_i w_i \left(\frac{\partial f_i}{\partial \alpha_1} \right)^2 \Delta\alpha_1 + \sum_i w_i \frac{\partial f_i}{\partial \alpha_1} \frac{\partial f_i}{\partial \alpha_2} \Delta\alpha_2 + \cdots$$

$$+ \sum_i w_i \frac{\partial f_i}{\partial \alpha_1} \frac{\partial f_i}{\partial \alpha_n} \Delta\alpha_n = \sum_i w_i \frac{\partial f_i}{\partial \alpha_1} (y_i - y_i^0) \tag{10}$$

where by $\Delta\alpha_j$ we mean $\alpha_j - \alpha_j^0$ and by $\partial f_i/\partial \alpha_j$ we mean $\partial f/\partial \alpha_j$ evaluated at $\alpha_1 = \alpha_1^0, \ldots, \alpha_n = \alpha_n^0, x = x$. To make the notation more compact, we define

$$A_{jk} \equiv \sum_i w_i \frac{\partial f_i}{\partial \alpha_j} \frac{\partial f_i}{\partial \alpha_k} \tag{11a}$$

$$h_j \equiv \sum_i w_i \left(\frac{\partial f_i}{\partial \alpha_j} \right)(y - y_i^0) \tag{11b}$$

Then Eq. (10) takes the form

$$
\begin{aligned}
A_{11} \Delta\alpha_1 + A_{12} \Delta\alpha_2 + \cdots + A_{1n} \Delta\alpha_n &= h_1 \\
A_{21} \Delta\alpha_1 + A_{22} \Delta\alpha_2 + \cdots + A_{2n} \Delta\alpha_n &= h_2 \\
&\vdots \\
A_{n1} \Delta\alpha_1 + A_{n2} \Delta\alpha_2 + \cdots + A_{nn} \Delta\alpha_n &= h_n
\end{aligned}
\tag{12}
$$

Equations (12), constituting a set of n simultaneous linear equations in n unknowns, are known as the *normal equations*. They can be solved to yield a set of equations in the form

$$\Delta \alpha_1 = B_{11}h_1 + B_{12}h_2 + \cdots + B_{1n}h_n$$
$$\Delta \alpha_2 = B_{21}h_1 + B_{22}h_2 + \cdots + B_{2n}h_n$$
$$\vdots$$
$$\Delta \alpha_n = B_{n1}h_1 + B_{n2}h_2 + \cdots + B_{nn}h_n$$

(13)

The details of the solution process do not concern us here. The quantities A_{ij} are the elements of an $n \times n$ square matrix **A,** and the B_{ij} are the elements of another $n \times n$ square matrix **B** that is the *inverse* of matrix **A.** The inversion of a matrix is a routine task with a computer.

Now a new and improved set of α's can be obtained:

$$\alpha_1^1 = \alpha_1^0 + \Delta \alpha_1$$
$$\alpha_2^1 = \alpha_2^0 + \Delta \alpha_2$$
$$\vdots$$
$$\alpha_n^1 = \alpha_n^0 + \Delta \alpha_n$$

(14)

If the function f is linear in the parameters α_j, the fitting procedure is now complete and the values α_j^1 are the best values in a least-squares sense. If f is nonlinear in terms of any of the α_j values, it is usually necessary to improve the α_j values by carrying out another "cycle" of minimization, in which α_j^1 plays the role previously played by α_i^0. The iteration process is repeated as many times as necessary to obtain convergence of the α_j values to some predetermined level of accuracy.

The normal-equations algorithm described here is generally the fitting method of choice when (1) the errors in the observations conform to a normal distribution (see discussion of this distribution in Chapter II), and (2) the observational equations are linear in the adjustable parameters. *As a matter of convenience, this algorithm is often used (especially in spreadsheet and other least-squares computer programs) when one or both of these conditions is not fulfilled.* This is not always bad practice, but one should be aware of the hazards discussed below.

The occasional large deviation of a y_i from its population mean can have a large effect on the refined parameters α_j because the quantity $[y_i - f(\ldots \alpha_j \ldots, x_i)]$ enters X^2 as the square. This is tolerable with a normal distribution because deviations become more and more unlikely as they increase in magnitude. However, if the error distribution has tails containing a higher-than-normal proportion of the total probability, the normal equations algorithm is significantly less reliable. Modifications of the normal-equations algorithm for cases of abnormal distributions have been proposed by Tukey and Andrews.[3]

Major difficulties that are often encountered with the normal-equations approach for *nonlinear* least squares are poor convergence to the "best" values of the parameters and possible convergence to a set of bad values of the parameters corresponding to a false minimum in X^2. A number of other algorithms are available that have advantages over the normal-equations algorithm for many circumstances.[4] In the "steepest-descent" algorithm, the parameter shifts follow the negative gradient of the function X^2 in parameter space to maximize the rate of decrease in X^2 in the early stages, but final convergence with this algorithm is often very slow. A mixed algorithm due to Marquardt[4-6] combines the virtues of the normal-equations and steepest-descent algorithms and is widely used. The Marquardt algorithm is equivalent to replacing Eq. (11a) by

$$A_{jk} = \sum_i (1 + \delta_{jk}\lambda)\frac{\partial f_i}{\partial \alpha_j}\frac{\partial f_i}{\partial \alpha_k}$$

(15)

where δ_{jk} is the Kronecker delta (equal to unity when $j = k$ and equal to zero otherwise) and λ is a parameter that is adjusted after each cycle to obtain optimum performance. When λ is large, as is typical in the early cycles, the procedure resembles that of steepest descents; when it is small, as it typically becomes in the later cycles, it resembles the normal-equations procedure. This adjustment of λ is done automatically in computer programs that utilize this algorithm.

Simple Examples. A trivial example of a least-squares calculations is the calculation of the arithmetic mean. In this case the independent variable x does not appear and there is only one parameter α, which is the desired average value of y. The m observational equations are of the form

$$y_i = \alpha = \Delta\alpha$$

since we may take $\alpha^0 = y_i^0 = 0$. The single normal equation becomes

$$\sum_i w_i(1)^2\alpha = \sum_i (1)w_iy_i$$

whence

$$\alpha = \bar{y} = \frac{\sum w_iy_i}{\sum w_i} \tag{16}$$

When the weights are all taken as unity, this becomes

$$\alpha = \bar{y} = \frac{1}{m}\sum_i y_i \tag{17}$$

A more significant example is that of a linear relationship

$$y = f(\alpha_0, \alpha_1; x) = \alpha_0 + \alpha_1 x \tag{18}$$

This is the equation of a straight line, where α_1 is the slope and α_0 is the y intercept. Here again, because of linearity, we may take $\alpha_0^0 = \alpha_1^0 = 0$. The m observational equations are of the form

$$\alpha_0 + \alpha_1 x_i = y_i$$

and the two normal equations are

$$\sum_i w_i(1)^2\alpha_0 + \sum_i w_i(1)x_i\alpha_1 = \sum_i w_i(1)y_i$$
$$\sum_i w_ix_i(1)\alpha_0 + \sum_i w_ix_i^2\alpha_1 = \sum_i w_ix_iy_i \tag{19}$$

Solving, we obtain

$$\alpha_0 = \frac{1}{D}\left(\sum_i w_ix_i^2\sum_i w_iy_i - \sum_i w_ix_i\sum_i w_ix_iy_i\right) \tag{20a}$$

$$\alpha_1 = \frac{1}{D}\left(-\sum_i w_ix_i\sum_i w_iy_i + \sum_i w_i\sum_i w_ix_iy_i\right) \tag{20b}$$

where

$$D = \sum_i w_i\sum_i w_ix_i^2 - \left(\sum_i w_ix_i\right)^2 \tag{21}$$

is the determinant $|A|$ of the matrix

$$\mathbf{A} = \begin{pmatrix} \Sigma \, w_i & \Sigma \, w_i x_i \\ \Sigma \, w_i x_i & \Sigma \, w_i x_i^2 \end{pmatrix} \qquad (22)$$

(If unit weights are employed, all w_i are deleted and $\Sigma \, w_i$ is replaced by m.) These equations may be useful when a simple straight-line fit ("linear regression") is being done with a spreadsheet program or a pocket calculator. Many calculators accumulate most or all of the sums required in Eqs. (20) to (22); some complete the calculation and offer both the refined parameters and their estimated standard deviations.

The student may find it illuminating to derive expressions for the three-parameter linear least-squares case with two independent variables:

$$z = \alpha_0 + \alpha_{1x} x + \alpha_{1y} y$$

Note that, if the independent variable y is replaced by x^2, we have a fitting function that is nonlinear in x but that can be treated with linear least squares since it is linear in the adjustable parameters. Similarly a general polynomial can be fitted by linear least squares:

$$y = \sum_{j=0}^{n-1} \alpha_j x^j$$

WEIGHTS

In general, a weight w_i must be assigned for each measurement y_i. If estimates of the standard deviations σ_i are available, the true weights $w_i^T = \sigma_i^{-2}$ can be used. If σ_i are not known but it is manifest that all measurements should have the same uncertainty, the weights are all equal and may be set equal to unity for convenience in finding the best fit. There may be other circumstances in which the standard deviations are not known but in which *relative weights* w_i^R can be assigned on an arbitrary scale by judgment based on experience or by commonsense criteria. (For example, a value of y_i that is the mean of two or three measurements of y at the same x_i has a weight two or three times that of a single measurement at that x_i.) Such relative weights are related to the true weights by Eq. (6).

In some cases the quantity measured is known to follow a Poisson distribution and the weight of a measurement of value y_i is simply $1/y_i$. Examples include radioactive counting experiments or fluorescence or Raman intensity measurements at low light levels, where the photon-counting rates are subject to statistical variation. Figure 1 shows how such a weighting choice would be implemented in a spreadsheet analysis of Raman intensities as a function of a solute concentration. As can be seen from Eq. (7), the weights are incorporated into the X and Y arrays by simply forming new columns of $w_i^{1/2} x_i$ and $w_i^{1/2} y_i$ values. These columns are then used in the normal linear least-squares operation of the spreadsheet program. In this example the parameters a and b differ only slightly for the choices w_i equal 1 and $1/y_i$, but the latter would be preferred because it has a theoretical basis.

The best basis for assigning weights is to make several measurements y_{ik} $(k = 1, \ldots, N)$ at the same x_i and determine the experimental variance S_i^2 of each measurement with Eq. (II-5). Since the mean of those N measurements will enter the least-squares calculations as y_i, that variance divided by N will constitute the first approximation to σ_i^2. Since N is likely to be small (less than 6), σ_i^2 will be rather uncertain and consequently a certain amount of pooling or smoothing of the σ_i^2 values may be appropriate. This may be done by plotting the σ_i^2 against x_i and drawing a smooth curve to obtain the values to be used in the least-squares

Example of a Weighted Least-Squares Regression

Raman intensity I as a function of solute concentration C

Assume $I = a + bC$

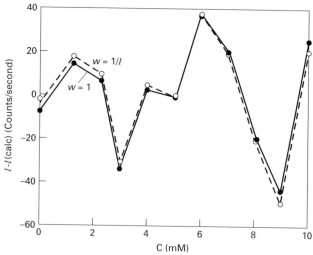

Weights $w = 1$

Unit column	C (m/M)	I	I (calc) counts/sec	$I - I$ (calc)
1	0	20	28.6	−8.6
1	1	160	147.2	12.8
1	2	271	265.7	5.3
1	3	352	384.2	−32.2
1	4	506	502.7	3.3
1	5	622	621.3	0.7
1	6	778	739.8	38.2
1	7	880	858.3	21.7
1	8	957	976.9	−19.9
1	9	1048	1095.4	−47.4
1	10	1240	1213.9	26.1

Regression output:

Constant	0
Std. error of Y est.	27.009
R squared	0.996
No. of observations	11
Degrees of freedom	9

	a	b
X coefficient(s)	28.64	118.53
Std. error of coefficient	15.24	2.58

Note: In each case, the X array encompasses the two columns at the left and the Intercept Equal Zero regression option is chosen

Weights $w = 1/I$ (Poisson statistics)

Sqrt(w)	Sqrt(w)C	Sqrt(w)I	I(calc)	$I - I$ (calc)
0.224	0.000	4.47	22.0	−2.0
0.079	0.079	12.65	141.6	18.4
0.061	0.121	16.46	261.3	9.7
0.053	0.160	18.76	381.0	−29.0
0.044	0.178	22.49	500.6	5.4
0.040	0.200	24.94	620.3	1.7
0.036	0.215	27.89	740.0	38.0
0.034	0.236	29.66	859.6	20.4
0.032	0.259	30.94	979.3	−22.3
0.031	0.278	32.37	1098.9	−50.9
0.028	0.284	35.21	1218.6	21.4

Regression output:

Constant	0
Std. error of Y est.	1.095
R squared	0.988
No. of observations	11
Degrees of freedom	9

	a	b
X coefficient(s)	21.98	119.66
Std. error of coefficient	4.56	1.83

Note: "True" parameters used to generate data are	a 22	b 117

FIGURE 1

Spreadsheet example illustrating the procedure for doing a weighted least-squares regression. Unit ($w = 1$) and $w = 1/I$ weighting choices are compared. The intensity data were generated using the spreadsheet random number generator for a Poisson distribution. This requires a mean value for each datum, which was calculated from $I = 22 + 117 * C$. If this analysis of such random data is repeated many times, the parameters obtained using Poisson weighting will generally be closer to the "true" values than parameters obtained with $w = 1$.

calculation. Alternatively, the σ_i^2 may be averaged in blocks of data, having in each block uniform conditions of measurement. We may summarize by writing

$$\frac{1}{w_i} = \sigma_i^2(\text{est}) = \left[\frac{1}{N(N-1)} \sum_{k=1}^{N} (y_{ik} - y_i)^2 \right]_{\text{smoothed}} \qquad (23)$$

where y_i is the mean of the y_{ik}.

Since y is functionally dependent on x, the uncertainty in y must contain a contribution from any uncertainty in x. If in the above-described procedure, the variable x is set separately and independently to its assigned value x_i for each of the N measurements y_i, then presumably the contribution of the uncertainty in x_i will automatically be reflected in the uncertainty of y_i and Eq. (23) will apply directly. However, if x is set only once to x_i for the N measurements y_i, so that the error contributed by x_i is the same for all N measurements, then the uncertainty in both must be reflected in the weight

$$w_i = \frac{1}{\sigma_i^2} = \frac{1}{\sigma_{iy}^2 + (\partial f/\partial x)_{x_i}^2 \sigma_{ix}^2} \qquad (24)$$

where σ_{iy} and σ_{ix} are the independent standard deviations in y and x, respectively, and $(\partial f/\partial x)_{x_i}$ is evaluated with $\alpha_j = \alpha_j^0$ (the trial values of the parameters). Equation (24) can also be used in other cases where the uncertainties in y and x are independent and have been estimated separately. In principle, the least-squares method assumes that all the uncertainty is in the y_i value at some exactly known x_i value, but Eq. (24) is a good approximation for adjusting the weights to reflect the experimental fact that both x and y are usually subject to experimental error.

It should be stressed that, in any experiment for which there is to be a least-squares refinement of parameters, intended to yield not only the best possible parameter values but also respectable estimates of their uncertainties *and* a test of the validity of the model, it is vital to take the trouble to analyze the methods and the circumstances of the experiment carefully in order to get the best possible values of the a priori weights.

REJECTION OF DISCORDANT DATA

There may be one or more y_i values that deviate markedly from the trend of the others. Not only may y_i values be in error because of the kinds of occurrences discussed in Chapter II (reading errors, transposition of digits, power-line transients), there may also be deviations occurring at discrete x_i values arising from physical effects that are not taken into account in the model. (Examples: in spectroscopy, an unexpected resonance happening at a certain frequency ν_i, enhancing or reducing the spectral response relative to that envisaged by the model; in x-ray crystallography, multiple reflection or thermal diffuse scattering producing abnormal intensity for a particular Bragg reflection).

Sometimes visual examination shows that the deviation is so gross that the offending measurement can be discarded at once. More often it is necessary to carry out one or more cycles of least squares and to obtain *weighted residuals* δ_i based on *properly scaled weights* [defined below by Eqs. (27a) and (27b)]. It must then be decided by inspection of the δ_i values whether the apparently discordant y_i should be rejected and one or more additional cycles of least squares carried out. This question is seldom discussed adequately, and there seem to be no well-defined and generally accepted criteria. We present some suggestions based on the magnitudes of the weighted residuals.

For values of m no larger than 10, one can use a Q test criterion (see Chapter II) for considering the rejection of *one* outlying datum y_i. In the present context, the quantity Q is defined by

$$Q = \frac{\delta_{max} - \delta_{near}}{\delta_{max} - \delta_{far}} \tag{25}$$

The quantities appearing in this equation may be best understood with the aid of an example. Let the unweighted residuals for a set of 10 measurements be arranged from the largest negative to the largest positive:

$$-1.0 \quad -0.9 \quad -0.7 \quad -0.7 \quad -0.3 \quad 0.0 \quad 0.3 \quad 0.4 \quad 0.4 \quad 2.0$$

Here δ_{max} (the δ with the largest magnitude) is 2.0, δ_{near} (the δ nearest to δ_{max}) is 0.4, and δ_{far} (the δ farthest from δ_{max}) is -1.0. Thus $Q = (2.0 - 0.4)/[2.0 - (-1.0)] = 0.53$. From Table II-1, the critical value Q_c is 0.41. Therefore the y_i point that yields the residual 2.0 should be rejected.

For values of m ranging from 11 to 50, we suggest that a seemingly discordant datum for which the weighted residual $\delta_i > 2.6$ be rejected. (A value of 2.58 corresponds to an a priori estimate that a given measurement has about a 1 percent chance of being valid; see Table II-2.) For larger values of m it is difficult or impossible to distinguish data having large weighted residuals because of faulty measurement from those having large weighted residuals because they happen to be in the tail of the normal distribution. However, data with "wild" values of weighted residuals δ_i (i.e., exceeding 4) may always be rejected.

GOODNESS OF FIT

The χ^2 Test. Having completed the least-squares computations (with enough cycles to obtain convergence if the function is nonlinear in the α_j) and having determined the best values α_j^* of the parameters, it remains to determine how good they are and how good the model is. In this and the following sections, a number of equations are presented without proof; detailed developments can be found in the books referred to under General Reading at the end of this chapter.

At this point the observed y_i are not in perfect agreement with the calculated y_i^* given by the model with the refined parameters

$$y_i^* = f(\alpha_1^*, \ldots, \alpha_n^*; x_i) \tag{26}$$

There remain residuals $y_i - y_i^*$, which we may subject to statistical analysis. To reduce these to the same statistical population we define a *weighted residual* δ_i:

$$\delta_i \equiv \frac{y_i - y_i^*}{\sigma_i} \tag{27a}$$

where the σ_i are the a priori standard deviations. In practice, the weighted residuals are approximated by

$$\delta_{i0} = \sqrt{w_i}(y_i - y_i^*) \tag{27b}$$

where w_i are the a priori estimated weights. If the true weights were used, these two quantities would be identical; however, in the case where the weights are estimated on an arbitrary relative basis [see Eq. (6)], it is necessary to make a distinction between them.

Analogous to Eq. (II-5), there is the following expression for the *variance of an observation of unit weight:*

$$S_{(1)}^2 = \frac{\Sigma \, \delta_{i0}^2}{m - n} = \frac{X_{\min}^2}{m - n} \tag{28}$$

where m is the number of observations and n is the number of adjustable parameters. The square root of this quantity, $S_{(1)}$, is an *estimate* of the standard deviation of an observation of unit weight. If $S_{(1)}$ were indeed equal to the standard deviation of an observation of unit weight, it would have the value unity. Presumably such a value would be obtained if there were an infinite number of observations, the model were rigorously correct, and the correct values $w_i = \sigma_i^{-2}$ were used in Eq. (27b). In general $S_{(1)}$ differs from unity. For an error-free model, $S_{(1)}$ may be greater than or less than unity, depending on whether the weights are overestimated or underestimated.† If the weights are correct, errors in the model tend to make $S_{(1)}$ greater than unity. Thus we see the reason for the emphasis on assigning a priori weights realistically and accurately. If one has confidence in the weights, the proximity of $S_{(1)}$ to 1 can provide a real test of the validity of the model. On the other hand, if weights have been assigned only on a relative basis, $S_{(1)}$ does not afford an effective test of a single model but may be of value in comparing two models as discussed in the next section; see Eq. (34).

The test of a fit can be put into the form of the following question: "At a specified level of confidence, is the value of $S_{(1)}^2$ consistent with the assumption that the residuals δ_{i0} are representative of a normal distribution as they would be if the model were correct and the weights properly assigned?" Here we make use of the fact that in principle the minimized quantity X^2, known as χ^2, should conform to the so-called *chi-square distribution.* The probability distribution function for χ^2, with $\nu = (m - n)$ degrees of freedom, is[1]

$$P(\chi^2, \nu) = \frac{(\chi^2)^{(\nu-2)/2} e^{-\chi^2/2}}{2^{\nu/2} \Gamma(\nu/2)} \tag{29}$$

where $\Gamma(\nu/2)$ is the gamma function. The probability that χ^2 will exceed a certain limiting value is

$$P_{\text{int}}(\chi^2, \nu) = \int_{\chi^2}^{\infty} P(x^2, \nu) \, dx^2 \tag{30}$$

where x^2 is a variable of integration corresponding to χ^2.

At this point it is convenient to make a change in notation. It has become common practice to cite the quantity "reduced chi square," χ_ν^2:

$$\chi_\nu^2 \equiv \frac{\chi^2}{\nu} \tag{31a}$$

$$\chi_\nu^2 \equiv \frac{1}{\nu} \sum_i \frac{(y_i - y_i^*)^2}{\sigma_i^2} \simeq \frac{1}{\nu} \sum_i w_i (y_i - y_i^*)^2 \tag{31b}$$

where

$$\nu = m - n \tag{31c}$$

is the number of degrees of freedom when m data points are fitted with a form involving n adjustable parameters. The quantity χ_ν^2 is defined by the first expression in Eq. (31b) and,

---

†One can rescale the weights a posteriori on the approximate basis that $w_i(\text{new}) = w_i(\text{old})/S_{(1)}^2$ if one is confident that the model is correct.

TABLE 1 Limiting values of $\chi_\nu^2 \equiv \chi^2(\nu)/\nu$, $F(1, \nu)$, and $F(\nu, \nu)$ for stated probability of exceeding these values^a

ν	Limiting χ_ν^2		Limiting $F(1,\nu)$	Limiting $F(\nu, \nu)$
	$P_{\text{int}} = 0.95$	$P_{\text{int}} = 0.05$	$P_{\text{int}} = 0.05$	$P_{\text{int}} = 0.05$
1	0.0039	3.84	161	161
2	0.0513	3.00	18.5	19.0
3	0.117	2.60	10.13	9.28
4	0.178	2.37	7.71	6.39
5	0.229	2.21	6.61	5.05
6	0.273	2.10	5.99	4.28
8	0.342	1.94	5.32	3.44
10	0.394	1.83	4.96	2.98
12	0.436	1.75	4.75	2.69
15	0.484	1.67	4.54	2.40
20	0.543	1.57	4.35	2.12
30	0.616	1.46	4.17	1.84
40	0.663	1.39	4.08	1.69
60	0.720	1.32	4.00	1.53
120	0.798	1.22	3.92	1.35
∞	1.000	1.00	3.84	1.00

^{a} Adapted from Refs. 7 and 8.

in practice, is calculated with the second expression. Thus, in the literature, χ_ν^2 is the symbol often given to what we have been calling $S_{(1)}^2$. One can refer to published tables[1,7,8] of χ_ν^2 for given probability P_{int} and number of degrees of freedom ν. An abridged table for $P_{\text{int}} = 0.95$ and $P_{\text{int}} = 0.05$ is given in Table 1, and spreadsheet formulas for these are given in Chapter III.

In using $S_{(1)}^2$, two limiting cases can be distinguished. *If the model is known in advance to be above reproach,* the value of $S_{(1)}^2$ can be used to answer the question: At a specified level of confidence, is the value of $S_{(1)}^2$ consistent with the assumption that the weighted residuals δ_{i0} represent a normal distribution with mean zero and standard deviation unity? This could be crudely translated into the question: Is the value of $S_{(1)}^2$ consistent with the assumption that the weights have been properly assigned? Here $S_{(1)}^2$ can be either less than or greater than unity; for 90 percent confidence limits, $S_{(1)}^2$ should lie between the *limiting* χ_ν^2 values corresponding to $P_{\text{int}} = 0.05$ and $P_{\text{int}} = 0.95$. For example, when $\nu = m - n = 30$, we find from Table 1 that $S_{(1)}^2$ should lie between 0.616 and 1.46 if the above question is to be answered in the affirmative. An affirmative answer is *not* a guarantee that the weights have been assigned correctly but only an assertion that they cannot be criticized on the basis of the statistics available.

If the assignment of weights is known in advance to be above reproach, the value of $S_{(1)}^2$ can be used to answer the more interesting question: At a specified level of confidence, is the value of $S_{(1)}^2$ consistent with the assumption that the data set conforms to the assumed model function? For the above example of $\nu = 30$ and 90 percent confidence limits, $S_{(1)}^2$ should lie between 0.616 and 1.46 if the question is to be answered positively. That is, if the model is valid, there is only a 5 percent a priori statistical probability that $S_{(1)}^2$ would lie below 0.616 and the same probability that it might lie above 1.46.

Good fits require χ_ν^2 values that lie between the limiting values specified (in Table 1 for example) for a given confidence level and ν value. To continue with our example of a fit to a data set with $m = 32$ and $n = 2$, let us assume that χ_ν^2 calculated with Eq. $(31b)$ has the value 1.7. At the 90 percent confidence level, the validity of the model, the assignment of weights, or both are suspect. However, one must be quite certain about the weights before this χ_ν^2 value can be used to reject the model (within the specified confidence limits). Let us now assume that the fit to our data set had given $\chi_\nu^2 = 1.2$. For this χ_ν^2 value, the validity of the model and the assignment of weights cannot be challenged on a statistical basis at this confidence level. However, this does not guarantee that the model is sound unless one can show that artificially low weights have not been used.

As a warning that unrealistic weights can yield misleading results, consider the implications if one chooses unrealistically small relative weights w_i^R. The resulting χ_ν^2 value can then be very small—smaller than the $P_{int} = 0.95$ value in Table 1. *This does not indicate a good fit;* χ_ν^2 values less than the $P_{int} = 0.95$ values are as statistically unlikely as those larger than the $P_{int} = 0.05$ values. For realistic weights w_i that are close to $1/\sigma_i^2$, a very small χ_ν^2 value is as statistically unlikely as an overly large value. Once again this stresses the key importance of making your estimates of w_i^R as carefully as possible if you wish to use statistical arguments to support a theoretical model fit to your data.

If the model function is a nonlinear function of the parameters α_j and the trial values α_j^0 are not quite close to the true values, the least-squares treatment may converge on a false minimum for χ_ν^2. Usually when this happens, $\chi_\nu^2 \simeq S_{(1)}^2 \gg 1$. However, in problems where m and n are both very large (as in x-ray crystallography), there are a great many false minima, some of which are not far from the true minimum in n-dimensional parameter space and yield χ_ν^2 values that are not greatly different from unity. This is one additional argument for making careful a priori estimates of σ_i. When the weights for the observations are properly estimated, it is unlikely (although occasionally possible) that a false minimum will satisfy the χ^2 test within the appropriate confidence limits.

Distribution of Residuals. When the number of measurements is large (preferably more than 100), one can carry out a χ^2 test of the frequency distribution of the δ_i values.[2] This frequency test can be more instructive than the χ^2 test of $S_{(1)}^2$, since it may allow a diagnosis of defects in the model or defects in the weight distribution apart from a mere scaling error in the weights.

Some workers have introduced a more detailed statistical test of the least-squares fit and the weighting scheme by using *normal probability plots*.[9,10] This test compares the actual distribution of the observed weighted residuals δ^{obs} to the ideal values δ^{ideal} expected for a normal distribution of mean zero and standard deviation unity. A plot of δ^{obs} versus δ^{ideal} should yield points close to a straight line with slope unity that passes through the origin, and most of the scatter should occur at the ends. If the plotted points look linear but with a slope considerably different from unity, the weights may be relatively correct but either overestimated or underestimated. If the plotted points exhibit a pronounced deviation from linearity, either the model is defective or erroneous relative weights have been assigned.

In the spirit of the above, it is helpful to make a qualitative inspection for trends in the residuals even if a formal analysis is not undertaken. In some fitting situations, one can see that dropping a few points will allow a change in the adjustable fitting parameters that will appreciably decrease all the remaining δ_i values and thus significantly lower χ_ν^2. Such a procedure may or may not be defensible on purely statistical grounds, but it can at least lead one to carefully inspect the experimental validity of possibly errant points. Even when there are no experimentally suspect points with peculiar residuals, it is important to be aware of the trend in the residuals across the data set. Models that give a best fit with systematic trends in the residuals (say a block of negative residuals at each end of the data set with a central block of positive residuals) are suspect even if the χ_ν^2 values appear to be satisfactory.

COMPARISON OF MODELS

It often happens that a decision has to be made between two somewhat different models for describing the data. Both models have physical significance, but they are based on different physical hypotheses. If both appear to provide reasonably good fits to the data, can preference be given for one model over the other?

The judgment may be based on the reduced chi-square values χ_ν^2 for the fits with the two respective models. However, even if fitting with the two models involves the same number of degrees of freedom ν, it should not be said that model 2 is significantly preferable to model 1 just because χ_ν^2 for model 2 is smaller than χ_ν^2 for model 1. One must answer the question: Is the difference *significant?* Here we need to look at the probability distribution for the ratio of reduced chi squares. Such a ratio, $\chi_{\nu1}^2/\chi_{\nu2}^2$, should conform to a distribution known as the F distribution, for which the probability distribution function is[1]

$$P_F(F, \nu_1, \nu_2) = \frac{\Gamma[(\nu_1 + \nu_2)/2]}{\Gamma(\nu_1/2)\Gamma(\nu_2/2)}\left(\frac{\nu_1}{\nu_2}\right)^{\nu_1/2} \frac{F^{(\nu_1-1)/2}}{[1 + F(\nu_1/\nu_2)]^{(\nu_1+\nu_2)/2}} \tag{32}$$

The test is made with the integral probability

$$P_{\text{int}}(F, \nu_1, \nu_2) = \int_F^\infty P_F(f, \nu_1, \nu_2)\, df \tag{33}$$

(where f is an integration variable corresponding to F), which expresses the probability that F obtained from a random data set exceeds a certain value. We make use of the fact that

$$F \equiv \frac{\chi_{\nu1}^2}{\chi_{\nu2}^2} = \frac{\chi_1^2/(m - n_1)}{\chi_2^2/(m - n_2)} = \frac{S_{(1)1}^2}{S_{(1)2}^2} \tag{34}$$

and look up in Table 1 or published tables[7,8] the value of

$$P_{\text{int}}(F, \nu_1, \nu_2) = P_{\text{int}}\left(\frac{S_{(1)1}^2}{S_{(1)2}^2}, m - n_1, m - n_2\right)$$

or else look up the limiting value of $F(\nu_1, \nu_2)$ for a given probability level, say 5 percent. In the latter case,

$$F(\nu, \nu) = \left\{ \frac{\left[\left(1 - \frac{2}{9\nu}\right)^2 + x_\alpha\left[\frac{4}{9\nu}\left(1 - \frac{2}{9\nu}\right)^2 - \frac{4}{81\nu^2}x_\alpha^2\right]^{1/2}\right]^3}{\left[\left(1 - \frac{2}{9\nu}\right)^2 - \frac{2}{9\nu}x_\alpha^2\right]} \right\} \tag{35}$$

where $x_{0.05} = 1.64485$ for $\alpha = 0.05$ (95 percent confidence level);[8] see Table 1.

For example suppose that there are two models, both with the same number of adjustable parameters. An example would be $y = a + bx + cx^2$ versus $y = a + bx + cx^3$. Each fit will involve the same number of degrees of freedom, say 20. Reference to statistical tables gives an $F(20, 20)$ value of 2.12 for a probability level of 0.05. That means that if $\chi_{\nu1}^2/\chi_{\nu2}^2 > 2.12$, it can be said at the 95 percent confidence level that model 1 may be rejected in favor of model 2. Note the important feature that an accurate value for the ratio

$\chi^2_{\nu 1}/\chi^2_{\nu 2}$ does *not* require a priori knowledge of the σ_i values and true weights. If good relative weights are used [see Eq. (6)], the reduced chi-square ratio will be correct.

This F test can also be used when the number of degrees of freedom is different for the two models. The most frequently encountered circumstance of this kind results from least-squares fitting with two models, one having n parameters and the other having the same n parameters plus p additional parameters associated with an elaboration of the first model. An example is $y = a + bx$ versus $y = a + bx + cx^2$. As a further simple example, consider a unimolecular gas-phase kinetics experiment such as Exp. 24. Strict first-order kinetics predicts that the partial pressure of decaying molecular species A will vary as

$$\ln p_A = \ln p_{A0} - kt \quad \text{(model 1)} \tag{36a}$$

However, at low pressures, owing to decreased probability of collisional deactivation of the activated molecule, the behavior may be better represented by

$$\ln p_A = \ln p_{A0} - kt + \frac{b}{p_{A0}}(e^{kt} - 1) \quad \text{(model 2)} \tag{36b}$$

Equation (36a) has two adjustable parameters: p_{A0} and k. Equation (36b) has three: p_{A0}, k, and b. The experimental data consist of measurements of p_A as a function of t over a range in which p_A decreases by a factor of about e^2. Least-squares parameter determinations are carried out separately with each of the two models. They give slightly different sets of values of p_{A0} and k. It is found that $S^2_{(1)}$ for model 2 is less than $S^2_{(1)}$ for model 1. Is model 2 to be preferred? That is, does a nonzero coefficient b really arise from a physically better model, or does it merely provide a cosmetic improvement in the fit by adding an extra and nonphysical adjustable parameter?

The F test in this case can be cast in a somewhat different form:[1]

$$F_\chi(p, m - n - p) \equiv \frac{\chi^2(m - n) - \chi^2(m - n - p)}{p(m - n - p)^{-1}\chi^2(m - n - p)} \tag{37}$$

Substituting in the $S^2_{(1)}$ values, we have

$$\frac{(m - n)S^2_{(1)}(m - n) - (m - n - p)S^2_{(1)}(m - n - p)}{pS^2_{(1)}(m - n - p)} = \frac{\nu_1 \chi^2_{\nu 1}}{p \chi^2_{\nu 2}} - \frac{\nu_2}{p} = F_\chi(p, \nu_2) \tag{38}$$

In the present case $n = 2$, $p = 1$; let us suppose that $m = 20$. For model 1, $\nu_1 = m - n = 18$; for model 2, $\nu_2 = m - n - p = 17$. Therefore

$$\frac{18S^2_{(1)1} - 17S^2_{(1)2}}{1 \cdot S^2_{(1)2}} = 18\frac{\chi^2_{\nu 1}}{\chi^2_{\nu 2}} - 17 = F_\chi(1, 17) \tag{39}$$

It can be shown that F_χ follows the F distribution. At $P = 0.05$, the limiting value of $F(1, 17)$ is found from Table 1 to be 4.5. The corresponding limiting ratio of reduced chi squares, $\chi^2_{\nu 1}/\chi^2_{\nu 2}$, is found from Eq. (39) to be $(4.5 + 17)/18 = 1.19$. Suppose in our hypothetical example we find that the actual ratio is 1.24. Then, at the 95 percent confidence level, model 2 is to be preferred, and the hypothesis that decay of the activated species is competing significantly with deactivation is confirmed. If on the other hand this ratio is found to be only 1.12, the "better" fit with model 2 is not statistically significant. We might have strong theoretical reasons for preferring model 2, but this set of data on this particular reacting system does not allow us to reject model 1. An example of an F test in a spreadsheet regression is given in Fig. III-3.

More detailed comparison of two models could involve the use of normal probability plots[9,10] or at least inspection of residuals for systematic trends as discussed earlier. If two models produce fits of equivalent quality in the sense of an F test, one might still prefer the model that gave the more random sequence of residuals (or put differently be suspicious of a model that gave blocks of residuals with alternating signs). Another way to compare models that yield almost equivalent fits is *range shrinking,* in which the data set is systematically truncated from either end of the range in the variable x and the stability of the parameter values is inspected.

Finally, one should pay considerable attention to the physical reasonableness of the adjustable parameters α_j before choosing one model over another. To embrace model 2 and reject model 1 on a purely statistical analysis of one set of data is dangerous. If possible one should vary the experimental conditions, analyze several sets of data, and look at the behavior of the α_j values. In our gas-kinetics example, one could vary the temperature at which the rate of decomposition is studied and test the temperature dependence of the k values obtained from least-squares fits at each temperature. What if model 2 were statistically better for fitting the data at each T but gave k (and probably b) values that were erratic functions of T, while model 1 gave k values that smoothly varied with T in a way that was theoretically pleasing? One might then prefer model 1 but report the statistical problems associated with the data. In particular, one should then look carefully for a posteriori evidence that the relative weights might be reassigned. Unsuspected systematic errors present over part of the range could have influenced the quality of the overall fit.

UNCERTAINTIES IN THE PARAMETERS

We now assume that the least-squares refinement has converged satisfactorily, that any necessary rejection of discordant data has taken place before the final cycles were carried out, and that statistical tests on the weighted residuals have given reassuring results. It is now appropriate to estimate the uncertainties in the determined values of the adjustable parameters α_j.†

A propagation-of-error treatment of Eq. (13) yields for the estimated standard deviation in parameter α_j the expression

$$S(\alpha_j) \;=\; B_{jj}^{1/2} S_{(1)} \tag{40}$$

where B_{jj} is the jth diagonal element in the inverse normal equation matrix $\mathbf{B} = \mathbf{A}^{-1}$ and $S_{(1)}$ is the estimated standard deviation of an observation of unit weight given by Eq. (28). In the unusual case where (1) the number m of observations is so small (less than 6 or 8 say) that $S_{(1)}$ cannot be determined with great reliability from Eq. (28), and (2) the reliance that can be placed on the a priori σ_j values is uncommonly high, $S_{(1)}$ can be simply set equal to unity. Equation (40) is the basis of the calculation of estimated standard deviations that are automatically given by packaged least-squares computer programs (and all too often are uncritically accepted by the users of such programs without much concern for the validity of the weighting system or of the model).

---

†If the results of statistical tests were *not* reassuring, only qualified confidence in the estimated uncertainties is justified; if they are to be quoted at all, it may be appropriate to multiply them by a factor of at least 2 or 3. In any such case, the circumstances should be fully reported along with the parameter values and the quoted uncertainties, if any.

For the trivial one-parameter case in which the parameter to be determined is the arithmetic mean,

$$B_{11} = \frac{1}{\sum_i w_i}$$

Making use of Eqs. (28) and (40), we obtain

$$S(\bar{y}) = \left[\frac{\sum \delta_i^2}{(N-1)\sum w_i} \right]^{1/2}$$

If all weights are taken as unity, $\sum_i w_i = m = N$ and

$$S(\bar{y}) = \left[\frac{\sum (y_i - y_i^*)^2}{N(N-1)} \right]^{1/2} \tag{41}$$

which corresponds to the result cited in Chapter II, Eq. (II-8).

For the two-parameter case of the linear relationship we may apply Eqs. (28) and (40) to the estimation of the standard deviations in the intercept α_0 and the slope α_1 of the corresponding straight line. From Eq. (22) we have

$$B_{00} = \frac{1}{D} \sum_i w_i x_i^2 \qquad B_{11} = \frac{1}{D} \sum_i w_i$$

where D is given by Eq. (21). Thus

$$S(\alpha_0) = \left(\frac{1}{D} \sum_i w_i x_i^2 \right)^{1/2} \left(\frac{\sum \delta_i^2}{m-n} \right)^{1/2} \tag{42a}$$

$$S(\alpha_1) = \left(\frac{1}{D} \sum_i w_i \right)^{1/2} \left(\frac{\sum \delta_i^2}{m-n} \right)^{1/2} \tag{42b}$$

As in the case of Eqs. (20) to (22), when unit weights are employed, the w_i are deleted and $\sum w_i$ is replaced by m.

Knowledge of estimated standard deviations is not by itself very useful, especially when they are unthinkingly accepted without regard to the validity of the model and of the weighting scheme. Assuming that these aspects are assuredly satisfactory, we are usually more interested in confidence limits for the parameters at a stated probability level P, as in the one-parameter case of the algebraic mean discussed in the section on confidence limits in Chapter II. Equation (II-29) for the one-parameter case applies also to the n-parameter case; for 95 percent confidence,

$$\Delta_{0.95, j} = t_{0.95, \nu} S_j \tag{43}$$

but here we must take as the number of degrees of freedom $\nu = m - n$.

This method is usually satisfactory for routine work, particularly when the model function is linear in the adjustable parameters. However, there may arise circumstances in nonlinear least squares for which more care is needed. Consider a situation in which one parameter, say α_k, is of special importance. It may be that it was the whole point of the experiment to determine that one parameter, while the other parameters are of minor importance apart from their being necessary parts of the model. Let the "best" value of this parameter, determined by a convergent least-squares procedure, be α_k^*. Let us step away from α_k^* in both directions by a constant interval to obtain several values of α_k for the procedure to be described. [The interval chosen might be, for example, S_k as calculated with Eq. (40), and the number of steps might be three or more on each side.] We then *fix* α_k at each of these trial values in turn, refine the remaining parameters to convergence, and evaluate χ_ν^2 with

Eq. (31*b*), using in each case the current fixed value of α_k and the refined values of the other parameters. Let us now plot χ_ν^2 against α_k and draw a smooth curve through the points. In a linear least-squares case, the plot may be expected to be parabolic around α_k^*, but in a nonlinear case, it may deviate considerably from parabolic shape and is likely to be unsymmetrical. We now use the *F* test to establish a ratio,

$$F(\nu, \nu) = \frac{\chi_{\nu max}^2}{\chi_{\nu min}^2} \tag{44}$$

that corresponds to a probability level of say 5 percent. For example, with 15 degrees of freedom, we find from Table 1 that $F_{0.05}(15, 15) = 2.40$. We multiply the observed $\chi_{\nu min}^2$, evaluated at α_k^*, by the appropriate value of *F* to obtain a value for $\chi_{\nu max}^2$, and we mark the two opposite points on the curve where $\chi_\nu^2 = \chi_{\nu max}^2$. The corresponding values of α_k are the maximum and minimum limiting values, and their differences from α_k^* may be quoted as the confidence limits at the stated probability level, say 95 percent. Note that the two limits may not necessarily be equal; instead of quoting a result in the customary form 3.154 ± 0.021, we may have to use the form

$$3.154^{+0.027}_{-0.015}$$

Note also that the above procedure can be used to deduce parameter errors for nonlinear fits done with the Solver operation in spreadsheets.

A plot of χ_ν^2 against a parameter that is fixed at selected values as described above is useful also for diagnostic purposes. It may for example indicate that the value obtained for α_k^* is at a false minimum of X^2. If this situation is suspected, the plot should cover a much wider range than is necessary to determine confidence limits of error.

It frequently happens that one wishes to obtain an estimated standard deviation for a quantity *F* that is calculated from two or more of the α_j according to a functional relationship of some sort, say $F(\alpha_1, \ldots, \alpha_n)$. It would *not* be correct to estimate separately the $\sigma(\alpha_j)$ with Eqs. (28) and (40) and apply Eq. (II-41). The reason is that the errors in the α_j are usually not independent; in general they are *correlated*. A simple example will show how errors can be correlated. Imagine three points *a*, *b*, *c along a straight line*. Let the distance *ab* be α_1, and let the distance *bc* be α_2. If the estimated standard deviation (e.s.d.) in the position of any of the points is *independently* σ_p, the e.s.d. of α_1 is $\sqrt{2}\sigma_p$ and the e.s.d. of α_2 is also $\sqrt{2}\sigma_p$. What is the estimated standard deviation of $F = \alpha_1 + \alpha_2 =$ the distance *ac*? If we use the propagation-of-errors treatment without allowing for correlation, we get the answer that $\sigma(F) = [\sigma^2(\alpha_1) + \sigma^2(\alpha_2)]^{1/2} = 2\sigma_p$, while on going back to first principles we see that the true answer is $\sqrt{2}\sigma_p$. The two distances α_1 and α_2 are correlated: when point *b* moves toward *a*, it is moving away from *c*. For a general discussion of this problem and methods for handling it with the correlation matrix, see Refs. 1 and 6. The correlation matrix may also provide useful information for the design of an experiment. Many commercial software packages perform nonlinear regressions and give parameter errors and other statistical information. These include Curvefit, IGOR Pro, Origin, SigmaPlot, and others.

SUMMARY OF PROCEDURES

This section will provide a brief "road map" of the steps recommended in carrying out a least-squares fit:

1. Establish the weights w_i to be assigned to each observed value y_i. It is vital to have good relative weights and preferable to have true weights based on an experimental determination of the standard deviations σ_i. Think carefully about your measurements before taking the easy way out and adopting equal weights (see Weights). It may be

necessary to smooth the observed variances over the entire data set (see Sample Least-Squares Calculation).

2. Reject any obviously bad points (see Rejection of Discordant Data).

3. Decide on the model function to be used. This choice may be guided by a theoretical prediction or by a rough empirical assessment of the type of dependence likely.

4. Choose a reasonable set of trial values α_j^0 for the adjustable parameters of the model if nonlinear fitting is to be done. A good choice of α_j^0 is very important to reduce computational time and to avoid false minima (see Goodness of Fit, toward the end of the discussion of χ^2 tests).

5. Carry out the least-squares minimization of the quantity X^2 in Eq. (7) according to an appropriate algorithm (presumably normal equations if the observational equations are linear in the parameters to be determined; otherwise some other such as Marquardt's[4,5]). The linear regression and Solver operations in spreadsheets are especially useful (see Chapter III). Convergence should not be assumed in the nonlinear case until successive cycles produce no significant change in any of the parameters.

6. After convergence has been achieved, check the residuals $y_i - y_i^*$ to see if any data points should be deleted from the data set (see end of the section on Goodness of Fit). If so, rerun the least-squares fit on the new, edited data set.

7. Obtain the reduced chi-square value χ_ν^2 given by Eq. (31b) and carry out the appropriate statistical tests for goodness of fit, including inspection of the weighted residuals for systematic trends (see Goodness of Fit).

8. If either theory or clear systematic trends in the residuals suggests an alternative model, carry out a new least-squares fit to the same data set with the new model. Make a statistical analysis (F test) to help decide whether one of the models can be rejected in favor of the other. Be careful to consider the physical reality of the adjustable parameters in the model that you finally choose (see Comparison of Models).

9. Evaluate the statistical uncertainties in the adjustable parameters obtained from the best fit (see Uncertainties in the Parameters).

10. *Think about what you are doing. Do not treat least-squares fitting as a magical mathematical game,* and do not accept without searching examination the results of packaged least-squares computer programs.

SAMPLE LEAST-SQUARES CALCULATION

We will illustrate here as many as possible of the principles and techniques discussed in this chapter with a sample "curve fitting" by least squares.

For values of x_i ranging from 2.1 to 5.0 ($i = 1$ through 30) in intervals of 0.1, measurements of y were made, four y_{ik} for each x_i value. The measured values are listed in Table 2 and plotted in Fig. 2. For each x_i, the four y_{ik} were averaged to give y_i, and the variance S_i^2 was calculated. Since the measurements were made under uniform conditions and since the variances, when plotted against x, showed (apart from their natural statistical scatter) a smooth variation suggestive of quadratic behavior, the variances were smoothed by least-squares fitting with a quadratic using unit weights. The variance function obtained was $S_{sm}^2 = 0.162 - 0.0879x + 0.01357x^2$. The values of $S_{sm_i}^2$ calculated with this equation are also listed in Table 2.

No measurements were rejected as being discordant. For example, let us look at the measurements at $x_i = 2.2$, where the observed y_{ik} values were

$$16.62 \qquad 16.46 \qquad 16.62 \qquad 16.60$$

TABLE 2   Sample least-squares calculation: data and least-squares fits with two and three parameters

i	x_i	$y_{ik},\ k=1,\dots,4$				The data set (Mean) y_i	S_i^2	$S_{\theta m_i}^2$	$S_{\theta m_i}^2/4$	w_i	Two parameters[a] y_i^{calc}	δ_i^2	δ_i	Three parameters[b] y_i^{calc}	δ_i^2	δ_i
1	2.1	17.15	16.73	16.80	17.24	16.98	0.0638	0.0372	0.0093	107	17.13	2.41	−1.55	16.85	1.81	1.34
2	2.2	16.62	16.46	16.62	16.60	16.57	0.0060	0.0342	0.0086	116	16.81	6.68	−2.58	16.58	0.01	−0.11
3	2.3	16.33	15.94	16.35	16.26	16.22	0.0363	0.0315	0.0079	127	16.48	7.32	−2.70	16.30	0.81	−0.90
4	2.4	16.14	15.70	15.77	15.97	15.89	0.0396	0.0291	0.0073	137	16.16	9.99	−3.16	16.03	2.69	−1.64
5	2.5	15.64	15.87	15.90	15.61	15.75	0.0228	0.0270	0.0068	147	15.83	0.94	−0.97	15.74	0.01	0.12
6	2.6	15.66	15.84	15.55	15.62	15.62	0.0303	0.0251	0.0063	159	15.50	2.29	1.51	15.46	4.07	2.02
7	2.7	15.10	15.28	15.20	15.32	15.22	0.0094	0.0235	0.0059	169	15.18	0.27	0.52	15.16	0.61	0.78
8	2.8	14.70	14.97	14.67	15.06	14.85	0.0355	0.0222	0.0056	179	14.85	0.00	0.00	14.87	0.07	−0.27
9	2.9	14.64	14.40	14.44	14.58	14.52	0.0129	0.0212	0.0053	189	14.53	0.02	−0.14	14.57	0.47	−0.69
10	3.0	14.23	14.11	14.12	14.18	14.16	0.0031	0.0204	0.0051	196	14.20	0.31	−0.56	14.27	2.37	−1.54
11	3.1	14.03	14.06	14.06	14.04	14.05	0.0002	0.0199	0.0050	200	13.88	5.78	2.40	13.96	1.62	1.27
12	3.2	13.66	13.75	13.52	13.57	13.63	0.0103	0.0197	0.0049	204	13.55	1.31	1.14	13.65	0.08	−0.29
13	3.3	13.33	13.28	13.05	13.28	13.24	0.0158	0.0197	0.0049	204	13.23	0.02	0.14	13.34	2.04	−1.43
14	3.4	12.81	13.11	12.88	13.16	12.99	0.0293	0.0200	0.0050	200	12.90	1.62	1.27	13.02	0.18	−0.42
15	3.5	12.79	12.48	12.55	12.74	12.64	0.0221	0.0206	0.0052	192	12.58	0.69	0.83	12.69	0.48	−0.69
16	3.6	12.13	12.62	12.67	12.25	12.42	0.0718	0.0215	0.0054	185	12.25	5.35	2.31	12.37	0.46	0.68
17	3.7	12.27	12.08	11.91	11.97	12.06	0.0250	0.0226	0.0056	179	11.93	3.03	1.74	12.04	0.07	0.27
18	3.8	11.67	11.66	11.81	11.75	11.72	0.0050	0.0240	0.0060	167	11.60	2.40	1.55	11.70	0.07	0.26
19	3.9	11.34	11.42	11.66	11.47	11.47	0.0185	0.0257	0.0064	156	11.28	5.63	2.37	11.36	1.89	1.37
20	4.0	11.26	11.24	11.03	11.28	11.20	0.0135	0.0276	0.0069	145	10.95	9.06	3.01	11.02	4.70	2.17
21	4.1	10.67	10.53	10.64	10.98	10.70	0.0372	0.0298	0.0074	135	10.63	0.66	0.81	10.67	0.12	0.35
22	4.2	10.39	10.40	10.08	10.33	10.30	0.0225	0.0323	0.0081	123	10.30	0.00	0.00	10.32	0.05	−0.22
23	4.3	9.96	10.00	9.65	9.93	9.88	0.0254	0.0351	0.0088	114	9.98	1.14	−1.07	9.96	0.73	−0.85
24	4.4	9.31	9.68	9.21	9.84	9.51	0.0893	0.0381	0.0095	105	9.65	2.06	−1.43	9.61	1.05	−1.02
25	4.5	9.09	8.85	9.16	9.25	9.09	0.0294	0.0414	0.0104	96	9.33	5.53	−2.35	9.24	2.16	−1.47
26	4.6	9.07	8.86	8.58	8.53	8.83	0.0498	0.0450	0.0112	89	9.00	2.57	−1.60	8.88	0.22	−0.47
27	4.7	8.71	8.39	8.17	8.34	8.50	0.0294	0.0488	0.0122	82	8.68	2.66	−1.63	8.50	0.00	0.00
28	4.8	7.94	8.52	8.04	8.04	8.17	0.0641	0.0529	0.0132	76	8.35	2.46	−1.57	8.13	0.12	0.35
29	4.9	7.74	7.88	7.71	7.84	7.79	0.0065	0.0573	0.0143	70	8.03	4.03	−2.01	7.75	0.11	0.33
30	5.0	7.01	7.37	7.70	7.62	7.42	0.0963	0.0620	0.0155	65	7.70	5.10	−2.26	7.37	0.16	0.40

$\sum \delta_i^2 = 91.33 \div 28$
$S_{(1)}^2 = 3.262$
$S_{(1)} = 1.806$

$\sum \delta_i^2 = 29.23 \div 27$
$S_{(1)}^2 = 1.083$
$S_{(1)} = 1.040$

[a] $y_i^{calc} = 23.957 - 3.251 x_i$
[b] $y_i^{calc} = 21.618 - 1.850 x_i - 0.2000 x_i^2$

FIGURE 2

Plot of experimental points
and least-squares lines:
(*a*) straight-line fit (two
parameters), (*b*) quadratic fit
(three parameters). A filled-in
circle represents two or more
nearly coincident points.

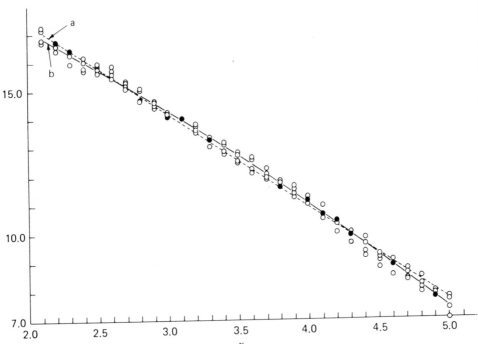

The value of Q is $24/26 = 0.923$, which exceeds $Q_c(90\%)$ for $N = 4$. Should the value 16.46 be rejected? If these four measurements were the only ones, one should certainly reject that value. Here however we have the benefit of much more information, and a glance at the rest of the data show that the range of y_{ik} values at $x_i = 2.2$ is reasonably representative. In applying rejection criteria or other statistical criteria, one must always take into account the context in which the criteria are given, and be careful to make the best possible use of *all* the data at hand.

There being four measurements averaged, the value of the smoothed variance for y_i, the mean of the y_{ik}, was taken as $S_{sn_i}^2/N$ where $N = 4$. This now is our best estimate of σ_i^2. Its reciprocal, rounded to an integer for computational convenience, was taken as the weight w_i. In many other circumstances, there will not be as many as four measurements of each y_i to form a basis for estimating w_i values. Often only one y_i value is available at each x_i. *If this is the case, it is important to carry out multiple measurements at least for some x_i values to provide a feel for "typical" σ_i values.* Also, one should use all one's general knowledge and pay careful attention to the range of y_i, x_i values in estimating the best weights.

Since the plot in Fig. 2 appears to confirm roughly a theoretical expectation of linear behavior, a two-parameter least-squares fit to the function

$$y = \alpha_0 + \alpha_1 x$$

was undertaken. This fit yielded $\alpha_0 = 23.9567$, $\alpha_1 = -3.2507$. Values of y_i^{calc} calculated with these parameters are given in Table 2. The value $\chi_\nu^2 \simeq S_{(1)}^2 = 3.262$ is quite different from unity; for $P_{\text{int}} = 0.05$ and $\nu = m - n = 30 - 2 = 28$, the limiting χ_ν^2 is 1.476. Therefore, at the 95 percent confidence level, our χ_ν^2 value is *not* consistent with the assumption that the residuals δ_i are representative of a normal distribution. Either our weighting scheme is wrong or the data are not well represented by a linear fit.

Examination of the fit of the straight line to the data points (as characterized by the residuals) shows that the points tend to be a little below the line near the ends and a little above near the middle, as if the correct fitting function should have a small amount of curvature. Let us now consider a quadratic model, which may be justified by a more refined theory or may be purely empirical. Accordingly a new fit was made with the function

$$y = \alpha_0 + \alpha_1 x + \alpha_2 x^2$$

This fit yielded $\alpha_0 = 21.6183$, $\alpha_1 = -1.8505$, $\alpha_2 = -0.2000$. The value of $\chi_\nu^2 = S_{(1)}^2$ was 1.083 for this three-parameter fit, clearly within the bounds of the χ^2 test. A decrease in χ^2 from that for the two-parameter fit was to be expected; χ^2 never increases and practically always decreases when an additional parameter is added to an existing model (provided the new model reverts to the previous one for some value of the added parameter). While the fact that $S_{(1)}^2$ is so close to unity is gratifying, we may apply an additional test to confirm that the improvement resulting from the addition of the quadratic parameter is statistically significant at the 95 percent confidence level. With the F test in the form given by Eq. (38), we obtain

$$F(1,\ 27) = \frac{28S_{(1)2}^2 - 27S_{(1)3}^2}{1 \cdot S_{(1)3}^2} = \frac{28 \times 3.262 - 27 \times 1.083}{1.083} = 57.34$$

At $P = 0.05$, the limiting value of $F(1,\ 27)$ is 4.22; thus the improvement is significant at the 95 percent confidence level; indeed it is still significant at about the 99.99 percent confidence level.

Having satisfied ourselves of the statistical adequacy of the fit, we are now entitled to estimate the uncertainties in the refined parameters. We apply Eqs. (28), (40), and (43) to obtain estimated standard deviations, given in parentheses, and 95 percent confidence limits for both the two-parameter and three-parameter determinations:

Two-parameter model ($\nu = 30 - 2 = 28$)

$$\alpha_0 = 23.96(13) \quad \text{or} \quad 23.96 \pm 0.22$$
$$\alpha_1 = -3.25(4) \quad \text{or} \quad -3.25 \pm 0.06$$

Three-parameter model ($\nu = 30 - 3 = 27$)

$$\alpha_0 = 21.6(3) \quad \text{or} \quad 21.6 \pm 0.5$$
$$\alpha_1 = -1.85(18) \quad \text{or} \quad -1.85 \pm 0.30$$
$$\alpha_2 = -0.20(3) \quad \text{or} \quad -0.20 \pm 0.04$$

The uncertainty values calculated for the two-parameter model are quoted only for illustrative purposes, since we have good reason to believe from our statistical tests that they are not to be relied upon. Note that they are much smaller than the uncertainty values calculated for the three-parameter model. Note also that the differences between the two models in the values of α_0 and α_1 are far outside the limits of error. Had we lazily accepted the two-parameter fit and its estimated error limits and attached physical significance to either of the two parameters, we would have badly deceived ourselves. On the other hand, in view of the excellent statistical performance of the three-parameter fit, we may accept the parameter values and the estimated uncertainty values of the three-parameter model with considerable confidence.

Coda. Now it can be told: The "experimental" y_i values in Table 2 were actually generated from *known* "true" values of the parameters and *known* standard deviations! Thus we have here a demonstration of how well the method of least squares works.

We started out with

$$y_i^{\text{true}} = 21.33 - 1.697x_i - 0.2232x_i^2$$

and the y_i values were assigned a standard deviation of

$$\sigma_i^{\text{true}} = 0.8 - 0.4x_i + 0.06x_i^2$$

Numbers from a random-number table were used to generate normal deviates. These were multiplied by σ_i^{true} as given above and the resulting synthetic "errors" were added to the y_i^{true} to produce the "experimental" data.

Let us see how well the least squares worked. The smoothed variance came out to be $S_i^2 = 0.162 - 0.088x_i + 0.136x_i^2$. This cannot correspond exactly with $(\sigma_i^{\text{true}})^2$, which is a quartic. The least-squares smoothing does not look particularly good, especially near the ends of the range. In the following comparison, results of a completely independent visual smoothing (not used) are also included:

x_i	$(\sigma_i^{\text{true}})^2$	$S_i^2(\text{l.s.})$	$S_i^2(\text{visual})$
2.1	0.0504	0.0372	0.041
3.0	0.0196	0.0204	0.017
4.0	0.0256	0.0276	0.024
5.0	0.0900	0.0620	0.061

The least-squares smoothing in this case cannot be said to be really superior to visual smoothing; although neither the least-squares nor the visual variances are particularly satisfying compared to the true variances, they are the best we have. Certainly either of them is better than equal variances and very much better than the individual variances for the purpose of assigning weights. (For example the variance determined for the four measurements at $x_i = 3.1$ is 0.0002, corresponding to a weight of 5000, about 25 times the maximum weight eventually assigned.) However, the least-squares method is not very sensitive to modest errors in the weights. If such errors are quite large, statistical tests should indicate the existence of problems with the assigned weights.

Now let us look at the parameter values with their 95 percent confidence limits, tabulated below.

	Two-parameter	Three-parameter	True
α_0	23.96 ± 0.22	21.6 ± 0.5	21.33
α_1	3.25 ± 0.06	-1.85 ± 0.30	-1.697
α_2	—	-0.20 ± 0.04	-0.2232

The values from the three-parameter determination agree with the true values well within the 95 percent confidence limits quoted for the former. However, the values from the two-parameter determination deviate from the true values by 12 and 26 times the confidence limits! This provides further illustration of the danger of relying on estimated uncertainties when the model may be defective.

REFERENCES

1. P. R. Bevington and D. K. Robinson, *Data Reduction and Error Analysis for the Physical Sciences*, 3d ed., McGraw-Hill, New York (2003).

2. A. I. Khinchin, *Mathematical Foundations of Statistical Mechanics,* p. 166, Dover, Mineola, NY (1949).

3. E. Prince, *Mathematical Techniques in Crystallography and Materials Science,* 2d ed., Springer-Verlag, New York (1994).

4. F. S. Acton, *Numerical Methods That Work,* rev. ed., chap. 17, Mathematical Assn. of America, Washington, DC (1990).

5. W. H. Press, S. A. Teukolsky, W. T. Vetterling, and B. P. Flannery, *Numerical Recipes—the Art of Scientific Computing,* 2d ed., Cambridge Univ. Press, New York (1992). [Available in C, Pascal, and FORTRAN versions.]

6. J. L. Stanford and S. B. Vardeman, *Statistical Methods for Physical Science,* Academic Press, San Diego, CA (1994).

7. W. H. Beyer (ed.), *CRC Handbook of Tables for Probability and Statistics,* 2d ed., pp. 296–298, 306, CRC Press, Boca Raton, FL (1968); R. D. Lide (ed.), *CRC Handbook of Chemistry and Physics,* 89th ed., CRC Press, Boca Raton, FL (2008–2009).

8. K. V. Mardia and P. J. Zenroch, *Tables of the F- and Related Distributions with Algorithms,* Academic Press, New York (1978).

9. S. C. Abrahams and E. T. Keve, *Acta Crystallogr.* **A27,** 157 (1971).

10. W. C. Hamilton and S. C. Abrahams, *Acta Crystallogr.* **A28,** 215 (1972).

GENERAL READING

P. R. Bevington and D. K. Robinson, *Data Reduction and Error Analysis for the Physical Sciences,* 3d ed., McGraw-Hill, New York (2003).

R. de Levie, *J. Chem. Educ.* **63,** 10 (1986).

E. Whittaker and G. Robinson, *The Calculus of Observations,* 4th ed., Blackie, Glasgow (1944).

J. H. Zar, *Biostatistical Analysis,* 4th ed., Prentice-Hall, Upper Saddle River, NJ (1998).

Appendix

A

Glossary of Symbols

Listed below are the most common meanings of those symbols that occur frequently in this book; special usages of these symbols and the meanings of any unlisted symbols are defined in the text wherever they occur. Symbols used to represent units for physical quantities are given in Appendix B. A more complete listing of symbols is given in I. Mills et. al., *Quantities, Units and Symbols in Physical Chemistry,* 2d ed., published for IUPAC by Blackwell, Oxford (1993).

Symbol	Meaning
a	Activity
c	Concentration, molecular speed, speed of light
a, b, c	Crystal unit cell dimensions
d	Diameter (molecular), density, Bragg lattice-plane spacing
e	Electronic charge, base of natural logarithms
f	Force, function, frequency, fugacity, formation (subscript)
g	Acceleration due to gravity, gas, degeneracy
h	Planck constant, height
i	$\sqrt{-1}$
k	Boltzmann constant (also k_B), rate constant, force constant, scalar wave vector
i, j, k	Indices of array elements (in vectors, matrices)
l	Liquid, length
h, k, l	Miller indices for crystal planes
m	Mass, mass of atom or molecule, molality
n	Number of moles, index of refraction
p	Pressure
p_B	Partial pressure of B
q	Heat absorbed by the system, molecular partition function
r	Radius, distance
s	Solid
t	Celsius (centigrade) temperature, time, parameter in Student distribution
u	Root-mean-square speed
v	Scalar velocity (speed), vibrational quantum number, volume adsorbed
w	Work done upon the system
x, y, z	Cartesian coordinates
z	Valence of an ion

A	Helmholtz free energy, area, absorbance
B	Second virial coefficient, scalar magnetic flux density, rotational constant
C	Heat capacity, capacitance, number of components
C_p	Heat capacity at constant pressure
C_v	Heat capacity at constant volume
D	Diffusion constant, centrifugal distortion constant
D_0, D_e	Dissociation energy referenced to ground state and potential minimum respectively
E	Energy, scalar electric field strength
E_a	Activation energy
F	Rotational term symbol
G	Gibbs free energy, vibrational term symbol
H	Enthalpy, scalar magnetic field intensity, Hamiltonian
I	Intensity of radiation, moment of inertia, ionic strength, electric current
J	Rotational quantum number, flux
K	Equilibrium constant, thermal conductivity
K_f	Molal freezing constant
L	Length
M	Molar mass ("molecular weight"), molarity, radiant excitance (emitted radiant flux)
N	Number of particles (molecules, atoms, ions), normality
$\bar{N}$	Concentration in molecules per unit volume
N_0	Avogadro's number
P	Number of phases, power, scalar polarization
P_M	Molar polarization
Q	Electric charge, generalized thermodynamic quantity, canonical partition function, vibrational coordinate
R	Gas constant, resistance, radius
R_M	Molar refraction
S	Entropy, estimated standard deviation
T	Absolute temperature, time period, spectroscopic term value, transmittance
U	Potential energy, ionic mobility
V	Volume, voltage (potential difference)
W	Weight
X	Mole fraction, reactance
Z	Collision frequency, atomic number, number of formula units per unit cell, compressibility factor, impedance

a, b, c	Crystal lattice vectors	⎫
p	Momentum	
k	Wave vector	
r	Position vector	
v	Velocity	
B	Magnetic flux density	⎬ vector quantities
D	Electric displacement	
E	Electric field strength	
H	Magnetic field strength	
M	Magnetization	
P	Polarization (dielectric)	⎭

$\mathscr{A}$	Pre-exponential factor in Arrhenius expression
$\mathscr{E}$	Electromotive force (emf)

$\mathscr{F}$	Faraday constant
$\mathscr{R}_\infty$	Rydberg constant
α	Thermal-expansion coefficient, degree of dissociation, polarizability, angle of optical rotation, adjustable parameter (least squares)
α_e	Vibration–rotation coupling constant
α_0	Distortion polarizability
β	$1/kT$, parameter in Morse potential function, hyperpolarizability constant, critical exponent
γ	Activity coefficient, surface tension, ratio C_p/C_v
δ	Deviation, chemical shift (NMR)
ε	Molecular energy, permittivity, molar absorption coefficient (absorptivity), emittance
ε_0	Permittivity of vacuum
ϵ	Error
η	Coefficient of viscosity
θ	Surface coverage, angle (e.g., Bragg angle)
κ	Electrical conductivity, relative permittivity (dielectric constant)
κ_S, κ_T	Adiabatic, isothermal compressibility
λ	Wavelength, ionic equivalent conductance, mean free path
μ	Chemical potential, Joule–Thomson coefficient, scalar dipole moment, reduced mass, permeability
μ_0	Permeability of vacuum
μ_B, μ_N	Bohr magneton, nuclear magneton
$\nu, \tilde{\nu}$	Frequency (in Hz), wavenumber (in cm^{-1})
$\tilde{\nu}_e x_e$	Anharmonicity parameter (in cm^{-1})
ξ	Magnetizability
ρ	Density, resistivity
σ	Molecular area or dimension, standard error, order parameter, shielding constant (NMR), symmetry number, cross-section
τ	Relaxation time, time constant, lifetime
ϕ	Apparent molal volume, angle
χ, χ_e	Magnetic susceptibility, dielectric susceptibility
χ^2	Distribution of goodness of fit
ψ	Wavefunction
ω	Circular frequency $(2\pi\nu)$
Γ	Surface concentration, gamma function (math)
Δ	Limit of error, difference
Θ	Characteristic temperature (e.g., Θ_{rot}, Θ_{vib}; Θ_D for Debye temperature)
Λ	Equivalent conductance
Π	Osmotic pressure
Ω	Solid angle
aq	Aqueous solution
ln	Natural logarithm
log	Logarithm to the base 10
pH	$-\log(a_{H+})$
°C	Degree Celsius (centigrade)
K	Degree kelvin
Q^0	Any thermodynamic property Q of a substance in its standard state
$\tilde{Q}$	Molal quantity Q
$\tilde{Q}_A$	Partial molal quantity Q for component A

Appendix

B

International System of Units and Concentration Units

The International System of Units (SI, for Système International) is based on seven dimensionally independent quantities:

Physical quantity	Name	Symbol
Length	meter	m
Mass	kilogram	kg
Time	second	s
Electric current	ampere	A
Thermodynamic temperature	kelvin	K
Amount of substance	mole	mol
Luminous intensity	candela	cd

There are also SI *derived units* with special names and symbols; some of these are listed below. More details on this and other matters are given in I. Mills et al., *Quantities, Units and Symbols in Physical Chemistry,* 2d ed., published for IUPAC by Blackwell, Oxford (1993).

Physical quantity	Name	Symbol	Definition
Frequency f (*not* $\omega = 2\pi f$)	hertz	Hz	s^{-1}
Force	newton	N	$m\,kg\,s^{-2}$
Pressure, stress	pascal	Pa	$N\,m^{-2} = m^{-1}\,kg\,s^{-2}$
Energy, work, heat	joule	J	$N\,m = m^2\,kg\,s^{-2}$
Power, radiant flux	watt	W	$J\,s^{-1} = m^2\,kg\,s^{-3}$
Electric charge	coulomb	C	$A\,s$
Electric potential, electromotive force	volt	V	$J\,C^{-1} = m^2\,kg\,s^{-3}\,A^{-1}$
Electric resistance	ohm	Ω	$V\,A^{-1} = m^2\,kg\,s^{-3}\,A^{-2}$
Electric capacitance	farad	F	$C\,V^{-1} = m^{-2}\,kg^{-1}\,s^4\,A^2$
Magnetic flux density	tesla	T	$V\,s\,m^{-2} = kg\,s^{-2}\,A^{-1}$
Magnetic flux	weber	Wb	$V\,s = m^2\,kg\,s^{-2}\,A^{-1}$
Inductance	henry	H	$V\,A^{-1}\,s = m^2\,kg\,s^{-2}\,A^{-2}$
Luminous flux	lumen	lm	$cd\,sr^{-1}$
Plane angle	radian	rad	—
Solid angle	steradian	sr	—

SI derived units exist for other physical quantities, but many of these do not have special names or symbols. Most of these are obvious—volume (m^3), mass density ($kg\,m^{-3}$), velocity ($m\,s^{-1}$), molar volume ($m^3\,mol^{-1}$), molar energy ($J\,mol^{-1}$), etc. A few important but less obvious cases are listed below as a convenient illustration.

Physical quantity	Definition
Heat capacity, entropy	$J\,K^{-1}$
Surface tension	$N\,m^{-1}$
Heat flux density, irradiance	$W\,m^{-2} = kg\,s^{-3}$
Thermal conductivity	$W\,m^{-1}\,K^{-1} = m\,kg\,s^{-3}\,K^{-1}$
Kinematic viscosity, diffusion coefficient	$m^2\,s^{-1}$
Dynamic viscosity	$N\,s\,m^{-2} = Pa\,s = m^{-1}\,kg\,s^{-1}$
Electric displacement, polarization	$C\,m^{-2} = m^{-2}\,s\,A$
Permittivity	$F\,m^{-1} = m^{-3}\,kg^{-1}\,s^4\,A^2$
Permeability	$H\,m^{-1} = m\,kg\,s^{-2}\,A^{-2}$
Electric field strength	$V\,m^{-1} = m\,kg\,s^{-3}\,A^{-1}$
Magnetic field strength	$A\,m^{-1}$
Luminance	$cd\,m^{-2}$
Exposure (X and γ rays)	$C\,kg^{-1} = kg^{-1}\,s\,A$

Note that the symbols for SI units and derived units are all represented by roman letters, with an initial capital letter used when the unit is named after a person. These symbols are never expressed as plurals (with an added s) and are never followed by a period except at the end of a sentence. A sequence of symbols should alternate with spaces as shown above. The format a/b can be used in place of a b^{-1} (e.g., J/K instead of $J\ K^{-1}$ for entropy), but the exponent format is preferable since it is unambiguous for compound units (e.g., there is some possibility of confusion with W/mK for thermal conductivity but none with $W\ m^{-1}\ K^{-1}$).

It is often convenient to handle decimal fractions and multiples of the above SI units by attaching prefixes to the symbols. The commonly used prefixes are listed below.

Fraction	Prefix	Symbol	Multiple	Prefix	Symbol
10^{-1}	deci	d	10	deca	da^a
10^{-2}	centi	c	10^2	hecto	h^a
10^{-3}	milli	m	10^3	kilo	k
10^{-6}	micro	μ	10^6	mega	M
10^{-9}	nano	n	10^9	giga	G
10^{-12}	pico	p	10^{12}	tera	T
10^{-15}	femto	f	10^{15}	peta	P^a
10^{-18}	atto	a	10^{18}	exa	E^a

[a] These prefixes are rarely used.

The concentration units and symbols defined below are widely used in physical chemistry.

Name	Symbol	Definition
Weight percent	%	(Grams of solute per grams of solution) $\times$ 100
Mole fraction[a]	X_A	Moles of A per total number of moles
Molarity	M	Moles of solute per liter of solution[b]
Normality	N	Equivalents of solute per liter of solution[b]
Formality[c]	F	Formula weights of solute per liter of solution[b]
Molality	m	Moles of solute per kg of solvent
Weight formality[c]	f	Formula weights of solute per kg of solvent

[a] The symbol Y_A is often used for the mole fraction of A in a gas phase that is in equilibrium with a liquid solution.

[b] Note that $1\ L \equiv 1\ dm^3$. The SI concentration unit of $mol\ m^{-3}$ is not very attractive for most work in chemistry, and non-SI concentrations based on the liter are widely used.

[c] These units are infrequently used but are of great convenience in expressing the overall composition of a solution when the solute is partially associated or dissociated.

Appendix

C

Safety

Given below is a short reminder list of basic safety principles and specific warnings about hazards that can occur in a physical chemistry laboratory. Items 1 and 2 are crucially important. Knowing how to respond effectively in the unlikely event of a serious accident is essential. **Safety information should be displayed prominently in the laboratory—** find it and read it before beginning any experimental work.

1. Review experimental procedures and identify possible safety hazards *before* beginning laboratory work.
2. Learn the location and proper use of all safety equipment available in the laboratory as well as the fastest method of obtaining emergency medical assistance.
3. Never work alone in the laboratory.
4. Beware of high-voltage electricity.
5. Wear safety glasses (or goggles) in the laboratory at all times.
6. Never pipette chemicals by mouth; use a pipetting bulb.
7. Dispose of any waste chemicals in a proper manner.
8. Never look directly into any laser beam; even a reflected part of such a beam that enters the eye can cause permanent damage.
9. Limit the use of open flames, and never use them in the presence of flammable materials.
10. Beware of possible explosions due to gas overpressures, especially in glass systems.
11. Secure all high-pressure gas cylinders and use proper reducing valves.
12. Do not eat or drink in the laboratory.

The remainder of this appendix contains discussions of several safety issues pertinent to physical chemistry and a brief description of safety equipment. Such information serves as a starting point for safe laboratory practices. The authors do not claim that this treatment includes a discussion of every conceivable hazard or establishes legal standards for safe behavior in the laboratory. The primary literature and specialized sources of safety information should be consulted. An extensive list of such sources is given at the end of this appendix.

Electrical Hazards. Several experiments make use of 110-V ac electrical power and employ apparatus in which exposed metal parts are "live." If the laboratory table has a metal surface, cover it with an insulating mat or sheet of plywood or other material before assembling an electrical circuit. Remember that metal fixtures of all kinds and pipes or tubes of any kind that carry water are usually grounded. All electrical equipment should have a conveniently located switch so that the power source can be quickly disconnected in an emergency. Proper wiring in compliance with National Electrical Code (NEC) standards must be used, and the equipment should be grounded and properly fused to protect the user from electrical shock. Electrical equipment, especially items involving high voltage, must be isolated or shielded in a manner that protects against accidental contact with live circuits. Turn off all electrical apparatus before altering circuits, and disconnect the power source. Check that capacitors are discharged and shorted before working in their vicinity (normal discharge may be slow or not take place due to a broken circuit or defective bleeder resistor). If it is necessary to check or adjust an electrical apparatus while it is turned on, use properly insulated test probes and work with only one hand, keeping the other hand in your pocket or behind your back. *Never carry out any electrical work if doubt exists about your personal safety.*

Naturally 220 V represents a greater hazard than 110 V. It should be kept in mind that the laboratory is often served with 220 V in a three-wire system, with 110 V each side of ground. If 110-V outlets are supplied with a ground wire and one side of the 220-V line, as much as a 220-V difference can be obtained in accidental contact between circuits plugged into outlets serviced by opposite sides of the 220-V line.

Shock, if it does occur, can be a serious matter; medical help should be summoned at once. Keep the victim quiet and comfortable; administer no stimulants of any kind.

Chemical Hazards.   Chemical hazards are many and varied. It should be taken for granted that any chemical substance taken by mouth or inhaled is toxic until and unless definite assurance has been given to the contrary. Reactions that produce toxic fumes or vapors or entail risk of fire should always take place in a fume hood. As a matter of standard safety practice, never pipette any liquid or solution by mouth; use a rubber pipetting bulb. Another insidious hazard is that of vapors from organic solvents. Such solvents should not be used indiscriminately for cleaning purposes, and spills should be avoided. Good ventilation is important.

Environmental exposure to chemical hazards is currently a subject of concern and awareness. Zero exposure or zero risk of exposure is impossible in practice, either in the chemical laboratory or elsewhere. Part of the professional role of chemists is to acquire knowledge and to develop judgment as to which precautions are necessary to limit these risks. One should not be blindly afraid of every chemical in the laboratory, nor should one be foolishly fearless. Many chemical hazards can be avoided by simply not eating or drinking in the laboratory or breathing large volumes of vapors. Most chemicals are eliminated from the body, so that the effects of exposure gradually diminish. However, some substances are not eliminated completely and they accumulate, usually in particular tissues.

It is important to distinguish between infrequent, perhaps one-time, research use of a potentially hazardous chemical and chronic exposure in an industrial setting. Many of the substances specified as hazardous under the Occupational Safety and Health Act (OSHA), the Annual Report on Carcinogens from the National Toxicology Program (NTP), and the International Agency for Research on Cancer (IARC) are ones that are harmful after prolonged exposure but represent rather low risks from brief use in an undergraduate laboratory course or even in research.

However, *chronic* exposure to low levels of certain chemicals has been shown to increase significantly the incidence of cancer. Such chemicals are referred to as *select carcinogens*. Another class of risk is to pregnant women, especially during the first trimester. These chemicals are called *teratogens*. A third group consists of acutely toxic compounds such as Cl_2 gas, hydrazine, and HCN. Compounds in these three categories are classified as *particularly hazardous substances.* No strongly carcinogenic or strongly teratogenic substances are used in the experiments described in this book nor are acutely toxic chemicals. In the limited cases where hazardous chemicals are involved, safety warnings are given in the text. If any doubt should exist about the safety of a chemical, the Materials Safety Data Sheet (MSDS) should be consulted. Such information, provided by the supplier of the chemical, will be on file in the departmental safety office. Two common organic chemicals deserve mention here: benzene and carbon tetrachloride. In the past these were widely used as solvents, but each has toxicity hazards if ingested or inhaled at a high vapor pressure. Their use in this book is limited, and detailed warnings are given in Exp. 12 for CCl_4 and in Exp. 29 for benzene.

Special mention is also appropriate for **mercury** and mercury compounds, which can cause death if swallowed or inhaled as a vapor at high concentrations. Mercury compounds are not used in this book, and liquid mercury is used only as part of an amalgam in Exp. 18. The likelihood of ingesting liquid mercury is remote, but one should be alert about exposure to its vapor. Chronic inhalation of mercury vapor is destructive to mucous membranes and the respiratory tract. Although the greatest danger of such exposure arises from the vapors of hot liquid mercury, mercury vapor can attain appreciable concentration from the presence of exposed surfaces of mercury at room temperature. The vapor pressure is 1.85 mTorr at 25°C, which corresponds to a concentration of 20 mg m^{-3} in the very unlikely case that saturation were to occur. This means that *chronic exposure* can be a concern. The threshold limit for long-term, continuous, 40-hour-per-week exposure is 0.025 mg m^{-3}. The occasional use of mercury in laboratory amounts in ventilated areas is not a serious hazard, but certain precautions are necessary. Mercury spills must be cleaned up (a capillary tube attached to a suction flask is convenient for this), and inaccessible droplets in floor cracks and hard-to-reach places should be covered with a light dusting of powdered sulfur. Any mercury recovered from spills or broken apparatus should be placed in thick-walled high-density polyethylene bottles for proper disposal by the laboratory staff.

Mercury-in-glass laboratory thermometers are convenient for many applications and are still frequently used. The only safety concern is the possibility of a broken thermometer bulb. If breakage occurs, the thermometer must be given to the laboratory staff for disposal and any mercury spillage must be promptly cleaned up. Safety concerns about mercury should be kept in perspective. There is no appreciable danger from the normal use of mercury thermometers or other laboratory devices containing mercury (such as manometers or barometers) if they are used properly.

Waste Disposal.   The general rule for waste disposal is that only dilute inorganic solutions and a few benign organic chemicals such as alcohol and acetic acid can be put down the drain. Materials such as organic solvents, concentrated acids and alkalis, or toxic chemicals (cyanides, arsenic, lead, and heavy-metal compounds) must be put into proper containers, securely capped and labeled, for disposal by the safety office.

Chemical Burns.   Strong acids (particularly oxidizing acids such as chromic acid cleaning solution) and bases may cause severe burns to the skin. Rubber gloves must be worn if one is handling such materials. If skin contact is made, wash copiously with water. If the exposure is to a strong acid, washing with a very dilute weak base (ammonia) is helpful; for a strong base use a very dilute weak acid (acetic acid). Particular attention should be directed to eye protection: *Safety glasses,* safety goggles, or a face shield must be worn in the laboratory at all times. Prompt and effective action is essential if any chemical agent gets into the eyes; a strong base such as sodium hydroxide can permanently destroy the cornea in a few seconds. *Speed is all-important* in getting the exposed individual to an eyewash fountain or other source of copiously flowing (but low-pressure) water and *thoroughly* bathing the eyeball. The eyelids should be lifted away from the eyeball to facilitate effective washing. Use nothing but water. *Get medical help promptly.*

Fire and Explosion.   Any flammable substance provides a potential fire hazard. In experiments that make use of hydrogen gas or other flammable gases, not only open flames but also cigarettes and sparking electrical contacts provide the possibility of explosion. If they are used in large quantities, such gases must be vented into the open air outside the building. A direct exhaust line from the experiment to a nearby window is best but an exhaust fan and cross ventilation will serve. The distillation of flammable liquids must be carried out in the absence of open flames; use a steam bath or electrical heating mantle. If an experiment involves an irreducible risk of fire or explosion, arrange for an adequate barrier. *Safety goggles* are required in all circumstances in which fire or explosion is a possible eventuality. Know the location of the nearest water supply and fire extinguishers (use water only on paper or cloth fires). In the event of serious burns, do not apply ointments or medications; get medical help.

Vacuum Apparatus and Liquid Nitrogen Hazards.   All-metal vacuum systems present essentially no hazards of implosion or explosion, and glass vacuum systems are also quite safe. However, certain precautions must be taken in the use of glass systems. The danger of injury from the implosion of large evacuated glass bulbs arises since atmospheric pressure is great enough to generate a hail of small (and sharp) glass fragments. Thus any bulb more than 1 L in volume should be surrounded by a fine-mesh metal screen or have its surface covered by multiple strips of plastic electrician's tape. Do not use large-size flat-bottom vessels in vacuum systems, although small optical or spectroscopic cells with flat windows can be evacuated without concern.

Another danger is the bursting of a system due to excessively high internal overpressures (explosion rather than implosion). This might happen in three distinct ways. If a substantial quantity of a gas has been liquefied or adsorbed on a high-area adsorbent at low temperatures, an explosion can result when the system warms up, if adequate venting has not been provided. Second, if a gas is being added to a vacuum system from a cylinder of compressed gas, there is the obvious danger that the pressure inside the system could significantly exceed atmospheric pressure if the gas cylinder does not have a suitable reducing valve. Both of these explosion hazards can be eliminated, if necessary, by including a safety valve in the vacuum system that will prevent the overpressure from exceeding some small preset value. The third, and least likely, explosion hazard arises if an oil diffusion pump (especially a glass one) is inadvertently vented to the ambient air atmosphere while the oil is still hot. *Never vent a hot oil diffusion pump.* Indeed, avoid exposure of hot pump oil to air at pressures greater than about 0.1 Torr.

A different set of safety issues arises from the use of liquid nitrogen, which is often used for vacuum cold traps as well as a general low-temperature bath fluid. Nitrogen is obviously nontoxic and is very safe, and liquid nitrogen is used in large quantities in most physical chemistry laboratories. The only hazard from liquid nitrogen is skin "burns" due to its low temperature. Thus *avoid direct contact of liquid nitrogen with the hands or other exposed skin* and wear insulating gloves if handling cold transfer tubes delivering liquid nitrogen. If liquid nitrogen is being poured from a storage Dewar to a smaller Dewar, pour it *slowly* to avoid messy and wasteful boil-off that might cause a spray of small liquid droplets. When an object (e.g., sample bulb or cold trap) is being inserted into a Dewar of liquid nitrogen, proceed *slowly* in order to reduce boil-off and limit thermal shock to the sample.

When using a liquid nitrogen cold trap for a vacuum system, there is a danger of condensing liquid oxygen inside the trap, since the vapor pressure of liquid oxygen is 157 Torr at 77.35 K (the normal boiling point of nitrogen and thus the approximate temperature of the trap) and air at a total pressure of 1 atm has an oxygen partial pressure of 179 Torr. Thus, if an appreciable quantity of air were pumped through a cold trap, liquid oxygen could accumulate. *Do not refrigerate a trap with liquid nitrogen until the system pressure is below a few Torr.* The concern about liquid oxygen is the possibility of an explosive reaction with organic substances that might condense in the trap during normal use of the system. Liquid oxygen alone can be the cause of excessive internal pressure and a possible nonchemical explosion should a trap containing liquid oxygen be warmed up without venting (as mentioned earlier).

Radiation Hazards.   Ultraviolet light from a mercury lamp or carbon arc can be highly damaging to the eyes. Ordinary glasses give some protection, but the experimental arrangement should be well shielded so as to decrease the possibility of accidental exposure to a minimum. Prolonged exposure of the skin to such radiation can produce a severe "sunburn."

Exposure to strong radiofrequency or microwave fields can "cook" tissue and produce deep internal burns. Exposure to X rays and to the radiation from radioactive materials must be carefully guarded against in experiments dealing with them. Any such experiments should be done under the direct supervision of an experienced research worker, who will assume personal responsibility for all required safety measures, and under an appropriate license if radioactive materials are involved.

Laser Hazards.   Lasers are used in several experiments in this book and in a variety of research and commercial applications. There are two significant hazards associated with lasers; one is an electrical risk due to the high voltages and currents used in many of them, and the second is the possibility of eye damage due to inadvertent exposure to a direct or reflected beam. Of these, the electrical risk is greater, and there have been several fatalities due to contact with laser power supplies or capacitors, usually during maintenance operations that are unlikely to be performed by students. No fatalities due to direct exposure to a laser beam are known, although a number of cases of eye damage have been documented, for example, by D. C. Winburn in the book *Practical Laser Safety.* Only for very powerful lasers is the risk of burns on exposed skin of any concern.

Lasers vary greatly in their power output and risk rating. At one extreme are low-power CW lasers used in laser printers, compact-disk players, and supermarket scanners. These do not require special labeling and precautions and are generally termed Class 1 devices in the classification scheme recommended in the American National Standards Institute report ANSI Z136.1. The precise definitions of the classifications are somewhat involved but are spelled out in this report, which is reproduced by Winburn and in other books on lasers. In general, Class 1 lasers are "incapable of producing damaging radiation levels." Class 2 CW lasers include helium-neon lasers with a power of 1 mW or less, and for these the blink reflex of the eye is normally sufficient to provide protection. The use of such a laser for the interferometric determination of the index of refraction of gases is described in Exp. 30, and such lasers are also employed in FTIR instruments to establish the position of the moving mirror. Visible CW lasers of powers up to 500 mW, such as obtained with argon- and krypton-ion lasers and dye lasers, fall in the Class 3 category, and safety goggles should always be used when working with these sources, as well as with similar Class 4 lasers that have CW powers in excess of 500 mW. Argon- or krypton-ion lasers, red diode lasers, 532-nm laser modules, and dye lasers serve as excitation sources for Raman or fluorescence measurements in Exps. 35 and 39. Special care should be taken in positioning samples in the laser beam in these experiments, and laser safety goggles capable of reducing the intensity by at least 10^4 should be used.

Pulsed lasers are more hazardous than CW lasers because the eye cannot react in the time of a laser flash. The ANSI Z136.1 safety classification scheme used for pulsed lasers is in terms of energy intensity ($J \; cm^{-2}$) per flash and is more stringent for pulse durations shorter than a nanosecond. For visible pulses of nanosecond duration, such as the 5 to 10 ns, 532-nm doubled Nd:YAG output used in Exps. 36, 40, and 44, total output energies in excess of a few microjoules are likely and would move the safety category from Class 1 to Class 3. (There are no pulsed lasers of Class 2.) Safety goggles are mandatory in working with any pulsed laser in Class 3, as well as in Class 4, which corresponds to energy intensities >31 $mJ \; cm^{-2}$. Suppliers of glass and plastic goggles include Bollé, Carl Zeiss Optical, Glendale Protective Technologies, Lase-R Shield, and other companies, and typical prices range from $100 to $400. One must specify the wavelength region to be blocked, since filters are chosen to pass other visible wavelengths so that one need not work completely in the dark. Safety goggles are especially important in dealing with invisible ultraviolet or near-infrared sources such as the 1060-nm output of a Nd:YAG laser. Most of the reported serious eye injuries have involved small reflected amounts of energy from this source, since it is widely used and can involve pulse energies of as much as a joule.

In working with lasers, a number of commonsense practices can reduce the possible exposure to reflections, the most common cause of accidents. When using intense sources, it is always wise to do preliminary alignment of beams and samples at reduced laser power before turning up the laser current. A check for stray reflections can be done, again at low power, by carefully removing your goggles while keeping your back to the source. If any are found, they should be eliminated. Avoid passing your eyes through any planes of reflection of beam-directing optics, and always close your eyes if you pass through such a plane, as might occur in picking up an item dropped on the floor. When working with lasers, take care to insert any windows or lenses so that reflections will go down instead of up toward your eyes. Avoid wearing shiny watches or jewelry that might cause reflections, and beware of reflections from your fingernails if you must insert your hands in the beam. *Do not look at a laser or likely reflecting objects without goggles and of course never stare directly into a laser beam no matter what its power level.*

Mechanical and Other Hazards. Most mechanical hazards are too clearly apparent to warrant mention here. The danger lies in forgetfulness or casual disregard of risks. The bursting of a container due to overpressure can be a cause of accident or injury. A compressed-air line (usual pressure of the order of 50 psi) should never be connected to a closed system containing rubber tubing or glass bulbs. No closed system, except a properly designed combustion bomb, should be attached to a cylinder of compressed gas (usual maximum pressure: about 3000 psi) unless a suitable reducing valve is attached; even then a relief valve should be provided to guard against accidental overpressure. Gas cylinders must be chained or strapped to prevent their falling over. The protective cap must be in place whenever a gas cylinder is being moved; cylinders should be moved with an appropriate hand truck, not dragged across the floor. *Mechanical pumps must have belt guards.*

Unattended operations must be planned with automatic safety switches that prevent serious damage (fire, flooding, explosion) in case of accidental equipment failure or interruption of utility services such as electricity, water, or gas supplies. Of special concern are the constant flow of cooling water and the operation of high-temperature baths. In the case of water flow, a device should be installed in the water line to (1) automatically regulate the water pressure (so as to avoid surges that might disconnect or rupture a water hose), and (2) automatically turn off electrical connections and water-supply valves in case of a total loss of water supply. In the case of hot thermostat baths or ovens, a sensor/control device should be installed that automatically turns off the electrical power to all heaters if the temperature exceeds some preset upper limit.

Caution must be taken in the proper handling and disposal of sharp objects that could puncture the skin: broken glass, jagged metals, razor blades, hypodermic or other needles. Safe disposal requires a container that is impervious to puncture and can be incinerated.

Safety Equipment. *Safety glasses* or *goggles* have been mentioned in this appendix and in other places in this book in connection with specific hazards. However, use of safety glasses equipped with side shields, or other approved means of eye protection (plastic goggles alone or over ordinary prescription glasses, plastic face shields), is usually *mandatory at all times in instructional laboratories,* just as it is in industrial research laboratories. The use of safety goggles is strongly recommended as standard laboratory policy. Their use is essential in all circumstances where there exists the possibility of fire, explosion, implosion, spattering of caustic chemicals, or flying fragments from machine-shop operations. Fortunately, very few of these risks exist in the experiments described in this book, and explicit warnings are given in each case where hazards do occur.

The laboratory should be equipped with a conveniently accessible safety shower and an eyewash fountain; there should be more than one of each in a large laboratory. Increasingly, the fixed type of eyewash fountain is being superseded by a spray nozzle at the end of an extensible hose; there should be one of these on each laboratory bench. In lieu of such devices—or in addition to them—2- or 3-ft lengths of rubber hose (*not* small-bore pressure tubing) attached with wire or clamps to water faucets are certainly better than nothing.

The laboratory should have convenient access to one or more fume hoods (with a face velocity of at least 100 ft/min) for any operations involving more than insignificant quantities of volatile chemicals in open containers.

An approved fire extinguisher [the "dry chemical" (bicarbonate) type is preferred, but the CO_2 type is satisfactory] should be mounted near at least one exit and refilled after every use, no matter how small. The laboratory should be arranged so as to provide two or more avenues of escape from any experimental setup in case of emergency. A first-aid kit containing Band-Aids, sterile gauze, adhesive tape, petroleum jelly, a mild antiseptic, sterile cotton swabs, tweezers, a set of sewing needles, a packet of razor blades, and a quick-reference first-aid manual will provide adequately for most minor emergencies.

The location of an inhalator, a stretcher, and other rescue equipment, if not in the laboratory itself, should be known. The telephone number of the nearest medical emergency room and the local ambulance service should be posted conspicuously. Instructions for emergency evacuation, including special procedures for evacuating physically handicapped persons, should also be posted. An evacuation drill held near the beginning of each academic term is recommended.

Under no circumstances should a person be allowed to work in the laboratory alone.

SOURCES OF SAFETY INFORMATION

M. A. Armour, *Hazardous Laboratory Chemicals Disposal Guide,* 2d ed., CRC Press, Boca Raton, FL (1996).

G. S. Coyne, *The Laboratory Companion: A Practical Guide to Materials, Equipment and Techniques,* rev. ed., Wiley, New York (1997).

A. K. Furr (ed.), *CRC Handbook of Laboratory Safety,* 5th ed., CRC Press, Boca Raton, FL (2000). Contains a wide variety of information.

L. B. Gordon, *IEEE Trans. Educ.* **34,** 231 (1991). Contains excellent general information on electrical hazards.

M. Grandolfo (ed.), *Light, Lasers, and Synchrotron Radiation,* Kluwer, Norwell, MA (1991). Contains a chapter discussing the American National Standard for the safe use of lasers.

J. Hecht, *The Laser Guidebook,* 2d ed., McGraw-Hill, New York (1992). A useful general text on lasers that includes discussion of safety concerns.

L. H. Keith and D. B. Walters (eds.), *Compendium of Safety Data Sheets for Research and Industrial Chemicals,* Wiley, New York (1985).

P. Patnaik, *A Comprehensive Guide to the Hazardous Properties of Chemical Substances,* Wiley, New York (1992). Contains a classification of chemicals by structure and functional groups with a discussion of general hazards.

Prudent Practices in the Laboratory: Handling and Disposal of Chemicals, National Academy Press, Washington, DC (1995).

D. Sliney and M. Wolbarsht, *Safety with Lasers and Other Optical Sources,* Kluwer, Norwell, MA (1980). An older but exhaustive review.

Standard on Fire Protection for Laboratories Using Chemicals, National Fire Protection Association (NFPA 45) (1991). Contains information on compressed gases and cryogenic fluids.

D. C. Winburn, *Practical Laser Safety,* 2d ed., Dekker, New York (1990). Contains ANSI Z136.1 classification scheme and a critique of this from the standpoint of a laser safety officer with much experience.

J. A. Young (ed.), *Improving Safety in the Chemical Laboratory,* 2d ed., Wiley, New York (1991). Contains information on high voltages, high pressures, vacuum, lasers, low temperatures, and cryogenics.

It is convenient to distinguish between two types of literature search: one concerned with a broad area of research, a method, a class of reactions, or a class of compounds; and another concerned with information about a single chemical species or reaction. In the former, methodology-oriented case, you should begin by checking the review literature—specialized monographs (consult the subject catalog in the library), reference books, encyclopedias, review journals, and annual series such as "Advances in _____." In the latter, substance-oriented case, a variety of options are now available. The print literature, especially a few key handbooks and encyclopedias, still is a valuable source of chemical and physical data. Alternatively one could make an online Internet search of computerized commercial databases using a wide variety of search interfaces, although this often proves to be an expensive option. In recent years, there has been a rapid growth in web-based data resources—both commercial and noncommercial ones like those maintained by the National Institute of Standards and Technology (NIST). Finally one can make an abstract search to find relevant journal articles.

REVIEW LITERATURE

Chemical literature of the review type consists of collections of articles or chapters on specific topics written by presumed experts. The introduction to these reviews usually specifies the period of time covered. Ideally the reviews include critical evaluations of the work and progress in a specific area of research together with a rather complete bibliography of the important original research articles pertinent to that area. However, it should not be assumed that every reference to a particular subject is given or that all of the author's critical interpretations are necessarily correct. A selection of the review literature that is of special value in physical chemistry is listed below.

Review Journals. Several journals specialize in articles reviewing the current status of experiment and/or theory in active research fields. The most valuable for physical chemistry are

Accounts of Chemical Research

Advances in Physics

Chemical Reviews

Chemical Society Reviews

Reviews of Scientific Instruments

Review Series. Review series are books that contain long chapters on specialized topics, in contrast to the generally shorter articles in review journals.

Advances in Chemistry Series, American Chemical Society, Washington, DC (1950–) (more than 220 books issued at irregular intervals).

Advances in Chemical Physics, I. Prigogine and S. Rice (eds.), Wiley-Interscience, New York (1958–) (more than 100 books).

Advances in Magnetic Resonance, Vols. 1–14 (1965–1989) and *Advances in Magnetic and Optical Resonance,* Vols. 15– (1990–), W. S. Warren (ed.), Academic Press, San Diego, CA.

Advances in Quantum Chemistry, P.-O. Löwdin (ed.), Academic Press, San Diego, CA (1964–).

Annual Review of Physical Chemistry, H. L. Strauss (ed.), Annual Reviews, Inc., Palo Alto, CA (1950–).

Solid State Physics, H. Ehrenreich (ed.), Academic Press, San Diego, CA (1955–) (more than 50 volumes).

Springer Series in Chemical Physics, F. P. Schafer (ed.), Springer-Verlag, New York (1978–).

Structure and Bonding, Springer-Verlag, New York (1966–). (more than 90 volumes).

Reference Books. The two series of books listed below describe important experimental techniques and apparatus for a variety of physical measurements. These books also deal with the phenomenological theory of the methods described.

Physical Methods of Chemistry, 2d ed., Vols. I–X, B. W. Rossiter and J. F. Hamilton (eds.), Wiley-Interscience, New York (1986–1993).

Methods of Experimental Physics (1959–1994) and *Experimental Methods in Physical Sciences* (1995–), Academic Press, San Diego, CA.

The *Physical Methods of Chemistry* is a multivolume series that includes Components of Scientific Instruments (Vol. I), Electrochemical Methods (Vol. II), Determination of Chemical Composition and Molecular Structure (Vol. III), Microscopy (Vol. IV), Determination of Structural Features of Crystalline and Amphorous Solids (Vol. V), Determination of Thermodynamic Properties (Vol. VI), Determination of Elastic and Mechanical Properties (Vol. VII), Determination of Electronic and Optical Properties (Vol. VIII), Investigations of Surfaces and Interfaces (Vol. IX), and Supplement and Cumulative Index (Vol. X).

PHYSICAL PROPERTIES OF SUBSTANCES

This section is devoted to numerical tabulations of various physical properties. They are convenient, but some are distinctly secondary sources of information. It is often difficult to judge the quality of the data listed, since references to the original sources are sometimes inadequate and transcription errors can occur. If at all possible, it is wise to confirm important information by consulting the original literature.

Numerical values of physical properties are available in many forms. The traditional print literature contains tabulations for a very wide range of properties. Although printed tables are being supplanted by computer databases, they are still of value, especially for older data, which is often of high quality. A brief selection of print tables is given below. It should also be noted that some data tabulations may be available in your library in the form of CD-ROMs. However, the principal replacements of print literature are databases that can be accessed over the Internet by *online* proprietary search interfaces for commercial databases or *web-based* searches, which include both simplified interfaces for commercial sources and access to noncommercial databases.

There has been a rapid proliferation of commercial search interfaces, such as DataStar and STN International. STN may deserve special mention since it is a science information service operated by Chemical Abstracts Service (CAS) in North America, Fachinformationszentrum-Karlsruhe (FIZ-K) in Europe, and Japan Association for International Chemical Information (JAICI) in Japan. It is available as online subscriber access to 220 databases (STN Express) and as web access (STN on the Web for experienced searchers and STN Easy for infrequent searchers). See www.cas.org/stn.html and/or www.stn-international.de/. Since Internet search techniques depend on the interface available, no detailed explanations are given here. There are several published guides to database searches for chemical and physical properties,[1–5] but the situation is evolving rapidly and it is recommended that you consult a reference librarian for information about the current situation concerning search interfaces at the library you use.

General Tables.   Of the three encyclopedic collections of chemical data—Beilstein, Gmelin, and Landolt-Bornstein—the last is the one of significant value in physical chemistry. Landolt-Bornstein's *Zahlenweke und Functionen aus Physik, Chemie, Astronomie, Geophysik und Technik,* H. Borchers et al. (eds.), 6th ed., Springer-Verlag, Berlin (1950–1979), comprises an 11-volume compilation of a vast array of numerical data that has been critically evaluated. Although written in German and containing no index, it is relatively easy to use since the contents are tables of numerical data arranged according to a scheme specified in a detailed table of contents. Recent additions to Landolt-Bornstein are published as volumes in the New Series, which is printed in English:

Landolt-Bornstein Numerical Data and Functional Relationships in Science and Technology. New Series, K.-H. Hellwege (ed.), Springer-Verlag, New York (1965–). Groups II–IV are especially useful for physical chemistry.

Other general tables that are less encyclopedic but often easier to use are listed below.

American Institute of Physics Handbook, D. E. Gray *et al.* (eds.), 3d ed., McGraw-Hill, New York (1972). Although old, this volume still contains some useful information that supplements that given in the *Handbook of Chemistry and Physics.*

CRC Handbook of Chemistry and Physics, D. R. Lide (ed.), 89th ed., CRC Press, Boca Raton, FL (2008–2009). This is a particularly useful reference for obtaining a wide variety of routine physical data. Although a new edition is issued every year, changes are introduced very slowly. Thus any recent edition is likely to be almost as useful as the newest one. Also available online by subscription; see www.hbcpnetbase.com.

Journal of Physical and Chemical Reference Data, published for Natl. Inst. Sci. Tech. (NIST) by American Chemical Society and American Institute of Physics, Washington, DC (1972–). Regular issues plus many supplements.

Physik Daten/Physics Data, Fachinformationszentrum, Karlsruhe, Germany (1976–). Volumes 3-1 to 3-4, 9-1, 21-1, and 31 are especially useful.

Thermodynamic Tables.

NIST-JANAF Thermochemical Tables, 4th ed., M. W. Chase, Jr. (ed.), Monograph 9 of *J. Phys. Chem. Ref. Data,* American Institute of Physics, New York (1998). Critically evaluated data for inorganic substances and organic substances containing one or two carbon atoms.

TRC Thermodynamic Tables: Hydrocarbons, NIST Standard Reference Data Program, Gaithersburg, MD. Fourteen volumes of critically evaluated data on physical and thermodynamic properties of more than 3700 hydrocarbons and some sulfur derivatives of hydrocarbons present in petroleum and coal.

TRC Thermodynamic Tables: Non-Hydrocarbons, NIST Standard Reference Data Program, Gaithersburg, MD. Thirteen volumes of critically evaluated data on physical and thermodynamic properties of selected nonmetallic inorganic compounds and organic compounds other than hydrocarbons and sulfur compounds.

NIST Resources.   The National Institute of Standards and Technology has a Standard Reference Data Program that maintains electronic databases on Analytical chemistry, Atomic and molecular physics, Biotechnology, Chemical and crystal structure, Fluids, Materials properties, Surface data, and Thermochemical data. Some of these databases are available at moderate cost as "PC products" (diskettes, CD-ROMs, or Internet downloads) and some are free online systems. Further information is available on the website www.nist.gov/srd/#begin.htm.

NIST also maintains a website called the *NIST Chemistry WebBook* (http://webbook.nist.gov), which provides access to a broad array of data compiled under the Standard Reference Data Program. This site allows a search for thermochemical data for more than 7000 organic and small inorganic compounds, reaction thermochemistry data for over 8000 reactions, IR spectra for over 16,000 compounds, mass spectra for over 15,000 compounds, UV/VIS spectra for over 1600 compounds, electronic and vibrational spectra for over 5000 compounds, spectroscopic constants of over 600 diatomic molecules, ion energetics data for over 16,000 compounds, and thermophysical properties data for 74 selected fluids. The site allows general searches by formula, name, CAS registry number, author, and structure and also a few specialized searches by properties like molar mass and vibrational energies.

ABSTRACT SEARCH

The use of *Chemical Abstracts* (*CA*) and *Physics Abstracts* (*PA* = Science Abstracts Series A) is the most effective way to make a detailed search of the physical chemical literature from 1907 to the present time. These abstract journals provide journal citations and brief summaries of the important contents of original research publications, often identical to the abstracts given at the beginning of such publications. Both of these abstract series maintain electronic bibliographic databases that allow searches over the past 40 or more years. In the case of *Physics Abstracts,* the database *Inspec* is produced by the Institute of Electrical Engineers and covers electronics and computing as well as physics. More than 3800 journals are scanned each year, and in 2007 this database contained 9 million records going back to 1969. *Inspec Archive* covers the period 1898–1968. *Inspec* is a web-based product that can be accessed via commercial web services like Ovid and Dialog as well as Elsevier's interface Engineering Village. More information can be obtained from the *Inspec* website [www.iee.org.uk/publish/inspec]. In the case of *Chemical Abstracts, SciFinder Scholar* is designed for academic access to *CA* and the *CAS Registry* file. Abstract searches can be made back to 1907 using a variety of sophisticated Explore search routines.

In addition to abstract searches, *SciFinder Scholar* allows browsing the table of contents for a group of journals and also citation linking. It should be noted that *SciFinder Scholar* is not web based but uses the client (local computer)/server (CAS computer) model, where the local library subscribes for such use. More information is available from the website [www.cas.org/SCIFINDER/SCHOLAR/index.html].

The material given below concerns the print versions of *CA* and *PA* since searches in these are not menu driven as with the computerized databases. The index structures of *PA* and *CA* are somewhat different. *Physics Abstracts* provides an author index and a subject index. The latter contains both general topic entries, such as "dielectric relaxation," with specific compounds arranged as subentries, and chemical substance entries such as "argon" or "Nickel compounds." *Chemical Abstracts* provides six different indexes: author, general subject, chemical substance, formula, ring systems, and patent. If one is searching *CA* for information on a specific compound, the chemical substance index is naturally the most useful. It is larger and more complete than the subject index; furthermore many convenient subheadings are used to display properties, reactions, and simple derivatives of the given compound.

Some practice is necessary to acquire a rapid and thorough search technique for using *CA* or *PA* index volumes. The considerations outlined below should be helpful as an introduction to the use of the print abstracts. First, it is usually advisable to begin this type of literature search with the most recent abstract indexes available. In this way one may find recent review articles or monographs that are sufficiently complete in coverage that a search of the older literature can be avoided. Second, search under both the chemical name and key subject entries for the property of interest. Third, it is of obvious importance to use the correct name in searching for the desired compound. To ensure that the name being used is consistent with *Chemical Abstracts* usage, it is advisable to find the compound in the formula index and then use the name listed there or to refer to the most recent *Collective Index Guide to the Chemical Abstracts*. Note also that each chemical compound has a unique Chemical Abstracts Service (CAS) registry number. In dealing with compounds with complicated structures and thus complex names, it may be useful to cite the registry number when referring to the compound. An excellent guide to the use of the *Chemical Abstracts* prior to the introduction of *SciFinder Scholar* is Schulz and Georgy, *From CA to CAS Online.*[6]

The various abstract indexes discussed above contain item numbers that refer directly to an abstract in one of the abstract volumes. If the abstract indicates that the research results are of interest, the original journal article should be consulted whenever possible, even if the abstract gives the desired information. If a cited journal article is especially pertinent but not available locally, a reproduction can be obtained through interlibrary loan services. The availability of all journals abstracted in *Chemical Abstracts* is listed in *Chemical Abstracts Service Source Index,* published by the American Chemical Society.

It is important to realize that there is a delay of about six months between the time a journal article appears and the time its abstract is printed in *Chemical Abstracts* or *Physics Abstracts.* This is not surprising when you consider that *CA* presents abstracts from 14,000 journals published in 150 countries. In addition, there is a delay after an abstract volume is complete before the subject and substance indexes become available. Thus it is wise to check the most recent literature by scanning the title pages of likely journals. A convenient method for doing this is to scan through *Current Contents: Physical, Chemical, and Earth Sciences,* a weekly compilation of the latest tables of contents for many journals. A list of specific journals of greatest interest to physical chemists is given in Appendix E.

Another search method involves the use of the *Web of Science,* a commercially licensed website (http://webofscience.com) maintained by the Institute for Scientific Information. This website provides access to the *Science Citation Index Expanded,* which allows a search from 1973 to the present of more than 3000 journals. One can search for articles that cite an author or an explicit reference, which is based on the idea that such articles are likely to contain information pertinent to that in the given reference. Broader searches can also be made by topic (using key words) or author's name (resulting in a bibliographic listing of the author's publications with the number of times each has been cited).

REFERENCES

1. A. Twiss-Brooks and F. Whitcomb, *Searching the Scientific Literature,* Appendix IV in R. S. Berry, S. A. Rice, and J. Ross, *Physical Chemistry,* 2d ed., Oxford Univ. Press, New York (2000).

2. S. M. Bachrach (ed.), *The Internet: A Guide for Chemists,* American Chemical Society, Washington, DC (1996).

3. R. E. Maizell, *How to Find Chemical Information: A Guide for Practicing Chemists, Educators, and Students,* 3d ed., Wiley, New York (1998).

4. D. Stern, *Guide to Informatlon Sources in the Physical Sciences,* Libraries Unlimited, Englewood, CO (2000).

5. D. F. Shaw (ed.), *Information Sources in Physics,* 3d ed., Bowker Saur, London (2000).

6. H. Schultz and U. Georgy, *From CA to CAS Online: Databases in Chemistry,* 2d ed., Springer- Verlag, New York (1995).

Appendix

E

Research Journals

Out of the hundreds of different research periodicals currently being published, those listed below describe research of special interest to physical chemists. A very complete tabulation of periodicals, the correct abbreviation of their titles, and their distribution in American and selected foreign libraries can be found in *Chemical Abstracts Service Source Index.* Issues after the mid-1990s of almost all these selected journals are available on the Internet under licenses to research libraries.

Abbreviated title	Full title
Acc. Chem. Res.	Accounts of Chemical Research
Acta Crystallogr.	Acta Crystallographica
Appl. Opt.	Applied Optics
Biochemistry	Biochemistry
Biophys. Chem.	Biophysical Chemistry
Biophys. J.	Biophysical Journal
Bull. Chem. Soc. Jap.	Bulletin of the Chemical Society of Japan
Chem. Phys.	Chemical Physics
Chem. Phys. Lett.	Chemical Physics Letters
Int. J. Quantum Chem.	International Journal of Quantum Chemistry
J. Am. Chem. Soc.	Journal of the American Chemical Society
J. Appl. Phys.	Journal of Applied Physics
J. Chem. Phys.	Journal of Chemical Physics
J. Chem. Thermodyn.	Journal of Chemical Thermodynamics
J. Magn. Reson.	Journal of Magnetic Resonance
J. Mol. Spectrosc.	Journal of Molecular Spectroscopy
J. Opt. Soc. Am. B	Journal of the Optical Society of America B
J. Phys. B	Journal of Physics B: Atomic, Molecular and Optical Physics
J. Phys. Chem. A, B	Journal of Physical Chemistry A, B
J. Phys. Chem. Ref. Data	Journal of Physical and Chemical Reference Data
J. Phys. Chem. Solids	Journal of Physics and Chemistry of Solids
J. Phys.: Condens. Matter	Journal of Physics: Condensed Matter

(continued)

Abbreviated title	Full title
J. Phys. Soc. Jpn.	Journal of Physical Society of Japan
J. Polym Sci. A, B	Journal of Polymer Science, parts A, B
J. Quant. Spectrosc. Radiat. Transfer	Journal of Quantitative Spectroscopy and Radiative Transfer
J. Ram. Spectrosc.	Journal of Raman Spectroscopy
J. Solid State Chem.	Journal of Solid State Chemistry
Nature	Nature
Opt. Lett.	Optics Letters
Philos. Mag. B	Philosophical Magazine B
Physica A, B	Physica, sections A, B
Phys. Chem. Chem. Phys.	Physical Chemistry Chemical Physics (merger of former J. Chem. Soc., Faraday Trans., and Ber. Bunsen Phys. Chem.)
Phys. Lett. A	Physics Letters A
Phys. Rev. A, B, E	Physical Review, sections A, B, and E
Phys. Rev. Lett.	Physical Review Letters
Rev. Sci. Instrum.	Review of Scientific Instruments
Solid State Commun.	Solid State Communications
Spectrochim. Acta A	Spectrochimica Acta A: Molecular and Biomolecular Spectroscopy
Surf. Sci.	Surface Science
Thermochim. Acta	Thermochimica Acta
Z. Phys. Chem.	Zeitschrift für Physikalische Chemie

A brief introduction to numerical methods for handling experimental data is given in Chapter II, and a detailed description of least-squares fitting procedures is presented in Chapter XXI. In addition, the use of spreadsheet programs for a variety of numerical operations is discussed in Chapter III. This appendix presents a concise review of several basic numerical methods that are widely used in the analysis of data—smoothing, series approximations (both polynomial and Fourier series), differentiation, integration, and root finding. References are given to more detailed treatments elsewhere.[1-3]

SMOOTHING

Before data can be interpreted, whether by analytic or numerical means, it is sometimes necessary to "smooth" the data so that differences from point to point do not show random fluctuations. This is the equivalent of "drawing a smooth curve" through the data points on a graph, but it can be done more objectively by numerical methods, several of which are described elsewhere.[1,3] We shall give here one method, based on fitting a third-degree polynomial to $2n + 1$ adjacent points. For $n = 2$, the "smoothed" value of y_i is

$$y_i' = \tfrac{1}{35}[17y_i + 12(y_{i+1} + y_{i-1}) - 3(y_{i+2} + y_{i-2})] \tag{1}$$

For $n = 3$,

$$y_i' = \tfrac{1}{21}[7y_i + 6(y_{i+1} + y_{i-1}) + 3(y_{i+2} + y_{i-2}) - 2(y_{i+3} + y_{i-3})] \tag{2}$$

For $n = 4$,

$$y_i' = \tfrac{1}{231}[59y_i + 54(y_{i+1} + y_{i-1}) + 39(y_{i+2} + y_{i-2})$$
$$+ 14(y_{i+3} + y_{i-3}) - 21(y_{i+4} + y_{i-4})] \tag{3}$$

This procedure is equivalent to the Savitsky–Golay method, algorithms for which have been included in computer software for scientific instruments such as the Fourier-transform infrared (FTIR) spectrometer. Alternatives to smoothing are weighted least-squares fitting or optimal (Weiner) filtering techniques.[2]

SERIES APPROXIMATIONS

For many purposes it is desirable to construct an arbitrary series representation of the function $y = f(x)$ over the available range of x. There are several ways of doing this, of which two are very common. In both methods smoothing is often important as a way to avoid erratic fluctuations that may necessitate artificially high-order series.

Polynomial Fits. The most straightforward method is to fit a polynomial,

$$y = f_1(x) = \sum_{j=0}^{N} a_j x^j = a_0 + a_1 x + a_2 x^2 + \cdots + a_N x^N \tag{4}$$

to the experimental data by determining the "best" values of the constant coefficients a_j.[4] If the number m of data points is exactly equal to the number of terms $N + 1$ (N being the degree of the polynomial), then the coefficients are determined by the solution of a set of $m = N + 1$ simultaneous linear equations (one for each data point x_i, y_i). The resulting polynomial will then fit each of the data points exactly, but between the data points the polynomial will provide only an approximate prediction of y. The predicted values of y', the first derivative of y with respect to x, are in general everywhere approximate and particularly so near the ends of the range. If $m < N + 1$, the system of equations is underdetermined, and a unique solution yielding the values of the coefficients is unobtainable. If $m > N + 1$, the system of equations is overdetermined, and the coefficients may be determined by any of several methods, chief among which is the method of least squares, described in Chapter XXI. The values of the coefficients a_j, as well as their standard errors, are easily obtained from the multiple linear regression operation of spreadsheet programs, as discussed in Chapter III.

In carrying out polynomial fits, one must be alert to possible pathological behavior. If the observed quantities y_i vary in a smooth but rapid and complex way with x_i, one will need a high-degree polynomial to fit the data. This leads to a potentially serious problem. Two high-degree polynomials of different order that each fit a limited data set very well may yield different interpolated values and derivatives. Another problem situation arises when the measured physical quantity represents a singular function; for example the heat capacity of a pure fluid at its critical density tends smoothly to infinity as T approaches the critical temperature T_c. In such a case, a polynomial fit will improperly round off the sharp physical feature.

In either of the above cases, the use of an empirical series fit can be dangerous. If polynomial fitting is desired, the best approach is to abandon the idea of using a single polynomial over the entire range. Instead one can join together a set of different cubic polynomials that fit adjacent groups of data points. This procedure, called *cubic spline interpolation,* is based on a mathematical analysis of the use of a draftsman's spline, the flexible plastic curve that can be used to draw pleasingly smooth curves by eye. The algorithm for calculating a cubic spline fit is quite straightforward; see Refs. 2 and 5.

Fourier Series Fits. The other widely used method of constructing the function $y(x)$ is by use of Fourier series:[6]

$$y = f_2(x) = a_0 + 2 \sum_{h=1}^{N} \left[a_h \cos\left(\frac{2\pi h x}{L} \right) + b_h \sin\left(\frac{2\pi h x}{L} \right) \right] \tag{5}$$

Here the coefficients to be determined are a_h and b_h. This function is periodic in x, with a period of arbitrary length L. The coefficients are determined by

$$a_0 = \frac{1}{L}\int_0^L y(x)\,dx \simeq \frac{\Delta x}{L}\sum_{i=1}^m y_i = \bar{y} \tag{6a}$$

$$a_h = \frac{1}{L}\int_0^L y(x)\cos\left(\frac{2\pi hx}{L}\right)dx \simeq \frac{\Delta x}{L}\sum_{i=1}^m y_i\cos\left(\frac{2\pi hx_i}{L}\right) \tag{6b}$$

$$b_h = \frac{1}{L}\int_0^L y(x)\sin\left(\frac{2\pi hx}{L}\right)dx \simeq \frac{\Delta x}{L}\sum_{i=1}^m y_i\sin\left(\frac{2\pi hx_i}{L}\right) \tag{6c}$$

If the period L is extended to infinity, we obtain a *Fourier transform pair:*

$$y = f_3(x) = \int_{-\infty}^\infty F(h)e^{2\pi ihx}\,dh \tag{7a}$$

$$F(h) = \int_{-\infty}^\infty f_3(x)e^{-2\pi ihx}\,dx \tag{7b}$$

where i is here the square root of minus one and $f_3(x)$ and $F(h)$ are in general complex quantities but may be real in special circumstances. In practical usage of course, the infinite limits are replaced by finite ones, taken as large as possible; generally the integrals are evaluated by numerical means, being approximated by sums with very narrow intervals between terms.

The Fourier series is particularly appropriate when the function $y(x)$ is known to be periodic and is much used in problems involving waves (sound, light, ocean waves, diffraction phenomena, etc.). It can however be used for general fitting of data, with L set at some value greater than the range of x. Often it is desirable that the range covered by L should extend somewhat beyond the range of x at both ends, since troublesome peaks may occur at both ends of the range covered by L owing to the fact that the number of terms in the series is not infinite. Generally the number of terms should be as high as possible but should not exceed a fraction (say a tenth to a quarter) of the number of data points. The use of the Fourier series for general data fitting is sometimes criticized as artificial because the trigonometric functions, on a computer, are themselves provided in the form of approximating polynomials; thus the Fourier series could itself be expressed as a polynomial constructed in a rather special way.

It is unrealistic to attempt the use of the Fourier series or of the Fourier integral transforms without the aid of a computer. In recent years a "fast Fourier transform" (FFT) algorithm for computers has become widely used.[1,5,6] This is particularly useful in certain kinds of chemical instrumentation, specifically nuclear magnetic resonance and infrared absorption spectrometers. In such instruments the experimental observations are obtained directly in the form of a Fourier transform of the desired spectrum; a computer that is built into the instrument performs the FFT and yields the spectrum (see Chapter XIX).

Having obtained a fitting function $y(x)$ in the form of a polynomial, Fourier series or integral transform, or other form, we may differentiate it or integrate it as desired.

DIFFERENTIATION

For a set of data x_i, y_i (which have been smoothed if necessary), we now consider methods for differentiation that do not require a fit to some analytic form or fitting function covering the entire range of x. *For convenience, we shall take the values of the independent variable x to be evenly spaced.* For differentiation, a *local* construction of the function $y(x)$ can be based on a general interpolation formula,[7] which utilizes successive differences obtained as follows:

$$
\begin{array}{llllll}
x_0 = x_0 & y_0 \\
& & \Delta y_0 \\
x_1 = x_0 + w & y_1 & & \Delta^2 y_0 \\
& & \Delta y_1 & & \Delta^3 y_0 \\
x_2 = x_0 + 2w & y_2 & & \Delta^2 y_1 & & \Delta^4 y_0 \\
& & \Delta y_2 & & \Delta^3 y_1 & & \Delta^5 y_0 & \quad (8) \\
& & & & & & \vdots \\
x_3 = x_0 + 3w & y_3 & & \Delta^2 y_2 & & \Delta^4 y_1 \\
& & & & & & \vdots \\
& & \Delta y_3 & & \Delta^3 y_2 \\
& & & & \vdots \\
x_4 = x_0 + 4w & y_4 & & \Delta^2 y_3 \\
& & & & \vdots \\
& & \Delta y_4 \\
& & \vdots \\
x_5 = x_0 + 5w & y_5 \\
\vdots & & \vdots
\end{array}
$$

where $\Delta y_0 = y_1 - y_0$, $\Delta y_1 = y_2 - y_1$, $\Delta^2 y_0 = \Delta y_1 - \Delta y_0$, etc. These differences are especially easy to generate with a spreadsheet. The degree of smoothness of the data can be seen by inspection of the differences; there is no point in continuing the table beyond the point where the differences cease to vary in a reasonably smooth and regular manner. An expression for y as a function of x in the neighborhood of x_0 is given by the Gregory-Newton interpolation expression

$$
y = y_0 + r\Delta y_0 + \frac{r(r-1)}{2!}\Delta^2 y_0 + \frac{r(r-1)(r-2)}{3!}\Delta^3 y_0 + \cdots \qquad (9)
$$

where

$$
r \equiv \frac{x - x_0}{w} \qquad (10)
$$

By differentiation we obtain

$$
\frac{dy}{dx} = \frac{1}{w}\left(\Delta y_0 + \frac{2r-1}{2!}\Delta^2 y_0 + \frac{3r^2 - 6r + 2}{3!}\Delta^3 y_0 \right.
$$
$$
+ \frac{4r^3 - 18r^2 + 22r - 6}{4!}\Delta^4 y_0
$$
$$
\left. + \frac{5r^4 - 40r^3 + 105r^2 - 100r + 24}{5!}\Delta^5 y_0 + \cdots \right) \qquad (11)
$$

$$
\frac{d^2 y}{dx^2} = \frac{1}{w^2}\left[\Delta^2 y_0 + (r-1)\Delta^3 y_0 + \frac{6r^2 - 18r + 11}{12}\Delta^4 y_0 \right.
$$
$$
\left. + \frac{2r^3 - 12r^2 + 21r - 10}{12}\Delta^5 y_0 + \cdots \right] \qquad (12)
$$

On referring to the display shown in Eq. (8), we see that the successive differences entering into Eqs. (9), (11), and (12) lie on a downward-slanting line. With as much justification, an upward-slanting line may be used and indeed must be used if x_0 is at or very near the end of the table. No changes in the above expressions other than a replacement of $\Delta y_0, \Delta^2 y_0, \Delta^3 y_0, \ldots$ by $\Delta y_{-1}(= y_0 - y_{-1}), \Delta^2 y_{-2}, \Delta^3 y_{-3}$ and a change of all signs in Eqs. (9), (11), and (12) to plus are required if we continue to define r with Eq. (10). Indeed a good procedure is to employ both downward-slanting and upward-slanting Δ's, and to restrict r to the range $-\frac{1}{2} \leq r \leq \frac{1}{2}$.

INTEGRATION

For evaluation of the definite integral

$$Y(a, b) = \int_a^b y \, dx = \int_a^b f(x) \, dx \tag{13}$$

it is often possible to utilize procedures that do not require smoothing of the data or fitting of the data.[8] Let us assume that there are $n + 1$ data points x_i, y_i.

Trapezoidal Rule. The simplest procedure, where one connects adjacent points by straight line segments, is illustrated in Fig. 1a. The area of a single trapezoid is half the product of the base times the sum of the heights. Thus

$$Y(x_0, x_n) = \frac{1}{2} \sum_{0 \leq i \leq n-1} (x_{i+1} - x_i)(y_i + y_{1+i}) \tag{14}$$

For an equally spaced set of points with $x_{i+1} - x_i = w$ for all i, Eq. (14) simplifies to

$$Y(x_0, x_n) = w\left(\frac{y_0}{2} + y_1 + y_2 + \cdots + y_{n-1} + \frac{y_n}{2}\right) \tag{15}$$

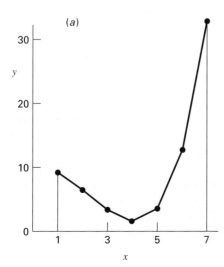

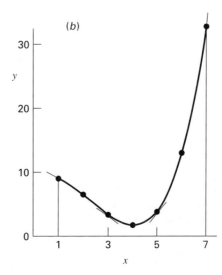

FIGURE 1

Numerical integration of a variable y measured at a set of equally spaced values of the independent variable x. The integral $Y(1,7) = \int_1^7 y \, dx$ is approximated with both the trapezoidal rule (a) and Simpson's one-third rule (b). In each case, the value of Y is given by the area under the heavy lines. The light lines in (b) represent extensions of the three parabolic sections that are used to construct this approximation.

Simpson's Rules. There are better procedures for approximating the integral of a function that make use of quadratic and cubic forms rather than linear segments. The simplest of these is *Simpson's one-third rule:*

$$Y(x_0, x_n) = \frac{1}{3} \sum_{0 \le i \le n-1} \left(\frac{x_{i+1} - x_i}{2} \right) \left[y_i + 4f\left(\frac{x_i + x_{i+1}}{2} \right) + y_{i+1} \right] \qquad (16)$$

Application of this equation to an unequally spaced set of data points requires the calculation of a y value at the midpoint of each interval $i, i + 1$. If a functional form $y = f(x)$ is known, this does not represent a problem and Eq. (16) is useful for the numerical integration of functions that cannot be evaluated in closed analytic form. For data sets with equally spaced points where $x_{i+1} - x_i = w$ for all i, no functional form is needed since Eq. (16) can be written in terms of successive pairs of intervals and simplifies to

$$Y(x_0, x_n) = \frac{w}{3}(y_0 + 4y_1 + 2y_2 + 4y_3 + 2y_4 + \cdots + 4y_{n-1} + y_n) \qquad (17)$$

This version of the rule requires that n be even, i.e., the number of values of y must be odd. It should be noted that the area under a parabola drawn through three equally spaced points is given exactly by $(y_0 + 4y_1 + y_2)w/3$. The piecewise application of this approach, which is shown in Fig. 1*b*, corresponds to Eq. (17).

Another rule, *Simpson's three-eighths rule,* is given by

$$Y(x_0, x_n) = \frac{3w}{8}[y_0 + 3(y_1 + y_2) + 2y_3 + 3(y_4 + y_5) + 2y_6$$
$$+ \cdots + 3(y_{n-2} + y_{n-1}) + y_n] \qquad (18)$$

for equally spaced points. This requires that n be divisible by 3 and is equivalent to fitting third-degree functions to four points at a time.

In cases where n is divisible neither by 2 nor by 3, the range of integration may be split into two parts, one for Simpson's one-third rule, the other for Simpson's three-eighths rule. Alternatively, if the curve is approximately linear in one or two intervals, the trapezoidal rule may be used in these intervals.

The errors resulting from the use of Simpson's one-third rule are much smaller than those associated with the use of the trapezoidal rule. This is illustrated by the example given in Fig. 1. The points shown in this figure were generated from the function $y = 10 - x^2 + 0.03x^4$ and are given below:

$x =$	1	2	3	4	5	6	7
$y =$	9.03	6.48	3.43	1.68	3.75	12.88	33.03

For this function we can easily evaluate $Y(1, 7) = \int y \, dx$ analytically:

$$Y = 10(7 - 1) - \frac{1}{3}(7^3 - 1) + \frac{0.03}{5}(7^5 - 1) = 46.502667$$

Application of the trapezoidal rule, Eq. (15), yields $Y(1, 7) = 49.250$, which is in error by 5.4 percent. Simpson's one-third rule, Eq. (17), yields $Y(1, 7) = 46.860$, which is in error by only 0.77 percent.

If the number of data points is large, it is essential to use a computer algorithm to carry out the numerical integration. Note that the trapezoidal rule as given by Eq. (14) can be used for an unequally spaced set of data points. If the data are equally spaced (or if a fitting form or interpolation scheme is available), then Simpson's rule is much better and a generalization of Simpson's approach called Romberg integration is even better and is recommended.[1]

ROOT FINDING

If a functional form $y = f(x)$ has been obtained, one may need to determine one or more zeros of this function.[9] The equation $y(x) = 0$ is in general nonlinear; analytic methods of solving such equations (other than polynomials of degree 1 or 2) are generally extremely complicated or nonexistent. However, solutions to such equations can be obtained routinely by numerical methods to any desired precision with the aid of a computer and the use of convergent iterative methods.[1] We will describe two methods here.

False Position. The equation $y(x) = 0$ may have several roots [if $y(x)$ is a polynomial of degree N, it will have N roots, but some of them may be complex]; we will seek only one root for this discussion. Assume that the function $y(x)$ varies slowly enough that it does not change sign more than once between two adjacent data points. Let us say that y_1 (at x_1) and y_2 (at x_2) are of opposite sign. Then, by linear interpolation between the two points, to the first approximation (first iteration) the root is given by

$$x_3 = \frac{|y_1|x_2 + |y_2|x_1}{|y_1| + |y_2|} = x_1 + \frac{|y_1|}{|y_1| + |y_2|}(x_2 - x_1) \tag{19}$$

The value of $y_3 = y(x_3)$ is now calculated. It is extremely unlikely that y_3 is zero, as it would be if x_3 is the root we seek. The zero in y will presumably occur at an x value on one side or the other of x_3. In the second iteration, we repeat the above procedure, interpolating either between x_1 and x_3 or between x_3 and x_2, depending on which range includes the zero in y. The procedure is repeated in subsequent iterations to obtain x_4, x_5, and so on until the change between iterations becomes insignificant and we can say that the process has effectively converged on the root to the desired degree of precision. This method is known as the "false position" method.

Newton–Raphson. Another method, closely related to the first, is the Newton–Raphson method. Here the derivative with respect to x, denoted y', is evaluated at x_1, a point judged to be close to the root. Then

$$x_2 = x_1 - \frac{y_1}{y_1'} \tag{20}$$

is the first approximation to the root (first iteration). This is equivalent, in graphical terms, to drawing a tangent to the curve at the point x_1, y_1 and extending it to intersect with the x axis. Then, for the second iteration, y_2 and y_2' are evaluated to obtain x_3, and so on.

Example. The data from an experiment are fitted by least squares with a third-degree polynomial:

$$y(x) = 0.331x^3 - 1.410x^2 + 2.050x - 4.015$$

We seek the value of a root of the equation

$$y(x) = 0$$

It is apparent from plotting the function that a zero lies between $x = 3.000$ and $x = 4.000$. From the equation for the function $y(x)$, it is found that

$$\text{for } x_1 = 3.000 \quad y_1 = -1.618$$
$$\text{for } x_2 = 4.000 \quad y_2 = +2.809$$

Application of Eq. (19) for the false position method gives

$$x_3 = \frac{3.000 \times 2.809 + 4.000 \times 1.618}{1.618 + 2.809} = 3.366$$

$$\text{for } x_3 = 3.366, \qquad y_3 = -0.469 \qquad \text{(Iteration 1)}$$

Since y_3 is negative, the zero must lie between $x_3 = 3.366$ and $x_2 = 4.000$. In the same manner we obtain

$$x_4 = 3.456, \quad y_4 = -0.107 \quad \text{(Iteration 2)}$$

Continuing we obtain

$$
\begin{aligned}
x_5 &= 3.476, & y_5 &= -0.023 & &\text{(Iteration 3)} \\
x_6 &= 3.480, & y_6 &= -0.005 & &\text{(Iteration 4)} \\
x_7 &= 3.481, & y_7 &= -0.003 & &\text{(Iteration 5)} \\
x_8 &= 3.481, & y_8 &= -0.003 & &\text{(Iteration 6)}
\end{aligned}
$$

Apparently the process has converged to a satisfactory stopping point. The failure of y to decrease below 0.003 in magnitude results from round-off error in x; keeping additional significant figures during the calculation would reduce this magnitude in the same number of iterations. (Try it and see!) Let us now apply the Newton–Raphson method to the same problem. Differentiation of the polynomial yields

$$y'(x) = 0.993x^2 - 2.820x + 2.050$$

$$\text{for } x_1 = 3.000, \quad y_1 = -1.618, \quad y_1' = 2.527$$

Using Eq. (20) we obtain

$$x_2 = 3.000 - \frac{(-1.618)}{2.527} = 3.640$$

$$x_2 = 3.640, \qquad y_2 = 0.730, \qquad y_2' - 4.943 \qquad \text{(Iteration 1)}$$

Continuing we obtain

$$
\begin{aligned}
x_3 &= 3.492, & y_3 &= 0.046, & y_3' &= 4.312 & &\text{(Iteration 2)} \\
x_4 &= 3.481, & y_4 &= -0.003, & y_4' &= 4.226 & &\text{(Iteration 3)} \\
x_5 &= 3.482, & y_5 &= 0.002, & y_5' &= 4.270 & &\text{(Iteration 4)} \\
x_6 &= 3.482, & y_6 &= 0.002, & y_6' &= 4.270 & &\text{(Iteration 5)}
\end{aligned}
$$

Satisfactory convergence has been obtained, somewhat faster than with the false position method. Usually the Newton–Raphson method is indeed the one that converges faster, but in some cases it can diverge in the early iterations. It may be helpful to use the false position method for the first one or two iterations, and then switch over to the Newton–Raphson method for further iterations leading to satisfactory convergence. This, and other root-finding procedures, are commonly used by the Solver or Optimizer operation of spreadsheet programs, whose use is described in Chapter III.

REFERENCES

1. W. H. Press, S. A. Teukolsky, W. T. Vetterling, and B. P. Flannery, *Numerical Recipes—the Art of Scientific Computing,* 2d ed., Cambridge Univ. Press, New York (1992). [Available in C, Pascal, and FORTRAN versions. CD-ROM disk versions of the text are available for C and FORTRAN90. *Numerical Recipes Example Book,* 2d ed., accompanied by disks containing the programs given in the books on C, Pascal, and FORTRAN77 are also available.]

2. L. F. Shampine, R. C. Allen, Jr., and S. Preuss, *Fundamentals of Numerical Computing,* Wiley, New York (1997).

3. R. W. Hamming, *Numerical Methods for Scientists and Engineers,* 2d ed., pp. 567–573, 597–599, Dover, Mineola, NY (1987) .

4. *Ibid.,* pp. 437–443.

5. R. Sedgewick, *Algorithms,* Addison-Wesley, Reading, MA (1998). [Also available as *Algorithms in C* and *Algorithms in C^{++}*.]

6. R. W. Hamming, *op. cit.,* pp. 503–545.

7. *Ibid.,* pp. 302–309.

8. *Ibid.,* pp. 577–581.

9. *Ibid.,* pp. 59–77.

GENERAL READING

K. Atkinson, *Elementary Numerical Analysis,* 2d ed., Wiley, New York (1993).

R. L. Burden and D. J. Faires, *Numerical Analysis,* 6th ed., Brooks-Cole, Pacific Groves, CA (1996).

C. Daniel and F. S. Wood, *Fitting Equations to Data,* 2d ed., Wiley, New York (1999).

J. Murphy and B. McShane, *Computational Mathematics Applied to Numerical Methods,* Prentice-Hall, Upper Saddle River, NJ (1998).

G. M. Phillips and P. J. Taylor, *Theory and Applications of Numerical Analysis,* 2d ed., Academic Press, San Diego, CA (1996).

F. J. Scheid, *Schaum's Outline of Numerical Analysis,* 2d ed., McGraw-Hill, New York (1989).

Appendix

G

Barometer Corrections

The entries Δ in the table below are calculated from Eq. (XVIII-1) on the assumption that the barometer has a *brass* scale graduated to be accurate at 0°C. See Chapter XIX for a description of the recommended procedure for reading a barometer. These corrections should be **subtracted** from the observed barometer readings; i.e., $p(\text{Torr}) = p_{obs}(\text{mm}, T) - \Delta$. (If the brass scale is accurate at 20°C, the appropriate corrections are approximately 0.3 mm greater than those given.) Once the barometer reading has been corrected to 0°C, the pressure is referred to as Torr rather than mm Hg.

t, °C	720 mm	740 mm	760 mm	780 mm	800 mm
16	1.88	1.93	1.98	2.03	2.09
17	1.99	2.05	2.10	2.16	2.22
18	2.11	2.17	2.23	2.29	2.35
19	2.23	2.29	2.35	2.41	2.48
20	2.34	2.41	2.47	2.54	2.60
21	2.46	2.53	2.60	2.67	2.73
22	2.58	2.65	2.72	2.79	2.86
23	2.69	2.77	2.84	2.92	2.99
24	2.81	2.89	2.97	3.05	3.12
25	2.93	3.01	3.09	3.17	3.25
26	3.04	3.13	3.21	3.30	3.38
27	3.16	3.25	3.34	3.42	3.51
28	3.28	3.37	3.46	3.55	3.64
29	3.39	3.49	3.58	3.68	3.77
30	3.51	3.61	3.71	3.80	3.90

Ethical Conduct in Physical Chemistry

Scientific research is an important social process. As such, it is based on "social contracts" as guidelines, and perhaps the most important of these is ethical behavior. Progress in science is based on mutual trust; and to achieve this, honesty is a crucial aspect of all scientific activity.

This brief statement of key elements in scientific ethics deals with both a general overview pertinent to all professional work in science and those particular issues of greatest concern to undergraduates in a course such as physical chemistry laboratory.

An excellent bibliography of information on scientific ethics is the website of D. G. Elmes[1], which lists books such as *Research Ethics* by Barnbaum and Byron[2] and a variety of ethics websites. The most significant of the latter for physical chemists are:

http://www.acs.org By using the American Chemical Society Google search engine with the entries "code conduct" and "professional ethics," one can access *The Chemist's Code of Conduct* and *Guidelines for the Teaching of Professional Ethics.*

http://www.aps.org/statements/02_2.cfm This is an American Physical Society site where *Ethics and Values Statement* for physicists can be found. This site provides more detailed information than the ACS sites.

GENERAL RESEARCH ISSUES

The summary given here is based closely on the APS website cited above and touches on three issues.

1. *Research results* must be recorded initially and maintained (archived) in a form that permits independent review and analysis of the data. Fabrication of data or even the selective reporting of some data and withholding of other data with the intent to mislead is completely unacceptable.

2. *Publication of results* often involves a paper with several authors. All who made a significant contribution should be included as coauthors. Others who may have contributed to the work in minor ways should be acknowledged but not listed as authors. All coauthors must have access to the entire range of results and an opportunity to review the manuscript prior to submission to a journal. If there are several authors, more than one author should verify the accuracy of the entire paper, and all authors share responsibility for the quality of the paper and must review the aspects of the work to

which they contributed. Plagiarism is unethical and a serious violation of scientific standards. Furthermore, the past work of others that is used in any research must be cited by proper references or acknowledgements.

3. *Peer review* of manuscripts submitted for publication and of research proposals is an essential part of scientific activity. Reviewers must have the necessary expertise to judge a manuscript or proposal and must be fair and objective in their evaluations. Potential conflicts of interest must be revealed, and in general it is best that persons with professional conflicts decline to act as reviewers of a given document.

UNDERGRADUATE LABORATORY COURSES

It is essential that students be exposed to high ethical standards as a way of learning scientific ethics as part of their professional development. In most cases, a laboratory experiment will be done by two students working together. However, each student should submit a personal laboratory report. Naturally, the data acquisition and the presentation of the "raw" data will be common elements, and the partner's name must be clearly stated in the report. *Copying of spreadsheets or text is plagiarism and is a very serious matter.* Indeed, plagiarized reports are usually given a grade of zero irrespective of the quality of the data or the text.

The data analysis is in general the independent work of each student. However, in group projects involving extensive data analysis, the instructor may assign distinct parts of the analysis work to each student in the group. As an example, in the HCl/DCl spectroscopy experiment (Exp. 37), if an analysis were to be made of the spectra of the four species $H^{35}Cl$, $H^{37}Cl$, $D^{35}Cl$, $D^{37}Cl$, then each student might be assigned the analysis of one of these species. In such cases, each student is responsible for reviewing the analysis work of other participants, and the contribution of each group member should be clearly identified. In any event, the presentation of results and conclusions must be the independent work of each individual, with appropriate references given to published work and any unpublished sources of information.

As with original research done for publication, data for undergraduate laboratories must also form a permanent record that can be archived (at least for the student's undergraduate career). Since no scientist routinely reports 100 percent of the data acquired, the crucial challenge is to include all the valid data and learn what can be discarded without artificially prejudicing the result. For undergraduate laboratory reports, it is recommended that any faulty or poor data be clearly labeled as such but included in the report so that the student's judgment can be checked by the instructing staff.

REFERENCES

1. D. G. Elmes, *Resources in Ethics and Responsible Research;* see http://psych.wlu.edu/ethics_and_responsible_research.htm.

2. D. R. Barnbaum and M. Byron, *Research Ethics: Text and Readings,* Prentice-Hall, Upper Saddle River, NJ (2001).

GENERAL READING

J. Kovac, *The Ethical Chemist,* Prentice-Hall, Upper Saddle River, NJ (2004).

On Being a Scientist: Responsible Conduct in Research, National Academy Press, Washington, DC. (1995). Available online at http://www.nap.edu/0309051967/html/index.html.

Responsible Conduct of Research Educational Committee, http://rcrec.org (click on Resources).

Index

A

Abbe refractometer, 615–617
Ab initio calculations, 82–84, 397, 405, 423, 435, 444–445, 453
Absorbance, 284, 396, 532–534, 631
Absorption coefficient, molar, 396, 631
  of iodine, 534–535
Absorption spectrum (*see* Spectrum)
Abstract searches, 10, 704–706
Accuracy, 29
Acetic acid, ionization constant of, 237–238, 257
Acetylene spectrum, 424–436
  ab initio quant-mech calcs, 435
Acid-catalyzed reactions, 263–270, 273, 277–278
Activated complex, 287–289
Activation energy, 264–265, 269, 281, 284, 319
  (*see also* Kinetics, chemical)
Activity, 189, 193, 249
Activity coefficient:
  from cell measurements, 248–253
  Debye–Hückel equation for, 193, 237, 250–251
  for hydrogen gas, 249
  of ionic species, 193, 237
  of neutral solute species, 193
  for a solvent, 189
Adhesives, 659–660

Adiabatic jacket, 150, 152–153
Adiabatic processes:
  in chemical reactions, 145, 148–151
  expansion of gas, 98–105, 109–114
Adsorbed molecules, area of nitrogen, 316
Adsorbents, 311–312
Adsorption:
  BET theory of, 307–308
  of gases, 308–317
    gravimetric method, 311
    volumetric method, 311–316
  isosteric heat of, 311
  isotherms, 308–309
  at liquid surface, 302–303
  multilayer, 309–310
  physical, of gases, 308–317
  (*see also* Isotherm)
Air:
  density of, 604
  dielectric constant, 352
  index of refraction of, 613–614, 620n
  viscosity of, 134
Aluminum, 653–654
Amici prism, 614–617
Amplifiers:
  frequency-selective (lock-in), 618
  instrumentation, 546–547
  operational, 542–547

Analog-to-digital conversion, 547–552
Anharmonic oscillator, energy levels of, 417, 438–439
Antibonding orbital, 463
Apparatus lists, 3
Apparent molar volume, 174–175
Area of adsorbed nitrogen, 326
Argon:
  Debye temperature of solid, 517
  dielectric constant, 352
  lattice energy of, 515–522
  lattice parameter of, 522
  Lennard–Jones potential parameters for, 518
  molecular diameter of, 143
  mutual diffusion constant for He–Ar, 143
  thermodynamic properties of, 516
  vapor pressure of solid, 515–522
  virial coefficient for, 97, 516
Arithmetical calculations (*see* Calculations)
Arrhenius activation energy (*see* Activation energy)
Ascarite, 140
Aspirator, water, 588
Atomic mass table (*see back endpaper*)
Autocorrelation function, 380–382, 386
Average deviation, 39
Azeotropes, 211

B

Balances:
  analytical, 601–605
  corrections and errors, 604–605
  Gouy, 365–368
Ballast bulb, 95, 203
Barometer, 605–606
  corrections, 718
Baths (*see* Temperature control)
Beattie–Bridgeman equation,
  101–102
Beer–Lambert law, 280, 284, 396,
  631
Benzene:
  compressibility factor for,
    200–201
  heat capacity of gas and liquid,
    202
  safety warning about, 346
  stimulated Raman spectrum,
    407–415
Benzenediazonium ion, rate of
  decomposition of, 283–286
Benzoic acid, specific energy of
  combustion, 157
Bernoulli's law, 132
BET (Brunauer, Emmett, and
  Teller) isotherm, 309–310
Bias, 6
Birefringence of nematic liquid
  crystals, 221–227
Birge–Sponer plot, 438, 444
Blackbody radiation, 574–575, 621
Bohr magneton, 363, 454
Boiling point, 199
Boiling-point diagrams, 208–211
Bolometer, 630
Bond energies, 158–161
Bonding:
  orbital, 463
  valence force model, 401–403,
    428–429
Bond type and magnetism,
  364–365
Bourdon gauges (*see* Gauges)
Bragg construction, 503
Bragg equation, 503–504
Bragg reflections, 503–507
  (*see also* Crystal structure; X-ray
    diffraction)

Brass, 654
  coefficient of thermal expansion,
    605
Bravais lattice, 502
Brillouin zone, 529–530
Brownian motion, 379
Bubblers, 141, 251–252, 646
Buffers, acetate, 257
Buoyancy correction, 604
Burette, 640–641
  gas, 311–314
Butyl peroxide, rate of
  decomposition of, 295–296

C

Cadmium electrode, 246–247
Calculations, 29–38, 709–716
  analytical methods, 32
  arithmetical, 31–32
  checking of, 32
  graphical methods, 34–37
  numerical methods, 33–34,
    709–716
    differentiation, 711–713
    integration, 713–714
    polynomial and Fourier series,
      710–711
    roots of equation, 77,
      715–716
    smoothing of data, 709
  precision, 31–32
  significant figures, 30
  (*see also* Least-squares method)
Callendar–van Dusen equation,
  560–561
Calomel electrode, 609
Calorimetry:
  adiabatic jacket, 150–153
  bomb, 152–166
  principles of, 145–151
  solution, 167–171
  (*see also* Heat)
Capacitance cell:
  for gas, 350–351
  for solution, 341–342
Capacitance manometer, 353,
  596
Capacitance measurements:
  bridge method, 342, 349

heterodyne-beat method,
  342–343
LC oscillator, 342–343,
  349–350
Capacitors, 540–541
Capillary rise, 304–306
Carbon dioxide:
  C_p of gaseous, 105
  critical point of, 229–234
  dielectric constant, 352
  molecular diameter of, 143
  van der Waals constants for, 101
  virial coefficients for, 97
Carbon tetrachloride:
  Raman spectrum of, 398–406
  safety warning about, 197
Carcinogenic chemicals, 694–695
Cassia flask, 176–177
Cathetometer, 240
Cell constant (conductance), 236
Cells (*see* Emf cells)
Celsius (centigrade) temperature
  scale, 92, 557
  (*see also* Temperature scale)
Cements, 659
Center of inversion, 425
Centrifugal distortion, 417, 419,
  423, 429–430
Charge-coupled device, 629, 634
Chemical Abstracts, 10, 704–715
Chemical equilibrium, 188–198,
  237–238, 422, 471–472,
  475–477
Chemical exchange reaction,
  263–270
Chemical potential, 174
  statistical mechanical calculation,
    524–530
Chemical shifts (NMR), 267, 270,
  373, 375, 466–468
Chi-square distribution, 673
  χ^2 test, 672–675, 681
Chlorobenzene, dipole moment of,
  345
Circuit elements, 538–542
Clapeyron equation (*see*
  Clausius–Clapeyron equation)
Classical probability statistics, 45
Clausius–Clapeyron equation, 180,
  200, 450, 515, 531, 535

Clausius–Mossotti equation, 338
Cleaning solution, 642–643
Clock reactions, 254–262
Coexistence curve, 230–234
Coherent anti-Stokes Raman
    Scattering (CARS), 409–410
Cold trap, 591
    safety, 696–697
Colligative properties, 179–183,
    188–190
    freezing-point depression,
        179–193
Collision frequency, 120–121
Combustion (*see* Heat)
Complexes, transition metal,
    364–365
Compressibility factor, 200–201
    of benzene, heptane, water, 201
Computer:
    databases, 703–704
    interfacing, 85–88
        serial/parallel/USB bus, 85
    LabPro system, 86, 224
    Lab VIEW, 87–88, 386–388
    software (*see* Software)
    spreadsheets (*see* Spreadsheets)
    virtual instruments, 87–88,
        386–388
Concentration, 692
    surface, 300–304
Conductance:
    of solutions, 235–244
        equivalent conductance,
            236–237
        Λ_0 for HCl, KCl, KAc, 244
        molar conductance, 236
        strong electrolytes, 236
        weak electrolytes, 236–237
Conductance bridge, 238–241
Conductivity:
    cell, 241
    electrolytic, 236–238
    of water, 242
    (*see also* Thermal conductivity
        of gases)
Confidence limits, 46–52
Constants, physical (*see front
    endpapers*)
Construction materials, 652–658
Contact angle, 305

Conversion factors (*see front
    endpapers*)
Cooling curves, 183–185
Copper, 654
    wavelength of $K\alpha$ radiation, 511
Correlation time, 380
Corresponding states, law of, 201
Counters:
    electronic, 636
    radiation, 613
Coupling constant (J) for spins,
    468–471
Coverage, surface, 309–310
Critical behavior, 215–228,
    229–234
Critical exponents, 219, 226, 233
Critical opalescence, 231
Critical point, 229–234
Critical temperature, 201, 233–234
Cross-section for quenching:
    foreign gas quenching of I_2,
        447–448, 451–452
    self-quenching of I_2, 447–448
Cryogenic:
    baths, 577–578
    thermometers, 575–576
Cryopump, 594
Crystal:
    partition function, 527–530
    vibrations, 527–530
    zero-point energy of, 515, 527,
        531
Crystal class, 502
Crystal-field complex, 364–365
Crystal lattice, 502–508
Crystal structure, 500–515
    of CsCl, 507
    deduction of, 507–508
    of I_2, 528
    lattice parameter for argon, 522
    Miller indices, 501, 504–507
    unit cell, 500–502, 528
    (*see also* X-ray diffraction)
Crystal systems, 501–502
Cumulants, method of, 389–390
Curie constant, 363, 368n, 377
Curie–Weiss law, 368n
Cyclohexane:
    density of, 187
    freezing point of, 183

molal freezing-point depression
    constant, 183
molar heat of fusion, 183
Cyclohexanone, refractive index,
    212
Cyclopentene, rate of decom-
    position of, 294–295

D

Data:
    collection (bias), 5–6
    database, 10
    recording of, 7–9, 23–24
    rejection of discordant, 42–43,
        671–672
    smoothing of, 709
    treatment of, 29–38, 42–43,
        51–52
    (*see also* Calculations; Error(s);
        Graphs)
Databases, 10, 703–704
Debye–Hückel theory, 193, 237,
    250–251
    limiting law, 250
Debye–Scherrer x-ray method,
    508–511
    (*see also* X-ray diffraction)
Debye temperature of argon, 517
Debye unit for dipole moment,
    341
Degeneracy:
    of d orbitals, 364–365
    electron spin, 455
    nuclear spin, 371–372, 431
    rotational, 431
    vibrational, 402, 427
Degrees of freedom for a molecule,
    107–108, 206, 527
Density:
    of air, 604
    of cyclohexane, 187
    of dibutyl phthalate, 113, 595
    of mercury, 113
    of NaCl solutions, 307
    pycnometers for determining,
        15–17, 176–177
    of water, 177
    Westphal balance for
        determining, 325, 369

Deshielding effects in NMR, 467–468

Deslandres table, 453

Detailed balancing, 264

Detectors, 612–613, 625–631
charge-coupled devices, 628
photodiodes, 626–628
photographic film, 626
photomultiplier tubes, 625–627
pyroelectric, 629–630
radiation counters, 612–613
thermal devices, 629–631

Deuterium chloride preparation, 421–422, 480–481

Deutero-acetylene preparation, 432–433

Dewar flask, 150, 168

Diamagnetism, 362, 373
of water, 377

Diameter (*see* Molecular diameter)

Diatomic molecule, energy levels, 416–417, 438–440

Dibutyl phthalate, density of, 113, 595

Dichlorobenzene, dipole moment of, 344–345

Dielectric cell for solutions, 341–342

Dielectric constant, 336–338
of gas, 349–350
of solution, 341–343
values for reference gases, 352

Differentiation:
graphical, 36
numerical, 711–713

Diffraction camera, 508–510

Diffusion:
Fick's laws of, 136–137
of gases, 124–127, 135–144
Loschmidt apparatus, 137–139, 142
of polystyrene spheres, 379–392
self-diffusion in gases, 126–127, 136

Diffusion constant for gases, 124–127, 135–136, 143–144

Diffusion pumps, 590–592

Digital multimeter (DMM), 550–552

Digital-to-analog conversion, 547–550

Dipole moment:
induced nonlinear of benzene, 407–408
of chlorobenzene, 344–345
of HCl gas, 347–360
quantum-mechanical calculation, 345–346, 359
of solutes, 336–346
of succinonitrile, 344–345

Dissociation energy, 438–439

Dissociation of weak electrolyte, 188, 236–238

Distillation, 210–212

Distribution function:
for molar masses $P(M)$, 322–324
normal, of errors, 44–46

Distribution ratio, 194–196

Distribution of residuals, 675

Dry Ice (*see* Temperature control)

Drying agents, 647

Duralumin, 654

Dyes, absorption spectra of, 393–397

Dynamic light scattering, 379–392

E

Effusion of gases, 120–121

Electrical heating, 146–147, 168–170

Electrical measurements, 542–556

Electrodes:
cadmium, 246–247
cadmium amalgam, 247
calomel, 609
glass, 609–610
hydrogen, 248, 251–252
platinum, 251, 651
silver–silver chloride, 249, 251–252, 651–652

Electrolytes, 188–192, 235–244, 248–252

Electromagnetic spectrum, 619

Electron g value, 454–456

Electron spin resonance spectrum (*see* Spectrum)

Electronic circuit elements, 538–542

Electronic spectrum (*see* Spectrum)

Electronic states, quantum-mechanical calculation, 397

Emf cells:
activity coefficients from, 248–252
electrodes for, 246–247, 251–262, 609, 651
standard, 554
temperature dependence of, 245–247
thermodynamics of, 245–247

Emission spectrum (*see* Spectrum)

Emittance, 575

Energy:
activation, 264–265, 270, 284, 319
bond, 158–161
conversion of units (*see front endpaper*)
of a crystalline solid, 517–518
dissociation, 438–439
strain, 158–166
zero-point, of crystal, 517, 526–527

Energy change in chemical reactions, 145–151, 157–165

Energy levels:
anharmonic oscillator, 417, 438–439
of conjugated dyes, 393–397
of diatomic molecules, 416–417, 437–440
of nuclei (NMR), 371–372, 466–470
of polyatomic molecules, 425–430
of unpaired electrons (ESR), 454–456

Enthalpy (*see* Heat)

Enthalpy change in chemical reactions, 145–151, 152, 158–160, 245
(*see also* Heat)

Entropy:
statistical mechanical, 206, 531
of sublimation, 531
of water, 206

Enzyme kinetics, 271–282
  Michaelis–Menton mechanism
    of, 271–273
Enzyme solution:
  preparation of, 279
  specific activity, 281
Epoxy resins, 659
Equation of state, 93, 100–102
  Beattie–Bridgeman
    equation, 101–102
  ideal-gas law, 93
  van de Waals, 100, 110–111, 233
  virial, 93, 97, 102
Equilibria:
  binary liquid–vapor phase,
    207–215
  chemical in gas phase, 475–483
  chemical in solution, 188,
    193–198, 237–238, 263–270
  coexisting liquid–vapor,
    199–206, 229–236
  helix-coil polypeptide, 327–334
  heterogeneous, 194
  homogeneous, 193–194
  isotopic exchange, 475–483
  keto–enol tautomerism, 466–474
  phase, 199–234
  physical adsorption of gas,
    308–316
  solution–solid solvent, 179–187
  vapor pressure of liquid,
    199–207
  vapor pressure of solid, 515–522
Equilibrium constant, 193–198
  for gas-phase exchange,
    475–483
  for hydrolysis reaction, 263–270
  for keto–enol tautomerism,
    466–474
  for weak electrolyte ionization,
    188, 237–238
Equipartition theorem, 108
Equipment, general laboratory, 3
Equivalent conductance, 236–237
Error(s), 38–65, 678–680
  in arithmetic mean, 39–40, 47
  confidence limits, 46–52
  error probability function, 43–48
  estimation of, 50
  examples, 43, 56–58

mistakes, 41–42
normal error function, 43–46
precision, 31–32, 38–40, 61–63
propagation of, 52–59
  sample treatment, 19–20,
    56–59
Q test, 42
random, 38–40
range, 39, 50
significant figures, 30–32
in slopes and intercepts, 37
in small samples, 48–50
standard deviation, 45–48
Student t distribution, 48–50
systematic, 40–41, 52
uncertainty in least-squares
  parameters, 678–680
variance, 39–40
(*see also* Least-squares method)
Euler's theorem, 173
Excel software, 69–72, 388
  Solver option, 77–78, 226
Exchange interactions, 458
Exchange reaction, 422, 475–483
Exciton, 495
Expansion of gas, 107–111
Experimental data (*see* Data)
Experimental execution, 4–5
Extensive variables, 172, 299–300
Extinction coefficient (*see*
  Absorption coefficient, molar)
Eye safety, 695, 697–698

F

F distribution, 676
F test, 75–77, 676–677
Fick's laws of diffusion, 137–138
Figures, 13, 34–36
Filters:
  interference, 633
  neutral density, 413, 415
Fitting procedures, 663–685,
  710–711
Fixed point (*see* Temperature
  scale)
Flowmeters, 647–649
Flow method for chemical kinetics,
  260–262

Fluorescence of iodine, 440–443,
  446–453
  decay lifetime, 448
Force constants:
  of acetylene, 426, 428–429
  of benzene, 411
  of CCl_4, 401–403
  of diatomic molecules, 419
Forepump, 588–590
Fourier series expansion, 710–711
Fourier transform:
  fast (FFT), 711
  FTIR, 634–636
  NMR (FT-NMR), 477–480
Franck–Condon factor, 80–82, 442,
  444
Freedom, degrees of, for a
  molecule, 107
Free-electron model of dye spectra,
  393–395
Free energy:
  changes in electrochemical cells,
    245–247
  for phase transition, 218–219
  surface, 301–302
Free radical anions, 456–458
Freezing-point depression, 179–193
  of electrolytes, 188–193
  molal, constant, of cyclohexane
    and water, 183
  for molar mass determination,
    179–187
  of nonelectrolytes, 179–187
Frequency doubling, 446, 486
Frequency of LC circuit, 342
FTIR (*see* Spectrometer)
Furnaces, 584–585

G

g value:
  of electron, 363
  of nucleus, 371
Gas burette, 314
Gas-handling procedures, 645–651
Gas law (*see* Equation of state)
Gas purification, 646
Gas regulators, 644–646
Gas thermometer, 91–98, 557

Gases:
adsorption of, 308–317
degrees of freedom in, 107
diffusion of, 124–127, 135–144
effusion of, 120–121
expansion of, 109–114
heat capacity of, 101, 105, 106–118, 202, 420
Maxwellian velocity distribution in, 119
partition function, 420, 475, 524–530
thermal conductivity of, 123–124
vibrational heat capacity of, 108, 420
viscosity of, 121–123, 128–136
water saturation of, 647
Gated integrator, 706
Gauges:
Bourdon, 649–650
ionization, 595, 597–598
McLeod, 596
manometers, 594–597
Pirani, 597
residual-gas analyzer, 598
strain-gauge pressure transducers, 596, 650
thermocouple, 597
(see also Manometers)
Gaussian computer program, 82–84, 397, 414, 423, 435, 453
Gaussian distribution, 45
Geiger–Müller counter, 613
Gel permeation chromotography, 320
Gibbs–Duhem equation, 173
Gibbs–Helmholtz equation, 246, 265
Gibbs isotherm, 301–302
Glass electrode, 609
Glasses, 652
thermal expansion of, 652
Glasswear:
calibration of, 641–642
cleaning of, 642–643
volumetric, 639–641
Glyptal, 660
Goodness of fit, 672–673
Gouy balance, 365–368
Graphical differentiation and integration, 36–37

Graphs:
errors in slopes and intercepts, 37
preparation of, 34–36
presentation of, 13
(see also Calculations)
Gratings, 412–413, 622–625
Guggenheim method, 293
(see also Kinetics, chemical)

H

Half-life, 294
(see also Kinetics, chemical)
Handbooks, 703–704
Harmonic oscillator functions, 81–82, 426
Hartree–Fock SCF calculations, 82–84
Hazards, safety, 693–699
HCl (see Hydrogen chloride)
HCN vibrational spectrum, 430
Heat:
of adsorption, isosteric, 311
of combustion, 152–166
for iron and benzoic acid, 157
of formation, 159–160
of fusion, 180–181
for cyclohexane and water, 183
of ionic reaction, 167–171
of reaction, 245–247
of sublimation, 515–522
of vaporization, 199–207
Heat capacity, 147
of benzene, heptane, water, 202
of gases, 101, 105, 106–118, 202, 420
vibrational contribution, 108, 420
Heat flow, one-dimensional, 123–124
Heating wire, 584
Helium:
C_p of gas, 105, 106–118
molecular diameter of, 143
mutual diffusion constant He-Ar, 143
van der Waals constants for, 101
virial coefficient for, 97

Helix-coil transition, 328–335
Heptane:
compressibility factor for, 201
heat capacity of gas and liquid, 202
Heterodyne-beat method, 342–343
Heterogeneous equilibrium, 194
Homogeneous equilibrium, 193–194
Hoses, 646
Hückel molecular orbital theory, 79–80, 461–465
Hund's first rule, 362, 364
Hydrochloric acid, Λ_0 value, 244
Hydrogen, van der Waals constants for, 101
Hydrogen bonding in polypeptides, 328–329
Hydrogen chloride:
ab initio quant-mech calc, 82–84, 423
dipole moment (gas), 347–359
equivalent conductance (soln.), 244
mean ion activity coefficient, 248–253
spectrum, 416–423
Hydrogen electrode, 248, 251–252
Hydrogen ion catalysis, 263–264, 273, 277–278
Hydrolysis reaction, 263–270, 274–281
Hyperfine splitting, 456–458

I

Ice bath, 96, 568, 578
Ice point, 92, 96, 559, 561–562
(see also Temperature scale)
Ideal-gas law, 93
Ideal solutions, 180, 208–209
Inconel, 654
Index of refraction, 614–617
of air, 614
of cyclohexanone-tetrachloroethane solutions, 212
Inductors, 541–542
Inertia, moment of, 416–417, 429, 526

Infrared spectrometers, 633–636
  detectors, 629–630
  gratings, 622–624
  prisms, 622
  sources, 618, 622
Infrared spectrum, 420–423, 424–435
  fundamentals, 402, 425
  gas cell, 421
  HCl, rotational structure of, 416–423
  isotope effect, 416, 419–420, 422–423, 426–435
  overtone and combination bands, 427–428
  quantum-mechanical calculation, 405, 435
  selection rules for diatomic molecules, 417–419
  vibrational modes of acetylene, 425
  (*see also* Spectrum)
Initial rates, method of, 254–262
Instrument control, 85–88
Instrumentation amplifiers, 547
Insulation materials, 657
Integration:
  graphical, 36
  numerical, 713–714
Intensive variables, 172–174, 297–300
  surface quantities, 300–301
Interfacial tension, 299–301
Interferometer, 354–356, 634–636
Interpolation, 712–713
Intrinsic viscosity (*see* Viscosity)
Inversion, center of, 425
Inversion of sucrose, 271–282
Inversion temperature, 101
Iodine-clock reaction, 254–262
Iodine crystal structure, 528
Iodine–iodide–triiodide equilibrium, 193–198
Iodine spectrum:
  absorption, 436–440, 443, 532–533
  emission, 440–444
  fluorescence, 446–452
Iodine sublimation pressure, 450, 532–535
Ion pump, 592–594

Ionic mobility, 236
Ionic strength, 194, 250
Ionization constant of acetic acid, 257
Ionization gauge, 595, 597–598
Ionization potential:
  values for various gases, 452
Iron, specific energy of combustion, 157
Isoelectronic ions, 508
Isoteniscope, 203–205
Isotherm:
  BET adsorption, 309–310
  Gibbs, 301–302
Isotope effect on spectra, 416, 419–420, 422–423, 429–435
Isotope exchange, 475–483

J

Joule coefficient, 105
Joule heating, 146–147
Joule–Thomson coefficient, 98–106

K

Kelvin temperature scale, 92, 557
  (*see also* Temperature scale)
Keto–enol tautomerism, 466–474
Kinetic theory of transport phenomena, 119–127
Kinetics, chemical:
  acid-catalyzed reaction, 263–270, 273, 277–278
  activated complex, 287–289
  Arrhenius activation energy, 264–265, 269, 281, 284, 319, 489
  chain mechanisms, 290–291
  continuous-flow method, 260–262
  dependence of rate constant on temperature, 264–265, 268, 281, 284, 319
  enzyme, 271–282
  exchange reaction, 265–267
  first-order reaction:
    half-life of, 294
    methods of testing, 284, 293–294

  gas-phase reaction, 287–298
    pressure method, 291–293
    wall effects, 291
  Guggenheim method, 293
  of hydrolysis reaction, 263–270, 271–282
  Lineweaver–Burk plot, 273
  method of initial rates, 256–258
  order of reaction, 255
  reaction mechanisms, 255, 258–259, 271–272
  reversible reaction, 263–270
  RRK and RRKM theories, 288–289
  specific rate constant, 263
  spectrophotometric methods, 274–275, 284–285
  steady-state mechanisms, 288, 290
  transition-state theory, 289n
  unimolecular reactions, 287–289
Knudsen flow, 120–121
Kovar seals, 654
Kundt's tube, 116–117

L

Laboratory notebooks (*see* Notebooks)
Laboratory reports (*see* Reports)
LabPro, 86, 224
LabVIEW, 87–88, 386–388
  virtual instrument (VI), 87, 386–388
Laminar flow, 128, 132
Landau theory of phase transition, 217–221, 383
Lasers, 354, 383, 399, 403–404, 407–414, 441–443, 446–453, 484–491, 618–619
  frequency doubling, 446, 486
  Nd:YAG, 441–442, 446–447, 485–486, 489
  ruby, 484–491
  safety precautions, 697–698
Lattice, crystal, 502–508
Lattice, reciprocal, 529
Lattice constants, 505–507
Lattice energy, 517–518
Lattice planes, 504–505

Lattice types, crystal, 505–507
Lattice vibrations, 527–530
LC circuit, frequency of, 342
Least-squares method, 33, 663–665
  chi-square (χ^2) test, 672–675,
    683–684
  degrees of freedom, 673
  estimated standard deviation of
    parameters, 678–680
  F test, 75, 676–678, 680, 684
  foundations of, 664–669
  goodness of fit, 672–676
  linear fits (linear regression),
    73–75, 666–669, 671–674
  nonlinear fits, 667–668
  normal equations, 666–667
  P test, 76
  sample calculation, 681–685
  spreadsheet examples, 71, 75,
    76–79, 670
  summary of procedures, 680–681
  uncertainties in the parameters,
    678–680
  weighting, 669–671, 679–680
Lennard–Jones potential, 101–102,
  105, 143, 518
  parameters for gases, 101, 518
Lifetime:
  fluorescence, 447–448, 451
  radiative, 488–489
Ligand-field theory, 364–365,
  377–378
Light scattering, dynamic, 379–392
Limits of error, 46–52
Linear regressions, 74–75
Liquid crystals, 215–229
  (*see also* Nematic liquid crystals)
Liquids:
  dielectric constant of, 336–346
  surface adsorption in, 302–303
  vapor pressure of, 199–207
Liquid–vapor coexistence, 229–234
Lissajous figures, 608–609
Literature searches, 704–706
Lock-in amplifier, 459
Lock-in detector, 618
Lorentz–Lorenz relation, 339
Lorentzian line shape, 265, 482
Loschmidt diffusion apparatus,
  137–143

Lubricants, 657
Lucite, 655–656

M

Machine tools, 661
Maclaurin series, 182
Macromolecules:
  intrinsic viscosity, 318–327
  light scattering, 320
  optical rotation and
    conformations, 327–335
Magnetic field, 361, 371n
Magnetic induction, 371n, 455
Magnetic moment:
  electron (orbital), 363
  electron (spin), 362–363, 374,
    377
  nuclear, 371, 477–479
Magnetic susceptibility, 361–370,
  372–374, 377
Manometers:
  capacitance, 353, 596
  flow, 648–649
  mercury, 594–596
  oil, 595
  open-tube, 111–112
  reluctance, 596
  temperature correction for
    readings, 595
Marquardt's algorithm, 667–668,
  681
Mass, reduced, 358, 416, 526
Mass action, law of (chemical
  equilibrium), 188–193,
  193–198, 237–238, 471–472,
  475–483
Maxwellian distribution of
  velocities, 119–120
McLeod gauge, 597
Mean, uncertainty in, 37, 47–50
Mean free path, 120
Mean speed, 120
Mechanical pumps, 588–590
Mercury:
  density of, 113
  safety hazards, 564, 695
  thermal expansion of, 595
  thermometers, 562–564

Metals for construction, 652–654
Michaelis–Menton mechanism,
  272–274
Miller indices, 501, 504–507
Mobility of ions, 236
Molality, 692
Molar mass of polymer:
  distribution function, 322–324
  mass-average, 323
  number-average, 322–323
  viscosity-average, 322–323
Molar mass of solutes, 179–187
Molar refraction, 339, 349
Mole fractions, 173
Molecular diameter, 120, 123,
  126–127
  values for Ar, CO_2, He, 143
  various gas values, 452
Molecular flow (effusion of gases),
  120–121
Molecular-orbital calculations,
  461–465
Molecular weight (*see* Molar
  mass)
Moment of inertia, 416, 429, 526
Monochromator, 631–633
Morse oscillator, 80–83
Morse potential, 70, 439
Mylar, 657

N

Nanocrystals, 492–499
Needle valve, 646
Nematic liquid crystals, 215–221
  Landau theory, 219–221
  nematic-isotropic transition,
    219–221
  order parameter, 216–227
Neoprene, 655
Nickel-chloride solution,
  susceptibility of, 369
Nicol prism, 611–612
NIST, 561n, 641, 704
Nitrogen:
  adsorbed, area of, 316
  dielectric constant, 352
  gaseous, C_p of, 105
  handling of liquid, 599, 696–697
  liquid, vapor pressure of, 521

pressure variation of boiling point, 97

van der Waals constants for, 101

virial coefficient for, 97

Nomenclature, 10

Normal equations, 666–669

Normal error probability function, 44

Normal modes of vibration, 402, 425, 429

of a crystal, 527–530

Normal probability plot, 675, 678

Notebooks, 7–8, 24

Nuclear magnetic resonanace (NMR), 263–270, 371–378, 466–474, 475–483

band-shape analysis, 265–267

chemical exchange, 266–267

chemical shifts in, 373, 375–376, 467–470

deshielding effects, 467–468

deuteron spectroscopy, 480

Fourier transform NMR, 477–480

free induction decay, 265–266, 269, 479–480

intensities in, 265, 269, 468–470

spectrometer, 375–376

spin–lattice (T_1) and spin–spin (T_2) relaxation time, 265–266, 269, 479–480

spin–spin splitting in, 468–470

Nuclear spin quantum number (I), 371

Null detectors, 553

Numerical methods, 33–34, 709–716

Nylon, 656

O

O-ring gaskets, 655–657

Ohm's law, 235

Onsager theory, 236, 244

Operational amplifiers, 542–547, 554

circuit configurations, 545

Optical density (*see* Absorbance)

Optical pyrometers, 574–575

Optical rotation, 223–227, 332–334, 611–612

Order of reaction, 255

(*see also* Kinetics, chemical)

Orthorhombic system, 501

Oscillators, LC, 342–344, 349–350

Oscilloscope, 605–608

Osmotic coefficient, 189–190

Ostwald viscosimeter, 320–321, 324–325

P

P test, 76

Palladium, 655

Paramagnetism, 362–363, 371–378

spin magnetic moment, 362–363, 374, 377

Partial molar quantities, 172–173

free energy, 174

volume, 172–178

Partition function:

configurational, for polymer segments, 329–332

rotational, 475, 526

translational, 475, 526

vibrational:

of a crystal, 527–530

of a gas, 420, 475, 527

Pascal's constant, 374

Pauli's exclusion principle, 362, 394

Perfect-gas law, 93

Permeability of vacuum, 363, 365, 372

Permittivity (dielectric) of vacuum, 336, 348

pH meter, 608–610

Phase diagram:

binary liquid–vapor system, 207–214

for one-component system, 179

Phase rule, 208

Phase transition:

Landau theory, 217–221

liquid-gas, 199–207

nematic-isotropic in liquid crystals, 215–229

order-disorder, 217–221

solid-gas, 515–522, 523–536

solution-vapor, 207–214

Phonon dispersion curves, 528–530

Phosphorescence, 437

Photodiodes, 627

Photography, 626

x-ray film, 512–513

Photomultipliers, 384, 625–627

Physical adsorption of gases, 308–317

Physical constants (*see front endpapers*)

Pipette, 640–641

Pirani gauge, 595–597

Planck radiation law, 558, 574–575, 621

Plastics, 655–657

Plating methods, 651

Platinum, 654–655

electrodes, 251–252, 651

resistance thermometers, 559, 568–571

temperature coefficient of resistivity, 569

Plexiglas, 655

Point groups, 401–402

Poiseuille's law, 130, 321

Poisson distribution, 670

Polar molecules in solution, 336–346

Polarimeter, 333–334,

Polarizability, 338, 399, 401, 407, 451

values for various gases, 452

Polarization, 337–338, 401–404

atomic, 348–349, 357–359

distortion, 336, 338–340, 349, 356–357

of electrodes, 238

molar, 338–341

of a gas, 348–349, 356–357

in solution, 340–341

nonlinear induced, 407, 409

orientation, 336, 338–339, 356

of solute at infinite dilution, 340–341

Polybenzyl glutamate, helix-coil transition in, 327–335

Polyethylene, 656–657

Polymers, 318–335
 statistically coiled, 322–323
 (*see also* Macromolecules;
  Plastics)
Polynomial, fitting with, 74–75,
 669, 681–658, 710
Polypeptides, 327–329
 hydrogen bonding in, 327–330
Polystyrene (latex) spheres, 385,
 388–389
Polyvinyl alcohol:
 chain linkage in, 318–327
 intrinsic viscosity of, 321–323
Potassium acetate, Λ_0 value for,
 244
Potassium chloride:
 Λ_0 value for, 244
 solutions, conductance of, 243
Potential energy:
 internuclear, 401, 417, 428, 437
 Lennard–Jones, 101–102, 105,
  518
 Morse functions, 70, 439
Potentiometer:
 basic circuit, 552–553
 null detectors, 553
Precision, 31–32, 38–40
 fundamental limitations on,
  61–63
Predissociation, 447
Pressure measurements (*see*
 Gauges; Manometers)
Pressure transducer, 353, 596–597
Pressure units, 587
Printers, digital, 613
Prisms:
 Amici, 614–617
 infrared, 622
 Nicol, 611
 visible, 622
Probability distribution, 44–46
Projects, special, 26–27
Propagation of errors, 52–61
 examples of, 19–20, 57–61
Proportional counter, 613
Pumps:
 cryopump, 594
 diffusion, 590–593
 direct drive, 588
 forepump, 588–590

ion, 592–594
 mechanical oil, 588–590
 Roots, 590
 turbomolecular, 590
 water aspirator, 588
Purification methods, 644–645,
 646
Pycnometer, 15–17, 176–177
Pyrex glass, 596, 651–653
Pyrometers, optical, 574–575
Pyruvic acid hydrolysis, 263–270

Q

Q-switched Nd:YAG laser, 411,
 413, 485
Q test, 42
Quantum-mechanical calculations:
 ab initio, 82–84, 397, 414, 423,
  444–445
 acetylene, 435
 dipole moments, 345–346, 359
 electronic states, 397
 Hartree–Fock self-consistent
  field, 82–84
 hydrogen chloride, 82–84, 423
 vibrational spectrum, 405–406,
  435
Quenching, 437, 446–453

R

Radiation, blackbody, 574–575,
 621–622
Radiation detectors (*see* Detectors)
Radiation sources, 618–622
Radical anions, 454–455
Raman spectrometers, 403–404,
 632–634
Raman spectrum, 398–406,
 407–415
 depolarization ratio, 401
 selection rules, 400–401, 410
 stimulated, of benzene, 407–415
 Stokes and anti-Stokes
  intensities, 400
 (*see also* Spectrum)
Random errors, 38–40, 43–52

Random walk, 89
Range shrinking, 678
Raoult's law, 180, 208–209
 deviations from, 209–211
Rate laws (*see* Kinetics, chemical)
Rayleigh scattering, 399
Reaction mechanisms, 255,
 258–259, 283
 (*see also* Kinetics, chemical)
Reciprocal space, 529
Reciprocity relation, 301
Recorders, 613
Rectilinear diameter, law of,
 229–230
Reduced mass, 126–127, 358, 416,
 526
Reducing valve, 644–646
References:
 abstracts, 10
 books, 702–704
 citation of, 23
Refraction, molar, 339
Refractive index, 354–356
 of air, 614, 620n
 of cyclohexanone-
  tetrachloroethane solutions,
  211–214
 of gas, 354–356
 of various optical materials, 623
Refractometers, 613–617
 Abbe, 615–617
 immersion, 614
 principles of operation, 613–614
Refrigerant baths, 577–578
 pumping on liquid nitrogen,
  518–520
 (*see also* Temperature control)
Regulators, gas, 644–646
Rejection of discordant date,
 42–43, 671–672
Relaxation times, 265–266, 269,
 380, 480
Reports, 10–25
 format of, 12
 presentation of figures in, 13
 sample, 14–25
 style of, 12
 word processors, 68–69
Resistance thermometers, 568–571
Resistors, 538–540

Resonance energy, 159
Reversible reaction, 263–270
Reynolds number, 132
Root-mean-square speed, 119
Rotation, specific, 332–333,
    611–612
Rotational structure:
    of IR spectra, 416–423, 424–435
    of vibronic spectra, 439–440
Rounding off, 30–32
Rubber, 654–656

S

Safety, 6–7, 197–198, 693–699
    chemical hazards, 694–695
    for vacuum systems, 599, 696
    reference sources, 699–700
    use of benzene, 346
    use of CCl_4, 197–198
    use of hydrogen and oxygen gas,
        251–252, 650–651
    use of lasers, 697–698
    use of liquid N_2, 696–697
Saran, 656–657
Science Citation index, 706
Second harmonic generation, 415,
    486
Seebeck coefficients, 565–566
Selection rules:
    electronic, 437
    ESR, 455
    IR, 400, 427–428
    NMR, 371, 467
    Raman, 400–401, 428
    rotational, 417–418, 430
    vibrational, 400–401, 417,
        427–428, 437
Self-consistent-field calculation,
    82–84
Self-diffusion (*see* Diffusion)
Semiconductor, parabolic band
    model, 494–495
Shopwork, 661
Signal averaging devices, 617–618
Significant figures, 30–32
Silver, 653–654
Silver–silver chloride electrode,
    249, 251, 651–652

Simpson's rules, 714
Smoothing of data, 709
Sodium chloride solutions, density
    of, 307
Sodium hydroxide, purification of,
    664
Software (computer):
    Excel, 69–72, 388
    LabVIEW, 87–88, 386–388
    mathematical programs, 79–84
    spreadsheets (*see* Spreadsheets)
    word processors, 68–69
Soldering, 657–659
Solids:
    crystal structure of, 500–515
    lattice energy of, 515–522
    sublimation pressure, 515–536
Solution:
    chemical equilibrium in, 193–198
    conductance in, 235–244
    dielectric constant of, 336–345
    freezing point of, 179–192
    ideal, 180, 208–209
    nonideal, 174, 209–211
    partial molar volume in, 172–178
    Raoult's law, 180, 208–209
    surface tension of, 299–308
Sound velocity, 114–118
Specific activity of enzyme, 280
Specific heat (*see* Heat capacity)
Specific rotation, 332–334,
    611–612
Spectrographs, 622
Spectrometer:
    diode array, 631–632
    ESR, 458–459
    Fourier transform IR, 421,
        431–432, 634–636
    infrared, 634
    NMR, 267, 375–376, 472,
        477–480
    Raman, 403–405, 413, 632–634
Spectrophotometer, 275–276, 284,
    395–396
    infrared, 634
    Raman, 632–634
    visible–ultraviolet, 631–633
Spectroscopic components:
    gratings, 622–625
    prisms, 622

radiation detectors, 626–630
radiation sources, 618–622
Spectrum:
    absorption:
        of I_2, 436–445, 532–534
        of nanocrystals, 495–497
    band heads, 440
    conjugated dyes, 393–397
    electronic, of I_2, 436–445
    electron spin resonance (ESR),
        454–465
    emission, of I_2, 437, 440–444
    fluorescence:
        of I_2, 446–453
        of nanocrystals, 495–498
    infrared (IR) vibrational:
        of acetylene, 424–435, 635
        of HCN, 430
    IR, rotational structure of,
        416–420, 429–430, 434
    isotope effect, 405, 411,
        419–420, 423, 431–435
    quantum-mechanical calculation,
        397, 405, 414, 423, 435,
        444–445, 453
    Raman vibrational spectrum of
        CCl_4, 398–406
    resonance fluorescence, of I_2,
        436–453
    rotational–vibrational:
        of acetylene, 424–436
        of HCl, 416–423
        of HCN, 430
    stimulated Raman, of benzene,
        407–415
    wavelength calibration, 440, 620,
        634
    (*see also* Infrared spectrum;
        Raman spectrum)
Speed (molecular):
    mean, 119–120
    root-mean-square, 119
Speed of sound, 114–118
Spin–lattice relaxation time (T_1), 480
Spin–spin relaxation time (T_2),
    265–266, 269, 479–480
Spin–spin splitting, 468–470
Spreadsheets, 31, 34, 69–79, 569
    data/formula entry, 70–71
    examples, 70, 72, 75–77

Spreadsheets—*cont.*
  exercises, 88–89
  functions, 71, 77, 680
  graphs, 72–73
  linear regressions, 73–75
  nonlinear regressions, 77–79
Standard cells, 553–554
Standard deviation, 45–50
  estimated, 40, 46–50, 678–680
  in least-squares parameters,
    678–680, 684–685
Standard electrode emf, 248–249
Standard emf (Weston) cell,
  553–554
Standard free-energy change, 247
Statistical mechanics:
  chemical potential of I_2 gas and
    crystal, 524–530
  classical equipartition theorem,
    108
  model for helix-coil transition,
    329–332
  vibrational partition function,
    420, 475, 527
Statistical thermodynamics, 420,
  475–476, 523–531
Steady-state chemical kinetics, 288,
  290
Steam point, 92, 96
Steel, stainless, 653
Stefan–Boltzmann equation, 621
Steric factor, 319
Stern–Volmer plot, 448, 451
Stokes-Einstein relation, 382
Strain energy, 158–160
Strain gauges, 596, 649–650
Student *t* distribution, 48–50
Succinonitrile, dipole moment of,
  344–345
Sucrose inversion, 271–282
Supercooling, 184–185
Surface adsorption in liquids, 302
Surface concentration, 300–304
Surface coverage, 309
Surface tension, 299–308
  capillary-rise method, 304–306
  of solutions, 299–308
  of water, 307
Susceptibility, magnetic, 361–370,
  371–378

Symbolic mathematics, 79–82
Symmetry:
  of normal modes of vibration,
    402, 425–427
  of vibrational levels, 425–428
Symmetry species, 401–402,
  425–428
Syringes, 641
Systematic errors, 40–41

T

t-butyl peroxide, decomposition of,
  295–296
t distribution, 48–49
t factor, table, 49
Tables, 12–13
Teflon, 656–657
Temperature, critical, 201, 233
Temperature control, 576–585
  controllers, 578–582
  fixed-temperature bath,
    577–578
    Dry Ice, 577–578
    ice, 561–562, 578
    liquid nitrogen, 577
    vapor, 578
  liquid baths:
    circulating type, 584
    oil, 583
    water, 582–583
  melting-point baths, 578
  ovens and furnaces, 584–585
  thermostat blocks, 584
Temperature scale, 91–92,
  557–561
  Callendar–van Dusen equation,
    560–561
  Celsius (centigrade), 91, 557
  characteristic, 517, 526
  fixed points, 91–92, 96, 558
    ice, 92, 96
    steam, 92, 96
    triple, of water, 92, 557, 561
  International Scale, 92–93, 558
  Kelvin, 92, 557
  perfect-gas, 93
Term value, 417, 425, 429, 438
Tetrachloroethane, refractive index
  of, 212

Thermal conductivity of gases,
  123–124
Thermal expansion:
  of brass, 605
  of glass, 97
  of mercury, 595
Thermal insulation, 657
Thermistors, 571–574
Thermochemistry, 145–171
  (*see also* Calorimetry; Heat)
Thermocouple gauge, 597
Thermocouples, 564–568
  amplification, 547
  differential, 103–105
  reference junction bath,
    565–566
  single junction, 565–566
  types of, 566–567
  use of, 567–568
Thermodynamics of cells, 245–247
Thermometer:
  calibration, 563, 567, 571,
    573–574
  calorimetric, 563–564
  comparison of different types,
    573
  cryogenic, 575–576
  cryoscopic, 183–184, 563–564
  gas, 91–98, 557
  mercury, 92, 562–564
    stem corrections, 563
  other devices, 576
  platinum resistance, 568–571
  quartz, 576
  special liquid, 564
  thermistor, 571–574
  vapor-pressure, 519–522, 576
  (*see also* Pyrometers, optical;
    Thermocouples)
Thermopile, 629
Thermoregulating circuits,
  578–582
Thermostats (*see* Temperature
  control)
Thyodene, 196
Timing devices, 636
Titration, 643
  of acids, 191
  of iodine, 196–197
Torr, definition of, 587
Transformer, 541–542

Transitions, vibrational:
  combination and difference, 428
  fundamental, 425–426
  overtones, 427
Transmittance, 631
Transport properties, kinetic theory of gases, 119–127
  (*see also* Diffusion; Viscosity)
Trapezoidal rule, 713
Triple point, 179
  cell, 561
  of water, 91, 557
Tubing connections, 661
Turbulent flow, 128, 132
Tygon, 654–656

U

Uncertainties (*see* Error(s)):
  in least-squares parameters, 678–680
Uncertainty principle, 63
Unimolecular reactions, 287–289
  (*see also* Kinetics, chemical)
Units:
  conversion factors (*see front endpapers*)
  international system (SI), 690–692

V

Vacuum techniques, 578–599
  O-rings, 655
  safety, 599, 696
  (*see also* Gauges; Pumps)
Valence bands, 494–495
Valence force model, 401–403, 428–429
Valves, needle and reducing, 645–646
van der Waals equation, 100, 110, 233
Vapor pressure:
  of argon (solid), 518–522
  empirical Antoine equation, 522
  of iodine (solid), 523–536
  of liquid, 199–206

boiling-point method, 202–203
  isoteniscope method, 203–205
  of nitrogen, 521
Vapor-pressure thermometer, 519–521, 576
Variance, 39–40, 208
Velocities, molecular, Maxwellian distribution of, 119–120
Velocity of sound, 114–117
Vermiculite, 657
Vibration, normal modes of, 402, 425, 429
Vibrational heat capacity of gases, 420
Vibrational partition function, 420, 475, 526
Vibrational Product Rule, 435
Vibrational spectrum (*see* Spectrum)
Vibrational wavefunctions, 426
Virial coefficients, 93, 102
  for He, Ar, N_2, and CO_2, 97
Virial equation, 93, 102
Virtual instrument (VI), 87–88, 386–388
Viscosimeter, 129, 131
  Ostwald, 321, 324–326
Viscosity:
  air, 134
  gases, 121–123, 128–136
  intrinsic, 321
    of macromolecules, 318–327
    of polyvinyl alcohol, 322
  kinetic-energy correction, 133
  several fluids, 128
  slip correction, 133
  solution of rigid particles, 321–322
  specific, 321
  water, 389
Viscosity coefficient:
  of gas, 121–123, 128–136
  of water, 326
Viscous flow, 128–134
Voltage measurement, 542–554, 605–608
Volume, molar:
  apparent, 174

partial, 172–178
  of water, 174
Volumetric procedures, 639–644

W

Waste disposal, 695
Water:
  compressibility factor for, 200–201
  critical temperature, 201
  density of, 177
  diamagnetism of, 369
  entropy of, 205–206
  heat capacity of gas and liquid, 202
  molal freezing-point depression constant for, 183
  molar heat of fusion, 183
  molar volume of, 174
  pressure variation of boiling point, 97
  purification of, 644
  surface tension of, 307
  triple point of, 91–92, 557
  viscosity of, 389
Water baths, 582–583
  (*see also* Temperature control)
Wave functions, 80–84, 426
Wavelengths:
  argon, 620
  copper $K\alpha$ x-ray, 511
  krypton, 620
  mercury, 620
  neon, 620
  selection, 622–625
  sodium D doublet, 611
  in vacuum, 441, 620n
  xenon, 620
Wavenumber, 416
Wave vector, 380–381, 528–529
Weighing errors, 604–605
Weight of an observation, 665–666, 669–671, 678–680
Weighted residual, 672
Weston cell, 553–554
Westphal balance, 325, 369
Wheatstone bridge, 238–241, 553–556, 569–570

ac circuit, 238–241, 554–556, 608
dc circuit, 553–555
Wiens law, 621
Word processors, 68–69
Work:
electrical, 147, 169–170
expansion (sign convention),

X

X-ray diffraction, 500–515
Bragg reflection, 503–505
camera, 511–512
film, 512, 514
intensities of reflections, 507–508

powder method, 500–514
(*see also* Crystal structure)

Z

Zeeman energy, 371, 455–457, 466–467
Zero-point energy of crystal, 517